W9-BMM-624

AUTOMOTIVE MICROCONTROLLERS

Other SAE books in this series:

Sensors and Transducers
by Ronald K. Jurgen
(Order No. PT-68)

Object Detection, Collision Warning and Avoidance Systems
by Ronald K. Jurgen
(Order No. PT-70)

Electronic Instrument Panel Displays
by Ronald K. Jurgen
(Order No. PT-71)

Navigation and Intelligent Transportation Systems
by Ronald K. Jurgen
(Order No. PT-72)

Electronic Engine Control Technologies
by Ronald K. Jurgen
(Order No. PT-73)

Actuators
by Ronald K. Jurgen
(Order No. PT-74)

For information on these or other related books, contact SAE by phone at (724) 776-4970, fax (724) 776-0790, e-mail: publications@sae.org, or the SAE website at www.sae.org

AUTOMOTIVE MICROCONTROLLERS

Automotive Electronics Series

PT-75

Edited by

Ronald K. Jurgen

Published by
Society of Automotive Engineers, Inc.
400 Commonwealth Drive
Warrendale, PA 15096-0001
U.S.A.
Phone (724) 776-4841
Fax: (724) 776-5760
www.sae.org

ISBN 0-7680-0396-2
Library of Congress Catalog Card Number: 98-88835

SAE Order No. PT-75

Introduction

Microcontrollers: Electronic Enablers

The first application of a microprocessor in a production automobile was the Delco-Remy MISAR (Microprocessed Sensing and Automatic Regulation) system used in the 1977 Oldsmobile Toronado as part of the spark-timing system.[1] This system controlled spark timing precisely for all conditions of load and speed consistent with driveability and emissions-control requirements. It was not obvious then that such systems—microprocessors and their sophisticated counterparts, microcontrollers or microcomputers—would be used in such large quantities as they are today in automobiles in all price ranges. Today, the microcontroller is one of the basic components of any automotive control system, and, with the aid of sensors and actuators, makes possible the multitude of vehicle control applications now available.

The *McGraw-Hill Electronics Dictionary*[2] defines a microcontroller as "A monolithic integrated circuit with a complete central processing unit (CPU) and enough semiconductor memory (RAM, ROM, EPROM or EEPROM) and input-output (I/O) capability to be considered as equivalent to a 'computer on a chip.'"
Because it has memory and I/O capability on the same chip with the central processing unit (a microprocessor), the microcontroller is versatile in a wide and ever-increasing range of automotive electronics applications. In addition to their early use to control emissions and fuel economy, microprocessors and microcontrollers now have applications in braking, transmissions, displays, body control, radios, navigation systems, and cruise control—and with time, this list grows longer.

The papers in this book have been placed in three categories: Microcontroller Design Considerations, Microcontroller Communications, and Examples of Microcontroller Applications. In each category, the papers appear in chronological order, making it easy to follow the trends in technological developments from ten years ago to the present.

Among the topics featured under Microcontroller Design Considerations are circuit design, software, fault-tolerant electronics, memory trends, packaging, reliability, and EMI (electromagnetic interference). In the Microcontroller Communications section, topics covered include distributed realtime processing, the CAN (controller area network) specification, OSEK (open systems and interfaces for distributed electronics in cars), TTP/A (time triggered protocol), and J1850. Examples of Microcontroller Applications include uses of microcontrollers in electronic engine control, ride control, anti-spin control, displays, radios, transmissions, and diagnostics.

The last paper in this book, "Future Developments in Automotive Microcontrollers" by Ross Bannatyne of Motorola Inc., provides an insightful view on what to expect in the future. In the words of the author, the object of this paper is "to discuss the changes that will occur in the use and implementation of automotive microcontrollers in the next generation of systems. The functions that will be required to be maintained by the microcontroller and the types of technology that will enable these function are addressed."

References

1. Trevor Jones, "Automotive electronics I: Smaller and Better," *IEEE Spectrum*, November 1977, pp. 34-35.
2. *McGraw-Hill Electronics Dictionary*, Sixth Edition, McGraw-Hill Inc., New York, N.Y., 1997.

* * * * *

This book and the entire automotive electronics series is dedicated to my friend Larry Givens, a former editor of SAE's monthly publication, *Automotive Engineering*.

Ronald K. Jurgen, Editor

Table of Contents

Introduction

Microcontroller Design Considerations

Microcontroller Communications

Typical Microcontroller Applications

What the Future Holds

MICROCONTROLLER DESIGN CONSIDERATIONS

890764

The Most Suitable Way to Design the LSI for the Engine Management Control System

H. Minamino and S. Matsubara
NEC Corp.
Tokyo, Japan

F. Bruce Gerhard, Jr.
NEC Electronics Inc.
Natick, MA

1. ABSTRACT

As electronics technology has made remarkable progress in recent years, 16-bit microcomputers have been increasingly employed in automotive engine control systems.

Initially, engine control systems which could control only fuel injection and ignition timing might have been sufficient, but today, systems are increasingly being required to cover many control, such as knocking and intake control.

In addition, demands for integrated engine control units that can also control other sub-units such as anti-lock braking control systems, traction control systems, and transmission control systems have been on a steady rise.

Against this background, the following requirements will be demonstrated by the engine control systems of today and the future:

1) Integration on a microcomputer chip of powerful real-time
control functions suitable for mechanical control
2) Miniaturization of unit size
3) Unification of control programs
4) Standardization of hardware

Some of these requirements can be satisfied by employment of a high-speed microcomputer and large-capacity on-chip memory. However, accelerating the operating speed of the microcomputer and increasing the on-chip memory capacity are likely to increase the workload and the chip size of the microcomputer. This will not necessarily improve the overall performances of the control system.

What microcomputer is the best for automotive engine control systems and how should the system itself be organized to meet all of the above requirements? In a bid to answer these questions, this paper discusses the LSI design techniques (especially microcomputers) to be used in engine control systems, placing a special emphasis on the advantages as well as disadvantages of multiple chip configurations for engine control systems.

This paper also presents some application examples of the 16-bit single-chip microcomputer μPD78322 and of the (ASIC) μPD71P301 (nicknamed "turbo access manager"), whose architecture is designed using LSI design techniques which will be introduced in this paper.

2. INTRODUCTION

Since the microcomputer became practical in the latter half of the 1970's as a device of the engine control system, the engine control system has made remarkable progress, and today's control system can control not only ignition timing and fuel injection but also other sophisticated control actions including anti-lock braking, traction, and transmission control actions. As the technology of the engine control system has advanced, the role that the microcomputer plays in the system has been steadily expanded. In addition, thanks to the recent development of the semiconductor technology, high-performance microcomputers are now available at low cost. As a result, more and more engine control systems, from those intended to control medium-size engines with displacement of 2,000 cc. to those made for compact engines having displacement of about 500 cc., have adopted microcomputers as their control elements. Additionally, the engine control systems for the so-called luxury models have been increasingly required to provide more sophisticated functions and higher performance. So now, many semiconductor manufacturers are making relentless efforts to develop microcomputers to meet these demands.

3. RECENT TREND OF ENGINE CONTROL SYSTEM AND MICROCOMPUTER FUNCTIONS

In recent years, automotive engine control systems, particularly medium and large-scale systems, have been required to provide more functions and to perform more control actions.

Especially for high performance engines, the number of items to be controlled and the number of control actions to be performed by control systems have been increasing in order to achieve high output power, low fuel consumption, and clean exhaust gas containing less harmful, toxic substances. These engine control systems, therefore, control many operations of an automobile, including:

1) Fuel injection
2) Ignition timing
3) Idling speed
4) EGR
5) Knocking
6) Supercharging pressure

Control techniques used to control these items have become increasingly complicated in recent years in order to control each item more precisely and delicately. Many control techniques have been invented and introduced so far. These include such techniques as independent sequential fuel injection in each cylinder and ignition control without a distributor that can also control the knocking of each cylinder are widely employed. In addition, many control systems are now provided with such functions as learning and self-diagnosis.

Moreover, many manufacturers have developed and put into practice integrated engine control systems that have functions not only engine control, but also communication with other control subsystem units such as transmission control, traction control, and anti-lock braking control systems.

To organize such high-performance engine control systems, it is expected that the microcomputers to be incorporated in the control systems will be increasingly required to provide more input/output functions, higher arithmetic operation speed, and more powerful real-time functions.

Specifically, the engine control systems for the future will need to satisfy these requirements:

1) Integration on a microcomputer chip of powerful real-time
functions suitable for mechanical control

The microcomputer employed in the engine control systems must provide powerful hardware as well as software functions in order to perform the increasing and diversified control actions.

Many control actions must be performed to control an automobile engines and most of these actions must be performed real-time. It is therefore important that the microcomputer be able to control its outputs in immediate response to input data that changes with time. Because today's 16-bit microcomputer can operate at very high speeds, quickly performing arithmetic operations and processing data, they can process, with an aid of accurate software, input data that changes with time, with only negligible delay time if any.

Therefore, 16-bit microcomputers have enormous potential in the construction of high-performance engine control systems. Yet, the problem to be solved is how the result calculated by the microcomputer should be output. Even if the microcomputer can operate at lightning speeds in quick response to inputs, unless the results of the microcomputer operations can be output at equally quick speeds, there will be no merit to using a microcomputer in an engine control system. It is therefore necessary to integrate on a microcomputer chip the hardware devices that can fully draw out the functions of the CPU and that can convey the results of the CPU operations as quick as possible. This includes items such as a powerful interrupt controller high-speed timers/counters with compare registers, and capture registers.

2) Miniaturization of unit size

The size of the control unit recently has been required to be compact so that the control unit can be installed in a very limited space, especially of a sub-compact car. In addition, in order to provide more passenger space, the mounting space available for engine control units has become increasingly small. For these reasons, the microcomputer to be incorporated in the control unit must be as small as possible. In reality, however, the number of I/O ports and pins of recent microcomputers have increased. This has resulted in larger size microcomputer chip and package as the performances of the microcomputers have been improved so that the more complicated and sophisticated control actions can be performed. To reduce the size of the control unit, the microcomputers and other peripheral devices housed in surface mount packages such as PLCC (Plastic Leaded Chip Carrier) and QFP (Quad Flat Package) should be employed to increase the package density. Photo 1 shows the LSI Package Variation.

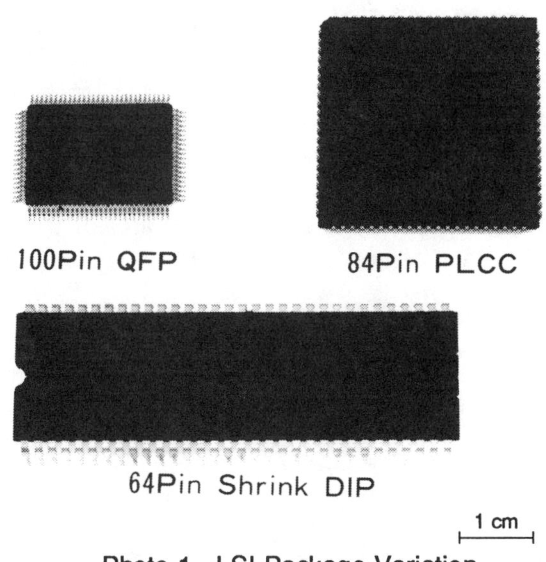

Photo 1 LSI Package Variation

3) Unification of control programs

As the number of things to be controlled and the number of control actions to be performed have increased, the number of software development processes have also increased. Nevertheless, an increase in the number of development processes itself does not pose a serious problem as the software development efficiency can be enhanced by several methods. For example, recycling the existing control programs to create a new, integrated program is one such possible methods.

Moreover, in recent years, high-level languages are being applied to design programs. High-level languages indeed bring about a significant increase in software productivity, but they also increase the number of object programs. The problem,however,is the translation efficiency of a high-level language codes into machine language codes.

Because of the ease of development and maintenance, high-level languages are expected to play an important role in standardization of control programs. However, this means an increase in the quantity of software and it can therefore be said that the microcomputer to be incorporated in an engine control system must have a large-capacity, on-chip memory.

4) Standardization of hardware

The development cycle of the engine control system has been progressively shortened year after year due to the expansion of the market and the progress of the electronic technology. Moreover, the time required for the development of a control system has been equally shortened. Consequently, the configuration of the LSIs to be adopted in the engine control systems of today should have a high degree of freedom so that if the functions of the existing system should need to be improved because of the progress of the engine control techniques, as many existing hardware devices as possible can be recycled to create a new system. Specifically, since memories such as RAM and ROM and some I/Os are necessary in any type of control system, these devices should ideally be developed as general-purpose LSIs.

4. DESIGNING MICROCOMPUTERS FOR ENGINE CONTROL SYSTEMS

As mentioned earlier, the number of roles an engine control system must be able to play has been steadily increased as the performance of the control system has improved. It is no exaggeration to say that the performances and functions of the control system are solely determined by the microcomputer employed in the system.

The questions are then what microcomputer is the best to improve the performance of the engine control system and, at the same time, to provide good cost-effectiveness, and how the control system should be organized.

An engine control system can be organized in three ways: 1) in a single chip configuration in which all the necessary control functions are integrated on a single chip of the microcomputer, 2) in a multiple chip configuration in which the control functions are distributed among several LSI chips, and 3) in multiple CPU configuration in which the control functions are divided into two blocks and each block is controlled independently. Of these, multiple CPU configurations can have many variations depending on the microcomputers to be used. It is difficult to compare these configurations with 1) and 2). In this paper, therefore, an emphasis is placed on comparison between the single chip configuration and multiple chip configuration. Figure 1 shows the Configuration of the Engine Management Control System.

1) Basic function of a model microcomputer

In order to form a basisfor discussing the relative advantages of single-chip and multiple chip configurations, we shall define a typical microcomputer system for engine control. The model microcomputer has the following specifications.

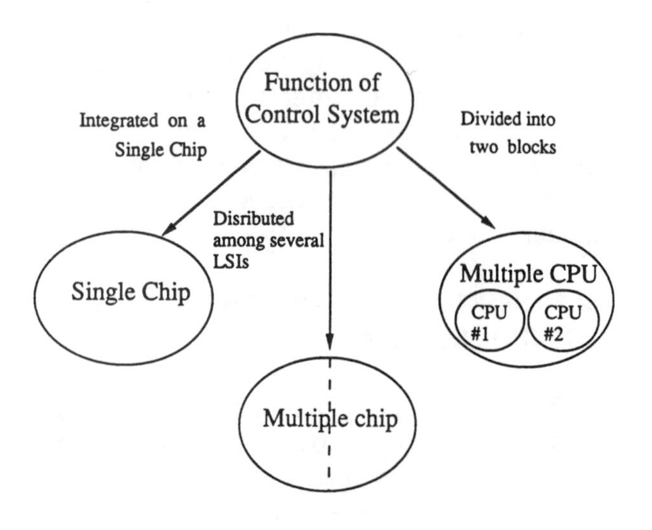

Figure 1 Configuration of the Engine Management Control System

Specifications:

- 16-bit CPU core
- EPROM: 32K bytes
- RAM: 2K bytes
- A/D converters: 10 bits, approx. 8 channels
- Timers: 16 bits, approx. 10 channels
- I/Os: approx. 80 lines
- Serial communication: 3 channels
- Others

2) LSI integration and design rule

Recent progress in LSI production technology is remarkable. For example, VLSI (very large scale integration) DRAMs containing 1M byte of memory are now available on the market, and it is hoped that it will soon be possible to produce chips using submicron design rules.

Generally, LSI production technology which has been established by mass production technology of memory chips is used in producing high-performance microcomputers for automobile instrumentation, and a design rule under 1.5-micrometer is the current mainstream of production technology. This means that the chip size of an LSI is generally determined by the functions and scale of integration of the LSI. Since the price of an LSI is closely related to the chip size, the functions to be integrated on an LSI must be carefully determined to provide the LSI with the greatest cost effectiveness. Of the hardware devices integrated on an LSI, on-chip RAM takes up a larger area on the LSI than the other memories. For example, if a mask ROM,

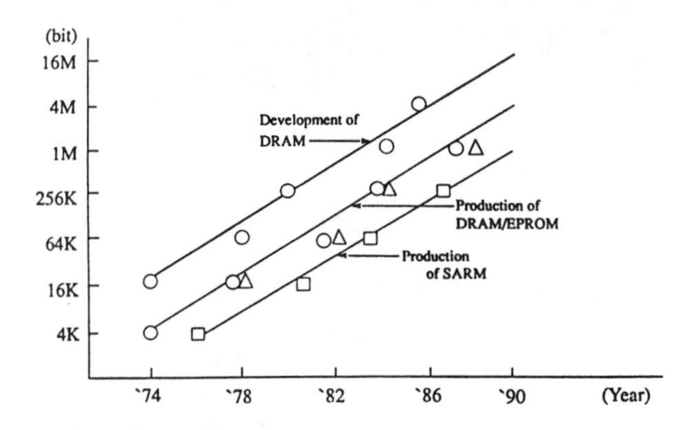

Figure 2 Trend of Memory Size

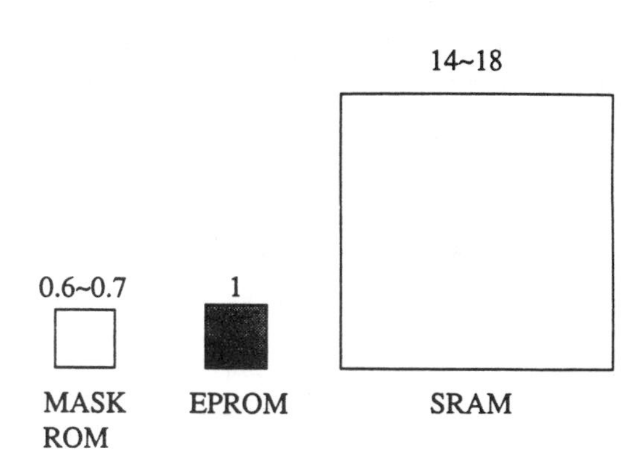

Figure 3 Comparison of Memory Chip Size

EPROM, and RAM each having the same capacity are created using the same design rule, the RAM requires an area 14 to 18 times larger than the area required by the EPROM, whereas the mask ROM takes up an area of six to seven tenths that of the EPROM. Figure 2 Shows the Trend of Memory Size. Figure 3 Shows the Comparison of Memory Chip Size.

3) Single chip configuration

What are the advantages and disadvantages of organizing an engine control system in a single chip configuration?

In a single chip configuration, the number of components required is naturally less than that in multiple chip configuration, because all the necessary hardware functions except the input/output circuit and special function logic (such as a supply voltage monitor circuit) can be integrated on a single LSI chip. Therefore, a single chip configuration can greatly contribute to reducing the size of the overall control system (though this does not mean that the LSI chip itself is small).

In addition, high operation speed and real-time input/output processing are distinct advantages of a control system configured on a single chip. In a single chip configuration, the hardware devices such as memory and peripherals can be placed on the chip at locations where the performance of the 16-bit CPU can be fully realized.

One of the disadvantages of the single chip configuration is that the chip size tends to be large because many functions are integrated on the same chip. The production cost of the chip is also high. Although the chip size varies depending on the architecture of the microcomputer to be produced, if a microcomputer having the specifications mentioned earlier is to be created using the 1.5-micrometer design rule, the chip size of this microcomputer will be larger than 10 millimeters square. In addition, the memories and peripheral I/Os including the timers will occupy about 60% of the chip area (30% each). Increases in cost are not tolerated these days, even if the functions and performances are improved. It is therefore important that the functions and performance of the microcomputer be improved while the production cost is held at a low level.Figure 4 shows Single Chip Configuration.

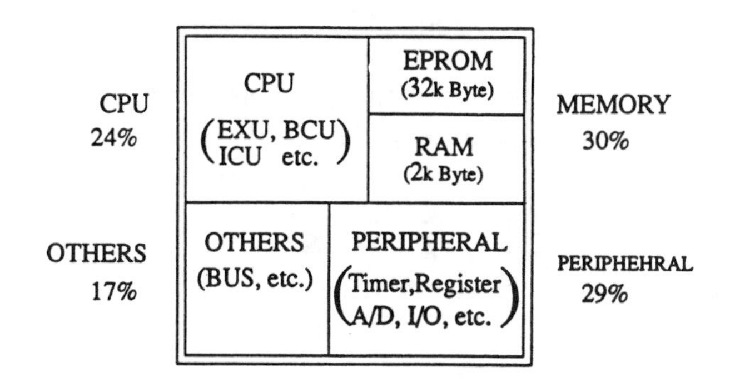

Figure 4 Single Chip Configuration

The next disadvantageous point of the single chip configuration is that the number of pins of the LSI package increase, making the package bulky. In order to achieve the specifications mentioned above, at least 100 pins are required, and at present, only QFP is available as a plastic package that can accommodate as many as 100 pins. Consequently, control unit manufacturers are forced to introduce surface mount technology. Furthermore, a high level of mount technology is required because the pin pitch decreases as the number of pins increases.

4) Multiple chip configuration

One of the advantages of a multiple chip configuration is a high degree of freedom.

LSIs can be broadly classified into two types: one of them contains memories and general-purpose I/Os and the other is an integrated CPU. The former type of LSI can be used in many standard products, not only in engine control systems. If these LSIs are incorporated in many products, the cost of the finished products can be substantially reduced. The CPU LSI can be produced by the latest semiconductor production technology, which is improved every year, to provide application systems with high performances. If a CPU is made up of two chips, the price of the CPU can be held at the current level.

If the model specifications are achieved in multiple chip configuration, a 32K-byte PROM, 2K-byte RAM, and 16 I/Os can be integrated on about 7.5 millimeter square chips. Therefore, if the functions of a microcomputer are divided into several groups with each functional group integrated on a chip, and if the overall engine control system is configured using these chips, the size of each chip can be reduced, and the cost of the overall system can be decreased. The functions of the system can also be improved.

In spite of these advantages, the multiple chip configuration also has disadvantages. One of them is that the multiple-chip engine control system takes up more space than the single-chip counterpart. In multiple chip configurations, many chips must be mounted on a printed circuit board (PCB), and therefore because wiring occupies a large part of the PCB, the size of the PCB itself increases. This runs counter to the recent trend toward miniaturization of the control systems.

Moreover, the performances of the CPU may not be fully realized in the multiple chip configuration. This is because, in the multiple chip configuration, the ROM storage instructions and programs, and the CPU are placed on separate chips. Although the processing speed of the CPU itself has been substantially improved as the CPU of today can operate on a much faster clock than before, separation of the CPU from the memory extends the CPU's memory access time. Consequently, the operating rate of the CPU is degraded. If the functional blocks, such as timers and interrupt controllers, that must be able to operate real time are separate from the CPU, the overall performances of the system may be degraded.

5) Single chip vs. multiple chips

As is evident from the above discussion, both the single chip configuration and multiple chip configuration have their own advantages and disadvantages. Nevertheless, the multiple chip configuration, especially two chip configuration is superior to the single chip configuration from the viewpoints of price and performance.

If an engine control system consists of two chips, it is considered that the functions should be divided between the chips as follows:

Chip A: ROM, RAM, and peripherals (mainly I/Os and outputs such as PWM)

Chip B: CPU and peripherals (mainly interrupt controllers, A/D converters, and timers)

Integrated on chip A are memories which can be mass produced at a low cost, input devices with relatively slow response, and PWM outputs and general-purpose output devices that operate in response to the outputs from chip B. The reason chip A is provided with all of these functions is to keep the size of chip B as small as possible.

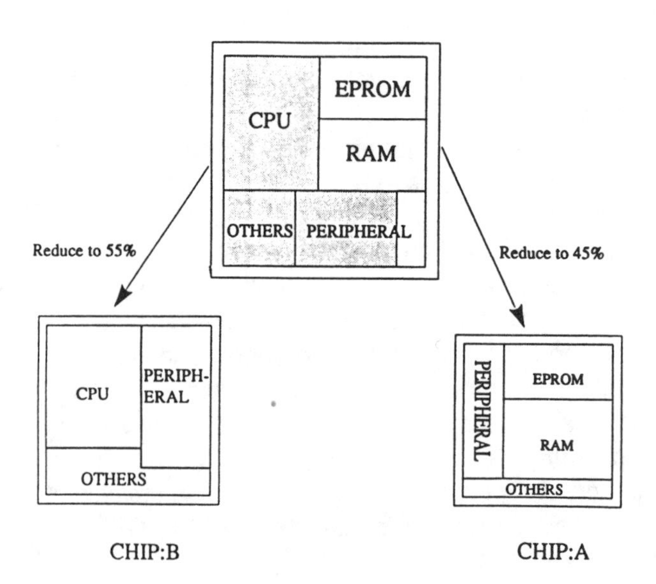

Figure 5 Multiple Chip Configuration

The functional blocks that have a significant influence on the overall performances of the system, such as an interrupt controller and timers must be integrated on chip B along with the CPU, so that these peripherals can operate as fast as the CPU. Figure 5 shows the Multiple Chip Configuration. Although the multiple chip configuration has disadvantages,such as increase of the PCB size,it gives us much more informative advantages.

One of the advantages of this two chip configuration is that the performance of the entire system can be improved by merely exchanging chip B with an upgraded chip. You can easily upgrade the system by only changing one of two chips. This can be done without much modification of the hardware. In a case of the single chip configuration,if you want to upgrade the system,you have to change the whole CPU. Because the performance of the CPU is subject to constant improvement as the semiconductor production technology makes progress, this provision for replacement is convenient when the currently employed CPU is superseded by a newly developed one.

ITEM	Multiple Chip Contiguration (2 Chip Set)	Single Chip Configuration
Relative Cost	0.7	1
Chip Size *1	A ; 7.5 mm square B ; 8.0 mm square	11 mm square
Package	A ; 68 pin PLCC B ; 68 pin PLCC	120 pin QFP
Program ROM Access Speed	2 ~ 3 (1.0)*2 [External ROM]	1 [On chip ROM]
Flexibilty for various System	Sufficient for present systems	Sufficient for small systems only

Note; * 1 Estimate on the paper

*2 μPD71P301

Figure 6 Single Chip vs. Multiple Chip Configuration

One more advantage of this multiple chip configuration is that the production cost is lower than that of the single-chip configuration,because smaller chip size makes production lower for increase of the yield. Figure 6 shows Single Chip vs. Multiple Chip Configuration.

5. ENGINE CONTROL SYSTEMS USING THE μPD78322 AND THE μPD71P301

This section discusses an application example of a medium-scale engine control system consisting of an application-specific integrated circuit, μPD71P301 (nicknamed a "turbo access manager"), and a 16-bit single-chip microcomputer, the μPD78322. This was designed based on the concept of the multiple chip configuration described in the preceding sections.

1) Features of the μPD78322

The μPD78322 is a 16-bit single-chip microcomputer designed specifically for automotive applications. It has 16K bytes of internal ROM and 640 bytes of RAM, and can directly access 64K bytes of memory space. In addition, the μPD78322 is provided with 109 powerful instructions and can execute an instruction in 250 ns (with a 16-MHz crystal oscillator).

Furthermore, the microcomputer has a general-purpose real-time pulse unit (RPU); timer/counter functions; an eight-channel, 10-bit A/D converter; and a powerful interrupt controller. It is also provided with DMA functions, called macro service functions, that

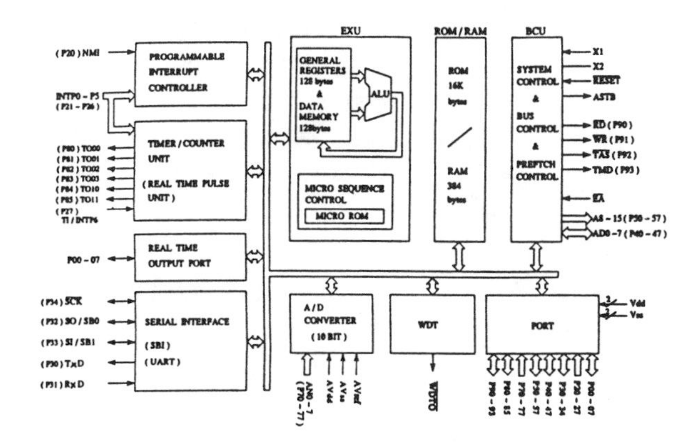

Figure 7 μPD78322 Block Diagram

include arithmetic operations such as addition and subtraction between memory and internal peripheral I/Os including timers.

This microcomputer also has a turbo access manager interface function with which to read the data in the internal PROM of the μPD71P301 (turbo access manager) at high speeds. Figure 7 shows the μPD78322 Block Diagram.

Features:

- 16-bit CPU (μCOM-78K/III)
- 109 powerful instructions,
  including:
  - 16-bit operation instructions
  - Bit manipulation instructions
  - Multiplication instructions
    (16 bits by 16 bits = 32 bits)
  - Division instructions
    (32 bits by 16 bits = 32 bits + 16-bit remainder)
  - String instructions
  - User stack instructions
- Minimum instruction cycle:
  250 ns at 16 MHz
- Memories
  - On-chip ROM : 16K bytes
  - On-chip RAM : 640 bytes
  - Memory space: 64K bytes
- General-purpose pulse input/output unit
  (real-time pulse unit)
  - 16-/18-bit free running timer: x1
  - 16-bit timer/event counter: x1
  - 18-bit capture register: x6
  - 18-bit capture register: x4
  - 18-bit capture/16-bit compare register: x2
  - Timer output: x6
- High-precision, 10-bit A/D converter
  (8 channels)
- Real-time output port: x8
- Serial interface (with baud rate generator): x2
  - Asynchronous serial interface (UART)
  - Serial bus interface (SBI)
- Interrupt controller
  - Vector interrupt function
  - Context switching function
  - Macro service function
- Turbo access manager control signal
  output function
- Watchdog timer
- Standby function (STOP/HALT)
- CMOS technology
- 68-pin PLCC package

2) Features of the μPD71P301 (turbo access manager)

The μPD71P301 is a peripheral LSI which includes 16K bytes of EPROM, 1K bytes of SRAM, and 16 I/O lines integrated on a single chip. This LSI also has high-speed memory access functions and bus interface functions.

The turbo access manager can be directly connected to a microcomputer equipped with a multiplexed address/data bus. When this LSI is connected to a microcomputer having turbo access manager (TAM) control signals such as the μPD78322, the memory can be accessed at high speeds. Figure 8 shows the μPD71P301 Block Diagram.

Features:

- High-speed memory access function
  Instruction code/data read speed:
  1 word max. per clock
- Direct connection to multiplexed address/data bus
- Internal memory space relocation function
  Any place in 1M byte.
- External expansive address output function
- Chip select signal output function

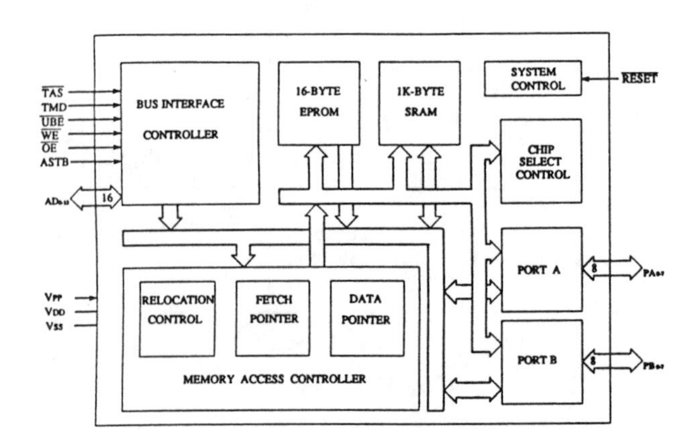

Figure 8 μPD71P301 Block Diagram

- Memories:
  - 16K-byte EPROM
  - 1K-byte SRAM
- I/O lines: 16
- CMOS technology
- 44-pin PLCC/LCC package

3) System consisting of the µPD78322 and the µPD71P301

An engine control system consisting of the 16-bit single-chip microcomputer the µPD78322 with the features mentioned above and ASIC µPD71P301 can provide following features . The biggest feature of the control system having the µPD78322 is that the general-purpose real-time pulse unit (RPU) of the microcomputer can be utilized. Figure 9 shows the Real-time Pulse Unit Block Diagram.

The RPU consists of two timers, TM0 and TM1, and registers connected to these timers.

TM0 is a 18-/16-bit free running timer that operates on the CPU system clock and is connected with four 16-bit compare registers, three 18-bit capture registers, and an 18-bit register that can be set in the compare or capture mode by software.

TM1 is a 16-bit timer/event counter that operates on an external input clock or CPU system clock. This timer is equipped with two compare registers.

In addition to these devices, the RPU is also provided with a capture register and a compare/capture register whose timer source can be selected by software, so that a flexible timer unit can be configured as the specifications of the system require.

By using this RPU, cyclical measurements of input signals, can be easily controlled. These are indispensable for engine control, fuel injection timing and injection period, and ignition timing can be easily controlled.

The RPU can also select up to six toggle outputs, which can be directly output by hardware in response to a signal indicating coincidence between the contents of a timer and those of a compare register, or a maximum of four set-reset outputs, so that highly accurate outputs can be issued in synchronization with the timer operations.

The µPD78322 also includes a real-time output port. Specific bits or groups of bits in this port can be either set or reset under timer control. Since this output port can be directly set or reset by the trigger signal output from the RPU, pulse control involving multiple phases, such as fuel injection control, can be performed in synchronization with the timer operation. Figure 10 shows the Block Diagram of the Real-time Output Port.

The powerful macro service function of the microcomputer can be cited as the third feature. This function allows not only DMA between the memory and timers or special function registers (SFRs) but also arithmetic operations such as addition and subtraction. Therefore, updating the value of a compare register connected to the free running timer (i.e., transferring the sum of data and old compare register value to the

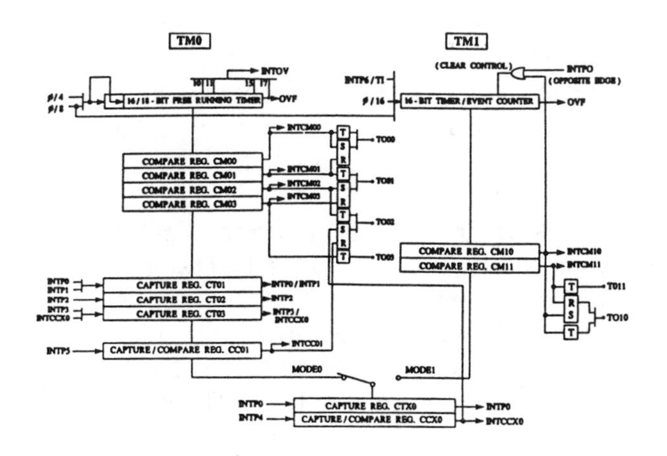

Figure 9 Real-Time Pulse Unit Block Diagram

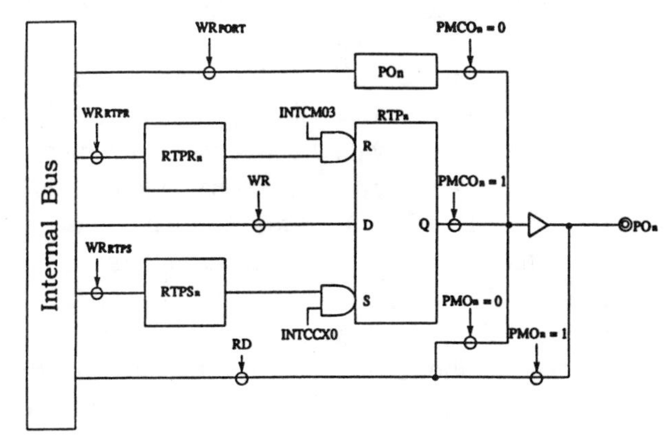

Figure 10 Block Diagram of the REAL-Time Output Port

compare register) and differential processing of a capture register (transferring the difference between the newly captured value and previously captured value to memory) can be started by an interrupt signal. This means the processing time can be substantially shortened as compared with the time required by conventional software processing.

Next, let's discuss the features of the ASIC μPD71P301, or "turbo access manager".

First of all, when the LSI's internal EPROM is used to expand the program storage capacity, the OP codes can be fetched from the EPROM of the LSI at the same speed as that of the on-chip ROM of the μPD78322. Recent advances in microcomputers have included increasingly faster CPUs, but their advantage is lost unless instruction fetch times can also be kept to a minimum. However, this cannot be done by the existing microcomputers because they are slow in accessing an external ROM. The μPD71P301 has a fetch pointer that operates in synchronization with the program counter of the CPU. This fetch counter makes it possible, especially when several OP codes are to be fetched continuously, to access the EPROM of the μPD71P301, which is connected to the CPU as an external device, at the same speed as that of the CPU's on-chip ROM. This function of the fetch pointer allows one byte of OP code to be read from the μPD71P301 to the μPD78322 in the duration equivalent to one system clock. Therefore by using the turbo access manager, degradation of the entire system's performance, which is inevitable with any other peripheral LSI, can be effectively prevented. Figure 11 shows the Instruction Fetch Bus Cycle.

The second feature of the turbo access manager is that it is provided with 16 I/O lines. When the external memory is expanded with a conventional microcomputer, an expansive I/O port device is required. This is because 8 or 16 I/O ports of the microcomputer are used as an address and/or data bus. As a result, the expansive I/O port device is needed to make up for the shortage of I/O ports. However, since the turbo access manager is provided as many as 16 I/O lines, use of this ASIC provides sufficient I/O ports. The third feature of μPD71P301 is that the turbo access manager is provided with special control registers that can be used to map the internal EPROM, RAM, and I/O spaces on any area of up to 1M byte of memory space. Therefore, if special external I/O devices are connected to the turbo access manager, no extra components such as address decoder is needed. Consequently, the increase in the mounting area, which is unavoidable in the multiple chip configuration, can be minimized.

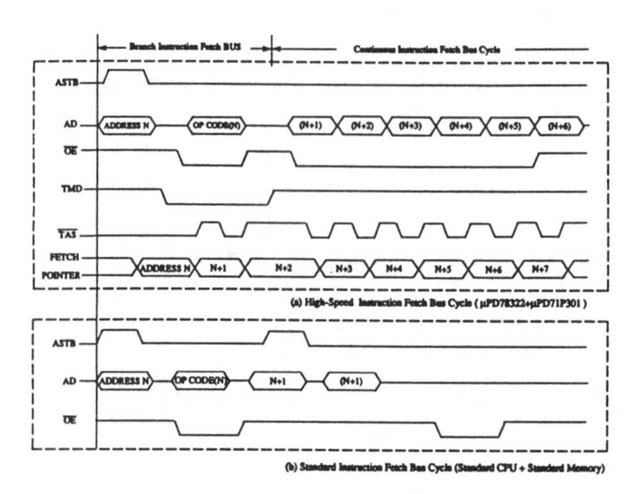

Figure 11 Instruction Fetch Bus Cycle

6. CONCLUSION

In this paper, we have discussed the effective design techniques utilizing a microcomputer for engine control systems It also discusses the advantages as well as disadvantages of single chip and multiple chip configurations of such engine control systems.

It is expected that in the future, engine control systems will need to control and process even more items. Accordingly, more functions must be integrated on the microcomputer to be incorporated in such a system, and thus, the scale of the control program becomes large.

It is considered that an engine control system should consist, presently at least, of two LSI chips in order to improve the overall performance of the systems including cost. One chip has integrated memories and peripherals (such as I/Os and PWM outputs) and the other chip contains the CPU and peripherals (such as timers, interrupt controller, and A/D converters).

In this paper,we have described that the multiple chip configuration is better than that the single chip configuration.But this idea depends on the scale of the system.

In a case of the small scale system,the basic function of the CPU would be more simple,so the chip size of the microcomputer would be smaller than that of the large scale system,and we do not have to divide the function of the CPU into two.It means that the cost of the single chip configuration would be cheaper than that of the multiple chip configuration.

The day may come when the semiconductor technology will make a breakthrough reducing the chip size. Then, should demands for engine control systems with more functions not increase from the present level, it may be possible in the future for the high-end engine control systems to be effectively configured of a single LSI chip.

And there are a few possible methods of configuring an engine control system, which are not covered by this thesis. Multiple CPU configuration and ASICs (custom ICs) have the potential to effectively create a high-performance engine control system. These configuration methods are the subject of future study.

900928

Impact of Microcontroller Design on Software Quality

John Suchyta
Microcontroller Div.
Motorola, Inc.

Abstract

Microcontrollers (MCUs) are found in all types of vehicle electronics, from powertrain control to interior features. The microcontroller suppliers who are successful in the vehicle electronics industry know that quality must be designed into their products and processes. The quality of the final product is just as dependent on the design of the software as it is on the design of the hardware.

How does the design of the microcontroller affect the task of the software engineer in developing quality software? Microcontroller design considerations to assure quality start with the processor architecture and the peripheral functions which the MCU must perform, and extend to processing, packaging, and testing. Motorola microcontroller architectures and peripheral functions are examined to show how designed-in quality reduces the risk of error for the software engineer developing code.

Introduction

Microcontrollers are microcomputer systems on a single chip, or a minimum number of chips, that perform tasks in applications where size, power, and cost all must be kept to a minimum. These microcomputer systems consist of a microprocessor, memory, and parallell and/or serial I/O, and also often include a timer and other peripheral subsystems.

Motorola's MCU designers consider customer requests for CPU performance, peripheral features, memory mix, power dissipation, packaging, and cost when they generate the specification for a new product family or derivative. They are also required to consider wafer process technology, testability, development support, and reliability of the process and product. These items are straightforward to specify, design, measure, and/or guarantee.

Quality, on the other hand, is not easy to specify and measure. Of course, outgoing quality can be measured in the factory, but overall quality is determined by the end user and is measured by whether or not the product does what it was intended to do and doesn't do what wasn't intended. While this may not be easy to quantify, we can consider the lack of surprises in functionality and performance to be a basis for MCU quality.

The above items concern what is done to guarantee hardware quality, but what do we do to guarantee that the programmer does not get surprised by unexpected quirks in the product's design? Let's look at the basic system architectures of Motorola's MCUs and comment on the aspects that relate to software quality in a microcontroller.

8-Bit MCU Architectures

Motorola 8-bit microcontrollers (MCUs) consist mainly of the versatile M68HC05 and high performance M68HC11 families which are based on the original 8-bit MC6800 microprocessor core introduced in 1974. The MC6800 microprocessor features two 8-bit accumulators for arithmetic operations or data manipulation, a 16-bit index register for effective address creation or data storage, a 16-bit stack pointer for interrupt and subroutine service, a 16-bit program counter that contains the address of the next instruction, and a 6-bit condition code register for interrupt masking and results of the last arithmetic or data manipulation operation. Enhancements were made to this CPU core to aid MCU system flexibility and increase throughput in the M68HC05 and M68HC11 families. High density CMOS processing is now used to combine high speed with low power.

Since the M68HC05 family is targetted for low cost single chip MCU applications, the CPU register set was reduced to one 8-bit accumulator, an 8-

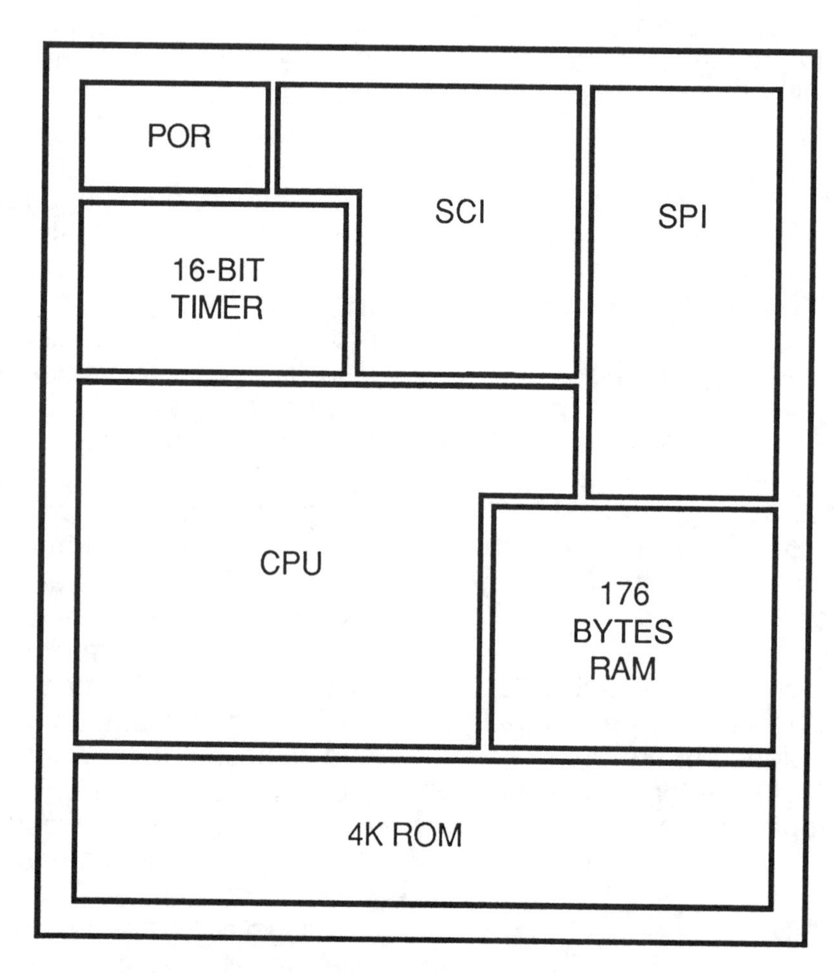

MC68HC05C4 BLOCK DIAGRAM

bit index register which can also be concatenated with the accumulator for unsigned multiplies, and reduced length stack pointer and program counter registers. (See M68HC05 programming model below.) The M68HC05 intruction set deleted or modified some M6800 instructions and incorporates new bit test/manipulation and other instructions for more byte-efficient program storage.

The M68HC11 family is targetted for both single chip and external memory based applications which require enhanced 8/16-bit arithmetic capabilities and larger memory sizes. To this end, the CPU register set was expanded from the MC6800 to include an extra index register, and the two 8-bit accumulators are concatenable for 16-bit operations. (See M68HC11 programming model below.) Many new instructions were added to take advantage of the 16-bit arithmetic capability and the extra index register.

Motorola intentionally started and stayed with the M6800 CPU architecture to provide upward and downward source code compatibility to system designers. Similarity in instruction set, code execution, and addressing modes provides a level of consistency and quality to the software engineer developing code for these machines. Likewise, the peripheral features integrated onto the M68HC05 and M68HC11 families are similar in operation, but can differ greatly in sophistication, even among family members.

M68HC05 And M68HC11 Hardware Features

The size and type of memory on board a single chip MCU directly affects system quality. Adequate program and data storage reduces the need for shortcuts or slick programming techniques which could compromise system integrity. Naturally, for simple control applications requiring few hardware features, a large amount of on-board ROM and RAM would be excessive and expensive. The M68HC05 family offers ROM to RAM ratios ranging from 12:1 to 45:1. This ratio is adequate for the M68HC05 family since the instruction set is tailored for 8-bit data and only five bytes of register information are stacked in RAM during interrupts. The M68HC11 family offers a maximum ratio of 32:1 of ROM to RAM, mainly because more RAM is necessary for 16-bit operations, and stacking operations store 9 bytes of register information at a time.

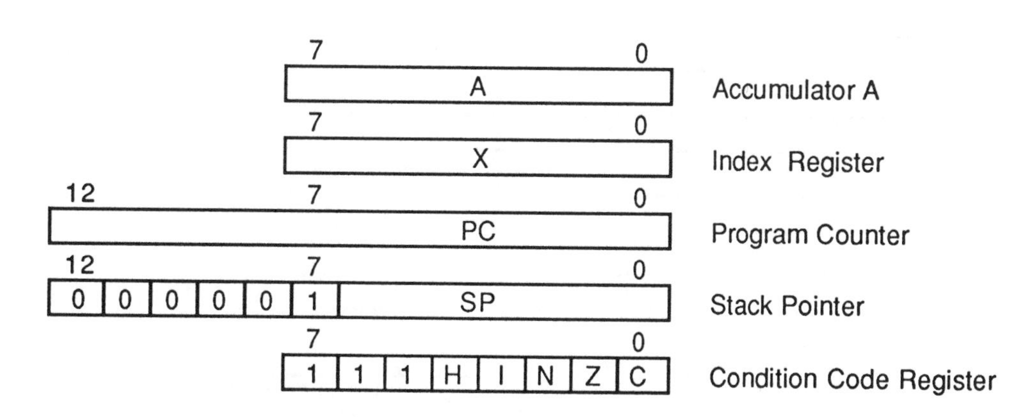

M68HC05 FAMILY PROGRAMMING MODEL

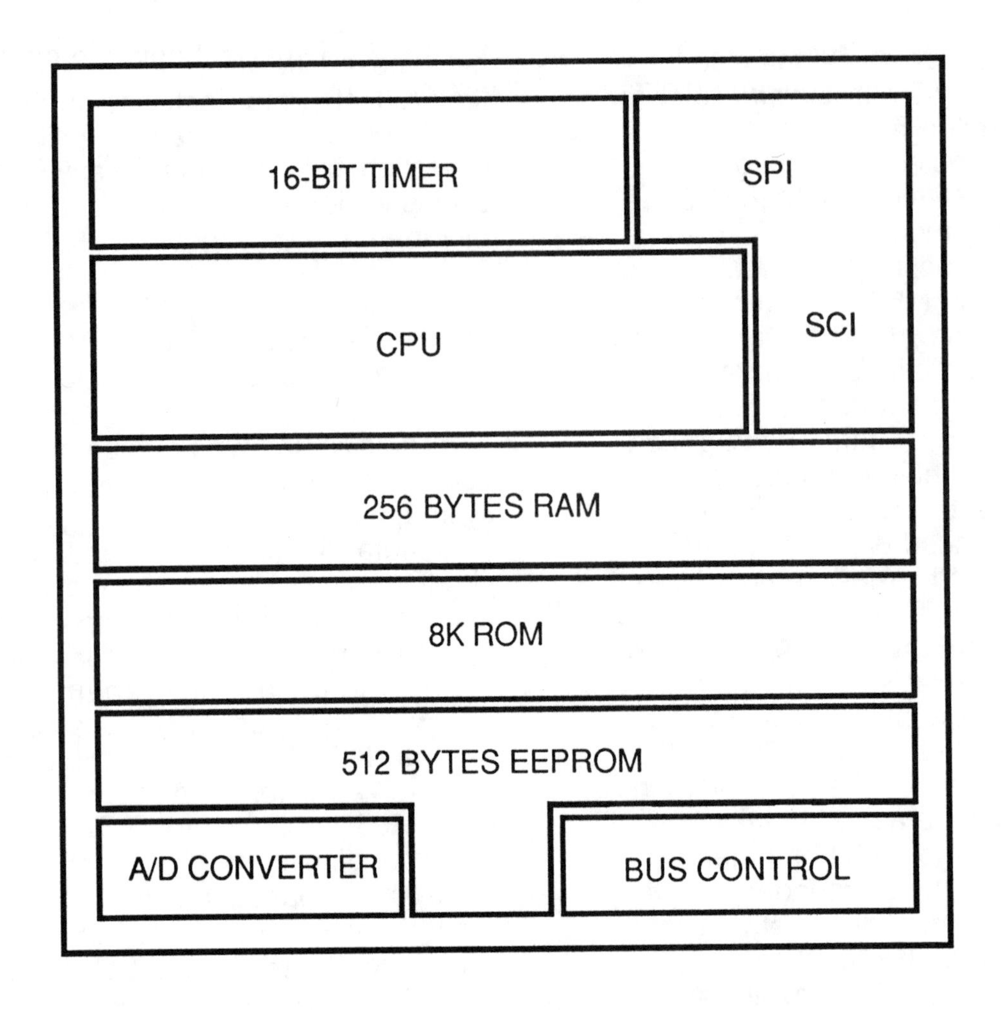

MC68HC11A8 BLOCK DIAGRAM

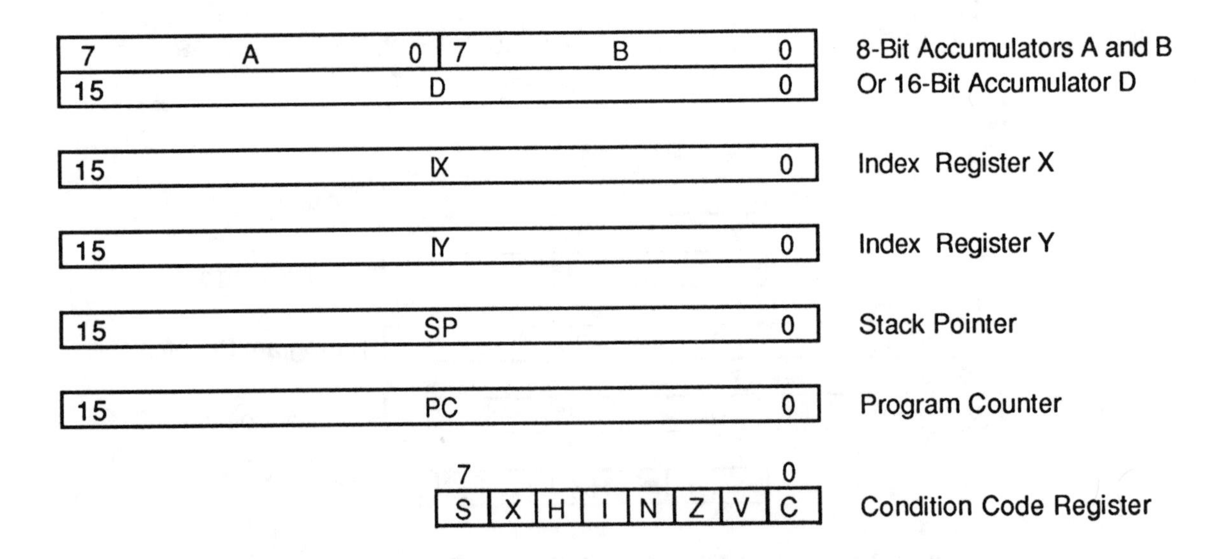

M68HC11 FAMILY PROGRAMMING MODEL

Generally, MCU subsystems such as timers, analog to digital and digital to analog converters, display drivers, EEPROM, watchdog timers, serial I/O, and some parallel I/O are implemented using control and status registers on chip which initiate, control, monitor, and terminate peripheral events. The M6800 CPU architecture treats these registers as memory locations, so special instructions are not necessary to activate MCU subsystems.

Control registers contain bits for subsystem enables, interrupt enables, rate or frequency control, and option enables. Status registers contain bits to monitor subsystem activity, so that the subsystems can be operated in either polled or interrupt driven systems. When using these registers in an interrupt driven system, interrupt vectors dedicated to the subsystem or even to individual subsystem features provide for a well structured software implementation. For example, separate interrupt vectors for each of the multiple output compares and input captures on the MC68HC11 obviate the need for polling the timer resources to determine the source of the interrupt.

Several versions of M68HC05 and M68HC11 derivatives offer on-board EPROM in place of ROM so that system development can be done on the actual hardware before committing code to a masked ROM product. These EPROM emulator devices are useful for low volume production and even initial production while a masked ROM device is being processed.

Motorola offers on-chip EEPROM on most M68HC11 and some M68HC05 derivatives for data storage and program execution. System personality information, user option settings, code patches, and self-calibration information can be stored in this non-volatile memory for added versatility, even after the system leaves the factory. Special bootstrap modes are available to load data or code during board assembly or in the field.

A watchdog timer is available on some M68HC05 and all M68HC11 devices. The watchdog, when enabled, resets the MCU in the event of software runaway. Programmable timeout periods, ranging from 16 milliseconds to several seconds, restart the M68HC05 machine, while a bidirectional reset pin pulls the reset line low on the M68HC11 and restarts the machine from a special reset vector.

The M68HC11 also includes an illegal opcode trap in the event software runs away. The trap vectors to a separate interrupt routine which can be set up to monitor illegal opcodes during development or in the application.

M68300 Family 32-Bit Microcontrollers

The M68300 family of 32-bit microcontrollers was recently introduced by Motorola. The M68300 family consists of a CPU core based on our 16/32-bit M68000 microprocessor family, a standardized internal bus, and modular, intelligent on-chip peripherals that plug into the bus. The goal of the M68300 family is to provide a high performance microcontroller by combining a powerful CPU and semi-autonomous peripheral features which require little intervention from the CPU. This combination enhances overall system performance since the CPU will spend much less time servicing interrupts.

The first member of the M68300 MCU family, the MC68332, starts with the 32-bit CPU core, called the CPU32, which is connected to the central Inter-Module Bus (IMB). MC68332 peripheral features include a powerful Time Processor Unit (TPU), an advanced Queued Serial Module (QSM), 2K bytes of RAM, and a highly integrated bus interface module called the System Integration Module (SIM) that eliminates virtually all glue logic for expanded memory systems. The MC68332 was designed for maximum performance at minimal power.

The CPU32 is a fully static 32-bit HCMOS CPU core based on the popular 16-bit MC68000 microprocessor and enhanced with features from the MC68010 and 32-bit MC68020 microprocessors. Multiply and divide execution times were reduced. A "Loop Mode" speeds execution of simple loops. Two new instructions, Low Power Stop (LPSTOP) and table interpolation (TABLE), were added to support microcontroller applications. The CPU32 maintains source code compatibility with the M68000 family, allowing the system designer to draw upon a broad base of available user software. High-level languages are efficiently supported by the M68000 family, so software development is faster, the software is more accurate, and is portable to other systems. Virtual memory is economically supported by an error recovery method called "instruction restart". Development support features implemented in the CPU32 include a trace on instruction execution and a breakpoint instruction which are used by traditional development support equipment, a

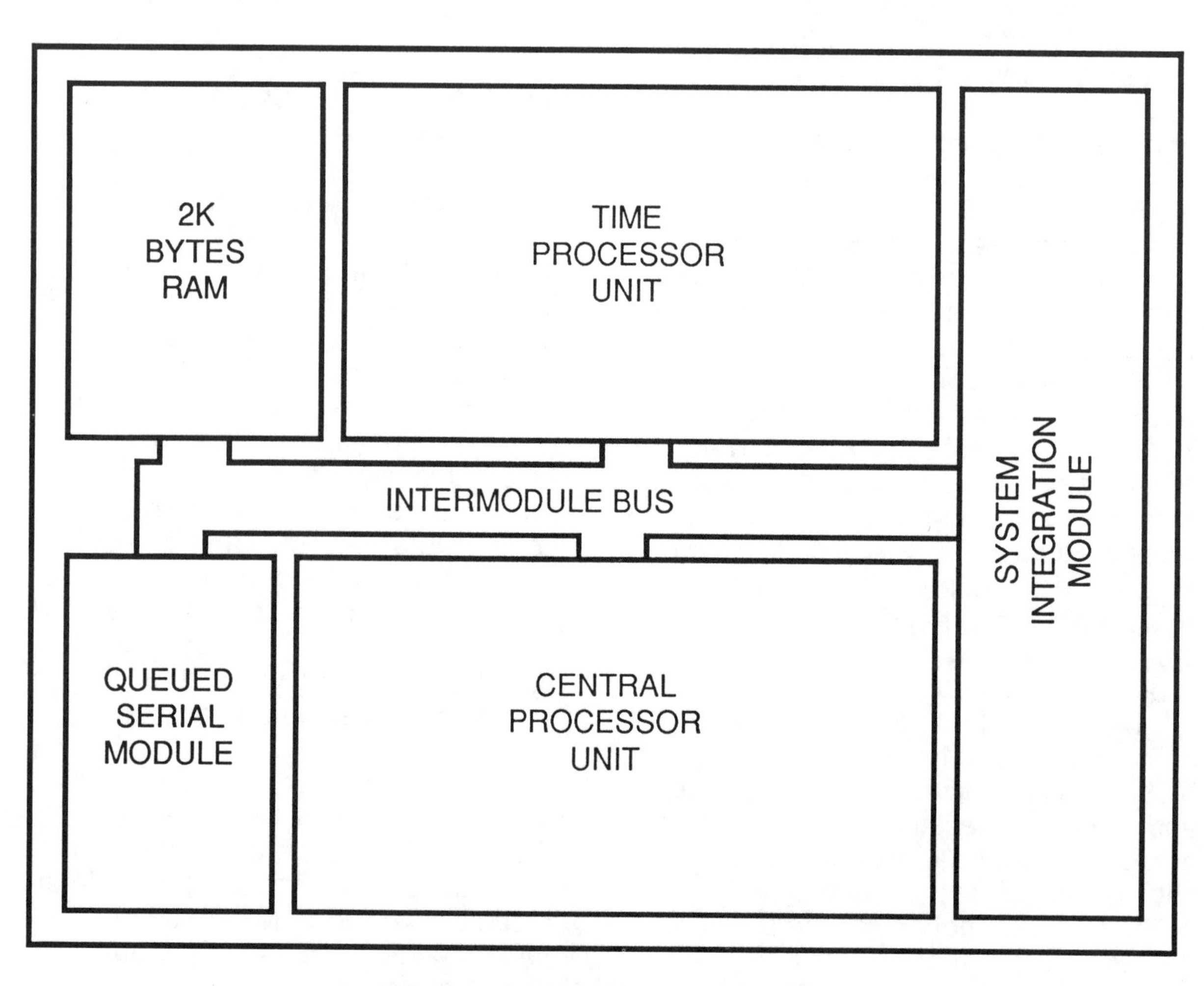

MC68322 BLOCK DIAGRAM

unique background debug mode which allows serial communication with a bus state analyzer to interrogate and modify processor activity without interfering with the bus, 2 pins which can be monitored to provide deterministic opcode tracking, and a pin and on-chip hardware which allow a breakpoint trap on any memory access, whether internal or external.

Two CPU32 programming models are shown below: one for the execution of non-privileged user programs and one for the privileged supervisor level system software. These two levels may exist in the same system to provide a memory map for an exclusive operating system which can control sensitive system features if necessary, and a separate memory map for normal application software. The CPU32 has 16 32-bit general purpose registers, a 32-bit program counter, a 32-bit supervisor stack pointer, a 16-bit status register, 2 alternate function code registers, and a 32-bit vector base register.

The register set of the CPU32 is different than that of the M68HC05 and M68HC11, but still performs similar functions. The 8 data registers (D0 - D7) are used in place of accumulators and can be used for bit, byte, word, long word, and quad word (64-bit) operations. The 7 address registers (A0 - A6) and user and supervisor stack pointers (A7 and A7') can be used as software stack pointers, base address registers, or for word and long word operations. All 16 data and address registers may also be used as index registers. The program counter (PC) is used to show the address of the next instruction to be executed. The status register (SR) contains the standard condition code register and, in the privileged supervisor mode, additional priority and control bits. The vector base register (VBR) contains the base address of the exception vector table (256 separate vectors) so that this table can be placed conveniently in memory. The 2 3-bit alternate function code registers (SFC and DFC) are used to extend the 24-bit address bus to define user and supervisor program and data spaces.

The Time Processor Unit (TPU) subsystem of the MC68332 is a sophisticated timer system which contains its own dedicated microprocessor to control time functions with little intervention by the CPU32. This 2 time base, 16-bit intelligent timer is initialized by the CPU through a host interface unit which contains registers to configure, control, and monitor the timer resources. The host interface also contains a dual ported parameter RAM which contains control and data information for each of the 16 channels. The host interface drives a scheduler and microengine which synthesize the requested time functions and provide interrupts to the CPU32 only when necessary. The microengine has an execution unit that can perform timer event calculations which would otherwise burden the CPU32. Each timer channel can perform any of the microcoded time functions (input capture, input transition, output compare, pulse-width modulation, period measurement, position synchronized pulse generation, stepper motor control, and period & pulse-width accumulation). Channels can be linked to perform complex time functions based on events occurring on another channel. Custom time functions can be developed and emulated by the software engineer by using the emulation mode of the TPU and the 2K byte RAM module which is connected to the TPU's microengine while in this mode.

The TPU's capability to execute repetitious, cascaded, and event

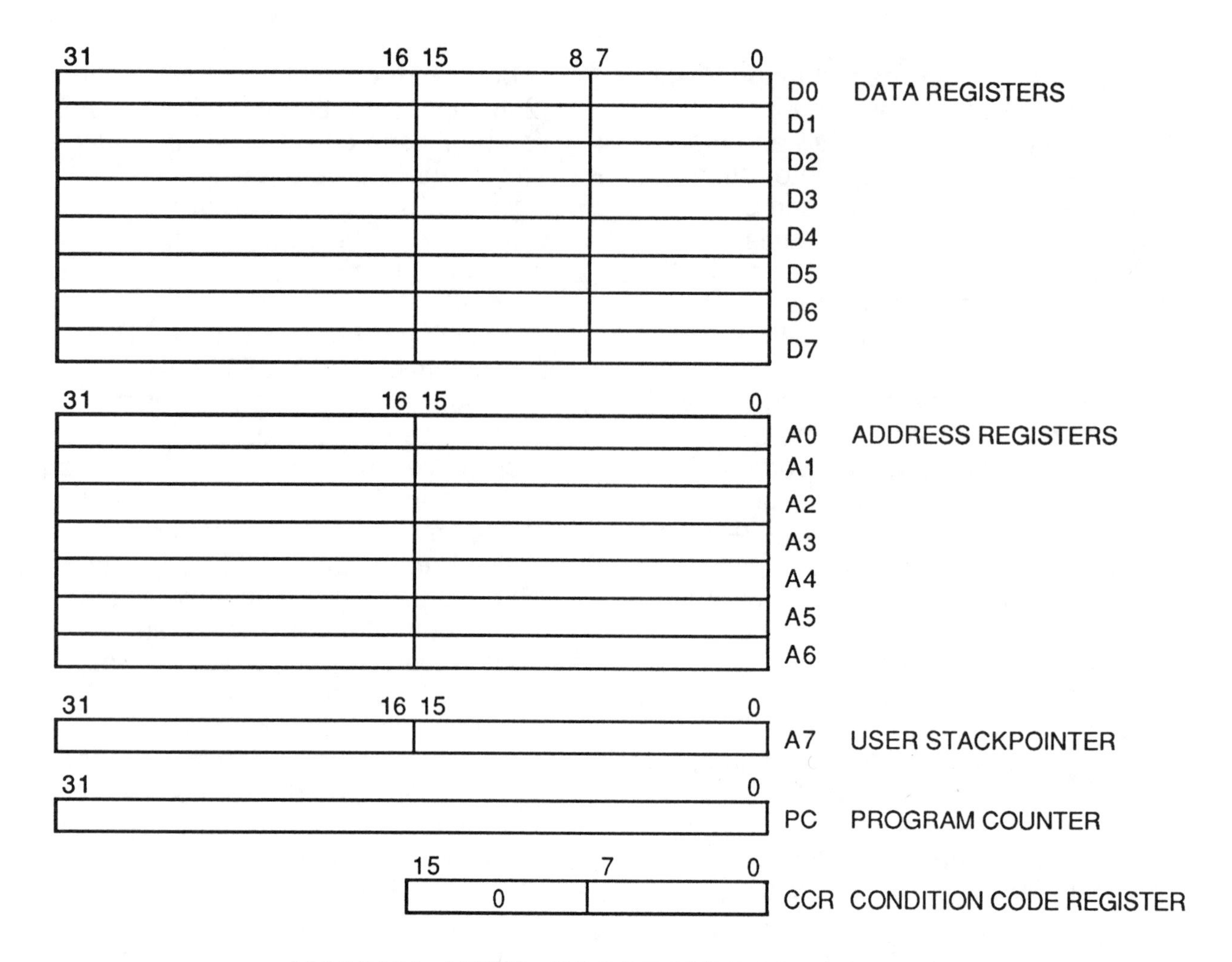

MC68332 USER PROGRAMMING MODEL

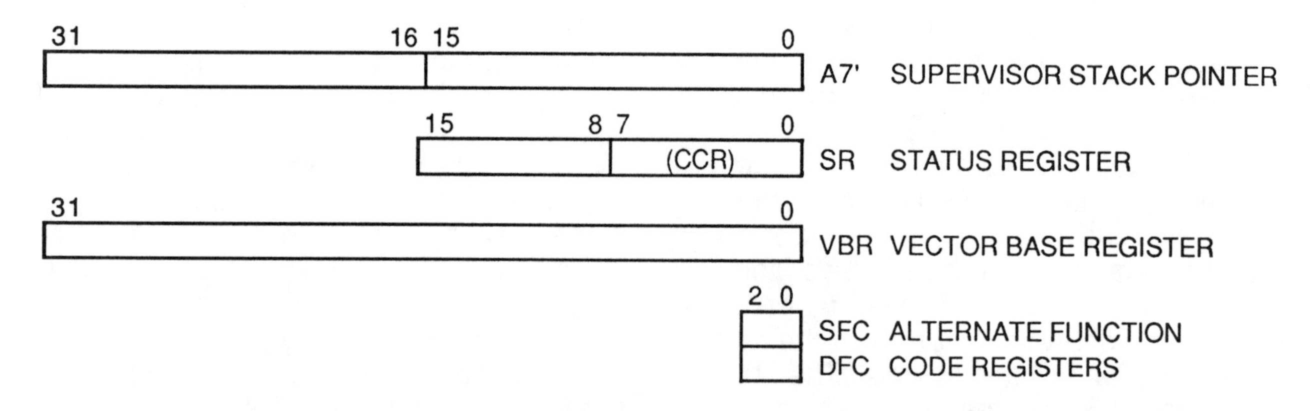

MC68332 SUPERVISOR PROGRAMMING MODEL SUPPLEMENT

triggered timing signals independent of the CPU allows the software designer to concentrate on other system throughput issues.

The Queued Serial Module (QSM) of the MC68332 is another advanced peripheral subsystem which contains an enhanced asynchronous Serial Communications Interface (SCI) and a programmable, high speed, synchronous Queued Serial Peripheral Interface (QSPI). The SCI offers a 13-bit programmable baud rate generator for data rates up to 524K baud, software selectable 8 or 9 bit word length, full and half-duplex operation, parity generation and detection, an idle line "wake-up" feature, and an advanced signal noise detection circuit to aid in filtering data in a noisy environment. The QSPI communicates over a 3-wire bus and includes 4 peripheral select pins to choose 4 (or 16 in a decoded mode) external synchronous serial peripheral devices.

The purpose of a "queued" SPI is to provide a programmable buffer so that up to 16 serial transfers or a data stream transmission of up to 256 bits can be accomplished with little or no CPU intervention. As with the TPU, after CPU initialization of control registers, the QSPI can operate without interrupting the CPU as each new data is transferred. The CPU can update new values and read the latest values in the buffer at any time without servicing QSPI resources. This semi-autonomous feature of the QSM reduces the throughput restrictions inherent in conventionally implemented serial subsystems.

The QSPI also operates in a slave mode to facilitate interprocessor communications.

The MC68332's System Integration Module (SIM) provides the external parallel interface to memory and peripherals, a programmable system clock, programmable chip selects, and system protection features. The external bus interface consists of 16 data bus lines, 24 address lines, address and data strobes, and other bus control signals to allow reading and writing of dynamically sized (byte, word, long word) data. The system clock feature is unique in that a 32 KHz watch crystal can be used to synthesize the 16.7 MHz (or slower) system clock. The 12 programmable chip selects feature a bootstrap chip select for initiating program execution out of reset, programmable block sizes, byte and word wide memory enables, read and/or write capability, synchronization to address and data strobes, and programmable wait states for slower memories and peripherals.

The system protection features in the SIM include a standard software watchdog timer to reset the chip if the device is lost in a loop. Also on-board is a hardware watchdog which initiates exception processing if an external device fails to respond in time to bus requests. A spurious interrupt monitor initiates a bus retry sequence when none of the external peripherals responds to an interrupt acknowledge. A periodic interrupt timer is part of the SIM, rather than the TPU, so that a real-time clock function can be implemented independent of the powerful TPU.

The MC68332's System Integration Module eliminates the need for glue logic in expanded memory systems. The software engineer can switch between byte and word wide data, on the fly. The ability to partition blocks of data space for read and write access reduces the likelihood of data corruption in runaway conditions. Programmable wait states and self-generated output and write enable signals allow the system designer to interface to peripherals with wide ranging access times without external logic.

As testament to the utility of chip selects and external bus interface, Motorola offers a low cost development system for the MC68332 called the Business Card Computer (BCC). The BCC is a complete computer system that contains the MC68332, a monitor program stored in a 128K EPROM, 64K bytes of RAM, and a serial communications line driver on a printed circuit board just a little larger than a standard business card. No other logic gates are necessary. That's a total of 5 chips for a complete system including memory!

The on-chip 2K byte static RAM can be read or written as byte, word, or long word data. The RAM would typically be used for stack space or data storage. Program execution from the internal RAM is fast since CPU accesses only take 2 clock cycles (normally 3 clock cycles per bus cycle externally except for fast termination cycles). The RAM has a battery backup pin for those applications which need to save data during power-down.

Summary

Motorola's microcontrollers are designed to provide the system designer a wide range of solutions for practically any control application. From the simplest serial I/O task to complex time functions, Motorola has an MCU with the appropriate processing power, memory mix, and I/O capability.

The components evaluated in this discussion, from CPU registers to peripheral subsystem features, show that great consideration is given to the design of the hardware functions of the microcontroller families. In all cases, this same consideration is given to how the user effectively and efficiently utilizes the functions from a software standpoint. The efficient instruction sets, similarity of CPU and peripheral system operation, adequate on-board memory types and sizes, orthogonality of I/O and memory, and versatile system protection features of the M68HC05 and M68HC11 families are the key to designing low risk system software. The M68300 MCU family combines a popular and powerful CPU with low maintenance, intelligent peripherals to eliminate throughput bottlenecks. Advanced development support tools, built into the M68300, further reduce the chance of error during system design verification.

References

Greenfield, Joseph D. and Wray, William C. *Using Microprocessors and Microcomputers: The Motorola Family, Second Edition* , Wiley, 1988

Motorola Inc. *MC68332 System Integration Module User's Manual,* 1989

Motorola Inc. *Microprocessor, Microcontroller and Peripheral Data, Volumes I & II,* 1988

Motorola Inc. *M68HC11 Reference Manual,* Prentice Hall, 1989

Peatman, John B. *Design With Microcontrollers,* McGraw-Hill, 1988

901123

Sensing and Systems Aspects of Fault Tolerant Electronics Applied to Vehicle Systems

Milt Baker
Motorola Inc.
Automotive and Industrial Group

Randy Frank
Motorola Inc.
Semiconductor Products Sector

Larry C. Puhl, Ezzy A. Dabbish, and Michael Danielsen
Motorola Inc.
Chicago Corporate R&D Center

ABSTRACT

New approaches to the design of automotive electronic systems can be used to achieve reliability and cost objectives for future vehicles. Present systems can benefit by applying fault tolerant design concepts. This paper is in two sections. The first section discusses the application of fault tolerant concepts to ECU design. The second portion covers the important role played by sensors in fault tolerant system designs.

SYSTEM ASPECTS OF FAULT TOLERANT ELECTRONICS

From its inception automotive electronics has achieved an outstanding growth rate. For the decades of the 70's and 80's, automotive electronics achieved wide-scale use. The power of microprocessing was harnessed to control basic internal combustion engine parameters and thereby increase efficiency and reduce pollution. Automotive electronics has also been employed to increase driver comfort and convenience in display, communications, and other functions.

The decade of the 90's may well be the decade where automotive electronics is utilized in high volume to improve the safety of the vehicle. Air bags and anti-lock braking systems are predicted to achieve high penetration rates in the 90's. Traction control will further enhance driver safety in the mid-90's. Toward the end of the decade collision avoidance systems may be possible. Into the new century automated headway designs will require systems that achieve very high levels of integrity.

With the introduction of high volume safety systems, designers of these systems must pay significant attention to what happens when a system malfunctions.

Reliability the Primary Focus

The prime focus for all automotive systems will continue to be to increase system reliability and strive toward a goal of zero failures for the life of the vehicle. However, as long as practical systems exist, malfunctions will occur. How these systems sense and then respond to malfunctions will always be an important design consideration.

Diagnostics Today's Focus

Today's approach to system malfunction is to build a high level of self-diagnostics into automotive electronic systems. If a malfunction occurs, the system is managed into a "fail safe" condition, which disables the control function, and provides some form of warning to the driver. Today's diagnostics systems are extensive; in some systems, hardware and software for diagnostics can approach 50% of total system cost. Some systems employ dual microprocessors to check each other and the entire system for faults.

For this significant investment in hardware and software the system benefit may not be obvious to the driver. If the system fails, the operator is without critical safety features.

A New Paradigm - Fault Tolerance

With automotive systems, uptime is a critical parameter as well as reliability. Uptime is defined as the amount of the time the system is operational. Although failure rates are low for today's automotive systems, a driver experiencing a malfunction will be without the feature until repairs can be made. A simple illustration of this concept as applied to aircraft is the single engine verses the twin engine aircraft analogy. Although the single engine aircraft may have high engine reliability, when the engine fails, actions must be taken immediately. A twin engine aircraft may have engines with the same reliability as the single engine aircraft but achieve higher uptimes during critical maneuvers.

The notion of systems that can continue to perform even though a malfunction has occurred is called fault tolerance. Fault tolerance is not a new concept. Fault tolerance is really nothing more than the concept of redundancy applied to systems. When a component fails, another takes its place and the system continues to perform. The typical implementation of fault tolerance is with three computer systems (see Figure 1) capable of switching from a main to a back up computer in a failure mode [1].

Fault tolerance is not new to automotive controls. For example, some manufacturers' engine control, control spark from a main ECU, but contain a back up spark control ECU in the distributor in the event the main processor fails.

Fault tolerant systems have generally been considered too costly for widespread system application in automotive. However, new technology discussed later may alter this conclusion.

Fault Tolerance - The Next Generation

It is the contention of this paper that systems that can perform even in the presence of a failure offer significant advantages over systems which only have a diagnostic warning.

FAULT TOLERANCE PRACTICAL IMPLEMENTATION

The challenge of fault tolerance is cost effective implementation. Technology changes may alter the perception that fault tolerance is too expensive for widespread automotive application. Three changes may cause a reexamination of the cost premise: system integration, semiconductor technology advances and natural system redundancy.

System Integration

The trend in systems is to reduce stand alone controllers to system controllers. An example of this is the integration of engine control and transmission control into one unit. As the number of controllers consolidate, fewer higher power microprocessors will be handling integrated functions. This sharing of electronics may significantly lower cost. Fault tolerant electronics may provide cost effective solutions for highly integrated control functions.

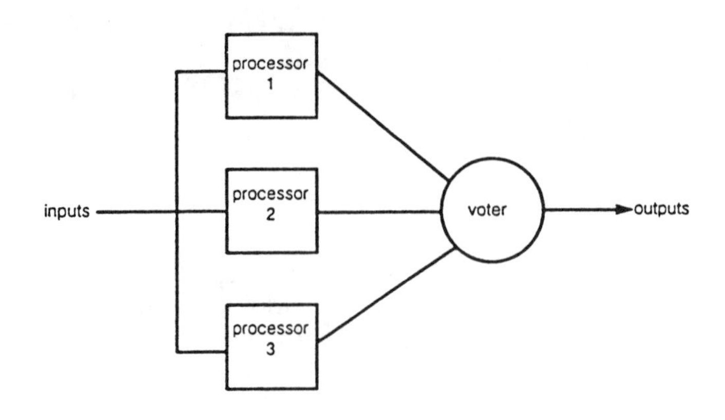

Figure 1 - Tripole Modular Redundancy System.

Semiconductor Technology

It is widely understood that the number of circuits that can be put on a semiconductor chip is increasing every year. As an example, today's one (1) megabit DRAM memory chip is expected to evolve to 64 megabits by 1995. As the number of circuits/chip increase, there is a corresponding reduction in the cost per function.

This decreasing cost function can be used to build redundancy into chips used for automotive control. New technology applying fault analyzing hardware on chip can contribute to fault tolerant ECU's. Motorola has developed technology for a self-diagnostic microprocessor which is a building block in a fault tolerant system.

Natural System Redundancy

The third trend which causes re-examination of the cost assumptions associated with fault tolerance is the natural system redundancy built into today's safety systems. Take for example anti-lock braking or active suspensions. Each system may have a sensor and an actuator deployed at each wheel. This redundancy can be utilized to operate the system on less than the full complement of sensors or actuators in the event of a malfunction. A four-wheel ABS system could be made to perform as a three or two-wheel ABS. Of course, the safety of this configuration would have to be carefully evaluated. Today, however, two-wheel anti-lock systems are in high volume production.

The Fault Tolerant Microprocessor

A key to helping a multiple sensor/actuator system remain operational is to assure that the microprocessor is operational. Although a system may be able to function with a defective sensor or actuator, a malfunctioning CPU will take the system down.

Although the microprocessor may be the highest reliability component in the system, the consequence of failure is high. No operation can take place once the microprocessor has failed.

Motorola has focused its efforts in this area on the technology for a fault tolerant microcomputer system to assure that the microprocessor is always operational. The fault tolerant microprocessor is essentially two self-diagnostic microprocessors. This technology is a derivative of Motorola's considerable expertise in secure communication systems where no failure of the encoding can be tolerated for security reasons.

FAULT TOLERANT MICROCONTROLLER REQUIREMENTS

A fault tolerant microcontroller must be designed to continue safe operation after a single failure. A microprocessor is deemed to be safe if the hardware and software are correctly executing the intended function. Maintaining safe operation requires both reliability and fault detection. Motorola has significantly improved the reliability of microprocessor hardware and software through their "6-sigma" quality control program. The goal of this program is to make the statistical probability of a failure to less than 3.4 parts per million. Safety is provided by detecting the failures that still occur and maintaining safe operation in the presence of a single failure.

When a failure occurs in most of the current automotive systems, continued operation occurs in a safe but degraded manner. An example of this is a fault in the electronic engine controller causing the system to operate in a limp home mode. Often, ensuring safety results in lower reliability due to the increase in complexity. Fault tolerance can be used to improve the safety and reliability at the same time.

The classic example of a fault tolerant system is one that incorporates triple modular redundancy and is called TMR for short. A TMR system contains three identical processors executing the same function. An additional piece of hardware called the voter, determines the correct output by comparing the outputs of all three processors. If a single processor fails, the two good processors will still out vote it allowing non-degraded system performance after a single failure. A TMR system with a single failure is not safe, however, because if a second failure occurs, the bad processors could out vote the single remaining good processor. A TMR system can be made safe by increasing the complexity of the voter to make it adaptive. When the voter detects a mismatch between the three processors, it removes the faulty processor from the system and reverts to a dual processor comparison system. While the classic TMR system hides errors from the system through voting, the adaptive system removes the faulty unit from the system resulting in higher system reliability and an increased meantime to failure [1].

The TMR approach is a traditional method to provide fault tolerance because it does not rely on any special error detection hardware. Another approach is to use two special processors which contain extensive error detection capabilities. In this system, shown in Figure 2, only two processors are needed, because the arbitration hardware can determine which processor is bad after a single failure. After removing the faulty processor, the system is still safe because the error detectors will detect a fault in the remaining processor (allowing a safe shutdown of the system).

At this point, one must make a decision as to which approach towards fault tolerance will be taken. The TMR approach is simple but expensive because it requires three processors. The dual processor approach is more involved due to the complexity of the error detectors. Error detection is difficult because of the many ways in which errors can be created. Errors can be caused by transient or permanent hardware faults and design errors. Design errors occur when the system does not work as it was intended. This can be hard to detect since the system is working correctly according to its specification. Since removing all bugs from a complex system is still a formidable task, many systems incorporate reasonability checkers to detect unintended behavior of the system. Reasonability checking is normally performed by a second processor in the system called

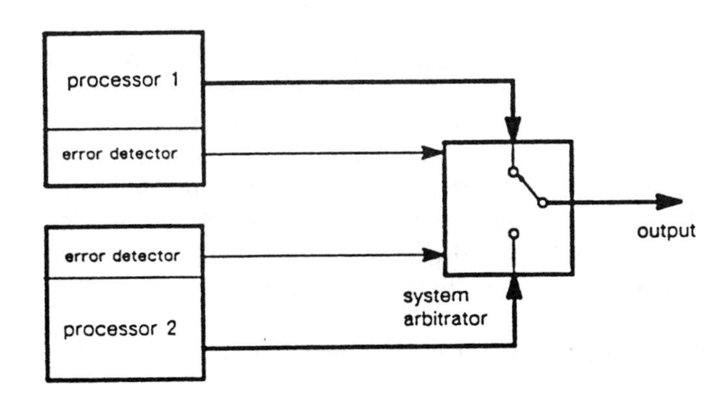

Figure 2 - A Dual Processor System.

the watchdog processor [2]. It should be noted that the TMR system only checks for hardware errors. To be safe, watchdog processors would have to be added to the TMR system. For this reason, we will continue with the dual processor design approach.

Before we examine the specifics of our dual processor architecture, it would be helpful to mention some of the problems that can occur in a dual processor system. Since the system reliability is dependent on the ability to detect errors and correctly choose which processor controls the system, the design of the error detectors and arbitration hardware is very important. A fault in the error detection circuitry could cause the system to be unsafe if it went unnoticed. The fault in the error detectors could cause other errors to go undetected or prevent the indication of error conditions to the system arbitrator. In order to provide a safe system, the arbitrator needs to know: Is the system OK and can the system tell if it is OK?

Fault Detection Using the Lock and Key Concept

A lock and key concept has been developed to address the requirements given above. Fault detection circuits and software are used to determine that a microcontroller is working correctly. Information that proves the microcontroller is working correctly is used to generate a continuous "key". The design is such that the key can only be generated if the microcontroller is working correctly. A separate "lock" function determines that the "key" is valid and allows the microcontroller to control the system. The lock function is designed to be simple and therefore very reliable. Since the lock function is critical to safe operation it should be designed using both fault detection and fault tolerance concepts. The simplicity of the lock circuitry considerably simplifies meeting these requirements.

Figure 3 shows a high level diagram of a lock and key system. In this system a microcontroller has to provide a key to the lock in order to control the output. The keys reflect the integrity of the microcontroller, proving to the lock circuit that the microcontroller is sane. If one microcontroller goes bad, it can no longer provide correct keys and therefore cannot control the output. The other microcontroller can continue to affect the outputs of the system. After the first microcontroller fails, the second microcontroller must continue to prove its sanity to the lock mechanism, before it can write the output.

Full Fault Tolerant Architecture

Figure 4 shows the architecture recommended for fault tolerant operation. The architecture consists of two fault detecting microcontroller and one lock

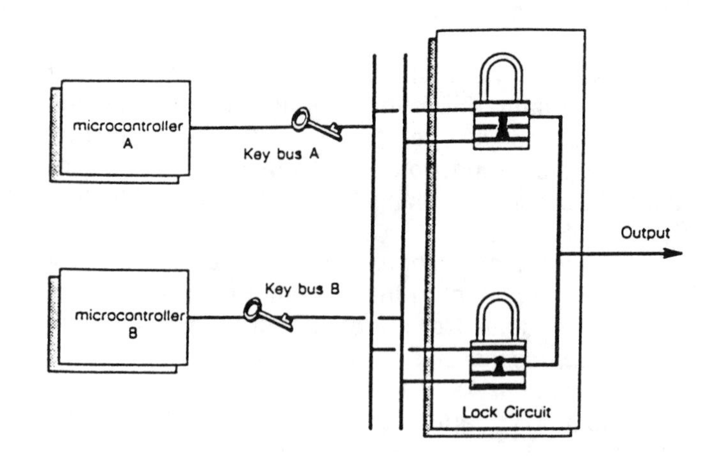

Figure 3 - A Lock and Key System.

circuit. The lock circuit will only allow a working microcontroller to control the system. This architecture will maintain safe operation after any single failure

Each microcontroller circuit consists of two processor systems: a main processor system and a watchdog processor system. The main processor system performs the main algorithm calculations, some self-checking, and output circuit checking. The

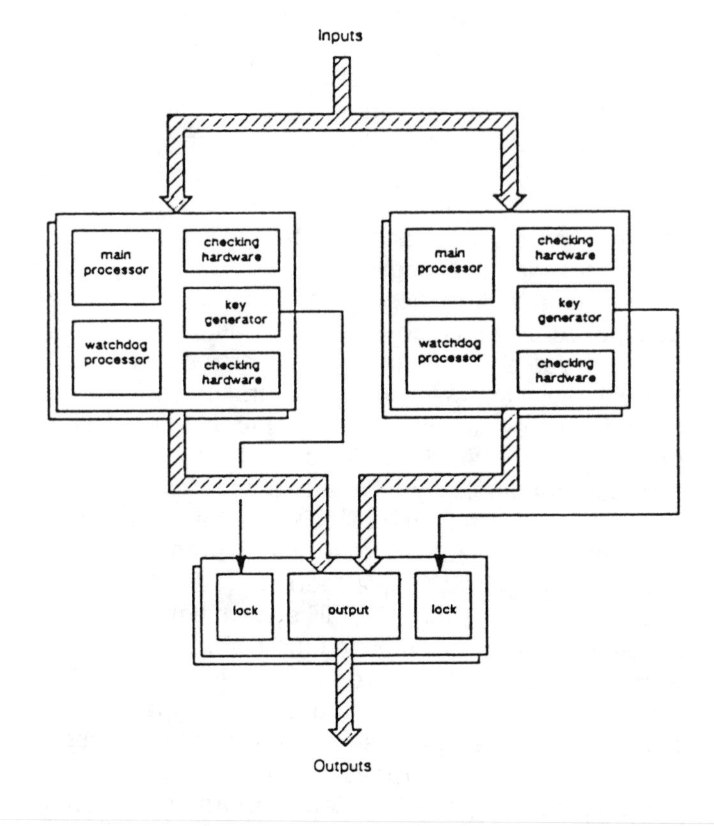

Figure 4 - Fault Tolerant Architecture for Automotive Applications.

second processor is used to check the main processor. This processor would be smaller than the main processor and have less memory associated with it. Both processors have error detection circuitry associated with them and both sets of error detectors feed into the key generator. The key generator, therefore, would indicate the sanity of both processors.

A more economical system could be built using one of the fault tolerant microcontrollers and a lock chip, as shown in Figure 5. This system would provide the same margin of safety and reliability as today's diagnostic system using fewer parts. Although this system would not provide fault tolerance, it would ensure safe operation. The lock and key concept provides a more economical solution by allowing the main processor and watchdog processor to be on the same integrated circuit.

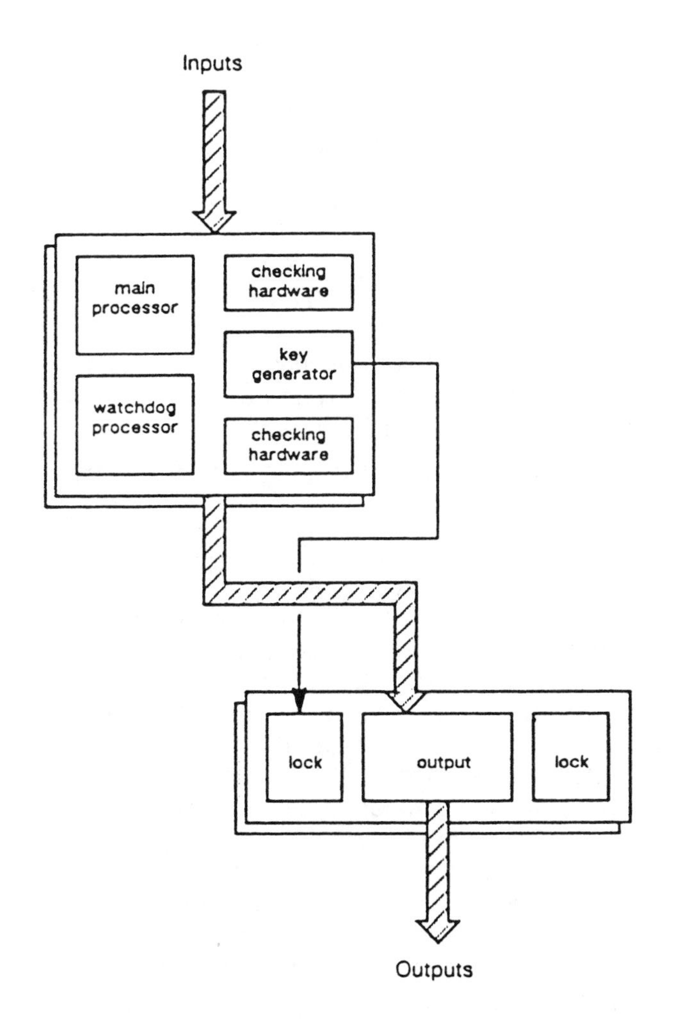

Figure 5 - A Diagnostic Architecture for Automotive Applications.

FAULT TOLERANCE - THE FOUNDATION FOR DESIGNING NEW SYSTEMS

By designing systems that continue to perform when faults occur, automotive manufacturers make full usage of the potential that electronic controls offer and provide their customers a significant improvement over a limp-home mode. With the foundation of fault tolerant systems approach, the addition of other system components that provide an improved level of performance and integration of additional functions can mean a significant change in the total system performance and reliability that can be achieved. The input that semiconductor sensing technology can provide to a fault tolerant system will be the focus of the remainder of this paper.

SENSING ASPECTS OF FAULT TOLERANT ELECTRONICS

One of the keys to fault tolerant design is the sensing of critical system parameters. Sensors provide information necessary for proper system functioning and can provide diagnostics to allow fault mode operating strategies, driver indication of a fault condition as well as improve vehicle servicing and repairs.

A variety of sensing technologies are available and new technologies are being developed to meet existing and future automotive system needs. Some sensors are even integrated into other semiconductor devices such as control IC's or smart power transistors. This section will examine: the sensing requirements of the air bag system and the integration of sensing in power devices for fault detection.

SENSING REQUIREMENTS IN AN AIR BAG SYSTEM

An inflatible restraint (air bag) system is one of the more important new vehicle systems designed to address the need for improved vehicle safety and satisfy the legislated requirements of FMVSS208. Air bag systems utilize as many as five accelerometers (crash sensors) to provide information to an ECU (electronic control unit). The crash sensors are positioned at strategic locations in the front of the vehicle and a low-g arming sensor is located in the passenger compartment. When a crash occurs, the series arming sensor is closed and the rapid deceleration produces a short duration signal from the crash sensor. If this signal is above the calibrated indication of a crash event (equivalent to 12 mph into a wall) it is allowed to pass the trigger current necessary to ignite the propellant that inflates the bag.

Present systems utilize electromechanical and piezoelectric sensors to provide the air bag deployment signal. These system have an excellent record, however, the system costs about $400 to $500 for the driver side only system and is estimated to cost from $800 to $1000 when passenger side units are added in 1992. A significant portion of the cost is attributed to the sensors. This situation is surprisingly similar to the history of silicon pressure sensors that measure manifold pressure (MAP) in automotive engine control systems. Micromachined silicon sensors that are used extensively to perform the sensing function today replaced technologies such as LVDT (linear variable differential transformer) and ceramic capacitive sensors that were used in some of the first engine control systems. In addition to meeting the system performance levels, reliability requirements and cost objectives of automotive applications that were achieved by MAP sensors, semiconductor accelerometers are being pursued for the capability to provide a self-test (diagnostic) feature, reduce the number sensors in the system and their potential for increased integration.

Mechanical Crash Sensor

One version of the crash sensor that is used on production vehicles is based on a design for artillery munition fuzes [3]. The sensor (Figure 6) consists of a ball, as the sensing mass, in a cylinder to produce a single axis for motion. The ball is positioned against a spring-loaded lever. Sufficient vehicle deceleration allows the ball to compress the spring, moving the lever which releases a spring-loaded firing pin. The firing pin activates a percussion cap detonator.

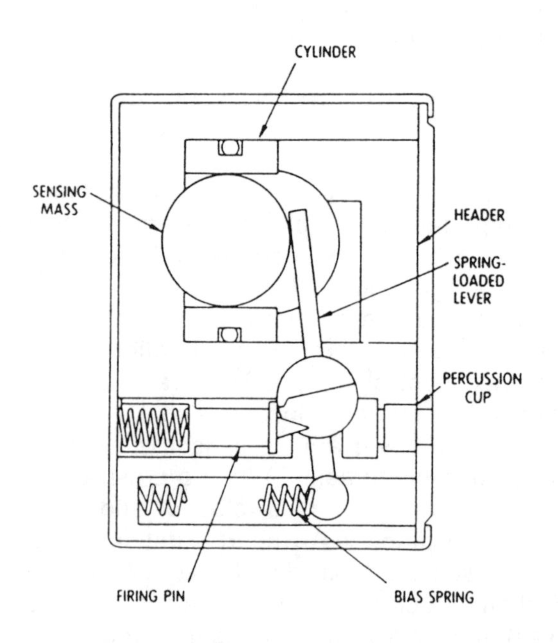

Figure 6 - Mechanical Accelerometer Currently Used as Crash Sensors.

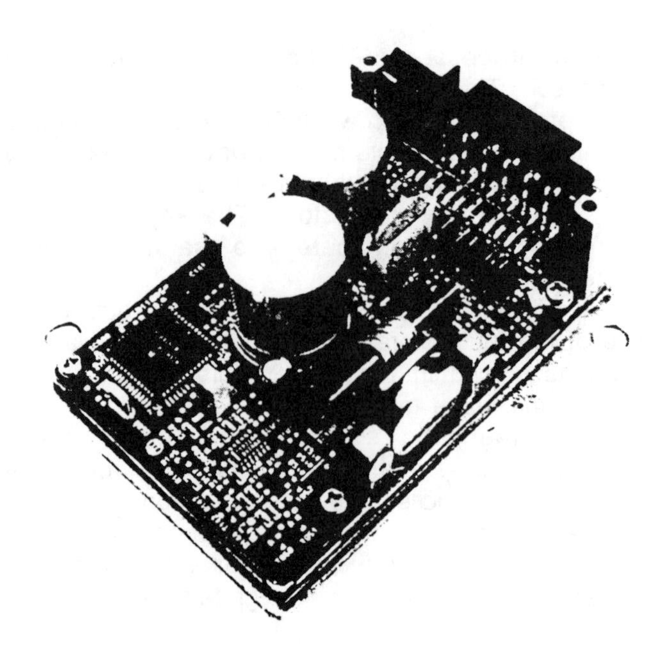

Figure 7 - Piezoelectric (Ceramic) Accelerometer.

Proper sensor operation requires precision mechanical components and minimal friction. The action of this sensor is a single-shot mechanism which requires a reset operation and firing pin replacement once it is activated.

Piezoelectric Accelerometer

Ceramic and thin film piezoelectric sensors overcome some of the cost and operational problems associated with mechanical crash sensors. They are analog sensors and not switches like the mechanical units. They can operate through a crash level input signal without requiring a reset. Figure 7 show a piezoelectric sensor currently used on production vehicles [4].

Silicon Accelerometers

Micromachining techniques utilized in the manufacturing of piezoresistive pressure sensors such as the MAP sensor have lead to the development of silicon accelerometers. Micromachining is a chemical etching process that is consistent with semiconductor processing techniques. Several wafers in a wafer lot can be fabricated at the same time and lot-to-lot control can be maintained by controlling a minimal number of parameters. Key parameters are crystallographic orientation, etchant, etchant concentration, semiconductor starting material, temperature and time. Figure 8 shows the similarities in the two structures. The piezoresistive accelerometer with beams and suspended mass is considerably more complex than the simple diaphragm used for most pressure sensors. A three

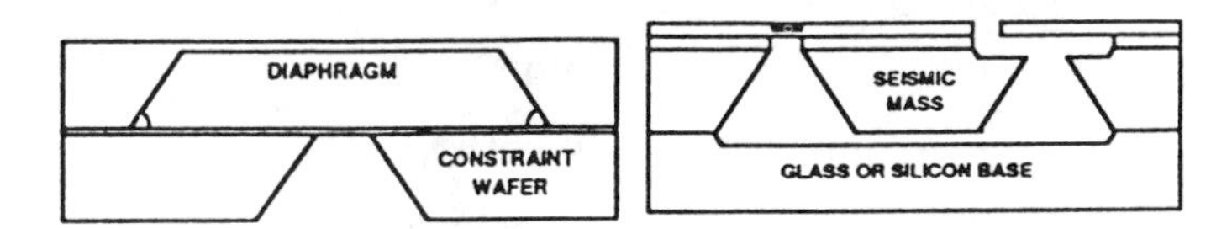

Figure 8 - Micromachined Piezoresistive Accelerometer Compared to Micromachined Piezoresistive Pressure Sensor.

layer structure is necessary to achieve over-range stops, bending beam and air damping.

Figure 9 [5] shows a piezoresistive silicon accelerometer with micromachined beams and suspended mass. Micromachining allows a very precise mass and support structure to be consistently produced in a batch process with several thousand device in a single wafer lot. The piezoresistive element(s) of the strain gage are ion implanted into the suspension arm for maximum sensitivity. The output is in mV/g and the resistive elements are temperature sensitive so additional circuitry is required to provide calibration, temperature compensation and allow interface to external circuits.

Capacitive approaches for semiconductor accelerometers are also possible such as the structure shown in Figure 10 [6]. Electroplated metal sense elements are fabricated partially on the top of a sacrificial layer. Selective chemical etching is used to produce two capacitors in asymmetrically shaped structures. Acceleration perpendicular to the substrate surface causes the sense elements to rotate around the torsion bars causing the capacitance of one side to increase and the capacitance of the other side to decrease. The output (in fF/g) is converted into a frequency that is proportional to acceleration that can be directly interfaced to a microcontrol unit.

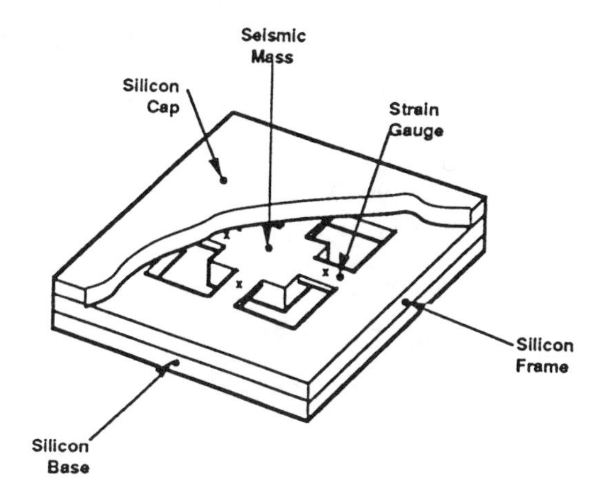

Figure 9 - Piezoresistive Accelerometer.

One of the major advantages of micromachined silicon accelerometers is the ability to integrate self test features. As shown in Figure 11, an electrostatic field applied to an integral capacitor built into a silicon accelerometer can be used to deflect the beam and provide an indication that the accelerometer is working properly and even check the calibration.

The ability to check the calibration can be used to test the accelerometer in wafer form, packaged form and in the final assembly. In the application, the device can be tested for drift and lifetime measurements without having to apply external forces or the occurrence of a crash. A comparison of the piezoelectric technology to piezoresistive and capacitive semiconductor units is shown in Table 1.

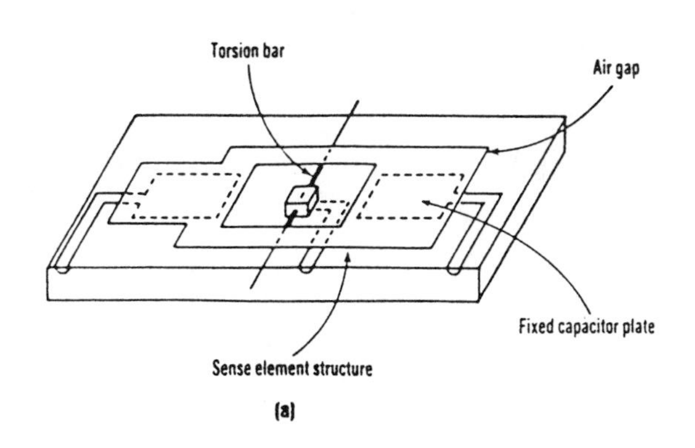

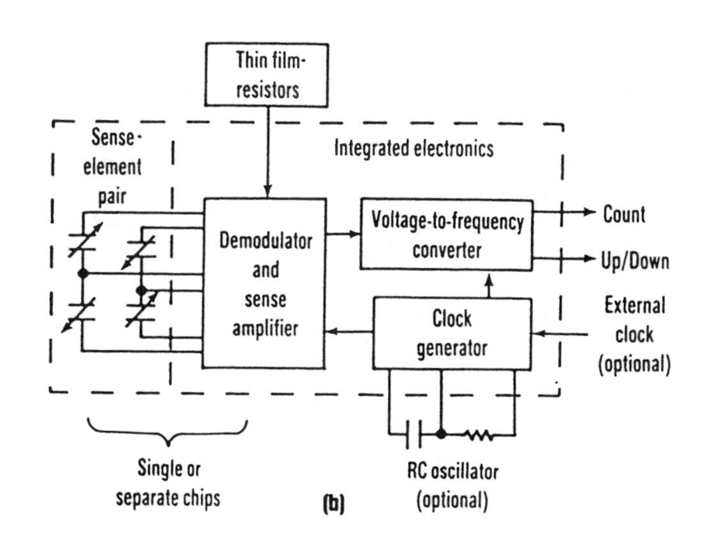

Figure 10 - Capacitive Accelerometer and Sensing Circuit.

Characteristic	Piezoelectric	Piezoresistive	Capacitive
Loading Effects	high	low	high
Size	small	medium	small
Temperature Range	wide	medium	very wide
DC Response	no	yes	yes
Sensitivity	medium	medium	high
Shift Due to Shock	yes	no	no
Self-test Features	no	capacitive	yes
Damping Capability	no	yes	yes

Table 1 - Comparison of Piezoelectric, Piezoresistive and Capacitive Accelerometers.

Fault sensing in an inflatible restraint system can provide information on 1] a failed accelerometer that is not capable of providing any signal in the event of a crash and 2] an extremely out of calibration accelerometer that is incapable of providing the correct firing point. Either event, inadvertent deployment or failure to deploy, has an impact far greater than just a dissatified customer. An indication to the driver of a system problem and the driver's reaction to the warning by obtaining service to correct the fault are essential.

SENSING IN POWER SEMICONDUCTOR DEVICES FOR FAULT CONDITIONS

Sensing for fault conditions is an integral part of power IC's or smart power transistors. The action that is taking based on sensing a fault condition can vary depending on the control system. Sensing excessive temperature in a power device may mean that the device can turn itself off to prevent failure in one case and in another situation a fault signal provides a warning but no action is taken. An example of a sense signal only situation will be considered.

Power FET with Integrated Thermal Sense

A very low on-resistance power FET that incorporates a temperature sensing diode is shown in Figure 12. The power device achieves an on-resistance of 5 milliohms (including packaging resistance) at 25°C utilizing a production power FET process [7]. The power FET process was chosen instead of a power IC process because a large die size was required to meet the power dissipation requirements of a very cost sensitive application. A

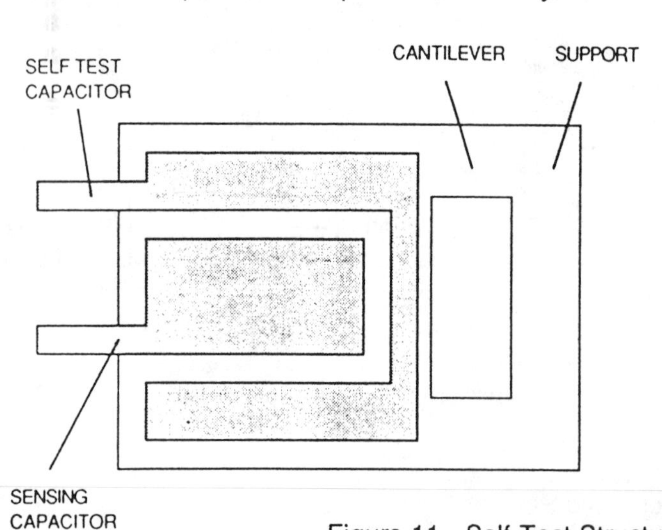

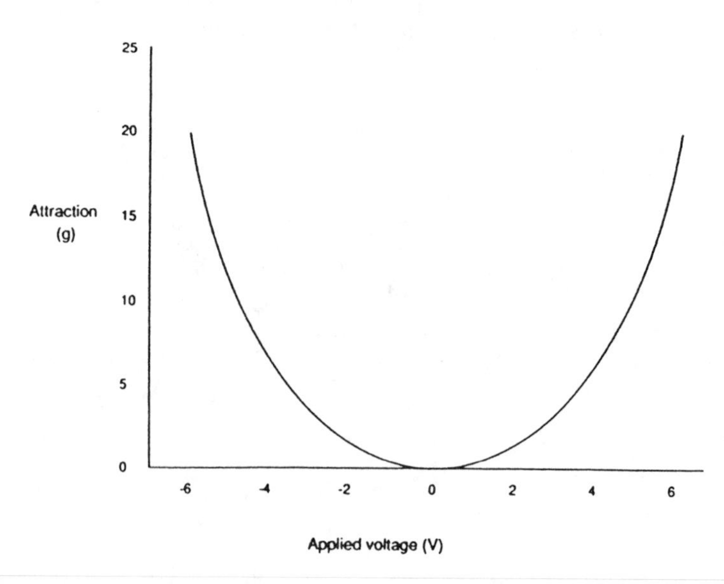

Figure 11 - Self-Test Structure and Applied Voltage.

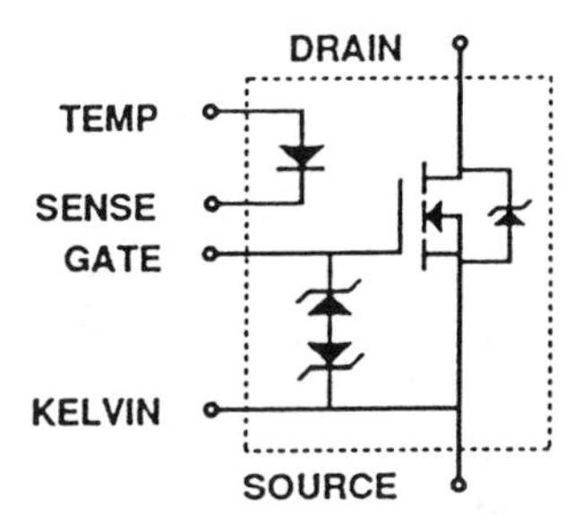

Figure 12 - Power FET with Integrated Temperature Sensor.

minor modification (the addition of a single masking layer) to a production power FET process allows both sensing and protection features to be integrated. Devices that are produced using the SMARTDISCRETE™ technology, as this approach is called by Motorola, and are one way to provide more cost effective smart power semiconductors, especially for high current or low on-resistance power applications.

The thermal sensing is provided by integrated polysilicon diodes. By monitoring the output voltage when a constant current is passed through the diode(s) an accurate indication of the maximum die temperature can be obtained. A number of diodes are actually provided in the design. A single diode has a temperature coefficient of 1.90 mV/°C. Two or more can be placed in series if a larger output is desired. For greater accuracy, the diodes can be trimmed during the wafer probe process by blowing fusible links made from polysilicon. The response time of the diodes is less than 100 μseconds which has allowed the device to withstand a direct short across an automobile battery with external circuitry providing shutdown prior to device failure. The sensing capability also allows the ouput device to provide an indication (with additional external circuitry) if the heatsinking is not proper when the unit is installed in a module or if a change occurs in the application which would ultimately cause a failure.

An over voltage protection circuit has also been integrated in the structure which allows the thin gate oxide of the logic level device to withstand higher levels of ESD (electrostatic discharge) for improved reliability.

The thermal sensing capability has been integrated into a chip designed to achieve very low on-resistance (less than 0.010 Ohms) utilizing a high volume production power FET process. It is intended for very high current applications that require high reliability and would otherwise be addressed by electromechanical relays. The 40 Volt, logic level FET is fully on with 5 Volts on the gate and has conducted several hundred Amperes in a short circuit condition prior to being shut-off by external circuitry. One of the potential applications is in an ABS system.

Temperature and current sensing capability, as well as, over-voltage detection and protection are possible in a low cost production power FET process. This allows semiconductor devices to provide sensing features normally associated with more complex smart power transistors to be manufactured at a considerably reduced cost. In addition a simple diagnostic output can also be provided to indicate that a sensed parameter has initiated a protection feature or that a fault condition has occurred.

SENSING, DIAGNOSTICS, SELF-TEST & THE FUTURE

The requirements for sensors in future vehicle systems include improved performance (sensitivity) and reliability in spite of operating at higher temperatures, lower cost, space and weight reduction, greater functionality including diagnostics and self-test, ease of interfacing with an MCU for adaptive control and interchangeability. The capability of semiconductor technology to provide

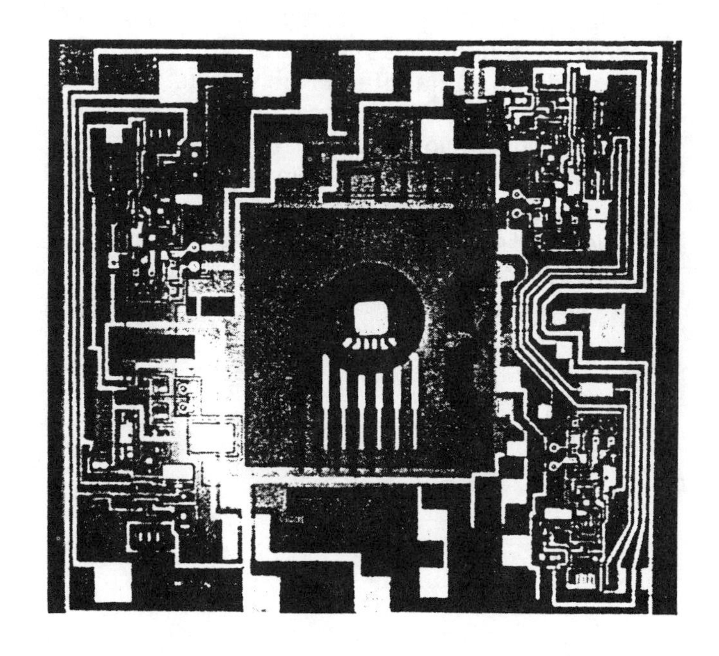

Figure 13 - MAP Sensor with Integral Signal Conditioning.

sensing information for failure detection provides a means for increased reliability in several vehicle subsystems. The present state-of-the art MAP sensor (Figure 13) demonstrates the capability that can be achieved by integrated semiconductor sensing technology applied to solving an automotive sensing problem. The sensor and circuitry necessary to achieve calibration, temperature compensation and amplification are all included in one monolithic, micromachined structure.

Micromachining is used to adjust the diaphragm to nominal targets and microelectronics is used to provide the precision for the fully integrated semiconductor sensor. The precision output is obtained by laser trimming of zero TCR (temperature coefficient of resistance) thin film metal resistors that are integrated around the periphery of the diaphragm.

A "smart sensor" that contains memory and can interface with a host microcomputer with bi-directional communication is the goal for future integration. By considering the transducer with integrated signal conditioning and amplification as one block (for example, the MPX5100D of Figure 13) and the A/D, memory and MCU as a separate building block (for example Motorola's MC68HC05B6), a totally custom smart sensor can be developed. The sensor is comprised of two pieces but can be achieved at a very low cost utilizing technology that is available today. The possibility of a fully integrated sensor-microcomputer is within the capability of today's technology. The extent and speed that more complex sensors are developed and designed into production automobiles will be determined by semiconductor manufacturers investment in research; automotive manufacturers implementation of systems integration and requirements for reduced packaging dimensions and diagnostics; and the ability of auto makers to define their requirements to provide the direction for semiconductor research.

Semiconductor technology will be part of the continuing evolution of sensors to provide the precision, reliability, cost reduction and features, including self-test and diagnostics, that auto manufacturer's have come to expect from other semiconductor devices.

CONCLUSION

This paper advocates designing systems that continue to perform in the presence of a failure. The first part of the paper focused on designing ECU's with fault tolerance. The second part of the paper focused on sensor aspects of system fault tolerance.

Obviously, total redundancy cannot be provided; however, a blend of diagnostic and fault tolerant approaches to ECU designs and sensors can be used to make practical higher uptime systems which will benefit both customers and manufacturers.

References:

[1] Siewiorek, D. P., and R. Swarz, The Theory and Practice of Reliable System Design, Digital Press, Bedford, Mass. 1982.

[2] Joel Garnault, "Third Generation Antiskid: Safety Diagnostic Features of the E.C.U.", SAE paper #890093.

[3] W. R. Iverson, "Why Electronics is no Easy Winner in Battle for Air-Bag Controller Sales", Electronics, July, 1988, p. 64.

[4] Phil Frame, "Restraints Rules Ignite Solid State Sensor Debate", Automotive Electronics Journal, July, 1990, p. 9.

[5] W.C. Dunn, "Accelerometer Design Considerations", Microsystems ICC Berlin, September 10-14, 1990.

[6] J.C. Cole, " A New Capacitive Technology For Low-Cost Accelerometer Application", Sensors Expo International, 1989, 106B: 1-5.

[7] Steve P. Robb, Judith L. Sutor and Lewis E. Terry, SMARTDISCRETES™, New Products for Automotive Applications, IEEE Workshop on Automotive Power Electronics, August, 1989.

901131

The Role of Microprocessors in Future Automotive Electronics

Murray A. Goldman, Stan E. Groves, and James M. Sibigtroth
Microprocessor Products Group
Motorola, Inc.
Austin, TX

ABSTRACT

After the introduction to automotive electronics in the mid 1970s, microprocessor-based automotive technology matured quickly in the 1980s as semiconductor manufacturers and automotive designers learned to work together to solve ever more challenging automotive problems. As we enter the 1990s, it is clear that microprocessor technology has become an essential technology for future automotive electronics. As advances in semiconductor technology allow more functions on a single chip, issues of reduced design cycle times have become a serious challenge. This paper explores these and other problems and discusses how semiconductor and automotive manufacturers can work together to develop solutions.

Microprocessor technology enables automotive designers and engineers the opportunity to solve a great many design problems and the unprecedented flexibility to innovate beyond mechanical limitations. At the same time, the number of microprocessors and the diversity of microprocessor applications in the vehicle create new challenges for the automotive manufacturer. In particular, the complexity of microprocessor components demands a close working relationship between the semiconductor manufacturer and the automotive manufacturer.

The microprocessor, or MPU, is the quasi-intelligent portion of a microcontroller, or MCU. The architecture of the MPU has changed from a simple 4-bit unit controlling fuel/air mixture to 8-bit, 16-bit, and now 32-bit MPUs that will soon appear in production vehicles. Semiconductor designers have improved performance by adding more buses and more features, all operating with greater parallelism. Engine control was one of the first automotive applications and is still one of the most complex. Microcontrollers like the Motorola MC68332, called by Morgan Stanley Research "the ultimate programmable engine controller," rival the computing power of todays most sophisticated personal computers and workstations, and perform engine control functions with reserve computational ability.

Newer architectures such as the reduced instruction set computer, or RISC, and the digital signal processor, or DSP, promise higher levels of performance and, thereby, open the doors for new applications. DSP microprocessors silence road noise by detecting sound waves and producing offsetting sounds that exactly cancel the road noises. Similar application to the exhaust will soon enable more efficient operation by cancellation of the exhaust output pressure wave. In the future, microprocessors could alter engine mount characteristics in real time to reduce the coupling of engine vibration to the vehicle chassis. DSP technology already enables sophisticated dynamic suspension controls for high-performance cars and will soon be practical for most vehicles.

Instead of using simple table look-up algorithms, these advanced controllers now execute real-time control algorithms which compute fuel allocation and spark advance, including avoidance of pre-ignition, for each individual cylinder. The increased computing power allows one MCU to perform multiple simultaneous tasks. A present example is the combination of engine and transmission control. An emerging example is the combination of antilock brakes and traction control. Current 32-bit microcontrollers combine all four functions and more into a single MCU. Considerations such as proximity to sensors, shared sensor data, or redundancy, not processor performance, will determine what functions will be combined within each electronic control module (ECM).

Processors can eliminate the carburetor by direct control of fuel injection, the distributor contact points by use of a magnetic flux sensor, and the distributor itself by direct control of spark advance for each cylinder. A next logical step will eliminate the cam and all of its associated bearings, rods, and adjustments by electronic control of valve openings and closings, producing an engine fully controlled by electronics.

Microprocessor technology makes it possible to explore engine technologies that were previously considered impractical for cars. Reasonable mechanical control systems could not produce adequate fuel economy or pollution reduction from two-cycle engines. Using leading edge microprocessor technology, an Australian company has produced a practical two-cycle engine that is suitable for a passenger car. This engine is much smaller than a traditional four-cycle engine of comparable horsepower. Several automotive manufacturers are currently evaluating two-cycle engines with expectation of prototype vehicles in a few years.

On-board navigation systems have been proposed for some time but technology limitations have prevented all but experimental implementations. These systems require powerful microprocessors and huge amounts of storage for map data. New 32-bit microprocessors meet processing demands, and recent developments in compact disk storage technology offer practical solutions to map storage requirements. Although exciting in themselves, these developments are merely the seed for a whole new field of applications. The presence of a comprehensive electronically accessible map in the car is a key requirement for several artificial intelligence applications.

Route planning could become a matter of entering information and letting an on-board computer suggest the route. The driver could select various judgement criteria such as timing of stops or minimum route time or minimum distance. Typical and current actual traffic conditions could also be used to improve the route planning process. The navigation map would also be useful in rerouting around an accident or traffic congestion.

Many changes to semiconductors occur outside the MCU itself. Each year semiconductor manufacturers fabricate denser chips using smaller geometries than before, shrinking existing MPUs and MCUs in order to reduce their cost. Each year we process larger die, and assemble them in packages with higher numbers of more densely spaced pins. Motorola today ships MCUs in packages ranging from the 20-lead SOIC to the 132-lead PQFP (Figure 1).

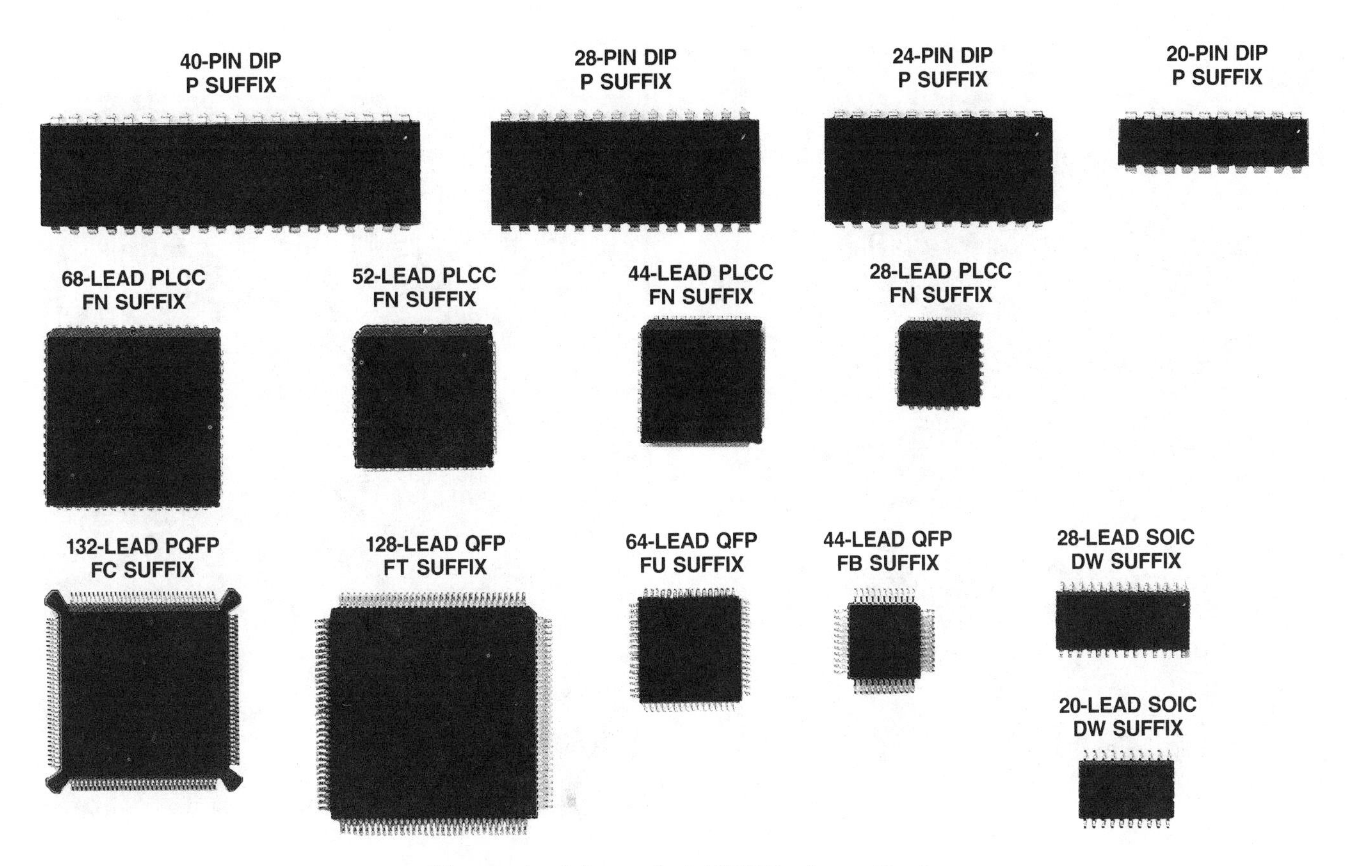

Figure 1—Motorola MCU Package Options

The use of smaller geometries together with larger die and denser packages permits the inclusion of more peripheral functions on the MCU chip, and the CPU and peripherals can have greater complexity and generality. For the automotive designer, this results in a continued reduction in chip count for a given feature, adaptation of a given MCU to more tasks, and consideration of a greater complexity of tasks for the MCU than earlier. Perhaps not as obvious, this added flexibility and capability allow easier accommodation of last-minute changes, or inclusion of feature enhancements.

Intel introduced the first MCU with EPROM* and Motorola the first MCU with EEPROM.* Continuous research is underway to improve the write/erase endurance and to reduce the geometry of these mem-ory structures, in order that the MCU may include larger arrays of non-volatile memories. These memories today provide for initial calibration and limited in-vehicle recalibration. Motorola's new MC68HC711E9 combines RAM, ROM, EPROM, and EEPROM together on one versatile HCMOS MCU chip (Figure 2).

Such larger arrays will allow retention of more diagnostic data for service bay analysis, and performance-correcting learning algorithms that mimic human learning by adapting to a changing environment such as wear or atmospheric conditions. For instance: as a car that is tuned for operation at low altitude is driven up a mountain, air density changes. Peak engine performance requires a continuous change in the fuel mix as the altitude increases. By storing the best fuel mix parameters in an EEPROM location, the engine will start more easily at the new altitude because of this localized tuning process. The operational adaptation occurs automatically as conditions change and thus driveability is improved.

* Electrically (Erasable &) Programed Read-Only Memory.

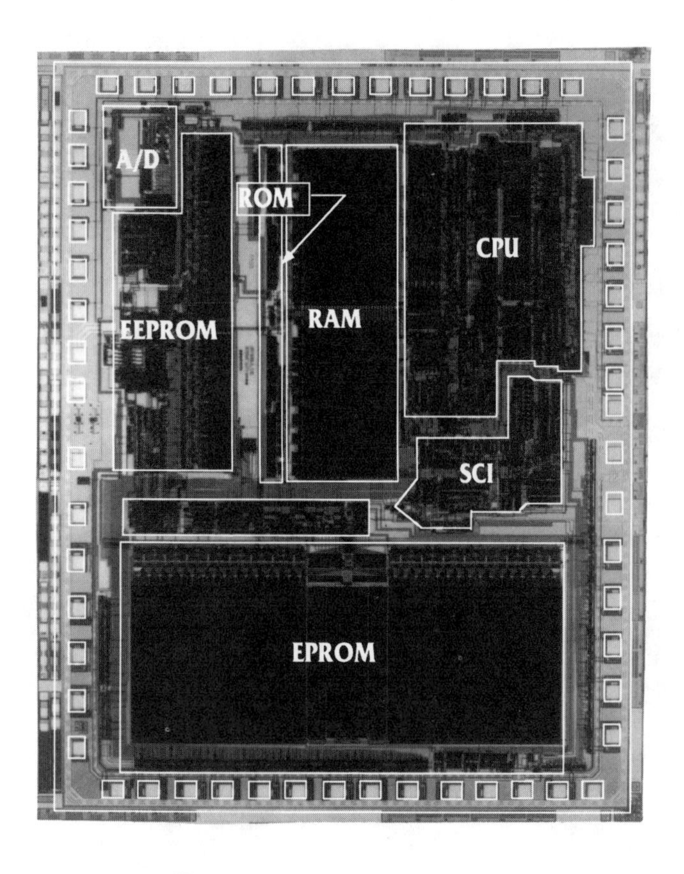

Figure 2—MC68HC711E9

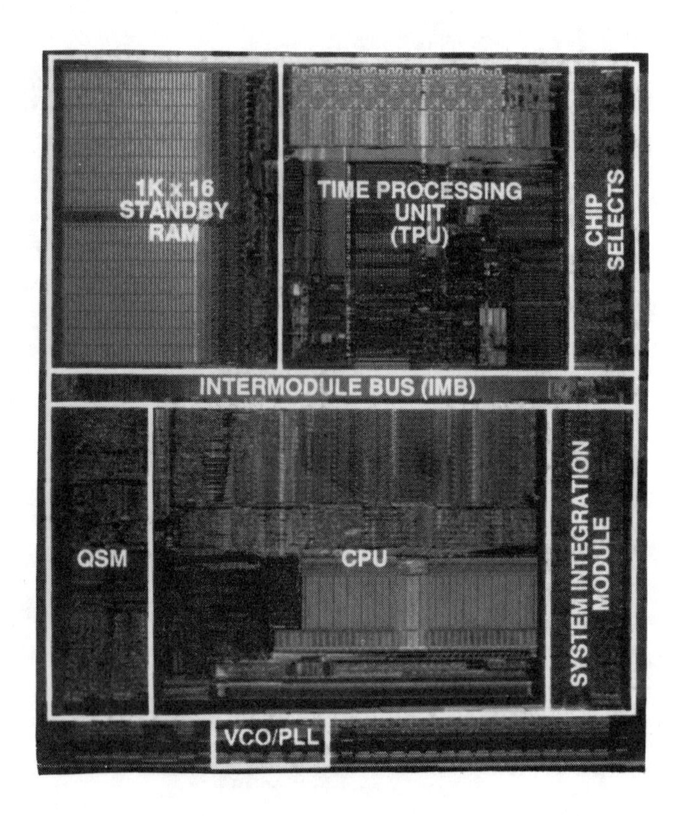

Figure 3—MC68332

The trend will continue in which the peripheral functions on the MCU need more intelligence and autonomy of operation. Motorola developed this idea to a great extent in the MC68332 (Figure 3).

In the MC68332 the serial interface can manage a block transfer unattended, and the timer is a stand-alone Time Processing Unit. The CPU is thus freed from routine interruptions for data transfers or waveform generation and analysis, allowing enhanced throughput for execution of control algorithms. This greatly magnifies the effectiveness of the CPU.

Until recently, the requirements in chip area and processing difficulty made the cost inappropriate to include circuitry on the MCU to allow connection to the raw battery in the vehicle. Reduction of the noise on the power source, together with new innovative circuit and processing techniques, make this direct battery connection a feasible future possibility. These new capabilities allow the designer to mount an MCU with integrated drivers directly to the glass substrate of an instrument panel with a direct power connection to the panel.

Power driver technology is developing in two areas. Smart Power devices are very high-current power transistors with a small amount of digital logic fabricated on the same silicon die. Today, an MCU can include on the same die medium power drivers designed to drive hundreds of milliamperes. In the future these developments will lead to MCUs that can directly drive heavy loads such as headlamps and motors. In the shorter term, multiplexed wiring systems can include both of these developments and thereby drastically reduce the amount of wire in the vehicle's wiring harness.

To most automotive engineers, a *module* is a box containing several integrated circuits mounted on a printed circuit board, or PCB. The semiconductor designer divides a larger IC into several modules, each configured with one standard dimension and a standard interface to a bus internal to the IC. A designer may easily re-use a module adhering to such standards on a later IC design. The I/C using modular design methodology is larger due to the techniques used; however, it offers the best cost and design-time trade-off for families of high-end MCUs. Motorola has created this modular technique for its new M68300 and M68HC16 families. Although it is certainly possible to use this technique on smaller I/Cs, the size penalty for an internal bus and standard dimensions is contradictory to the goal of lowest-cost applications.

Much effort has gone into the development of application-specific integrated circuits, or ASICs. Giving the automotive designer the tools and a library of predefined cells with which to customize his own IC promises an IC with little wasted silicon. Such a device in theory could be cost competitive with a general-purpose IC.

With regard to microcontrollers, this approach was not as practical as hoped, primarily because these custom ICs do not benefit from the high-volume learning curve associated with standard ICs. Also, because of the computer-generated routing to standard cells, the ASICs tend to be less silicon-efficient than hand-packed high-volume ICs. Motorola also found that a non-IC designer may easily miss subtle problems and is typically less efficient in his result. Another pragmatic problem is the testing of these ASIC designs. Test time is a fair part of the cost of a high-volume IC, especially when test-grading of fault coverage is required. In the future, when more sophisticated CAD tools including efficient computer routing programs are available, the ASIC dream may be fully realized.

Motorola has developed a pragmatic low-cost design methodology called Customer Specified Integrated Circuits (CSIC) using the popular M68HC05 MCU family. The CSIC methodology, recently extended to the M68HC11 family, starts from a base chassis composed of a CPU with core peripheral functions and then adds various combinations of memory and semicustom peripheral functions. Unlike the ASIC methodology, the semiconductor manufacturer takes responsibility for circuit design and layout activities. Engineers from the primary customer and from the semiconductor manufacturer form a team. The specification for a CSIC comes primarily from the customer with advice as needed from the semiconductor engineers. The expert assistance of one customer toward a design intended for public sale to an entire class of customers allows the full benefits of economy of scale to all users of the new part.

Most MCUs developed as CSICs, such as those in Figure 4, become standard products, making pricing more attractive than that of an ASIC device. Because of the large number of existing peripheral function blocks available to CSIC design teams, a typical CSIC can be completed in less than six months from specification to samples.

One consequence of the CSIC methodology is the proliferation of small MCUs to all corners of the vehicle. Very specialized small MCUs will be used at switches, lamps, and other loads to intelligently interface these devices to a standardized multiplexed wiring bus. It is already feasible to design an MCU for the door of a car that can multiplex all of the switches and loads in the door, so that only three wires need to pass through the hinge, instead of more than 50 as presently occurs. Eventually, very simple MCUs could be built into each switch and load.

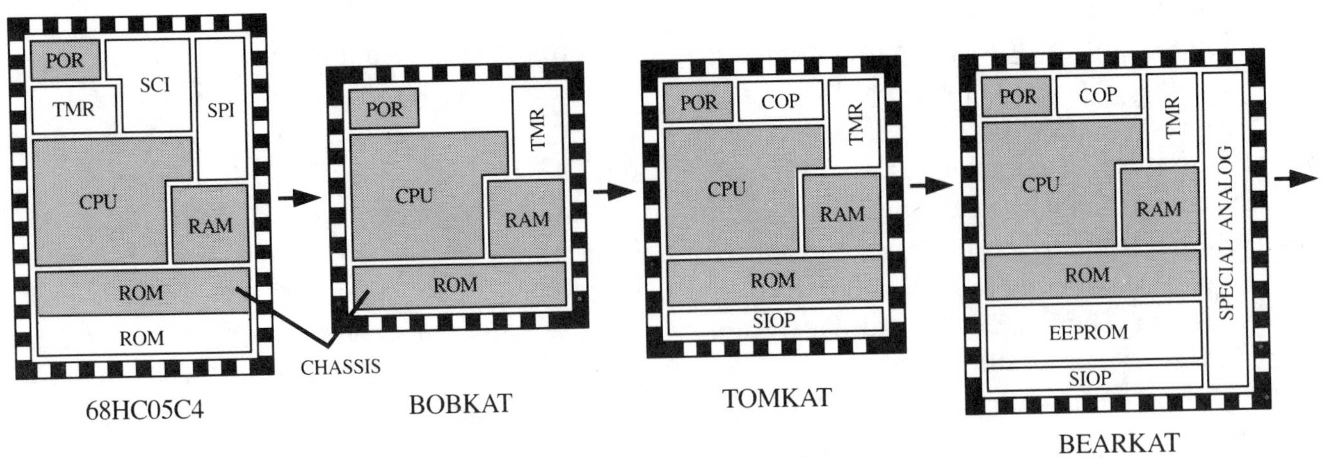

Figure 4—CSIC Examples

Microprocessors record data during deployment of the air-bag in newly standard inflationary restraint systems. They appear in keyless-entry and anti-theft systems, climate control systems, reminder lights, cruise control, memory seats, and in speed-proportional power-steering assist. In the area of entertainment, they are used in the majority of automotive radios for digital tuning, station memory, noise reduction, and equalization; cellular radios, CD, optical disk, and digital audio tape players all include microprocessors. Both digital and analog instrument panel displays now make use of microprocessors, and often include trip and navigation computers, digital clocks, and a display of engine diagnostics.

To a casual observer it may appear that these MCUs drive the cost of automobiles higher, but a closer look reveals many ways in which microcontrollers actually simplify automobiles and thus reduce their cost. For example, the wiring harness in a car has grown to contain miles of copper wire, which adds direct cost for the wire and connectors and indirect costs due to the design difficulty, lowered reliability, and extra weight caused by routing so many wires into a door or through a firewall connector block. Once an MCU installation is committed, the presence of the MCU allows flexibility and feature fringe benefits never before possible.

In a few cars now, and in most future cars, microcontrollers will pass command signals over high-speed serial data links using standardized communications protocols such as J1850 or J1939. Multiplexed wiring uses microprocessor technology to drastically reduce the amount of copper wiring, interconnection, and mechanical hardware in the car.

Multidimensional benefits ensue: reduced cost and weight of wire and connectors, added flexibility to add options without adding wires, ability to use the bus for diagnostics, ability for formerly dumb loads and switches to report malfunctions to a central body computer which can then temporarily reconfigure the system around a fault. At the dealership, multiplexed wire systems will allow rapid diagnosis of all lamps, switches, etc., as well as single-point communication with all on-board systems, including full electronic recalibration of engine, transmission, brake, traction, and body computers.

As in the past, when the federal government mandated the CAFE requirements, some of the more significant changes occur outside of the vehicle itself. The recently enacted On Board Diagnostics (OBD) required by the California Air Resources Board (CARB) may soon become a federal standard for all vehicles in the US. The use of a higher-performance 16-bit MCU to detect and analyze the special failure and diagnostic patterns involved will allow system designers to incorporate these extra diagnostic functions into the software in the engine control computer. Motorola sees a different approach on the technological horizon.

The human brain operates on different computing principles than today's digital computers. Although digital computers are very good at some kinds of control problems, they are no match for the human brain in subjective decision-making. In the future there will be new types of microprocessors that operate on new computing principles including some not yet conceived. Artificial neural networks are one such computing technology, but others are likely to appear.

Although the CARB OBD requirements appear to be a complex computational problem, they are actually a relatively simple case of pattern recognition, as that used in speech recognition and speaker identification. Pattern recognition problems such as speech recognition and the OBD requirements of CARB are not easily solved using traditional digital computing techniques. Traditional MPUs easily make decisions by comparison between numbers, objects, or arrays which must either be alike or not alike, greater-than or less-than, etc. Newer computing technologies such as artificial neural networks (ANNs) are better suited to handle problems involving recognition of a pattern that is *nearest* the reference pattern. ANNs will do more than make a simple interface for the CARB-required OBD system. ANNs will bring about the feasibility of true collision-avoidance and driverless vehicles because of the ability to learn in a real-time environment. The "Smart Highway" will become a property of the vehicle where it belongs, rather than an attribute of a very small number of interstate highways.

Today one can find anti-skid braking systems and traction control systems on many product lines. Higher-performance processors now becoming available will improve performance and response of these systems, and allow them to be easily integrated into a single system. Current anti-skid systems attempt to prevent skid and lockup. In the future these systems will be able to evaluate a potential skid and compute a recovery sequence. The complexity of the patterns of sensor information involved and the rapid response required are, again, a potential application of artificial neural networks.

As mentioned earlier, microprocessors presently record data during deployment of some air-bags. The exact data probably includes the vehicle speed and odometer values. With the existence of a full multiplexed wiring system, one can easily include data from the digital clock to include time of day, from the braking-traction system to determine wheel skid, from sensory wipers to indicate presence of rain (but not a wet or oily road surface), and automatic headlamp systems can indicate night conditions. Better resolution or additional sensors, with other automatic features of tomorrow's highways, could allow the inclusion of road surface condition, traffic light position, and force of impact. Further progress, such as the use of pulse-encoded IR in collision avoidance systems, can provide the ability to capture the license number or other vehicle identification. Like the famous "black box" on an airplane, such data could help resolve the cause of accidents, and prove useful in court.

There are many ways microprocessor technology can make cars more reliable, yet many mechanics, dealers, and owners think otherwise. Why do they believe microprocessors are less reliable than the mechanical systems they replace? What must happen to change their minds?

When microprocessors were first introduced to automobiles, the technology was foreign to almost everyone. There is a great deal of suspicion with anything unfamiliar, and especially with the microprocessor. To add fuel to this suspicion, there was a learning curve with the introduction of microprocessors and sensors into the harsh environment of the car. The large number of components required to build an early electronic control module (ECM) resulted in reliability problems.

The service shop was poorly prepared to deal with an ECM failure. The standard service action was complete replacement of a complex, expensive ECM assembly. In reality, few returned ECMs had a malfunction within the microprocessor itself; most returned ECMs were in fact operational and improperly diagnosed as defective.

Rapid advances in semiconductor fabrication technology result in much higher levels of integration. Newer ECMs require fewer components. At the same time, the increasing capability of the microprocessor allows it to perform diagnostic functions in addition to primary control functions. These diagnostic functions provide valuable information which helps the service technician make a faster and more accurate diagnosis.

As prejudice fades with familiarity, an objective look at microprocessors in the car will reveal their advantages over their mechanical predecessors. Electronic control systems have few if any moving parts. This simplifies assembly and eliminates potential for wear and failure. Mechanical systems constrain results by physical limitations; only the imagination of the designer constrains the microprocessor system.

Consider the traditional distributor ignition compared to a modern microprocessor-based engine controller. Spark timing in the mechanical system is only as accurate at idle speeds as the eye of the mechanic. Optimization at higher RPM depends upon manifold vacuum, which is more or less proportional to throttle position, and some flying weights which crudely track engine speed. By comparison, the microprocessor system can accurately modify spark timing to optimize engine performance at all engine speeds. Dwell time, which determines spark energy, should ideally increase for higher speeds, but the mechanical system inherently decreases dwell time. In an attempt to overcome this limitation, older high-performance engines used dual-point systems. By contrast the microprocessor controller can easily behave as if a separate set of distributor points existed for each spark plug, even though the points and the distributor no longer exist.

A complete discussion of future automotive vehicles must include mention of electric cars. One of the most serious hurdles for electric cars is the expense and the reliability of the batteries. MPUs can play a major role in controlling the charging of the batteries, managing the load to cover more distance with each charge, and to make batteries last longer between replace–ments. As part of the load management, Motorola expects that braking and deceleration will become energy-recovery activities, instead of today's dissipation of heat. This energy recovery technique is also applicable to hybrid gas/electric vehicles.

The manufacture of microprocessors requires a very high capital investment. Motorola is currently contructing a new facility which will cost about $500 million before fabricating the first wafers; this capital outlay requires very high volume for amortization. Testers for these higher-performance, multi-module MCUs cost around $2 million each; again, high volume is required to amortize this investment. Motorola currently ships the MC68040 containing over

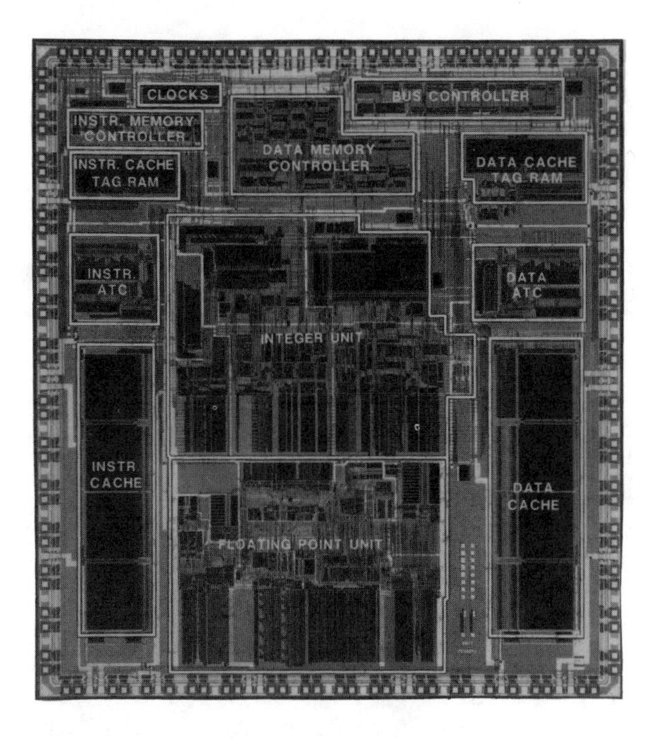

Figure 5—MC68040

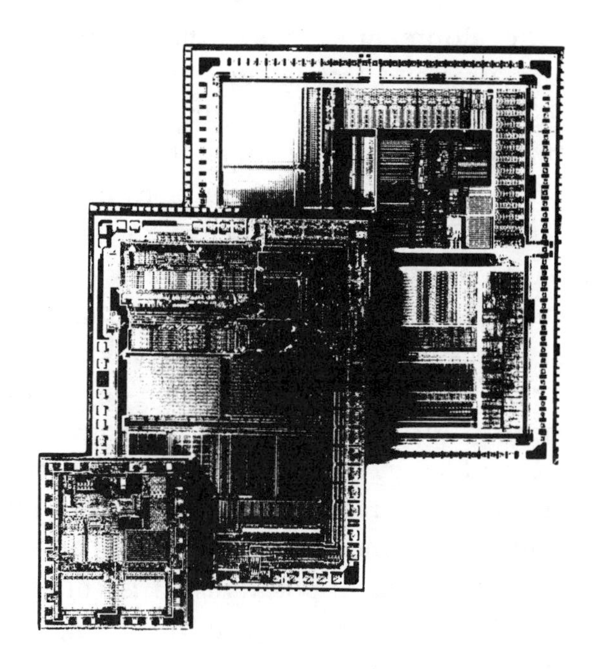

Figure 6
MC68HC05P1, MC68HC11E9, MC68332

1 million transistors (Figure 5) and the computing power of the mainframe computers of yesterday. We will be shipping single I/Cs with over 10 million transistors by 1995! Yet the time required to test the large number of functions and transistors must be kept short to keep costs competitive. To do this, we design testability into every MCU, and test-grade the fault coverage for highest quality and reliability.

MCU designers and manufacturers are justly proud of the capability of the newest 16- and 32-bit advancements in their art; however 8-bit MCUs will fill the majority of applications through the end of this century, just as at present. As the leading supplier of 8-bit MCUs in the world, Motorola's vision is to provide the best combination of features and best-in-class quality production, while providing a software compatible evolution to 16-bit MCUs, and a migration path to 32-bit MCUs when needed for future performance. (Figure 6) Microcontroller suppliers must also provide enhanced development support equipment to enable generation and testing of software. Motorola believes it is essential to supply these products from regional factories located at multiple sites around the globe, to facilitate regional support for both design and assembly of products by the OEM.

The microprocessor is like a raw material used to produce a great variety of finished goods. Like other raw materials, the microprocessor is of little use to the consumer until an OEM such as the automotive manufacturer molds it into a functional component by writing a software program, designing a printed circuit board, and performing a multitude of tests.

The day has arrived where virtually every department of every automotive product line will require some expertise in microprocessors. Motorola and other microcontroller manufacturers must provide better documentation at an earlier point in the design cycle, and provide training expertise to the vehicle manufacturer to accomplish such prolific applications with speed efficiency, and reliability.

The development of new microprocessor architectures and technologies requires teamwork between automotive engineers and semiconductor engineers, combining their respective talents to produce practical solutions to design problems. The automotive engineer must focus on the needs, both present and future, of the car buyer, while the semiconductor manufacturer must focus on the needs, both present and future, of the automotive engineer. Semiconductor designers must concentrate on the broad view of the total automotive semiconductor requirement and trust the automotive engineers to concentrate on the details of specific applications. It is counterproductive for either group to attempt to accomplish the whole task alone.

Automotive engineers and semiconductor engineers depend upon one another in that each has a different set of competencies to combine in a team effort to advance the overall technology. If an automotive engineer develops a new system without the benefit of inputs from the microcontroller manufacturer, or if a microcontroller manufacturer develops a new microprocessor architecture without significant input from the automotive manufacturer, the result will be a less-than-ideal match between the microcontroller and the application.

Automotive engineers and microcontroller engineers must combine very different points of view to arrive at mutually beneficial results. The ideal microprocessor for an application may not be practical to produce, and the easiest microprocessor to produce may not be the best at performing the application. Only through thoughtful consideration of conflicting requirements can the best cost/performance compromise result.

SAFETY AND CONVENIENCE

Climate Control
Cruise control
Keyless Entry
Light Reminder
Memory seat
Sensory wipers
Auto door lock
Headlight dimming
Traction Control
Antiskid braking
Load-sense braking
Window Control
Air Bag restraints

POWERTRAIN

Dynamic Engine Mount
Electronic "Camshaft"
Ignition timing
Spark distribution
Fuel delivery control
Turbo control
Emissions monitor
Voltage regulator
Alternator
Transmission shift
On-Board Diagnostics
Operational Adaptation
Energy Recovery

ENTERTAINMENT

Noise-Reduction Systems
Cellular Radio/Telephone
CD & Optical Disk Players
CB Radio
Digital Audio Tape

DRIVER INFORMATION

Digital & Analog Guages
Engine diagnostic display
Service reminders
Digital clock
Trip Computers
Navigation Computers
Intelligent Highways
Collision Avoidance
Drowse/DWI Alert

BODY CONTROL

Multiplexed wiring
Inter-module network
Body System Diag.
Smart power drivers
Dynamic ride control
Active Suspension
Load Leveling
Anti-theft devices
Electronic steering
Electronic muffler

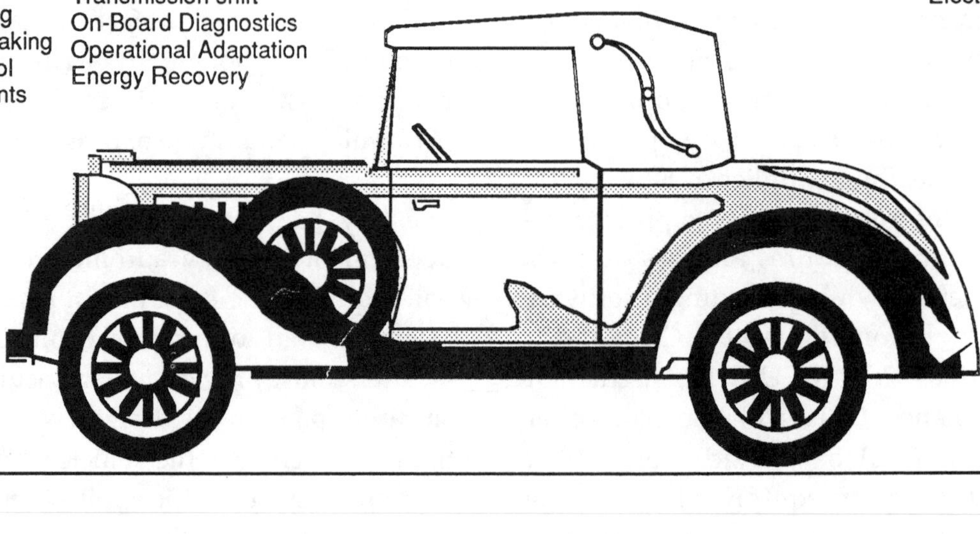

901133

Programmable Memory Trends in the Automotive Industry

Pradeep Shah and Gregory Armstrong
Texas Instruments

ABSTRACT

Semiconductor memory has served as a driving force behind integrated circuit chip technology over the past three decades at the same time as it has revolutionized the computer, consumer electronics, and industrial electronics industries. Recent developments in low cost reprogrammable memory technologies have added new dimensions of flexibility, serviceability, customization, and intelligence to electronic subsystems. These features are of particular value to automobiles of the future. Uses of UV-erasable EPROM's, electrically erasable EEPROM's, and related Flash erasable EEPROM's are already beginning to revolutionize powertrain and embedded controller applications, and it is anticipated that they will rapidly pervade additional safety, driver information, and comfort applications in the coming decade. Reprogrammable memory is also expected to enable the "intelligent automobile" of the future, capable of self diagnosis, repair, and navigation, as well as revolutionizing the driver vehicle interface.

INTRODUCTION

Electronic content in automotive systems, both in the vehicle itself, as well as in the automobile support and service industries, is projected to grow very rapidly in the next decade. Automotive electronics is now recognized to be the only electronic equipment segment whose growth outpaces the growth of the end equipment market it is serving. This growth is driven by emerging automotive requirements of enhanced system functionality and flexibility. These needs are currently well served by traditional semiconductor component technologies such as discrete components, linear IC's, microcontrollers, ASIC's, and standard function memories. Emerging products with increasing functional capabilities such as digital signal processors, intelligent power devices, smart sensors, distributed intelligence for system network applications, and reprogrammable Flash EEPROM's are expected to provide superior solutions to automotive systems. Programmable memories are playing multiple roles in this developing environment: first, as traditional stand-alone memory components to store data and programs; and second as an integral part of the increasing number of applications for embedded controllers, application specific processors, and distributed intelligence requiring programmable memory bits for product customization, adaptability and in-system serviceability.

AUTOMOTIVE SYSTEM CHALLENGES

The automobile industry's challenge to semiconductor manufacturers extends well beyond the component level to include supporting automotive system requirements such as:

1. VEHICLE - Systems for vehicle control such as powertrain, braking, and suspension. Engine efficiency, performance, comfort; electromechanical component replacement for cost reduction and fuel economy, reliability, and overall system integration.

2. SAFETY/ENVIRONMENTAL CONTROL Vehicle safety enhancement; compliance with government clean air, emission, safety and regulatory requirements.

3. DRIVER INFORMATION - Vehicle fuel, trip, and serviceability status; maps and navigation.

4. DRIVER COMFORT/ENTERTAINMENT - Audio, video displays, vehicle /driver interface, inter-system communication through multiplexing, networking.

5. FOR THE AUTOMOBILE AND SYSTEM MANUFACTURER

   a. Overall reduction of development cycle time and flexibility of program changes during system evolution.

   b. Manufacturing flexibility and streamlining.

   c. In-vehicle alterations, testability, diagnostics for avoiding recalls.

6. FOR THE SERVICE/DEALER INDUSTRY
   a. Diagnostics, repair, testability for enhanced maintainability.

   b. Personalization for performance versus economy.

   c. Geographic, government regulation-related changes at the local level.

7. FOR THE CONSUMER
   a. Personalization for optimum ride, control and safety.

   b. Communication from vehicle to home/office and highway systems.

   c. Improved environment, comfort and entertainment systems.

   d. Vehicle economy and performance.

Figure 1 shows various automotive system categories and typical electronic applications for each system. Thanks to the immediate need for fuel economy and emission control, power train systems have already begun a significant incorporation of micropro-

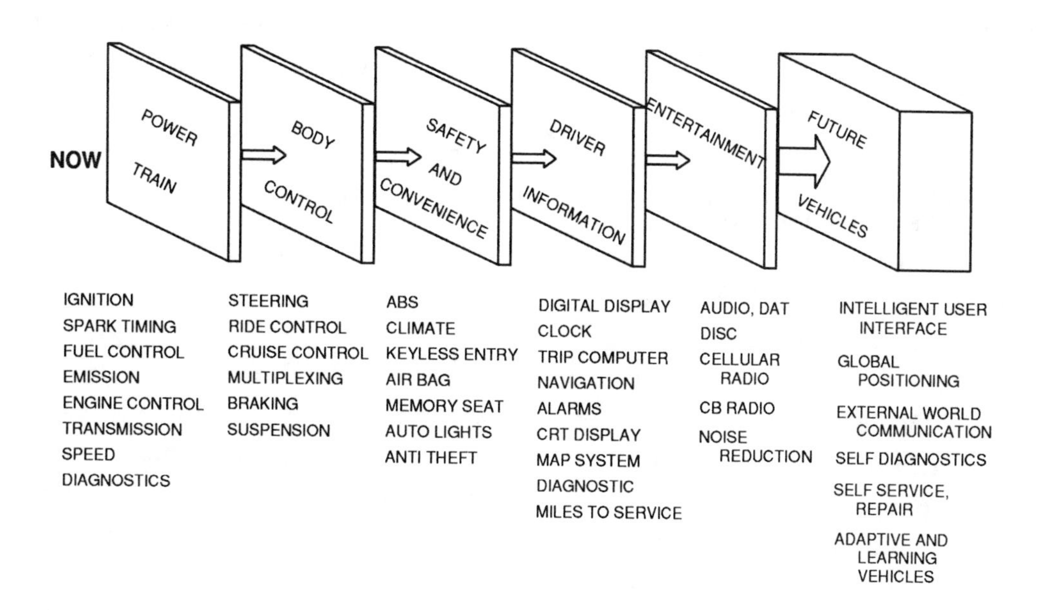

FIG. 1 EVOLUTION OF APPLICATION OF ELECTRONICS IN AUTOMOTIVE SYSTEMS

cessors, linear IC components, and of course programmable memories. This trend is expected to continue over the next decade, with 100% of engine control and 80% of transmission systems projected to be electronic by the end of the 90's. Programmable memories, which traditionally have employed UV erasable technology for storage of codes, lookup tables, and engine and transmission parameters, are projected to make a transition to flash electrically erasable reprogrammable memory during this time. It is also expected that automotive microcontrollers will integrate larger and larger amounts of such memory on-chip for improved system level integration.

Increasing uses of programmable memory in the areas of body control, safety and convenience are expected to follow closely the path set by the powertrain systems. This is enhanced by programmable memory's ability to provide added flexibility to embedded controllers and dedicated application processors, as well as providing highly reconfigurable stand-alone program memory. In parallel, user-friendly driver interfaces, displays, trip computers, and entertainment systems will use increasing amounts of user reprogrammable memory. Further uses of smaller numbers of reprogrammable bytes of such memory for distributed intelligence applications such as keyless entry, instrumentation cluster applications and personalization of systems are expected to accelerate this trend.

II. SEMICONDUCTOR MEMORIES IN STEP TO MEET THE AUTOMOTIVE CHALLENGE

The emergence of 16Mb DRAM and 16Mb Flash EEPROM's by the mid nineties represents a 10,000X growth in memory bit density in a mere two decades (Figure 2). Memories have not only provided the fuel to revolutionize the electronic equipment market in such areas as personal computers and industrial systems, but have been a driving force behind the evolution of the semiconductor industry in terms of device processing technology, manufacturing equipment capability, reliability, and chip cost learning. Emergence of high density, low cost reprogrammable memories such as UV erasable EPROM's, and more recently Flash EEPROM's, have added new dimensions to the impact memory technologies expected to have on the evolution of electronic systems: flexibility, differentiation, and intelligence (Figure 3).

PROGRAMMABLE MEMORY SPECTRUM

Figure 4 displays the current spectrum of programmable and reprogrammable memories. Starting with the oldest and lowest bit cost approach, the mask programmable ROM's, the figure shows a progression towards increased flexibility (with a corresponding increase in bit cost). This cost versus flexibility trade-off moves from ROM's through one time programmable PROM's (or OTP's), UV-erasable EPROM's, Flash EEPROM's, full EEPROM's and finally to true non-volatile RAM's.

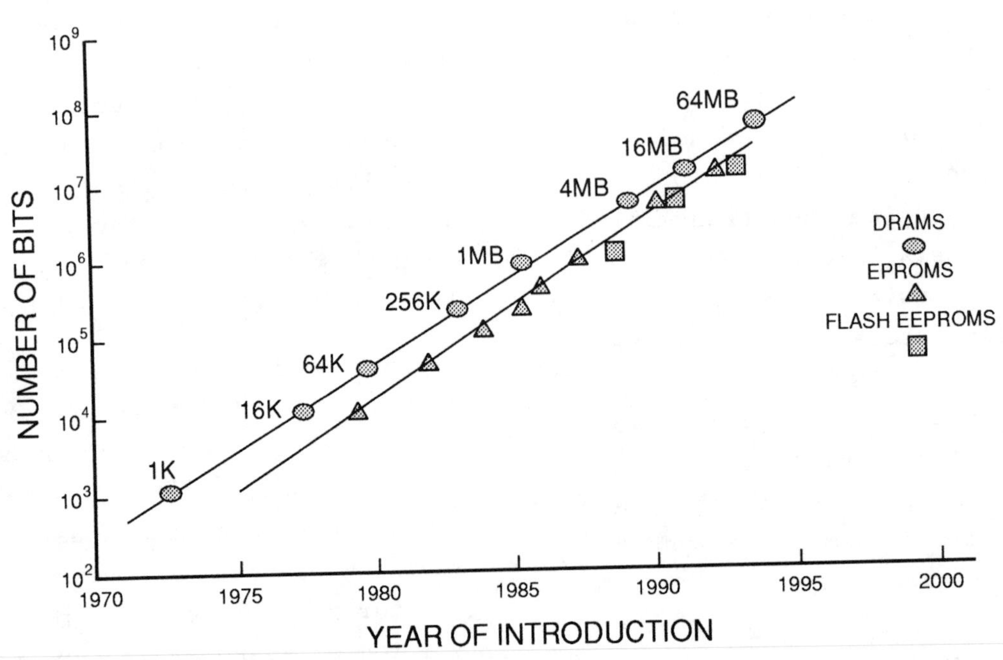

FIG. 2 EVOLUTION OF DRAM/EPROM/FLASH EEPROM MEMORY BIT DENSITY OVER PAST TWO DECADES

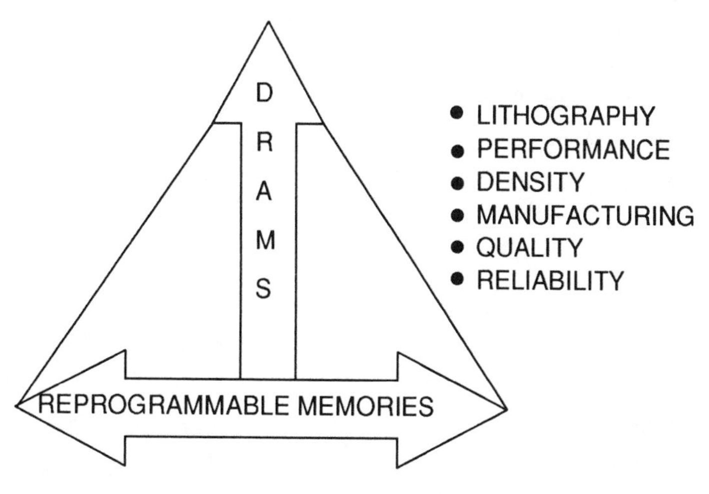

FIG. 3 TWO DIMENSIONS OF IMPACT OF DRAM AND REPROGRAMMABLE MEMORIES ON SEMICONDUCTOR COMPONENT AND SYSTEM TECHNOLOGY.

MOS NON-VOLATILE MEMORIES

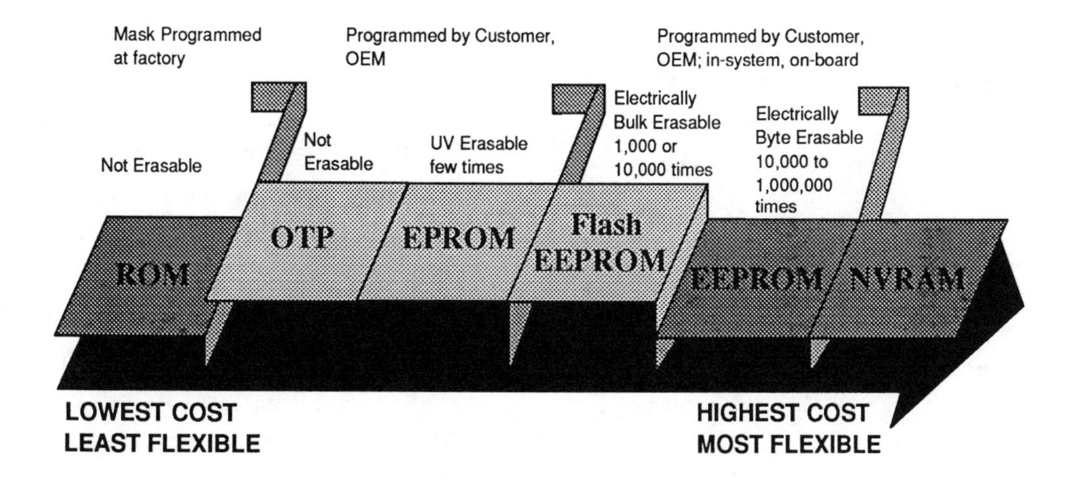

FIG. 4 SPECTRUM OF MOS NON-VOLATILE MEMORIES WITH INCREASING COST/FLEXIBILITY.

In cases where erasability is important, UV-erasable EPROM's are currently the most commonly used reprogrammable memory due to their reprogramming and erase capabilities at costs that are much lower than full EEPROM's. However, two new product types are emerging to change this. The first is the plastic-packaged EPROM used as one time programmable PROM and having functionality similar to a ROM, but with the added flexibility of being programmable at the chip level at the user's manufacturing line. High density ROM and EPROM applications where multiple reprogamming cycles are not required can take advantage of the relatively lower cost per bit for OTP's made possible by the elimination of the windowed ceramic EPROM package.

The second emerging non-volatile memory product is the Flash EEPROM. In this device a bulk erase of the chip is possible electrically, rather than requiring a labor and time-consuming separate UV erase procedure. It is projected that this product will revolutionize wide ranging applications within automotive electronics and beyond due to its added feature of permitting electrical erasing of the chip at a per bit cost closer to that of a UV-EPROM than to that of a relatively expensive full feature EEPROM.

Currently two versions of Flash EEPROM's are offered. One approach uses two power supplies (12V or 21V for programming and erase and 5V for read). This approach is taken by Intel, AMD, Hitachi, Toshiba, and SEEQ. This device thus programs much like a traditional EPROM and erases electrically much as would a full EEPROM (hence the high voltage and current requirements). The other approach to Flash EEPROM's developed by Texas Instruments and ATMEL requires a single 5V power supply for all operations (program, erase, read). This approach is particularly adapted to true in-system programming operation since it provides system simplicity without need for extra power supplies, regulation circuitry as well as low power for portable battery power system.

Rapid evolution in densities of flash memory technology has already surprised most memory analysts and is projected to compete with DRAM in terms of bit density and with full EEPROM for write/erase cycle endurance, offering the promise of the ideal cost-effective reprogrammable read only memory. Significantly, the added features of in-system erase and reprogramming offered by the Flash EEPROM addresses most of the automotive system challenges listed in the section above.

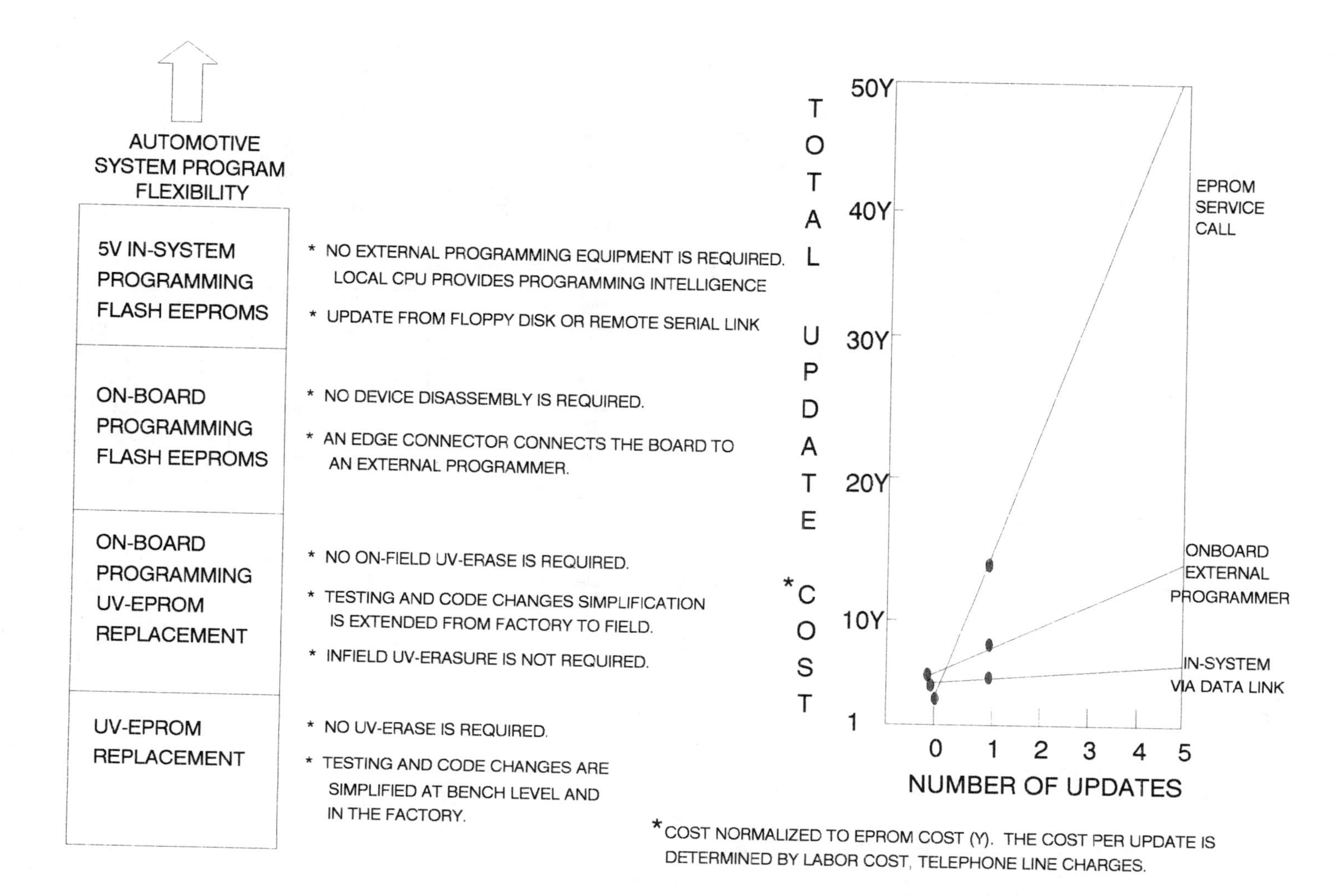

FIG. 5 COST OF CODE UPDATES BASED ON SERVICE CALL, EXTERNAL ONBOARD, AND OVER DATA LINK IN-SYSTEM PROGRAMMING

Fig. 5 shows a comparison of the cost of program updates at the system level for various options of UV-EPROM replacement. Total code change costs not only includes chip cost, but also system support components and overhead cost for each update. In an true in-system programming application, reprogramming over a data link with only a telephone line charge for code transfer is the most cost effective method for multiple code changes.

APPLICATIONS OF PROGRAMMABLE MEMORIES

The features of non-volatile data storage and increasing flexibility of alteration of the data through in-system reprogrammability make these class of memories suitable for wide ranging applications in the segments of computing, consumer, industrial and office equipment, mass storage, telecommunication and military systems. The extended features of differentiation, personalization, and reconfigurability at densities comparable to traditional low cost volatile semiconductor memories such as DRAMs make them even more attractive in newer class of application for mass storage in personal portable computers, image storage and processing, archival of large volume of larger data such as books, maps, personal records, and pictures. Fig. 6 shows evolution of non-volatile and reprogrammable memories from 1987 to 1995 with typical applications...Flash EEPROMS, One-Time programmable ROMS appearing as emerging products.

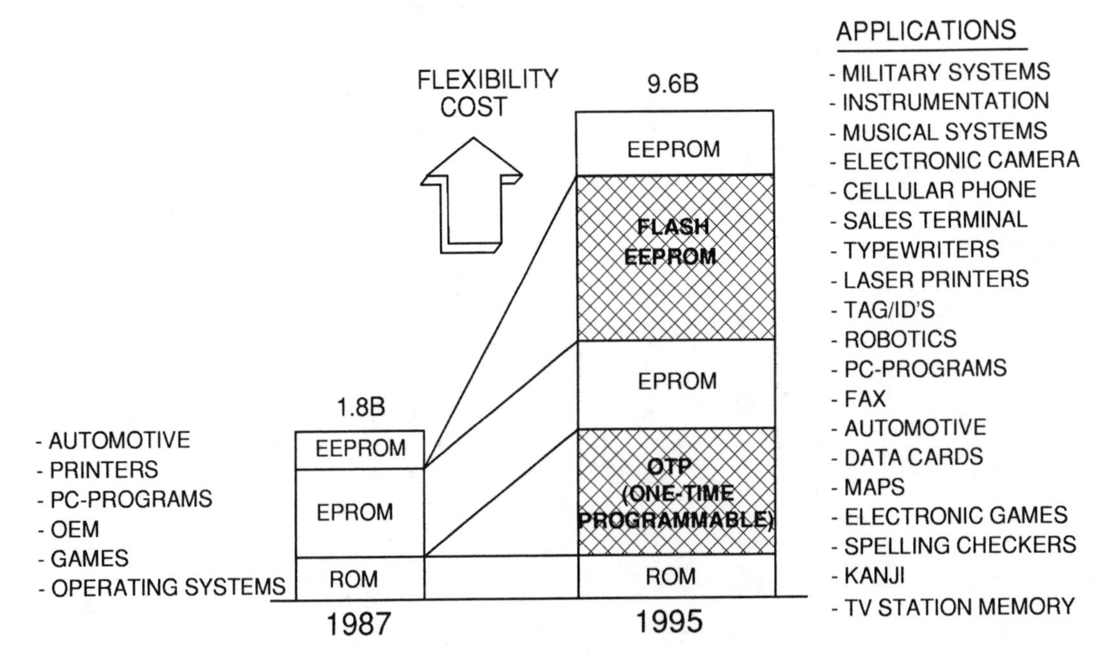

FIG. 6 EVOLUTION OF PROGRAMMABLE MEMORY SEGMENT AND APPLICATIONS FROM 1987 TO 1995.

This added flexibility of reprogrammable non-volatile memories is of particular value to automotive application. Reprogrammability makes available a number of key features to the system designer, vehicle manufacturer, and after-sales service groups:

1. In-system code change for development and optimization without hardware changes such as in powertrain systems.

2. JUST-IN-TIME manufacturing flexibility to select appropriate codes and programs for particular modules for vehicles.

3. INVENTORY CONTROL - Minimization of part numbers.

4. In-vehicle selection of features - engine, transmission, ride options.

5. Testability using a test bus and system service diary chips to store service history and codes.
6. Continuous system optimization through vehicle life cycle by real time system programmability.

7. In-field repair/alteration, maintainability, recall avoidance, by in-vehicle alteration at dealer.

III. EMBEDDED MEMORIES

Embedded programmable memories integrated on the same chip as a processor or controller have been used in the past for system development when program code changes were made often, but read-only memory typically would be converted to a cheaper ROM configuration when codes stabilized and product volumes increased. However, with an increasingly competitive environment, shortening product development cycle times, and increasing chip development costs, a trend is beginning to emerge in which controllers are beginning to have high density, plastic package compatible EPROM and/or EEPROM, as well as SRAM's, embedded, within the final controller chip. Additional EEPROM bits or larger density EPROM's make these microcontrollers flexible and capable of customization in a variety of automotive applications. An example of this is the Texas Instruments TMS370 8-bit microcontroller family with on board EEPROM versions. This sort of teaming of microcontroller and reprogrammable memory can be very powerful in applications such as electronic climate control and entertainment systems (Figure 7-A and 7-B).

TMS370 ELECTRONIC CLIMATE CONTROL

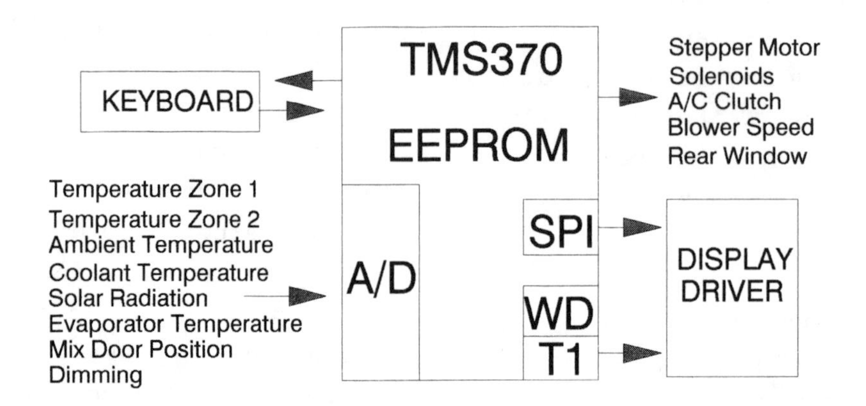

Automatic Controls	Programmable Response to blower speed, vehicle speed, ambient conditions, and solar load.
Passenger Comfort	Control multi-zone temperature, humidity, and air quality.
Personalization	EEPROM storage of heating/cooling sequence/blower speed.
Manufacturing Configuration	EEPROM selection of Climate Control System functions and calibration.
Power Consumption	Ignition-off sequence can use two low-power modes during engine cool-down. EEPROM storage enables complete power-off.

FIG. 7-A

TMS370 DISTRIBUTED AUTOMOTIVE ENTERTAINMENT SYSTEM

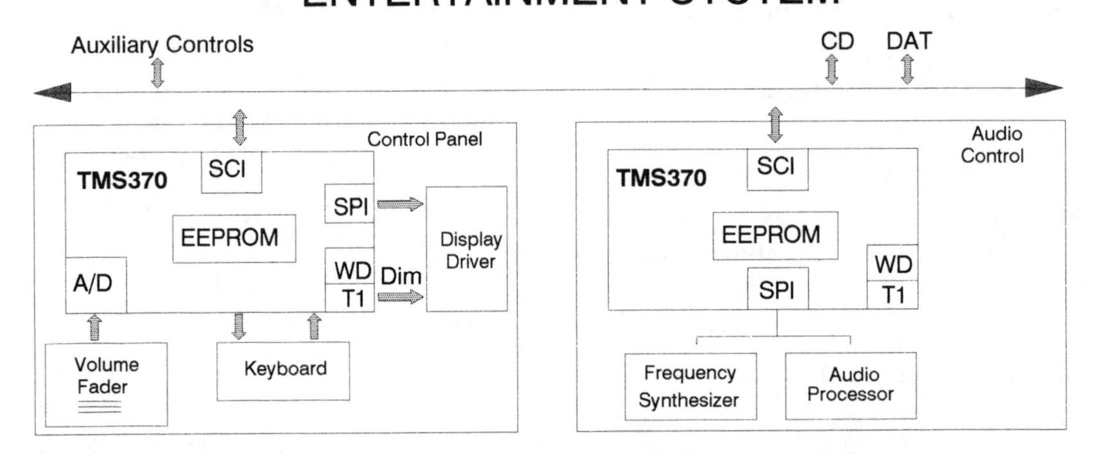

Distributed System	Enables system expansion/frees up dashboard area Paritions software and hardware development
Power Consumption	Two power-saving modes with multiple wake-up
Personalization	EEPROM storage of AM/FM presets, graphic equalizer, etc. EEPROM storage of anti-theft security codes
Manufacturing Configuration	EEPROM selection of entertainment system functions
Cost/Performance/Reliability	Integrated Features: A/D, EEPROM, SPI, SCI

FIG. 7-B

IV. DISTRIBUTED INTELLIGENCE

Another class of system applications for reprogrammable memory elements is storage of system intelligence, typically in a few bytes of EEPROM, distributed in various parts of the system. Using such an approach, for Example, one can place system security codes, fault information, system service diaries, manufacturer's system ID's, code ID's for overall system testability, etc., throughout the system. The Joint Testability Action Group (JTAG) has, for example, defined a system bus which can include serial flash EEPROM memory to store information related to board service history (Figure 8). As the electronic content within an automobile expands, such testability features will take on additional urgency during development, manufacturing, and maintainability & field service environment. The EEPROM element can also be used in applications for system identification, vehicle location (as on toll roads, parking lots, garages), and eventually will be used as part of intelligent sensors for remote diagnosis and servicing.

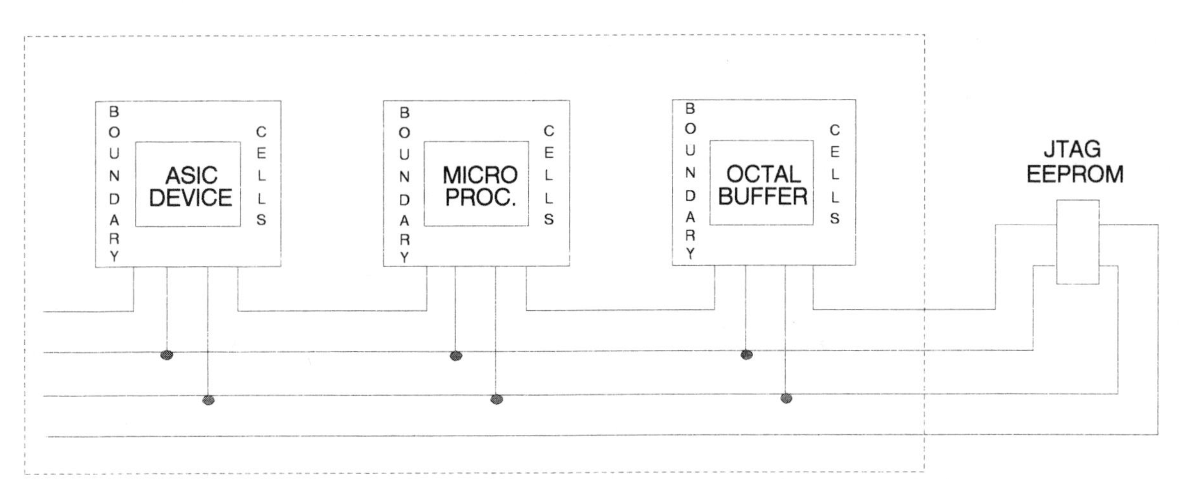

FIG. 8 The circuits within the dotted lines already exist in a computer or on any logic board using devices that are too dense to test any other way but with self diagnostics, or SELF TEST. The JTAG EEPROM simply takes advantage of the existing circuits which eliminates the need for additional hardware or or the need for relayout of the board for additional data paths to the EEPROM.

FUTURE

The steady trend in memory towards increased density and higher performance at lower cost per bit will clearly be complemented in the future by the added flexibility requested by the system designer. A "one size fits all" approach to electronic system design is becoming increasingly less acceptable. Application specific IC (ASICS), programmable gate arrays reprogrammable memory, and in particular, Flash EEPROM's, are expected to address both the cost/performance and flexibility requirements of the system designer in the years ahead.

Automotive applications have been projected to evolve towards vehicles that are not only more efficient and safe, but also more user friendly in terms of on-board vehicle intelligence such as trip computers, intelligent service warnings, providing the driver with information for decision making, and access to the world beyond the windshield in terms of communication links to households, offices, highways, tourist centers, and so on. Personalization of vehicles for driving efficiency, comfort and safety will require designers to begin using increasing electronic content in terms of processing, memory, and interface and communication systems. Thanks to intelligent memories, controllers, and computers, by the early 21st century we can imagine cars that drive, analyze, diagnose, participate in their own repair, talk to the drivers, warn of road hazards, communicate with the highway, find themselves on a map, and adapt themselves to the unique driving habits of each family member all requiring in-system reconfiguration, adaptation realizable with high density reprogrammable semiconductor memories such as Flash EEPROMS.

901134

Current Status and Future Trends of Electronic Packaging in Automotive Applications

Hitoshi Minorikawa and Seiji Suda
Sawa Works,
Hitachi, Ltd.

ABSTRACT

Since the late 1970's the microcomputer has been introduced and rapidly expanded to various kinds of vehicle electronics applications. This technology has been utilized to provide automobiles which not only have higher performance but also run more smoothly and cleanly.

Microcomputer technology has also entered vehicle entertainment systems such as TV, mobile phone, VCR, and many other applications. Vehicle electronic packaging problems have developed as a result of this rapid expansion in vehicle microcomputer usage. Such problems include; size limitations, wire harness weight, wire harness complexity, connector size, electronic module packaging, and numerous other problems.

This paper provides historical packaging technology issues including system integration, hybrid IC module technology, and CAE reliability analysis for extreme conditions.

INTRODUCTION

Recently customer personal tastes in vehicle design and equipment has been increasingly expanding. At the same time, social needs for living environments are directed towards clean air, safety and amenities. Car manufacturers are pursuing systems and products which will meet both the customer personal tastes and social needs through state of the art electronics or electromechanical technology. Automotive system and component suppliers have presented various kinds of electronics products including engine control, powertrain control, automatic transmission control, chassis control, and entertainment equipment. Those electronics products have been required to show reliable and cost effective packaging technology under harsh underhood and passenger compartment conditions.

This paper will review the past 20 years of electronic packaging technology and progress to study some future packaging trends ranging from printed circuit to hybrid IC, from electronics components to components designed for packaging versatility.

AUTOMOTIVE ELECTRONICS AND PACKAGING TRENDS

Fig. 1 illustrates the electronic packaging areas that were improved in order to match the expanding vehicle requirements.

In 1970's, most countries strengthened the anti-pollution laws to decrease total HC, CO_2, NOx exhaust gas emissions. Most drivers, however, were concerned with fuel economy due to the sudden rise in fuel cost and the fuel shortage.

All car manufacturers rushed to introduce newly developed engine control systems, especially in areas of fuel and emission control. New state of the art sensors, actuators and electronic control units were needed. Most of the new electronics sensors and actuators were installed in the engine compartment. This location created very harsh operating conditions for these components. They were new and required to endure high temperature (-40 ~ +125°C) high vibration (30G, 20 ~ 2000Hz) exposure to the elements and EMI (Electro Magnetic Interference).

Fig. 1 - Outline of Automotive Electronics and Packaging Trends

Item \ Year	1960-1970's		1980'S	1990'S
Vehicle and Engine Needs	Anti-pollution Law Low Emission	Fuel Shortage Fuel Economy	Prosperity High Performance Versatile Tastes	Green-house Effect Fuel Economy Low Emission High Performance
Trends for Electronics	Unitization of Small-Scale Electronics into Mechanical Equipment	Total Engine Control (Medium-Scale Electronics)	Total Engine/Powertrain Control (Large-Scale Electronics)	Unitization of Medium/Large-Scale Electronics to Mechanical Equipment
Electronics Products	IC Voltage Regulator IC Ignition Module Pressure Sensor Mass Air Flow Sensor	Engine Control Unit (Engine)	Multi-Function Control Unit (Engine/Powertrain)	Total-Function Control Equipment.....LAN Engine/Powertrain Chassis, Information
Installation	Engine Compartment	Passenger Compartment	Passenger/Engine Compartment	Passenger/Engine Compartment
Electronic Packaging Technology	Surface Mount Technology on HIC	Pin Insertion Technology on PCB	Surface Mount Technology on PCB	High Density Surface Mount Technology on PCB
Electronic Component	Small-Scale Hybrid IC Module ASIC-Bare IC Chip Chip Capacitor	Single-layer Printed Circuit Board DIP/SDIP-LSI, IC	Two-layer Printed Circuit Board PLCC/QFP - LSI SOP - IC Chip Component Vertical Mount HIC	Multi-layer Printed Circuit Board Fine-pitch Device
Problem	Thermal Stress Vibration Water		EMI Vibration at Engine Water Compartment	Noise Thermal Stress

Additionally, some units were forced into small confined spaces because they were integrated to existing mechanical components. In the 1960's, Hybrid ICs were providing high reliability in applications such as alternators with built-in IC voltage regulator and distributors with unitized IC ignition module. Therefore, most new applications started with hybrid ICs.

Microcomputers were also introduced in engine control units to attain the accurate and precise air fuel ratio and ignition timing control

In those days, the hybrid IC packaging designs concentrated on eliminating thermal stress damage and providing waterproof package construction.

In the 1970's and 1980's, the concerns of packaging turned to space utilization.

World-wide prosperity required high performance vehicles as well as diversified equipment not only in engine control but also vehicle control and entertainment, such as, Automatic Transmission Control, A.B.S(Antilock Braking System), diagnostic displaying, TV, mobile phone, and VCR.

Installation space limitation and wire harness complexity became the most pressing problem as the amount of electronics equipment increased.

Meeting this new challenge required electronic control unit size reduction, integration of sensors to control unit, underhood control unit installation, and LAN(Local Area Network).

For packaging technology, multi-layer substrates (HIC or Printed Circuit Board), electronic component mounting, small waterproof connectors, and high temperature resistant (125°C max) components (especially LSI) were studied and developed.

ELECTRONIC PACKAGING IN THE 1970'S

Requirements for electronic packaging in 1970's are outlined in Figure 1. The hybrid IC was the principal technology to respond to those requirements.

Environmental conditions for underhood installation are shown in Table 1.

Table 1-Comparison of Underhood and Passenger Compartment Environmental Conditions

Item	Underhood	Passenger Compartment
Temperature Range	-40 ~ +125°C	-40 ~ +85°C
Vibration	30G, 20 ~ 2000Hz	4.4G
Humidity	85°C, 85%RH	40°C, 95%RH

Photo 1 shows an alternator IC regulator module and Fig. 2 is the cross-sectional view showing bare chip and wire bonding technology. Photo 2 shows an IC ignition module and Fig. 3 is the cross-sectional view showing flip chips and solder bonding. The typical HIC manufacturing process flow chart is shown in Fig. 4.

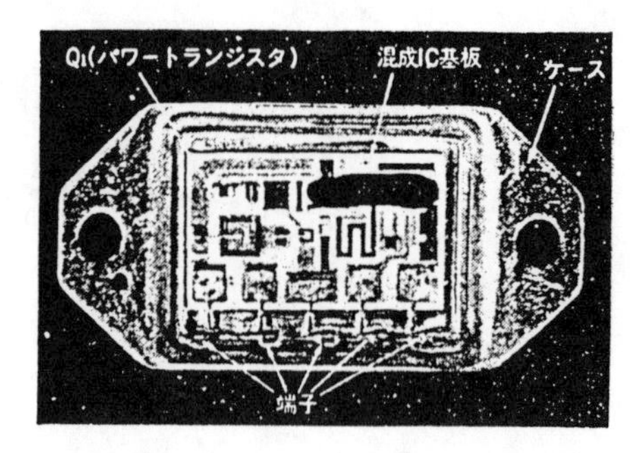

Photo 1-Alternator IC Regulator Module

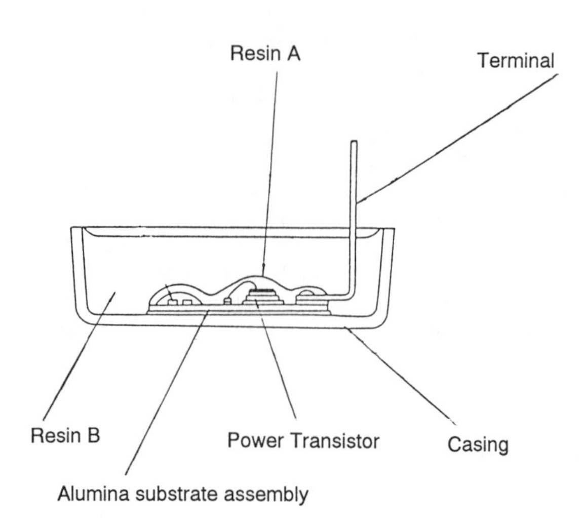

Fig. 2-Cross-Sectional View of IC Regulator

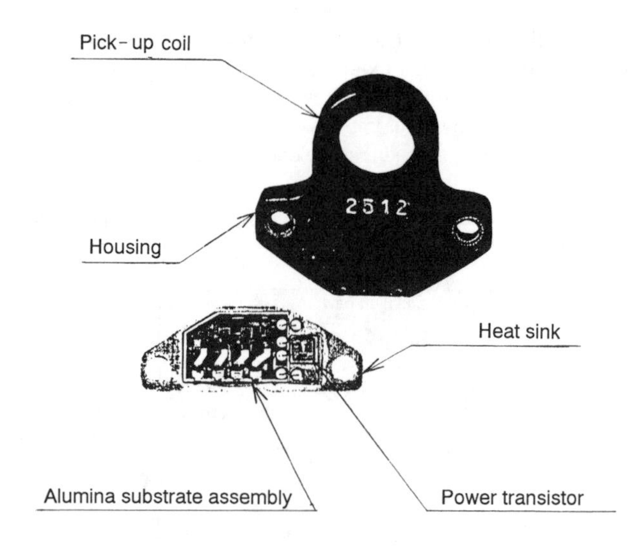

Phto 2-IC Ignition Module

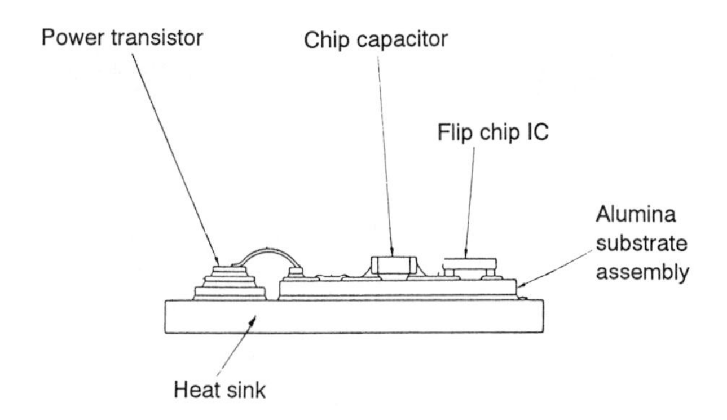

Fig. 3 - Cross - Sectional View of IC Ignition Module

Manufacturing process	Quality control item
△ Receiving Inspection (Component, Material)	
○ Printing and Firing	Appearance, Thickness of Thick Film
○ Resistor Trimming	Resistance, Appearance
○ Solder Printing	Appearance
○ Parts Mounting	Appearance
○ Solder Reflow & Cleaning	Appearance, Characteristics
○ Heat Sink Adhesion	Appearance
○ Wire Bonding	Pull Test
○ Housing Adhesion	
○ Silicone Gel Potting	Appearance
○ Lead Frame Welding	Appearance
○ Cover Adhesion	
◇ Outgoing Inspection	Characteristics

Fig. 4 - Manufacturing Process of Hybrid IC

In these typical applications, all of the semiconductor chips and capacitors mounted on an alumina ceramic substrate are bare chips, not plastic or ceramic package types. This style of mounting decreases thermal stress and outline dimensions. The current package technology utilized hermetic sealed packaging for industrial application. A more cost - competitive packaging type was needed to enter into the vehicle market. For the vehicle market, a plastic housing and cover were adopted. New technology was required to make the packaging reliable. Specially developed passivation technology like Si_3N_4 and PSG was applied on the semiconductor surface to provide moisture protection. Electrode material of capacitors and thick film material on the ceramic substrate had to be carefully selected to avoid electro - chemical migration. Lead rich solder was selected not only to gain sufficient flexibility to handle wide temperature ranges but also to resist corrosion.

Specially designed flexible lead - frames or wires were utilized for interconnecting the substrate and terminals. Finally, silicone gel was applied over electronics components to provide further moisture protection.

ELECTRONIC PACKAGING IN THE 1980'S

(1) EMI Protection

Due to world - wide prosperity, the 1980's was a time for customer satisfaction. Higher quality and higher reliability standards than had been previously reached were pursued in order to satisfy the car customer. Three major areas of packaging improvements were addressed: EMI protection, thermal stress analysis, and system integration.

The most significant improvement was required in the area of EMI protection. Increasing amounts of EMI were being generated within the vehicle and also from external equipment. Various kinds of on - vehicle electronics equipment started to interrupt each other with their pulse signals and switching signals as the use of vehicle electronics expanded. The use of high frequency microwave transmissions from off - vehicle equipment was increasing.

As a first step, most electronic sensors like semiconductor pressure sensors and mass air flow sensors utilized the sensing element and the signal conditioning circuit as shown in Fig. 5.

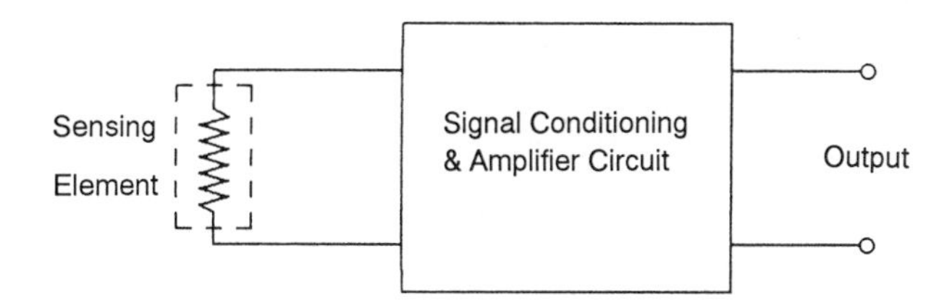

Fig. 5 - Electronic Sensor Circuit

Even with this change, it is difficult to reach acceptable levels of EMI protection because of necessary high levels of signal amplification (100 or more) in the signal conditioning circuit.

Further improvements in EMI protection required careful design and analysis of the circuit itself, amplifier IC design, component and pattern layout, distributed stray capacitance, ground points, additional internal metal shielding, and feed - through capacitors between substrate assembly and terminals. In the case of a control unit using printed circuit board, EMI filters are placed on the board to reduce the electromagnetic wave.

As an example of improvements that can be realized, Photo 3 and Fig. 6 show the results for a mass air flow sensor up to 1000 MHz.

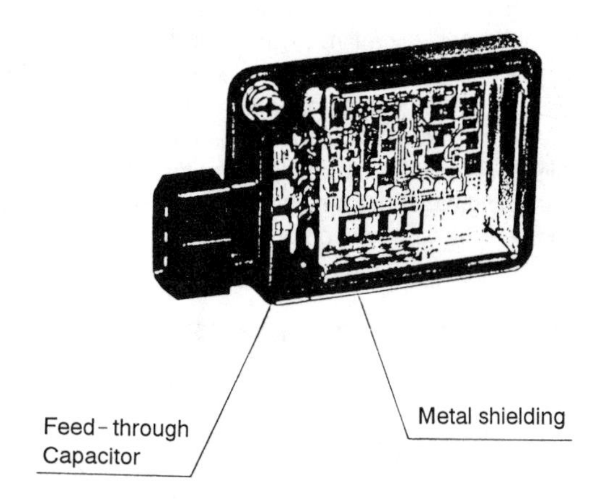

Photo 3 - Mass Air Flow Sensor Module

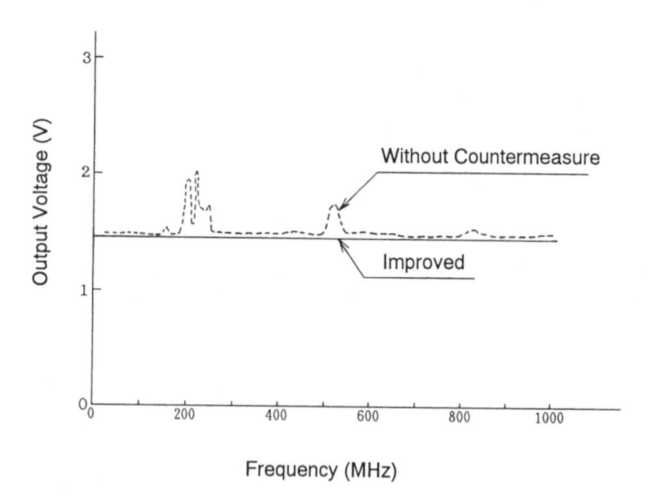

Fig. 6 - EMI Countermeasure

(2) Thermal Stress Analysis by CAE

Recently, specially developed thermal stress analysis methods are being normally used to increase durability for both hybrid IC and printed circuit packaging. As Already shown in Fig. 3, most automotive hybrid IC products use flip chip ICs. This flip chip IC is the custom - designed packageless construction which has high temperature melting point solder bump electrode on the active function surface protected with Si_3N_4 as shown in Fig. 7.

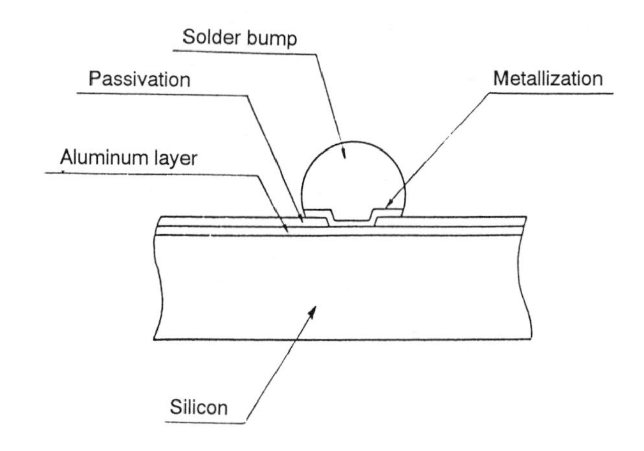

Fig. 7 - Cross - Sectional View of Flip Chip IC

Fig. 8 (a) (b) show a thermal fatigue analysis model and deformation of a flip chip solder joint. Its thermal stress can be calculated by modeling the diagonal - positioned solder bumps and then using three dimensional non - linear construction analysis methods developed by our laboratories. The deformation is made towards the chip center.

Fig. 8 (c) shows the equi - stress distribution of the three dimensionals. The highest portion is on the boundary of the silicon chip side.

Fig. 8 (d) is the sectional SEM photo after accelerated life test (- 55 ~ +150°C temperature cycling). Small cracking due to partial deformation by periodic thermal stress is growing gradually. Its thermal stress life cycle Nf is calculated using equation (1).

$$Nf = C.f^{m}(\Delta\varepsilon_{P})^{-n} \exp \frac{Q}{KT} \quad (1)$$

Where
- C : constant
- f : thermal cycling frequency
- $\Delta\varepsilon_P$: deformation strain width
- Q : energized energy
- k : Boltzman constant
- T : temperature
- m,n : index

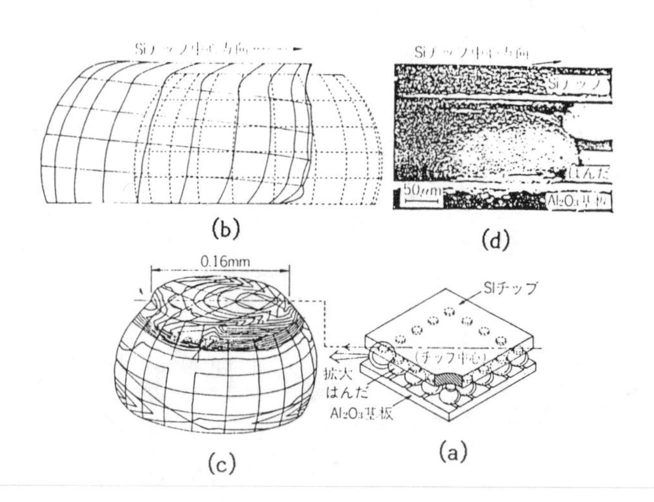

Fig. 8 - Thermal Fatigue Analysis of Flip Chip Solder Joint

Surface mounting technology has been developed to reduce packaging size for printed circuit type packaging. One of the most typical examples is the PLCC (Plastic Leaded Chip Carrier) 68 pin LSI soldered on a glass epoxy substrate. The solder joint becomes stronger as its solder volume and thickness increase. The modeling for such a package is shown in Fig. 9 (a)(b). Fig. 9 (c)(d) show the thermal stress strain curve from 20°C → 150°C → - 55°C.

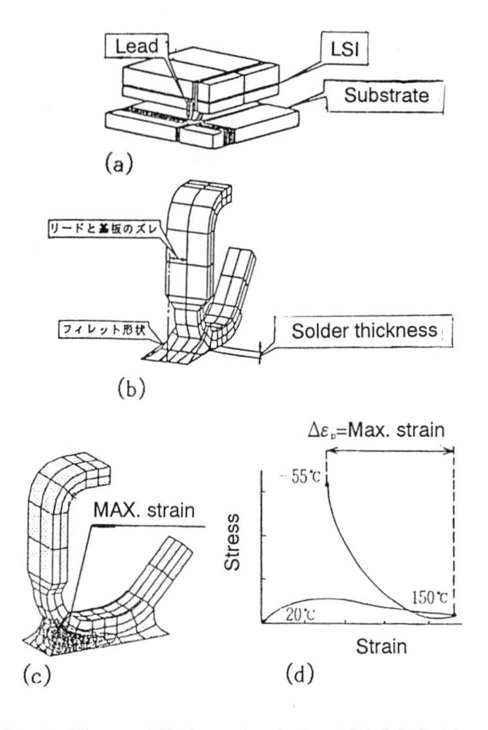

Fig. 9 - Thermal Fatigue Analysis of PLCC Solder Joint

Table 2 uses non-linear and linear analysis methods to make a life comparison between varying thicknesses (1, 2, 3mm) solder joint which can normally occur in production. The 5% error in the linear method is sufficient to study life reliability with regard to parametric effects.

Table 2 - Life Versus Solder Thickness of PLCC

Analysis method \ Solder thickness (mm)		t 1	t 2	t 3
Non-linear (Thermal stress 20 → 150 → - 55°C)	Equi-strain ratio	100	101	137
	Life ratio	100	98	62
Linear (Thermal stress 20 → 150°C)	Equi-strain ratio	100	103	142
	Life ratio	100	95	59

(3) Integration of Control Functions in Control Unit.

An adaptive control unit is necessary which can provide the best vehicle control in regards to emissions, fuel economy, and driveability. Such a unit would be required to monitor and control all areas of engine, powertrain, chassis, and vehicle controls including diagnostics.

As can be seen in Fig. 10 best emission, best fuel economy, and best driveability are not compatible areas of engine operation. This is particularly true in the area of engine control and transmission control. In order to realize such an adaptive control unit a packaging construction such as in Table 3 should be considered from the standpoint of control performance, cost effectiveness, smaller package size, etc. At this time case "A" provides the best construction method from the point of view of cost, performance, and compact size.

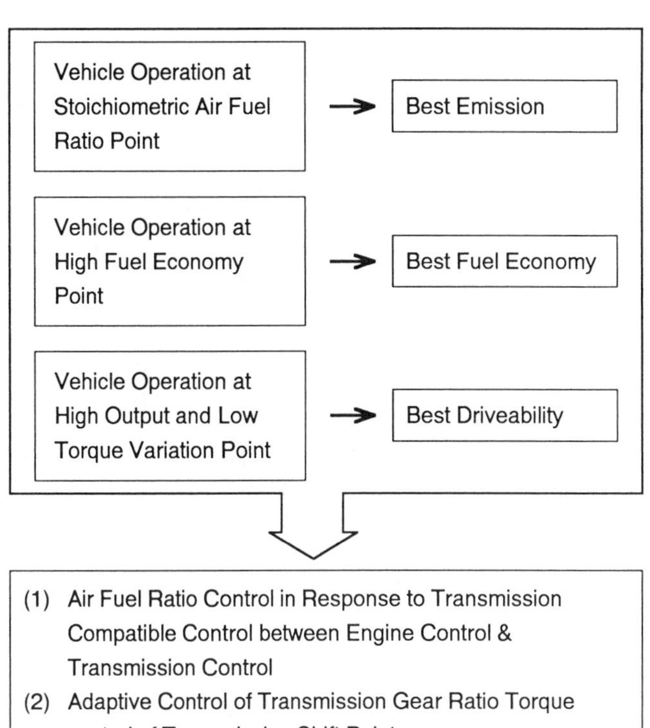

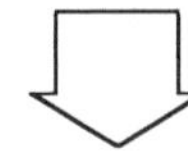

Fig. 10 - Control System Integration

Table 3 - Comparison of Packaging Construction

Item / Case	Construction	Size	Independence of development	Cost	Controllability
A	Case, LSI, PCB	○	○	○	○
B		△	○	△	○
C		△	○	△	△
D	IC	△	○	△	○

⟺ Parallel ; ←→ Serial

Data Communication

FUTURE TRENDS

In the next decade, some organizations forecast electronic automatic transmissions, antilock braking, traction control, multi-plexing, air bags, and active suspension as the most prevailing automotive electronics applications.

As automotive electronics are predicted to become more and more sophisticated and complex, the following items in Table 4 are most important in considering the future total electronic packaging problem.

Table 4 - Remaining Items in Advance

(1) Centralization of Control Unit Function
(2) Mechanics - Electronics Interaction
(3) Appearance of Large - Scale High Density Hybrid IC or Printed Circuit Board
(4) Simplification of EMI protection
(5) Future IC Packaging Trends

(1) Centralization of Control Unit Function

Currently, centralization is being clearly restricted by packaging size, pin size, pin pitch, and insertion force of connector, development time constraints, serviceability, cost, and so on.

Innovation in packaging technology and the development of smaller multi - pin connectors will promote the integration of some functions due to a decrease in control unit size.

On the other hand, over - integration of functions will probably increase required R&D activity, difficulties for product improvement through design modification, and lack of serviceability in the market.

Therefore, adequate centralization of control unit functions will be pursued in all applications including engine powertrain controls, chassis controls, vehicle controls, and information systems.

In the near future, interactive control of engine control with automatic transmission and traction control most probably will become an effective answer to achieve the strict fuel economy requirements and the smooth differential - gear shift requirements by control of the engine output torque during starting, accelerating, and deceleration as shown in Fig. 11.

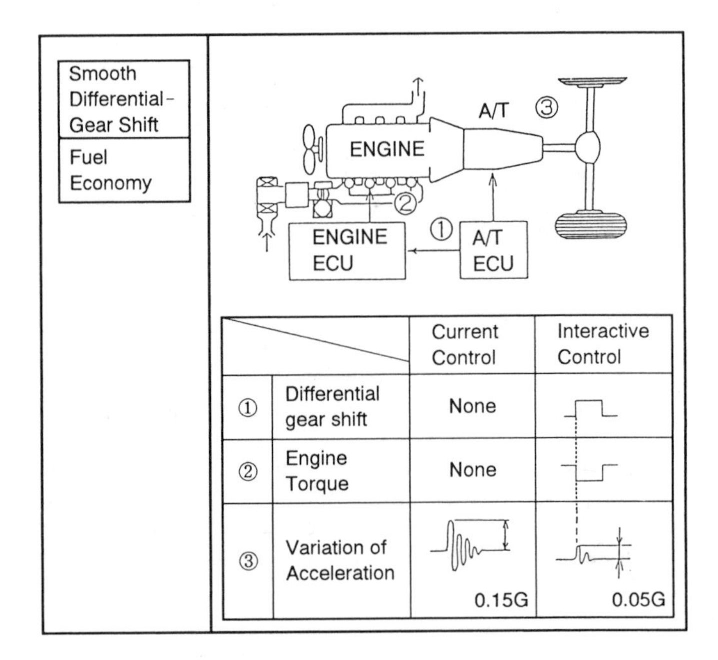

Fig. 11 - Interactive Control of Engine Control with Automatic Transmission

Furthermore, the idea of multi - plexing and/or LAN (Local Area Network) will be gradually introduced to resolve the above problem, and be useful for the improvement of fuel economy accompanied by the wire harness weight reduction as well as the simplification of market service.

(2) Mechanics - Electronics Interaction

As already mentioned, the comparably small HIC electronics have been integrated to the mechanical components, like IC ignition modules or air flow sensors.

In the future, large scale functional (LSI) electronic units will be introduced in conjunction with a sophisticated mechanical products. It is possible that two of the most likely products for this break-through are multipoint fuel injection throttle body and multi - function ignition controller.

These items are shown in photo 4.

Photo 4 - Multi - Point Fuel Injection Throttle Body

(3) Appearance of Large - Scale High Density Hybrid IC or Printed Circuit Board

Multi - layer printed circuit boards (organic or ceramic material HIC) and multi - layer mounting (similar to surface mount technology) on both sides of circuit boards should become increasingly prevalent due to the requirements for the function/volume ratio of control units.

Fig. 12 shows the dramatic size reduction that is possible in engine control units. Table 5 shows similar size reductions for principal components.

The large - scaling of HIC and PCB may induce increase thermal stresses on the units and solder joints. Careful consideration must be given to this point during the design and manufacture to avoid conductor and solder deterioration.

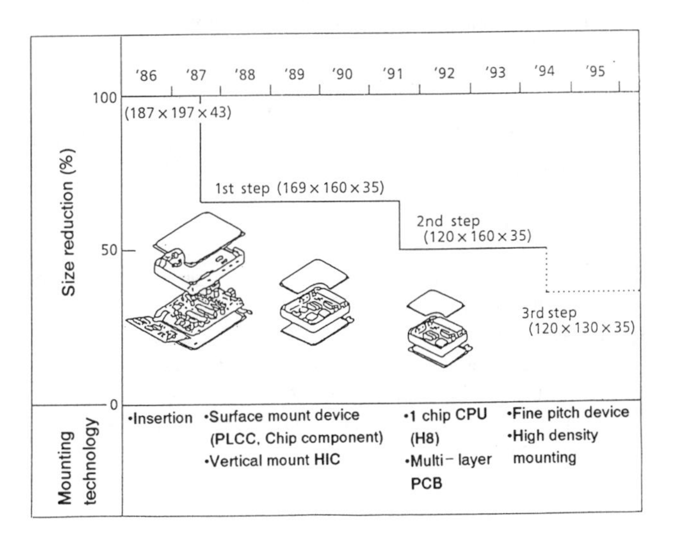

Fig. 12 - Size Reduction in Engine Control Units

Table 5 - Size Reduction for Principal Components

a) Components *as Discrete Parts Size=1

Items	Package	Area Ratio*	Remarks
Resistor	Discrete → Ceramic Chip	0.46	3.2 × 2.6(1/4W) 2.0 × 1.25(1/10W)
Capacitor	Discrete → Ceramic Chip	0.46	
	Discrete → Chemical Chip	0.83	
IC	DIP(19.3x6.3) → SOP	0.46	
CPU	DIP → PLCC	1.65	
	DIP → QFP	0.20	

b) Multi - layer PCB

Element
•Size reduction for ECU •Size reduction for electric parts •High density wiring

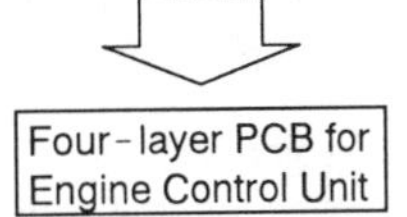

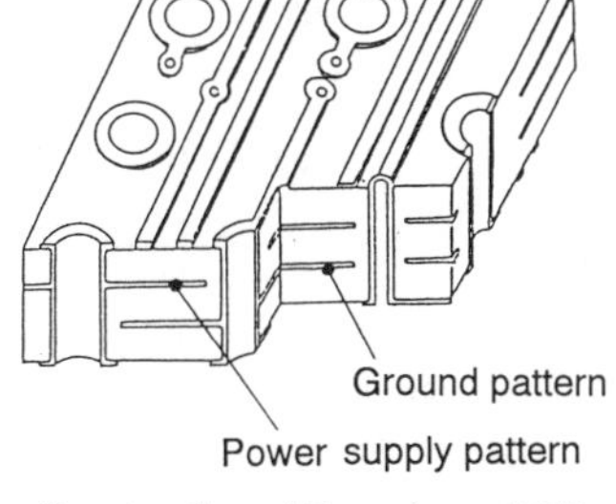

Construction of Four - layer PCB

Comparison between PCB and Four - layer PCB

	Two - layer PCB	Four - layer PCB
Mounting density	1	1.6
Area ratio	1	0.75
Cost ratio	1	1.8

c) Hybrid IC

	Discrete	Hybrid
Construction		

(4) Simplification of EMI protection

It is inevitable, I think, that unique circuits or semiconductors will be developed to withstand harsh electrical noise induced inside and/or outside of the control units. The circuits will not only simplify the packaging of their control units but will probably provide higher temperature resistance.

For example, there will be strong demands for semiconductors which will restrain and/or attenuate the high frequency noise entering from the input circuit and also decrease the output noise generated by it.

Some of the bipolar transistors and bipolar IC's have low gain x frequency band products. Also noise reduction will be attained through the use of noise suppression plastic containing some metal material. At this moment the noise suppression plastic performance is limited, so further work is needed on material. Simplified and lower cost shielded wire will also be strongly desired to block microwave radiation.

(5) Future IC Packaging Trends

Table 6 - IC Packaging Trends

Electronic equipment trends		IC packaging trends
Performance	•High functions	•Larger chip •Multi-pin
	•High speed	•Smaller size •Lower heat resistance •Lower noise
External configuration	•Small size •Thin Type	•Narrow pitch •Thinner type
Packaging	•High density packaging •Automation	•Surface mount •Waterproof packaging •High precision •Tape automated bonding
Application	•Area expansion	•Diversification •Standardization

Table 6 shows the future trends of IC packaging versus electronic module or equipment. All the car makers request higher performance, smaller size and lower cost electronic equipment to their assembly makers.

So, year by year, electronic equipment assembly makers pursue higher density packaging and highly automated assembly line. Therefore, the future of IC packaging is expected to realize the items shown in the above table.

The transfer of IC packaging from pin insertion type to surface mounting type is rapidly on its way.

With the increase of circuit density in IC chips, the pin numbers are increasing gradually and at the same time the pin pitch of the IC package is becoming finer and finer to the extent of 0.25 mm as shown in Fig. 13.

Accompanying with this trends multi - pin flat package QFP (Quad Flat Package) and TAB (Tape Automated Bonding) package must be expanded shown in Fig. 14.

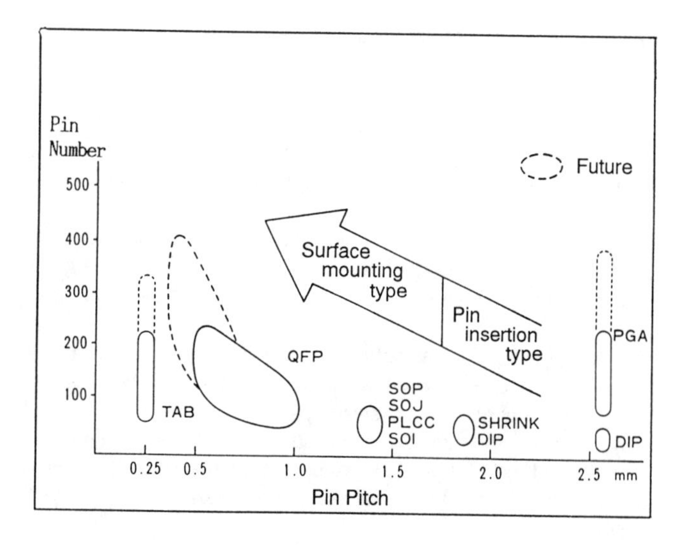

Fig. 13 - IC Packaging Trends (Pin Pitch & Pin Number)

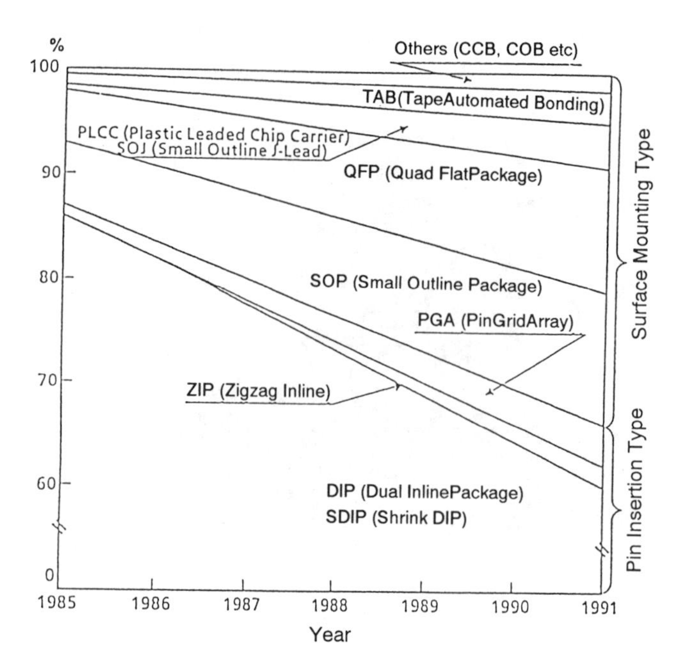

Fig. 14 - Usage Ratio for IC Packaging

CONCLUSION

Based upon the analysis of automotive electronics trends in the past twenty years, electronic packaging trends were clarified. At the same time, the problems were given to electronic packaging and their responding solutions were disclosed concerning EMI countermeasure, thermal stress analysis by CAE, integration among functions of control units.

Next to the above, future trends of electronic packaging were studied from components to control units. Interactive control of engine control with automatic transmission and multi - plexing will be gradually introduced for the improvement of fuel economy. Electronic module interaction with sophisticated mechanical components will be rapidly adopted in the underhood like multi - point fuel injection throttle body integrated mass air flow sensor module.

In order to realize the above mentioned electronic modules, large - scale high density hybrid IC or printed circuit with multi - layer mounting is needed due to the size reduction of modules and components.

Future IC packaging trends including packaging configuration, pin pitch, pin number and usage ratio of configuration were guided.

Surface mounting technology with surface mounting components like multi - pin flat package, QFP must be expanded.

ACKNOWLEDGEMENT

We want to thank our co - development engineers who have contributed to the above achievements;
particularly Dr. Y. Ohyama of Hitachi Research Laboratory, Mr. A. Yasukawa of Mechanical Engineering Research Laboratory, Mr. R. Sato of Production Engineering Research Laboratory, Mr. M. Tanimoto of Head Office, Mr. T. Hasegawa and Mr. H. Tatsumi of Sawa Works, Hitachi Ltd.

REFERENCES

1. D. Bergfried et al., "Engine Management Systems in Hybrid Technology" SAE Paper 860593, 1986

2. K. Nihei et al., "Current Situation and Development of Hybrid Microelectronics" American Electrochemical Society, Proceeding of the symposium on Electrochemical technology in Electronics, 1987

3. A. Yasukawa et al., "Simulation for Designing Semi-conductor Packages with High Reliability under Thermal Cycling" ISPSD '88

4. S. McIntyre Sr. et al., "AUTOMOTIVE ELECTRONICS: THE FUTURE?" Automotive engineering, Vol. 97., No.8, pp 27 ~ 32, 1989

5. Y. Ohyama, "Overall View and Future Trends in Fuel System" Journal of the Society of Automotive Engineers of Japan, Vol. 44, No. 1, pp 100 ~ 105, 1990

6. T. Oguro, "Impact of Advanced Semiconductor Technology on Automotive Electronics" Journal of the Society of Automotive Engineers of Japan, Vol. 43, No. 1, pp 25 ~ 29, 1989

7. T. Inui, "Automotive Electronics in Japan" Proc. of IMECH, 7th International Conf. of Automotive Electronics, 1989

901145

LSI Technology for Meeting the Quality Goals for Automotive Electronics

Hajime Sasaki, Shigeki Matsue, and Yukio Maehashi
NEC Corp.

ABSTRACT

The continuously increasing integration level and resultant "system-on-silicon" and customization trends in VLSI technology will have a significant impact on future automotive electronics. The microcomputer, which is the kernel semiconductor device in automotive electronics, reflects the tremds described above. As the VLSI integration level increases, reliability or quality issues will become more and more important, because of the increased impact of a possible device failure. This is particularly so in VLSIs for automotive electronics. In this paper, VLSI technology trends and ways for meeting reliability or quality goals will be reviewed. Also it outlines a future look at automotive electronics in the 21st century, based on a system-on-silicon microcomputer chip, in which several processor units with different functions are integrated together by ULSI technology, where more than ten million device elements are integrated within a single silicon chip with low submicron feature size.

AUTOMOTIVE ELECTRONICS SYSTEMS began common in the 1970s with engine control systems for exhaust gas pollution suppression and fuel economization. Since then, progress has been marked, and today, sophisticated systems like anti-locking brake control systems to suppress skidding and in-vehicle information network systems, are being used. It is no exaggeration to say that automobiles cannot be discussed without considering these complex erlectronic systems. The role of automotive electronics will continue to increase in importance, aiming at higher levels of safety, fuel economization and comfort.

Semiconductor integrated circuits, which are the key component of today's electronics, have progressed continuously from ICs to LSIs and to VLSIs, such as the microcomputers of today, which contain about one million device elements per silicon semiconductor chip. This evolution is based primarily on technological efforts aiming at higher integration with finer device pattern and higher functionality. The high integration level for VLSIs has enabled them to be " system-on-silicon ", i.e., an electronic system function can be integrated on a single silicon chip. Such VLSI capability has dramatically expanded their applications, both volume wise and function wise, bringing about almost the " Electronics Revolution Era ".

With the progress in the " system-on-silicon " trend, the system engineering role has been becoming more and more important. Microcomputers are now indispensable for automotive electronics. Through sophistication of the system functions, various advanced control schemes, such as learning control, predictive optimization control, fuzzy control, or AI control, will become common in engine control or brake control systems of the future.

In the 1990s, where C&C, or unification of computers and communications, will be fully implemented, on-vehicle systems that can exchange a great volume of information at a high rate with external systems, such as GPS (Global Positioning System), mobile communication systems and so on, will be installed in automobiles. Also, safety systems, such as collision avoidance systems or airbag systems, and comfort systems such as surround audio systems, will be installed in the automobile to make it a " mobile information and comfort environment ".

The environmental conditions that semiconductor devices encounter in automotive electronics are in general considerably more severe than those in other applications. The devices must be durable to withstand higher ambient temperature, severe temperature cycling, and stringent mechanical and electrical environmental conditions. In addition, the semiconductor devices must be highly reliable, since device failure would threaten both passenger and bystander safety as well as probable damage to both the vehicle and traffic facilities. Therefore, careful provisioning for reliability and quality assurance must be implemented.

The following description will be devoted to: (a) VLSI trend in the 1990s focusing on "system-on-silicon" VLSI technology, (b) semiconductor products reliability for automotive electronics, (c) future generation ULSIs and related technology issues, and (d) ULSIs contribution to the automotive electronics of the 21st century.

VLSI TREND IN THE 1990s - "SYSTEM-ON-SILICON"

The marked VLSI advance in integration level and speed performance has enabled accomodating a system level function on a single silicon chip. This situation is called " system-on-silicon ", and it will be one of the most important VLSI technology trends in the 1990s. The integration level has increased with a rate of 4 times per 3 years in the case of DRAM, from 1kbit DRAM in the early 1970s to today's 4Mbit DRAM, which is now in commercial production (see Fig.1). The data processing capability of microprocessor chip has advanced to a level of as high as 10 MIPS, which is almost equivalent to thatof a small mainframe computer as shown in Fig.2. These advances make it possible to integrate functions which otherwise are distributed among plural silicon chips. This integration of function realizes the "system-on-silicon" concept.

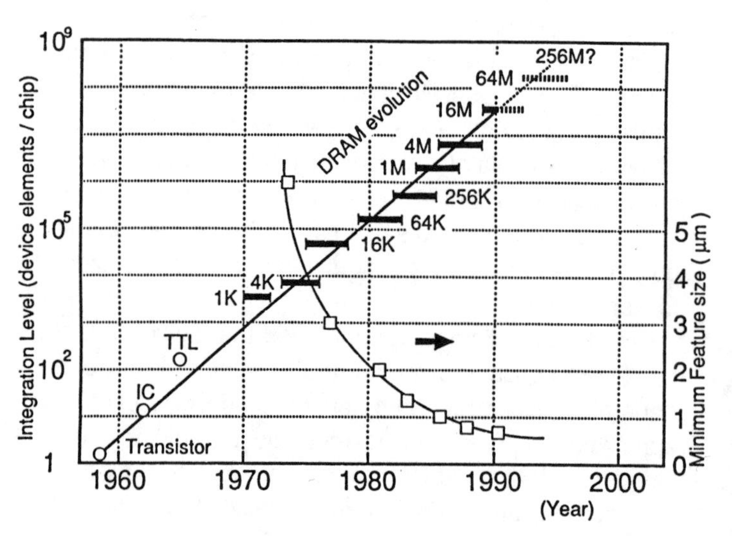

Fig. 1 Integration level and feature size in DRAM vs. year

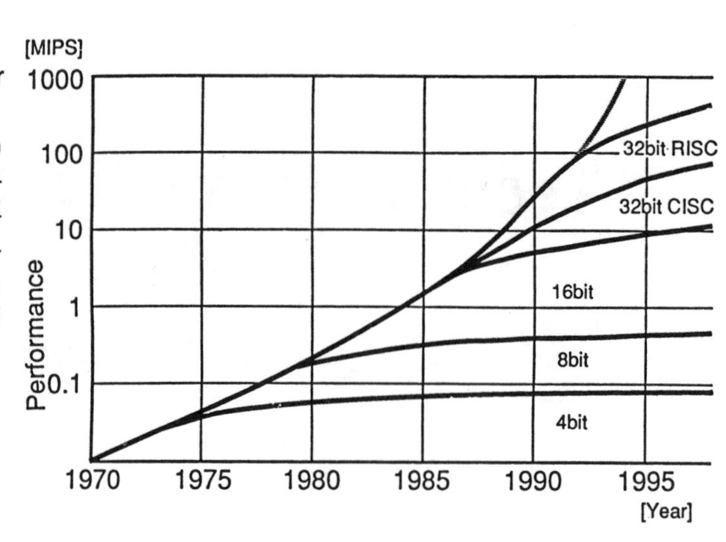

Fig. 2 Microprocessor performance evolution

Another important VLSI trend is the popularization of "customization", such as seen in recent marked popularization of gate array technology. Gate array is a semi-custom VLSI design approach, in which the customer, or system manufacturer, can implement customized functions on the chip by defining final interconnection for a half-made chip wafer containing an array of logic gates.

The "system-on-chip" and "customization" trends are implemented in VLSI microcomputer chips for automotive electronics in the following way.

Until recently, conventional microcomputers have been implemented using a standard hardware architecture that is capable of handling various applications, while customization for individual applications has been carried out by software. Recently, ASSP, or application-specific standard product approach, in which hardware that is specific to a certain application is designed, has become common. One typical example is a high-performance 16 bit microcomputer (Photo.1) designed for engine control, in which a high-performance CPU that can quickly execute 32 bit calculations, an 8 channel 10 bit A/D converter and several timer units are provided to implement full sequential control of an eight-cylinder engine. This is a " system-on-chip ", application-specific standard product.

As an effective approach for customization of microcomputer chips, " Custom Micro ", shown in Fig.3, has been proposed. In this approach, CPU and peripheral functions hardware are implemented using optimized combination ofstandard " macros ". Customized information with usergenerated "macros" is incorporated " user's macro ". Furthermore, combinations of " macros " for various subsystem functions are stored

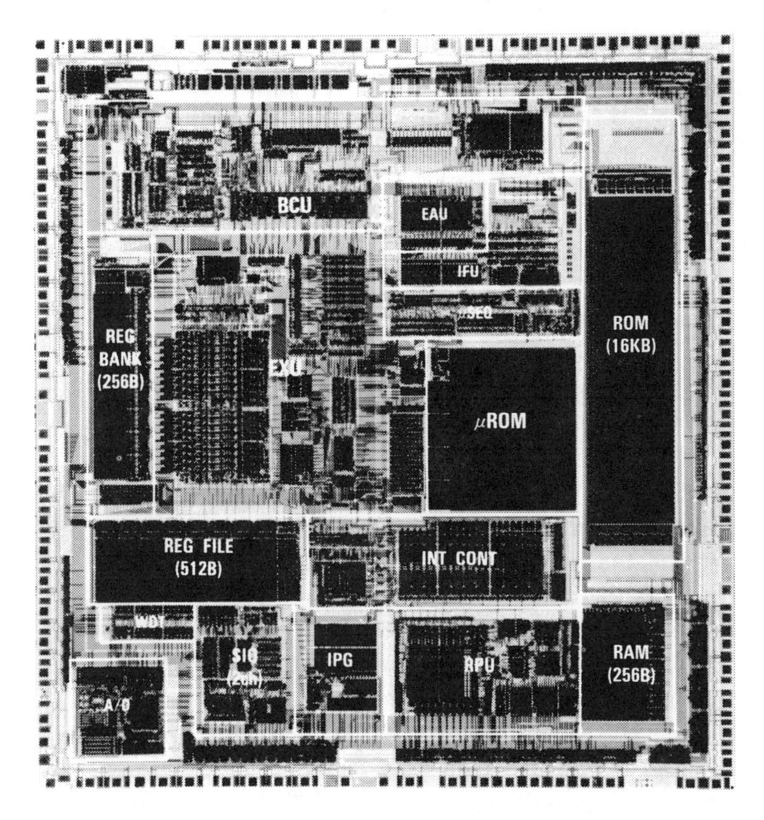

Photo. 1 μPD78602

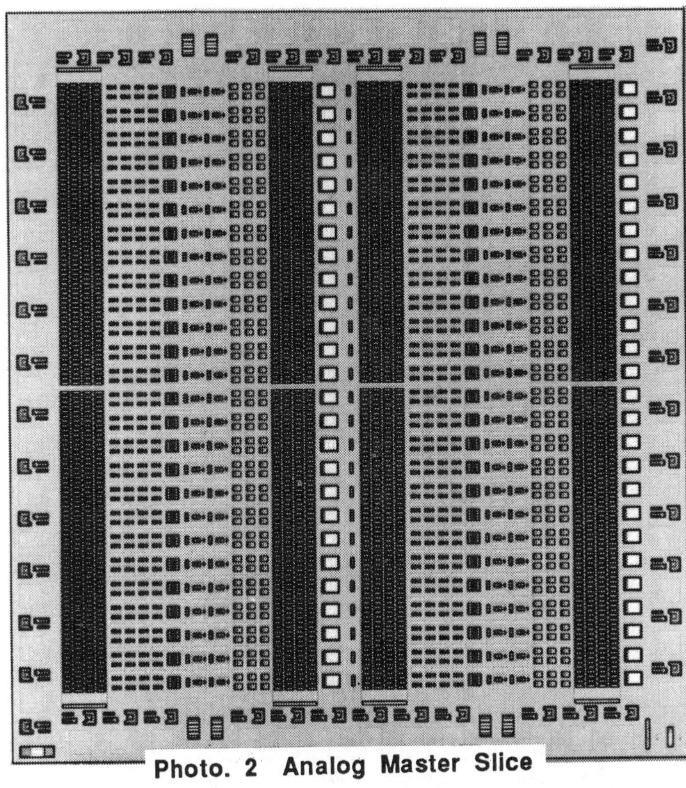

Photo. 2 Analog Master Slice

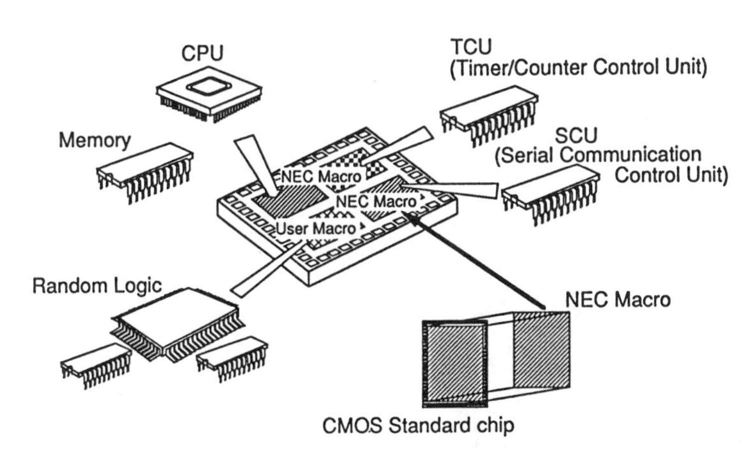

Fig. 3 Custom Micro

as" mega-macros ". Using several selected mega-macros and by adding some new circuit blocks, a system-on-silicon chip of higher system integration can be prepared quite efficiently.

Customized analog integrated circuit technology, which have not advanced as rapidly as the digital technology, is now progreassing with the advent of the " analog master slice " approach. In this approach, analog circuit elements such as transistors, capacitors and resistors, are placed on a chip beforehand(see Photo.2). The customized function is realized as in the gate array case by means of final interconnection on the wafer. In the not too distant future, highly advanced analog and digital mixed system-on-silicon custom products will be commercially available, in which BiCMOS technology optimized for analog-and-digital mixed circuits, and the " analog macro " approach, similar to the aforementioned approach, are utilized.

In the power integrated circuits field, various " intelligent power MOS ICs ", such as for power driver IC for actuator driving, have been developed. Control circuits have been integrated into a single chip, while the power loss in the driver circuitry has also been reduced significantly by adoption of power MOS technology.

SEMICONDUCTOR PRODUCTS RELIABILITY for AUTOMOTIVE ELECTRONICS

In the course of the semiconductor device progress, a number of technological issues have been overcome so far, such as problems in material crystal technology, fine pattern process technology, interface stabilization technology and so forth. From now on, even more difficult tasks must be solved, to meet the goals demanded from the application markets. The reliability or quality issue for meeting severe use point environments in automotive electronics is one of their major items.

The initial period of semiconductor device application to automotive electronics was beset by many difficulties, such as the wide operation temperature ambient, constantly repeating temperature cycles, severe noise environments and so on. However, by patient and close cooperation between automotive and semiconductor industries has resolved the problems steadily. For example, the microcomputer, first announced in 1971, had a -10℃~ +70℃ operation temperature range. This certainly could not meet the -40℃ ~ +85℃ dash board environment. It took five years for the microcomputer to meet the requirement. Today, microcomputers that are usable in a -40℃~ +125℃ engine compartment environment, are available commercially. To reach this goal, key technologies, such as improvements in semiconductor chip surface passivation film technology, in molded package material technology, reliability assurance technology for automotive electronics environments, technology for initial failure rate reduction, and so forth, have been developed and improved intensively.

In the " system-on-silicon " era, the reliability issue becomes much more important, because of the higher reliance on electronics systems in automotive applications. This trend is also enhanced by the more enhanced function and system complexity, higher circuit integration levels, finer patterned device and circuit elements, and associated manufacturing technology complexities. Designing, manufacturing and testing of larger scale integrated circuit require more complicated works. Failure analysis needs more sophisticated techniques. Software development also becomes more commplicated. A higher level of total reliability management approach is necessary. The essential philosophy should be to incorporate the actions to provide product reliability in every phase of the product preparation, from planning to final shipment, and even to after-sales services(see Fig.4). Among these actions, reliability incorporation in the design phase and manufacturing phase is the most vital.

In the design phase, CAD (Computer Aided Designing) is indispensable to accomplish " perfect " designing. In order to design complicated circuit systems without failure, CAD is vital. While designing a circuit, it is important to design the circuit so that it will operate correctly and reliably under rated operating conditions while keeping a reasonable harmonization with hardware layout on the chip and with device and process technologies. Logic designing, logic verification, automatic place and routing, layout designing, layout verification, test designing etc. are utilized to establish reliable design of VLSIs. Another important phase is to implement reliable device and process designing by utilizing simulations. Device simulations and process simulations have been becoming useful and important tools for design of reliable VLSIs. CAD environments, equipped with advanced computer systems and communication networks (LANs) are utilized to efficiently implement such designs.

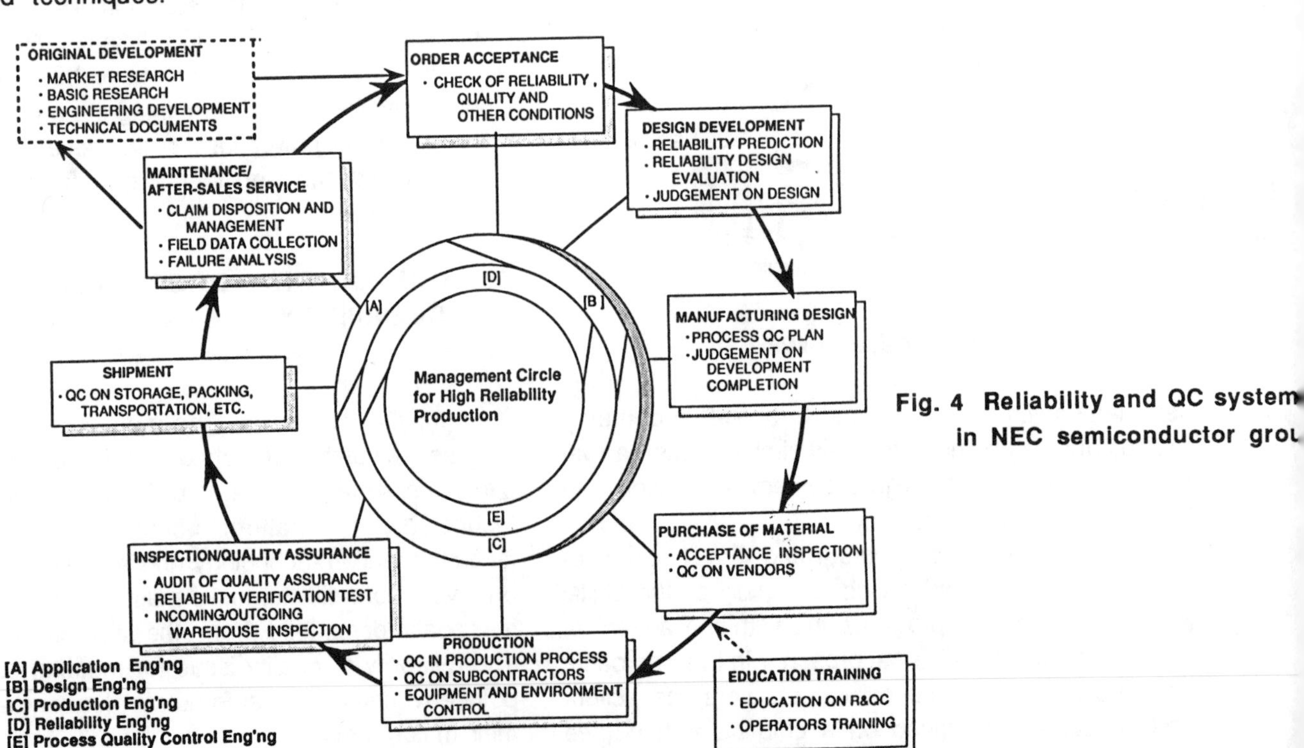

Fig. 4 Reliability and QC system in NEC semiconductor grou

In the manufacturing phase, quality control for materials, processes, inspections, testing, equipment, and manufacturing facilities and environments, are vital. Stringent as well as cost-effective controls are being conducted in VLSI manufacturing. Recent modernized VLSI manufacturing plants are equipped with automated systems, controlled by extensive computerized production management systems. Therefore, quality control on manufacturing equipment and systems is very important. " TPM ", or " Total Preventive Maintenance ", is performed to prevent equipment and systems failures by making programmed maintenance. As the pattern feature size becomes smaller and smaller, from micron to submicron level, extremely stringent control must be carried out as the barrier to for dust and contaminants in the manufacturing process. More and more stringent controls are becoming indispensable for VLSI wafer manufacturing.

In the field of packaging and assembly, "SMT", or surface mounting technology has recently become common. Soon after this technology was introduced, chip and package cracking, due to thermal stresses, was often encountered. However, those problems were solved by close cooperation between the manufacturers and the users to develop a number of new techniques, such as infrared reflow, laser soldering, VPS (Vapor-Phase Soldering), and temperature profile monitoring, to reduce the thermal stress applied to the package. Moisture absorption or penetration in the package material may also cause the cracking. Therefore, " dry-pack " packing has become common, in which packaged LSIs are kept dry up to the moment of mounting process

In the testing phase, it is important to design test pattern that maximizes the failure detection rate. Ideally, the best way is to completely test all of the functional operations of the circuits on the chip. However, this is very difficult to realize and is becoming more so with increasing levels of integration and functionality. For this reason, the emphasis is on data-basing test patterns used for storing accumulated know how, increasing commonality, or providing built-in dedicated test circuits within the chip itself to improve efficiency and reduce time required for the testing. Cost effective screening procedures for rejecting initia failure parts are also being improved.

Along with th e increased level of VLSI integration, failure analysis requires more and more sophisticated techniques. It is no longer enough to use conventional optical microscope, SEM (Scanning Electron Microscope) and LSI testers, but employing much more sophisticated techniqques , like an E-Beam (Electron Beam) tester that can measure real time electrostatic potential distribution of the chip surface while under the dynamic operation of the circuits, a FIB (Focused Ion Beam) system which can dig out a very small precisely defined portion of a silicon chip to reveal a sharp cross-section, or SIMS (Secondary Ion Micro Spectroscope) which can reveal chemical constituents in the fine patterned and thin layered structures. By fully utilizing such modernized tools and techniques, appropriate corrective actions for removing the failure causes can be taken place quite effectively as well as efficiently.

Reliability and quality of semiconductor devices for automotive electronics have been improved significantly, by close cooperation between the semiconductor and automotive industries. In the future, even more closer cooperation between the two industries will be neccessary, to allow the automotive industry to take full advantage of the " system-on-silicon " technology.

VLSI FUTURE PROGRESS

VLSI technology are now progressing continuously as aforementioned. In the 1990s, the level of integrationl will increase to in excess of 10^7 device elements per chip. This is called ULSI (Ultra Large Scale Integrated circuit). The design rule, or device feature size, will be reduced to below the half micron range ($<0.5\mu m$), as shown in Fig. 1.The chip size will be increased up to around an inch square, while the wafer diameter will be increased to 10 inches for efficient production. Multi-layered interconnection, with more than five layers and with highly reliable, low resistance aluminum-based alloy material, will be realized. Operation speed would be around 0.3ns gate propagation delay time for CMOS circuits, and 10ps for bipolar circuits. As for the integration level, DRAM of 256Mbits or more, and more than ten-mega gate logic ULSIs, shoud be developed by the year 2000.

In order to evolve such high ULSI levels, a number of barriers must be overcome. Extreme reduction in device size will bring about physical barriers, such as extremely thin gate oxide breakdown, degraded transistor operations due to the so-called hot electron effects,enhanced statistical fluctuation in device parameters, and so on. Extremely thin interconnect wirings will suffer from stress migration and electromigration, which may cause disconnection failure for interconnects. Process technologies will need significant advancements to become capable of low submicron level device circuit fabrication.

Lithography technology, which is the heart of the process technology, will use advanced photolithography, employing shorter wavelength ultraviolet than currently used. X-ray may even be used for low submicron processing. Molecular or atomic level control will become necessary to accomplish the goals. These situations also require sophisticated micro-analysis and diagnosis tools. Key materials, like photoresist, must be developed further. Devices will be much more sensitive to contaminations, and the materials used for the processing as well as the processing environments, must be ultra clean. Intensive, as well as extensive research and development are now conducted aiming at the realization of ULSIs for the future.

Another barrier to be considered for the development of ULSIs is the shortage of engineering human resources. While the trend is for ever increasing integration scale and functionality, VLSI product development cycle continues to shorten in terms of time and expand in terms of variety. The number of available engineers as well as their talents, are limited. Therefore, it will be necessary to provide CAD tools that enable efficient designing of VLSI or ULSI hardware, while for sofware design, efficient and flexible high level language environments are necessary.

Through overcoming the barriers mentioned above, the semiconductor industry can go into the 21st century or the ULSI era. In the next section, the ULSI impact on automotive electronics will be described.

AUTOMOTIVE ELECTRONICS IN ULSI ERA

Let us examine a ULSI microcomputer chip to be used for automobile in the year 2000. From an architecture point of view, it contains five processor units; i.e., for Power Train Control, Chassis Control, Vehicle Communication Network, Diagnosis, and Back-up processors. Each processor unit consists of individual ROM, RAM and several peripheral functional blocks. Individual processor units operate independently from each other, within the Control ULSI chip (see Fig. 5). In designing the Control ULSI chip, individual processor units to be contained with in the chip will be designed and developed independently from each other. As a consequence, a large amount of accumulated design data or know how can be utilized repeatedly, saving efforts of new development. By adopting such integration of several processor units, some common functions can be commonized within the chip. Individual processor units are connected to an internal communication bus via a shared memory. They can communicate with other processor units and can exchange data with each other.

The Power Train Control processor manages engine and transmission systems, which are its typical control targets. Major control targets of the Chassis Control processor are the active-suspension and 4WS(four wheel steering).The Vehicle Communication Network processor supports low-rate to high-rate data transmissions between microcomputer systems within the vehicle.

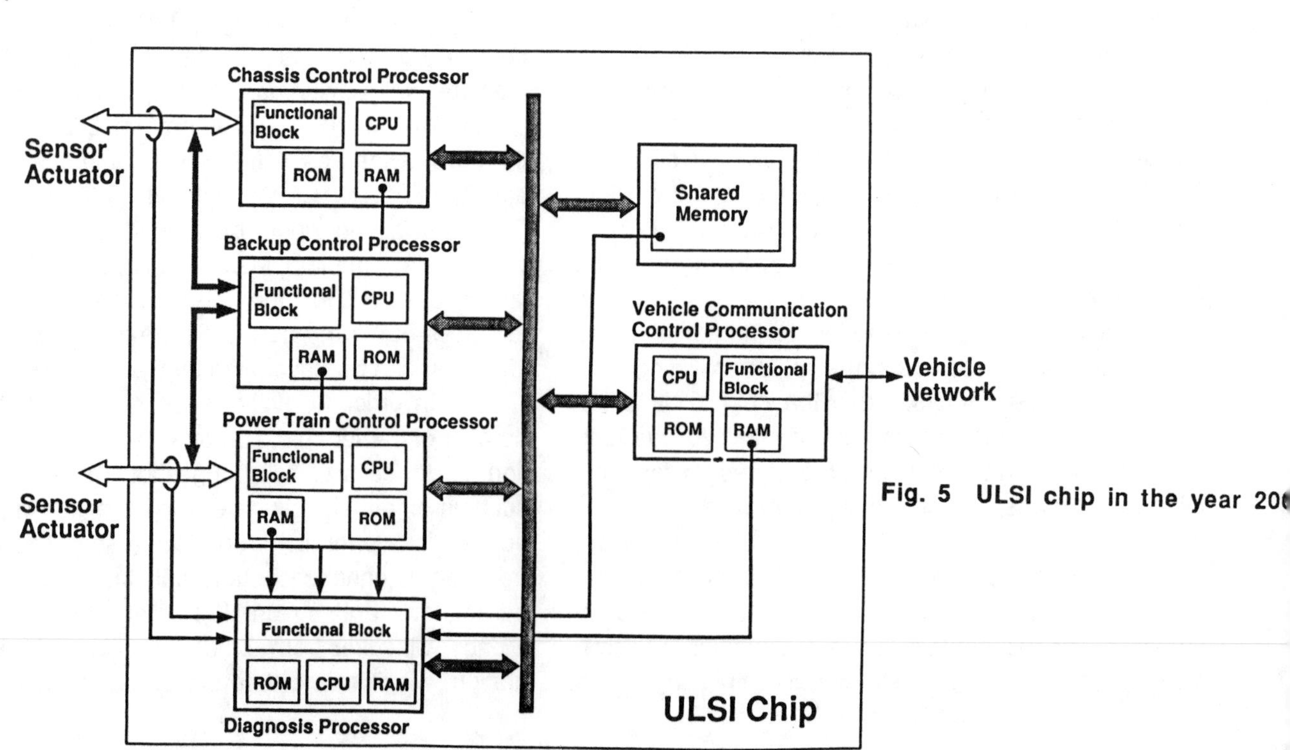

Fig. 5 ULSI chip in the year 20

The Diagnosis processor unit monitors other processor units operations and overall system operations. In case any anomaly in operations is detected, the Back-up processor unit memorizes previous operation data, and takes over the major role of the malfunctioning block.

The cores of the five processor units and communication buses are standarized and provided by the semiconductor manufacturer. The peripheral blocks, however, will be selected from standard options offered by the semiconductor manufacturer or customer-specific circuit blocks designed by the user.

Many of the new technologies, which are currently under reserach and development, will become commercialized to realize such ULSIs. Currently, automobiles are driven by their human drivers. However, in the 21st century, there would be many technologies and systems that will assist automatic driving and navigation. A car could be expected which is driven automatically, through utilizing AI (artificial intelligence) control technology, where the automobile is equipped with pattern recognition system utilizing fuzzy and neuron theory for finding roads and signs.

The driver only needs to instruct destination point at the beginning of driving, and a fully automatic navigation system will take the automobile to the destination, utilizing a mobile communication systems to determine the cars location (auto locating system) and refering to traffic information provided via this mobile communication system.

In addition, an automotive-tuned personnal computer with multi-media capability will be standard equipment on an automobile for incoming data acquisition from all control processor units(see Fig. 6). This computer will function as a navigation system in addition to reporting the vehicles current state. Using this system, the driver can perform real-time monitoring of engine conditions and others on a display panel. It also indicates to the driver procedures to cope with trouble, as it happens, via an easy-to-understand voice and visual display interface. This computer can also serve as a part of an office automation system, and can communicate with a host computer at a main office via a satellite. The automotive environment described above would be a "comfortable moving office".

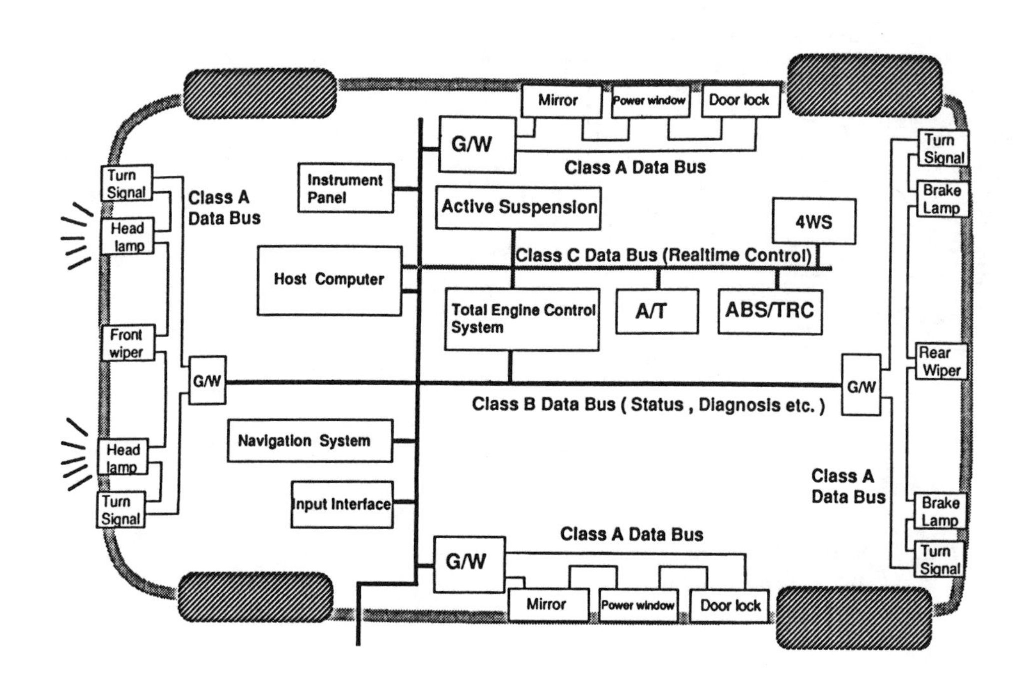

Fig. 6 Vehicle communication networks

CONCLUSION

Automobiles in the 21st century will provide a means of transportation of enhanced safety, energy economization and comfort. The driver enviroment will be easy to operate for driving. On board computers will assist the driver in decision and control operations, and provide comfortable cruising. The automobile will not be merely a transportation means from one place to another, but will have become a mobile information terminal, which will be serve as one of the key nodes in main-trunk network systems(see Fig. 7).

To achieve the goal stated here, cooperation will be necessary, not only between the semiconductor and automotive industries, but also among the communications and office automation systems industries.

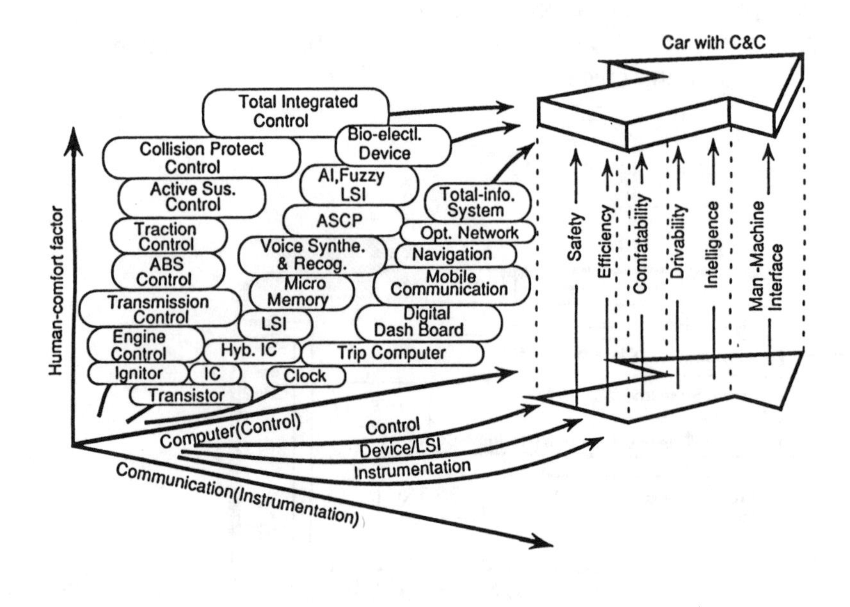

Fig. 7 C&C technology in automotive electronics

901708

A New 4-Bit Microcomputer with Highly Reliable Architecture for Peripheral Circuits of ECU

Hiroshi Nagase, Yoshihisa Harata and Mitsuharu Takigawa
Toyota Central Res. & Dev. Labs., Inc.
Keiji Aoki, Shinichiro Tanaka and Tokuta Inoue
Toyota Motor Corp.

ABSTRACT

Electronic Control Units (ECUs) for automobiles are usually composed of a main single-chip microcomputer and peripheral circuits with some standard and/or custom ICs. The peripheral circuits vary with the kinds of control or models of automobiles.

When the peripheral circuits are replaced with a single-chip microcomputer, the ECU becomes compact and low in cost. This is because the ECU is constructed with only two LSIs and can be used for various kinds of control and various models of automobiles only by changing the program of the microcomputer. The microcomputer, however, requires many I/O functions and high reliability.

We have developed a new 4-bit microcomputer suitable for these requirements. The new microcomputer has two remarkable features. One is powerful I/O functions such as high speed I/O, serial I/O, parallel I/O, analog I/O, and default output that is generated in place of the calculated output by the main CPU when it fails. The other is highly reliable architecture avoiding program runaways to the utmost with the following characteristics: (1) The instruction set is simplified. (2) All instructions are 1-byte long including a parity bit and an operand field. (3) No operation is executed when a parity error occurs. (4) Branch instructions on which the address goes back are limited to only one. (5) A power-on and supply voltage-drop reset circuit is included. The new microcomputer has been fabricated on an LSI chip using a standard CMOS process technology with a 2-μm design rule. The chip contains about 47,000 transistors and measures $5.4 \times 4.6 mm^2$.

WITH THE PROGRESS OF SEMICONDUCTOR TECHNOLOGY, electronic control systems within an automobile have been increasing rapidly and their application area have been spreading. Nowadays, some automobiles contain more than 20 electronic control systems based on the microcomputer. An automotive electronic control system usually has an Electronic Control Unit (ECU). Most of the ECUs use an 8-bit or 16-bit microcomputer as the CPU. These microcomputers, which are mostly custom LSIs developed for automotive use, are different from the standard microcomputers for consumer product use in the I/O functions, the instruction set, the architecture and so on. However, these microcomputers are not subject to wide variations because the cost for developing a custom microcomputer is still very high.

In oder to realize various ECUs for many kinds of control and models of automobiles, therefore, ECUs are sometimes composed of the same microcomputer as that for automotive use and the different peripheral circuits with some standard and/or custom ICs. Future development of an ASIC microcomputer with easily changeable I/O functions, instruction set and architecture makes it possible to produce an ECU composed of only the ASIC microcomputer with the function of peripheral circuits, resulting in simple structure, small size and low cost.

Therefore, we considered to replace the peripheral circuits with a single chip microcomputer and made various ECUs with the same composition. The microcomputer enables the ECU to be compact and inexpensive because the ECU is made of only two LSIs and is used for various kinds of control and models of automobiles only by changing the program of the microcomputer. Moreover, the microcomputer makes it easy to change the design of electronic control systems and reduce the time for development. However, the microcomputer which is substituted for peripheral circuits of ECUs needs powerful I/O functions and high reliability.

We have developed a new 4-bit microcomputer suitable for these purposes. This paper describes the new microcomputer designed so as to be used not only as the slave CPU mentioned above, but also as a main CPU of simple control systems such as a cruise control system and a knock control system. These systems do not need high computation ability.

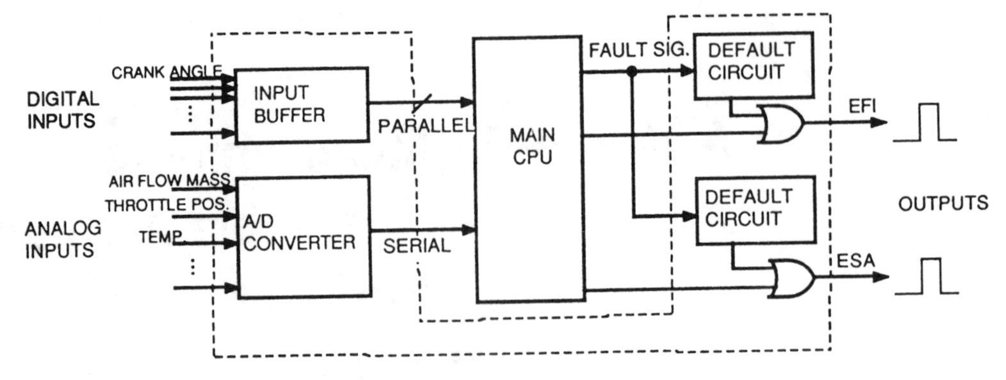

Fig. 1 - ECU for engine control

REQUIREMENTS

An example of ECUs for engine control is shown in Fig.1. We intended to replace the peripheral circuits enclosed by a dotted line with the new microcomputer. In this case, the new microcomputer needs powerful I/O functions such as serial I/O, parallel I/O, analog I/O and default output because it works as an I/O controller. The default output function sends default output signals instead of the calculated output signals from the main CPU when it fails.

We also intended to use the new microcomputer as a main CPU of small scale ECUs. In this case, it needs special functions such as the high speed I/O required for automotive controls. Of course, in both cases, it is very important to make the new microcomputer highly reliable so that it prevents easily illegal addressing and program runaways.

I/O FUNCTIONS

The new 4-bit microcomputer was designed with the following I/O specification in order to satisfy the above-mentioned requirements.

DEFAULT OUTPUT - Default output circuits are sometimes included in the peripheral circuits of ECUs which need the backup function when the main CPU fails. The default circuits send the output signals with the default value instead of the calculated value by the main CPU.

We integrated 2-channel default circuits in the new microcomputer. Utilizing this circuits or not depends on the program. This circuits use some pins of the parallel I/O port as their inputs and outputs. Fig.2 shows the 1-channel default output circuit. The calculated output signal such as EFI or ESA from the main CPU goes through the new microcomputer when the main CPU operates normally. However, it is exchanged to the default signal when the fault signal from the main CPU is received. The default signal is specified by the program.

ANALOG I/O -One of the features of the new microcomputer is that it contains not only an A/D converter but also a D/A converter as the powerful analog I/O functions.

The A/D converter is a successive approximation type with 0-Vcc input voltage range, 8-bit resolution, 7-bit accuracy and 72-μs conversion time(at 2 MHz). It includes a resistor string type 8-bit D/A converter and a comparator. When Vcc is supplied to the reference voltage pin, the conversion is done in proportion to Vcc. Analog inputs have 7 channels.

The D/A converter is the same as the resistor string type used in the A/D converter. The output voltage range is 0-Vcc and the accuracy is 7 bits when the output current is below 100nA. The analog output can be used as a variable reference voltage signal for an external comparator or as an analog control signal for an actuator through an external power amplifier.

HIGH SPEED I/O - The microcomputer for automotive applications requires a powerful timer-based high speed I/O function. This function is used to measure the input and output timing of pulse signals with a high degree of accuracy without software assistance.

We gave this function to the new microcomputer so that it could also be used as the main CPU for relatively simple automotive control systems. The high speed I/O section consists of a 8-bit free run timer, a compare register and a capture register, and six kinds of time resolution from 8μs to 512μs are available to be selected by the program. The default output function and the high speed output function cannot be used simultaneously.

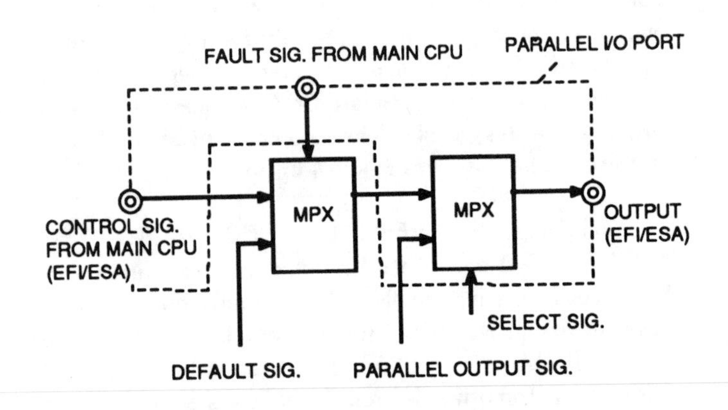

Fig.2 - Default output circuit (1CH.)

SERIAL I/O - As the serial I/O which communicates with a main CPU, the standard Universal Asynchronous Receiver Transmitter (UART) is adopted. Timing clock is supplied from the external, and the maximum transmission speed is 62.5Kbits/s. The data frame consists of 10bits : a start bit "0", 8 data bits and an end bit "1". And "1" is being transmitted during the idle state.

PARALLEL I/O - There are 16 lines in all as the parallel I/O. All of them are used as input or output and this selection is available with 4-bit unit or every bit. Some pins for the parallel I/O are shared with the analog I/O pins and the high speed input pin.

HIGHIY RELIABLE ARCHITECTURE

In the design of the new microcomputer, we applied some ideas to avoid the occurrence of software upsets and to obtain the stable and reliable program run [1]. The highly reliable architecture is described in this section.

MEMORY ACCESS - 4K-byte ROM and 128-word($\times$ 4bits) RAM are included as the program memory and the data memory, respectively. We used the following memory access method to obtain highly reliable addressing.

The ROM is divided into a main program area (2K bytes), an interrupt program area (1K bytes) and a data table area (1K bytes) as shown in Fig.3. The memory access to these areas is completely separated by the hardware architecture. When the main program is being executed, the program counter is limited from 000_H to $7FF_H$ and is automatically reset to 000_H after it reached $7FF_H$. When the interrupt program is being executed, the program counter is limited from 800_H to BFF_H and after it reached BFF_H, it automatically returns to the address of the main program just before the interrupt. The data table area is directly accessed only by the table read instruction without using the program counter.

Moreover, address buses of RAM and ROM are also separated.

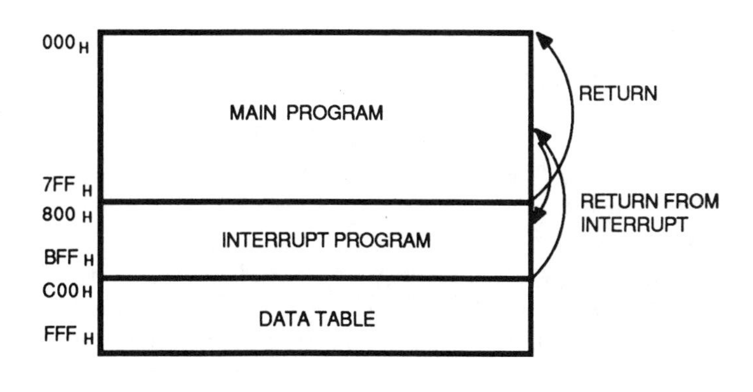

Fig. 3 - ROM map

SOFTWARE DESIGN - We considered that simplifying the instruction set provided the following advantages: (1) The structure of hardware is also simplified, miniaturizing the chip. (2) The instruction cycle becomes short. (3) The software error is reduced, and the high reliability is attained. These considerations are based on the new computer architecture called Reduced Instruction Set Computer (RISC).

Our microcomputer aims exclusively at automotive applications. Therefore, we investigated some microcomputer programs for automotive controls and studied to reduce the number of instructions to the utmost by omitting the instructions that were rarely used.

As a result, we picked up only 29 instructions for the new microcomputer, and decided that all instruction codes were 8 bits, 7 bits of which were assigned to opcode and operand and the rest to parity bit. These are all 1-cycle instructions except only the table read instruction which needs two cycles to be executed due to the indirect addressing by using the internal register.

The 1-byte instruction codes with a parity bit lead to adoption of a new method for avoiding program runaways and getting high reliability instead of a watch-dog-timer [2] in the new microcomputer. Parity check is always done at the time of instruction fetch. When a parity error occurs at that time, NOP(No OPeration) instruction is executed and a parity error signal is sent to external.

This method prevents the program runaways caused by an exchange between the opcode and the operand due to an address error, which sometimes occurs in the conventional microcomputer, since the opcode and the operand are included in 1-byte code. When an access to non-programming area is suddenly done for some reason, a parity error is certain to occur. When it occurs, NOP instruction is executed with no damage and the program counter is counted up. After these procedures, the access come to the program area and finally the program sequence returns to the normal loop. When a parity error is also induced by other causes, the program sequence returns to the normal operation soon after the same procedures.

Since subroutine call and branch instructions sometimes cause program runaways, we omitted subroutine call instructions and reduced the branch instructions to only two. Moreover, we restricted forward-jump instructions to only one, namely the absolute jump instruction.

RESET CIRCUIT - Since supply voltage largely changes in automobile environment, we designed the new microcomputer so as to operate in the supply voltage range from 3V to 5.5V.

Furthermore, we integrated a reset circuit into the microcomputer in order to ensure its operation start and to avoid program runaways due to the supply voltage drop. The reset circuit generates a reset pulse when the supply voltage exceeds a certain voltage level at power-on or when the supply voltage becomes lower than another certain voltage level. These voltage levels were set at about 3V.

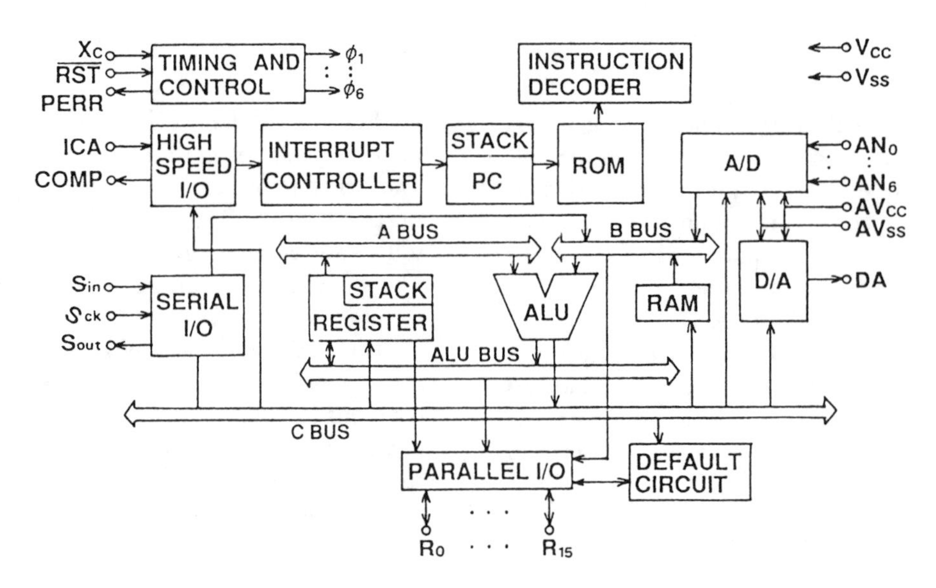

Fig. 4 - Block diagram

Fig. 5 - Microphotograph of the new 4-bit microcomputer

Table 1 - Features of the new microcomputer

ROM	4Kbytes
RAM	64bytes (128 × 4bits)
Analog I/O	8-bit A/D (0 ~Vcc), 7CH. 8-bit D/A (0 ~Vcc), 1CH.
High speed I/O	1CH. each (8-bit timer)
Serial I/O	1CH. each
Parallel I/O (External data bus)	4 × 4CH. or 16 × 1CH.
Default output	2CH.
Interrupt	3sources, 1level, 1vector
No. of instructions	29 (1-byte length in all)
Instruction code length	1byte with a parity bit
Instruction cycle	2~4μs at $f_{CLK} = 2MH_z$
Internal data bus	4bits
Process technology	2μm CMOS
No. of pins	28
Operating supply voltage	3-5.5V
Chip size	5.4mm × 4.6mm
No. of transistors	47,000

FABRICATION RESULT

We fabricated the new 4-bit microcomputer using the standard CMOS process technology with the 2-μm design rule. The block diagram and microphotograph of the new microcomputer are shown in Fig.4 and 5,respectively. About 47,000 transistors are integrated within a chip, measuring 5.4× 4.6mm^2. The features of this microcomputer are summarized in Table 1.

ASSEMBLER

It is generally hard to develop application programs for the RISC type microcomputer which contains small number of instructions. Therefor, we realized a new assembler which has the following characteristics to facilitate the development of application programs for our microcomputer.

(1) Users can define hierarchy of macro operations with arbitrary depth, and some intrinsic macro operations are prepared beforehand. These make up for the inconvenience that our microcomputer has no subroutine call instruction.

(2) Free expression with real constants and multiple parentheses is available in operand fields. This makes it easy to fill operand fields.

(3) Free format description is possible.

(4) Symbolic expression with arbitrary length for labels and variables is possible. This makes it easy to understand a flow of application program.

(5) Visible and substantial assembly list shown in Fig.6 is printed out. Macro expansions are indented in proportion to the depth of hierarchy. CDOWN in Fig.6, for example, is a hierarchic macro operation and it is expanded to a macro operation LAT1 and three executable instructions LY1, SETC and SBC. LAT1 is further expanded to two executable instructions LY1 and XAY. This assembly list makes debugging very easy. Characteristics of the assembler are shown in Table 2.

APPLICATIONS

We have studied to apply the new microcomputer to two kinds of ECUs. One is for engine control and the other is for Exhaust Gas Recirculation (EGR) control of diesel engine.

Fig.7 shows the structure of the former, which employs the new microcomputer as an I/O controller in place of the peripheral circuits. The main CPU is Toyota's 8-bit single chip microcomputer [3] that is a custom microcomputer developed for automotive use. High speed input signals are supplied to the main CPU and the new microcomputer. However, the high speed output function of the new microcomputer is not utilized because that the default output function is utilized. The analog output of the new microcomputer is used as the threshold voltage to the knock signal. The knock signal and the analog output signal are supplied to a comparator and its output signal supplied to a parallel I/O pin of the new microcomputer. When the knock signal level becomes higher than the analog output signal, that is, the threshold voltage, the ECU controls knocking. The threshold voltage is set by the program of the 4-bit microcomputer or is changed according to engine condition or circumstances by the microcomputer operation. This example shows that using the new 4-bit microcomputer, the ECUs are constructed of two LSIs and a few of circuits such as voltage regulator, quartz crystal clock generator and comparator.

```
line#  address  code  label  operation  operand
100    A0E      E3           XD         NO
101                          LXI        RAMO      : a macro
       A0F      40             LYI      $0
       A10      76             XAY
       A11      75             XX
102                          CDOWN      C0,MEMC0  : a hierarchic macro
                               LAI1     $0        : a macro
       A12      40               LYI    $0        : a executable instruction
       A13      76               XAY
       A14      40             LYI      $0
       A15      F7             SETC
       A16      FE             SBC
103    A17      02           JMP        SKIP
104    A18      F1           ST
105    A19      45    SKIP   LYI        5
```

Fig.6-An example of assembly list showing macro expansions

Table 2 - Characteristics of the new assembler

		The new assembler	Typical assemblers for 4-bit microcomputers
Kinds of Operation	Executable instruction	29	More
	Pseudo operation	11	Similar
	Intrinsic macro operation	9	None
	User's macro operation	Available	Unavailable
Expression in Operand	Operation procedure	1.( ) 2.×, ÷, 3.+, -	In described order
	Parentheses	Usable	Unusable
	Nearest integer function	Usable	Unusable

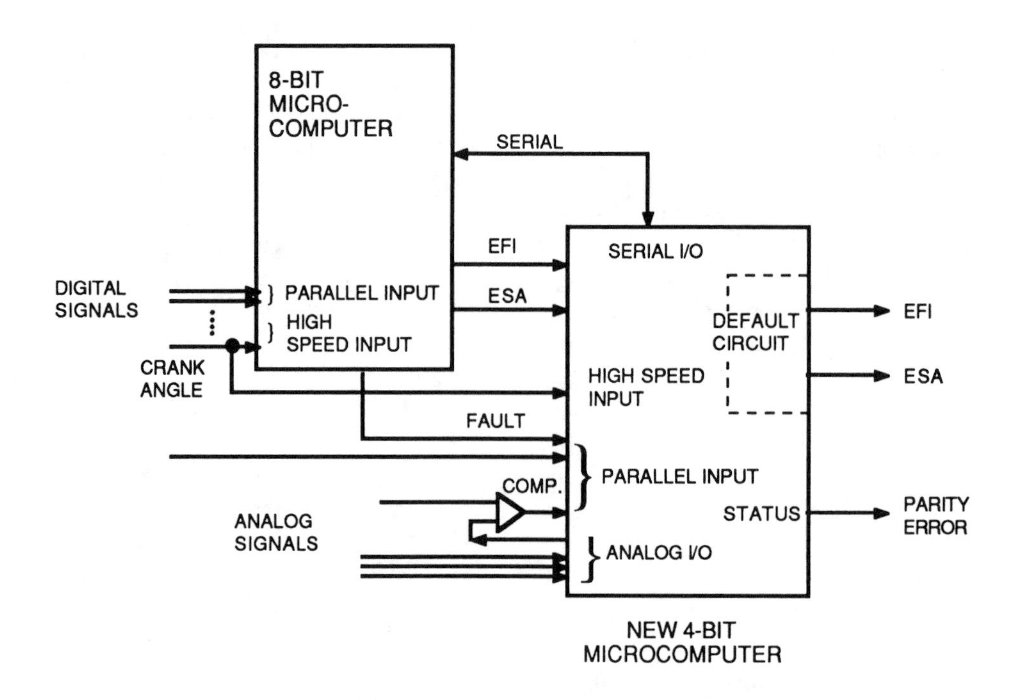

Fig.7 - An example of ECU using the new microcomputer as an I/O processor (Engine control system)

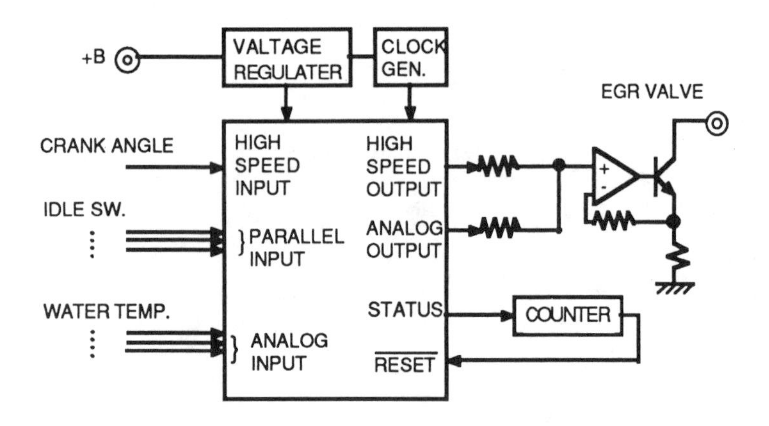

Fig. 8 - Another example of ECU using the new microcomputer as a main CPU (EGR control system)

Fig. 9 - Photograph of the ECU using the new microcomputer

Fig.8 shows the architecture of the ECU for EGR control of diesel engine. In this ECU, the new microcomputer is used as a main CPU, and many sensor signals are supplied directly to the analog inputs and the parallel inputs. The high speed I/O function is utilized, and its output is EGR valve control signal. Fig.9 shows a photograph of the ECU. The new 4-bit microcomputer is available to be used as a main CPU as shown in this example.

CONCLUSION

We have developed a new single-chip 4-bit microcomputer for automotive applications. This microcomputer aims particularly at a slave CPU in place of peripheral circuits of ECUs. It has been designed so as to have multifunctional I/Os including high speed I/O, serial I/O, parallel I/O, analog I/O and default output and to obtain highly reliable architecture such as separated memory access, the simplified instruction set, the instruction code being 1-byte with a parity bit, no operation at a parity error and an on-chip reset circuit.

Using this microcomputer instead of peripheral circuits of an ECU, the ECU becomes compact, applicable to various controls only by changing its program and easy to change the design of control systems, and then it becomes low in cost.

REFERENCES

[1] T.Yasui, M.Kato, "The One Bit Microcontroller", SAE860569.

[2] D.F. Hagen, D.F. Wilkie, "EEC-IV Tomorrow's Electronic Engine Controls today", SAE820900.

[3] T.Kawamura, M.Kawai, K.Aoki, "Toyota's New Single-Chip Microcomputer based Engine and Transmission Control System", SAE850289.

911658

The Impact of Higher System Voltage on Automotive Semiconductors

Ben Davis, Randy Frank, and Richard Valentine
Motorola Semiconductor Products Sector

ABSTRACT:

Doubling or quadrupling the present 12V electrical system provides advantages and disadvantages to many components, and to systems that utilize these components. The effect that this increase has on vehicle electronics starts at the semiconductor component level. Several design implications including technology tradeoffs must be evaluated, and the cost impact on silicon area, processing complexity, and current process limitations considered. This paper focuses on the impact of increasing the system voltage to 24V or 48V with and without centralized load dump transient suppression. The semiconductor devices principally affected are transient suppression devices, power semiconductors, smart power ICs and linear drivers. In addition, the effect of a change in system voltage on MCUs, DSPs, memory devices, and other interface circuits will be discussed.

INTRODUCTION

Semiconductor manufacturers, as well as other component manufacturers, continue to develop more power efficient devices. Semiconductor devices of yesterday were predominantly bipolar structures consuming large amounts of power compared to the MOS technologies of today. Bipolar structures inherently consume relatively high operational power by virtue of their required base currents in contrast to MOSFET structures. High demands for more electrical power are prompting automotive engineers to consider raising the battery voltage in order to make those systems more practical.

STANDARD 12 VOLT SYSTEM

The 12V system currently used in passenger cars and light trucks has several additional conditions that must be taken into account when using electronic components. Normal operating charging systems regulate the output of the alternator to provide sufficient voltage to keep the battery charged under various temperature and load conditions. This charging voltage can range from 15+ volts when very cold (-40°C) to below 13V during extreme underhood temperatures (>125°C), and below 12V with a heavy load at idle. Electronic assemblies on the car must be able to withstand a wide range of electrical supply conditions: Reverse battery hookup (-13V), jump start from tow trucks (24V), short duration voltage transients that can easily exceed ± 100V, and long duration (≥ 400 ms) load dump transients that can exceed +80V. ***Table 1*** gives a summary of these extremes using the 12V battery system as a reference and scaling up for the 24V and 48V systems. Preventing semiconductor component damage due to excessive voltages normally requires devices with sufficiently high breakdown voltage and/or protective circuits.

CONVERSION FROM 12 TO 24 OR 48 VOLTS

Some of the disadvantages of changing to a 24V or 48V battery system include poor lamp life (filaments tend to be longer and more fragile), higher leakage currents, higher voltage rated switching devices such as power semiconductors, and probably the most significant disadvantage, retooling/redesign costs of all the vehicle's electrical/electronic components. Part of these costs include many of today's semiconductor products when they make a transition from low voltage (30V to 85V) to high voltage (100V to 250V) processing. Design rules and costs change considerably when

	12 V *	24 V	48 V
Battery Cells	6	12	24
Nominal battery voltage	12.6V	25.2V	50.4V
Charging voltage range	12.6V to 16V	25.2V to 32V	50.4V to 64V
Cranking Voltage (minimum)	5.6V	11V	22V
Jump Start	12-24V	24-48V	48-96V
Reverse Battery	-13V	-26V	-52V
Semiconductor $V_{(BR)}$ **	85V	170V	340V
Semiconductor $V_{(BR)}$***	60V	120V	240V

* derived from SAE J1113a and SAE J1211 reports, 24V and 48V are estimates

** typical non-protected electrical system; depends upon each OEM spec.

*** for a protected electrical system; depends upon each OEM spec.

Table 1 **- Voltage Extremes on the Automotive Primary Power Supply**

this occurs, in some instances requiring more expensive processing. In other cases, the process extends into higher voltage ranges that require redesign of some existing automotive products, such as bipolar analog ICs or smart power devices. It has taken the semiconductor industry several years and considerable investment to develop design processes specifically for the 12V system. To design and manufacture higher voltage semiconductors that meet the automotive electronics cost performance requirements requires large up-front investment, as well as significant research and development.

TRANSIENT SUPPRESSION

Unprotected alternator load dump transients can exceed 100V under worst case conditions (1)*. The duration of this transient ($e^{-t/0.188}$) may cause excess junction temperatures even in the most rugged semiconductor devices when exceeding their voltage ratings. However, by designing avalanche rectifiers into the 12V alternator circuit, its maximum voltage can be held to 40 volts even with 90 amps of reverse current at 175°C junction temperatures for a duration of 80 µ seconds (2).

Without centralized load dump protection in the alternator an avalanche rectifier or MOV (Metal Oxide Varistor) device can clamp the load dump voltage transient in the automotive electronics module. These devices limit the load dump transient to a level above a double battery voltage condition. As a result the maximum voltage limit is generally between 32V to 40V for the 12V battery system. A zener device usually exhibits a tighter clamping voltage range over wide current levels than a MOV device. Increasing the system voltage will subsequently require increased resistivity and epi thickness in the clamping device to obtain higher voltage capability. This is not a significant processing change but will require time to fabricate new material and qualify the resulting product.

*** Numbers in parentheses designate references at end of paper.**

CENTRAL PROTECTION NEEDED

Many 12V automotive systems require every device or system module connected to the battery to be capable of sustaining the level of protection cited in ***Table 1.*** Each module design and any external semiconductor device tied directly to the battery bus incurs a cost penalty for this localized protection. With increased usage of electronics in automotive equipment (anti-skid brakes, air-bags, etc.) the cost penalty multiplies many times over. To gain overall cost savings a 24V or 48V electrical system design must address the central voltage protection issue. Avoiding development of new processes and silicon design rules capable of supporting the substantially higher voltages brings additional savings. Furthermore, if clamp voltages maintain their present level, current processing (60V to 85V) would apply with minimal modifications required.

LOAD PROTECTION

The fast turn off of high current inductive loads,

such as air conditioning compressor clutches, generate high voltage transients. The polarity of high voltage inductive kickback spikes depends upon the control switch configuration. A clamping network located at the terminals of inductive load can lower the inductive kickback voltage to a safe level. A wide variety of small zener or MOV transient suppression devices are available for suppressing lower energy transients in 24V or 48V systems.

As battery voltage increases for a given inductive load function (a motor, solenoid, etc.), the ampere-turns value of the inductor will remain constant with its ability to deliver the same force. This being the case, the current through an inductor will decrease allowing smaller gauge wire to be used but having a greater number of turns to maintain the ampere-turns constant. The increased number of turns will cause the inductance to increase according to the square of the turns increase ($L \sim N^2$). In light of the same force requirements, it is no surprise that the total inductor energy, in this case, will remain the same ($E = 1/2\ LI^2$). The transient voltage (V_t) can be expressed by $V_t = L\ di/dt$. The inductor resistance will increase directly with the turns increase and inversely with the wire cross-sectional area. Transient voltages will increase proportionately with increases in battery voltage unless inhibited by some other means.

REVERSE BATTERY

Reverse battery protection also needs attention. Many semiconductor devices will fail with -24V or -48V connection (reversed battery hookup). A single protection unit at the battery location instead of at the individual device or module level would simplify the system. This approach would solve the problem associated with MOSFET type structures inherently passing reverse current to a load via the MOSFET body diode. When the MOSFET structure is controlling an inductive load having a flyback diode across the inductor, reverse battery cannot be tolerated. A common method of protection against reverse battery conditions is to incorporate a blocking diode in series with the load. This introduces a diode forward voltage drop in series with the load reducing the applied voltage to the load. This technique has not proven to be satisfactory for many 12V battery applications; 1V drop out of 12.6V represents an 8.6% voltage loss. This technique, however, may be effectual with a 24V or 48V system, where a 1 volt drop may be tolerated for 2A or less current loads (based upon a 2W rectifier power dissipation constraint). Again, this method adds additional cost on a per unit basis, whereas a single reverse protection unit, ***Figure 1***, near the battery connections protects the entire electrical system, with the exception of, perhaps, the starter motor.

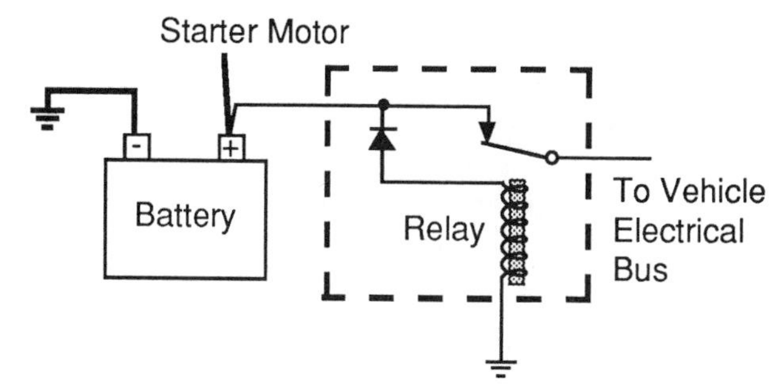

Figure 1. **Reverse Battery Protection**

LOAD CONTROL

To implement the automotive control strategies made possible by the MCU (microcontroller unit) and DSP (digital signal processing), actuators (solenoids and motors) usually interface to the controlling element by a power control device. Power FET devices or smart power ICs generally receive consideration for interfaces in new systems. The power FET in high volume automotive usage is an enhancement mode, N-Channel device that becomes fully enhanced with 10 (or 5) volts of gate bias. This permits current levels of several amperes to flow easily from the FET's drain-to-source. By selecting a die size and package to minimize power dissipation, the need for additional heat sinking can be eliminated in certain instances. The most common and readily available automotive qualified product at this date is in the TO-220 package with a 60 volt / 0.028 ohm on-resistance die.

SUPPLY VOLTAGE IMPACT ON FETS

Figure 2 compares MOSFET die sizes for a 12V, 24V and 48V system with a constant load wattage and semiconductor power dissipation. In this example, the 24V and 48V system offer about a 20% die size reduction over the 12V system. Note how the higher voltage semiconductor process design rules and the higher $R_{DS(on)}$ temperature coefficients counteract the higher voltage system's reduced load current values.

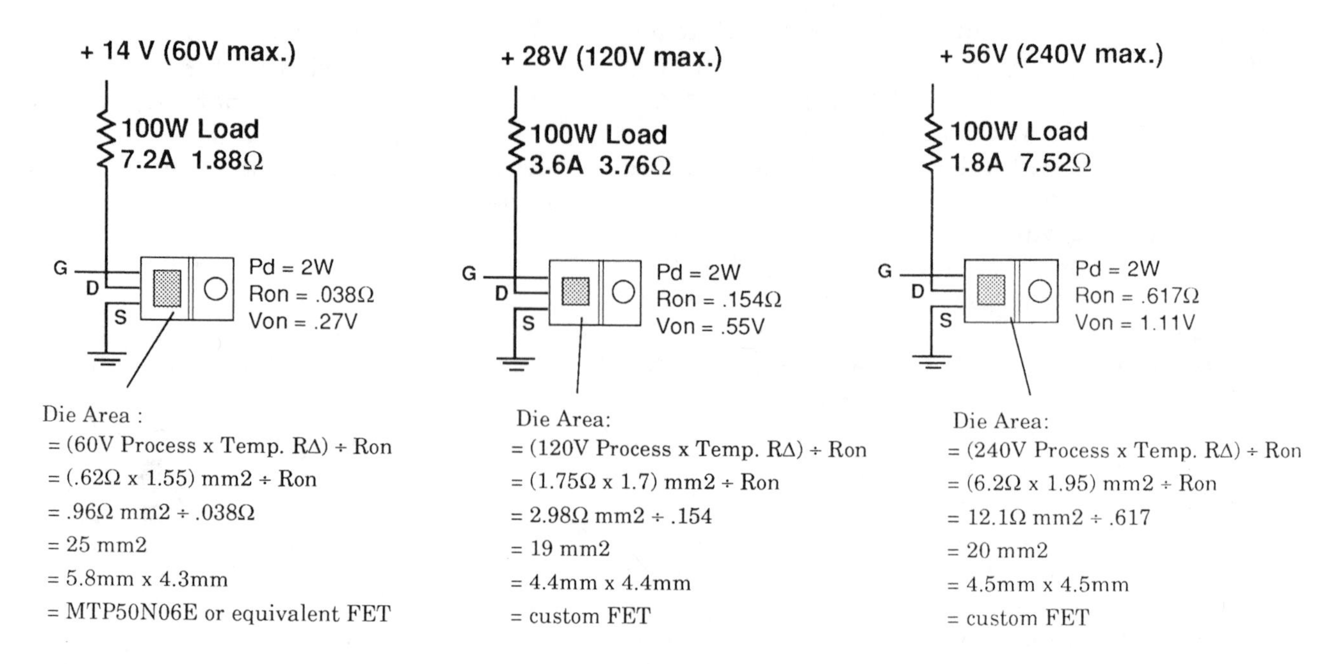

Figure 2. **MOSFET Size Comparison (Fixed 2W Dissipation at 125°C)**

Theoretically, the specific on-resistance of power FETs increases with the 2.5 power of the breakdown voltage. The theoretical limit of bulk silicon shown in ***Figure 3*** indicates a .0001 ohm cm^2 limit at 50 volts and a .1 ohm cm^2 limit at 1,000 volts.

From a circuit design perspective, reduced $R_{DS(on)}$ for same die size can be achieved by utilizing power FETs with a lower breakdown voltage as shown in ***Figure 4***. Note the 10x $R_{DS(on)}$ increase from 60V to 250V for the same 1 mm^2 die area. This has a direct impact on cost versus performance factors.

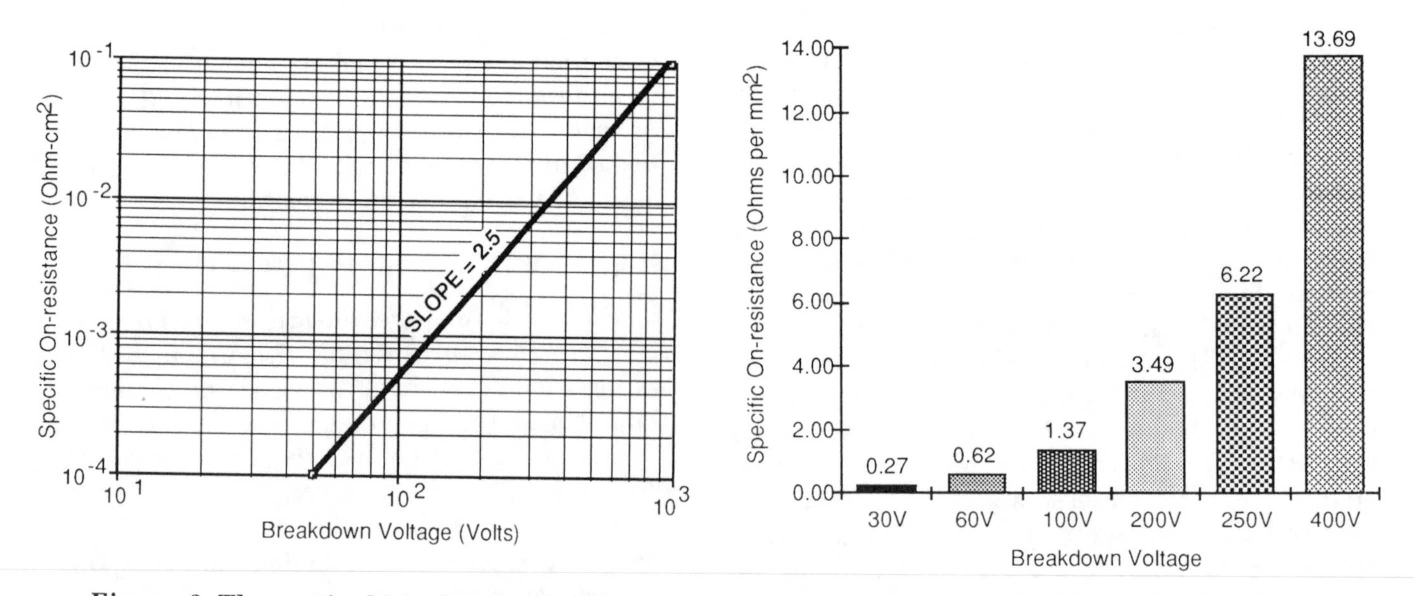

Figure 3. **Theoretical Limit of Bulk Silicon**

Figure 4. **FET Specific ON Resistance**

In the low voltage range (approximately 50 volts) the specific on-resistance (area x $R_{DS(on)}$ product) is continuously being improved. Semiconductor manufacturers have several levels of low voltage design rules to provide high yielding, high cell density, reduced on-resistance, and low cost power FET products.

Both cell density and epitaxial starting material have a significant effect on products having breakdown voltages below 100 volts (3). Above 100 volts the epitaxial material thickness contribution is a dominant term. This can be seen from the equivalent circuit model for on-resistance components of a vertical power FET shown in ***Figure 5***. In high voltage devices, the bulk and JFET contribution are about 90% of the device on-resistance. As a result, available high voltage power FETs are close to the theoretical limit as previously shown. In low voltage devices, the channel resistance (R_{CH} and R_D) is the dominant factor.

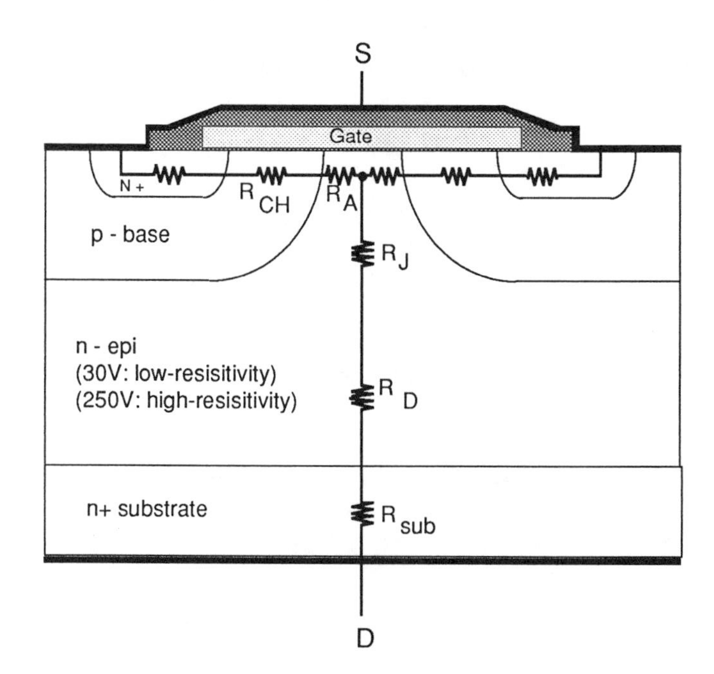

Figure 5. **Ron Components of a Vertical FET**

The vertical power DMOSFET cross-section (difussed MOSFET) shows the various resistance components in the on-state. The cross section shown is not to scale. The active area, top surface to the bottom of the P-well, is actually less than 5 microns. The epitaxial thickness ranges from 8 to 45 microns with a breakdown variation from 50 to 500 Volts. Growing thicker epi layers results in increased processing cost and consequently higher prices for a comparably sized high voltage FET versus a low voltage FET.

For a 30 Volt power FET the on-resistance components that make up 98.5% of the total on-resistance of the silicon die are (4):

R_{CH} = 32%	R_D = 28.1%
R_A = 4.5%	R_{SUB} = 13.6%
R_J = 20.3%	

For a higher voltage system, the effect of on-resistance is more than offset by the reduced current. This assumes the use of a protected or limited supply voltage. As shown in ***Table 2***, which utilizes available power FETs of about the same die size, a 60 volt power FET (**Case 1**) controlling a 225 watt load, such as a 16A blower motor fan, in a 12V system normally dissipates 12.5 watts and requires a heatsink with a ØSA (thermal resistance sink to ambient) of 3°C/W ØSA. This is an above average sized aluminum heatsink and is equal to about 80 square inches (516 cm^2) of 1/8 inch (.32 cm) bright aluminum. Limiting a 48V system to over-voltages below 100 volts, the same size power device with a 100 volt rating (**Case 2**) and 4 amps runs considerably cooler dissipating only 2.5 watts and requires a 25°C/W $Ø_{SA}$ rated heatsink or about 1 square inch (6.5 cm^2). Reducing the FET die size (higher $R_{DS(on)}$) for cost savings is somewhat offset by increased heatsink size. The cost trade-off is between $R_{DS(on)}$ (FET size) and heatsink size. On the other hand, there is essentially no change in the power dissipated and no cost savings relative to the power switching devices (**Case 3** for the 48V and **Case 4** for the unclamped 24V) if the existing transients and power FET breakdown voltage rating scale up to handle the increased voltage transient requirements (for example, 240V for the 48V system).

The cost factor and availability of power FETs in higher voltage ranges are important considerations. To withstand the 12V system voltage extremes shown in ***Table 1***, power devices must have 60 to 100 volt breakdown voltages. Unless the system voltages are controlled or the way power FETs are specified changes, the choice of standard products and processes available for the 150 to 400 volt range is presently limited. Of additional significance is the fact that on-resistance of higher voltage power FETs increases at higher temperatures faster than that of low voltage devices (***Table 3***). The on-resistance of a 400 volt device operating at 125°C junction temperature is approximately 2.1 times higher than its room temperature on-resistance value. A low voltage device is only 1.55 times the 25°C value.

	CASE 1: 12V System, 60V FET (MTP50N06E or IRFZ44)			
$R_{DS(on)}$ @ 25°C	0.028Ω	0.028Ω	0.028Ω	0.028Ω
$R_{DS(on)}$150°C factor	1.75	1.75	1.75	1.75
$R_{DS(on)}$ @ 150°C Tj	0.049Ω	0.049Ω	0.049Ω	0.049Ω
Ø-JC (°C/W)	1	1	1	1
Ø-CS (°C/W)	0.7	0.7	0.7	0.7
Ø-SA (°C/W)	**5**	**4**	**3**	**2**
Id (AMPS)	16	16	16	16
Ta (°C)	85	85	85	85
Tj (°C)	169.0	156.5	144.0	131.4
Vdrop	0.8	0.8	0.8	0.8
P	12.5W	12.5W	12.5W	12.5W
	Power Efficiency ≈ 94%			

	CASE 2: 48V System, 100V FET (MTP25N10E or IRF540)			
$R_{DS(on)}$ @ 25°C	0.077Ω	0.077Ω	0.077Ω	0.077Ω
$R_{DS(on)}$150°C factor	2	2	2	2
$R_{DS(on)}$ @ 150°C Tj	0.154Ω	0.154Ω	0.154Ω	0.154Ω
Ø-JC (°C/W)	1	1	1	1
Ø-CS (°C/W)	0.7	0.7	0.7	0.7
Ø-SA (°C/W)	**25**	**15**	**10**	**5**
Id (AMPS)	4	4	4	4
Ta (°C)	85	85	85	85
Tj (°C)	150.8	126.1	113.8	101.5
Vdrop	0.6	0.6	0.6	0.6
P	2.5W	2.5W	2.5W	2.5W
	Power Efficiency ≈ 99%			

	CASE 3: 48V System, 250V FET (MTP10N25E or IRF644)			
$R_{DS(on)}$ @ 25°C	0.28Ω	0.28Ω	0.28Ω	0.28Ω
$R_{DS(on)}$150°C factor	2.3	2.3	2.3	2.3
$R_{DS(on)}$ @ 150°C Tj	0.644Ω	0.644Ω	0.644Ω	0.644Ω
Ø-JC (°C/W)	1	1	1	1
Ø-CS (°C/W)	0.7	0.7	0.7	0.7
Ø-SA (°C/W)	**5**	**4**	**3**	**2**
Id (AMPS)	4	4	4	4
Ta (°C)	85	85	85	85
Tj (°C)	154.0	143.7	133.4	123.1
Vdrop	2.6	2.6	2.6	2.6
P	10.3W	10.3W	10.3W	10.3W
	Power Efficiency ≈ 95%			

	CASE 4: 24V System, 100V FET (MTP25N10E or IRF540)			
$R_{DS(on)}$ @ 25°C	0.077Ω	0.077Ω	0.077Ω	0.077Ω
$R_{DS(on)}$150°C factor	2.	2.	2.	2
$R_{DS(on)}$ @ 150°C Tj	0.154Ω	0.154Ω	0.154Ω	0.154Ω
Ø-JC (°C/W)	1	1	1	1
Ø-CS (°C/W)	0.7	0.7	0.7	0.7
Ø-SA (°C/W)	**5**	**4**	**3**	**2**
Id (AMPS)	8	8	8	8
Ta (°C)	85	85	85	85
Tj (°C)	151.0	141.2	131.3	121.5
Vdrop	1.2	1.2	1.2	1.2
P	9.9W	9.9W	9.9W	9.9W
	Power Efficiency ≈ 96%			

	CASE 5: 48V System, 500V IGBT (custom IGBT)			
$R_{DS(on)}$ @ 25°C	0.13Ω	0.13Ω	0.13Ω	0.13Ω
$R_{DS(on)}$150°C factor	1.1	1.1	1.1	1.1
$R_{DS(on)}$ @ 150°C Tj	0.143Ω	0.143Ω	0.143Ω	0.143Ω
Ø-JC (°C/W)	1	1	1	1
Ø-CS (°C/W)	0.7	0.7	0.7	0.7
Ø-SA (°C/W)	**5**	**4**	**3**	**2**
Id (AMPS)	4	4	4	4
Ta (°C)	85	85	85	85
Tj (°C)	100.3	98.0	95.8	93.5
Vdrop (FET)	0.6	0.6	0.6	0.6
P (FET)	2.3W	2.3W	2.3W	2.3W
Vdrop (diode)	0.7	0.7	0.7	0.7
P Total	5.1W	5.1W	5.1W	5.1W
	Power Efficiency ≈ 91%			

	CASE 6: 48V System, 400V FET (MTP10N40E or IRF740)			
$R_{DS(on)}$ @ 25°C	0.55Ω	0.55Ω	0.55Ω	0.55Ω
$R_{DS(on)}$150°C factor	2.4	2.4	2.4	2.4
$R_{DS(on)}$ @ 150°C Tj	1.32Ω	1.32Ω	1.32Ω	1.32Ω
Ø-JC (°C/W)	1	1	1	1
Ø-CS (°C/W)	0.7	0.7	0.7	0.7
Ø-SA (°C/W)	**5**	**4**	**3**	**2**
Id (AMPS)	4	4	4	4
Ta (°C)	85	85	85	85
Tj (°C)	226.5	205.4	184.3	163.1
Vdrop	5.3	5.3	5.3	5.3
P	21.1W	21.1W	21.1W	21.1W
	Power Efficiency ≈ 81%			

Table 2 **Specific FET Devices vs. System Voltage and Heatsink Size**

Furthermore, low voltage power FET devices are specified to operate up to 175°C in automotive applications. Automotive rated power FETs for 175°C and above 100 volt operation will require intensive reliability testing to insure that their silicon design and processes meet the increased voltage stress levels.

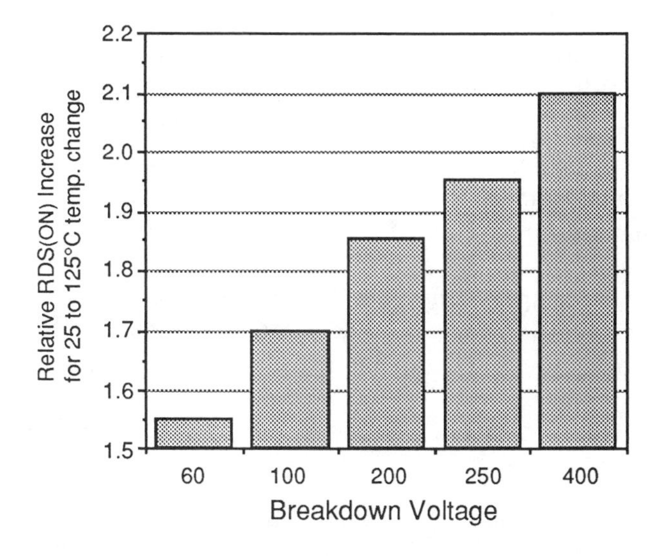

***Table 3.* Effect of Voltage on $R_{DS(on)}$ Temp. Coefficient**

Another FET circuit design issue is high side switching. High-side switching designs (which are frequently used in 12V systems) probably are even more desirable with a 24V or 48V system because of the removal of the voltage at the load when the load is inactive or switched "off". This reduces the chance for accidental contact with the higher supply voltages and minimizes leakage paths from loads subject to water/salt/dirt exposure. The 12V system typically uses either an N-channel FET with a gate voltage multiplier scheme (charge pump) or a P-channel FET to control high side loads. The charge pump design with a 12 volt supply provides a 30 volt level from gate-to-common that insures that the N-channel remains fully enhanced. ***Figure 6*** compares 12V to 48V high side designs using standard FETs. P-channel MOSFETs offer easy high-side design, but their silicon die has to be twice as large as a similarly rated N-channel MOSFET. Die sizes are about 25 mm^2 for both 60V and 200V P-Channel units shown. A voltage multiplier allows the N-channel gate to be biased above source voltage. Die sizes are about 14 mm^2 for both 60V and 200V N-channel units shown.

Linear IC charge pump designs are generally limited to 24V system voltage. The maximum

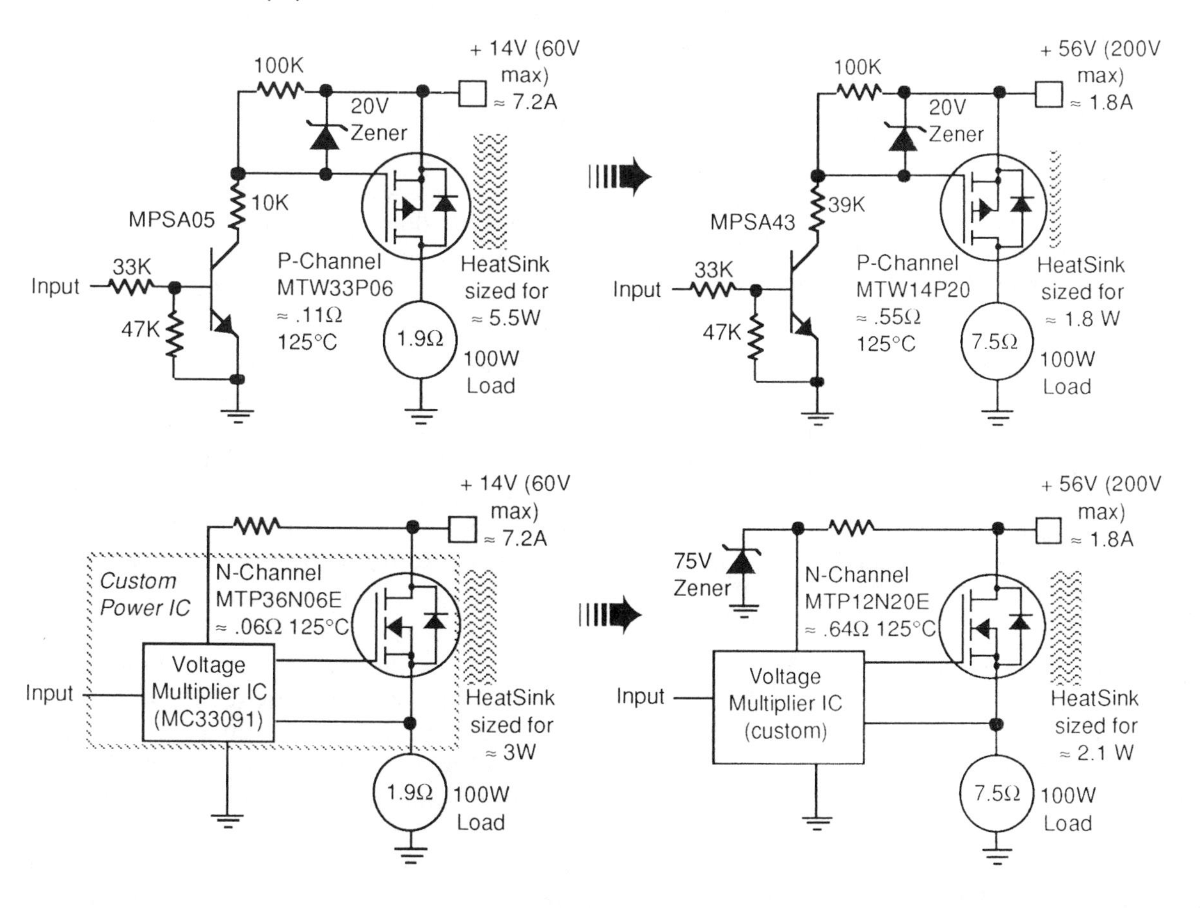

***Figure 6.* High-side Switching Designs**

possible charge pump output voltage is about 85V when integrating the charge pump design onto a present day smart power IC. This may suffice for a 24V system limited to 60V maximum. Note that the present 12V N-channel high-side design, using available technology, can be an integrated monolithic circuit, whereas the 48V system cannot.

IGBT's in Automotive Applications

IGBTs (insulated gate bipolar transistors) use the power FET cell structure and an additional P+ layer to achieve a power semiconductor device that can be used with higher supply voltages (typically above 250 volts) and typically achieve a 1/3 to 1/4 size reduction in silicon area when compared to a high voltage (above 250V) vertical power FET. For automotive usage the interest in IGBTs is primarily in ignition systems. The IGBT allows cost savings in the MCU or logic interface, when compared to a Darlington design. **Table 2, Case 5** and **6** shows a 500V IGBT compared to a 400V MOSFET with similar die area. The 500V IGBT can be operated at a safe temperature even with the extra dissipation of its intrinsic series diode. The 400V MOSFET cannot be used based on excessive Tj.

The IGBT's additional P+ layer, in effect a series diode with a 0.7 volt drop, causes its power dissipation to be much greater than a power FET in the low voltage range. However, the combination of higher system voltage and additional protection (a precision voltage clamp to limit the voltage across the IGBT that can be integrated during the device design) makes the IGBT an attractive alternative.

A fuel injector circuit, ***Figure 7***, uses an IGBT driver with an integrated 150 volt collector-gate clamp, and ESD protection (gate-emitter clamp). The design of the clamps allows the voltage to be selected in approximately 8 volt increments. The variation of the clamp voltage over 40°C to 150°C is within ±2% for a given device.

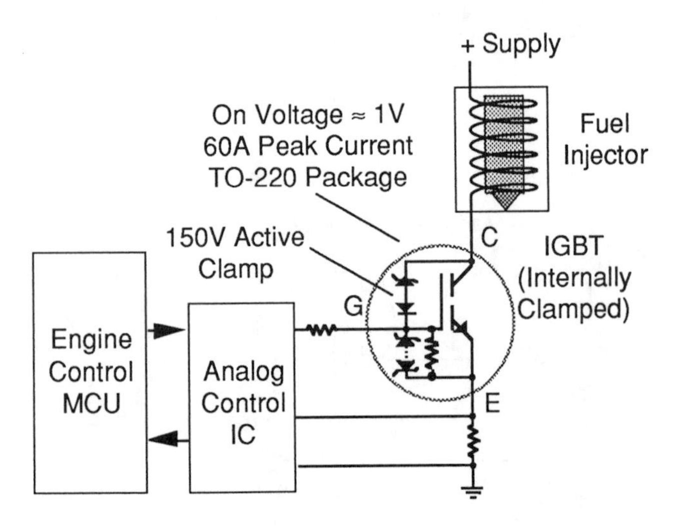

Figure 7. **Injector Circuit Using IGBT**

BIPOLAR DARLINGTONS

Darlington power transistors have found wide acceptance in low speed applications that require high voltage switching with reasonable base current drive levels. The darlington and IGBT devices appear to merit consideration for a 48V system, as their intrinsic 1 to 1.5V on voltage becomes less of an issue with the reduced current levels. ***Table 4*** compares specific devices for the power technologies reviewed. Other power device technologies (SCRs, Triacs, etc.) find usage mainly for AC power control.

FAULT PROTECTION AND HIGHER VOLTAGES

Higher battery voltage systems will probably require more attention to protecting the power semiconductors against load fault conditions. Due to the increased battery voltage, the power transistor's junction temperature may rise beyond safe limits in a short time during a shorted load condition. The transistor will tend to limit the current, and the power supply impedance will set a maximum shorted load current value. Generally, when the load shorts, the power transistor goes into a linear mode due to its gain limitation, overheats in a few milliseconds, and fails due to excessive junction temperature per the device's power dissipation (P=EI).

It is possible to add short circuit protection to power FETs with a limited amount of integration. The monolithic circuit shown in ***Figure 8*** uses a standard power FET process and one additional masking step to provide integrated current limiting of approximately 1.2 amps at 150°C. In addition, it incorporates output and input voltage clamps similar to the IGBT, discussed previously. The current limit provides short circuit protection as long as the package (TO-220 with a maximum of 30 watts) has an adequate heatsink to dissipate the power when the device is in current limiting. If the supply voltage increases to 64V, the 1.2A current limit alone may be insufficient to provide short circuit protection because of the increased power dissipation (77W), and more complex (higher cost) approaches will be required. Reducing the current limit to 0.3A diminishes the manufacturing economics of using this simple level of integration on power FETS (most power FETs are designed for more than 1A).

SMART POWER, LINEAR DRIVERS AND INTERFACE IC'S

Power integrated circuit development has advanced to a level that allows a mix of technologies such as high density MOS logic functions combined

Vbat=14V (60V max) 100W Load

Device Type	Part Number	Die Size*	Driver Loss	Device ON Loss	Package Style
NPN	MJF3055	5	10.4 W	3.6 W	TO-220
NPN Darlington	BDW40	9	.33 W	13.5 W	TO-220
PNP	MJF2955	5	10.4 W	5.0 W	TO-220
PNP Darlington	BDW45	9	.33 W	17.8 W	TO-220
N-Channel FET	√ MTP50N06E	25	.01 W	2.0 W	TO-220
IGBT	MPPD2021	26	.01 W	10.7 W	TO-220
P-Channel FET	MTW33P06	25	.01 W	5.5 W	TO-220
High-side IC	MC33391	15	.01 W	6.3 W	TO-220
Low-side IC	custom	15	.01 W	4.7 W	TO-220

√ Optimal Device Specs: *mm2

Vbk≈60V, I≈ 7.1A, Von≈.28V, Ron≈.04Ω, Pd≈ 2W, TO-220

Vbat=28V (120V max) 100W Load

Device Type	Part Number	Die Size*	Driver Loss	Device ON Loss	Package Style
NPN	MJF15030	9	9.72 W	1.1 W	TO-220
NPN Darlington	BDX53C	6	.14 W	4.7 W	TO-220
PNP	MJF15031	9	9.72 W	.7 W	TO-220
PNP Darlington	BDX54C	6	.14 W	5.8 W	TO-220
N-Channel FET	√ Custom	19	.03 W	2.0 W	TO-220
IGBT	MPPD2021	26	.03 W	5.0 W	TO-220
P-Channel FET	Custom	38	.03 W	2.0 W	TO-218
High-side IC	none	-	-	-	-
Low-side IC	none	-	-	-	-

√ Optimal Device Specifications:

Vbk≈120V, I≈ 3.6A, Von≈.56V, Ron≈.15Ω, Pd≈ 2W, TO-220

Vbat=56V (240V max) 100W Load

Device Type	Part Number	Die Size*	Driver Loss	Device ON Loss	Package Style
NPN	2N6497	8	19.69 W	.4 W	TO-220
NPN Darlington	MJE5740	16	4.92 W	1.4 W	TO-220
PNP	MJE5851	23	19.69 W	.4 W	TO-220
PNP Darlington	MJH11021	26	.98 W	1.6 W	TO-218
N-Channel FET	√ Custom	20	.06 W	2.0 W	TO-220
IGBT	MPPD2021	26	.06 W	2.5 W	TO-220
P-Channel FET	Custom	40	.06 W	2.0 W	TO-218
High-side IC	none	-	-	-	-
Low-side IC	none	-	-	-	-

√ Optimal Device Specifications:

Vbk≈240V, I≈1.79A, Von≈1.1V, Ron≈.63Ω, Pd≈ 2W, TO-220

Table 4. **Specific Devices vs. System Voltage and Heatsink Size**

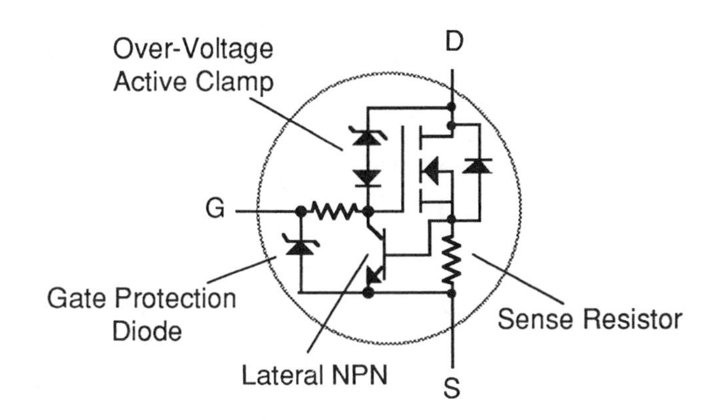

Figure 8. **Internally Clamped and Current Limited FET**

with power MOS functions on the same monolithic chip. Continual development is underway to meet the automotive system designer's demands for power IC devices with more complex control logic and explicit output fault diagnostic reporting as shown in ***Figure 9*** (5). This mixed IC semiconductor technology is attractive for the numerous smart power IC load control applications encountered in today's automobile. Fabrication of multiple outputs on a monolithic chip are common place using this technology. In addition, this technology affords relatively low power dissipation while being capable of switching medium current loads (.5A to 10A).

The source of power dissipation in these types of smart power IC structures is fundamentally due to $R_{DS(on)}$ of the power MOS structure. In order to

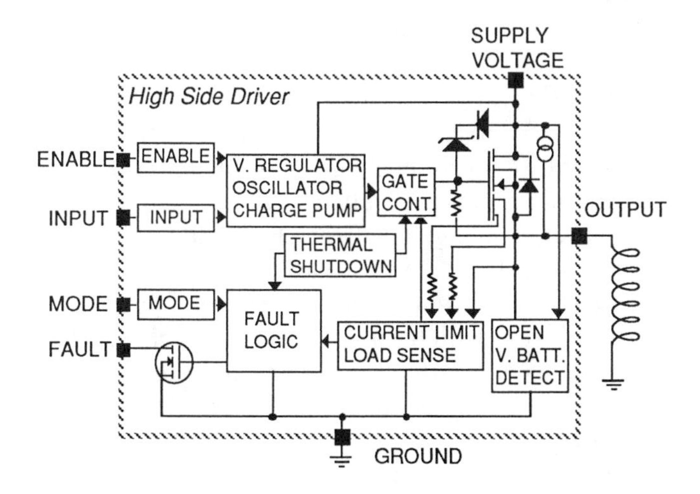

Figure 9. **Mixed Technology Power IC**

obtain low device power dissipation it is necessary to have low values of $R_{DS(on)}$. Therefore, the power MOS device occupies the majority of the chip area, and the associated logic functions occupy a relatively small portion of the chip, operate at low current levels (less than 1mA), and contribute relatively little in the way of power dissipation. The principle parameters controlling the power MOS devices $R_{DS(on)}$ are epi resistivity, epi thickness, and source/drain area. Many power IC processes maximize the IC's performance to obtain a specific voltage breakdown capability while minimizing $R_{DS(on)}$. As in the case of the power MOSFET device, the value of $R_{DS(on)}$ for a constant die area increases with breakdown voltage.

Power IC devices operated from a 12V system may have breakdown voltage safety margins of 5 times, 60V, the battery voltage (6). Many power IC devices manufactured for automotive electronic systems comply with this requirement and exhibit typical $R_{DS(on)}$ values of 0.5Ω (max) at 25°C at nominal battery voltages. Having the same breakdown voltage safety margin (5x) applied to new power IC devices operating from a 48V system would require 250V rated devices. 250V devices require higher resistivity epi material, which in turn demands new design rules to be developed calling for increased lateral spacing (pitching) of source-to-source cells of the power device structure. Increased spacing accommodates the enlarged depletion layer spreading (associated with the higher resistivity epi material) and significantly increases $R_{DS(on)}$ per unit area. For mixed technology power ICs this would mean approximately a 2 times increase epi thickness coupled with a 7 times increase in resistivity to meet the presently imposed 5 times breakdown safety margin. This would amount to about a ten fold increase in the on-resistance area product $R_{DS(on)}$ x A). This requires larger chip sizes (2 or 3 times larger for 48V system). The larger chip sizes require larger packages even though the power dissipation could be handled by an TO-220. Present IC package designs reflect the needs of current device processes. In association with this, present package leadframes have limited "flag" area available on which to mount an IC chip. New leadframes and possibly new packages may have to be developed to accommodate the physically larger chip size.

A 24V or 48V battery system requires higher voltage processes to be developed, particularly for power ICs that are commonly specified for operation in the present 12V system electronics. Increasing to a 48V system, in and of itself, will require higher voltage IC processes to be developed. Common requirements for an IC device operating from a nominal 12V system include survivability of double battery, reverse battery, external battery chargers, as well as load and field dump transients.

Requiring higher voltage ICs to be capable of the projected 48V system operation protection level of ***Table 1*** would present a severe burden for IC manufacturers. Typical automotive ICs are designed for relatively low voltage operation (60 to 85V max.) and are optimized for cost and performance. Few higher voltage IC processes currently exist for the IC designer to use. To develop new high voltage production ready processes would take an estimated three to five years lead time to develop. This added onto the average two years for device design and qualification time presently encountered for automotive devices would mean five to seven years lead time required in order to have a higher voltage automotive part available in production quantities.

MICROCONTROLLERS

Present MCUs use a 5V power source but there is developmental work underway to reduce this voltage to 3V to reduce cost. In any event, they will continue to require a separate regulated power source. Some development is also underway to integrate into the MCU, higher power elements that tie directly to the battery. For these "high power MCUs" to be practical to produce in high volumes and operate directly from a 24V or 48V system, it will require another era of silicon process inventions.

The 5 volt supply currently used to provide the power for MCU's, DSP's, memory devices and sensors will continue to be required for higher voltage systems. A typical 5 volt regulator device, LM2931T for example, is only rated at 40V continuous operation. To operate this device from a 24V or 48V system would require a pre-regulator to step down the supply voltage to under 40V. A simple 39V Zener shunt regulator and series resistor may suffice for a pre-regulator. The cost trade-off is between the 39V zener with resistor versus developing a new higher voltage 5V regulator IC. One other aspect of the higher voltage system is that the supply voltage remains well above the 5V regulator dropout minimum during engine cranking; the MCU's operation will not be affected by the low battery voltage.

CONCLUSIONS

Changing to a 24V or 48V system in the automobile without controlling the resultant worst-case voltage conditions will increase the cost of many semiconductor components. The variety of automotive rated semiconductor products available to system designers also will be restricted. By controlling the over-voltage conditions, especially limiting the load dump voltage, a cost benefit can be realized in the area of power semiconductor devices performing load control functions.

The 50% to 75% reduction in load current levels may slightly decrease the physical size of semiconductor wire bond and pad design rules, but in general, will have little cost impact. Changing from one packaging system to another, such as TO-220 to a TO-218, to accommodate larger higher voltage silicon chips will raise costs. If the 5X voltage factor is reduced to 3X, then the need for larger higher voltage chips is minimized, and larger packages may not be required.

While the failure rate of semiconductor devices is inherently low, some consideration will have to be given to the fact that increased voltage is one of the factors that affect semiconductor reliability over time (7). Higher voltage reliability stress testing will be required to verify that the semiconductor device failure mechanisms remain under control and are at acceptable rates...

ACKNOWLEDGEMENTS

The authors thank Sam Anderson, Gary Beaudin, Cliff Peterson, John Pigott, John Suchyta, Keith Wellnitz, John Wertz and Randy Wollschlager for their contributions in the preparation of this paper.

REFERENCES

(1) **Recommended Environmental Practices For Electronic Equipment**, SAE J1211, Table 3

(2) Mike Mihin and Richard J. Valentine, **New Rectifiers Clamp Alternator Voltage Levels,** SAE 872002

(3) Paulo Antognetti, **Power Integrated Circuits,** 1986, Mc-Graw-Hill, pp 3.35-3.40

(4) Krishna Shenai, **Performance Potential of Low-Voltage Power MOSFET's in Liquid-Nitrogen-Cooled Power System**, IEEE Transactions on Electron Devices, Vol. 38, No. 4, April, 1991, pp. 934-935.

(5) J.M. Himelick, J.R. Shreve, G.A. West, **Smart Power For The 1990s In General Motors Automobiles**, ISATA 22nd Int. Symposium on Automotive Technology & Automation, Florence, Italy, May 14-18, 1990

(6) James C. Erskine, **Power Dissipation in MOS Power Transistors For Automotive Applications: System Voltage Considerations,** 1991 SAE FTT Conf, August' 91

(7) **Reliability Data Analysis,** Motorola DMTG Reliabilty Audit Report, Q490, pp 1.11-1.16

911750

Cost and Time Effective Development of Microprocessor Based Control Systems

Adrian G. Kallis
Phoenix International Corp.

ABSTRACT

Through the use of modular electronics designs and specially designed electronic development tools, the development time and cost of electronic control systems, such as an engine controller or transmission controller, can be reduced. The control system electronics are based on a circuit board which includes the required control system components, including SAE J1708 compatible communications, high-current and low-current outputs, analog inputs, and a number of digital inputs. By using the same electronics, all development tools can be used across all development programs. Several such tools include a remote programmer, serial link development system, and a data acquisition system.

Although the use of vehicle electronics is greater than ever before, the use of microprocessor-based electronics has been limited to higher volume products, such as components for monitoring and display. Complex stand-alone electronics for engine control or transmission control are becoming more common, yet more sophisticated control systems linking engines, transmissions, and other hydraulic systems are still limited (1).* This is partially due to the high cost of development, and the risk involved in designing a new, unproven system for each engine-transmission combination. The purpose of this paper is to describe a method of achieving cost and time benefits through the use of a proven, modular system which can be used for a variety of control system and monitoring tasks.

* Numbers in parentheses designate references at end of paper.

PRE-PROTOTYPE DEVELOPMENT

Much of the development and testing of the mechanical portions of an electronic control system can be started without the design of any custom electronics or software by using a control card which communicates with a personal computer. This card interfaces to several high-current drivers and is controlled by PC software which is easily modified to provide open-loop control of the system by controlling the drivers in a user-programmable sequence. When used in the development of a transmission controller, for example, the PC based controller can first be used in the selection and testing of the electro-mechanical valves used to control the hydraulics in the transmission. From that point, it can be interfaced to a dynamometer, and initial shift patterns can be tested and modified from the computer keyboard. When a prototype vehicle becomes available, a lap-top personal computer can be used in the vehicle for continued development until any custom software which may be required is developed. In addition to this initial development, the PC controller on the dynamometer can be used for continued durability testing by utilizing a program which cycles through desired gear patterns in a continuous user-programmable sequence, and records the number of shift cycles completed.

Throughout the selection and testing of the mechanical components, an evaluation of the required system electronics is completed, and the design of any custom software which may be required can begin as soon as the input and output parameters are defined.

ELECTRONIC HARDWARE DEVELOPMENT

Most control systems have similar electronic requirements for inputs and outputs. By using a circuit which handles most of these requirements, most systems can use common, proven electronics (See Fig. 1). Each electronic module can be configured to operate on a 12 Volt or 24 Volt system with or without an auxiliary power input, and has the following input and output features:

1. Asynchronous serial communications port which complies with SAE J1708. This provides a method of communication with any other components in the vehicle which have this capability, including other controllers, displays, and "smart" switches and sensors.

2. Four high current outputs which may be pulse width modulated. The modulated frequency range is 0 Hz through 1000 Hz, and the duty cycle range is 0% through 100% in 0.5 microsecond increments. This gives a minimum duty cycle increment of 0.01% for the frequency range up to 200 Hz. Each output has the capability of sourcing 3 Amps at 12 Volts DC, or 1.5 Amps at 24 Volts DC. These drivers can be used, for example, to control electro-mechanical valves used to engage a clutch. By varying the duty cycle of the output (at the proper frequency based on the mechanics of the valve) the clutch engagement rate can be controlled to provide smooth engagement.

3. Four high current outputs which are on/off outputs. Each output has the capability of sourcing 3 Amps at 12 Volts DC, or 1.5 Amps at 24 Volts DC. These outputs can be used to drive solenoids or other coils which do not require pulse width modulation.

4. Three lower current digital outputs which can be configured to be open-collector outputs. These outputs can be used to drive any number of indicator lamps or displays by using them for synchronous serial communication with a standard display driver.

5. Three frequency inputs which can be configured for active sensors, if required, or for other sensors such as variable reluctance or reed type. These inputs can measure frequency from 1 Hz to 9.6 kHz, depending on the type of sensor being used. This allows the measurement of vehicle parameters such as engine speed, ground speed, or the speed of any other shaft on the vehicle.

6. Two analog voltage inputs, two regulated analog voltage reference outputs, and two analog voltage signal ground connections. These six lines are used to connect up to two position sensor potentiometers which can be used to determine engine throttle position, clutch pedal position, or the position of any other operator controlled lever or knob.

7. Seven current sinking and two current sourcing digital inputs which require a signal higher than 10 Volts DC at 10 milliamps to activate each input. These inputs can be used for any operator or control input switches required.

8. Two input lines used for remote programming of the control system software.

The control of the system is based on a Motorola 68HC11 microprocessor which can have up to 512 bytes of RAM and up to 2 Kbytes of EEPROM on chip. The operational code for the device is in external memory, which can be an EEPROM or EPROM up to 32Kx8, or a masked ROM up to 8Kx8. During the development of a control system, the EEPROM option is used so that the operational code can be reprogrammed by remote means (described below) without disconnecting the control system from the vehicle.

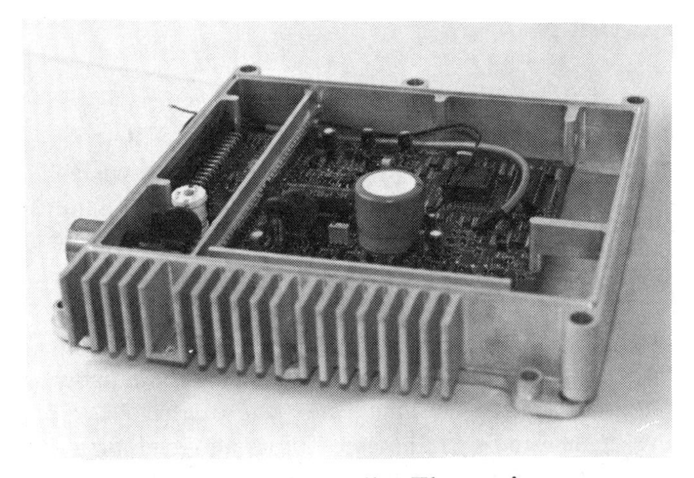

Figure 1: Controller Electronics

For systems which require less capability than this standard module, the standard module can still be used in a depopulated form by not placing all of the components at assembly time. For systems which require additional quantities of the same input and output configurations as the standard module, several identical modules can be layered together with a high-speed (up to 1 Megabit per second) synchronous internal data bus for data transmission between layers. The base module is enclosed in a die cast aluminum housing with removable top and bottom covers. Any additional modules are then stacked between the top and bottom covers using an identical housing for each module (See Fig. 2).

If a particular system requires additional features which are not addressed by the base module, a new module is designed which will function as its own layer and communicate with the other layers as if it were a standard module. By designing a new layer dedicated to the non-standard requirements, any risk involved in changing the circuit board to accommodate the new feature is removed. The new layer is designed to be the same size and shape as the base module, and uses as much of the same circuit as possible, such as the asynchronous and synchronous serial communication hardware, and any other common microprocessor connections. All additional modules, once designed and tested, can be used as an add-on layer for any other systems.

Figure 2: Modular Aluminum Enclosure

SOFTWARE DEVELOPMENT

Just as most control systems have similar electronics requirements, many software requirements are also similar. The use of identical hardware modules magnifies the amount of software which can be shared. By designing software with a modular concept in mind, the same benefits of decreased cost, decreased risk, and shorter development time can be further achieved. Each module contains the following standard software:

1. Power-up routines and module self-diagnostics routines. Each module will initialize the microprocessor internal registers, perform a complete functional check of RAM, and do a checksum calculation of the operational code before any additional software can be executed. After initialization, internal and external watchdogs must be serviced to allow the microprocessor to operate. Software is designed to monitor the external watchdog circuit to ensure that it is functional in case of a microprocessor failure. Internal registers and RAM locations are checked throughout software execution to diagnose any failed cells. Error handling routines for diagnostics failures are custom designed, depending on the effects of the system being controlled.

2. Standard time-based loops. Routines which are designed to be executed on a constant time basis can be inserted in the corresponding time-based loop desired.

3. Subroutines for mathematical operations and input/output control.

4. Asynchronous serial communications software which complies to SAE J1708. All software is designed to interface with the system development tools which utilize the serial communications port.

5. Common output driver routines for controlling the high current and low current output drivers. A routine for pulse width modulation of the four high current drivers allows the frequency and duty cycle to be easily modified, and verifies the requested parameters are within the operating range of the software.

6. Output driver diagnostics, which can sense shorted or open drivers, and shorted or open solenoid coils. The criteria for failing the diagnostics can be easily modified based on the resistance of the coil being driven. Often in the development of a new system, diagnostics routines are not developed until near the end of development. By having the diagnostics module included at the beginning of development, many problems can be found and corrected more efficiently.

7. Frequency measurement routines and interrupt routines, used with the frequency inputs. The software noise rejection, filtering, and scaling parameters can be modified based on the period range of valid inputs and the response times required.

8. Analog to digital conversion routines, used for position sensor inputs and output driver diagnostics. The scaling parameters for the position sensors can be either calibrated through a special calibration routine, or updated using a learning algorithm if input switches are available to indicate full travel positions of the sensors.

9. Switch input and debounce routines.

10. Multiple layer synchronous serial communications software with communications error checking.

Additional routines which may be required can often be derived from other custom routines which have already been written, once again decreasing the development time and risk involved in generating new software.

PROTOTYPE DEVELOPMENT TOOLS

In order to eliminate the time required to design custom development tools, and to decrease the time spent testing software changes, several development tools were created. One of these tools is the remote programmer, mentioned above, and another is a hand-held monitor and calibration development system. Both of these devices operate using the asynchronous serial communications lines.

REMOTE PROGRAMMER - The remote programmer provides a method of reprogramming the software that the system uses without removing the system from the vehicle. It is a hand-held unit powered from vehicle ignition and connected to the serial communications interface and the two input lines dedicated to remote programming (See Fig. 3). The programmer is designed around a plastic encapsulated EEPROM made by Datakey. The 8Kx8 EEPROM device is shaped like a key, and is placed into a slotted receptacle where contact with the EEPROM terminals is made. New software is transferred from a personal computer to the portable key-shaped EEPROM using a PC interface card. The key and a field remote programmer can then be taken to the vehicle, and the system reprogrammed.

Figure 3: Remote Programmer

A typical development process could occur as follows: While prototype vehicle testing and calibration is in process, software development would continue. After new software is developed and bench-tested, the vehicles at remote locations are updated by sending the assembled software via modem to a personal computer at the test sites. The file received is then programmed to the EEPROM key using the menu-driven PC software. The key and the remote programmer are then taken to each vehicle at that test site, a programming connector is plugged into the target system, the key is inserted into the socket and twisted, and the start button is pushed. A "Programming" indicator light is illuminated while the programming is in progress, and after the 8 Kbytes of software are transmitted to the target system, it is read back to the programmer for verification. Four LED indicators are used to display any error condition which may have occurred during programming. If no errors are encountered, the programming light is turned off, and a "Ready" indicator is illuminated. If the new software for the target system is larger than 8 Kbytes, up to four keys may be used in the same manner. Once programmed, the vehicle ignition can be cycled off then on, and the new program is executed upon power-up. The key can then be taken to the next vehicle for programming.

SERIAL LINK DEVELOPMENT SYSTEM - By using the SAE J1708 communications capability of the standard circuit board, the time required for vehicle calibration and system diagnostics can be reduced. The serial link development system is connected to the target system through the same connector as the remote programmer, and is powered from vehicle ignition. Data from the target system is sampled over the serial link by requests from the development system and displayed on a four-line, eighty character dot matrix LCD (See Fig. 4). The data received by the development system can be displayed in a variety of formats, including hexadecimal, decimal, and binary as one-byte or two-byte values.

In addition to displaying data from the control system, the development system can send requests to change the contents of internal RAM and EEPROM in the target system. By designing software with system calibration values in the microprocessor internal EEPROM, changes can be made while driving the vehicle and will be retained even after vehicle ignition is switched off.

At times it is desirable to experiment with different values, yet be able to return them to a known state by cycling ignition. To allow this type of development, software can be designed to read the most recent calibration values from EEPROM into RAM on power up, and use the RAM values which can be changed over the serial link. If a better calibration is achieved and needs to be saved, the serial link development system can issue a command to then store the modified values from RAM to EEPROM as the starting point for the next ignition cycle.

In the development of a transmission controller,

for example, the serial link development system would request a standard data stream from the vehicle system which would include engine speed, vehicle speed, powershift output shaft speed, clutch pedal position, throttle position, and the contents of any user-specified address from the microprocessor. With this standard data stream requested every 66 milliseconds, the displayed data is effectively a real-time window into the addresses of the microprocessor.

Another useful example of the serial link development system is in system diagnostics, especially in finding intermittent occurrences, which are often difficult to diagnose. Diagnostic error codes and any related data can be stored in the control system EEPROM when an error condition or other event of interest is sensed by the system. At any time after the error occurs, the development system may be connected to the control system and any error codes can be retrieved. The system is also useful for debug of new installations. By requesting the address which contains any switch input information and displaying it as a binary value, each switch input can be tested completely, including the microprocessor interpretation of that input. As each switch is toggled off and on, the corresponding bit displayed can be seen to toggle between 0 and 1.

Another feature of the development system is two digital-to-analog converters which provide two analog voltages which correspond to any of the data received over the serial link. This allows an oscilloscope or an XY plotter to be connected to the development system and the values of any two addresses in the target system can be plotted versus time or one value versus the other.

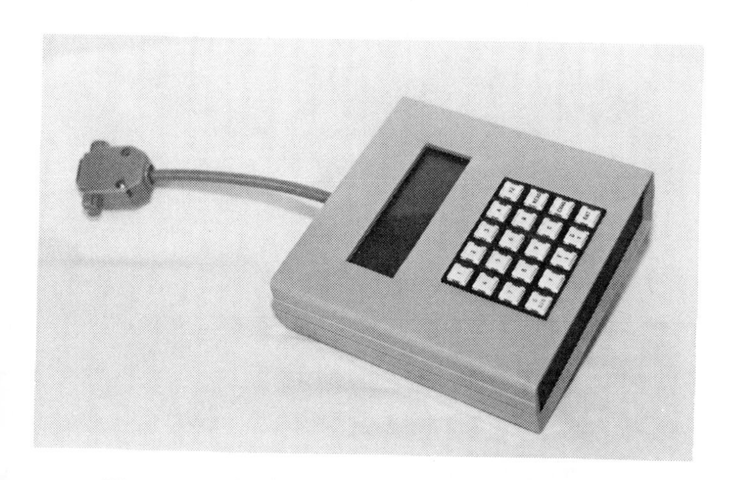

Figure 4: Serial Link Development System

DATA ACQUISITION SYSTEM - By combining the capabilities of the remote programmer and the serial link development system, the necessary hardware for a data acquisition system is already in place. The serial link portion samples data from the target system, and the data received can be stored in the same EEPROM device which is used in programming with the remote programmer. The serial link system can be user programmed to request the data of interest from the target system at the desired sampling rate. The data received is spooled to RAM and each sample is given a time stamp and corresponding sample number. During the time when data is not being sent or received from the target system, the development system is writing the data from RAM to the remote programmer portion of the data acquisition system via an internal high speed synchronous communications bus. The data that is written to the EEPROM key is then taken to the remote programmer personal computer interface and uploaded to the PC in a comma-delineated ASCII file. This file can be imported into most spreadsheets and manipulated and plotted as needed. Up to 8Kx8 bytes can be stored, which includes a header for each group of samples taken as well as the data received. The amount of data taken is limited only to the number of EEPROM keys available, and the amount of data lost during the time required to remove a "full" key and insert another one.

CONCLUSION

To decrease the risks involved in the design of a complex control system, proven, modular electronics can be used for nearly all control systems. By using this electronic hardware, development time and development costs are also reduced. Existing development tools can be used for system calibration, as opposed to designing new ones for each new system. System cost is also reduced through increased production volume. By using this system, common, proven software can also be used.

REFERENCES

1. Mike Osenga, Cost Effective Vehicle Electronics for Small OEMs?, Diesel Progress, February 1991

931810

S-Parameter Characterization for Oscillator Design with Intel Microcontrollers

David W. Elting
Intel Corp.

ABSTRACT

S-Parameter characterization techniques are used to aid oscillator circuit design with embedded microcontrollers. Off-chip oscillator circuit design techniques are improved using these characterization methods. S-parameter characteristic data is taken and used to form an equivalent circuit model. The equivalent circuit model is then used to perform Spice simulations of the on-chip inverter and the off-chip circuitry that models the functionality of the oscillator circuit design.

INTRODUCTION

Due to the increased frequency range, broad temperature spectrum, and the electrically noisy environments to which automotive electronics are subjected, a need has surfaced for more robust design with respect to oscillator circuits. In order to remove the guess work from designing the oscillator circuit Intel provides its customers with additional information, in the form of S-Parameters. The purpose of this paper is to describe how to use S-parameter characterization data to enhance oscillator circuit design with embedded microcontrollers.

S-PARAMETER NETWORK BASICS

Circuit designers who understand S-parameters are able to develop more stable oscillator circuits. Basic information about S-parameters needs to be understood before exploration of the oscillator circuit is continued.

Multi-port networks may be characterized by network parameters, such as S- and Y-parameters, measured at the ports with little or no regard for the contents of the network. From the network parameters the external environmental behavior of a system may be accurately predicted.

Equivalent circuit models may be developed from the network parameters to characterize the actual circuit contents. The equivalent circuit model allows a semiconductor manufacturer to disclose system behavior without disclosing the proprietary contents of the system design.

Network parameter characterization information, in conjunction with the equivalent circuit model, gives the design engineer an adequate amount of information for designing the oscillator circuit.

S-parameters and Y-parameters are commonly used to characterize 2-port networks. S-parameters, or *scattering* parameters, are used in high frequency systems, above 20-30 MHz, while Y-parameters, or *admittance* parameters, are used in the lower frequency systems. Both S-parameters and Y-parameters are useful for characterizing the "Black Box" values for a 2-port network. From the S-parameters or Y-parameters an equivalent circuit model for the network is derived. The equivalent circuit model is then used for design applications to accurately represent the 2-port network. There are a few things to consider when dealing with equivalent circuits. First, an equivalent circuit is only as good as the derived model. Second, an equivalent circuit is only an approximation. Although generally accurate, an equivalent circuit model only assures that the printed circuit board, PCB, design is essentially correct.

S-parameter characterization was chosen by Intel over Y-parameter characterization for two main reasons. The first reason was due to the trend in microcontroller design toward higher frequencies. In the not so distant future microcontrollers may be operating in the 50 MHz region, which is beyond the testing abilities of most Y-parameter testers. So using S-parameters at this early stage allows the microcontroller manufacturer the opportunity to develop the knowledge base for future product testing. This will then prevent the microcontroller manufacturer from having to change from Y-parameter characterization to S-parameter characterization and lose the information in the process. The second reason for using S-parameters is S-parameter testing is more compatible with MOS devices. Y-parameter testing requires shorting the two-port network, which poses some obvious issues with MOS devices. Due to these issues S-parameter testing was chosen to characterize oscillator circuits.

S-parameters are used to characterize systems that operate at higher frequencies, above 20 to 30 MHz. This 20 to 30 MHz limitation is due to tester issues rather than S-parameter limitations. S-parameter analysis represents an arbitrary 2-port network as a simple equivalent real and imaginary component. If the network is not trivial, S-parameters are useful in demonstrating how network parameters change with frequency. Since S-parameter measurements do not require the ports to be shorted during characterization of the circuit, they are useful for characterization of MOS devices.

CONVERTING S-PARAMETERS TO Y-PARAMETERS

To transform network parameters from S to Y or from Y to S is relatively simple. The conversion formulas may be found in many text books. However, when dealing with the frequency range that high-performance CMOS microcontrollers such as the Intel C196Kx operate, the transformation requires some correction factors, due to frequency limitations of the testing equipment. When the correction factors are used in the transformation equations an explanation will be provided. When utilizing the transformations one must also remember that network parameter values are all normalized to Zo, the characteristic impedance used in the S-parameter test set-up.

The basic formula set for transforming S-parameters to Y-parameters is as follows:

$$Y11 = \frac{(1+S22)(1-S11) + S12S21}{(1+S11)(1+S22) - S12S21}$$

$$Y12 = \frac{-2S12}{(1+S11)(1+S22) - S12S21}$$

$$Y21 = \frac{-2S21}{(1+S11)(1+S22) - S12S21}$$

$$Y22 = \frac{(1+S11)(1-S22) + S12S21}{(1+S11)(1+S22) - S12S21}$$

The Y-parameter values attained from these calculations are all normalized to Zo, which is 50Ω for the test set used for the data presented in this paper. To convert the Y-parameters to the actual Y'-parameters one needs to divide Yxx by Zo.

S-parameter testing typically begins to be valid in the frequency range of 20 to 30 MHz, depending on which tester is used for data collection. A correction factor had to be introduced into the transfer equations to account for tester limitations and to produce more accurate Y-parameters. The following equations include the correction factors being introduced into the above equations. An explanation of the correction factors is given following the equations.

Y11' = Y11 x (8E6/*frequency*)/50

Y12' = (Y12 / 2) x (8E6/*frequency*)/50

Y21' = (Y21 / 2) x (8E6/*frequency*)/50

Y22' = Y22 x (8E6/*frequency*)/50

EXPLANATION OF CORRECTION FACTORS

S-parameters, as in s=jω, are valid from DC to high frequency. Formulas for converting parameters from S to Y are also accurate at any frequency range, which is why the original conversion formulas have no compensation for frequency. However, S-parameter testing is accurate when the test frequencies are above 20 to 30 MHz. Since our S-parameter data was collected at frequencies of 8, 12, 16 and 20 MHz the factor of 8E6/*frequency* is required to correct for the low frequency range. The Hewlett Packard 8751A network analyzer and test system used for data collection has properties of a bandpass filter. At the frequencies used for our data collection we are operating near a zero for the bandpass filter that the tester represents. The tester does not correct for the zero, which is then incorporated into the characterization data. The 8E6/*frequency* is used to cancel the effect of the zero on the characterization data. The correction factor was determined from characterization data taken at several test sites and from information from Hewlett Packard.

The characterization data showed the 8E6/*frequency* correction to be valid over the temperature and the frequency ranges used for data collection.

The correction factor of removing the 2X from the numerator in the Y12 and Y21, also is used to correct for tester idiosyncrasies. The HP tester incorporates a power splitter. At the frequencies of operation it is necessary to compensate for the effects of the power splitter by dividing by two, or in this case removing the 2X multiplier from the numerator in the Y12 and Y21 terms.

Dividing by 50 is not a correction factor but is used to un-normalize the Y-parameters so they yield the equivalent circuit parameters.

OSCILLATOR EQUIVALENT CIRCUIT

When using 2-port network parameters such as S-parameters the black box is often represented with an equivalent circuit. An equivalent circuit is used to represent the contents of the black box without having to divulge information that is proprietary. This is particularly relative for oscillator circuits on microcontrollers. An equivalent circuit model was derived to provide information for AC analysis without jeopardizing the proprietary circuit information. The equivalent circuit model for the oscillator on Intel's 16-bit microcontroller family, the 8xC196Kx, is shown in Figure 1 below.

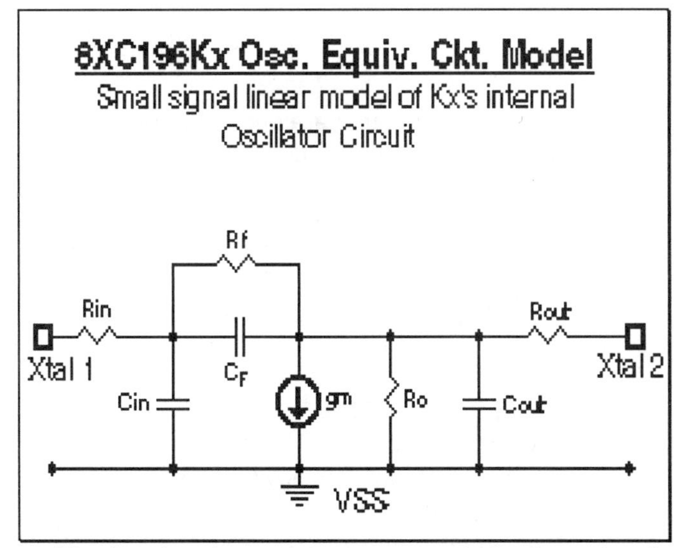

Rin Input series resistance accounts for the network parameter S11. This resistance is small. This is the resistance of the pin and metal line going to the inverting amplifier. Semiconductor makers hold Rin to a minimum since this is a parasitic resistance that is present in CMOS devices.

Cin Imaginary part of the network parameter S11. This capacitance is in the range of several picofarads. This is the gate to ground capacitance of the microcontroller and is added to the external load capacitance.

Rf This is the feedback resistor used in the inverting amplifier. Its purpose is to allow for start-up of the oscillator circuit. The value for this resistor is in the range of 750kΩ to 1.2MΩ. This value would combine with an external Rf, if used, as a parallel combination. Rf should be kept to a minimum; however, Rf must be relatively high to avoid high current drains through the oscillator.

C_F Feedback capacitance, used in the Miller effect. Cf in parallel with Cin creates high input capacitance values. Co for the crystal and board and Cf for the on-chip circuitry should be kept as small as possible since these are parasitic capacitances that have negative effects on the performance of the oscillator circuit. As Co and Cf increase the instability region of the oscillator decreases, which implies that the operating region is reduced as Co and Cf values increase.

gm Transconductance of the CMOS device. gm is kept relatively high by microcontroller manufacturers, but the PCB design must have ample margin to allow for process variations on the circuit. A further explanation of the gm margin will be given in a later section. gm is generally in the range of a few mmho's.

Ro Finite output resistance of the MOSFET in the pinch-off region. In the range of 10 to 100 kΩ and is inversely proportional to the D.C. bias current. Ro is kept in the range of hundreds of ohms. It is important for the Ro value to be relatively large to maintain optimal oscillator performance.

Cout Imaginary part of the network parameter S22. This capacitance is in the range of a couple of picofarads. This is also known as the gate to ground capacitance of the circuit.

Rout Output series resistance to account for real part of S22. This is a small value. Rout, like Rin, is kept as small as possible. Again, this is a parasitic resistance present in CMOS logic and is unavoidable.

The values for the passive circuit elements will be derived from the S-parameter data that is collected.

DERIVING THE EQUIVALENT CIRCUIT

The following formulas are used to convert Y-parameters to the equivalent circuit model. Values such as Rf are not converted from the characterization data but were tested using other techniques, in which case the value is presented without a transformation.

ω = 2 * π * *frequency*
Rf = approximately 750kΩ to 1.2MΩ
Cf = -*Imaginary* [Y12'] / ω
Cin = (*Imaginary* [Y11'] /ω) - Cf
Rout= Rin = *Real* [1 / Y11']
Ro = (*Real* [1/Y22'])
Cout= (*Imaginary* [Y22'] / ω) - Cf
gm = *Real* [Y21] + 1/Rf

ANALYSIS TOOLS

From the equivalent circuit some analysis tools are used to determine the stability of the oscillator circuit. Two commonly used analysis tools are the loop-gain measurement and the AC Nyquist analysis. From these two analysis tools the relative stability of the oscillator circuit may be determined. Both of these analysis tools will be used to provide a further understanding of the oscillator circuit, which will allow for a more robust design.

LOOP-GAIN ANALYSIS

Loop-gain analysis is a good first pass test for an oscillator circuit. For proper oscillation of a circuit two criteria must be met. First, the gain of the circuit must be greater than one. For cold start-up the gain of the circuit must be greater than 1.5 and it is preferable that the gain is at least 2. The second criterion is that the total phase shift is 360 or 0 degrees. The loop-gain analysis will determine if we meet the first criterion.

To perform the loop gain analysis the following calculation will be used:

$$\text{loop gain} = \frac{gmR_0}{1 + R_0[\omega_o{}^2C_1C_2R_{eff}]}$$

The terms for the equation are defined as follows:
gm Transconductance of the oscillator circuit.
Ro On chip oscillator circuit resistance.
ωo Oscillator frequency in rad/sec.
C1 Total capacitance from Xtal1 pin to ground, includes the load capacitance, the internal capacitance from the pin, and the parasitic capacitance of the PCB.
C2 Total capacitance from Xtal2 pin to ground, includes the load capacitance, the internal capacitance from the pin, and the parasitic capacitance of the PCB.
Reff The effective resistance of the crystal circuit. Determined from the following equation.

$$R_{eff} = R_s[1+\frac{C_o}{C_L}(1+\frac{|C_L-C_P|}{C_L})]^2$$

The terms used for Reff are as follows:
Co The capacitance measured from the Xtal1 pin to the Xtal2 pin.
Cp The parallel combination of C1 and C2.
CL The load capacitance of the crystal, as specified by the manufacturer.

It is obvious from the definition of the terms that there is some work involved in determining all the parameters. Although this process is not simple it may provide new insight when designing oscillator circuits.

LOOP-GAIN EXAMPLE

An example is provided to show the use of the loop-gain analysis. For the example an Intel 87C196KR is the microcontroller used in the design.

Suppose a system has the following measured parameters:

16 MHz Crystal - Co = 5.4 - 5.6 picofarads;
Rs = 5.4 - 7.1Ω; C1 = 21.9fF; L = 4.5mH

8XC196KR- S11 = 0.95615641-0.07103j
5 Volt S12 = 0.018841 + 0.07322j
16MHz S21 = 3.795783 + 2.34936j
S22 = 0.946093 - 0.057404j

Parasitic PCB cap. - Cxtal1 = 4.7pF;
Cxtal2 = 4.9pF

Load Capacitors- Cl1 = 22pF; Cl2 = 22pF

Next the S-parameters are converted to Y-parameters, which yields the following:
Y11' = -3.2323E-4 + 1.04748E-3j
Y12' = -4.3196E-5 - 3.75489E-4j
Y21' = -1.4158E-2 - 1.72506E-2j
Y22' = -2.64069E-4 + 9.8687E-4j

Now the Y-parameters are converted to the equivalent circuit parameters:

ω = 2 * π * 16000000
= 100530965
Rf = approximately 750kΩ to 1.2MΩ
= 750000 to 1200000
Cf = -*Imaginary* [Y12'] / ω
= 3.735pF
Cin = (*Imaginary* [Y11'] /ω) - Co
= 6.684pF

Rout = Rin = *Real* [1 / Y11']
= 268

Ro = (*Real* [1/Y22'])
= 253

Cout = (*Imaginary* [Y22'] / ω) - Co
= 6.081pF

gm = *Real* [Y21] + 1/Rf
= -0.01416

The next step is to calculate the loop gain using the parameters we have listed above:

$$\text{loop - gain} = \frac{gmR_0}{1 + R_0\left[\omega_o{}^2 C_1 C_2 R_{eff}\right]}$$

$Rx = \omega_o{}^2\, C_1\, C_2 R_{EFF}$

Rx = 1.011E16 * 3.3384E-11 * 3.298E-11 * 6.7 * 1.25

Rx = 9.3194E-5

$$\text{Loop Gain} = \frac{-0.01416 * 253}{1 + [253 * 9.319E\text{-}5]}$$

Loop Gain = 3.5

The loop gain of 3.5 shows that adequate gain is present in this configuration for the oscillator design. This information determines that the design is in the correct range.

NYQUIST ANALYSIS

A second useful analysis tool is the Nyquist analysis. Since this analysis uses the SPICE simulation it is used after the design has passed the loop-gain criterion. The Nyquist analysis will determine the overall marginality of the oscillator design. An example will be used to clarify the use and model for the Nyquist analysis.

A Nyquist analysis on Spice will provide the valid range for the parameter gm. The valid gm range is determined from inverting the voltage magnitude at the 0° phase crossing points, in the example these crossings will be in bold and underlined. If the gm of the device is within the valid range the oscillator circuit will function properly.

NYQUIST EXAMPLE

Using the same values as the previous example for loop-gain analysis a Nyquist analysis will be performed.

```
****************** 06/07/93 08:39:18 ******************
Evaluation PSpice (January 1991) ***********************
16MHz NYQUIST PLOT OF AN OSCILLATOR*********
************  CIRCUIT DESCRIPTION******************
   IIN 4 0 AC=1
   .OPTIONS ACCT ABSTOL=10N VNTOL=10U
RELTOL=.001 NOPAGE NUMDGT=6
.WIDTH OUT=80
.OP
.TEMP 25
.PROBE
.AC LIN 2000 16.0MEG 16.2MEG
*Cin is the input capacitance of the 2 port network
Cin 2 0 6.684pF
*Rx1 is the open resistance of the Xtal1 pin
Rx1 1 0 150MEG
*Rin is the input resistance of the 2 port network
Rin 1 2 268
*Co is the Co of the Crystal
Co 1 4 5.5pF
*Rs is the series resistance of the crystal
Rs 1 5 5.5
*Cs is the series capacitance of the   crystal
Cs 5 6 21.9fF
*L1 is the inductance of the crystal
L1 6 4 4.5mH
*Cl1 is the board and load capacitance on Xtal1
Cl1 1 0 22pF
*Cl2 is the board and load capacitance on Xtal2
Cl2 4 0 22pF
*Ro is the Ro of the active device
Ro 3 0 253
*Cout is the output capacitance of the 2 port network
Cout 3 0 6.081pF
*Rf is the FEEDBACK RESISTOR
Rf 2 3 1000000
*Cf is the  feedback capacitance also called Co in
     equiv. ckt model
Cf 2 3 3.735pF
*Rout is the output series resistance of the network
Rout 3 4 268
*Rx2 is the output open resistance of Xtal2 pin
Rx2 4 0 150MEG
***********************************
.PRINT AC VM(1) VP(1)
.END
**** SMALL SIGNAL BIAS SOLUTION
TEMPERATURE =  25.000 DEG C
```

FREQ	VM(1)	VP(1)
1.604E+07	2.882E+02	4.760E+00
1.604E+07	2.805E+02	3.319E+00
1.604E+07	2.730E+02	1.940E+00
1.604E+07	2.656E+02	6.207E-01
1.604E+07	**2.583E+02**	**-6.413E-01**
1.604E+07	2.513E+02	-1.847E+00
1.604E+07	2.444E+02	-3.001E+00
1.604E+07	2.377E+02	-4.102E+00
.	.	.
.	.	.
1.605E+07	1.296E+01	-2.982E+00

1.605E+07	1.262E+01	-1.854E+00
1.605E+07	**1.228E+01**	**-6.759E-01**
1.605E+07	1.195E+01	5.549E-01
1.605E+07	1.163E+01	1.840E+00
1.605E+07	1.133E+01	3.181E+00

JOB CONCLUDED

The spice simulation shows that the gm of the system must be in the following range:

$1/12.28 > gm > 1/258.3$ or

$0.0814 > gm > 0.00387$

Since the gm of the 87C196KR is 0.01416 it fits the range and the design should be suitable for the operating conditions used in the simulation.

If the results indicated that the current design as specified in the oscillator circuit model was not able to meet the Nyquist analysis criterion, design changes would have to be investigated.

MODELING ACCURACY

Whenever a circuit designer encounters an equivalent circuit model the question of the model's accuracy is always present. To prove the usefulness and accuracy of the equivalent circuit model two simulations were run. The first computer simulation is a simulation of the proprietary circuit model. This model incorporates the active devices and the corresponding parameters for the devices. The second computer simulation is of the equivalent circuit model. Both simulations were run using the same characteristics for the crystal and PCB. The graphical results are shown below.

From the graphic results of a spice simulation it can be seen that the magnitude of the voltages from the two separate models track very closely. There is less than a 5% error between the two models. The main difference is in the phase of the equivalent circuit model. Although the overall shape of the phase curve is similar to that of the actual circuit phase curve, a phase difference of approximately 50 degrees exists.

Actual Circuit

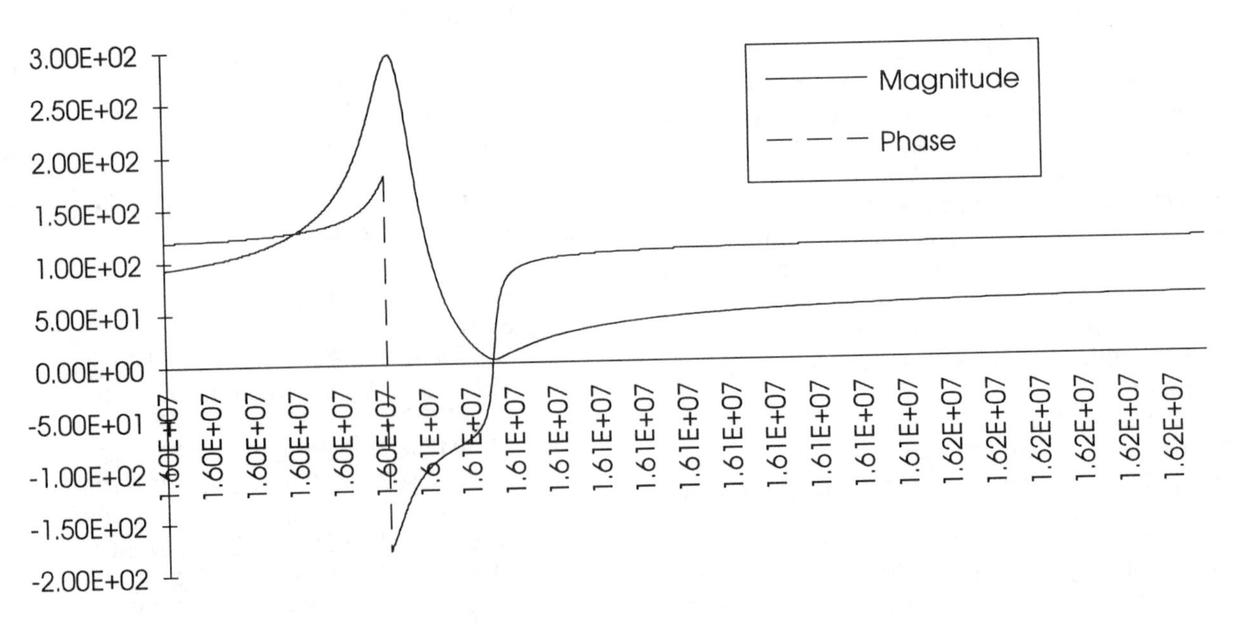

Equivalent Circuit Model

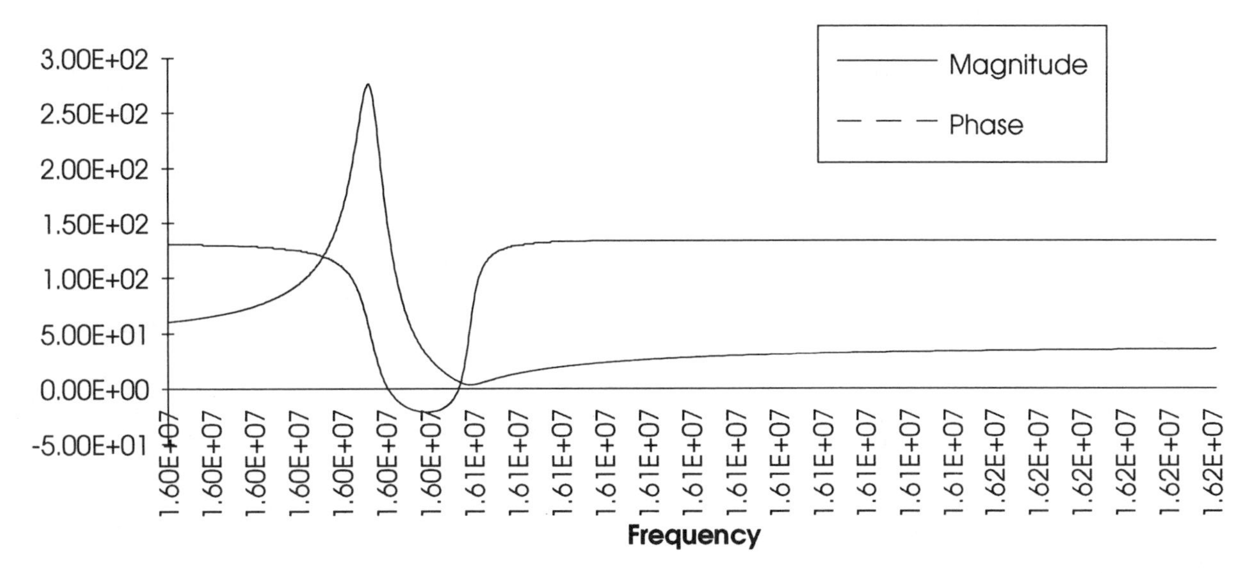

Currently an investigation is continuing to determine if the phase difference is due to the test set-up used to collect the S-parameter data or if the phase difference is due to the model being passive instead of active. Since the equivalent circuit model is more of a worst case scenario, the difference is not detrimental.

SUMMARY

From the S-parameter information that Intel provides and the equivalent circuit model contained within this paper, oscillator circuit designs may be optimized for a particular PCB design.

REFERENCES

1. R.G. Meyer, "MOS Crystal Oscillator Design," IEEE J. Solid-State Circuits, vol SC-15, no. 2, pp. 222-228, April 1980.
2. E.A. Vittoz, M.G. Degrauwe, and S. Bitz, "High-Performance Crystal Oscillator Circuits:Theory and Application," IEEE J. Solid-State Circuits, vol. 23, no. 3, pp. 774-783, June 1988.
3. Bottom, Virgil E. Ph.D., Introduction to Quartz Crystal Unit Design. New York: Van Nostrand Reinhold Company.
4. Matthys, Robert J., Crystal Oscillator Circuits. New York: Wiley-Interscience Publication.
5. Parzen, Benjamin, Design of Crystal and Other Harmonic Oscillators. New York: Wiley-Interscience Publication.
6. Intel Application Note AP-125 "Designing Microcontroller Systems for Electrically Noisy Environments"
7. Intel Application Note AP-155 "Oscillators for Microcontrollers"
8. HP Application Note 95-1 "S-Parameter Techniques for Faster, More Accurate Network Design"
9. HP Application Note 200-2 "Fundamentals of Quartz Oscillators"
10. HP Application Note 154 "S-Parameter Design"
11. NDK Application Note T_X-OZ-E004 "The Relationship Between A Crystal Oscillation Circuit and Load Capacitance"
12. NDK Application Note TN-X-2Z-E007 "For Proper Use of Crystal Units"
13. NDK Application Note T-X-9X-E002 "Frequency Jumps and Measures to Prevent Them"
14. NDK Application Note TN-X-19-E006 "The Frequency VS. Temperature Characteristics of Crystal Units and the Frequency Stability of Oscillation Circuits"
15. NDK Application Note TN-X-44-E009 "How to Oscillate Crystals Without Fail"
16. MURATA ERIE Catalog P-04-B "Ceramic Resonators for Timing Control"
17. Kimmel Gerke & Associates, LTD. "Practical Tools, Tips, and Techniques for High Performance Design." Presented April 1993.

931838

Multi-Chip Modules for Automotive Applications

Patrick J. McCaffrey
Lucas Advanced Engineering Centre

ABSTRACT

Increased functionality, increased reliability and reduced system costs have been the driving forces behind silicon integration. However, conventional packaging options are becoming a barrier both technologically and in terms of cost effectiveness to further silicon integration. Multi-chip modules (MCMs) provide a cost effective solution to handling the most advanced of silicon chips while increasing reliability. An example of an automotive system which merits the use of thin film (MCM-D) technology is described. The preferred type of interconnect for this circuit together with the alternatives is discussed, with emphasis on the multi-chip module substrate.

INTRODUCTION

AUTOMOTIVE ELECTRONICS - Electronic systems have been added sequentially to automobiles at an increasing rate over the past two decades. For further control interaction between them is required. Fast communication within and between systems is the key to the next stage of systems development. High speed communications to and from an increased range of actuator and sensor units and the real time processing of information from these systems will be necessary. Network systems have already been developed for vehicles and will soon begin to permeate through to the mass market. Intelligent sensors and actuators will be used to carry out the primary signal processing and facilitate direct links to the network. This will allow certain malfunctions to be taken care of by spare processor capacity on the network. Typically intelligent sensors and actuators will require the individual electronic units to be small and capable of being sited in a hostile environment.

The enhancement of automobiles with electronic systems is in its infancy. The potential for significant improvements in safety, road efficiency, reliability and communications is only now being developed. The work of programmes such as PROMETHEUS in Europe, IVHS in America and equivalent work in Japan envisages systems which extend beyond the confines of the automobile to the surrounding environment. Sensing systems such as laser and radar, as well as beamed-in information from ground stations mean that a massive increase in data processing power is necessary.

THE DEMAND FOR MCMS - Large electronic systems are the products of packaging and interconnect levels beyond the silicon integrated circuit. Although most of the processing functions are carried out within the silicon chips, they represent only a fraction of the size and cost of the system. Each integrated circuit has it individual package (wire-bonds, lead frame and solder joints) to contend with.

Even for plastic packaged parts, the silicon chip cost is typically less than half that of the packaged components. Next there is the interconnect level; this may consist of a printed circuit board (PCB) or hybrid (with the option of pasivating materials) which in turn may be connected to further boards via a range of joints (flexible and fixed).

The unit will then be supplied with an electrical connector to the outside world and boxed to provide mechanical and limited chemical protection as well as mounting arrangements. Therefore, any cost savings made through IC package integration are increased at each further level of packaging.

System failures can very often be attributed to joints between different packaging levels. Reliability is therefore dependant on the number of joints, the type of joints, the application and the environment.

By using MCMs the surface mount or through hole joints are reduced or dispensed with, and so an immediate reliability improvement is offered.

Speed of devices is increasing. Signal delays and distortions become more critical as frequency increases. The contribution of increasingly large packages and the necessarily large interconnect systems means that the performance of silicon devices is limited. The first move to reduce propagation delays and reduce signal distortion is to make the lengths of connections shorter. This can be achieved by the removal of the individual IC package, allowing components to be placed more closely together. The very high tracking capacity of MCMs allows tracking to occur beneath the component. Impedance can be controlled by systems such as silicon hybrids reducing signal distortions still further.

HARDWARE OPTIONS

SUBSTRATE TECHNOLOGY-Printed circuit boards (PCBs) with surface mounted components have largely displaced plated through hole technology to become the accepted standard electronic assembly technology. The range of different types of materials and design rules is wide, however, the mainstay substrate of automotive electronic systems is flat rigid PCBs with two to four conductor layers.

The main area of weakness in surface mount technology is in the actual solder joints themselves. There is an inherent drawback in the solder material, which even at room temperatures is relatively close to its melting point when compared to metals such as steel or copper. Solder joints are damaged by mechanical stresses which are often induced by thermal expansion differences between the components and the PCB. Thermal cycling is one of the severe conditions that solder joints can be subjected to, but even small stresses can lead to the solder creeping and eventually fatiguing and the joint failing. In automotive environments the additional hazard of the presence of aggressive chemicals assists the failure mechanisms through corrosion processes.

PCBs are used widely within the automotive industry for their simplicity and cost. However, as both IC pin counts and the complexity of systems functions are increasing, the ability of PCB technology to cope with the tracking requirements, and still keep cost down is limited.

Plastic package size is primarily determined by the number of I/Os and the I/O pitch. The I/O pitch of package pins is steadily being reduced but this makes packages more difficult to manufacture, (eg having to etch lead frames instead of stamping) and much more difficult to assemble onto interconnect boards.

DIRECT CHIP ATTACH -

Three options are available for direct chip attach, wire-bonding, flip chip and TAB.

Wirebonding is the most common type of direct chip attach. The silicon ICs are glued to the substrate surface with the active surface of the device facing upwards. Fine wire-bonds are then made between the chip metallization pad and the corresponding pad on the substrate. Chip attach materials vary widely depending upon the thermal, mechanical and electrical requirements, so that solders, epoxies, and polymers are all used. The selection of die attach material is critical and much effort has been spent on evaluating and qualifying materials. Both gold thermo-compression and aluminium ultrasonic wire-bonding are used with direct chip attach. Gold has the advantage of a faster bonding operation, but for use at elevated temperatures, aluminium wire-bonding is preferred. This is due to the fact that diffusion induced voids (Kirkendale) can form between aluminium and gold resulting in failed joints. Wire-bonding has two advantages as an electrical chip attach mechanism; firstly the set-up costs are relatively low, (no special processing of the devices is required) and secondly if aluminium is used as the wire and interconnect metal then a mono-metallic system between chips may be created, together with its accompanying integrity.

Flip chip is a method of electrical attach and mechanical chip attach using solder. Solder bumps are formed at the chip I/Os and the devices are attached directly to the substrate with the active face down, (hence the term flip chip). Solder wettable metal must be applied to both the chips and the substrate as well as various adhesion and diffusion barrier layers. All of the drawbacks of using solder are again present with this method, but for certain applications is has found favour due to the ability to form all connections in one operation and thus the saving of costs. Flip chip has the lowest impedance of the electrical chip attach methods and thus will find applications in the high speed data processing area.

A certain amount of redesign and processing of silicon devices is necessary to manufacture flip chip versions of devices, this, is very costly in the prototyping stage. Also the use of very fine pitch very fine diameter solder bumps is likely to be suspect particularly during power cycling of the modules - although this would be eased by use of silicon substrates. The drawback as with wire-bonding is that only limited testing of the devices will be available. Visual inspection of solder joint quality with peripherally bonded flip chip components is difficult; with area bonded flip chip it is generally not possible to inspect.

TAB (tape automated bonding) is a technology which has found favour in high volume consumer applications. In this technology thin foil lead-frames are formed on a polymer carrier film. The carrier film is formed into a reel, and the chips are connected to the lead-frame as the reel is unwound. Connections between the chip and the lead-frame are made by thermal-compression of gold bumps. The lead-frame is then cut out of the reel and the devices placed and connected. Connections between the outer lead and the substrate are typically solder although gold bumps are again an alternative. TAB has tended to become the desired chip attach technology for MCMs because it offers a route to providing tested die prior to assembly. However TAB is still restricted in its application due to set-up tooling and assembly charges. The cost of prototyping in TAB is very expensive and the cost of prototyping in TAB for multi-chip modules is prohibitive.

SUBSTRATE OPTIONS Choice of MCM substrate technology is determined by several factors. Key among these are reliability, system size, MCM performance requirements (power and speed), component and assembly costs, interconnect capacity and component handling ability. The substrates must also be easily interfaced with other packaging parts, such as macro-substrates and the unit box mounting.

Three general classes of substrate are available: (1) Laminated boards (MCM-L), (2) Ceramic substrates (MCM-C), (3) Deposited thin film substrates (MCM-D).

Laminated boards can be easily cut into a variety of shapes and may be populated on either one or both sides. Laminated boards are capable of containing a combination of plated through holes, surface mount and direct chip attach. For MCM-L substrates, very high tracking density boards are used. By virtue of their volume usage and the large area manufacturing processes, laminated boards are by far the lowest cost substrates and are likely to remain so.

Ceramic substrates are available in two groups; (1) Thick films, (2) Co-fired. Thick film substrates are the more common of the two and have been used for decades in the automotive industry with bare chips. Metallic inks are screen printed to form conductors. The ability to print resistors increases reliability by eliminating the component joints. Multi-layers of thick film hybrids are provided by printing conductors and the dielectric layers. The sequential nature of this process means that if one defect occurs during a print stage then the whole of the subsequent structure is failed, and thus the yield tends to go down as the number of layers goes up. Substrates can be produced with up to nine conductor layers, however, at this level the cost is too high for automotive applications.

Co-fired ceramic hybrids have been developed to help address the yield difficulties with thick film hybrids. Thin layers of ceramic have holes made in them for vias, these vias are filled and then a metal conductor pattern is printed and inspected when dry. Finally the ceramic layers are stacked together, laminated under pressure and fired to form one multi-layer substrate. The ability to check each layer before firing means that very good yields on substrates with many layers can be achieved. Multi-chip modules with over sixty conductor layers have been made using this process.

Thin film substrates have been available for decades, however the new wave of thin film substrates for multi-chip modules started in the mid 1980s. In essence this is a different approach to that of other MCMs, so that high tracking capacity is obtained by reducing line widths instead of increasing the number of layers. To some degree thick film hybrids are being reduced in track pitches and thin film hybrids are having more conductor layers added but the main approach to achieving tracking density is different. There are wide variations within the categories of thin and thick film hybrids, but on average the tracking pitch for thin film is about one tenth that of thick film. Thus, to achieve the same tracking capacity as one layer of thin film, ten layers of thick film would be required. This is the crucial advantage of thin film MCMs.

On the thin film substrate, the conductor patterns are fabricated from thin deposited metal films, (typically less than 3um) using standard IC photolithographic techniques. Normally the deposited films will be copper or aluminium. Dielectric layers either tend to be polymer films (such as benzocyclobutane or polyimide) or inorganic materials. The polymer films are normally spun on in precursor state and later dried and cured. Film thickness of up to 20um are easily achieved. Photoimagable versions of the polymers have been developed but patterning using a photo-resist can also be used. Silicon dioxide, silicon nitride, or silicon oxinitride can also be used for the dielectric layer. These materials are used routinely within the semiconductor industry and their compatibility with silicon processing has been established.

MULTI-CHIP MODULES FOR AUTOMOTIVE

All of the chip attach and substrate options can be used for automotive applications. A natural evolution will occur from chip on board to MCM-L and also from chip and wire thick film hybrids to MCM-C. The overall trend is for hardware to become more reliable and have increased functional capability.

This will see a move away from surface mount to bare chip, from MCM-L to MCM-C and from MCM-C to MCM-D. In many applications MCM-L and MCM-C shall provide adequate reliability and functionality at suitable cost, but for the more advanced applications and for high reliability requirements MCM-D will be necessary. An example of one such application is described below.

THE AICC (AUTONOMOUS INTELLIGENT CRUISE CONTROL)

The AICC system controls car performance via information from an intelligent radar sensor located at the front of the vehicle. The AICC system is comprised of a compact intelligent sensor which is linked through actuator control systems to the vehicles braking and accelerating systems. Vehicle and component manufacturers already have systems in place for cruise control, braking control and skid control and it is not required that the AICC duplicate these efforts. Therefore this work has been on the sensor, the vehicle recognition algorithms and the hardware to contain these algorithms.

Spectral analysis is used on the incoming radar signal to obtain an image of the surrounding environment. This image is then compared to a series of models stored as software, within the sensor unit. This processing function is computationally very intensive. The software models are very sophisticated being built-up over several years of development experience, (radar control for vehicles was started at LUCAS in 1968). When an image is identified as another vehicle it becomes a target, signals are sent to the driver and also to the throttle and braking ECUs. The vehicle then is maintained at a steady speed and controlled distance behind the target vehicle. Speed changes in the target vehicle are observed by changes in the vehicle image size. This requires the control module to store information on the target. Lateral changes in the target will be identified by the target moving out of view. By using three radar beams on each AICC unit it is possible to progress the movement of targets and the entrance of a new target into the field of view. Advanced control algorithms are used to prevent unsafe changes in speed when a change of target occurs, (for instance if a vehicle pulls in behind the target). The system has been demonstrated successfully on private roads in Europe and America in a range of different situations.

Ultimately, control of the whole transport system would provide the most comprehensive road safety and be most effective at reducing traffic congestion. This would be a dramatic change to the current systems. A move to a complete information network for vehicles would be a very costly step forward and is therefore likely to be delayed until a much later date when technology prices have fallen sufficiently. The likely scenario for the development of transport control systems is by the gradual addition of new functions and features to vehicles, the combining of these systems on the vehicle, and then the interlinking between systems. The aim of this work has therefore been to develop autonomous systems which provide vehicle control without the need to interact with similar systems on other vehicles.

THE REQUIREMENTS OF THE AICC - The computing power required for vehicle recognition is massive when compared to a typical automotive electronic control unit (ECU) available today. Modelling of a wide range of vehicle types and situations, and then processing real time radar data to fit these models requires a very large amount of signal processing and memory as well as analogue and digital filtering.

The system size must be relatively small. A space envelope of 10cm x 10cm x 3cm has been set by the systems designers. This allows location of the unit on a wide variety of sites and different models and does not impinge on the vehicle design. Likewise, the options for under-bonnet location must be considered and therefore, the hardware reliability must be appropriate. The unit is also likely to increase in complexity in the future; it is not useful to provide a first system which has no capacity for expansion. The success of this unit will lead to demands for the incorporation of further processing and communications with other ECUs and systems.

The current AICC circuit uses forty silicon ICs. The largest device has 150 I/Os and while the average number of I/Os (35) may not be as high as on some VLSI (very large scale integration) based computer MCMs the large number of ICs raises the average track length and in turn demands a high density interconnect system.

Parameter comparisons for each of the three AICC hardware technologies (surface mount, single layer MCM, double layer MCM), are shown in Table 1. The number of connections in brackets for the surface mount version includes internal connections within the plastic package. The number in brackets for the single layer hybrids includes the connections due to cross-overs.

THE MCM SUBSTRATE

The starting point for the chosen MCM-D substrate is a 2.54cm diameter silicon wafer. Silicon was selected as the base material because it is available in a suitably flat format. It's flatness allows easy processing to very fine feature sizes and it is compatible with standard photolithographic IC processes and handling. For MCMs the feature sizes are much larger than for ICs, and so processing can be carried out using older IC production facilities. Thus a ready supply of process equipment is available at low cost. Other base materials such as polished alumina or aluminium nitride could also be used for this type of substrate.

Once silicon has been selected, a number of other benefits are gained; these include high thermal conductivity, a low TCE, (Thermal Coefficient of Expansion, matched to the device), and the ability to include active circuitry in the substrate.

The structure of the two metal layer substrates is shown in Figure 1. The materials selected for this version of the silicon hybrid were aluminium conductors and silicon dioxide dielectric and passivation.

DESIGN - In order to evaluate the impact on cost and reliability of the number of conductor layers the substrate was partitioned into two parts. Initially a single metal layer version of the co-processor and memory was laid out. Then the main processor, memory and A/Ds were designed as a two metal layer hybrid. Although a single metal layer can provide a very high tracking capacity, sites at which tracks need to cross represent a difficulty. Cross-overs were accommodated with wire-bond jumpers over several tracks. This layout was mainly carried out manually as automatic routing tools tended to place far too many connections on the second metal layer which in this case represented the wire bonds. Current CAD tools do not have the capability to provide high efficiency, single layer, layouts. The advantages of a single metal layer substrate are in the simplicity of substrate manufacture. A very high yield was assured and visual inspection was sufficient to determine open and short circuit testing. The disadvantages of single layer hybrids are seen at the next manufacturing process, where for this application, the additional wire-bonds due to cross overs represented 20% of the total number of wire-bonds.

The size of this module was increased due to the area for crossovers, although by further design optimisation it was estimated that the module could be 20% smaller. The layout for the single layer substrate is shown in Figure 2.

Design time for the two layer silicon hybrid was about one quarter that for the single layer hybrid. Routing and placement of the devices took advantage of the experience from the layout of the single layer version. Almost all of the tracks were routed automatically with manual alterations required for a few tracks only. Both the single layer and two layer substrates incorporate test structures to monitor process quality.

PACKAGING - Metal or ceramic hermetic packages are the most obvious way to package silicon hybrid MCMs. .Automotive applications have previously not normally accepted the additional costs of hermetic packaging. For certain cases the replacement of several high pin count plastics IC packages with one low pin count hermetic package may actually provide a reduction in cost due to the package and assembly. However, generally for automotive applications non-hermetic packaging will be used.

The first stage of package development was therefore to access the robustness of the technology to humidity. Both 85°C/85%RH (relative humidity) and HAST (highly accelerated stress testing) environmental evaluations were carried out on a range of different substrate test modules. The results of this work enabled the selection of the substrate technology.

A ceramic carrier was used to support the silicon substrate. The ceramic was a standard alumina and contained thick film resistors.

Electrical contact to the hybrids could be provided by either wire-bonds from the alumina carrier to a moulded-in lead frame or soldered on lead-frame clips. These options allowed the silicon hybrid development to fit in with the various stages of the system improvements. A range of gel and glob topping materials were evaluated for corrosion protection. A silicone gel material was selected to coat the whole of the silicon hybrid. For mechanical protection during assembly a plastic cap will be attached by adhesive to cover the silicon hybrid. The option exists to replace the plastic cap with a ceramic one to provide a low cost hermetic package. The package construction is shown in expanded format in Figure 3.

TEST AND RELIABILITY - Sourcing of pre-tested known good dice is the major issue for multi-chip module manufacturers. Although certain tests are carried out at wafer level the majority of device tests have evolved to be performed after die packaging. Much research and development work is now underway to determine methods for ensuring bare dice quality and reliability. These include the development of special reusable test fixtures and testable electrical connectors. The design of devices for functional test and burn-in will also make an impact in this area.

Probe testing to check for open and short circuits is necessary for multi-layer substrates, but visual inspection is sufficient for single layer substrates.

In order to allow successful module build and to find defects early on each device is functionally tested after wire-bonding and prior to die attach adhesive cure. Rework equipment and processes and die attach materials are been developed as part of this project and the design rules for chip placement took these items into consideration. Each device was given the possibility of one rework attempt before the module was failed. Conventional wire-bond pads can be reworked but the quality of the reworked joint is suspect and difficult to prove. The wire-bond pads on the substrate were therefore sufficiently large to accommodate the easy placement of a second bond. The redundancy of these bonding sites and other possible redundancies such as extra ICs are issue closely linked with the provision of known good die.

Reliability testing of the modules will be performed to several automotive environmental test specifications. An example of one such test is shown in Table 2.

CONTINUED WORK

Electronics for moving systems suffer particularly harsh environments, especially when located close to a hot engine and in a non-hermetic package. However some of the performance demands such as signal speed are not currently as high as other electronic systems. In order to demonstrate the range of merits of silicon hybrids, demonstrator circuits will also be manufactured for other application areas. These are: aerospace, computing and telecoms.

CONCLUSIONS

The AICC (Autonomous Intelligent Cruise Control) system is a new step in safety and comfort. The rapid increase in automotive electronics heralded by the development of systems such as the AICC will push reliability requirements beyond those achievable with conventional electronic assembly technologies. The failure rate of today's technology can not be reduced enough to cope with the needs of large future systems. Solder joints are the weak link in the system and the use of this material should be maintained at a low level. Direct chip attach is the most suitable way of improving reliability. Various types of MCMs will be employed at different locations within the vehicle. For locations with high ambient temperature cycles thin film MCMs provide most benefits At this stage wire-bonding is the most appropriate method for chip attach for automotive multi-chip modules. The experience gained from design and manufacture of an automotive module has clarified the details of each stage in the process.

The reliability, performance and cost of silicon hybrids when compared with conventional technology will be used to further define the role of this technology.

ACKNOWLEDGEMENTS

This work has been carried out as part of the JESSI T9 project : Silicon Hybrids. The members of the JESSI T9 group are :-

Lucas Electronics, United Kingdom
MCE (Micro Circuit Engineering), United Kingdom
ES2 (European Silicon Structures), France
SOREP, France
SMS (Swedish Microsystems), Sweden
IMM (Centre for Industrial
Microelectronics & Materials Tech), Sweden
Saab-Scania Combitech, Sweden

BIBLIOGRAPHY

1. Automotive Electronics in The Year 2000
   Rivard J G
   Paper No. 861027, Convergence '86
   Detroit, Michigan (1986)

2. New Horizons
   L Brooke
   Automotive Industries
   Page 52-55 August 1991

3. Electronic Packaging For the Automotive Environment
   IEMT Proceedings 1986, page 251-253

4. Automotive Electronics: Where Are We Going?
   R. Dell'Acqua and F. Forlani
   Hybrid Circuits No 15 January 1988 page 8-12

5. Technology Comparison For Multichip Modules With Emphasis On Substrate Alternatives
   P Collander, J-O Andersson, E Martensson
   Proceedings of the 1992 International Conference on Multichip Modules
   page 46-51

6. Application Of Embedded MCM Technology To Automotive Electronics
   R A Fillion, S Eisenhart, R Rauerson
   Paper No 930011 SAE International Congress and Exposition 1993

7. Autonomous Intelligent Cruise Control Incorporating Automotive Braking
   P Martin
   SAE Paper No 930510
   International Congress, March 1993

8. Automotive Application of Microwave Radar
   R Tribe
   Proceedings of IEE Colloquium; Consumer Applications of Radar and Sonar
   25 May 1993, Savoy Place, London, UK

Table 1 Parameter Comparison for AICC Hardware Versions

System	Construction	No of ICs	Number of connections on-module	Substrate Size/cm^2	Max Power Dissipation/W
Total System (Parts 1 & 2)	Surface Mount PCB	35	1070 (2100)	20.0 x 11.5	20.3
Co-processor and Memory (Part 1)	Single Layer Thin-Film on thick-film carrier	14	410 (490)	4.5 x 5.8	8.5
Main Processor Memory and A/Ds (Part 2)	Dual Layer Thin-Film on thick-film carrier	21	660	4.3 x 5.0	11.8

Table 2 Environmental Stress Tests for AICC

Test	Level	Nominal Duration
Thermal Shock	-40°C to 150°C	100 shocks
Temperature Cycles	-40°C to 150°C	1000 cycles
Vibration	Sinusoidal, up to 35g up to 2000Hz	8.5 hours per plane
Salt Spray	5% at 35°C	500 hours
Humidity	-40°C to 40°C 93% RH	10 cycles

FIGURE 1 : SILICON HYBRID STRUCTURE

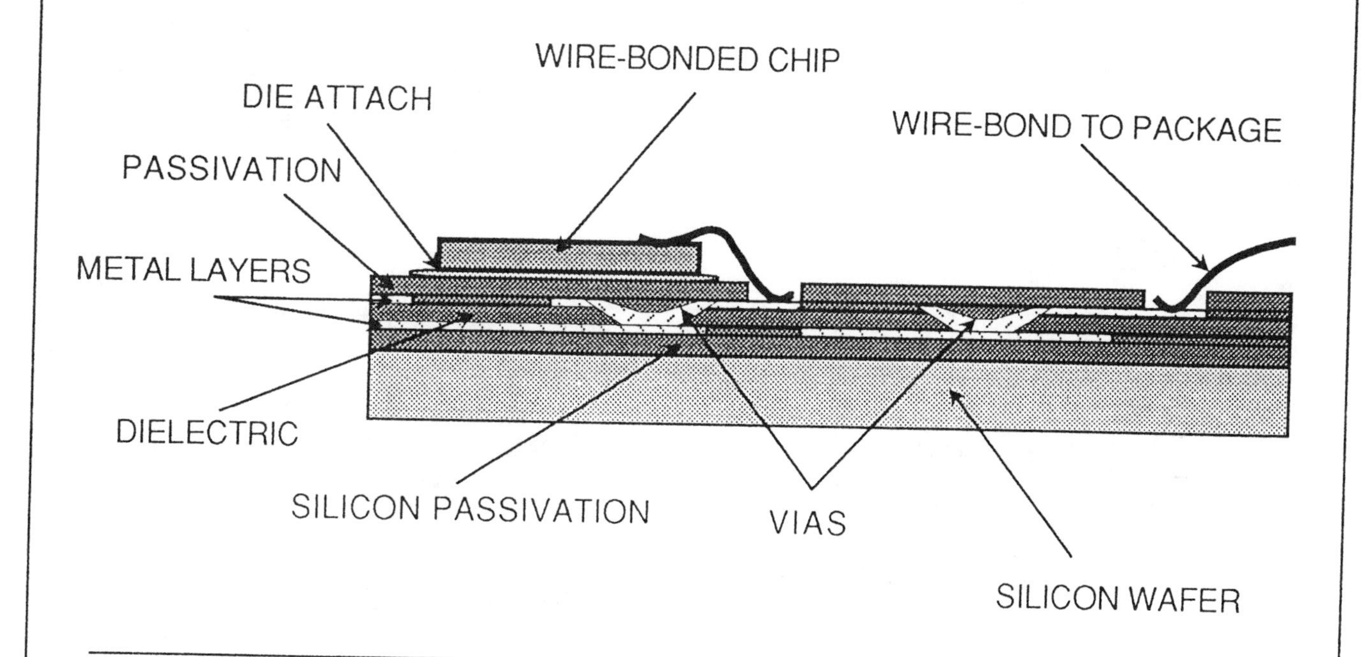

FIGURE 2 : LAYOUT FOR COPROCESSOR

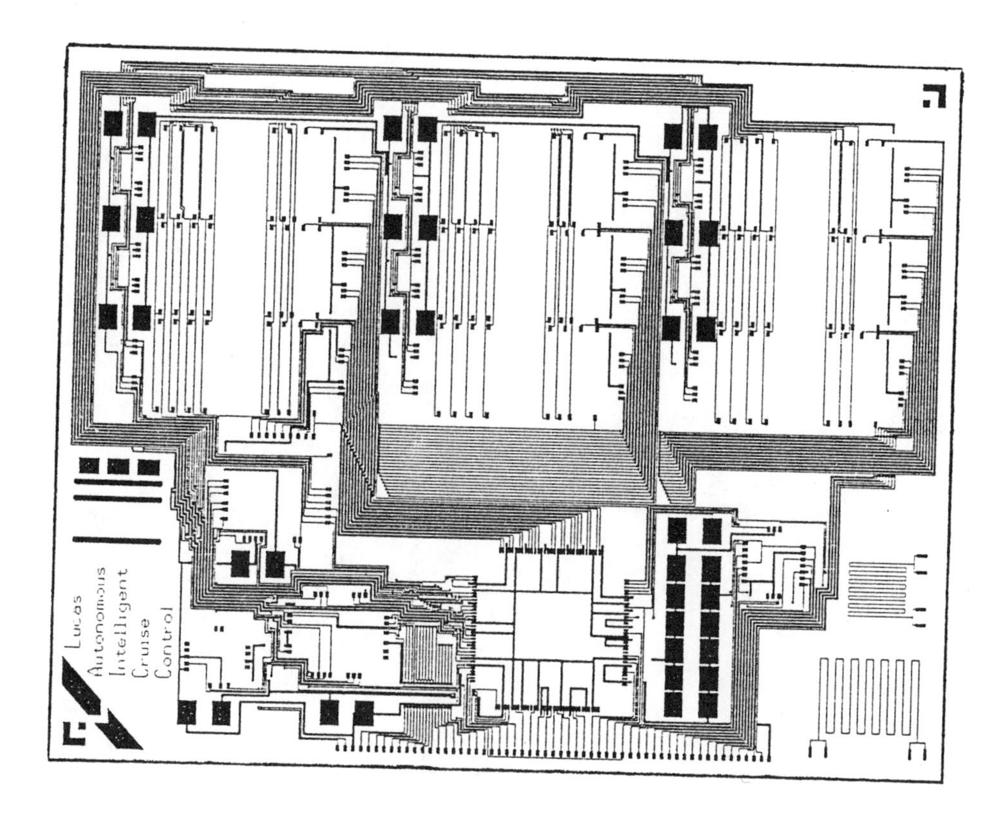

FIGURE 3 : SILICON HYBRID PACKAGING

PLASTIC CAP

SILICONE GEL

SILICON HYBRID

WIRE-BONDED CHIPS

SURFACE MOUNT CAPACITORS

CERAMIC CARRIER

THICK-FILM RESISTORS

WIRE-BONDS TO LEAD FRAME OR

SOLDERED CONNECTORS

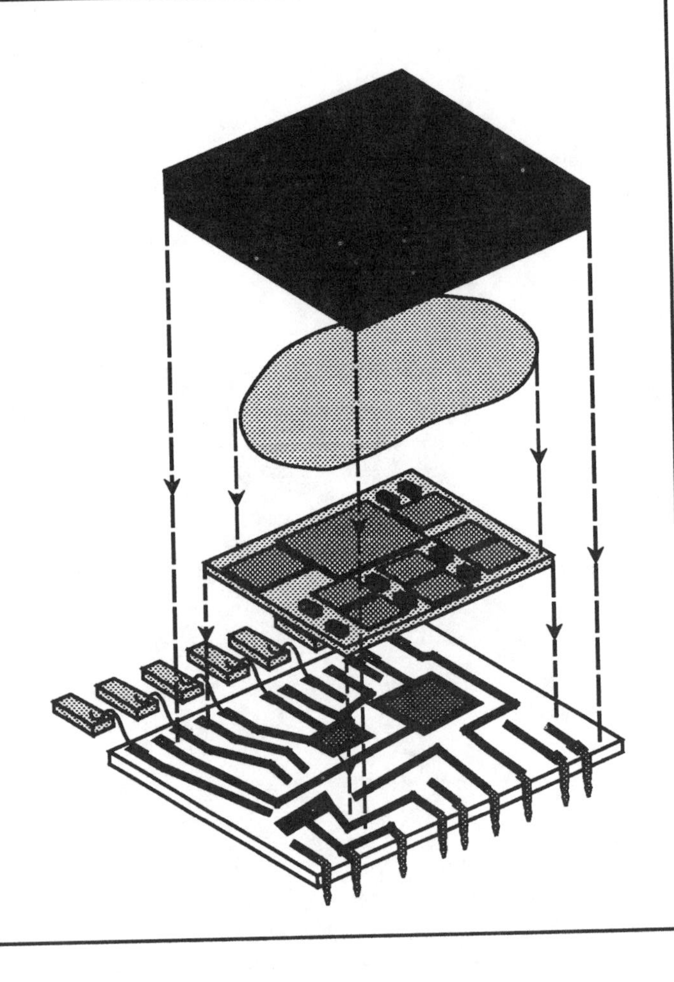

950431

Electronic Control Systems in Microhybrid Technology

Richard Schleupen, Walter Reichert, Peter Tauber, and Gerold Walter
Robert Bosch GmbH

ABSTRACT

The vast majority of automotive electronic control systems utilize printed circuit board technology. Hybrid substrate technology is successfully applied to smaller high volume systems for applications with high demands with respect to temperature range and vibration. In the past few years, the introduction of microcontroller based systems in hybrid technology has succeeded for high volume applications with limited model variety, as for example antiskid (ABS) systems. The newly developed microhybrid technology alleviates the shortcomings of standard hybrid technology regarding complexity, design flexibility and cost of production. This paper discusses two microhybrid systems, an engine management system mounted onto the engine and an antiskid ECU attached to the hydraulic unit.

INTRODUCTION

Thick film hybrid technology has proven to provide products that are able to withstand the sometimes harsh environmental conditions electronic systems face in an automobile. Wherever the neccessity for an application exists to withstand extreme temperatures, humidity and strong vibrations in coincidence with the demand for high reliability, reduced weight and compact size, a hybrid solution has advantage in comparison to other techniques [1] [2].

The production of printed standard substrates is a sequential process. Screenprinting, drying and firing pastes for bondpads, circuit lines, resistors and isolating layers are processes taking place one after the other. Cyclic repetition of isolation layer and circuit line prints allows multilayer designs. The sum-up of all yield losses of the single production steps ends up in an obvious tradeoff in cost growing with the number of layers.

Replacing the alumina substrate by a stack of low temperature cofired ceramic (LTCC) sheets, each of them serving as an individual substrate for the different layers, parallelizes substrate production. At the same time, a miniaturization of the characteristic dimensions of the hybrid is achieved while avoiding the drawbacks of a sequential production flow. This microhybrid technology, specially developed for use in rough environments, permits further reduction of area needed for representation of a given circuit, while keeping all the advantages of standard hybrids (Figure1).

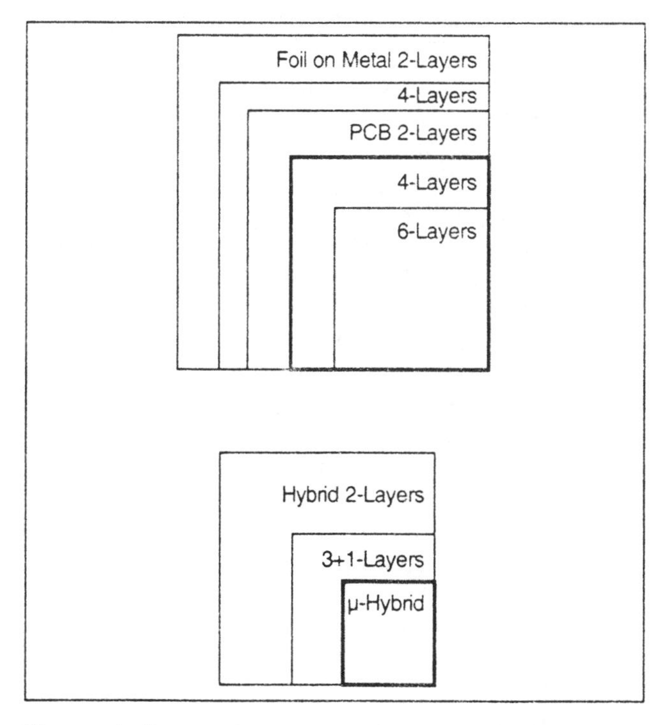

Figure 1: Comparison of substrate areas needed for same functionality

Specific synergetic benefits result when the electronic system is integrated in a mechanical assembly group. We will demonstrate this with two examples from different applications of automotive electronics, an ABS system and an engine management system.

ENGINE MANAGEMENT SYSTEM

The ECU discussed here is a microcontroller system designed to be mounted onto the engine. This design is based on the experience gathered with former systems [1] and their successors (Figure 2) as there are an ignition and a MOTRONIC system.

IGNITION SYSTEM - Module with knock-control and CAN interface with the following specification.

Main Features

- Ambient temperature range - 40/+110° C
- Mounted in the engine compartment
- Hybrid size 50.8 x 101.6 mm²
- Control box size 80 x 130 x 45 mm³
- 11 ICs (CPU, 32 K EPROM, 256 Bytes µC-internal RAM) 3 power stages with voltage regulator
- Integrated ignition output stages (11A,max.2) and pressure sensor
- Two 8 pin, one 3 pin and one coaxial connector.

Functions

- Controls 6- or 12-cyl. engines with conventional distributor (12-cylinder:2 ECUs per engine) or 8-cylinder with two distributors
- Load detection by internal manifold pressure sensor
- Adaptive knock control (cylinder selctive) with two knock-sensors
- Cylinder identification by inductive camshaft sensor
- Cylinder selective closed loop dwell angle control
- Missing spark detection by primary spark duration and voltage monitoring
- CAN communication .

MOTRONIC SYSTEM - Module with distributor-less ignition (DIS) and specification shown below.

Main Features

- Manifold pressure controlled MOTRONIC for 4-cylinder engines
- Mounted in the engine compartment
- Ambient temperature range -40/+110° C
- Hybrid size 76.2 x 101.6 mm²
- Control box size 139 x 130 x 57 mm³
- Highly integrated system: 6 ICs (CPU, 64K EPROM, RAM ,7 power stages with voltage regulator included
- Integrated ignition output stages (DIS) and pressure sensor.

Functions

- Multipoint injection (2 groups)
- Distributorless ignition system
- Adaptive lambda-control
- Idle speed control with digital throttle actuator position control .

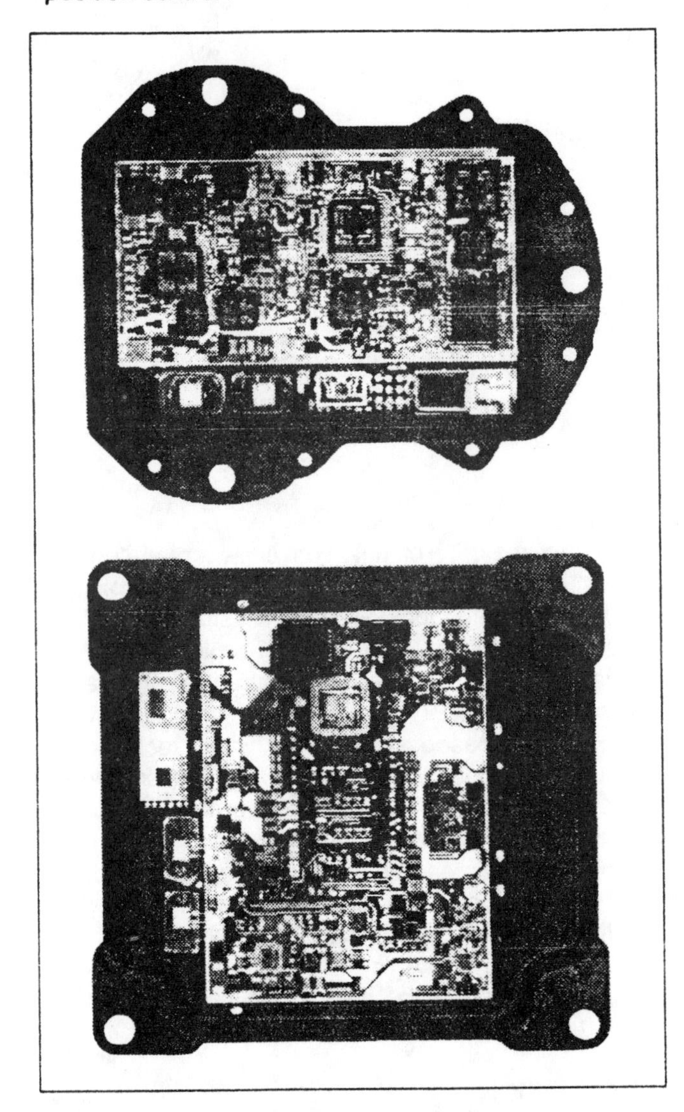

Figure 2: Microcomputer controlled electronic spark timing module (top) and engine management system (bottom)

Both systems are mounted in the engine compartment but not on the engine. They use conventional hybrid 3-layer substrates for the ignition system and 2-layers for the MOTRONIC system. A PLCC-EPROM contains the program and has to be mounted first during production. Therefore any exchange of software during the production process or after results in the replacement of the module.

ENGINE MOUNTED ECU - To overcome the restrictions mentioned above, the Motronic system presented here uses flash memory. For ease of production LTCC technique is used. Every layer can be tested before cofiring. The reduction in size for μ-hybrids, shown in Figure 1 allows mounting onto the engine (Figure 3).

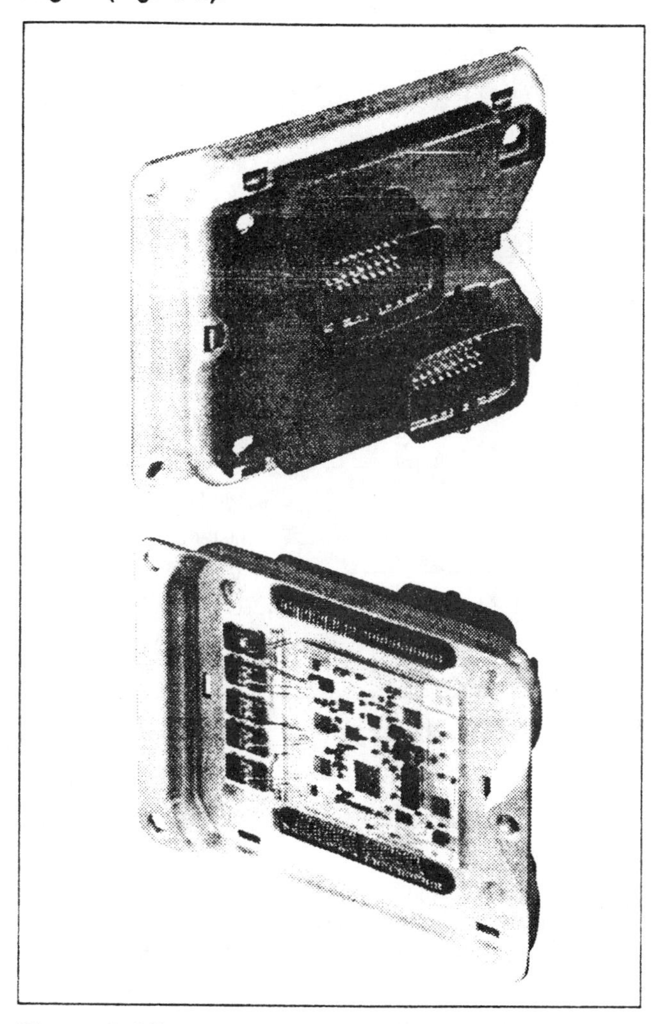

Figure 3: Microcomputer controlled engine management module in μ-Hybrid technology for application on the engine

CIRCUIT DESCRIPTION - The module is built with an 8bit flash-microcontroller, programmed to handle injection, ignition with knock control, idle speed control and closed loop exhaust gas recirculation (EGR). The microcontroller has 64 or 128k flash memory on board together with 3k RAM. Update of program can be done through serial link protected by security code with the ECU connected. Therefore no specially constructed application module to allow for change of software is neccessary. Load is detected from an external manifold pressure sensor. A maximum of four ignition output stages four distributorless ignition with single ended coils are mounted on the baseplate. The ECU is grounded on the engine.

Additional features -
- Ambient temperature range -40/+110° C
- μ-Hybrid size 50.8 x 50.8 mm^2
- 13 ICs (8bit CPU with flash memory)
- 4 ignition output stages for 7,5 A each
- Two 38-pin connectors.

Additional functions -
- 3/4 cylinder increment triggering from the crankshaft
- 60-2 teeth with inductive pick-up
- Multipoint/sequential fuel injection with phase sensor
- Idle speed control with digital throttle actuator position control
- Adaptive lambda-control
- Adaptive knock-control
- E^2PROM 256 byte
- Immobilizer link
- Pulse width modulated communication with external gearbox-control module
- CAN-communication optional
- 8 low power output stages.

A block diagram of the circuit is shown in Figure 4.

Mechanical design - The LTCC substrate is mounted onto the base plate of the housing consisting of a sheet aluminum deep-drawn part . Besides being neccessary for the mechanical fixture of the LTCC substrate and the power strip, bearing the driver stages, the base plate serves as a heatsink. On the base plate a connector housing with two 38 pin connectors is mounted. This housing is an injection moulded part. Connection between microhybrid substrate and connectors is performed by wire bonding. A metal lid seals the inner volume hermetically.

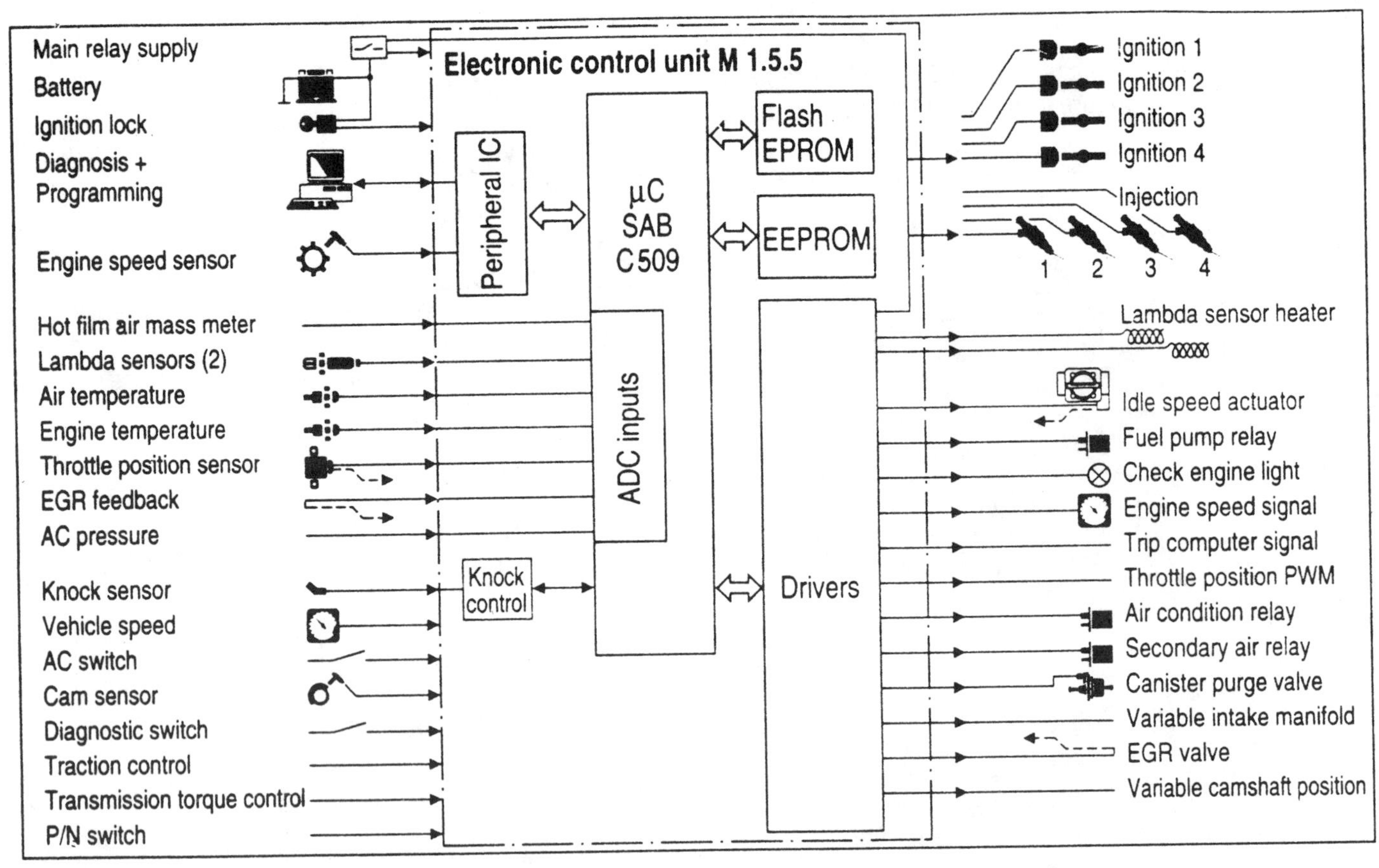

Figure 4: Engine management system in microybrid technology for application on the engine.
Block diagram

ANTILOCK BRAKING SYSTEM

As for the engine management system the antilock braking system has developed from printed circuit board technique (PCB) through conventional hybrids to microhybrid technology [3] [4]. The step from PCB to conventional hybrid technology simplified cable harness as the hybrid ECU is directly attached to the hydraulic unit (Figure 5). The further evolution of development from standard hybrid technique to microhybrids allowed more compact designs of the hydraulic unit and supported the efforts to reduce weight. This is demonstrated in Figure 6 where a comparison is made between three generations of hydraulic units with attached ECUs. Relative weight for the most advanced system scales down to about half the value of the first generation. The reduction in volume is even more pronounced.

CIRCUIT DESCRIPTION - The reliability and proper function of an ABS or traction control system mainly depends on the correct information of the rotational speed of the vehicles wheels and the vehicle speed itself. The wheel speed information is gathered by inductive wheel speed sensors (WSS). These signals are filtered and buffered by an application specific integrated circuit also monitoring disconnection of the sensor wiring. In the event of a WSS interruption, the coded information on the non-functional input channel is transferred to the microcontrollers. There the information is logically filtered, checked for non-plausibility and the vehicle speed is calculated. Using this information the ABS and traction control algorithm actuates the return pump motor and the hydraulic valves to individually reduce, maintain or increase the pressure in each of the wheel brake cylinders. The hydraulic valves are driven by low-side power stages with integrated monitoring functions to detect any irregularity in the actuation circuit, e.g. overload, short circuit or disconnection. In the event of an external failure the system returns immediately into the non-ABS state and the brakes remain fully functional as in vehicles that are not equipped with ABS. At the same time detailed information about the failure that has occurred and the environmental conditions at the time of the occurrance is stored in a non-volatile memory.

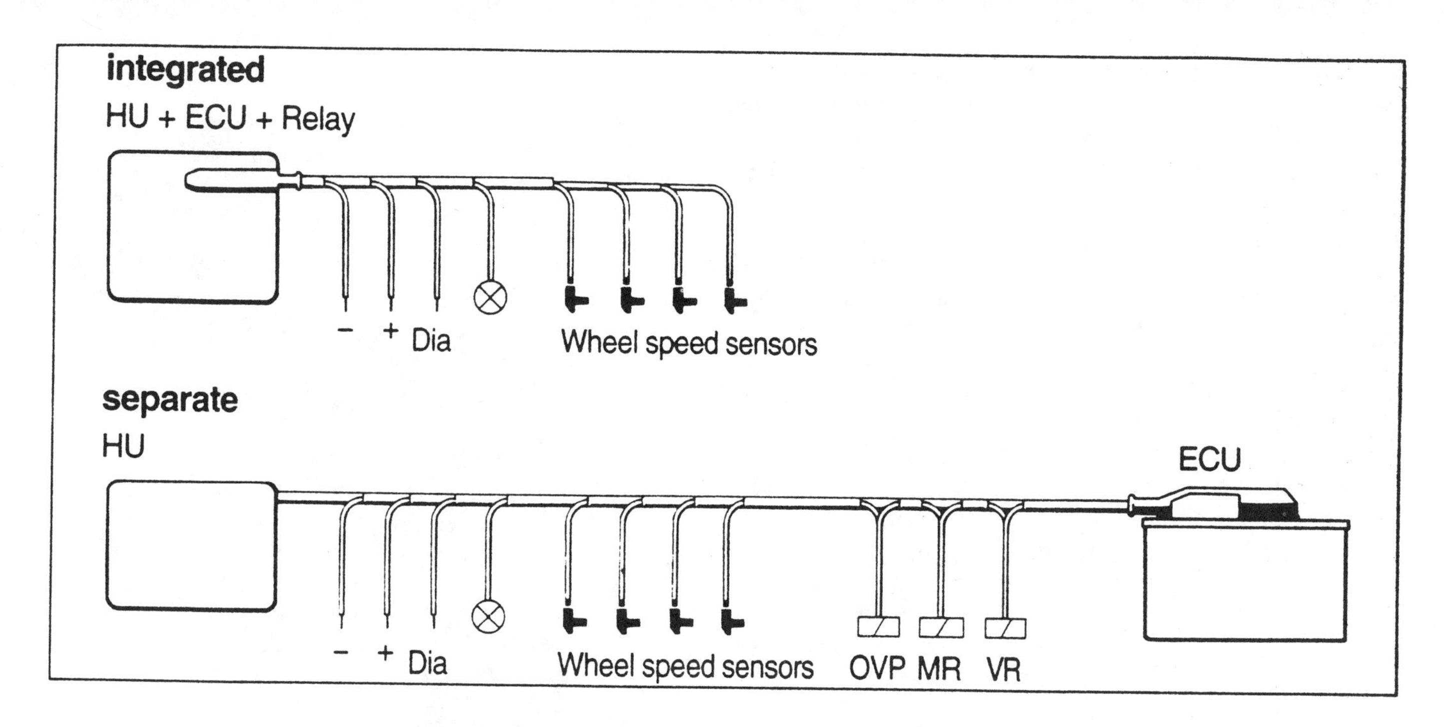

OVP, MR and VR stand for overvoltage protection, motor relay and valve relay, respectively. Dia signifies the line required for diagnosis, the system control lamp while + and - designate battery and ground connections.

Figure 5: Simplification of cable harness for integrated hydraulic unit (HU), electronic control unit (ECU) and relays versus seperate mounting of the components.

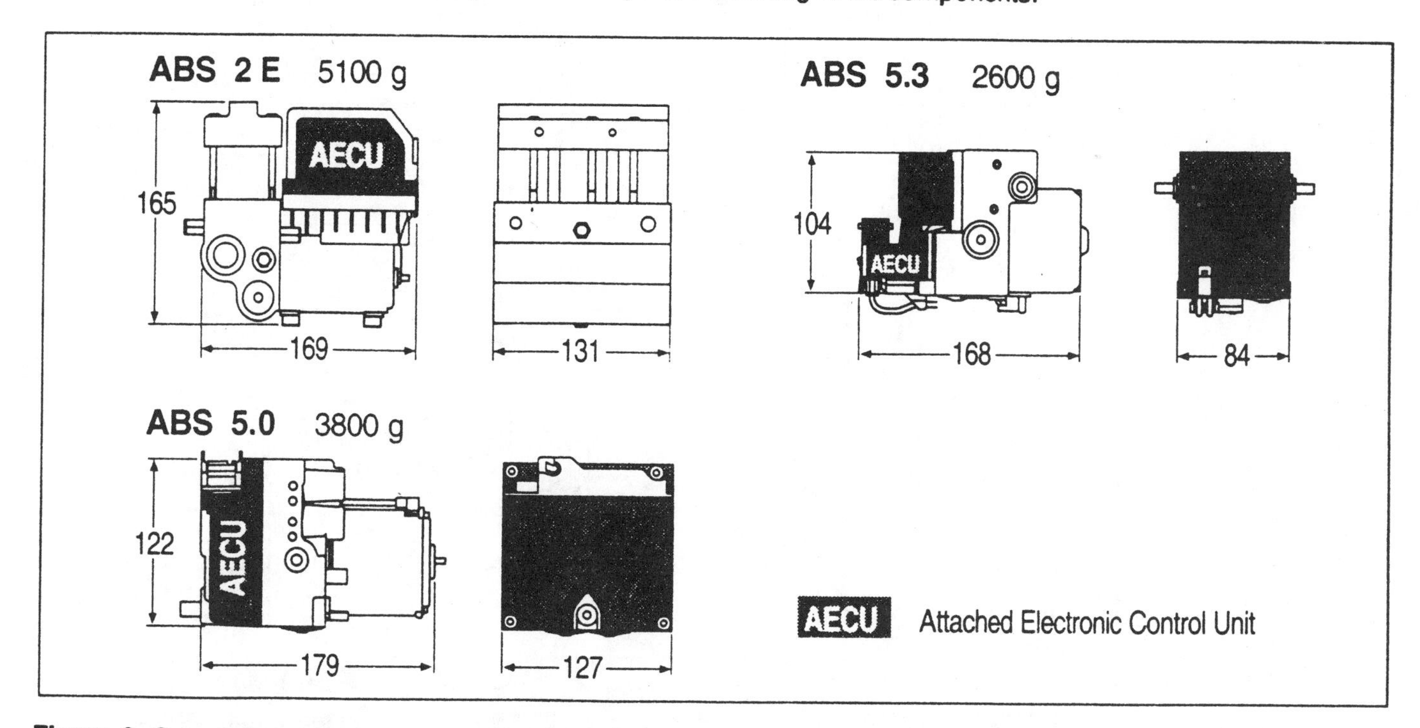

Figure 6: Comparison of size and weight of hydraulic units with attached ECUs of three different antilock braking systems. On the left two versions of hydraulic units with conventional hybrid AECUs are shown. On the right the system using a microhybrid AECU is depicted. All dimensions are given in millimeters.

An application specific diagnostic IC allows implementation of the interactive diagnosis functions according to a broad range of customer requirements. Being part of a redundant microcontroller safety concept, the ABS/traction control and safety logic functions are implemented in two microcontrollers supplemented by an intelligent safety circuit independent of the microcontrollers. Varying the number of input/output interface ICs, this concept allows an easy and flexible expansion from ABS to traction control systems with various engine management interfaces.

Front wheel drive vehicles can be equipped as well as rear wheel drive and four wheel drive vehicles with all kinds of ABS or traction control systems including vehicle bus systems like CAN (Controller Area Network). In Figure 7 a block diagram of an ABS/ASR system is given. A comparison of the same ABS circuit in standard hybrid technique with the microhybrid solution is shown in Figure 8.

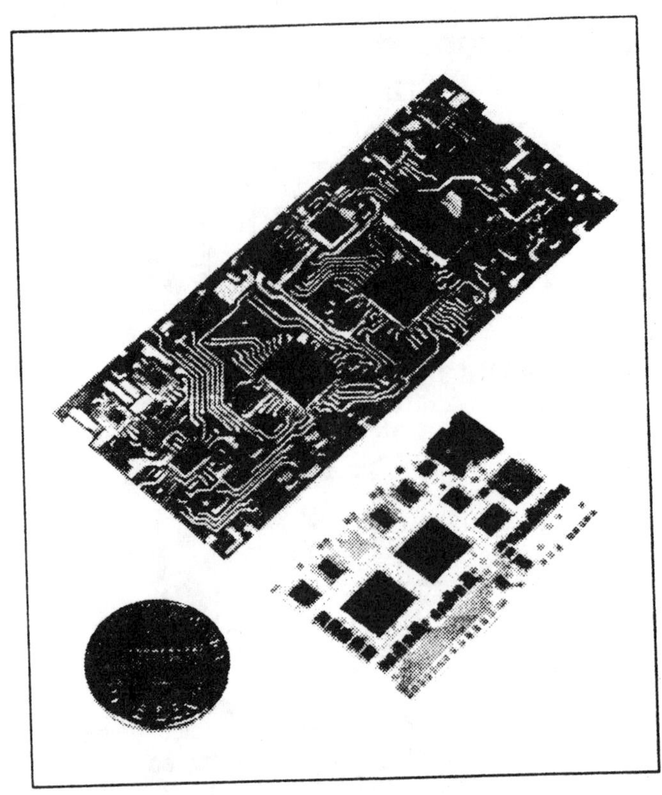

Figure 8: Comparison of the same ABS circuit in standard hybrid and microhybrid technique

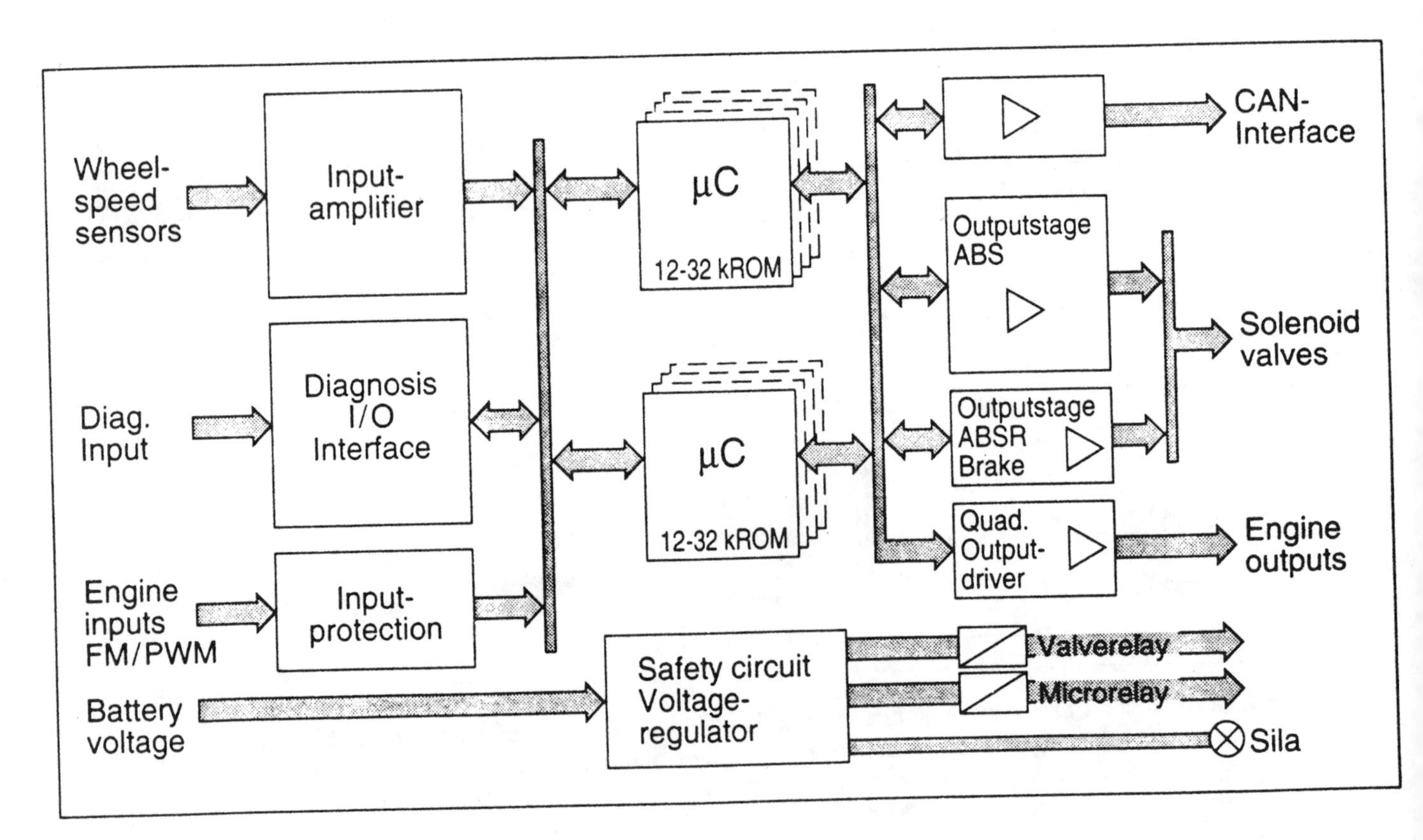

Figure 7: Block diagram of an ABS/ASR system with CAN interface

CONSTRUCTION - The design of the ECU is modular. It closly combines electronic and mechanical functions (Figure 9).

Housing - The housing is an injection moulded part containing connectors as well as a punched out grating. So, the whole interconnection system is an integrated part of the housing.

Relay Plate - Not all electronic components are well suited to be placed on the microhybrid substrate. It was neccessary to mount these components, e.g. relays, big capacitors and power diodes, seperately. The solution selected here, was to weld them to an autonomous punched out grating, the relay board. This board, on the other hand, is welded to the punch grid of the housing. The accommodation of the relay board is closed by a leakproof plastic lid.

Coils - The considerable reduction in size of the whole system made it possible to integrate the coils of the magnetic valves into the ECU housing. The coils are mounted, comparable to the relay board, in a cavity of the housing body and are welded to the punch grid of the housing, too.

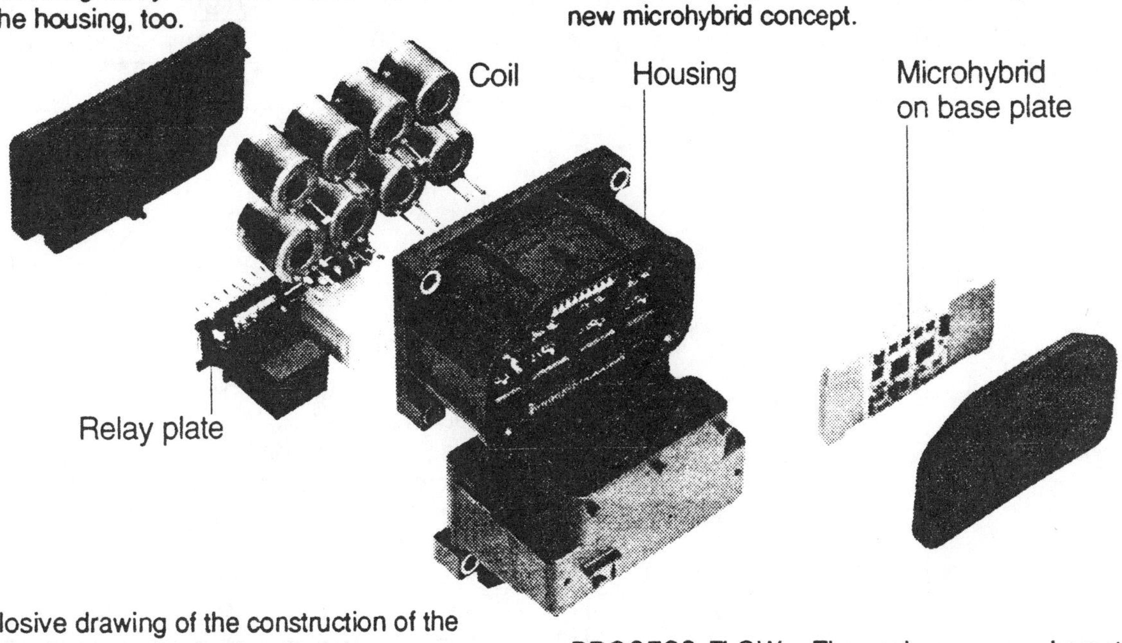

Figure 9: Explosive drawing of the construction of the attached electronic circuit unit of the antilock braking system using microhybrid technology.

Microhybrid on base-plate - Containing all the substantial parts of the electronic circuit the microhybrid is adhered to an aluminium base plate serving as a heat sink. Connection from the substrate to actuators and sensors is achieved by bonding to the punched out grating. Closing of the microhybrid mounting cavity is performed in the same way as for the relay board. Constructed as described, a very compact assembly group is formed which is directly attached to the hydraulic unit.

LTCC - PROCESS DESCRIPTION

Thickfilm technique is used to build up electronic circuits (standard hybrids) which are proved to have a high reliability for automotive applications [1] [2] [5]. The new microhybrid concept allows all advantages of standard hybrids to be kept while adding additional superior properties and capabilites. The high reliability at extreme environmental conditions (temperatures, humidity, vibration), the possibility to form resistors by screen printing and the capability to mount components by adhesive die attachment or soldering is a characteristic of both concepts. The capability of microhybrids to build multilayer circuits with high number of layers combined with the superior possibility for printing conductors with fine line structures will allow the drastic reduction of the size and weight per function. On the other hand it is possible to increase the number of functions per area to a much higher and more cost effective level. Table 1 shows a comparison of standard hybrids with the new microhybrid concept.

PROCESS FLOW - The main process characteristic of a standard multilayer hybrid is a sequentiel process. Each thickfilm printing is performed onto an alumina substrate followed by a drying and firing process step. For e.g. a three layer design the first conductive layer has to be printed, dried and fired onto an alumina substrate. The second layer will be built up by printing, drying and firing of an isolating layer followed by printing, drying and firing of a conductive layer. For the third layer this sequence has to be performed again (Figure 10). Screen printing, drying and firing of resistors completes the standard multilayer hybrid production flow. In case of open silver conductive lines on top an additional protective over coat has to be applied, still.

Table 1 : Capabilities of Standard Hybrid and Microhybrid

Capability	Standard Hybrid	Microhybrid
Multilayer	medium	high
Reliability		
– extreme temperatures	good	good
– high humidity	good	good
– strong vibrations	good	good
Printed resistors	standard	standard
Component Mounting		
– adhesive attach	standard	standard
– soldering	standard	standard
Wire bonding	standard	standard
Thermal conductivity	good	lower without thermal vias good with thermal vias
Size per function	1	0.25
Printed fine line structures	medium	high
TCE match to silicon	good	nearly same as silicon
Weight per function	medium	low

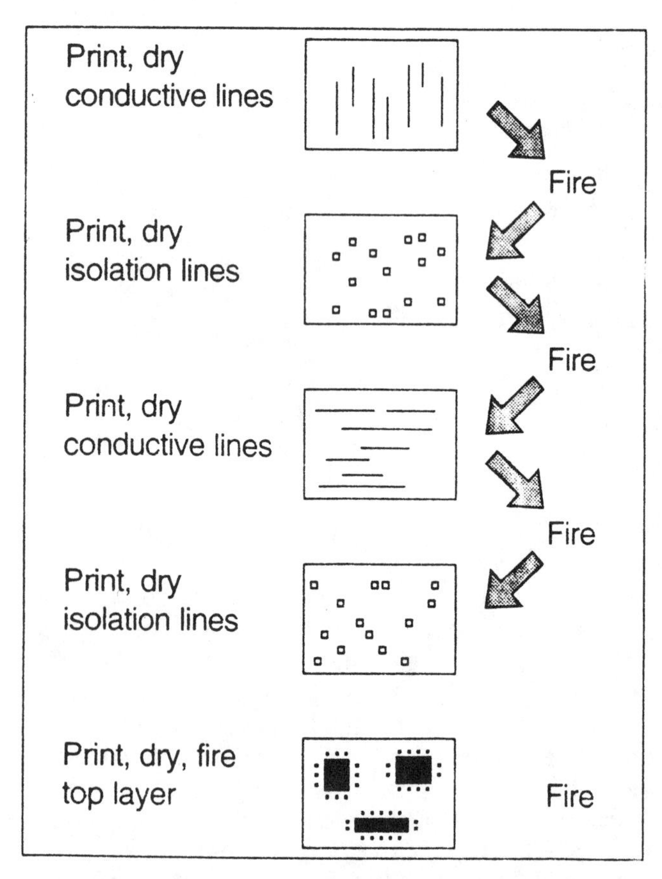

Figure 10: Schematic three layer production flow in standard hybrid technique

In contrast to the flow described above microhybrid technique avoids the repeated sequentiel use of the triad printing, drying and firing. For building up microhybrids each layer of the multilayer circuit is processed individually. Instead of a fired alumina substrate as for standard hybrids, microhybrids use low temperature cofirable ceramic foils. Punching vias, via filling and conductor printing is performed for each layer of the mulitlayer circuit separately. After processing and checking each layer individually, the layers are stacked one over another followed by lamination and a sintering step. Result is a sintered glass ceramic substrate with completed internal wiring (see Figure 11). To reduce the tolerances after sintering a process for minimizing dimensional tolerances is applied. Screenprinting external wiring, including resistors, completes the manufacturing process similiar to standard hybrids. The processing of each layer in parallel gives the obvious advantage to test each functional layer separately and to combine only defect-free layers for further processing. This characteristic results in an improvement of yield while at the same time manufacturing safety is increased. Compared to standard hybrids the parallel processing has the capability for more flexible prototyping. Additionally, the time for preparing a customer sample will be significantly reduced.

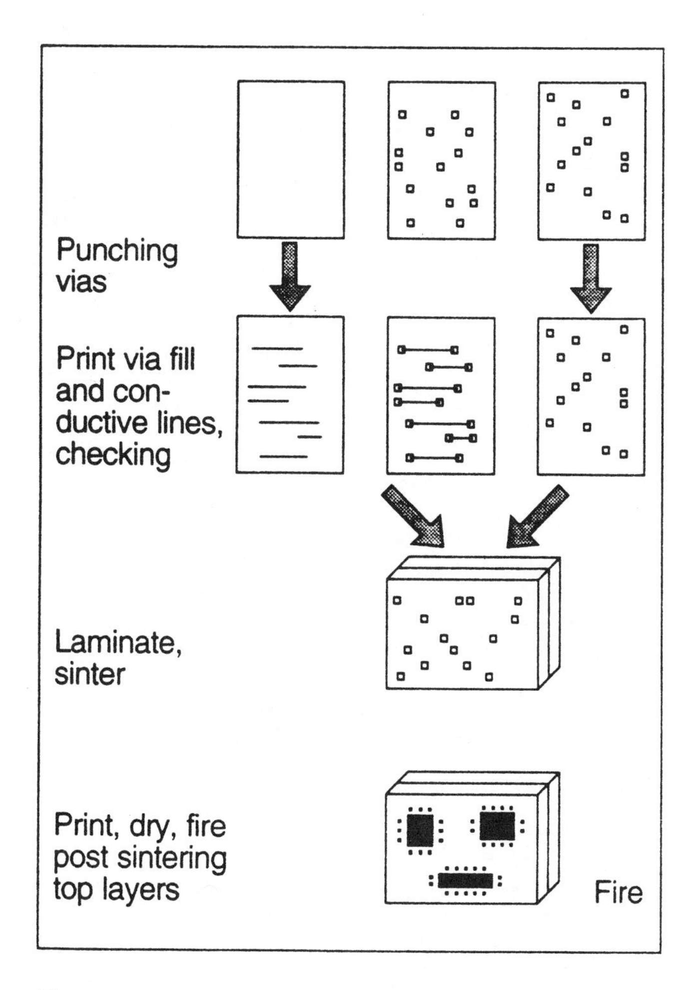

Figure 11: Schematic microhybrid substrate production flow

SUMMARY

LTCC substrates offer new perspectives for low-cost direct-mount ECUs. Their inherent multilayer capability allows for extremely dense packing of bare ICs and other mounted components, thereby reducing ECU size and weight significantly. The capability of this technology for automotive applications has been demonstrated with direct-mount ECUs for engine management and for an ABS system.

REFERENCES

[1] D.E. Bergfried, U. Mayer, R. Schleupen and P. Werner, "Engine Management Systems in Hybrid Technology" SAE-Paper 860593 (1986)

[2] S. Mitsutani and T. Ohtake, "The Strategies of Engine Control ECU at Nippondenso" Proceedings of the 17th International Symposium on International Technology & Automation, Florence, Italy, 11th-15th May 1987, pp. 145-157

[3] N. Rittmannsberger, "Antilock Brake System and Traction Control" Proceedings of the International Congress On Transportation Electronics, Dearborn (Michigan), USA, 17th-18th October 1988, pp. 195-202

[4] W. Maisch, W.-D. Jonner, R. Mergenthaler, and A. Sigl, "ABS5 and ASR5: The New ABS/ASR Family to Optimize Directional Stability and Traction" SAE-Paper 930505 (1993)

[5] H. Reichl (ed.) "Hybridintegration" Dr.Alfred Hüttig Verlag, Heidelberg (Germany), 1986 (in German)

950836

Designing with Microcontrollers with Low EMI

Chris Banyai
Intel Corp.

ABSTRACT

This paper describes the results of Intel Automotive's recent experiments for designing systems for low electromagnetic interference (EMI) using the MCS® 96 microcontroller family. The experiments focused on several configurations of a "representative" ABS module. The objectives were to study a layout technique called "micro-island", various grounding techniques, and the EMI benefit of 4 layer printed circuit boards (PCB) boards over 2 layer PCB boards. The tests showed: the wiring harness is a large contributor to the system emissions, the "micro-island" technique provides significant EMI reduction, and 4-layer PCB designs perform significantly better than 2-layer PCB's. In addition, it was demonstrated how poor layout can lower the performance of a 4-layer PCB to that of a 2-layer PCB.

INTRODUCTION

In the automotive environment, electronic design for low electromagnetic interference (EMI) is a difficult challenge. The design goal is to make sure that electronic systems do not interfere with each other. FM and cellular receivers are the most difficult to protect because their very function is to detect low level electromagnetic emissions. Microcontroller based electronic control units (ECU's) have long been recognized as an EMI source. Synchronous switching in the modules can act as tiny broadcast stations. Interference may be by direct radiation from the module or indirect radiation by conducting noise to "effecient" antennas such as the module wiring harness.

Microcontroller based ECU's have been essential in improving the safety, performance, and comfort of todays vehicles. A few examples include: antilock brakes (ABS), traction control, front and side airbags, suspension control, transmission control, engine management, and passenger comfort control. The proliferation and sophistication of these systems has driven the need for higher performance microcontrollers.

The first engine management and anti-lock brake systems were based on 8-bit archatectures operating at 12MHz.. Today, the 16-bit microcontroller is the standard data path and 16-20Mhz the standard operating frequency. As frequencies and edge rate increase, the wavelengths for efficient EMI radiation start to approach the length of the wiring harness and of PCB traces. In short, at higher frequencies the modules become better transmission antennas of EMI. Also, an increase in data path from 8 to 16 bits increases the switching current. As the demand for higher integration and performance has grown to support safety, performance, and comfort, inevitably so have the major EMI source contributors: frequency, edge rate, and current.

One potential solution, the elimination of the conventional wiring harness with a fiber optic network has had limited vehicular use due to cost. Such a solution would have the side benifit of increased fuel efficiency due to reducing vehicle weight.

Low EMI design is tricky due to both the extreme cost sensitivity and difficult design requirments of these applications. Extreme cost reduction pressures have driven the elimination of many traditional EMI protective measures such as metal enclosures and shielded cabling. So while the sources of EMI have increased, cost constraints have decreased the ability for electronic module manufactures to employ traditional EMI protective measures.

Given these difficulties, what can a module designer to do to minimize the system contribution to EMI? Unfortunately, application of good system design for low EMI is often difficult due to the large gap between textbook examples featuring Maxwell's equations and simple practical methods. The results of recent experiments by the Intel Automotive operation in partnership with EMI experts from Kimmel-Gerke, Ltd. provide a basis for some useful system design recommendations and practical guidelines. An overview of the EMI mechanism is presented followed by a description of a typical ABS module, the "representative" ABS modules used for testing, the test configurations and methods. The results and a low EMI design checklist are covered last.

EMI MECHANISM

Every EMI problem has a source, path, and victim. The source and victims are easily identified in the automotive environment. The high frequency component sources are usually attributable to the main and fail-safe computing devices and the victims are typically the FM radio receiver and cellular phone. Frequency components, are generated by a source, conducted or radiated by a path, and cause interference to a victim circuit. The next section describes the mechanism of microcontroller based sources and how system design impacts the EMI path.

MICROCONTROLLER FREQUENCY SOURCES - The high frequency components sourced by CMOS microcontrollers are most strongly dependent upon frequency, edge rate, amplitude, and current. While 3V devices have potential for reducing EMI (reduced amplitude and current), to date they have not been widely used in automotive applications due to their smaller tolerance for noise.

Narrow current spikes are an inherent characteristic of CMOS devices switching at high speed (see figure 1). These spikes of charge show up as noise radiated either directly from the chip or indirectly radiated off of PCB traces and the wiring harness. Most often the indirect noise gets to the PCB

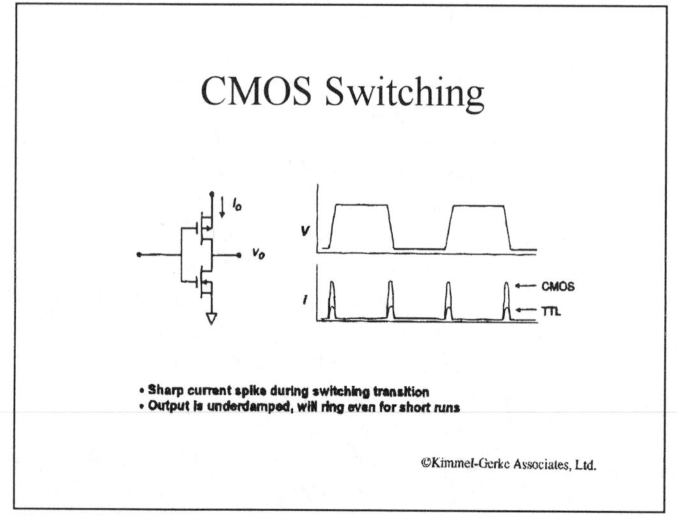

Figure 1: CMOS EMI Mechanism

on the Vcc and Gnd supply lines. Digital CMOS switching not only contains the fundamental oscillator frequency but also many harmonics. Thus a microcontroller operating at 16Mhz would be expected to generate harmonic noise at 32Mhz, 48Mhz, 64Mhz, ... What is the upper limit for frequency components due to harmonics?

Using a trapezoidal approximation for digital switching Kimmel-Gerke has developed what they call the extremely fast fourier transform (EFFT) to quickly predict the upper limit at which harmonics are significant as a function of the fundamental frequency (see figure 2). The first breakpoint at which the harmonics start to significantly fall off (20dB/decade) is related to the frequency by $1/(\pi*T)$. Doubling a 16MHz clock doubles the point at which the 20dB/decade point occurs ~10Mhz. The same concept can be applied to predicte how an emission level will change with frequency. Using the EFFT we would expect a 10dB increase in emissions for a 3x increase in frequency (see figure 3). A second breakpoint (40dB/decade) can be found at $1/\pi*t_r$ where t_r is the switching risetime. Frequency components in the FM and cellular frequency bands are mainly generated by the on-chip switching edge rate. According to the EFFT a switching edge rate of 10nS places the 40dB/decade breakpoint at ~32 Mhz. A faster edge rate of 3nS places this breakpoint at ~95Mhz.

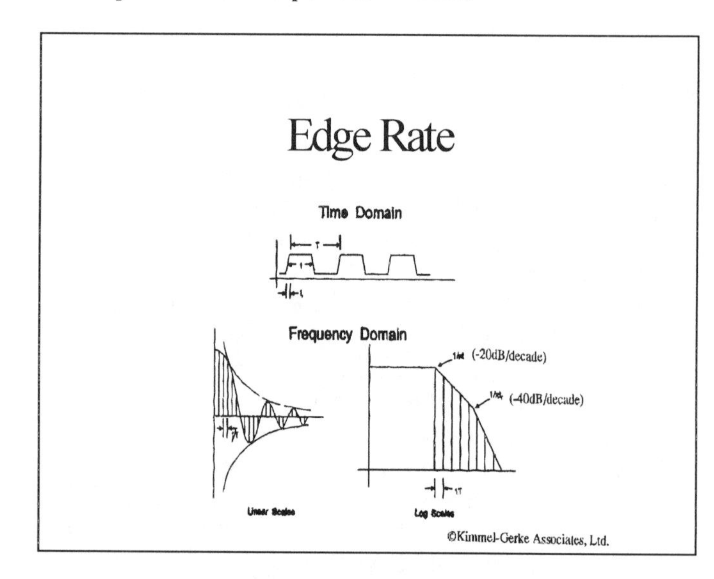

Figure 2: Extreamly Fast Fourrier Prediction

Three classical reasons why chip designers do not typically limit on-chip switching rates to 10nS or greater are size, speed, and temperature. Small feature size processes are used to meet cost targets. Along with the small size of these processes comes higher switching bandwidths. So, slower switching means bigger chips and higher cost. Speed is the second consideration. For the computer to work properly combinatorial logic must resolve between clock pulses. Finally, the automotive temperature extremes effect the on-chip edge rates. The same device designed to work under worst case propagation conditions (125 C°) (see Figure 3c) will switch faster under typical (figure 3b) and cold temperature (25 to -40 C°) (figure 3a).

SYSTEM DESIGN AND THE EMI PATH

Automotive electronic modules are often located as close as 1 meter away from highly sensitive victim circuits such as the car's FM radio or mobile communications antenna. "The typical voltage sensitivity of such recievers is between 0.1 to 0.5uV. Assuming a quarter wave antenna, this translates to a field intensity sensitivity of approximatly 0.23uV/m to 3.5uV/m over the 30-46MHz frequency range*. It only takes a fraction of a uV/m to cause interference! Automotive EMI requirements are up to 60dB (1000 times) more stringent than FCC Class A requirements**. The primary sources of EMI frequency components are the microcontroller "clocks" and other highly repetitive signals. System design however, determines to what degree frequency components have a path to radiate. In automotive electronic modules, key paths are free space, intra module PCB traces, ground and Vcc supplies.

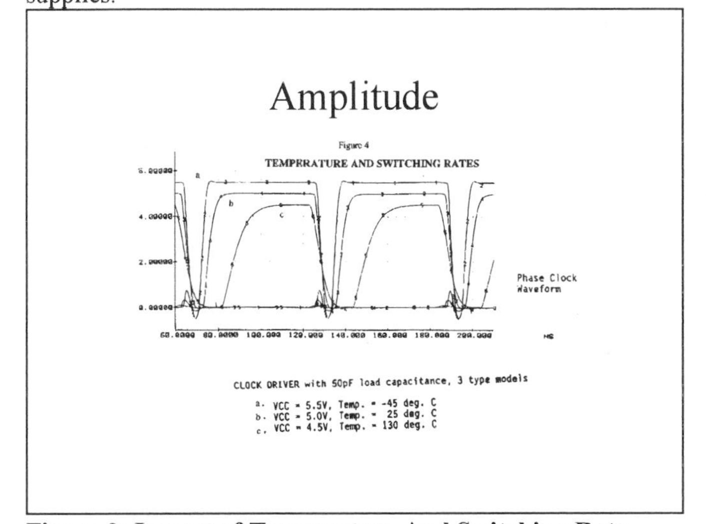

Figure 3: Impact of Temperature And Switching Rate

The degree to which a frequency source radiates efficiently has to do with how good of an antenna it has available. A conductor is a "good antenna" when its physical length is greater than or equal to the free space rise time of the signals it carries. Kimmel-Gerke defines this length as the critical length (Lcrit). Appreciable radiation is generated at fractions of Lcrit. Lcrit/2 is considered the typical length of rise time on a cable or PCB. Kimmel-Gerke recommends trying to keep dimensions less than Lcrit/20 for robust EMI design. Figure 4 summarizes the relationships beween frequency, signal rise time, and critical lengths. This table shows that a robust 32MHz design with 10nS edge rates would avoid structures exceeding the Lcrit/20 of 6 inches.

The goal of system design for low EMI is to minimize the EMI path. The path between source and victim is usually complex since it is dependent on a slew of variables including frequency, amplitude, time, impdance, distance, and the physical dimensions and of materials in-between and around the source and path. The path effects the efficiency to which the source noise couples to the victim circuits either by conduction or radiation.

Low EMI system design is achieved by minimizing the coupling of source EMI frequency components onto structures that are efficient antennas. Once on the wiring harness a change in the length or bend can significantly alter its EMI signature. This means EMI performance of an ECU will vary with the particular wiring harness and packaging of a specific car. Low EMI design strategies contain high frequency components before they get onto the wiring harness. The next sections present how a typical ABS system was implemented for testing in the form of a representative ABS system. Various configurations of this system were then measured for relative EMI performance.

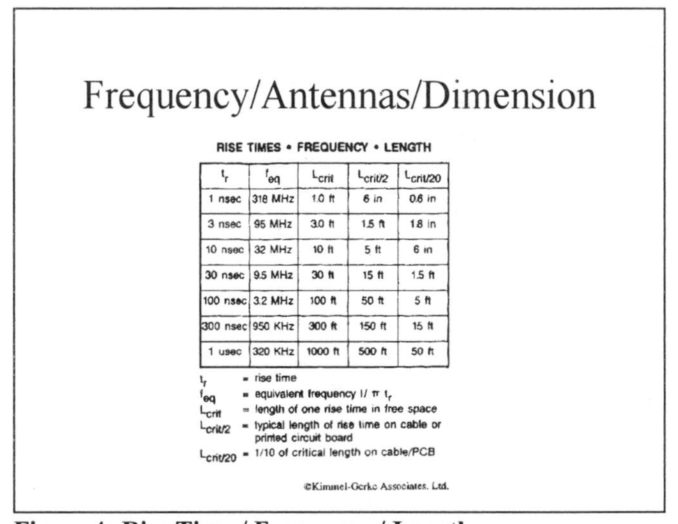

Frequency/Antennas/Dimension

RISE TIMES • FREQUENCY • LENGTH

t_r	f_{eq}	L_{crit}	$L_{crit/2}$	$L_{crit/20}$
1 nsec	318 MHz	1.0 ft	6 in	0.6 in
3 nsec	95 MHz	3.0 ft	1.5 ft	1.8 in
10 nsec	32 MHz	10 ft	5 ft	6 in
30 nsec	9.5 MHz	30 ft	15 ft	1.5 ft
100 nsec	3.2 MHz	100 ft	50 ft	5 ft
300 nsec	950 KHz	300 ft	150 ft	15 ft
1 usec	320 KHz	1000 ft	500 ft	50 ft

t_r = rise time
f_{eq} = equivalent frequency $1/\pi t_r$
L_{crit} = length of one rise time in free space
$L_{crit/2}$ = typical length of rise time on cable or printed circuit board
$L_{crit/20}$ = 1/10 of critical length on cable/PCB

Figure 4: Rise Time / Frequency / Length

TYPICAL ABS

The computing portion of a typical ABS system is comprised of a main and a fail-safe microcontroller. Typical inputs include wheel speed signals from each wheel generated by variable reluctance sensors in each wheel. These sensors generate a pulsed signal which varies in both frequency and amplitude as a function of wheel speed. The wheel speed signals may reach 150V p-p at a frequency of 6KHz. The typical outputs include power signals to drive the hydraulic pump and individual brake hydraulic actuators. The main and fail-safe constantly check each other for agreement. If the fail-safe or main processor sense a problem either component can take action ranging from a software adjustment to the complete shut down of the electronic braking module (leaving driver actuated braking). More sophisticated designs may not need to shut down. For example, if the module happens to be connected to other modules in the vehicle via CAN or some other networking scheme a wheel speed sensor failure may be handled by requesting the needed information from another module on the network (figure 5). Finally, ABS modules must endure the harsh automtive environment. These modules are usually mounted on or near the master cylinder

---

* Gerke, Daryl, PE, Microprocessors and VHF Radios - Mutual Antagonists Part II (Kimmel-Gerke Associates, Ltd., 1994), p. C6.19.

** Gerke, Daryl, PE, Designing for EMC, Practical Tools, Tips and Techniques for Bullet Proof Designs (Kimmel-Gerke Associates, Ltd., 1994), p. 2.

subject to moisture, electrical transients, and must endure the automotive temperature range of -40C to 125C.

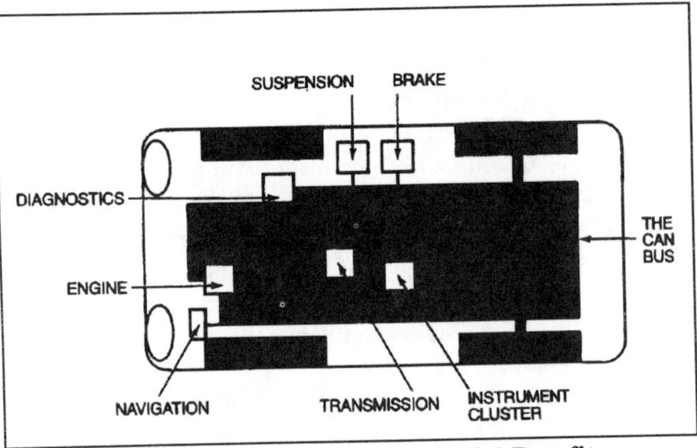

Figure 5: In-Vehicle Networking Potential Benefits

REPRESENTATIVE ABS

For our experiments we populated the representative ABS test boards with the 8XC196JT main microcontroller, regulator, load dump diode, and hex buffers (figure 6). The fail-safe load and other circuitry was duplicated by placing appropriate cap and resistor loads on the test board. The placement of these loads was approximately the same as a fully populated board. This was an important consideration since it imposed realistic PCB routing constraints and provided realistic paths for radiated and conducted emissions. Actual production type connectors and cabling was used. Jumpers and pads were added to easily reconfigure the test boards for various experiments.

Representative ABS Test Boards

- **Populated With 8XC196JT**
- **Failsafe And Other Loads Modeled**
- **Production Connectors/Cable**
- **Realistic Placement And Routing Constraints**

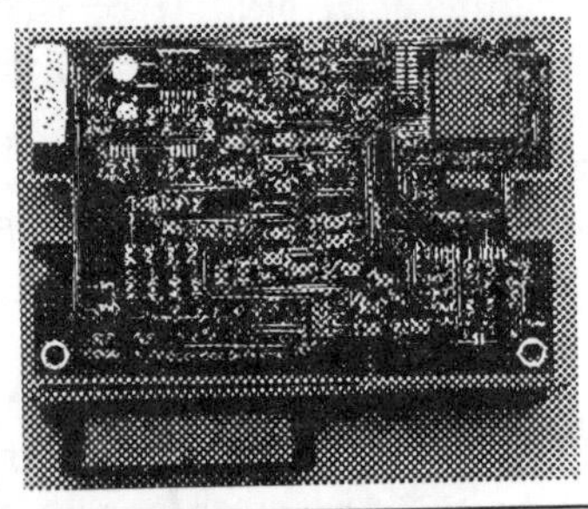

Figure 6: Representative ABS

EXPERIMENTS

The Intel experiments tested various configurations of the representative ABS system featuring the 8XC196JT 16-bit CHMOS microcontroller. The configurations varied in several design aspects including number of PCB layers, general layout technique, decoupling, and grounding. Several different software codes were run on the test boards to study their EMI characteristics. The characterization exercised the chip over a range of operating conditions typical of ABS operation. The test procedure was based on the GM9100 module level procedure which is a far field measurment technique. This procedure is used by General Motors to qualify electronic modules supplied by GM vendors.

MEASURMENT TECHNIQUE - The actual measurements were taken by using biconnical and log-periodic antennas at a distance of 1 meter from the test module. Signals picked up by the antennas were fed into a preamplifier and then a spectrum analyzer (figure 7). The analyzer scans were then corrected for the amplifier effect and antenna factors to produce the final data plots. The module was powered with a standard automotive battery through the wiring harness cable or directly at the module input. In addition, EMSCAN near field spatial measurements were taken of the test modules. The noise floor of the anechoic chamber used was slightly higher than the GM9100 proceedure specified but it was more than adequate for the purpose of comparative study.

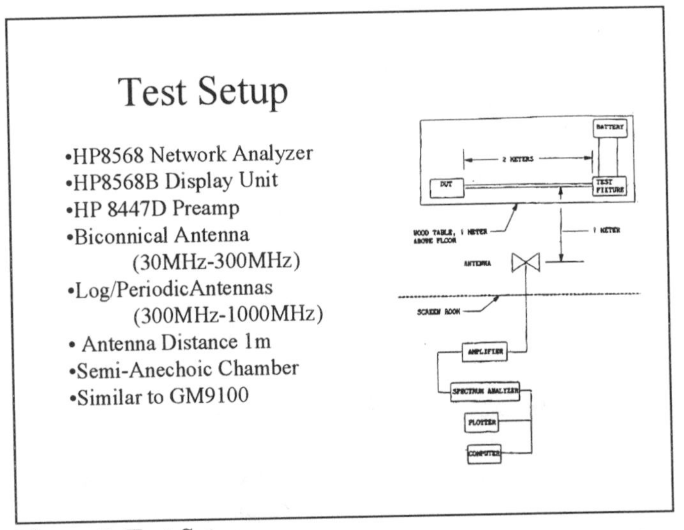

Test Setup
•HP8568 Network Analyzer
•HP8568B Display Unit
•HP 8447D Preamp
•Biconnical Antenna
(30MHz-300MHz)
•Log/PeriodicAntennas
(300MHz-1000MHz)
• Antenna Distance 1m
•Semi-Anechoic Chamber
•Similar to GM9100

Figure 7: Test Setup

CONFIGURATIONS - Five configurations were tested to study the impact of the micro-island technique and the relative effectivenes of 4-layer PCB's vs 2-Layer PCB's.:

1. 2LNMI 2-Layer PCB basic design
2. 2LMI 2-Layer PCB micro-island technique
3. 2LMIC 2-Layer PCB Same as #2 with capacitors
4. 4LNMI 4-Layer PCB Same as 1 with Vcc and Gnd plane
5. 4LMMI 4-Layer PCB micro-island technique, decoupling, best methods

Micro-Island - The key layout technique studied was "micro-island" as coined by EMI expert Daryl Gerke. While certainly not a new concept this method is a practical technique to isolate high frequency sources on only one portion of the PCB board. High frequency isolation was achieved by placing the main and fail-safe microcontrollers on a PCB "island" and decoupling all traces that bridge to the island with ferrites including the ground and Vcc (see figure

8 & 9). The intended effect was to eliminate the high frequency escape paths from the island to the rest of the PCB board. Depending on the frequencies involved resistors may be substituted for the ferrites. The mico-island prevents noise from coupling to the PCB and wiring harness. Once on the wiring harness noise radiates quite easily.

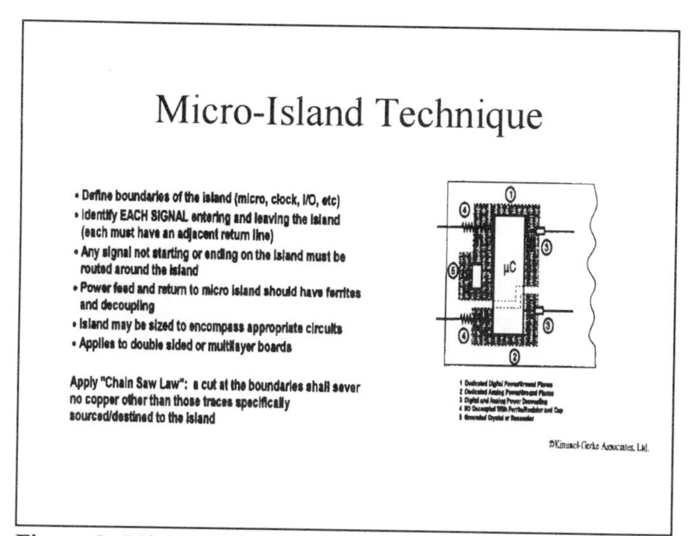

Figure 8: Micro-Island Technique

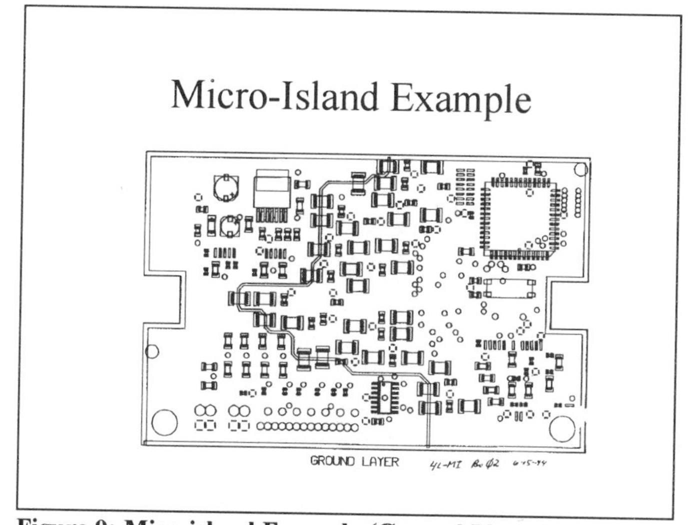

Figure 9: Mico-island Example (Ground Plane Layout)

Configuration Details - The first configuration named 2-Layer Non micro-island (2LNMI) was designed to be a good/typical 2-layer PCB design. The second configuration 2-Layer micro-island (2LMI) was designed to evaluate the micro-island technique. Due to layout routing difficulties the PCB designer was unable to implement the 2LMI design without compromising the micro-island technique. As is often the case in real designs, there simply was not enough space to route the needed traces without "cutting up" or "piercing" the ground fill with signal traces. A variation of the 2LMI board was made to see what difference it would make if the ground fill was solid instead of pierced. This variation, labeled 2LMIC, used decoupling capacitors to bridge where the solid plane was "cut up" (figure 10). While simple wires could have also been used, the capacitors had the effect of "sewing up" the cut up plane into a solid plane at high frequency. The fourth configuration took the exact same layout design from the 2LNMI and created a 4-layer PCB design by adding a dedicated ground and Vcc plane 4LNMI. By comparing the 2LNMI to the 4LNMI we could quantify the EMI benefits of 4 layer vs 2 layer PCB designs. The fifth configuration was designed to be the best design EMI consultant Daryl Gerke could design without using shielding techniques. This configuration, 4LMI again featured the micro-island technique and extensive use of surface mount ferrite decoupling devices. We constrained the PCB size and general part placement to be the same for all of the configurations. The only change in component placement was to make space for the decoupling devices such as in the 4LMI design.

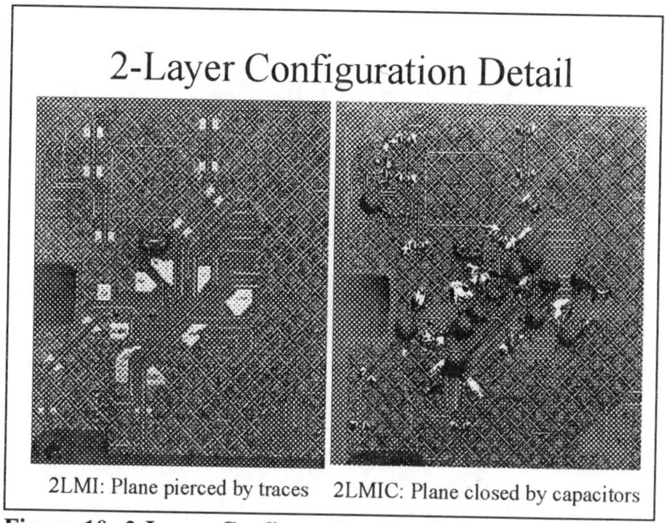

2-Layer Configuration Detail

2LMI: Plane pierced by traces 2LMIC: Plane closed by capacitors

Figure 10: 2-Layer Configuration Detail

SUMMARY OF FINDINGS

The various board designs yielded significantly different radiated emission profiles as described below. We noticed that many frequency components that appear in near field measurements such as strip line, loop probe, and EMSCAN were not measurable in our far field experiments. For typical module operation there was not enough detectable EMI generated for good comparisons.

By holding the microcontroller in a hardware reset we were able to generate sufficient emissons for good comparisons. The first scans were taken from 30MHz-1GHz. After analyzing these scans we determined the 30-300MHz range was adequate. Data is shown for the results from the primary tests. Only the results are included for the secondary tests.

The tests showed that while the source of the emissions originated on the module, most of the radiation actually occurred from the interconnecting cable. The primary sources were related to the microcontroller clock, which were then coupled to the cable.

PRIMARY TEST RESULTS - For typical ABS code, emissions were low (figure 11). Most of the radiation actually occurred from the interconnecting cable (figure 12). The addition of a ground plane (2-Layer to 4-layer) reduced emissions a minimum of 6dB with 20dB reductions typical above 100MHz (Figure 13). "Cutting up" or "piercing" a ground area with other traces can destroy the benefit of a ground area. (figure 14). Proper implementation of the micro-island is critical. The version with 1000pF capacitors (2LMIC) was up to 15dB lower than the non-capacitor (2LMI) version. Both the 2-layer and 4-layer "micro-islands" significantly reduced emissions above 100 Mhz. The MI version was about the same below 100MHz, but yielded additional reductions of up to 10dB above 100MHz.

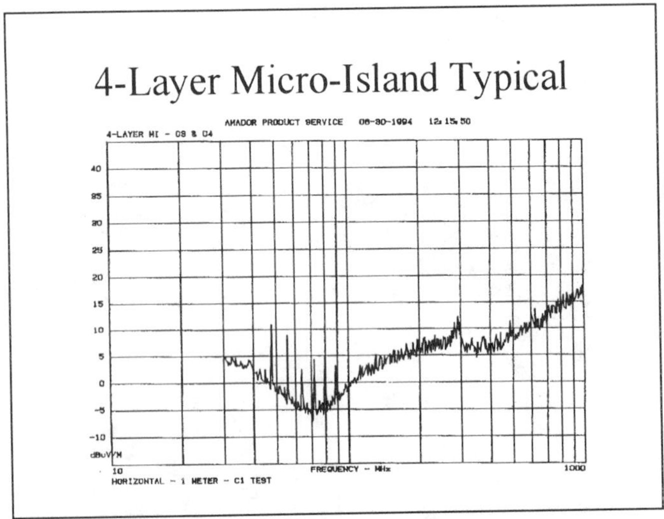

Figure 11: 4-Layer Micro-Island Typical

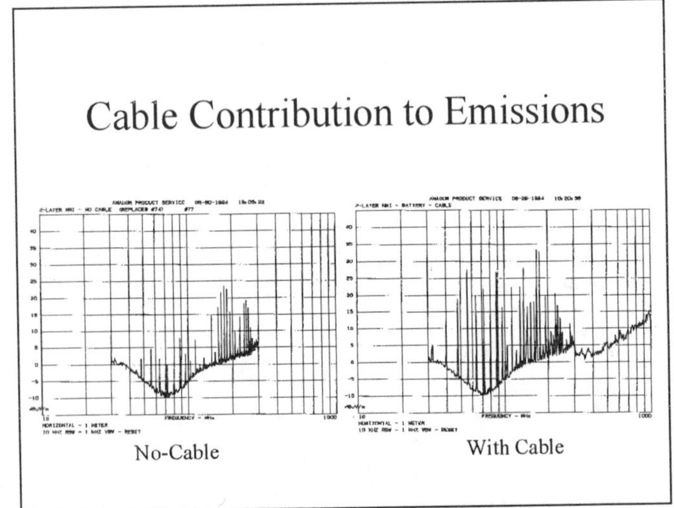

Figure 12: Cable Contribution to Emissions

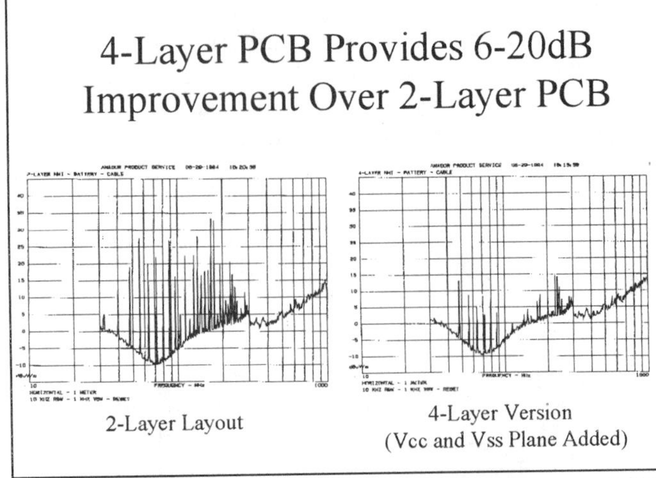

Figure 13: EMI Benefit of 4-Layer PCB Vs. 2-Layer PCB

SECONDARY TEST RESULTS - A "local shield" positioned over the microcontroller typically yielded 6dB reductions. There was an improvement when 5 unused pins were grounded vs not being grounded. There was little difference between grounding unused pins physically at the pin vs driving the pin to logical 0 via software (active ground). Ferrite decoupling was beneficial in the 150-200Mhz range. Differences were observed as a function of software. The more circuitry that was active the higher the emissions. As integration increases so does EMI.

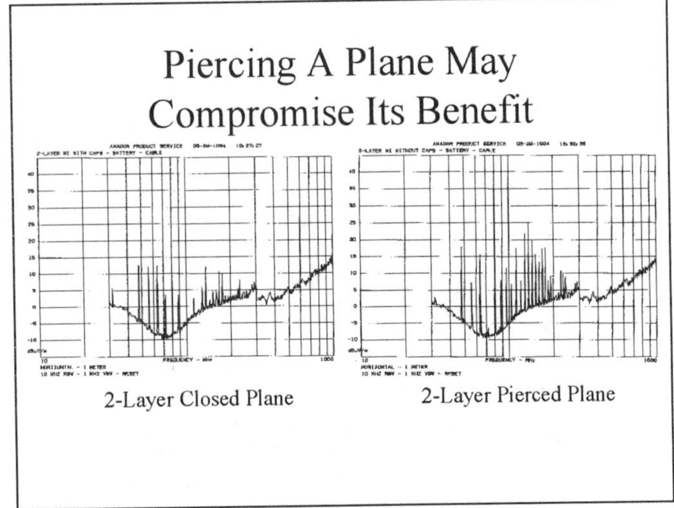

Figure 14: Effects Of Piercing A Plane May

DESIGN CHECKLIST

Based on the results from these experiments and the experience of EMI experts from Kimmel-Gerke, Ltd here are some practical suggestions to add to your design checklist for low EMI. The aim is to contain the high frequency components, minimize reflections, smooth high edge rates, and minimize the coupling paths to structures that can radiate efficiently. The checklist is divided into three sections: PCB Board Structure, Placement/Routing, and Decoupling.

<u>PCB Board Structure:</u>

1. 4-Layer Printed Circuit Board for designs 10Mhz or higher
2. Keep maximum ground plane area (do not "cut up" or "pierce" the planes with other traces)
3. Keep the power and ground plains aligned (avoid routing ground where you do not route Vcc) (figure 18).
4. Use micro-island technique for isolation of high frequency sources with surface mount ferrite or resistor "bridges" (ferrites 100 Ohm at 100Mhz)

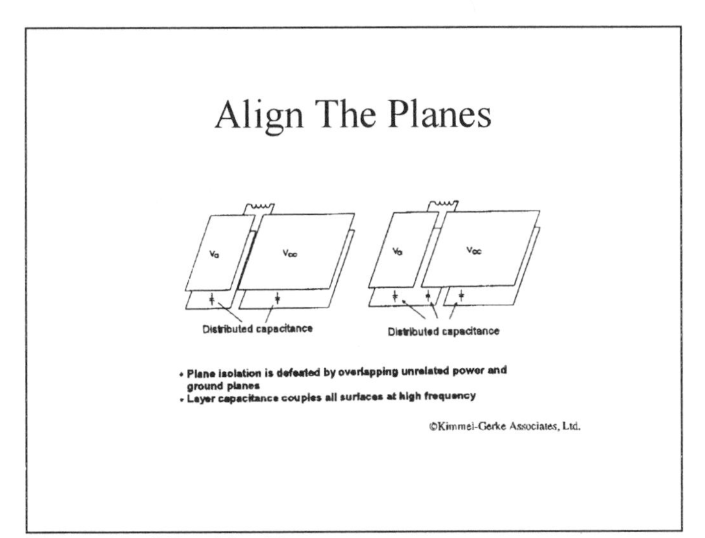

Figure 15: Align The Planes

Placement/Routing:

1. Mount crystals flush to the board and ground them
2. Minimize trace length of clocks and other periodic signals
3. Keep clock traces, buses and chip enables separate from I/O lines and connectors
4. Critical traces should be handled by:
   - Routing close to ground plane
   - Guardbanding (ground on each side of signal)
5. Avoid running traces under crystals and other hot circuits
6. Route only one continuous trace (avoid) stubs
7. If possible, route on 45 degrees to minimize reflections
8. Route balanced traces in parallel
9. Do not route signals under the microcontroller

Decoupling:

1. Ceramic capacitors at headers, connectors, and chip power pin (0.1uF up to 15MHz, 0.01 uF over 15MHz)
2. Bypass capacitors of similar sizing on all power feed and reference voltage pins for analog circuits
   - Bypass capacitors on fast switching transistors
   - Bypass capacitors on pull-up and pull-down resistors
3. Decouple power and ground supply at module entry with cap and ferrite
4. Shielded or decoupled Box and or Cable where possible

CONCLUSIONS

A key concern for the designers of automotive electronics is interference to on-board VHF radios, such as FM broadcast receivers or land mobile VHF communications receivers. Unfortunately, even very low levels of radiated emissions can cause problems. These EMI problems are aggravated by increasing clock speeds, higher chip integration, the shift from 8 to 16 bits and even 32 bit data paths and the widespread use of CMOS processes for high performance, low cost microcontrollers. The current clock speeds in the 8-20MHz range often result in interference well into the 200-300Mhz range. These problems can be solved through a combination of "clean" system design and "low noise" components. While many system designers would like to completely push the EMI responsibility back to the chip manufacture, that approach is not entirely practical with today's automotive requirements of low cost, high performance, and severe environmental extremes such as temperature. A joint effort is necessary, dealing with the components and the reset of the system (printed circuit boards, and wiring harness, etc.) The Intel Automotive operation is currently addressing the EMI problem on both fronts - through new "low noise" components and by providing "systems design" guidance.

Chris Banyai is an Sr. applications engineer with the automotive operation of Intel Corporation and holds a B.S. in computer engineering from the University of Michigan and a B.A. in Physics from Hope College.

Daryl Gerke is a principle of Kimmel-Gerke & Associates, Ltd. which is an engineering consulting and training firm based in St. Paul, MN that specializes in EMI troubleshooting, eductation, and computer technology.

REFERENCES

1. Gerke, Daryl, PE, Microprocessors and VHF Radios - Mutual Antagonists Part II, (Kimmel-Gerke Associates, Ltd.), 1994.

2. Gerke, Daryl, PE, Designing for EMC, Practical Tools, Tips and Techniques for Bullet Proof Designs, (Kimmel-Gerke Associates, Ltd.), 1994.

951909

Power Integrated Circuits for Powertrain Control Modules

Keith Wellnitz and Jeff Kanner
Motorola

ABSTRACT

This paper describes a series of power ICs designed specifically for powertrain control applications. The series includes low-side drivers and high-side drivers. The drivers are capable of fault detection and reporting to the Micro-Control Unit (MCU). Reported faults include short circuit detection, thermal limit, over-voltage protection, "on" open load detection, and "off" open load detection.

INTRODUCTION

Powertrain Control Modules (PCMs) use microcontrollers to control the system. The PCM environment requires Integrated Circuits (ICs) withstand extreme voltage transients. Without protection, these voltages are destructive to the MCU. MCUs need power ICs as protection from these voltages.

In addition, PCM modules must have extensive fault diagnostic capability to ensure reliability and meet emission requirements. These faults are detected and then reported to the motorist to indicate a potential need for servicing. Reliable designs off-load diagnostics from the MCU to the power IC to eliminate single point failures. Therefore, the power IC plays a critical role in the PCM system.

The ICs are fabricated using a mixed-mode process containing bipolar, MOS, and power FET technology. The power FET drives the load and is clamped to handle inductive flyback voltages. Bipolar and MOS devices allow both analog and digital functions. The choice of device technology depends on the type of control elements that must be integrated. Operational amplifiers, comparators and regulators work best with bipolar devices. CMOS devices handle logic, active filters, and current mirrors. This series of PCM ICs fully utilizes the diversity of available circuit technology.

QUAD INTEGRATED DRIVER

The Quad Integrated Driver (QID) is a four output low side switch with protection and diagnostics. The QID is designed specifically for PCM applications and utilizes the *SMARTMOS*™ process. The QID is capable of controlling fuel injectors, transmission solenoids, canister purge solenoids, fuel pump relays, and incandescent bulbs. Each of the four power outputs on the QID is comprised of a DMOS transistor. The control inputs and open drain fault reporting are CMOS compatible. Fault diagnostics include current limit, thermal limit, "on" and "off" open load detection, and over-voltage protection. The Rdson of each output is 200 mΩ. The IC is packaged in a high power 15 pin SIP.

A block diagram of the QID is shown in Figure 1. The key circuit blocks are Input, VREG, Output, and Fault Diagnostics. A description of each block follows.

INPUT - The input circuitry controls the gates of the power DMOS devices. It is CMOS compatible to receive direct MCU communication. There are 250 mV of hysteresis on each pin to provide double pulse suppression. The patented hysteresis cell takes advantage of unique characteristics of MOS devices and is insensitive to temperature variation. All inputs have active 10 μA pull-down current sources. This ensures that the power transistors do not engage the loads during an intermittent or lost connection. The input also contains a dual-select pin. This pin allows a fail-safe MCU to take control of outputs 1+2 and/or 3+4 in the event that the primary MCU is not functioning properly.

VREG - Battery is the only source of power available in automobiles. Most ICs are unable to operate from direct battery and require pre-regulation. The QID needs no pre-regulation because of the internal VREG circuit. VREG converts the battery (8-28 volts) to usable voltages and monitors the battery for an over-voltage condition. VREG uses high voltage analog transistors to withstand over-voltage conditions. Total quiescent current is less than 5 mA.

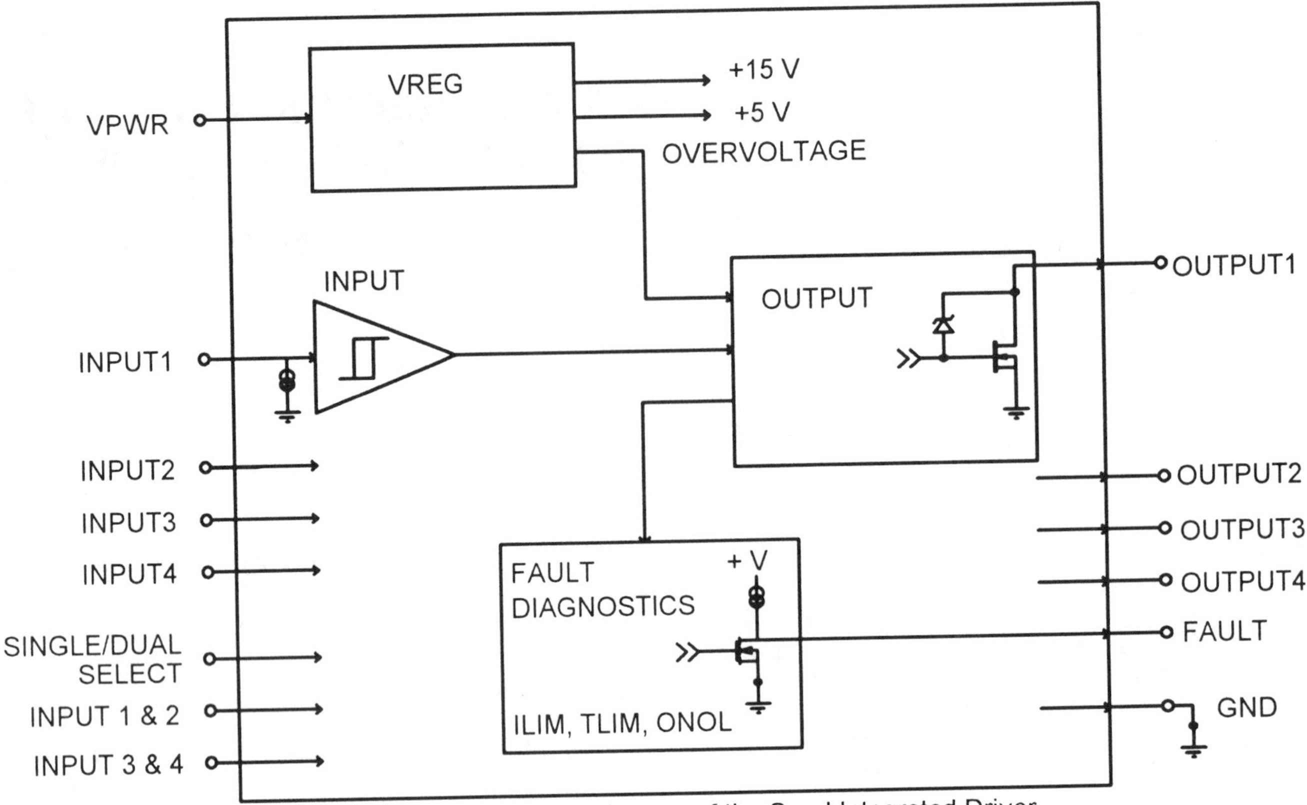

Figure 1: Block diagram of the Quad Integrated Driver

VREG contains a 5 volt regulator that powers the internal CMOS logic and bias circuitry. Its precision eliminates potential communication errors with the MCU due to supply mismatch. VREG also has a high voltage buffered shunt regulator. It produces a voltage one Vce_{sat} below the battery up to a 15 volt clamp. The 15 volt regulator drives the gate of the power FET. The ability to drive the gate to battery voltage levels lowers the RDS_{on} improving the power performance of the IC.

Over-voltage shutdown protects the QID during load dump. Since load dump overdrives the battery line, the power FET's drain experiences 80 volts. If a power device is on, excessive power dissipation causes a thermal overload. This extreme temperature condition puts unnecessary stress on the power device. Instead, VREG's over-voltage comparator trips and generates a fault flag. The flag signals the gate drive circuitry to turn off all power devices.

OUTPUT - Federal regulators place strict limitations on the amount of noise generated by automotive electronics. Fuel injection systems and full lock-up torque converters are prone to generate noise because the injectors and solenoids are pulse width modulated. Rapid switching causes high current slew rates at the output. The QID output circuitry controls gate charging current and limits the output current slew rate (dI/dt). Figure 2 shows the MCU's input voltage and the QID's output voltage and current. Note the drain current switches at less than 1 A/µs. External noise is mitigated.

Load dump, ESD, and the unclamped injector flyback voltage exceed the 80 volt breakdown of the DMOS device. To clamp the output to 60 volts, zener diodes are connected from drain to gate. When a disabled output is pulled to 60 volts, the zener clamp sources current into the gate enabling the output. This clamp enhances the robustness of the IC by turning on the DMOS and preventing avalanche breakdown. Avalanche breakdown is less robust because it increases the power density at the edges of the device. The clamp forces current to flow uniformly through the silicon and lowers the power density. At 25 °C, the energy capability of each output of the QID exceeds 100 Joules.

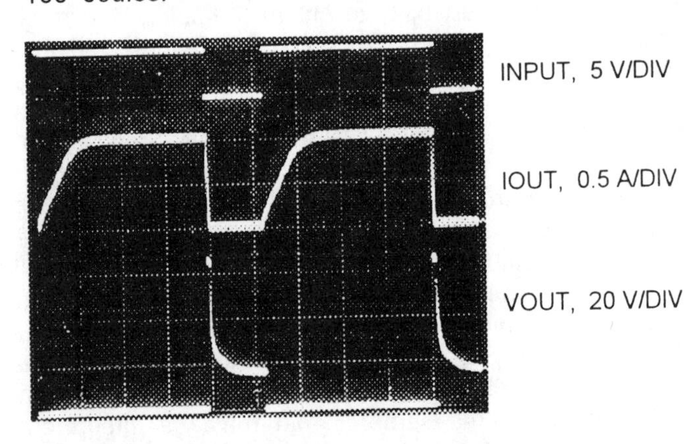

Figure 2: Oscilloscope photograph of the MCU input voltage and the QID's output current and voltage

The QID's DMOS device has a nominal RDS_{on} of 200 mΩ. The IC is packaged in a 15 pin SIP. Junction to ambient thermal resistance (Rθja) is 15 °C/W when mounted in still air with no heat sinking [1]. Proper heat sinking improves Rθja to 5 °C/W.

FAULT DIAGNOSTICS - Real time current limit (ILIM) protects the QID's outputs and external board traces during a faulted condition. Analog ILIM actively regulates the output DMOS to limit the current as shown in Figure 3. ILIM is implemented using SENSEFET™ technology to sample load current.

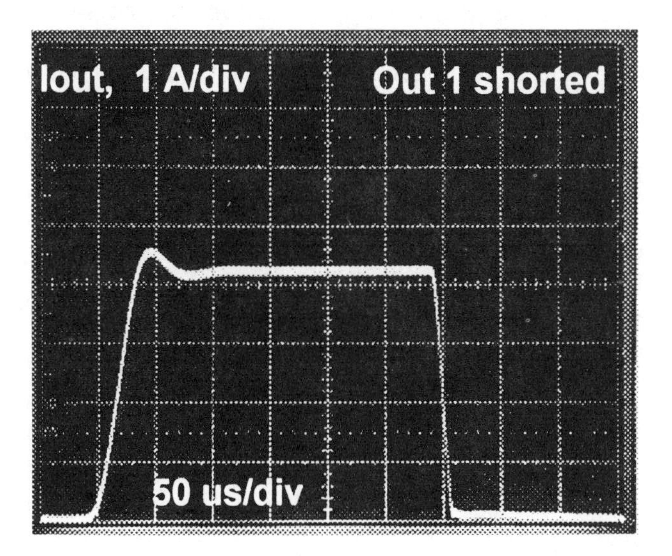

Figure 3: Analog current limit actively regulates the output current

The IC must protect itself from both hard and soft shorts. Hard shorts are direct shorts to battery. The ILIM circuitry detects hard shorts and turns off the outputs. Soft shorts are below current limit but exceed the power dissipating capability of the device. Soft shorts are difficult to detect. Soft shorts require over temperature sensing to protect the IC.

The QID includes over temperature detection and protection (TLIM). Faults are detected for each output separately and then the faulted devices are turned off. In a multiple output power IC it is highly desirable to turn off only the output device experiencing the faulted condition. Multiple output devices require independent local temperature sensing in place of a global temperature sense.

As shown in Figure 4, the four outputs of the QID turn off when the thermal limit of 170 °C is exceeded. All of the outputs were shorted to a 14 volt supply at 25 °C. A total current of 16 amperes initially flowed through the device. Note that each output turns off independently. Variations in turn off time of each output results from differences in current limit and thermal efficiencies. The shut-off temperature of each output is within 2 °C.

A PCM contains both Pulse Width Modulated (PWM) and DC switched loads. The injectors and torque converters are PWM loads. The fuel pump relay, canister purge solenoid, and incandescent bulbs are DC switched. For PWM applications, "off" open load detection is sufficient to verify the presence of the load. "Off" open load circuitry detects an open circuit between load and output when the input is off. "On" open load circuitry detects the same fault when the input is on. For DC switched applications, the activated load must be monitored to detect faults that occur during start-up.

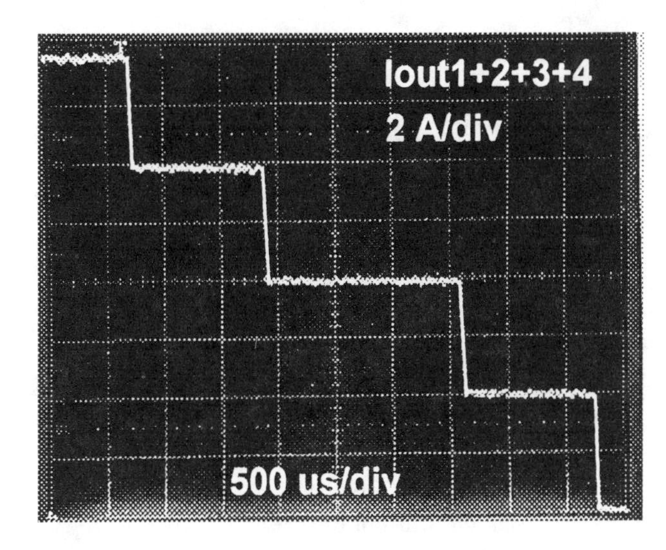

Figure 4: Independent thermal shutdown protects each output of the QID from a simultaneous short of all outputs to battery

For relay, solenoids, and lamps, the QID has "on" open load detection as a diagnostic feature. "On" open load circuitry must detect millivolt changes in the drain voltage. This requires precision analog circuitry. The QID uses a comparator network capable of detecting 10 mV changes on each output.

Figure 5 shows a die photograph of the QID. Each output has independent "off" open load, "on" shorted load, "on" open load, and over temperature detection and protection.

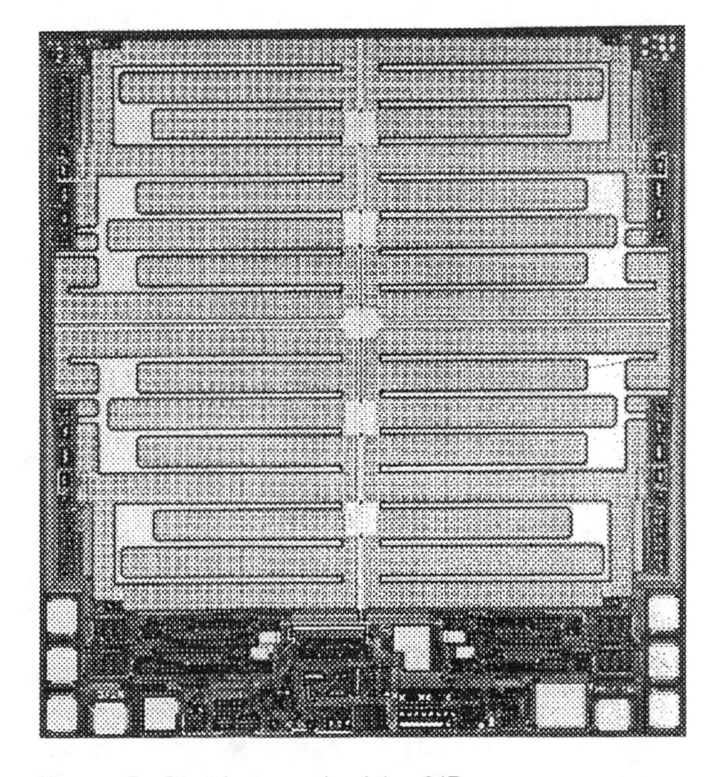

Figure 5: Die photograph of the QID

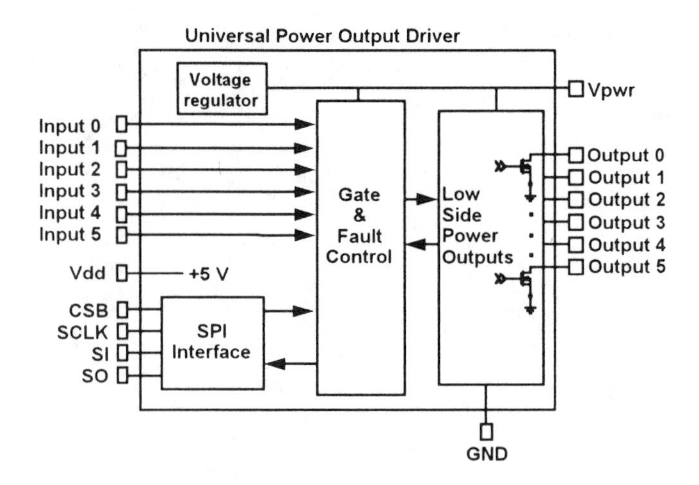

Figure 6: Block diagram of the UPOD

UNIVERSAL POWER OUTPUT DRIVER

The Universal Power Output Driver (UPOD) is a six output low side switch with fault protection, diagnostics, and a serial peripheral interface (SPI). Like the QID, the UPOD is designed for a variety of PCM applications from injectors to bulbs. Fault diagnostics include current limit, thermal detection, "on" and "off" open load detection, and over-voltage protection. The Rdson of each output is 200 mΩ. The IC is packaged in a high power 23 pin SIP.

High speed communications and increased flexibility are the trend in PCM ICs. The increase of ICs and diagnostic data in electronic modules increases the information available to the MCU. In order for the MCU to process this information, the data must be readily available in a high speed format. Parallel outputs are slow and require a separate MCU input pin for each output.

The UPOD utilizes an SPI to increase the communication with the MCU. The SPI is a two way communication port allowing the IC to receive instructions and transmit diagnostic feedback. With the SPI, the UPOD sends diagnostic information to the MCU in 8 bit words at frequencies up to 1.8 MHz.

Parallel control of the outputs is preferred for PWM applications. For DC switched applications, controlling the output through the SPI is sufficient. The outputs of the UPOD can be controlled by either the parallel inputs or an input command received on the SPI. Controlling non-PWM output via the SPI reduces the number of parallel ports required on the MCU. The UPOD block diagram is shown in Figure 6.

A die photograph of the UPOD is shown in Figure 7. The digital, analog, and power portions of this IC are labeled. The power DMOS outputs occupy sixty percent of the total die area. Even with a serial interface, the digital circuits require the least area because of the small geometry of CMOS.

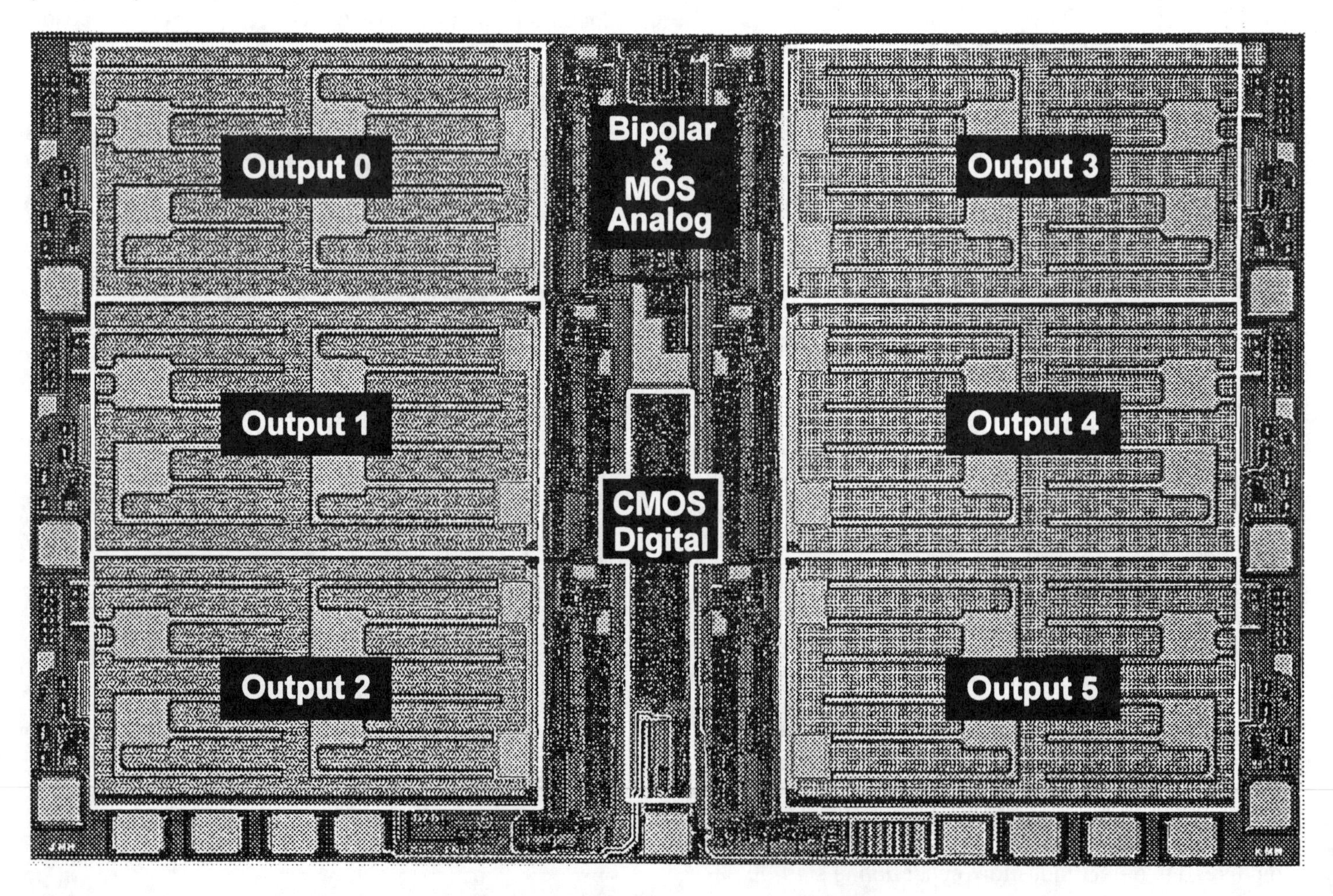

Figure 7: Die photograph of the UPOD

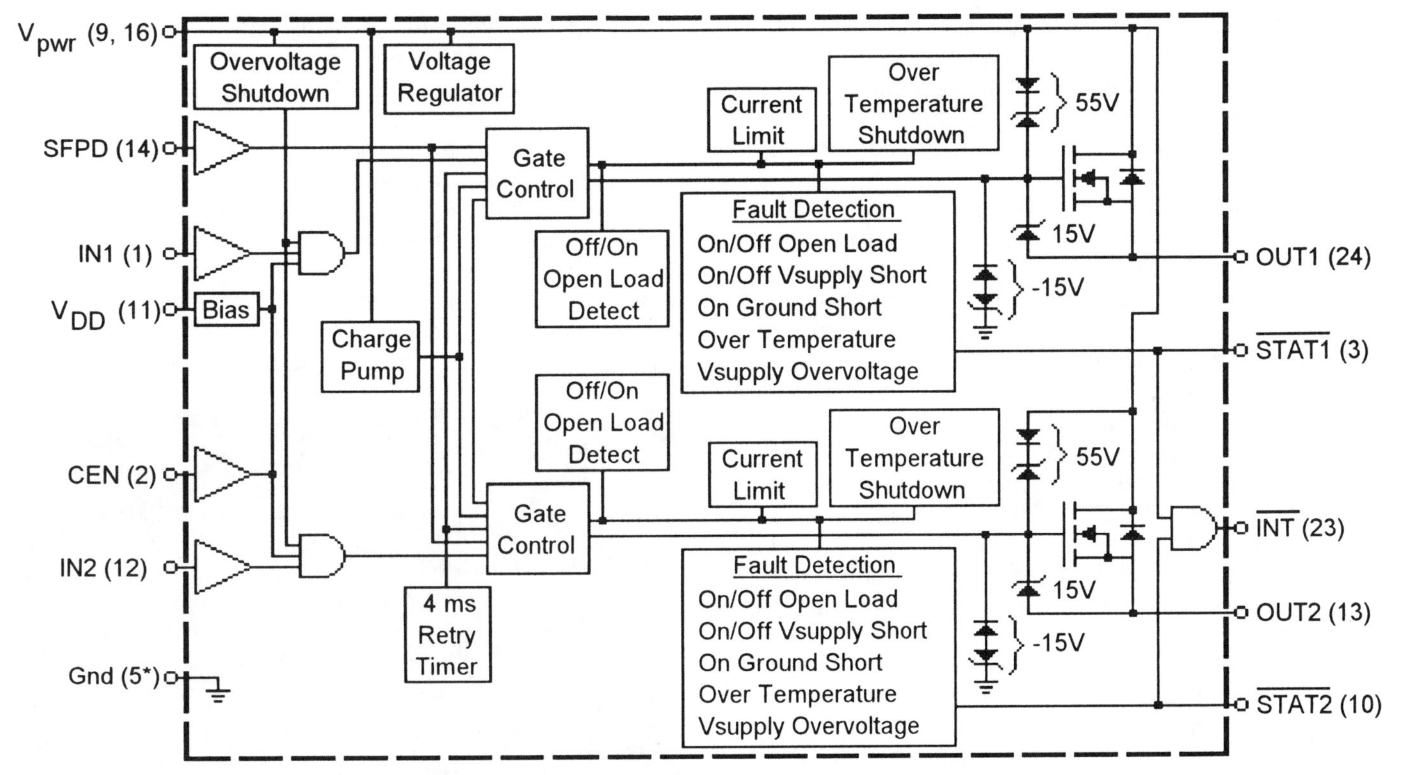

Figure 8: Block diagram of the DHSSA

DUAL INTEGRATED DRIVER

The Dual Integrated Driver (DID) is a dual output low side switch. It is designed specifically to control automotive loads such as electronic fuel injectors and transmission solenoids. The Rdson of each output is 150 mΩ and the IC is packaged in a 16 pin wide body SOIC. Fault protection capability includes ILIM, TLIM and over-voltage shutdown. The DID's diagnostic reporting strategy is different than the QID and the UPOD.

Rather than detect "on" or "off" open load conditions, the DID utilizes internal flyback detection circuitry for diagnostic reporting. At turn-off, a normally operating injector generates a flyback voltage. The DID's circuitry detects and reports the status of this flyback voltage. Under open load or shorted load conditions, no flyback voltage occurs. In this case, a fault is reported on the diagnostic pin.

The DID is packaged in a thermally enhanced 16 pin wide body SOIC. This standard outline package has eight leads connected to the flag of the leadframe for improved heatsinking. The heat is conducted from the die through the flag and leads to the circuit board. With an FR4 board the thermal resistance is 55 °C/W. When mounted on a polyimide board, the thermal resistance is 22 °C/W allowing 2 W of power dissipation with a board temperature of 105 °C. The use of standard outline SOICs for power IC packaging is an industry trend. These packages are cost effective for IC suppliers due to existing manufacturing equipment. Likewise, existing assembly line equipment makes these packages cost effective for automotive manufacturers.

Each DID output DMOS device contains 5200 active transistor cells with an area of 1,700,000 um^2. This is equivalent to 567,000 transistors on a standard 1 um CMOS process. To maintain a low threshold voltage, it is desirable to use thin oxide for the DMOS gate. However, the mean-time-to-failure decreases proportionally to oxide thickness due to particle defects in the oxide. The challenge for the manufacturer is to provide long term reliability of DMOS structures.

The DID offers a gate stress testing option to enhance the reliability of the IC. Since it is known that gate oxide rupture voltage is reduced by oxide defects, unreliable transistors can be eliminated by stress testing. A pad contacts directly to the DMOS gate so that the device may be tested prior to shipment.

Another reliability enhancement that the DID offers is the ability to measure the DMOS breakdown voltage. Since most IC DMOS outputs are clamped, it is not possible to measure their drain-source breakdown voltage. However, to ensure reliable operation, it is crucial that the DMOS breakdown remain above the clamp voltage. The DID has active circuitry that disables the clamp for testing purposes. Therefore, breakdown testing rejects DID's with low breakdown voltage.

DUAL HIGH SIDE DRIVERS

The DHSSA and DHSSB are Dual High Side Switches (DHSSs) with protection and diagnostics. The DHSSA is designed to control transmission solenoids, relays, and automotive lamps. The DHSSB is capable of driving small relays and solenoids. Faults protection and detection applies to "on" and "off" open loads, "on" shorted loads, current limit, over-voltage shutdown, and output over-temperature. The outputs are clamped 15 volts below ground for inductive flyback clamping.

The block diagram for the DHSSA high side switch is shown in Figure 8. A description of the key features of both DHSS follows.

ULTRA-LOW NOISE CHARGE PUMP - Either DHSS's charge pump radiates less noise than any other integrated High Side Switch (HSS). The charge pump provides the gate voltage necessary for the output DMOS switches. The charge pump is a voltage tripler with VPWR (battery) as the input. The tripler allows operation with a battery potential from 5.5 to 28 volts.

The three sources of noise in an HSS are the storage capacitor, VPWR and OUTPUT pins. Noise generated in a charge pump is typically routed beyond the module via the VPWR and OUTPUT wires. This noise will manifest itself in the form of conducted emissions. The DHSSs' charge pump and output control circuits are current source driven. This limits the current transients and the resulting conducted emissions.

The major source of noise in any HSS is the VPWR input because the charge pump runs directly off VPWR. The noise spectrum from the DHSS at the VPWR circuit board input is shown in Figure 9. Above 10 MHz, the noise is less than -60 dB. This is a 20 dB improvement over the industry standard HSS.

The internal charge pump is self oscillating, modulating the frequency based upon the current loading. This circuit further eliminates current spikes on VPWR by insuring that one phase of the charge pump is always charging. The dynamic response of this circuit eliminates the requirement for an external storage capacitor. This significantly reduces radiated emissions within the PCM.

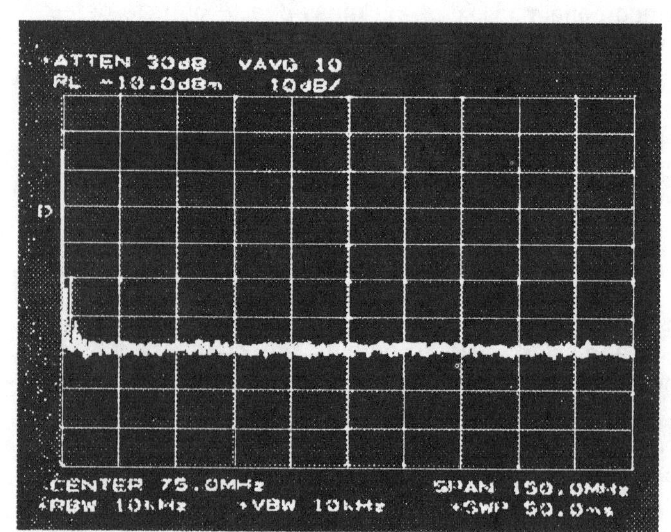

Figure 9: DHSS noise at the VPWR connector of the circuit board

The DHSS OUTPUT is coupled to the charge pump via internal current sources and the gate capacitance of the output DMOS. Figure 10 charts the OUTPUT noise spectrum for the ultra-low noise DHSS. In the AM, IF, and audio ranges (<15 MHz), the DHSS OUTPUT is 50 dB quieter than the industry standard and 20 dB quieter than a "Very Low Noise" HSS [2].

OVER CURRENT PROTECTION - The DHSS offer three layers of protection from over current conditions. The three layers are analog current limit, short fault protection, and thermal limit.

Analog current limit actively regulates the output current. Limiting the maximum current protects the printed circuit board and the IC. During current limit, the output control circuitry senses the presence of excessive current and reduces the gate voltage on the output DMOS transistor. Reducing the gate voltage forces the

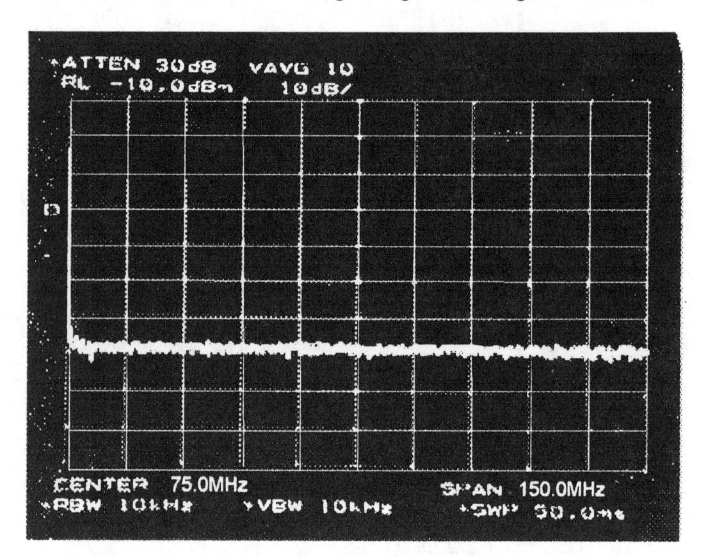

Figure 10: DHSS noise at the OUTPUT connector of the circuit board

DMOS transistor to operate in the MOS saturation region where the output current is a function of gate voltage. During current limit, the voltage drop from VPWR to OUTPUT is limited only by the external power source and load impedance. The resulting voltage across the output DMOS is larger than nominal and produces an increase in power dissipation. The increase in power produces a rise in the junction temperature of the IC. Left unchecked, the temperature rise is destructive.

The second layer of protection, Short Fault Protection (SFP), reduces the IC's temperature during analog current limit. SFP limits the duration of current limit to 50 μs to validate the fault condition. Then the output turns off for 4 ms to allow the IC to cool. The output is then reactivated for another 50 μs to monitor for fault condition removal. This cycle repeats itself until normal operation is achieved or until the output is commanded off via the Input. Figure 11 is a photograph of VPWR current for a DHSS with a short on Output 2. Figure 11 demonstrates current limit and SFP on Output 2 while Output 1 continues to operate normally.

The third and most significant layer of protection is Thermal Limit (TLIM). TLIM actively limits the temperature of the IC. TLIM deactivates the output when the junction temperature of the IC exceeds the over-temperature threshold, 180 °C. The over-temperature threshold is above the maximum operating temperature to allow the full range of operation. Seven degrees of hysteresis is provided to limit thermal oscillations. The output reactivates when the junction temperature drops seven degrees below the over-temperature threshold.

OUTPUT SPECIFIC TLIM - Output specific TLIM is essential for multiple output high side switches to insure that only the faulted output is disabled. Each DHSS output contains a dedicated TLIM circuit. Mismatch between over-temperature thresholds on the same

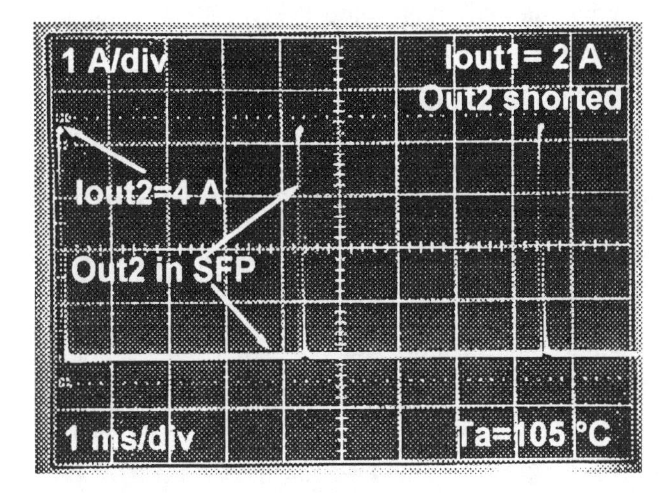

Figure 11: VPWR current with a short on Output 2 and normal operation of Output 1

device is less than 2 °C [3]. This matching ensures that the device deactivates the faulted output before causing an adjacent output to fail. Figure 12 is a photograph of output specific TLIM in the 140 mΩ DHSS. At 105 °C, Output 1 controls 2 amperes while Output 2 controls a 3 ampere load. Every 46 ms, Output 2 exceeds its over-temperature threshold and TLIM turns off the output. After cooling off, Output 2 turns on. Output 1 is never deactivated by the fault on the adjacent output.

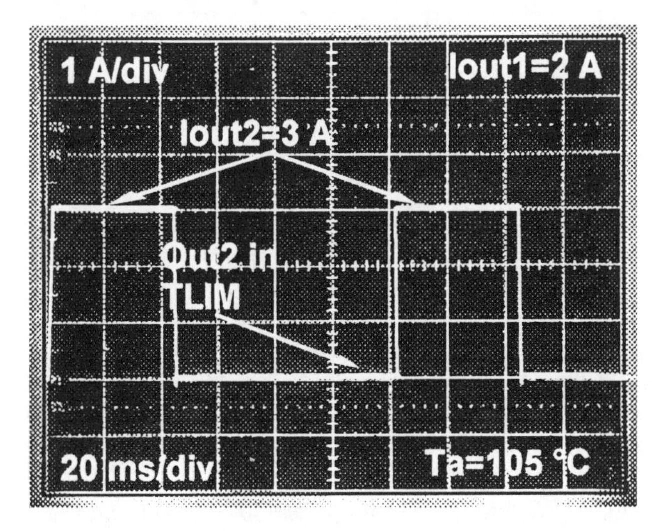

Figure 12: VPWR current with excessive power in Output 2 and normal operation of Output 1

The slow heating demonstrated in Figure 12 is a worst case test scenario. This slow heating of Output 2 allows the heat to propagate across the IC raising the temperature of Output 1. It would be easier to disable only the faulted output if a hard short were used for the fault condition. A hard short would have higher instantaneous power dissipation forcing a TLIM on Output 2 before the heat can propagate across the IC. As a result, the thermal gradient would be higher and easier to manage.

SHORT FAULT PROTECTION DISABLE - Short Fault Protection Disable (SFPD) is a selectable feature that extends the duration of current limit operation. SFPD is a CMOS input that disables the SFP circuitry. With SFPD active, a faulted output remains in current limit until attaining a normal current level or until TLIM deactivates the output.

Figure 13 is a photograph of the in-rush current into a brake lamp when driven directly from a supply. The current exceeds the DHSS current limit threshold, 4 amperes, for 30 ms. For the DHSS driving the same lamp with SFPD low, SFP activates the output 50 μs out of every 4 ms. The lamp would not incandesce with such a small duty cycle and on time. The MCU could incandesce the lamp by toggling the input every 50 μs. However, switching losses would increase the power dissipation and the MCU overhead requirements would degrade the system chronometrics.

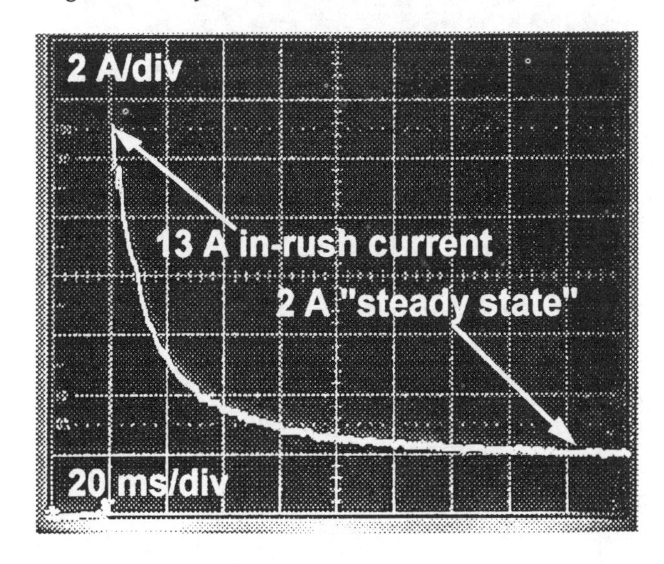

Figure 13: Brake lamp shorted to a voltage source

The SFPD feature permits incandescent lamp applications with in-rush currents in excess of the current limit threshold. To illuminate the lamp, the transient thermal response must keep the junction temperature below the over-temperature threshold. Figure 14 is a photograph of the in-rush current into a brake lamp driven by the DHSS with SFPD active. Having sufficient thermal capacity, the DHSS limits the current to 4 amperes until the bulb incandesces.

LOW OVERHEAD FAULT REPORTING - Fault reporting on the DHSS is software efficient. Monitoring the output status requires no background loops or word compares which are typical of SPI operated devices [3].

A fault on either output will induce a high to low transition on the INTB output. INTB is an open drain output with a current source pull-up of 40 μA. INTB is designed to be wired-or with other ICs to drive the MCU interrupt request input. INTB eliminates the need for the MCU to poll the DHSS for fault status. Because normal operation is the dominant mode of operation, a substantial saving of MCU "horsepower" is realized. When a fault occurs, individual status outputs, STATB1 and STATB2, indicate which output is faulted.

OPEN LOAD DETECTION IMPEDANCE - The load impedance that is reported as an "on" open is a function of the "on" open detection current (I_{oon}), VPWR, and R_{DSon}. This relationship is shown in

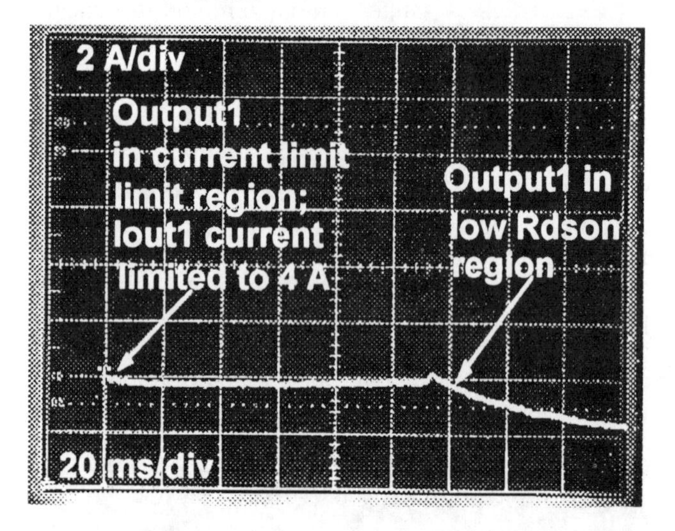

Figure 14: Brake lamp controlled by DHSS with SFPD active

Equation (1). Figure 15 plots the load resistance that is reported as an open by the DHSSA.

$$R_{on\text{-}open} \geq (VPWR/I_{oon})\text{-}R_{DSon} \quad (1)$$
$$R_{on\text{-}open} \approx (VPWR/I_{oon})$$

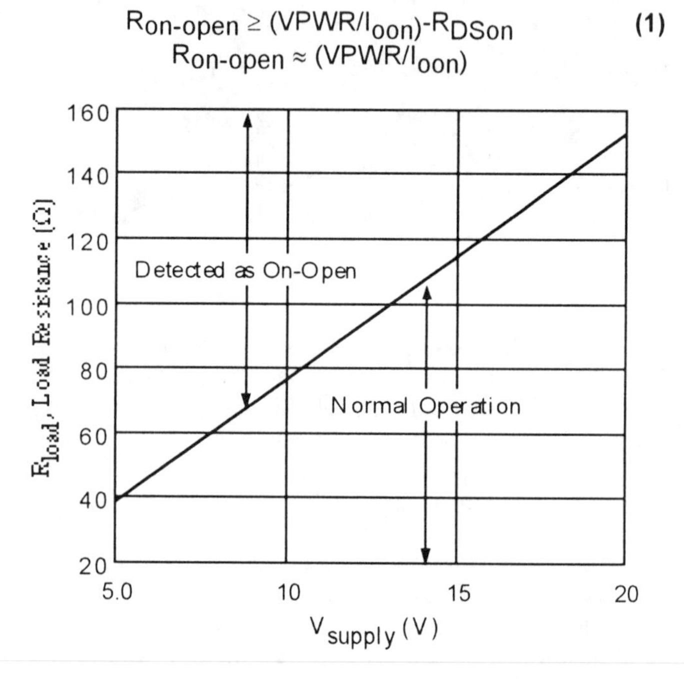

Figure 15: Load resistance reported as an "on" open for the DHSSA

The load impedance that is reported as an "off" open is a function of the output pull-up resistor (R_{opu}), VPWR, and the output fault threshold (V_{tho}). This dependence is shown in Equation (2). Figure 16 plots the load resistance that is reported as an open by the DHSSA with R_{opu} of 20 kΩ.

$$R_{off\text{-}open} = (VPWR\text{-}V_{tho})/(V_{tho}/R_{opu}) \quad (2)$$

"On" open load detection provides better resolution than "off" open load detection. The "on" open detection circuitry will detect loads in the 80 to 120 Ω range for VPWR in the range of 10 to 15 V. This impedance region is close to but above the normal load range of 12 to 40 Ω. Loads from 80 to 130 kΩ are detected as "off" opens for the same VPWR range. For "off" opens, the load impedance must change by two orders of magnitude before a fault is reported. "SLEEP STATE" - The DHSSs are designed to virtually eliminate VPWR current during periods of non-use. This circuit is called "Sleep State" (SS). SS is used in battery powered applications where the DHSS may not be active for extended periods of time and prolonged battery life is important.

The DHSSB controls 0.6 A per output in a 105 °C ambient (Tj ≤ 150 °C). Each output has 380 mΩ of resistance at 25 °C. The DHSSA delivers 1.6 A per output in a 105 °C ambient (Tj ≤ 150 °C). Each output has 140 mΩ of resistance at 25 °C. The die photograph of the DHSSA is shown in Figure 17.

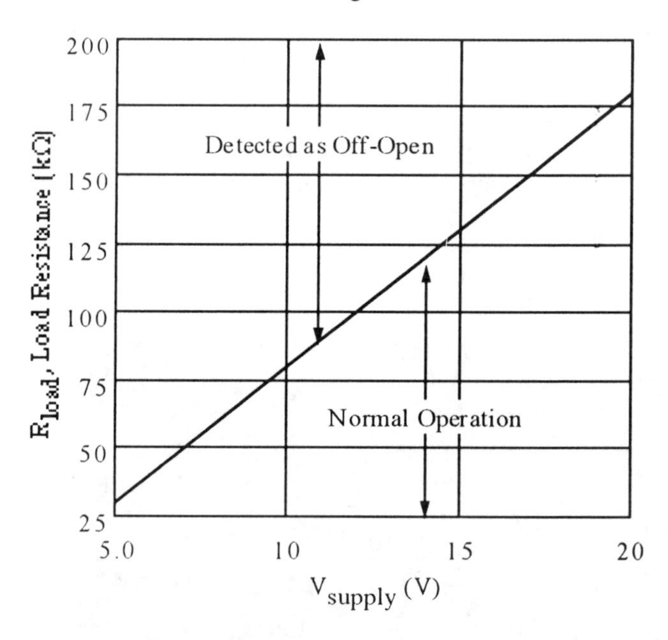

Figure 16: Load resistance reported as an "off" open for the DHSSA

SUMMARY

The Powertrain Control Module environment requires power ICs withstand extreme voltage transients and have diagnostic capability. The QID, UPOD, DID, DHSSA, and DHSSB utilize a mixed-mode process that enables them to meet rigorous PCM demands. Outstanding features of the ICs are reviewed in Table 1. All five of the ICs currently operate in PCM modules.

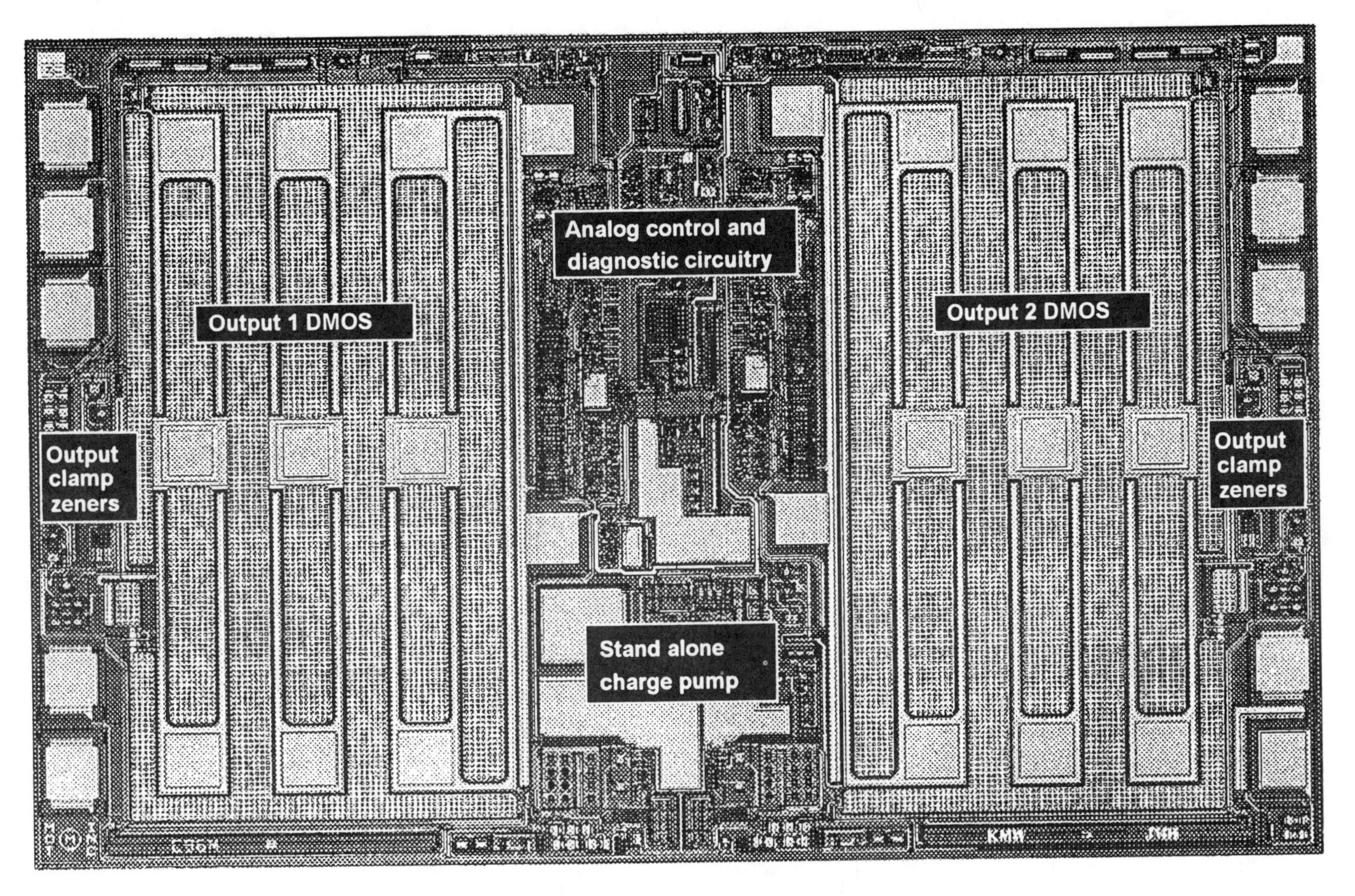

Figure 17: DHSSA die photograph

REFERENCES

[1] Laurie Carney, Randy Frank et. al., "Recent Developments in Surface Mount Power IC Packages", Power Conversion and Intelligent Motion Proceedings, October, 1993, p. 300.

[2] Mario Paparo, William Javurek, et. al., "A Very Low Noise, μP Interfaced, Multiple High Side Driver with Self Diagnostics for Car Radio Application" presented at the SAE International Congress & Exposition, Detroit Michigan, February 28-March 3, 1994

[3] Jeff Kanner and Keith Wellnitz, "Power IC for Controlling Electronic Fuel Injectors" presented at the PCIM/POWER QUALITY '94 with Mass Transit System Compatibility '94 CONFERENCE & EXHIBIT, Dallas/Ft. Worth, Texas, September 17-22, 1994

[4] Ben Davis, Keith Wellnitz, and Randall Wollschlager, "A New Automotive SMARTMOS™ Octal Serial Switch", presented at IEEE Workshop on Power Electronics in Transportation, Dearborn, Michigan, October 22-23, 1992

Integrated Circuit	PCM Application	Number of Outputs	RDSon per Output at Tj=25°C	Features
QID	fuel injectors transmission solenoids canister purge solenoids fuel pump relay incandescent bulbs	4	200 mΩ	current limit independent thermal limit on and off open load detection over-voltage protection high energy output clamp
UPOD	fuel injectors transmission solenoids canister purge solenoids fuel pump relay incandescent bulbs	6	200 mΩ	current limit independent thermal limit on and off open load detection over-voltage protection high energy output clamp two way SPI communication
DID	fuel injectors transmission solenoids canister purge solenoids fuel pump relay incandescent bulbs	2	150 mΩ	current limit independent thermal limit flyback detection over-voltage protection high energy output clamp
DHSSA	fuel injectors transmission solenoids canister purge solenoids fuel pump relay incandescent bulbs	2	140 mΩ	current limit independent thermal limit on and off open load detection over-voltage protection high energy output clamp low noise charge pump
DHSSB	engine relays transmission relays	2	380 mΩ	current limit independent thermal limit on and off open load detection over-voltage protection high energy output clamp low noise charge pump

Table 1: Outstanding characteristics of the QID, UPOD, DID, DHSSA, and DHSSB Powertrain Control Module ICs

960625

Cost Effective Input Circuitry for Texas Instruments TMS370 Microcontroller Family

Michael S. Stewart and David T. Maples
Texas Instruments

ABSTRACT

Severe voltage transient conditions and widely varying battery voltage levels place a premium on designing input protection and signal conditioning circuitry for microcontroller based systems. Various circuits have been implemented to provide the required robust designs. However, the costs associated with implementing some of these protection and conditioning circuits can excessively drive up system costs. Using TMS370 microcontrollers designed with internal diode protection circuitry and TTL specified logic levels allow designers to implement simple low cost input circuitry thus greatly reducing the total system cost.

INTRODUCTION

Today's microcontroller based systems are subjected to electrically harsh environments that require the existence of input protection circuitry. Depending on the embedded system environment and the design of the microcontroller, this external protection circuitry can add substantial system costs. Microcontroller based systems typically have a significant number of inputs and outputs (I/O). The I/O will be exposed to an environment that requires the use of discrete circuitry to condition input signals and to protect the microcontroller from high voltage transients. An opportunity for cost savings exists if the input circuitry of the microcontroller is designed with these automotive system requirements in mind.

The purpose of this paper is to outline the cost advantages resident with the Texas Instruments TMS370 microcontroller family when used in an automotive system with its 12V DC battery and potentially damaging transient noise spikes. The principles developed herein are applicable to other electrically harsh environments such as industrial, motor control, etc. This paper will discuss the following three areas:

1. Advantages and disadvantages of TTL and CMOS specified logic levels relative to required input circuitry.
2. Guidelines for designing external conditioning circuitry for TTL specified inputs.
3. Guidelines for designing external protection circuitry when using TMS370 microcontrollers with internal diode protection circuitry.

A brief cost analysis comparing various input protection and conditioning circuits is provided.

BASIC MICROCONTROLLER BASED REQUIREMENTS

A goal of the automotive system designer is to translate vehicle voltages to a voltage range that the microcontroller can recognize as a logic 1 or logic 0, outside of the indeterminate range, while not exceeding the maximum or minimum input voltage specification of the device. The following two typical conditions should be considered for the automotive environment:

1. Switching to battery voltage (V_{BAT}) as illustrated by figure 1.
2. Switching to battery ground as illustrated by figure 2.

One of the greatest difficulties in designing external input circuitry in both conditions is created by the wide fluctuations in the vehicle battery voltage. The battery voltage typically ranges from 9 to 18V during normal vehicle run conditions (not considering double battery or crank conditions). The vehicle ground may range from -2 volts to +2 volts due to vehicle ground offsets.

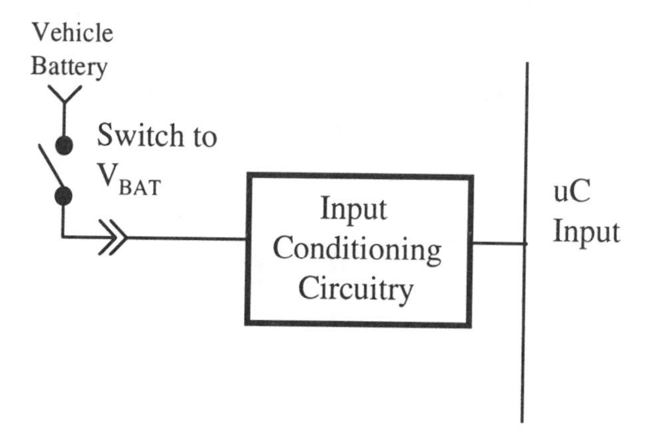

Figure 1. Switching to Vehicle Battery (V_{BAT})

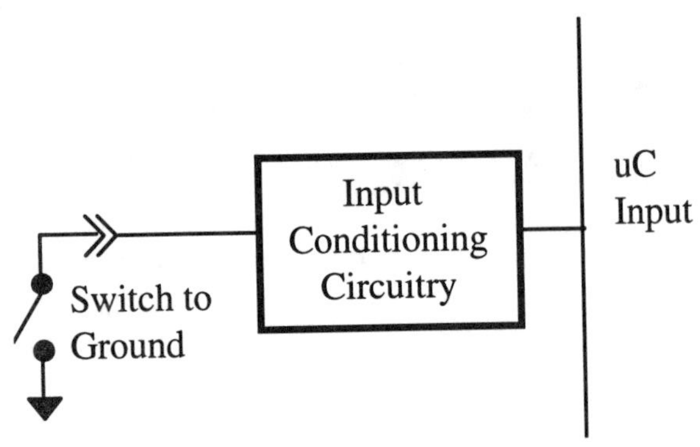

Figure 2. Switching to Vehicle Ground

To begin the analysis of designing required input signal conditioning and protection circuitry, consider the simple voltage divider circuit. Figure 3 illustrates the functionality of a simple voltage divider circuit with the TMS370 I/O buffer structures.(Figure 9 in the appendix illustrates actual buffer structures of the TMS370 device.)

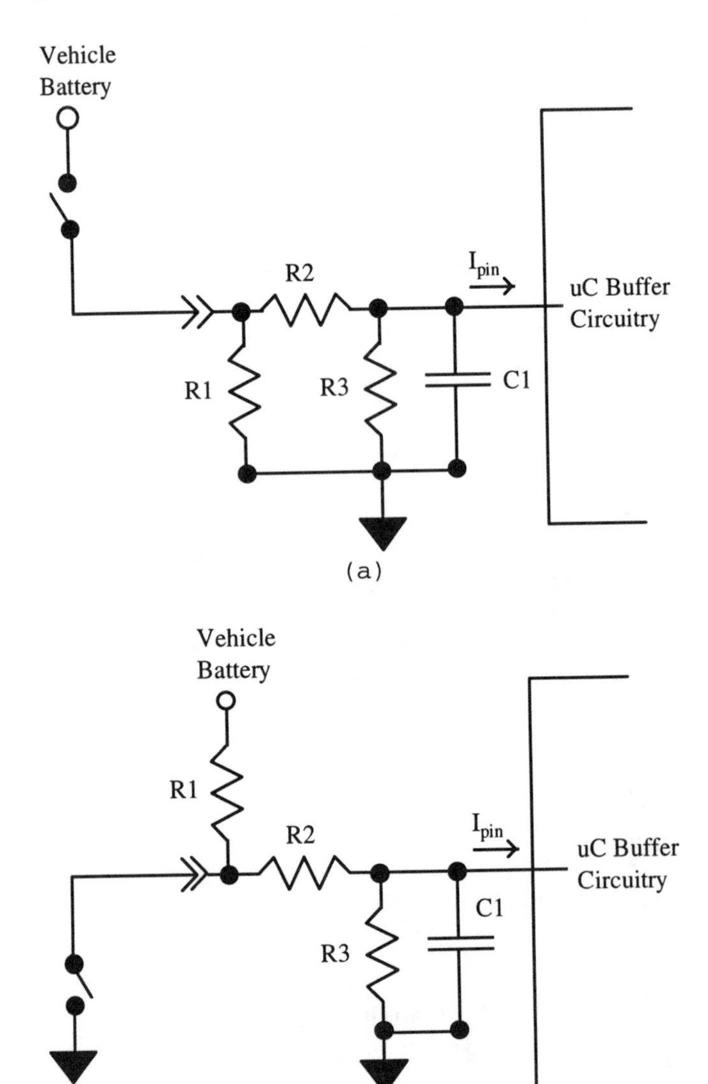

Figure 3. TMS370 microcontroller buffer circuitry with external voltage divider circuitry

In these figures, resistor R1 holds the input voltage at a known level in an open switch condition. Capacitor C1 in conjunction with resistor R2 make up a single pole low pass filter to minimize noise detected by the software and to assist in transient suppression. Resistors R2 and R3 make up a resistor divider network with the following familiar equation:

Equation 1:

$$\text{Input Voltage} = \frac{R3}{R2 + R3} \text{X (VBAT)}$$

Given system parameter with fixed values, a simple voltage divider/RC circuit could be developed for most applications. However, to effectively design within the automotive system of varying battery voltages, the advantages and disadvantages of different logic level specifications must be understood.

ADVANTAGES AND DISADVANTAGES OF TTL AND CMOS SPECIFIED LOGIC LEVEL RELATIVE TO REQUIRED INPUT CIRCUITRY

DESCRIPTION OF TTL AND CMOS SPECIFIED INPUT LOGIC LEVELS - Input levels of the microcontroller, commonly referred to as V_{IL} and V_{IH}, are the voltages required to guarantee that the microcontroller will interpret the voltages at the device input pin as a logic 1 or logic 0. Table 1 illustrates the input thresholds of industry standard microcontrollers.

Table 1. Industry Standard Microcontroller Input Thresholds:

Device	Minimum V_{IH}	Maximum V_{IL}
TMS370	2.0 V	0.8 V
HC11	0.7 Vcc	0.2 Vcc
HC05	0.7 Vcc	0.2 Vcc
80C51	0.2 Vcc+1.0V	0.2 Vcc-0.15V
COP888	0.7 Vcc	0.2 Vcc

As illustrated above, TMS370 input thresholds are specified at TTL levels while most competitor's devices are typically specified at CMOS voltage levels. The key difference in specification is that CMOS voltage levels have a wider indeterminate region than the TTL levels as illustrated in figure 4. Understanding the advantages of TTL logic levels in automotive applications is vitally important when designing cost effective input conditioning and protection circuitry.

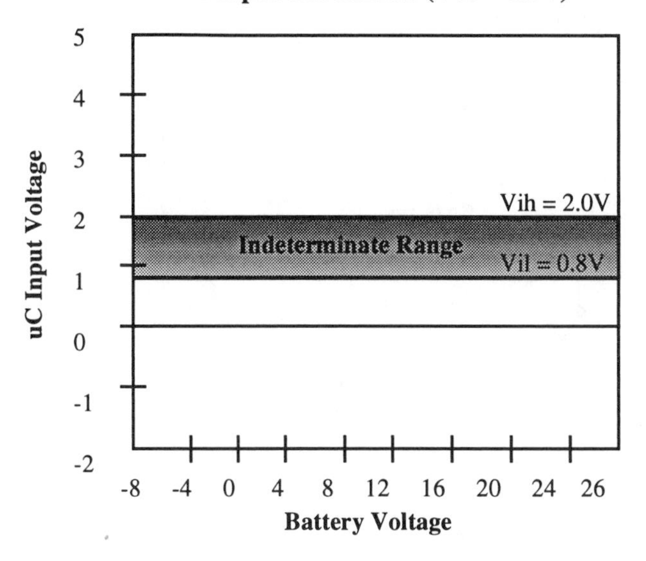

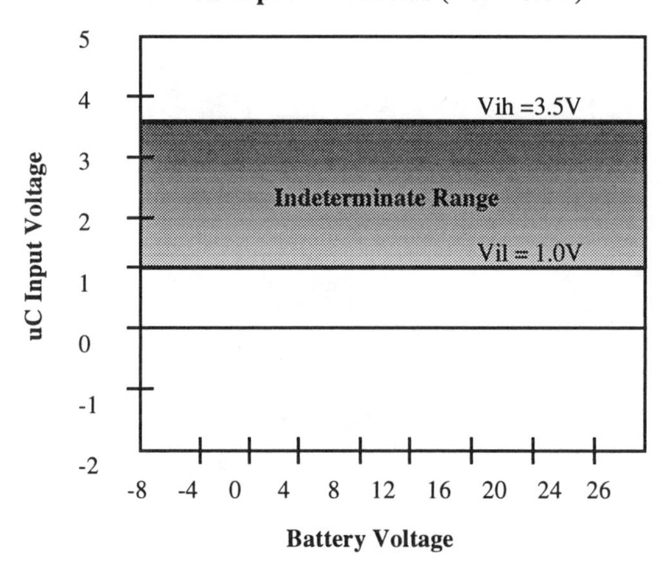

Figure 4. Indeterminate Range for TTL and CMOS Input Thresholds (Vcc = 5V)

Consider the CMOS input levels of most standard microcontrollers. Table 2 illustrates the conditions that the input conditioning circuitry will be exposed to and the requirements it must satisfy.

Table 2. Typical CMOS Parameters and System Conditions

Parameter	
Normal Battery Range (switch to V_{BAT} condition)	9V <= V_{IN} <=18V
Ground Range (switch to gnd condition)	-2V <= V_{IN} <= 2V
Vcc	5V+/-10%
Microcontroller Vih	0.7 Vcc
Microcontroller Vil	0.2 Vcc
Microcontroller absolute max. input Voltage range	7V
Microcontroller absolute min. input Voltage range	-0.6V

UNDERSTANDING DISADVANTAGES OF CMOS SPECIFIED INPUTS - Once the system and microcontroller specifications have been determined, an attempt can be made to find the resistor ratios necessary for the simple voltage divider network that will operate over the entire V_{BAT} range. Figure 5 located in the appendix plots the voltages seen at the microcontroller pin versus the battery voltage fluctuations for CMOS logic levels. Figure 5 also shows several resistor ratios that could be selected for a voltage divider circuit. The input voltage ranges of interest are noted to the right of the plot. Input voltages that exceed 5.7V (assumes Vcc = 5V) or are less than -0.7V will forward bias the input diodes found on most micorcontroller protection circuitry inputs. Good design practice dictates that V_{IN} conditions operate within the recommended operating conditions of the microcontroller. Therefore, conditions that would allow the diodes to conduct during normal operation are considered out of spec and should be eliminated.

> **Note**: The specifications for maximum and minimum VIN values is device and vendor dependent. These limits are primarily determined by the overvoltage protection circuitry. Each vendor has different protection circuitry and thus different absolute maximum and recommended operating range specifications.

The range between V_{IH} and V_{IL} is the digital indeterminate range. The microcontroller cannot be guaranteed to distinguish a logic 1 from a logic 0 across manufacturing process variations, voltage fluctuations, temperature range, etc. The other regions are the voltages that the microcontroller is guaranteed to recognize as a logic1 and logic 0. Therefore, for all valid voltages that the input conditioning is exposed to (i.e. 9V to 18V for an automotive switch to battery condition) the resistor curves must fall within the logic 1 or logic 0 to satisfy the design constraints.

A review of Figure 5 shows that all the design considerations cannot be met for CMOS inputs with a simple resistor divider. The switch to battery condition is shown between the two arrows on the right of the figure. Take the 1/4 ratio as an example. Battery voltages between 9 and 14 volts violate V_{IH}. The 1/3 ratio has better performance with respect to V_{IH} but battery voltages greater than 17 V exceeds the maximum input voltage of the device. Some type of active circuitry must be designed to satisfy all the design constraints adding to the total system cost. The switch to ground condition is shown between the two arrows on the left hand side of the figure. The design conditions can be met for a switch to ground with CMOS input levels for all three resistor ratios since V_{IN} falls within V_{IL} and the minimum input voltage of the device.

This analysis begins to illustrate several challenges inherent with designing with microcontrollers specified with CMOS levels.

UNDERSTANDING ADVANTAGES OF TTL SPECIFIED INPUTS - Designing with TTL level CMOS inputs are considered next. Table 3 shows the conditions that the input conditioning circuitry will be exposed to and the requirements it must satisfy. The design requirements are identical to the previous example except for the change in V_{IH} and V_{IL}.

Table 3. Typical TTL Parameters and System Conditions

Parameter	
Battery Range (switch to V_{BAT} condition)	$9V <= V_{IN} <= 18V$
Ground Range (switch to gnd condition)	$-2V <= V_{IN} <= 2V$
Vcc	5V+/-10%
Microcontroller V_{IH}	2 V
Microcontroller V_{IL}	0.8 V
Microcontroller absolute max. input Voltage range	7 V
Microcontroller absolute min. input Voltage range	-0.6 V

Figure 6 located in the appendix plots the voltages seen at the Microcontroller versus the battery voltage fluctuations for TTL logic levels. Again, several resistor ratios are plotted and the input voltage ranges of interest are noted to the right of the plot. A review of the figure shows that all the design considerations can be met with a 1/4 ratio for TTL input levels and a simple resistor divider. The switch to battery condition is shown between the two arrows on the right hand side of the figure. The Microcontroller input voltage is always greater than V_{IH} and less than the max. input voltage spec. for normal battery voltages between 9 and 18 volts.

The switch to ground condition is shown between the two arrows on the left of the figure. Again, the design conditions can be met for a switch to ground with TTL input levels. The microcontroller input voltage for all three resistor ratios fall with in V_{IL} and the minimum input voltage of the device

This analysis clearly illustrates the advantages inherent with TTL specified input logic levels versus CMOS specified input logic levels when designing input conditioning circuitry. The next section focuses on another area of design concern, the protection of input pins during transient conditions.

ADVANTAGES OF INTERNAL DIODE PROTECTION CIRCUITRY

The TMS370 family of microcontrollers has been designed with internal diode protection circuitry on all I/O pins. These diode protection circuits coupled with an external current limiting resistor can be used to successfully protect the microcontroller from excessive external high voltage spikes.

Typically, embedded microcontroller systems applications require the use of expensive external protective circuitry due to high voltage noise spikes present in the system. These high voltage spikes can easily exceed the absolute maximum specifications of CMOS microcontrollers. To protect the input pins from these high voltage signals, external suppression circuitry must be implemented. Figure 7 illustrates several common suppression circuitry methods including, the addition of external clamp diodes, zener diodes, and buffer circuitry.

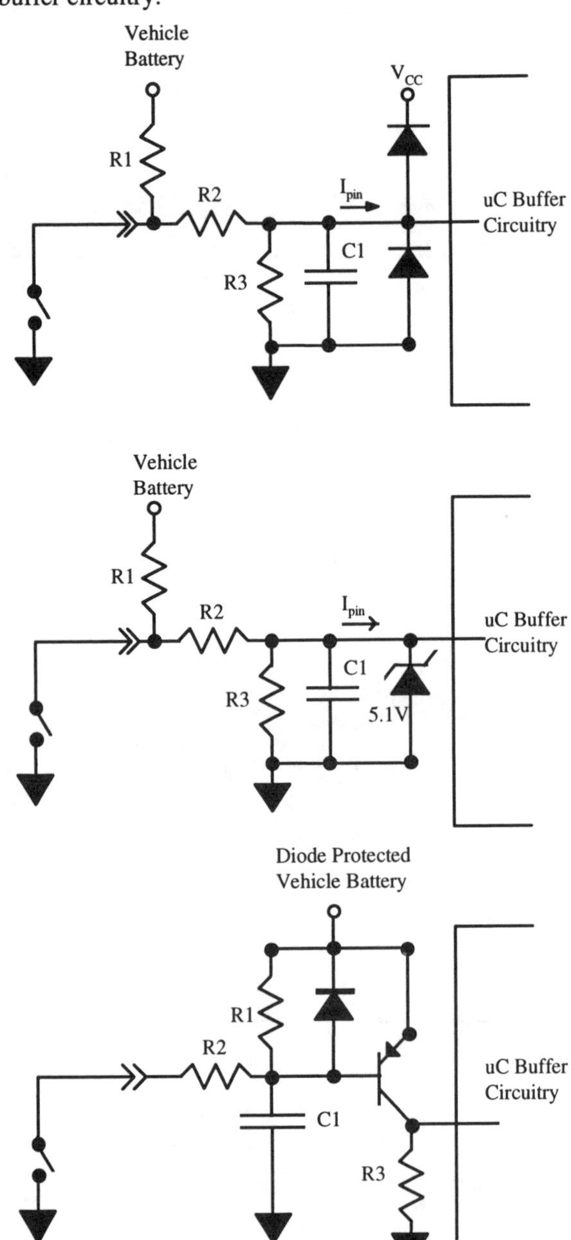

Figure 7. External electrical noise suppression circuitry

The external noise suppression circuits illustrated in figure 7, are necessary for over-voltage protection. However, the TMS370 microcontroller family has been designed with internal diode protection circuitry. A simple calculation can provide the necessary value for an external current limiting resistor that, coupled with the internal diode protection circuitry can adequately protect the TMS370 microcontroller from external high voltage spikes. Figure 8 illustrates the alternative low-cost circuitry required to protect TMS370 based microcontroller designs.

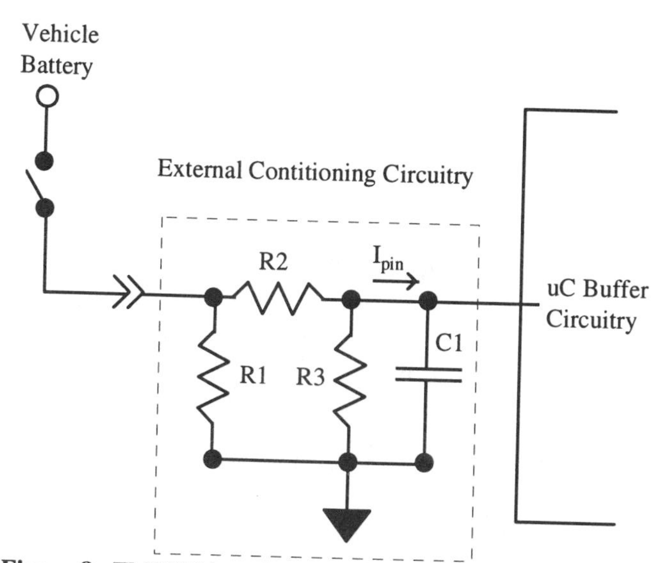

Figure 8. TMS370 based external noise suppression circuitry

The system cost advantages of designing with the TMS370 family of microcontrollers becomes quite evident when compared to competitive microcontrollers that do not contain internal diode protection circuitry or TTL input levels.

DESIGNING INPUT PROTECTION CIRCUITRY FOR TMS370 MICROCONTROLLERS -The next step in the cost reduction process is to design the input protection circuitry to meet the criteria for transient suppression and the TTL input thresholds. This section will provide an example for selecting the two external resistors (R2 and R3) required for a simple voltage divider protection circuit.

Using the external current limiting resistor (R2), you can limit the voltage and current seen on the I/O pins such that external protection diodes are not needed. There are actually two absolute maximum specifications that need to be considered. These are:

1. Input and Output clamp current: This specification is equal to + or - 20 mA when V_{IN} (or V_{OUT}) is less than V_{SS} or greater than V_{CCD}.
2. Input voltage range: This specification is equal to a minimum of -0.6V or a maximum of 7V on all pins except INT1. For INT1, the minimum is -0.6V and the maximum is 14V.

Continuous power dissipation should also be considered when selecting the external circuitry. Continuous power dissipation is dependent on package type and the maximum ambient temperature requirement. The real requirement is that the max. power consumption of the package not be violated during the transient.

> Note: Remember that transient suppression is designed to protect the microcontroller from over-voltage conditions and not for normal operation.

The TMS370 family has gone through several silicon "shrinks". Silicon "shrinks" are redesigns to use smaller silicon geometries. The TMS370 has gone through two shrinks commonly referred to as the 80% silicon and the 60% silicon. The original TMS370 was a 2 micron process (100%). The 80% shrink is a redesign for a 1.6 micron process. Likewise, the 60% shrink is a redesign for a 1.2 micron process. The 1.2 micron silicon is typically provided for new applications. The internal diode protection circuitry is identical for both 1.2 and 1.6 micron devices. However, the 1.2 micron devices have replaced most 'Fast' I/O buffers from the 1.6 micron devices with 'Slow' I/O buffers to help reduce EMI emissions.

Device symbolization for the 1.2 micron silicon will either have an 'A' or 'B' at the end of the device name. For example, the device name TMS370C056A would indicate a 1.2 micron silicon design. Device symbolization for the 1.6 micron silicon will not have either letter. For example, the device name TMS370C056 would indicate a 1.6 micron silicon design. Table 4 illustrates the different types of I/O pin buffer circuits used on TMS370 microcontrollers.

Table 4. TMS370 Microcontroller I/O Pin Buffer Types

I/O Pin Type	TMS370 pins (1.2 Micron Design)	TMS370 Pins (1.6 Micron Design)
Fast Input	INT1	INT1
Analog Input	AN0 - AN14	AN0 - AN14
Slow I/O	All Others	Reset-, D3, D6
Fast I/O	D3/CLKOUT	All others

Figure 9 located in the appendix illustrates the effective equivalent I/O pin buffer circuitry for both 1.2 micron and 1.6 micron silicon.

The current limiting resistance is not simply a matter of selecting a value that will limit the clamp current to +/- 20 mA. The external current limiting values need to be selected while keeping in mind the resistive characteristics of the internal protection circuitry. The goal is to limit the absolute maximum CLAMP CURRENT to less than +/- 20 mA and also at the same time limit the absolute maximum VOLTAGE to 7V (14V for INT1). With this in mind, the following example illustrates how to calculate the external current limiting resistor (R2) value necessary to adequately protect the input pins on the TMS370 device. Let's look at the following example.

EXAMPLE: What minimum external resistance value (R2) is needed on the AN0 pin to prevent damage to the TMS370 device during transient voltage spikes of +/- 150V?

Conditions: Limit the absolute maximum voltage on AN0 to -0.6V - 7V and the absolute maximum input clamp current to +/-20 mA. (Both conditions must be taken into account) Also, note that the resistance characteristics of the negative voltage protection diode circuitry is much smaller than the positive voltage protection diode circuitry. In this case, the example illustrates solving for both the positive and negative absolute maximum conditions.

I/O pin resistive characteristic value: The AN0 pin (Analog Input) has a resistive characteristic value of 2,000 ohms.

Solving for R2 to protect against a positive +150V voltage spike:

Given: V_{IN} = + 150V
V_{CCD} = 5.5V (Worst case for this example. A value of 4.5V would allow a larger voltage drop across the internal resistance)
V_{PAD} = 7.0V (Absolute Maximum value)

Solve for V_{RINT}: $V_{RINT} = V_{PAD} - V_{CCD}$
= 7.0V - 5.5V
= 1.5V

Solve for I_{RINT}: $I_{RINT} = V_{RINT} / R_{INT}$
= 1.5V / 2,000 ohms
= 750 uA

Solve for R2 $R2 = V_{IN} - V_{PAD} / I_{RINT}$
= 150V - 7V / 750 uA
= 143V / 750 uA
= 190.667K ohm minimum

Therefore, the value of R2 to adequately protect against a 150V transient condition would be ~ 191K ohms.

Solving for R2 to protect against a negative -150V voltage spike:

Given: V_{IN} = - 150V
V_{SSD} = 0V
V_{PAD} = -0.6V (Absolute Maximum value)

Solve for V_{RINT}: V_{RINT} = $V_{SSD} - V_{PAD}$
= 0V - (-0.6V)
= 0.6V

Solve for I_{RINT}: I_{RINT} = V_{RINT} / R_{INT}
= 0.6V / 20 ohms
= 30 mA

Since 30 mA exceeds the absolute maximum clamp current of 20 mA, the following equation will substitute the lower value of 20 mA.

Solve for R2: R2 = $V_{PAD} - V_{IN} / I_{RINT}$
= (-0.6V) - (-150V)/ 20 mA
= 149.4V / 20 mA
= 7.47K ohm minimum

To protect against a negative 150V transient condition a R2 value of ~ 7.5K ohms would be adequate. Select the R2 value that is highest. Since the minimum external resistance (R2) is larger for the positive external voltage spike, select a value of ~ 191K or greater for R2.

Now that R2 has been determined, calculate a value for R3. The first section of this document described the TTL inputs and the necessity that the resistor ratio between R2 and R3 be 1/4. Use this relationship to calculate R3.

$$\frac{1}{4} = \frac{R3}{(R3 + R2)}$$

$$R3 = R2 \; / \; 3$$

$$R3 = 191K\Omega \; / \; 3$$

$$R3 = \sim 64K\Omega$$

The TMS370 can withstand voltage transients and interpret vehicle battery variations as logic 1's or Logic 0's using a simple voltage divider. The series current limiting resistor (R2) limits the voltage and current seen on the I/O pins such that the internal diode protection circuitry can withstand the defined system transients. The addition of one additional pull down resistor (R3) creates a divider circuit with R2 and additional circuitry is not required to convert the vehicle battery levels to voltage levels recognizable by the microcontroller inputs.

Table 5 below and table 6 located in the appendix have been provided as a quick reference for the types of I/O pins that are available on both the 1.2 micron and 1.6 micron devices as well as a matrix to help select the minimum external resistance (R2) necessary over various external voltage conditions.

Table 5. Typical values of R2 required for 1.2 and 1.6 micron TMS370 silicon assuming an external +/- 150V spike

I/O Pin Type	TMS370 pins (1.2 Micron Design)	TMS370 Pins (1.6 Micron Design)	Min. R2 (Theory)
Fast Input	INT1	INT1	128K[1]
Analog Input	AN0 - AN14	AN0 - AN14	191K
Slow I/O	All Others	Reset-, D3, D6	29K
Fast I/O	D3/CLKOUT	All others	39K

Note 1. The absolute maximum V_{IN} value for the INT1 pin is 14V.

The values for R2 above coupled with the calculated value for R3 (1/4 ratio) satisfy the protection requirements for the TMS370 Microcontroller input. They limit voltage and current seen on the microcontroller I/O pins and ensure that TTL voltages thresholds are not violated across all normal operating voltages. A much more detailed analysis can be done for a specific transient specification. Since most transients are AC in nature, the low pass filter can be designed to ensure that a voltage transient with some frequency content will be attenuated.

The values calculated for R2 and R3 should be considered minimum values. Increasing the value of R2 and R3 may realize the following benefits.

1. Power consumption of the microcontroller will be reduced during a transient event. Quiescent current of the system will be reduced.

2. A greater R2 enables a lower value C1 for an equivalent low pass filter. Typically, lower value capacitors are less expensive.

> Note: The value of R2 has a direct effect on the A/D converter when used to limit current on analog input pins. There is a minimum sample time of 1 us per 1K ohm of source impedance. The system designer needs to determine the appropriate value to meet system requirements.

COST ANALYSIS

This paper establishes that designing with Texas Instruments' TMS370 family's TTL specified input circuitry can be more cost effective than designing with CMOS specified input circuitry. This advantage allows system designers to simplify their external conditioning and protection circuitry. The ultimate goal and the reason for this analysis is to minimize **COST** at the system level. This section establishs that there is a substantial system level cost savings associated with robust TMS370 TTL level specified input circuitry.

Several typical input conditioning circuits are shown in figure 11. This is by no means an exhaustive list, but it will serve the purpose of cost comparisons between different types of input circuits. Figure 11 illustrates the simple resistor divider input conditioning circuit for TMS370 TTL inputs as well as other external protection circuits such as external diodes, external zener, transistor level shifter, and a buffered hex-inverter used as a level shifter.

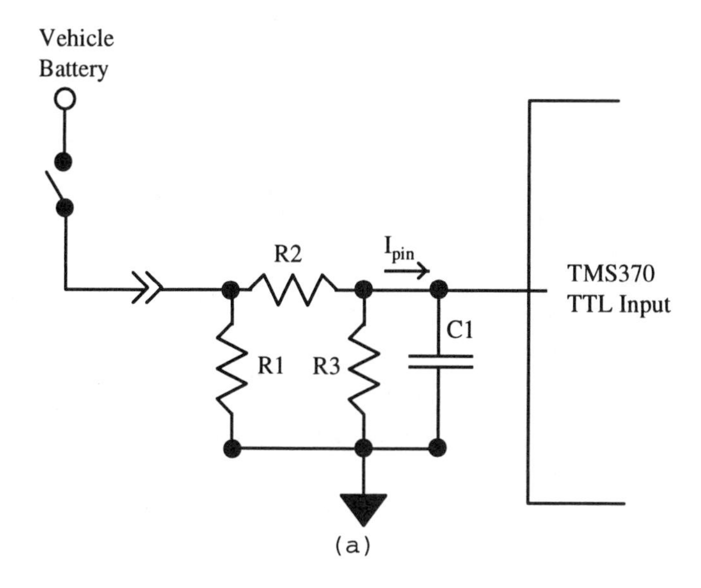

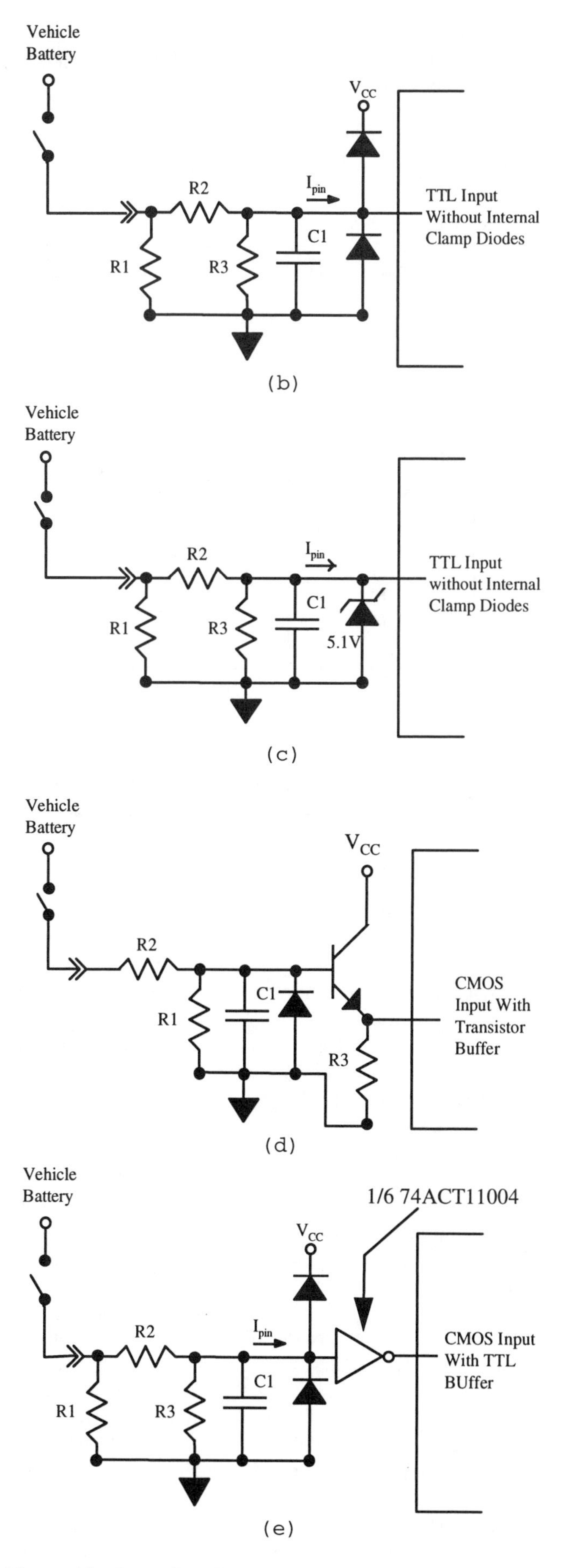

Figure 11. Examples of external protection circuitry

Table 7 illustrates a cost comparison between the five implementations shown in figure 11. The following component cost assumptions are used for comparison purposes only.

Resistor $0.01
Capacitor $0.02
Signal Diode $0.04 (assume dual SOT23)
Zener Diode $0.05
Small signal transistor $0.05
Inverter (74ACT11004) $0.05 (assume 1/6 cost of device)

The totals shown at the bottom of table 7 indicate that the simple resistor divider circuit used to condition TMS370 TTL inputs is the most cost effective using the fewest number of components. Additional cost savings are realized when parts count is minimized. Some of those costs are manufacturing cost (costs to insert extra parts), inventory costs, board space, test time, etc. These costs vary widely between manufacturers and therefore are not reflected in the example above. When multiplying the total costs of input pin conditioning and protection circuitry by the number of input pins implemented in the system, the advantages of TTL specified inputs can clearly be seen.

Table 7. Cost Comparison

Component	TI's TTL input	TTL Input Diode protection	TTL Input Zener Protection	CMOS transistor buffer	CMOS TTL buffer
R1	.01	.01	.01	.01	.01
R2	.01	.01	.01	.01	.01
R3	.01	.01	.01	.01	.01
C1	.03	.03	.03	.03	.03
D1	N/A	.04	N/A	.04	.04
Zener	N/A	N/A	.05	N/A	N/A
Q1	N/A	N/A	N/A	.05	N/A
1/6 74ACT11004	N/A	N/A	N/A	N/A	.05
Totals	$0.06	$0.10	$0.11	$0.15	$0.15

SUMMARY

In summary, microcontroller based automotive systems are exposed to electrically harsh environments. System module specifications dictate areas of electrical conditions that these systems must meet. Two areas of concern to the system designer have been discussed.

1. Input signal conditioning
2. Input transient noise suppression

It has been established that designing with TTL specified input logic levels simplify the circuitry required for the input signal conditioning. Also, designing with microcontrollers with internal protection diodes simplify the circuitry required for input transient suppression.

Designing with TTL levels allows a simple voltage divider network to be used to allow the microcontroller to recognize logic levels across a typical battery operational range of 9V to 18V. Selecting the proper values for this voltage divider network will also allow microcontrollers with internal diodes to adequately protect the input pins from potentially damaging transient noise conditions.

TMS370 family microcontrollers are designed with both these conditions in mind. Using simple passive circuitry provides an immediate system cost savings over active circuitry required by microcontrollers designed with CMOS logic levels or those without internal diode protection circuitry. Additional cost savings associated with board space, part count, inventory costs, and reliability will also be reduced by designing with the fewest and simplest components.

REFERENCES

1. Texas Instruments Inc., TMS370 Family Data Manual, pg. 16-18, 1993.
2. Motorola Corp., MC68HC11E9 Data Sheet, Appendix A, pg. 2, April 1992
3. Motorola Corp., MC68HC05C9 Data Manual, pg. 13-3, March 1992.
4. Phillips Semiconductor Corp.,80C51 Data Sheet, pg. 142. Jan 26, 1993
5. National Semiconductor COP888CF Data Manual, pg. 7, May, 1992

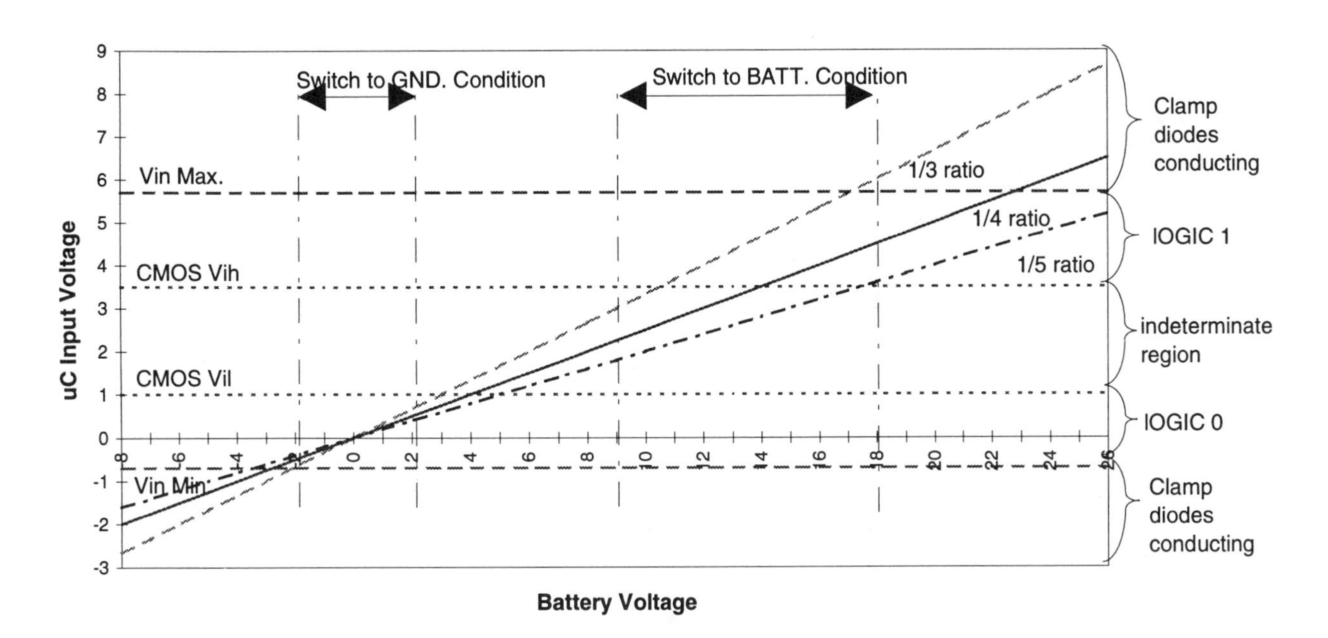

Figure 5. CMOS input levels over variations in V_{BAT}

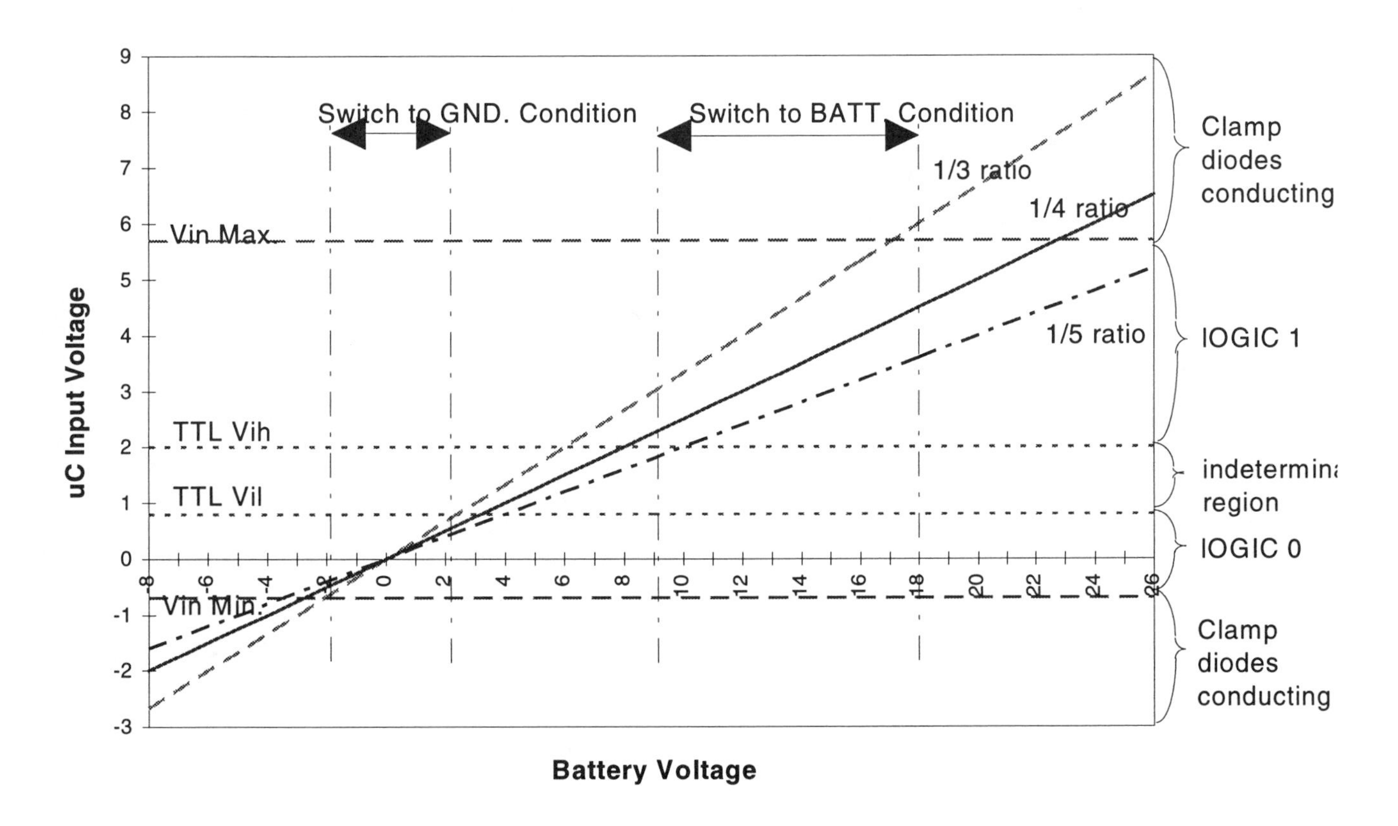

Figure 6. TTL input levels over variations in normal V_{BAT}

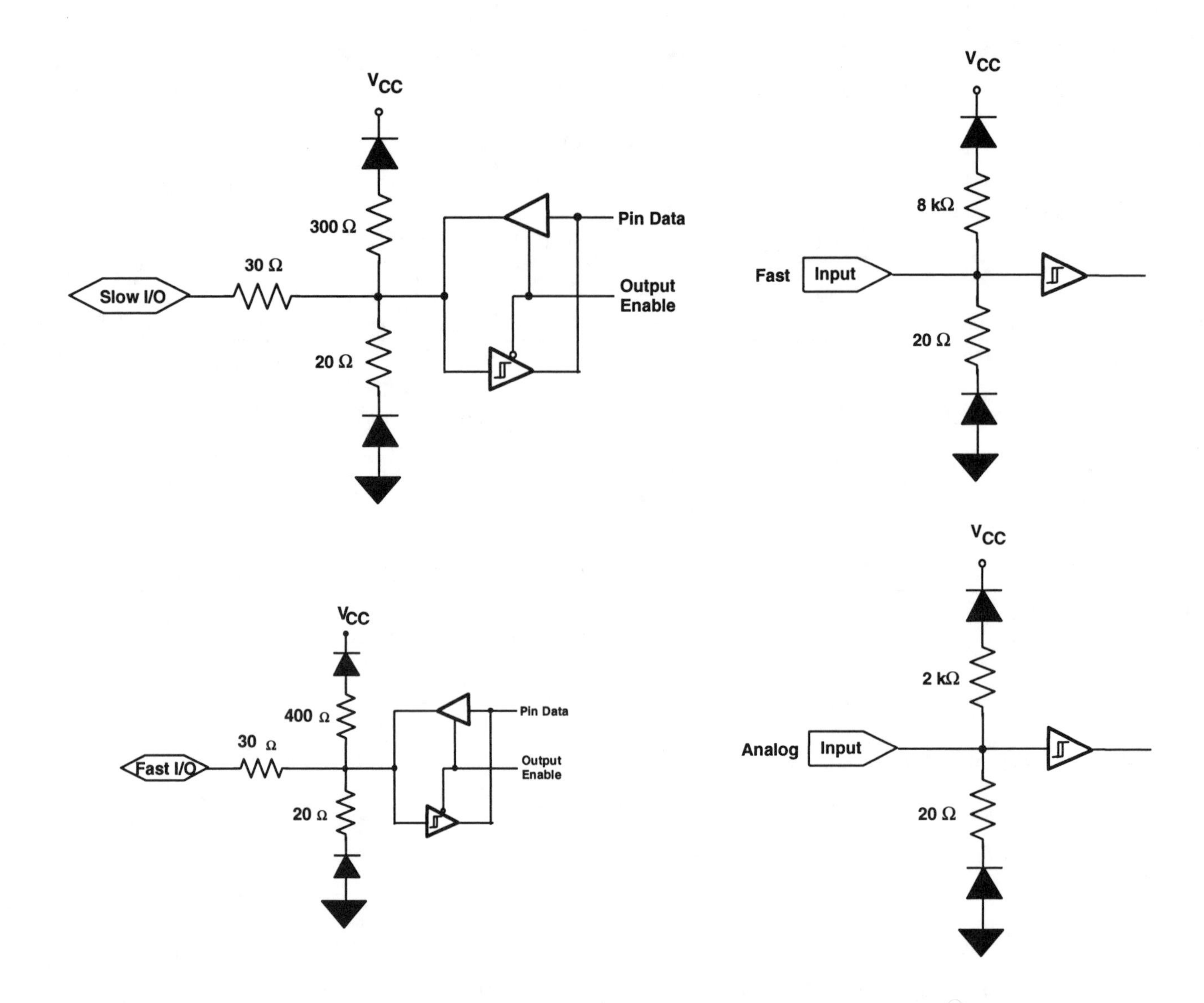

Figure 9. TMS370 Simplified 1.2 micron and 1.6 micron Silicon Buffer Circuitry

Table 6. External Resistance (R2) values for various external transient voltage conditions

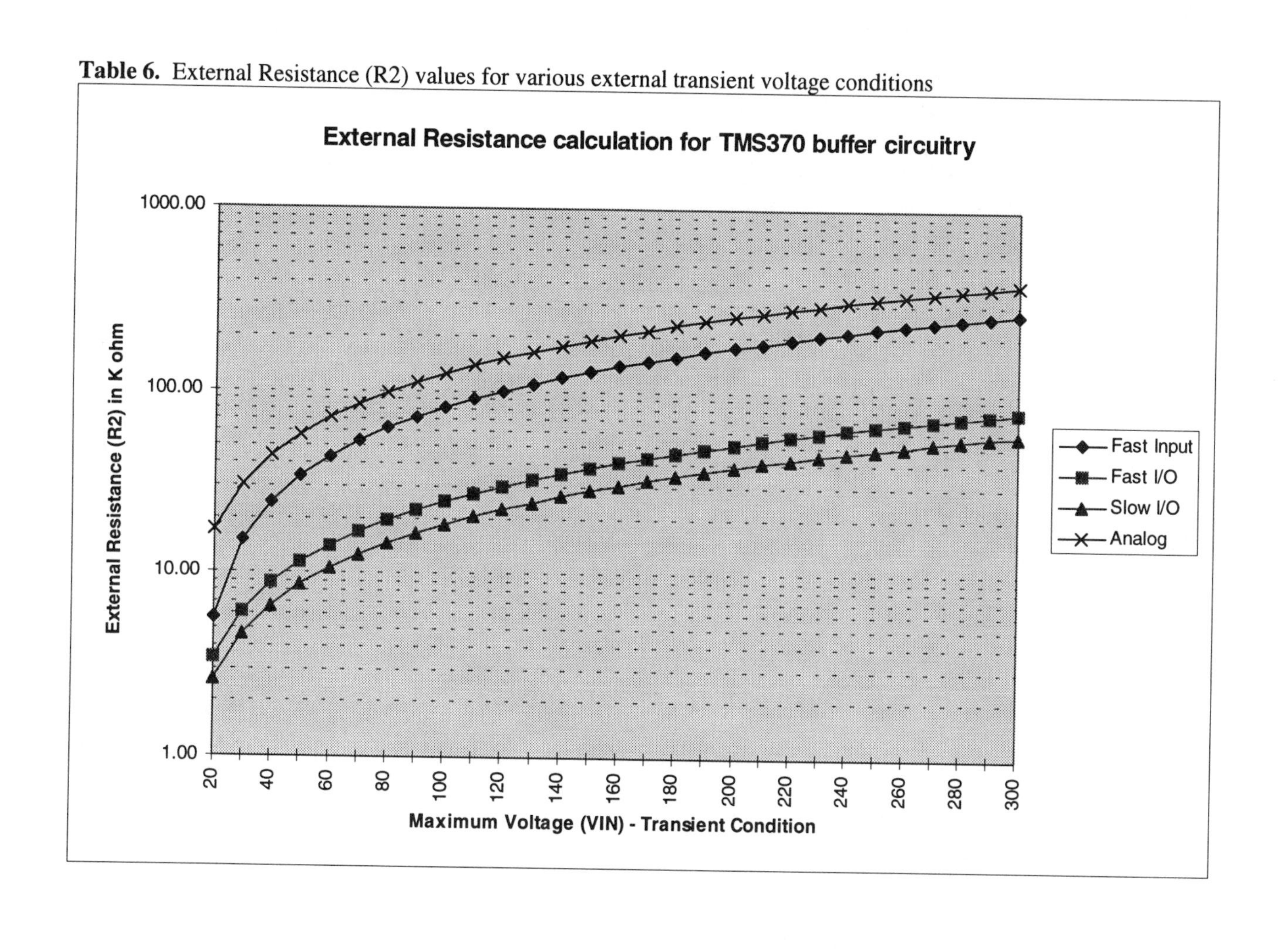

970863

A Single-chip RISC Microcontroller Boarding on MY1998

Akihiko Watanabe, Satoshi Tanaka, Masayoshi Kaneyasu, Michio Asano, and Shrio Baba
Hitachi, Ltd.

ABSTRACT

This paper presents a single-chip 32bit RISC microcontroller boarding on MY1998 dedicated to highly complicated powertrain management. The high performance 32bit RISC CPU provides the only solution to meet requirements of drastic CPU performance enhancement and integration.

Furthermore, a 32bit counter, based on a 20 MHz clock, and a 32bit multiplier make possible misfire detection and precise analysis of the engine management strategy, especially cylinder individual air-fuel ratio control.

INTRODUCTION

In the early 1990's, at single-chip 16bit microcontroller was introduced for engine control in place of an 8bit microprocessor and an external ROM. A historical summary of integration of microcomputers for powertrain control is shown in Figure 1. Several control items, e.g., automatic transmission(AT) control, functions to meet the California Air Resources Board(CARB) requirements for On Board Diagnosis phase II(OBD II), and processing to satisfy emission regulations, were also added in the early 1990's to basic control activities such as fuel injection and ignition timing control. On the other hand, engine control strategy itself has been progressing radically toward higher accuracy and speed to realize lean burn and low exhaust emission [1]. To streamline the integration, these control items must be centralized onto a single-chip microcontroller.

In this paper, first, the CPU performance of the 32bit RISC microcontroller for usual engine management procedures and the application propriety of the developed 32bit RISC microcontroller are described. Secondly, the development of the misfire detection and the cylinder individual air-fuel ratio control using the 32bit RISC microcontroller are described.

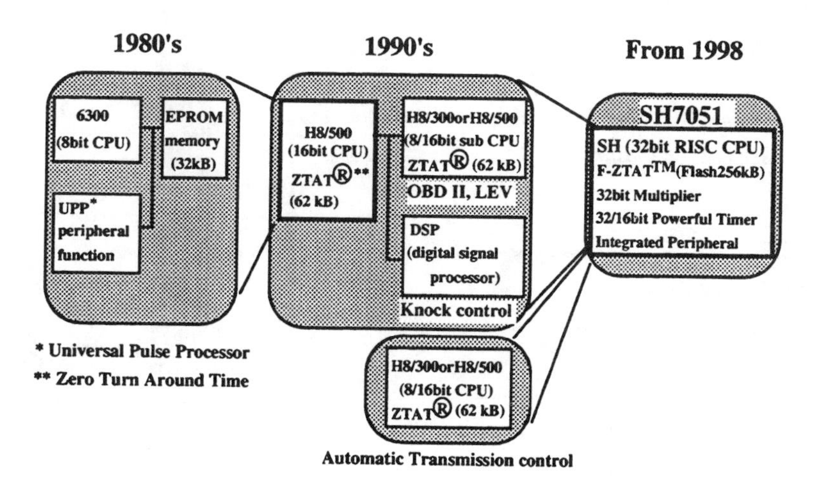

Figure 1. Integration of microcomputers for powertrain control

CPU PERFORMANCE OF 32BIT RISC

Generally, CPU performances are estimated by using standard benchmarks or actual application programs. This time, both were applied. The results of dhrystone benchmark are shown in Figure 2. The SH(32bit RISC CPU) has 10-fold higher performance than the H8/500(conventional standard 16bit CISC CPU).

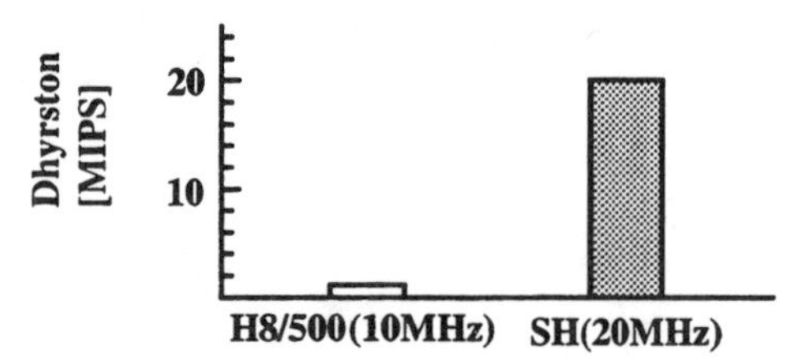

Figure 2. SH and H8/500 performance on Dhrystone benchmark

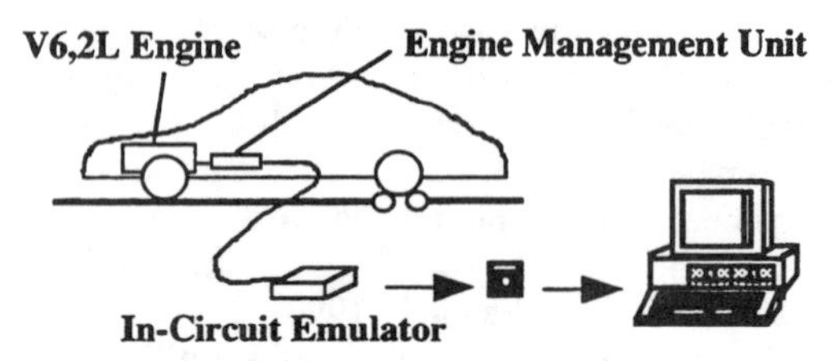

Figure 3. Configuration of the real time trace analysis system

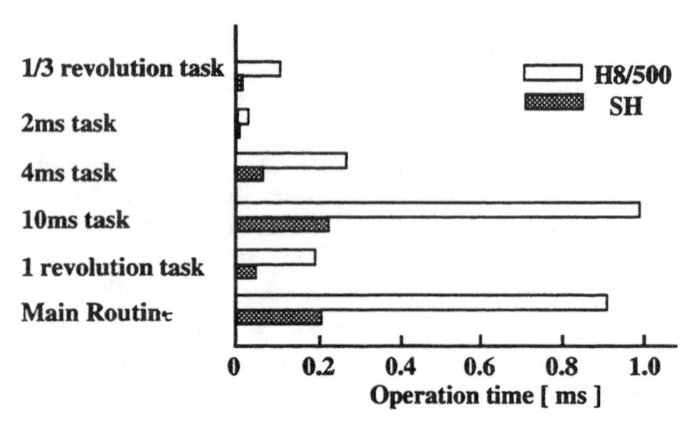

Figure 4. Task operation time of SH and H8/500 on 6-cylinder engine management program

Table 1. Performance comparison between SH and H8/500 on 6-cylinder engine management program

	H8/500(a)	SH(b)	a/b
1/3 revolution task	123 μ s	26 μ s	4.7
2ms task	32 μ s	7 μ s	4.6
4ms task	270 μ s	56 μ s	4.8
10ms task	1,036 μ s	236 μ s	4.4
1 revolution task	190 μ s	48 μ s	4.0
Main Routine	911 μ s	207 μ s	4.4

We also had the conventional V6, 2L engine management program employed for CPU performance estimation. The configuration of the real time trace analysis system is indicated in Figure 3. From the trace data, operation frequencies of tasks were measured. Part of the results are shown in Figure 4 and Table 1. The SH performance was 4 - 5 times higher than that of the H8/500. The program optimization for this SH benchmark, which was not done this time, would lead to more than a 5-fold enhancement. Furthermore, considering the change of the air-fuel ratio control method from a table look-up to a model based modern control approach, the 32bit RISC CPU is more advantageous for next generation powertrain management.

APPLICATION PROPRIETY OF THE DEVELOPED SINGLE-CHIP 32BIT RISC MICROCONTROLLER

The same analysis was adapted to other control items. The burden on 1998 powertrain management is shown in Figure 5. Estimation of knock control was on the premise of frequency analysis method. We concluded that the burden on the H8/500 was 83% for only engine management and 191% for 1998 powertrain management which included other control items' operation times. On the other hand, the total burden on the SH came to 37%. This means that the SH has sufficient performance for 1998 powertrain management.

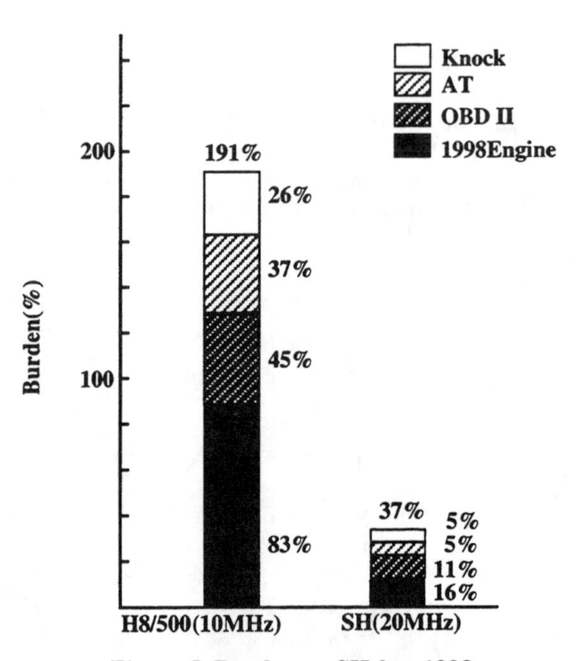

Figure 5. Burden on SH for 1998 powertrain management

Program sizes required for 1998 model vehicle engine management, OBD II and AT were predicted as 60kB, 50kB and 30kB respectively in terms of the H8/500 assembler just in case of an appropriate powertrain control system(Fig. 6). We have experience with code efficiency estimation that the length of the compiled SH C was less than 1.3 times that of the assembled H8/500 for the engine management program. The compiled SH C was expected to be 182kB long in all, which means SH7051 still has a memory allowance of 74kB which can be used for additional control items or supplement of system specifications.

The burden on the CPU and memory capacity indicates the developed 32bit RISC microcontroller can be one of single-chip solutions. Specifications of SH7051 are shown in Table 2.

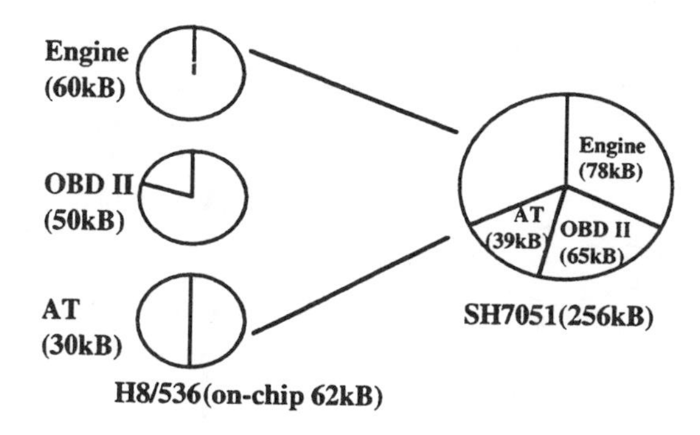

Condition: Compiled SH C = 1.3 x Assembled H8/500

Figure 6. SH7051 memory allowance for 1998 powertrain management

Table 2. Specifications of the SH7050 series

	SH7050 / 7051
CPU	SH
Multiplier	32bit
Operation Frequency	20MHz
Flash ROM	128kB / 256kB
RAM	6kB / 10kB
ATU(Advanced Timer Unit)	32bit IC* 4ch 16bit IC/OC** 18ch 16bit One shot 8ch 16bit PWM 4ch
APC(Advanced Pulse Controller)	Pulse Output 8ch
DMAC(Direct Memory Access Controller)	4ch
A/D Converter	16ch
SCI(Serial Communication Interface)	3ch
Input/Output	118ch
Package	QFP168

* Input Capture ** Input Capture or Output Compare

DEVELOPMENT REGARDING MISFIRE DETECTION[2]

Misfire detection is covered in the OBD II regulations. It is generally realized by several sensing operations, i.e., combustion presssure or knocking. Fluctuation analysis of engine revolution during each firing is another method. Fluctuation in misfiring, however, is usually less than a few percent. Resolution of upper 10 bits is necessary for engine rotation sensing. The 32bit free running counter is shown in Figure 7. It is based on a 20 MHz clock applied to the 12 degree crank angle pulse sensing system for engine speed monitoring. The counter becomes 6,666 per one crank pulse at 6000rpm, which is upper 10 bits resolution. The 20 MHz clock makes the counter sensitive enough to find fluctuation in the engine revolution with high resolution.

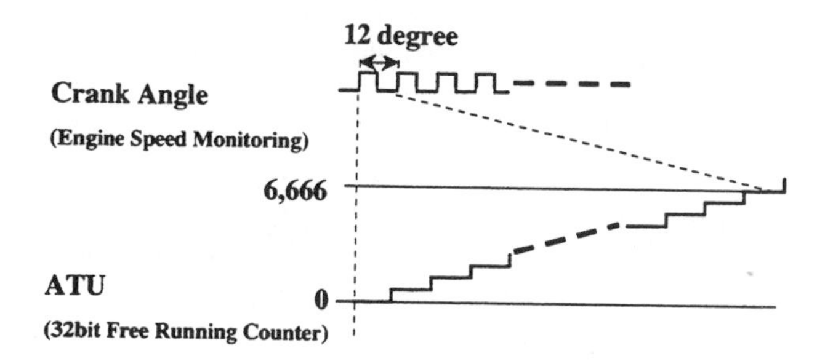

Conditions:Engine Speed = 6,000rpm
Counter Clock=20MHz

Figure 7. SH7051 ATU application for 12 degree crank angle pulse sensing system

The 32bit multiplier, takes 100 to 200ns to execute 32bit multiplication and accumulation, and allows frequency analysis using a Discrete Fourier Transform (DFT) in a short time. The spectrum of the crank angle speed ω_k are transformed to the following sequence of complex Fourier coefficients Ap.

$$Ap = \sum_{k=1}^{K} \omega_k (W_K)^{pk} \quad (1)$$

where

$$W_K = \exp\{2\pi j/K\}$$

In case of misfire detection, sufficiency of computation on the only fundamental Fourier component A1 was presented[2]. The above formula is as follows:

$$A1 = \sum_{k=1}^{K} \omega_k(W_K) = Re + j \cdot Im$$

$$= \sum_{k=1}^{K} \omega_k\{\cos(2\pi k/K)\} + j \cdot \sum_{k=1}^{K} \omega_k\{\sin(2\pi k/K)\} \quad (2)$$

Using the above equation, misfire detection was applied for 6-cylinder engine with magnitude computation and phase classification of A1. First, The fundamental Fourier component A1 is derived from multiplication and accumulation of ω_k and constants respectively as shown in Eq.(2). The items of $\{\cos(2\pi k/K)\}$ and $\{\sin(2\pi k/K)\}$ can be tabulated simply by definition of K. The case of a 12 degree crank angle pulse based measurement is shown in Table 3. In this case, K is 60. The memory necessary for this table is merely 480Bytes, which can be fully located in an on-chip ROM.

Table 3. $\{\cos(2\pi k/60)\}$, $\{\sin(2\pi k/60)\}$

	k = 1	2	3	···	60
$\{\cos(2\pi k/60)\}$	0.994	0.978	0.951	···	1
$\{\sin(2\pi k/60)\}$	0.105	0.208	0.309	···	0

For misfire detection, the magnitude of A1 is calculated and compare with a threshold value.

$$|A1|^2 = Re^2 + Im^2 \quad (3)$$

On the other hand, misfiring cylinder is classified by Re and Im in complex plane described below. The boundary lines to identify and regions of six misfiring cylinders are shown in Figure 8. Misfiring cylinder is defined by the combination of Re and Im on the fundamental Fourier component instead of phase analysis.

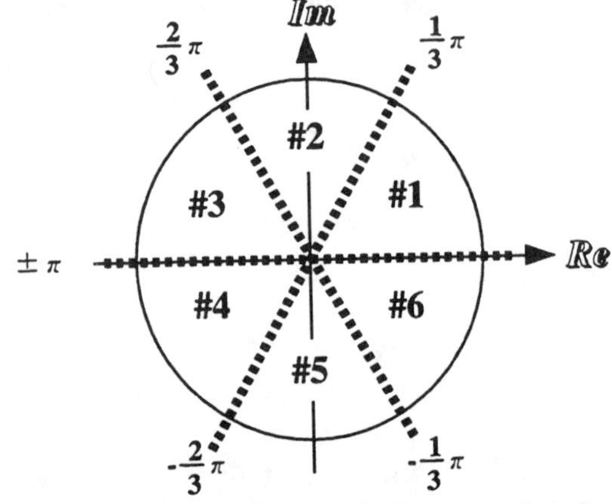

Figure 8. The boundary lines (dotted lines) and regions of six misfiring cylinders

Table 4 provides the conditions for the misfiring cylinder identification. Plus or minus judgements and easy comparison of Re and Im make possible the misfiring cylinder identification. Figure 9 gives an example flowchart.

Table 4. Conditions for misfiring cylinder identification

Conditions	Cylinder #
Re>0, Im>0, $|1.732 \times Re| > |Im|$	#1(0 to $\frac{1}{3}\pi$)
Im>0, $|1.732 \times Re| < |Im|$	#2($\frac{1}{3}\pi$ to $\frac{2}{3}\pi$)
Re<0, $|1.732 \times Re| > |Im|$	#3($\frac{2}{3}\pi$ to π)
Re<0, Im<0, $|1.732 \times Re| > |Im|$	#4($-\pi$ to $-\frac{2}{3}\pi$)
Im<0, $|1.732 \times Re| < |Im|$	#5($-\frac{2}{3}\pi$ to $-\frac{1}{3}\pi$)
Re>0, Im<0, $|1.732 \times Re| > |Im|$	#6($-\frac{1}{3}\pi$ to 2π)

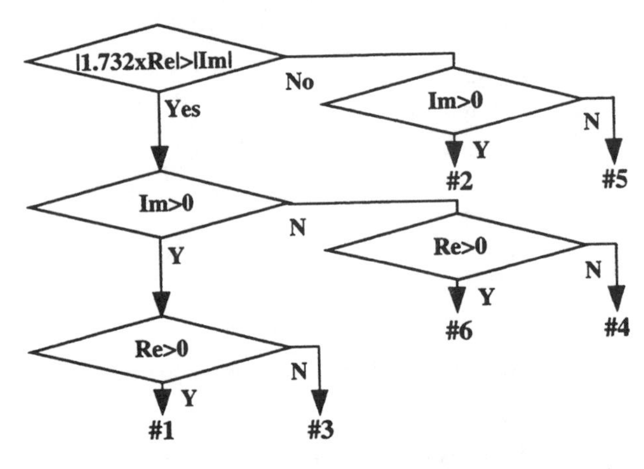

Figure 9. Flowchart on misfiring cylinder identification

COMPUTATION RESULTS The misfire detection program was prepared and operation times were measured(Table 5). Total operation time was 40.2 μs. One engine cycle (2 crankshaft revolution) took 20ms at 6,000rpm. Consequently, the burden on the CPU was 0.2%. The conventional 16bit CISC CPU takes more than 10-fold operation time for even 16bit multiplication and accumulation, and an additional processor, e.g., Digital Signal Processor (DSP) must be necessary for misfire detection. The SH and 32bit multiplier make possible realtime misfire detection.

Table 5. Operation time for misfire detection

Operation	Time
Re calculation	14.8 μs
Im calculation	14.7 μs
$Re^2 + Im^2$ calculation	6.0 μs
Cylinder identification	3.0 μs
Others	1.3 μs
Total	40.2 μs

CPU PERFORMANCE REGARDING CYLINDER INDIVIDUAL AIR-FUEL RATIO CONTROL[1]

The proper fuel amount must be supplied to a combustion cylinder to generate the desired air-fuel ratio within the cylinder under all engine driving conditions. The intake air volume is measured through an air flow meter or a manifold pressure sensor and used to calculate the basic fuel injection amount. The air-fuel ratios in the cylinders are managed to be close to the desired value based on lambda feedback control. However, in a conventional system, there are some problems caused by the response delay of the lambda sensor, transportation lag of the emission in the exhaust manifold, perturbation of the fuel amount injected, and differences of air amount introduced between the cylinders.

The cylinder individual air-fuel ratio control method was developed[1] which applies a state estimator and observer to eliminate differences between air-fuel ratios of each cylinder and desired value.

The operation time to perform cylinder individual air-fuel ratio control was simulated. The configuration of the control system is shown in Figure 10. The air-fuel ratio values(A/F) in cylinders of a L-4 engine are estimated based on the output of air-fuel ratio sensor (AFRS) mounted at the confluence position of the exhaust manifold using the AFRS model and cylinder individual air-fuel ratio estimator.

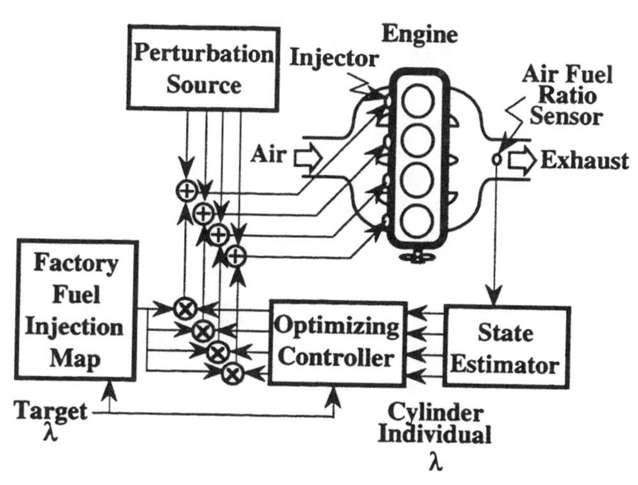

Figure 10. Schematic of cylinder individual air-fuel ratio control using state estimator and optimizing controller

The AFRS model was defined as Eq.(4) with first order lag to estimate actual A/F at the confluence point:

$$A/F(k-1) = t \cdot AF(k) + (1-t) \cdot AF(k-1) \tag{4}$$

where t is the weighting factor that works as the time constant of the first order lag system of the AFRS, k is the sample number and AF(k) means the A/F value at the confluence point that is converted based on the output signal of the AFRS.

The cylinder individual A/F estimator is described in a matrix equations as follows. Eq.(5) is a state equation in the exhaust manifold using state vector X and coefficient matrix A. The output function is defined in Eq.(6)

$$X(k+1) = A \cdot X(k) \tag{5}$$

$$y(k) = C \cdot X(k) \tag{6}$$

where

$$X(k) = \begin{bmatrix} F/A(k-3) \\ F/A(k-2) \\ F/A(k-1) \\ F/A(k) \end{bmatrix}, \quad y(k) = F/A(k) \text{ at confluence point}$$

(F/A means fuel/air ratio)

$$A = \begin{bmatrix} 0 & 1 & 0 & 0 \\ 0 & 0 & 1 & 0 \\ 0 & 0 & 0 & 1 \\ 1 & 0 & 0 & 0 \end{bmatrix}, \quad C = [\,0.05 \;\; 0.15 \;\; 0.30 \;\; 0.50\,]$$

An observer generates adaptive gain vector K to determine the fuel injection amounts for every fuel injector mounted around the suction port, based on air-fuel ratio values generated individually for a cylinder by the AFRS model and state estimator. The gain vector K is calculated in Eq.(7)

$$K = PC^{T}R^{-1} \tag{7}$$

where matrix P must be the unique positive solution that satisfies the following Riccati equation. An off-line calculation is done to determine the matrix P.

$$PA^{T} + AP - PC^{T}R^{-1}CP + Q = O \tag{8}$$

$$R = 1, \quad Q = \begin{bmatrix} 100 & 0 & 0 & 0 \\ 0 & 100 & 0 & 0 \\ 0 & 0 & 100 & 0 \\ 0 & 0 & 0 & 100 \end{bmatrix}$$

The elements of the weighting matrix Q are set at the above values to execute the state estimation in real time. The gain matrix K is calculated as follows:

$$K = \begin{bmatrix} 0.0436 \\ 0.2822 \\ 1.8283 \\ -0.2822 \end{bmatrix} \tag{9}$$

Eq.(10) is the observer to determine the A/F values of a cylinder individually. The state vector X(k) is fixed from the estimated A/F value y(k-1) and the former state vector X(k-1):

$$X(k) = (A - KC) \cdot X(k-1) + K \cdot y(k-1) \qquad (10)$$

where the 4x4 matrix (A-KC) is fixed because of the constant gains of the vector K.

The estimated A/F values of each cylinder were got using the observer for use in PID control for each cylinder. This cylinder individual air-fuel ratio control generated smaller A/F deviations between cylinders than the conventional lambda feedback control system.

COMPUTATION RESULT Table 6 shows the operation times needed for the cylinder individual air-fuel ratio control. Total operation time was 46.3 μ s. One firing engine cycle (half crankshaft revolution) took 5ms at 6,000rpm in a L-4 engine. Therefore, the burden on the SH was 0.96%, which means the SH and 32bit multiplier achieve the desired cylinder individual air-fuel ratio control.

Table 6. Operation times for cylinder individual air-fuel ratio control

Operation	Time
A/F estimation based on AFRS model	4.1 μ s
Cylinder individual A/F estimation	42.2 μ s
Total	46.3 μ s

CONCLUSIONS

A new single-chip 32bit RISC microcontroller was developed. Its performance was demonstrated by real time trace analysis of the engine management, computing the misfiring cylinder identification, and operating the cylinder individual air-fuel ratio control. The SH performance was 5 times higher than that of a 16bit CISC CPU .

The SH made the misfiring cylinder identification in 40.2 μ s , which meant that the burden on the SH was 0.2%. The operation time to perform the cylinder individual air-fuel ratio control was 46.3 μ s, and the burden on the SH was 0.96%.

The foregoing results show that the new single-chip 32bit RISC microcontroller provides the only solution to meet the requirements of next generation powertrain management.

ACKNOWLEDGMENTS

The high precision knock detection method was studied using real-time frequency analysis based on the SH and a piezoelectric accelerometer mounted on a V-6 engine. Knock indices were derived arithmetically for each cylinder using resonant frequency components from engine block vibrations [3]. The knock detecting program, which utilizes the multi-spectrum method, was evaluated through digital signal processing. The program worked well at high engine speeds in real time and made knock control possible throughout the entire engine operation range for all V-6 engine cylinders. The burden in case of introducing the multi-spectrum knock detection method on the SH was 5%.

The authors are grateful to Shigeki Morinaga, Junichi Ishii, Mitsuru Watabe, Mamoru Ohba, and Rika Minami for their help in building the real time trace analysis and useful discussions. Thanks are also expressed to Kiyoshi Matsubara and Fumio Tsuchiya for development of the SH7050 series.

REFERENCES

[1] Y. Hasegawa et al.,"Individual Cylinder Air-Fuel Ratio Feedback Control Using an Observer", SAE Paper No.940376.

[2] W.B. Ribbens and J. Park,"Road Tests of a Misfire Detection System", SAE Paper No.940975.

[3] M. Kaneyasu,"Engine Knock Detection Using Multi-Spectrum Method", SAE Paper No.920702.

980202

Floating-Point Number for Automotive Control Systems

Yoji Yamada
Mazda Motor Corporation

ABSTRACT

The software of automotive control systems has become increasingly large and complex. High level languages (primarily C) and the compilers become more important to reduce coding time. Most compilers represent real number in the floating-point format specified by IEEE standard 754. Most microprocessors in the automotive industry have no hardware for the operations using the IEEE standard due to the cost limitation, resulting in the slow execution speed and large code size. An alternative format to increase execution speed and reduce code size is proposed. Experimental results for the alternative format show the improvement in execution speed and code size.

INTRODUCTION

In recent years, the software of automotive control systems has become increasingly large and complex to meet the several requirements for system functionality and performance. It is very difficult to develop large program, especially for RISC, in assembly language. High level languages (primarily C) become more important and more realistic for development of such large control program. For high-level language program, a software, called compiler, is necessary and very important to generate high performance executable program. However, in general, compiler-generated program is not so fast as the program written by programmer since compilers on market which target many kinds of applications can not optimize program depending on the feature of the program. Especially, the execution of real number computation in the compiler-generated program is very slow since most compilers use very expensive IEEE standard format as real number representation.

Although some technique to improve performance by changing data structure depending on program features has been researched for numerical computation[1], there are very few approaches for control systems. This paper discusses the features of the IEEE standard from the viewpoint of automotive control systems. Alternative formats to increase execution speed and reduce code size are also discussed.

REPRESENTATION OF REAL NUMBER

Most automotive control systems use real numbers that are not integers. Real numbers that are not integers can be represented by two major methods in computer system. One is fixed-point representation and the other is floating-point representation.

In the fixed point representation, the binary point is imagined somewhere in the representation. Since the information about the location of the binary point is not included in the representation itself, programmers have to keep the information on record somewhere. The operations of fixed point numbers can be done using integer ALU (Arithmetic and Logical Unit) and general purpose registers, enabling the execution to be done in the same manner as integer operations. However, the range of fixed point number is very small like integer number.

In the floating point representation, a computer word is divided into two parts: an exponent and a mantissa[2]. Floating-point numbers greatly increase the range obtainable using the two parts. Since operations using floating-point numbers are much complicated than those using fixed-point numbers, most computers in office use special hardware called floating-point unit (FPU) for high-speed execution.

Most micro controllers on vehicle have no FPU due to the cost limitation. Therefore, fixed-point numbers are used in the control program. Physical numbers are converted into fixed-point numbers using some rule which is dependent on each physical number. The conversion rule for each fixed-point number must be kept on record somewhere. Furthermore, since the range of fixed-point number is very small, programmer must carefully control each binary point of all fixed-point numbers in program to perform operations in order to avoid overflow and underflow. These extra management not only reduces the productivity of software but also affects the reliability of software. It is desirable for programmers to design software using physical numbers without care about conversion, overflow and underflow.

An approach to resolve the fixed-point design issues is to use floating-point numbers. If floating-point numbers are used for automotive control systems, no more care is necessary for conversion, overflow and underflow. A standard format for floating-point numbers, known as the IEEE 754 standard [3][4][5], has been widely adopted in computer industry. Figure 1 shows the format, representation and characteristic of the single (32-bit) IEEE format. Most C compilers on market adopt IEEE standard format as floating-point representation.

IEEE format is very sophisticated and very effective in most applications, especially for numerical computation. However, it is somewhat complicated due to the sophistication. For example, the IEEE standard supports various unusual numbers such as overflow and NaN (Not a number), and use hidden leading bit in mantissa. Some extra processes are necessary for these features. On most office computers, these extra processes are done by special hardware or FPU. Since most automotive controllers have no FPU, these extra processes must be done by software in addition to actual operation such as add or subtraction. In other words, IEEE format is very expensive for the execution by software.

Table 1 shows the approximate execution cycles of primitive operations on a microprocessor without FPU using fixed-point numbers and IEEE floating-point numbers. In addition to these execution cycles, actual operations in program involve the overhead of function calls to operation library, requiring extra execution cycles.

Table 2 shows the approximate code sizes of operations using fixed-point numbers and IEEE floating-point numbers. These tables show the difficulties in the use of floating-point numbers for automotive control programs. A software-oriented floating-point format is necessary for microprocessors of automobiles in order to speed up execution cycles and reduce code sizes.

(a) Format

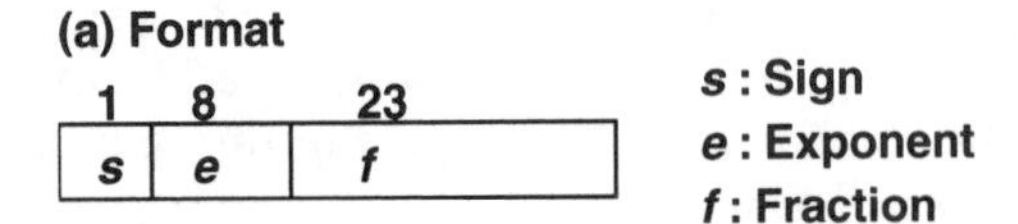

s : Sign
e : Exponent
f : Fraction

(b) Representation

e	*f*	Represents
0	0	0
Not 0	Not 0	$1.f \times 2^{e-127}$
0	Not 0	$0.f \times 2^{-126}$ ...underflow
All 1's	0	Overflow
All 1's	Not 0	Not number (0/0, etc.)

(c) Characteristics

Precision	24 bits
Decimal range	approx. 10^{-38} to 10^{38}
Denormalized smallest	approx. 10^{-45}

Figure 1: IEEE Floating-point Format (Single)

Table 1: Execution Cycles for Primitive Operations

Operation	Fixed-Point Number	IEEE Floating-Point Number
Compare	1	more than 20
Add / Subtract	1	more than 50
Multiply	2 ~ 4	more than 80
Divide*	approx. 70	more than 130

Note * : Use 1-bit divide instruction

Table 2: Code Sizes for Primitive Operations

Operation	Fixed-Point Number	IEEE Floating-Point Number
Compare	1	approx. 70
Add / Subtract	1	approx. 290
Multiply	1	approx. 180
Divide*	approx. 70	approx. 210

Note *: Use 1-bit divide instruction

FLOATING-POINT FORMAT FOR AUTOMOTIVE CONTROL SYSTEMS

There are some approaches to speed up operations for floating-point numbers. One of the approaches is to modify the floating-point format. According to the computational features of automotive control system, new floating-point format is proposed.

SIMPLIFICATION OF IEEE FORMAT

The first approach to speed up floating-point operations is to simplify the format based on the analysis of the computational features of control programs.

No Representation for NaN, Overflow and Underflow

As one of the features of IEEE standard format, the format has special representation for NaN (Not a Number) and overflow. However, it is rare that these special cases happen in control system. Figure 2 shows the decimal exponents of run-time operands for primitive operations of an engine control program. Unlike numerical computation, automotive control program uses limited and expected range of numbers. Unless there is a program mistake or system failure, the special cases such as NaN or overflow never happen. Even if the special cases happen, it is not necessary to continue the operations using the special representations, but enough to execute failure-mode routine instead. These special representations can be omitted for control systems.

Another special representation is underflow. However, unlike numerical computation, the precision of numbers which are close to zero is not so significant for normal control systems. Underflow can be regarded as zero.

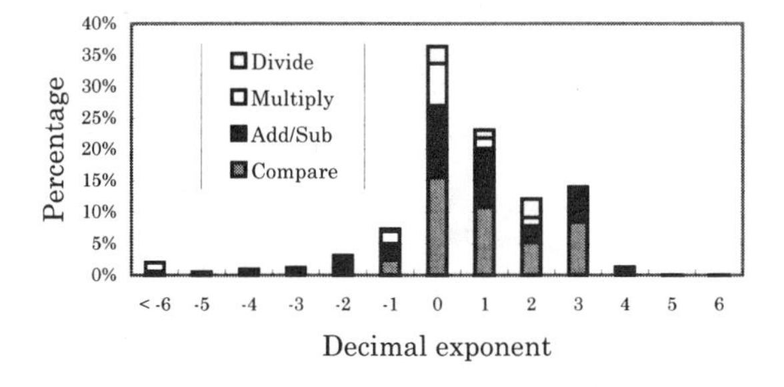

Figure 2: Distribution of Decimal Exponents

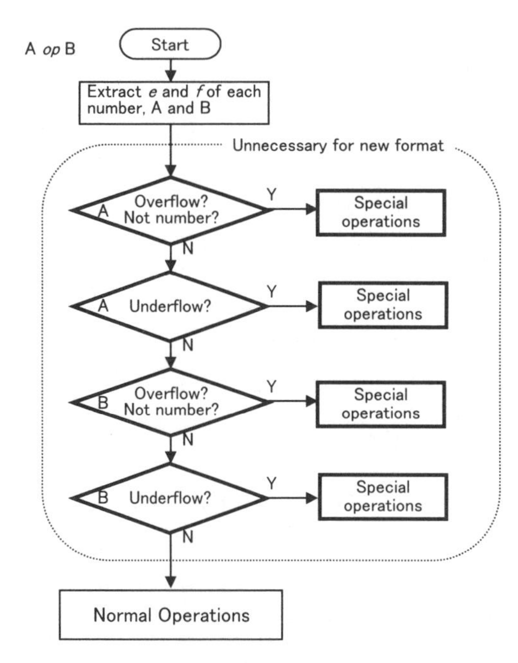

Figure 3: Extra Process for Special Representations

In IEEE format, it is necessary to distinguish the special representations from normal numbers before actual operations since special routines are used for the special representations. If floating-point format has no special representations for NaN, overflow and underflow, extra execution cycles and code for these special cases can be omitted, resulting in the speedup of operations for normal operands. (Figure 3)

Unification of Zero Representations

Another complicated matter is the representation of zero. In the IEEE standard format, zero can be represented in two ways: Positive zero and negative zero. Although it may be important from the view point of mathematics to distinguish negative zero from positive zero, two representations for zero make the operations complicated. For example, the operation to compare positive zero with negative zero should return true in control system. However, in the IEEE standard, the representation of positive zero is different from that of negative zero. Single compare instruction using integer ALU can not get the correct result for this compare operation. Another example is the operation to compare a variable with zero. In this operation, it is necessary to compare the variable with both positive zero and negative zero. For automotive control system, it is not necessary to distinguish positive zero and negative zero. By using single zero, the extra process due to two zeros is not necessary any more. (Figure 4)

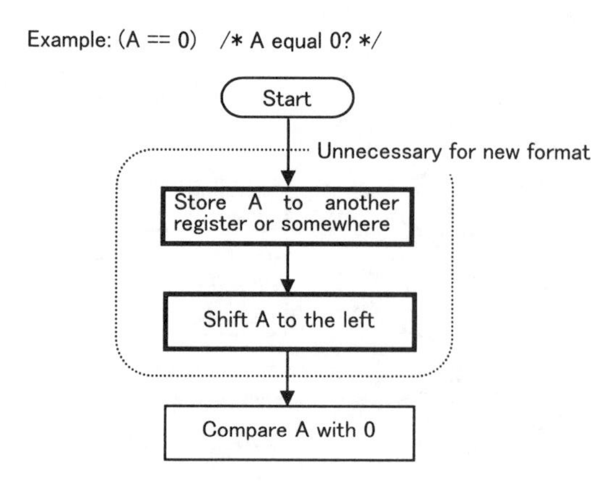

Figure 4: Extra Process for Two Zeros

Explicit Leading Bit of Mantissa

Another feature of IEEE format is the representation of mantissa. In the IEEE standard, the complete mantissa, called significand, is actually 1.M, where the 1 to the left of the binary point is an implicit or hidden leading bit which is not stored in the word. Therefore, although the number of mantissa bits for single precision is 23, the precision of actual mantissa or significand is effectively increased by 1 bit to 24 bits. This is very sophisticated, but complicated for operations by software. Before actual operation, the siginificand of each operand must be recovered by combining the hidden bit and mantissa, and after operation, the leading bit must be removed from the significand of the result. These processes need extra execution cycles as shown in Figure 5. In order to speed up the execution of operations by software, it is better to remove the extra process by using explicit leading bit rather than implicit leading bit. Using explicit leading bit for mantissa causes the one bit reduction of exponent bits, which is acceptable for current automotive control systems.

MONOTONOUS INCREASE OF FORMAT

The IEEE standard uses sign magnitude representation. The bits for exponent and mantissa are assigned in a word in this order. For positive numbers, a floating-point number with larger exponent is always larger than another floating-point number with smaller exponent. For numbers with the same exponent, it is also obvious that a number with larger mantissa is larger than another number with smaller mantissa. Therefore, the comparison of two positive floating-point numbers can be done using single integer compare operation. In other words, positive floating-point representation increase monotonously as shown in Figure 6-a. However, for negative numbers, the magnitude of floating-point numbers is reversed, meaning monotonous decrease of representation. Therefore, it is necessary to use different compare algorithm for negative numbers from that for positive numbers, causing extra execution cycles for compare operation.

If negative representation increases monotonously like positive numbers as shown in Figure 6-b, it is possible to use the same

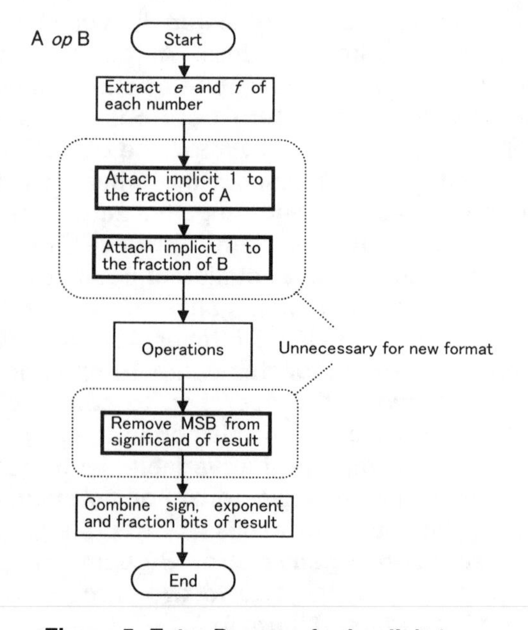

Figure 5: Extra Process for Implicit 1

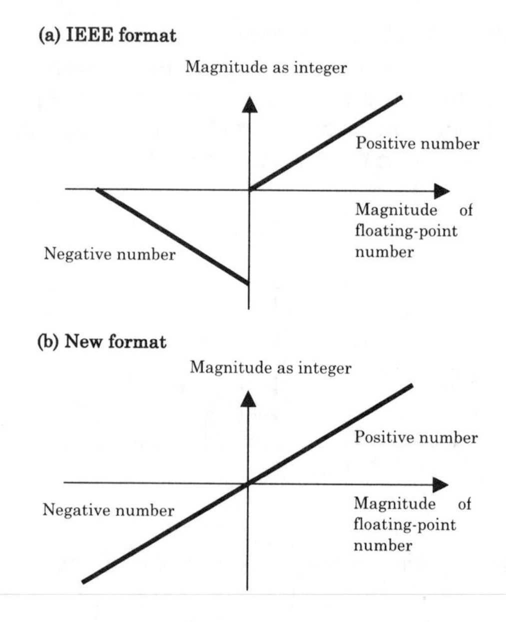

Figure 6: Magnitude of Floating-point Number

compare operation for both positive and negative numbers. Since new format uses single zero representation as described in previous section, it continuously increase from negative number to positive number. This means that the compare operation for new floating-point format can be executed using single compare instruction for integers.

NORMALIZATION BY MULTI BITS

One of the causes which make floating-point operations complicated is normalization. Normalization is done by shifting mantissa so that the MSB (Most Significant Bit) of mantissa becomes 1 in order to keep the same precision for all floating-point numbers. Since it is complicated to find the left-most 1 in mantissa by software, normalization takes several execution cycles. For example, five comparisons are necessary to find the left-most 1 in 32-bit mantissa using binary search.

Another issue due to normalization appears for add operations. It is necessary to equalize the exponents of two operands for addition if they are different. Since the IEEE standard normalizes every one bit, normalized numbers unlikely have the same exponents. Therefore, some extra process to equalize the exponents is necessary for add operations.

A solution to the issues above is to normalize every multi-bits rather than just one bit. It is necessary for this method to represent the MSB of mantissa explicitly, which is proposed in previous section. If the normalization is done every multi-bits, the process for normalization can be simplified. For example, if mantissa is normalized every 8-bits, only two comparisons are necessary for 32-bit mantissa. Also, since more numbers can be represented with the same exponent, operands for add operations likely have the same exponent with higher possibility, eliminating the extra process for equalization of the exponents.

IMPLEMENTATION OF NEW FORMAT IN C COMPILER

It is necessary to implement new floating point format and the operations in C compiler. Preprocessor [6] is one of the easiest ways to implement new floating-point format in an existing C compiler. The preprocessor reads original C source program, then generates new C program after replacing the IEEE floating-point format with new floating-point format. The preprocessed C programs can be converted into object modules using an existing C compiler. To generate load module, the object modules need to be linked with library functions which are newly created for operations of new floating-point numbers. The flow of processing code is shown in Figure 7.

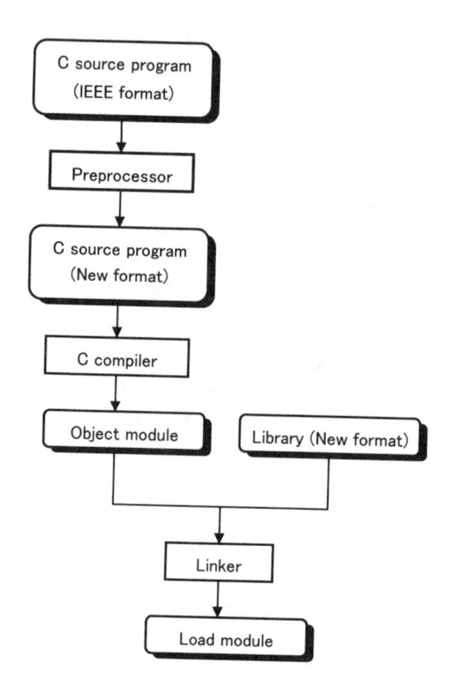

Figure 7: Code Process Flow

EXPERIMENTAL RESULTS

In order to show the effectiveness of new floating-point format, experimental results of execution time and code size are presented for primitive operations in Figure 8 and Figure 9 respectively. It is desirable to measure the actual execution time of each primitive operation while real control program is running on electronic control unit. However, it is not realistic since some interruptions may happen during operations. Therefore, in the experiment, run-time operands of floating-point numbers have been gathered when primitive operations were called in actual control program running on an electronic control unit. Then, using these numbers, test program which includes primitive operations was ran on emulator, and the execution time of each primitive operation was measured. The micro controller used for the experiment includes RISC style processor with multiplier.

Figure 8 shows the speedup of the execution time of operations using newly proposed format compared with that using the IEEE standard format. The compare operation using new format is approximately 20 times faster than that using IEEE format. The compare operation for new format can be executed by single instruction of integer ALU just like compare operation for fixed-pointer numbers. Figure 10 shows the frequency of primitive operations in an engine control program. Since current control system often uses maps as control parameters, many compare operations are performed in control program. Therefore, the large speedup for compare operations is significant. The speedup for add operations is approximately four. This is mainly due to the normalization by multi-bits.

Figure 9 presents the code size of operations for the IEEE standard format and newly proposed format. The code size of compare operation for new format is minimized because of the same reason for the speedup above. Since the compare operation is performed using single instruction, the library function for the compare operation is not necessary anymore and can be in-lined in the caller programs. Therefore, it is possible to reduce the execution cycles for the function call. The code size of add operation for new format is approximately halved.

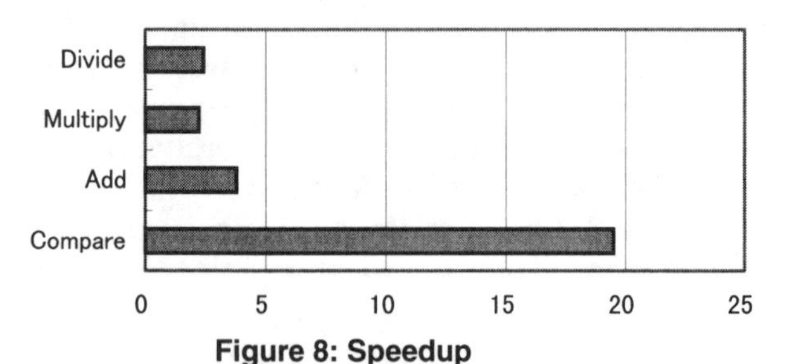

Figure 8: Speedup

Figure 9: Code Size (Byte)

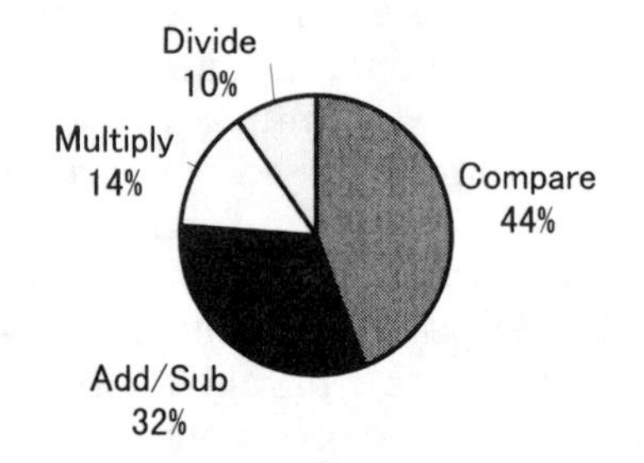

Figure 10: Frequency of Operations

CONCLUSIONS

New floating-point format for automotive controller which has no special hardware for floating-point operations has been proposed to speed up the execution of operations for floating-point numbers in control systems. Simplification such as explicit leading bit of mantissa rather than the hidden bit of the IEEE standard improves the performance of primitive operations in control programs. Monotonous increase of new format makes it possible to execute *compare* operation for floating-point numbers using single integer compare instruction. Normalization by multi-bits improves the execution speed of operations, especially for *add* operation.

CONTACT

Yoji Yamada <yamada.y@lab.mazda.co.jp>

REFERENCES

[1] Y. Yamada, Ph.D. thesis, Department of Computer Science, University of Illinois at Urbana-Champaign, 1995.

[2] S. Waser and M. Flynn, "Introduction To Arithmetic For Digital Systems Designers," CBS College Publishing, 1982.

[3] J. Hennessy and D. Patterson, "Computer Architecture A Quantitative Approach, Second Edition," Morgan Kaufmann, 1996.

[4] IEEE, "IEEE standard for binary floating-point arithmetic," SIGPLAN Notices 22:2, pp.9-25, 1985.

[5] A. Tanenbaum, "Structured Computer Organization, Third Edition," Prentice Hall, 1990.

[6] A. Aho, R. Sethi, and J. Ullman, "Compilers: Principles, Techniques, and Tools," Addison-Wesley, 1986.

980345

Integration of Simulation and Testing for Microelectronics Package Reliability Improvement

Quan Li, David Dougherty and Yong Li Xu
Semiconductor Products Sector (SPS), Motorola Inc.

ABSTRACT

Computer simulation has been used as a vital and powerful tool for evaluating stresses in microelectronics packages due to thermal-mechanical loading. Experimental measurement and reliability testing have been performed for modeling correlation and verification. In the past several years, the authors have been integrating computer simulation and testing to significantly improve package reliability at Motorola SPS. Several examples are presented for illustration and demonstration of the methodology.

INTRODUCTION

Microelectronics package reliability is becoming increasingly important when the trend for smaller packages to deliver more power continues. In general, package reliability issues are always associated with the stress in the package. For more than a decade, FEA (Finite Element Analysis) has been widely used for evaluating the thermal-mechanical stresses in the microelectronics packages[1-2]. With FEA, one simulates a physical system with a discretized mathematical model, or the so-called elements. All the elements are then assembled together to represent to the whole system in terms of internal energy, which turns to be an integral equation. One obtains the solution by minimizing the functional. FEA has been recognized to be a very powerful and effective method for assessing behaviors of many physical systems. However, in the process, several factors can significantly effect the simulation results, e.g., geometrical representation of the model, material properties, boundary conditions, and loading characteristics, etc. This is why every one of the commercial FEA software vendors has disclaimed its responsibility for the accuracy of any analysis results. Therefore, confirming the accuracy of any analysis or solution is the responsibility of the licensees or users.

Some work has been published in the area of correlation of simulation and experimental measurement[3-5]. At Motorola SPS (Semiconductor Products Sector), simulation and experimental measurement & testing have been working hand-in-hand on new product development (NPD) as well as reliability improvement of existing products. The methodology is presented in this paper along with several examples for illustration and demonstration.

CHALLENGES IN MICROELECTRONICS PACKAGE SIMULATION

Microelectronics packages are comprised of dissimilar materials with different CTE's (coefficients of thermal expansion). In general, a package has to go through several temperature excursions during it's manufacturing processes. In addition, it has to withstand some severe thermal loading conditions due to the required reliability tests, e.g., thermal and power cycle tests, high temperature storage, and thermal shock, etc. Finally, it has to withstand the thermal loading on the customer side during board mounting, power cycle, burn-in test, and thermal cycle, etc. Due to the CTE mismatch of different materials in the package, significant stress develops during various thermal loading conditions. The major task for package reliability simulation is to evaluate the stress in the package through out all the processes. By the way, the method of finite element analysis (FEA) is a very powerful simulation tool and thus most widely used. In order to achieve accurate simulation results, the FEA model has to closely represent the physical model and processes. All of these present several challenges to package reliability simulation to be summarized as the follows.

1. MANUFACTURING PROCESSES

Typical manufacturing processes (or package assembly processes) include die attach, interconnecting, and encapsulation. These are sequential processes. During each process, material may be added or removed. The FEA model should be able to simulate these sequential manufacturing processes.

2. GEOMETRICAL VARIATION

Firstly, some critical components in the package may have some very small dimensions causing significant element distortion and high element aspect ratio. Secondly, some of the dimensions are very difficult to control during the manufacturing processes. One typical example is the solder thickness. The die attach solder thickness may be designed to be 0.0010" in theory. But in reality, it can vary somewhere from 0.00015" to 0.00150" resulting in significant variation in die stresses. Therefore, a good FEA model has to take this variation into consideration.

3. MATERIAL PROPERTIES

Material property is the most critical issue of all. Due to historical reason and rapid development of material technology for microelectronics package, complete material property data seem to be always difficult to get. Some frequent requirements for material properties are temperature dependency, elasto-plasticity, and other non-linearity. In general, a non-linear stress FEA model is required for microelectronics package simulation.

Important note: the CTE given by a typical material database (e.g., CINDAS Database) is of either instantaneous CTE or based on a preset reference temperature (e.g., room temperature of 20 °C or simply 0°C). They have to be converted into the right format for a specific reference temperature, i.e., stress-free temperature.

4. LOADING

As mentioned earlier, the thermal loading is sequential. In many cases, the thermal loading can be represented by a uniform temperature excursion applied to the system. in some cases, a transient thermal loading needs to be considered, e.g., the thermal loading during a power cycle. Note also that the thermal loading is path dependent in case of elasto-plastic deformation.

It is also worth while to point it out that in many cases, there are more than one stress-free temperature or reference temperature.

5. OTHER CHALLENGES

There are some other challenges, e.g., interfacial characteristics, stress singularity along the boundary of an interface of two dissimilar materials[6], material creep, visco-elasticity, and visco-plasticity, intrinsic stress, etc. We may not have answers for all the challenges yet. However, these issues are some times critical to certain reliability problems and need to be accurately simulated .

6. REMARKS

Several advanced modeling options and techniques have been practiced at Motorola SPS to achieve simulation accuracy and to meet above challenges. The following are some examples.

- Element birth and death options are used to simulate material removal and addition during assembly process of microelectronics packages. These options are particularly useful in capturing the effect of sequential thermal loading during package assembly, reliability testing, and board mounting processes. These options are readily available with several commercial FEA codes, e.g., ANSYS and ABAQUS.

- Non-linear structural modeling techniques are used to address material eleasto-plasticity, visco-elasticity, visco-plasticity, etc. These material behaviors are critical for many microelectronics package materials.

- Submodeling and substructuring for effectively simulating the small dimension critical features of the system[7].

Note that package simulation is not a turn-key operation as many people may think. Even with the most sophisticated modeling techniques, one still needs to correlate the modeling results with experimental measurement or reliability testing. In the following several sections, some successful examples are presented.

ACCELEROMETER PACKAGE IMPROVEMENT VIA MODAL ANALYSIS AND TESTING

PROTOTYPE DESIGN

The Motorola micro-machined 50G accelerometer sensor and control chips are packaged in a SIP (Single In-line Package) unit. The prototype package consists of three major components, i.e., the SIP unit, the adhesive, and the plastic saddle, as shown in Figure 1. During the board assembly, the seven 0.01" thick Alloy 42 leads are plugged into the through-holes and soldered to the PCB. Note that the two snap-fit arms and four legs or "standoffs" on the plastic saddle enclosing the SIP unit are meant to provide adequate stiffness to the SIP unit as well as to position it perpendicular to the PCB. Since this accelerometer is a critical component in the airbag safety system, the fundamental resonant frequency of the assembled package is required to be above 2,000 HZ. In another word, any resonant frequency

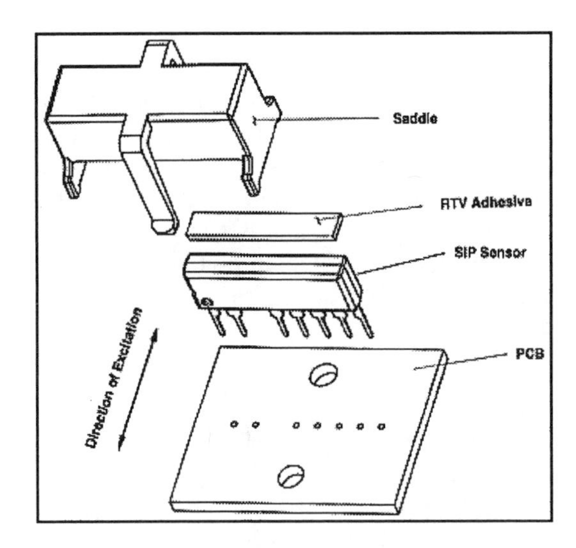

Figure 1. Motorola 50G accelerometer.

below 2,000 HZ could be detrimental to the sensor function since an erroneous sensor response may occur at the resonant frequency. Meantime, the accuracy of the sensor axis is equally important for this accelerometer to function correctly. With the prototype saddle design, the sensor axis was accurately maintained during the wave soldering process for board assembly. However, the resonance requirement was not met. During the electrical test of the prototype parts, a strong electrical signal at was observed on both SIP unit and the complete assembly. This presented a very serious reliability issue with the prototype design of this 50G accelerometer package. Subsequently, an extensive investigation and re-design effort was launched.

INVESTIGATION OF THE VIBRATORY BEHAVIOR

Under a great time pressure, the development team recognized the need for parallel efforts in simulation and experimental measurement. First, a modal analysis was conducted using ANSYS, a general purpose finite element analysis (FEA) computer software code, on the current prototype package design to understand its modal characteristics. Meantime, a modal test via time average holographic interferometry[5,8] was conducted to measure the resonant frequencies and the vibratory mode shapes of the prototype parts.

SIP Unit

First, the SIP unit was investigated. Since the seven leads on the SIP unit provide the basic support for the package and the early electrical testing was conducted on the SIP unit only, the modal analysis of the SIP unit should be a good starting point. The FEA model for the SIP unit is as shown in Figure 2.

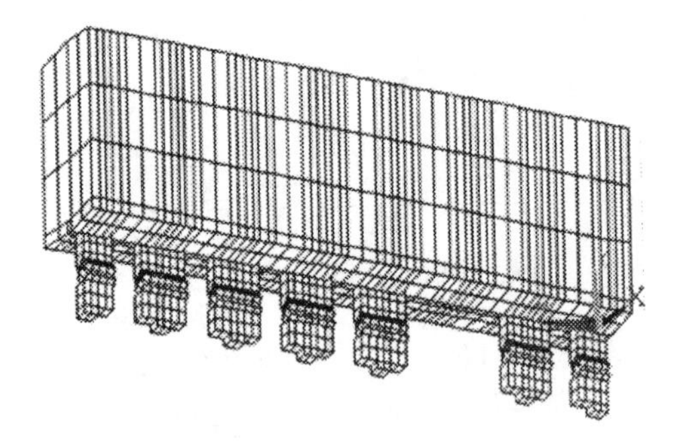

Figure 2. 3-D FEA model for SIP unit.

Note that the sensor chip, the control chip and the related structural details were represented by a homogeneous 3-D solid with an elastic modulus equivalent to that for the molding compound and an assumed bulk mass density that gives the total mass of the SIP unit of about 0.9 gram. Since the seven 0.01" Alloy 42 leads provide the major support for the SIP unit when it is soldered to the PCB, they were precisely modeled for both their embedded and exposed portions. From the modal analysis, the fundamental frequency was predicted to be about 797 HZ. Meantime, from the modal analysis, one can see that the fundamental mode is due to bending of the leads as shown in Figure 3. The predicted fundamental resonant frequency is close to the observed frequency of peak response (about 800 HZ) during the electrical test of the prototype parts, which validated the modal FEA model. Further more, the holographic modal testing independently conducted by Mechanical Engineering Laboratory (MEL) of Motorola Government and Space Technology Group (GSTG) showed some more details about the vibratory characteristics of the SIP unit. The holographic testing was done on six SIP

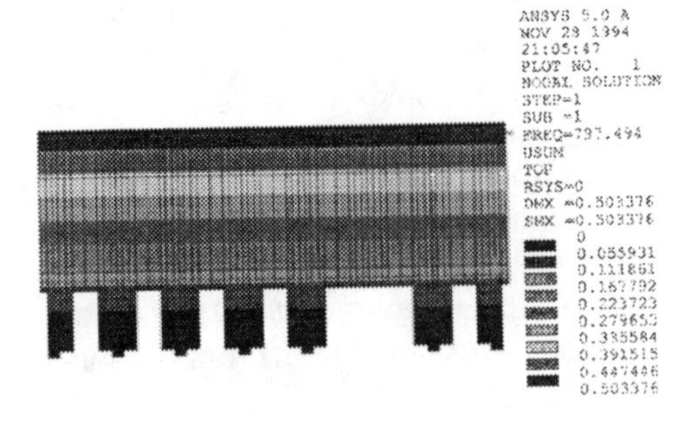

Figure 3. Fundamental mode shape of the SIP unit.

units soldered on a test PC board (Figure 1). The average fundamental frequency of the SIP unit was about 810 HZ. The typical fundamental mode shape is as shown in Figure 4.

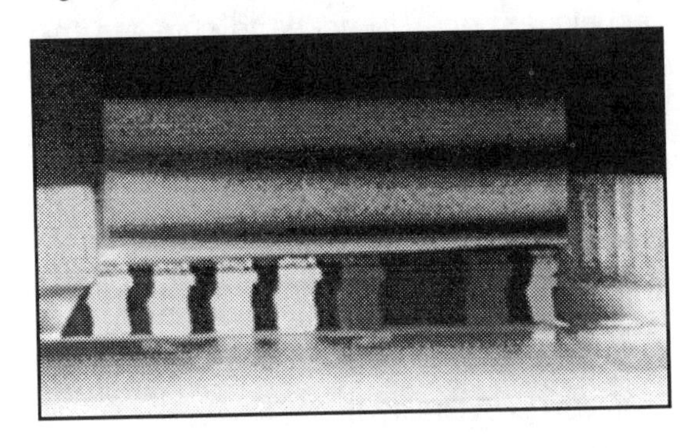

Figure 4. Fundamental mode shape of the SIP unit.

The correlation between the modal analysis prediction and holographic modal testing is excellent indeed.

Complete Package

After the FEA model for the SIP unit was verified, it was expanded to include the RTV adhesive layer and the plastic saddle to simulate the whole package as shown in Figure 5.

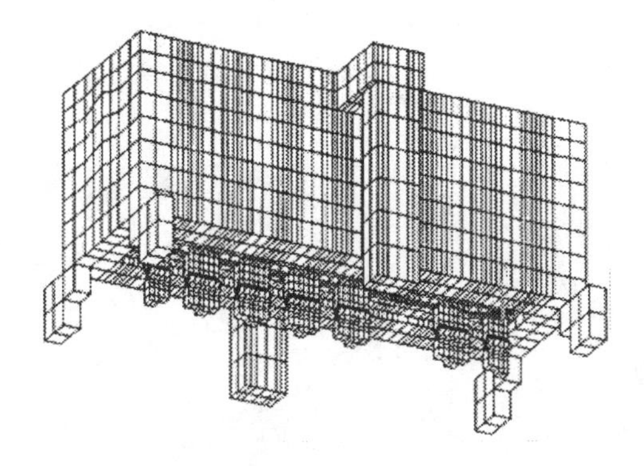

Figure 5. 3-D FEA model for prototype package assembly.

The fundamental resonant frequency was predicted to be 1,475 HZ, which is below 2,000 HZ and within the operating frequency range. Again, the holographic modal testing was performed on the complete package assembly. From the modal testing, the average fundamental frequency is found to be 1,427 HZ, which correlates with simulation prediction fairly well. Further more, the simulated mode shape (Figure 6) and that measured by holographic interferometry (Figure 7) correlate reasonably well.

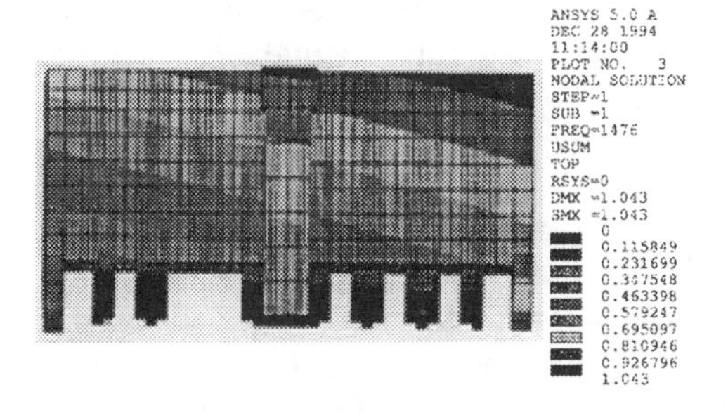

Figure 6. Simulation predicted fundamental mode shape of the prototype assembly.

Note that one can quantify the correlation between the simulation predicted mode and measured mode by digitizing the holographic fringe pattern and calculating the normalized vector correlation factor following the approach proposed by Li and Malohn[5].

Figure 7. Holographic fringe pattern depicting the fundamental mode shape of the prototype assembly.

DESIGN IMPROVEMENT

After the above investigation, a thorough understanding of the modal characteristics of the prototype package design for the 50G accelerometer was established. Due to the poor design of the saddle, the structure of the assembly is not stiff enough to push the fundamental frequency above the operating frequency range. With this validated FEA model, several modified saddle designs were evaluated. After many iterations, a new saddle design was recommended as shown in Figure 8.

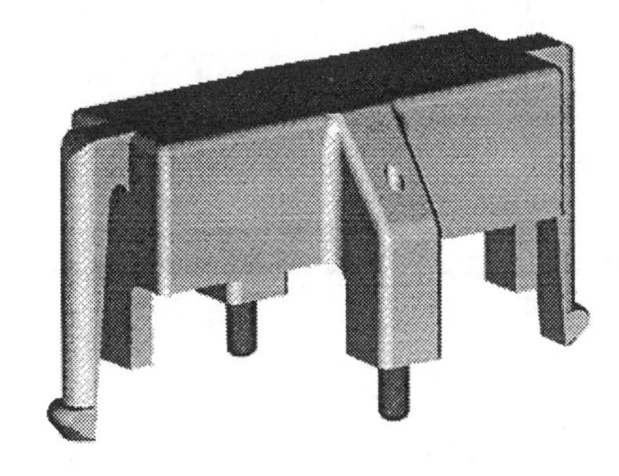

Figure 8. New saddle design for the 50G accelerometer package.

According to the simulation, this new saddle should push the fundamental frequency of the assembly above 3,000 HZ. Note that all these design iterations were performed on a computer. The cost saving and time reduction are significant. The new design has also incorporated manufacturability, assemblability and other reliability issues. With all the lessons learned, the new design worked for the first time. In fact, after the first part was made, its fundamental frequency was measured to be about 3,310 HZ, which correlates well with that the simulation predicted, i.e., 3,375 HZ.

SIMULATION & MEASUREMENT CORRELATION

During this investigation, the correlation of simulation and measurement was vigorously pursued. Specially the boundary conditions for modal analysis were found to be very important. However, some of the assumptions were found to be rather forgiving, e.g., representation of the SIP unit (excluding the leads) with one solid was well acceptable as far as modal analysis is concerned. The major simulation and measurement correlation data are as summarized in Table 1.

Table 1. Simulation predicted and measured fundamental frequencies (HZ).

Description	Holographic Testing	FEA Modal Analysis
SIP Unit	810	797
Prototype Assembly	1,427	1,475
New Saddle Design	3,310	3,375

CONCLUSION

- FEA modal analysis and holographic modal testing is an effective method for achieving robust semiconductor vibration sensor package design.

ROBUST PACKAGE DESIGN FOR PRESSFIT DIODES

BACKGROUND

From 1995 to 1996, the Motorola pressfit diodes (see Figure 9) were experiencing low ppm (parts per million) failures due to IR (reverse current) leakage at the Ford EFHD alternator plant in Detroit.

Figure 9. The Motorola pressfit diodes.

The package design for this diode has been through several major iterations for reliability improvement. Prior to 1995, the die stress was so high that the diodes were failing due to die fracture. After several design modifications, die stress was reduced and the die fracture issue was resolved. However, reliability issue remained to be a problem for not meeting the high reliability requirement, i.e., <10 ppm, set by Ford. The failures were due to unacceptable IR leakage after the diodes were pressed in. It was believed that the IR leakage to be still caused by high die stress due to mechanical-electrical behavior of the silicon die. In fact, it was observed that a few failed parts recovered after about two to three weeks of storage at room temperature, which was apparently due to the stress relaxation of the die attach solder layer. It was also noticed that all the failures were reported on the 40A diodes (die size 0.191'x0.191") whereas none reported on the 25A diodes (die size 0.161"x0.161"). This is because of the higher die stress for the bigger die. At the present time, we do not have a model to simulate as to how exactly the IR leakage is effected by the die stress. But we do know that the die stress is such a critical factor that if it is reduced to a certain level, no IR leakage should occur.

To determine causes of the failure and to identify this critical stress level for achieving a robust design quickly, a structural simulation effort started. Based on the simulation results, two major design improvements were proposed. Subsequently, parts were made and an accelerated reliability test was conducted on a total of 10 groups of 12 DOE (Design Of Experiment) parts based on a failure function developed by one of the authors[9]. The accelerated reliability testing has not only verified the structural models, but also provided a guide line for robust package design for the pressfit diodes.

DIE STRESS EVALUATION VIA FEA

Cross Section

The cross section of the Motorola 40A pressfit diode is as shown in Figure 10.

Figure 10. Cross-section of the Motorola 40A pressfit diode.

From the cross-section, one can see that this package consists of six major components, i.e., copper heat sink, copper lead, silicon die, mold compound, die attach solder layer, and polyimide passivation around the die. The big copper heat sink is pressed into the mounting plates of the Ford alternators as shown in Figure 11.

Figure 11. Pressfit diodes on a set of Ford alternator mounting plates.

The purpose of having such a big copper heat sink is to minimize the die stress during the press-in operation. However, the press-in force is so great (i.e., 1,000 ± 200 lbf) that it causes the heat sink to deform plastically resulting in high die stress.

The 2-D FEA Model

A 2-D axisymmetry structural FEA model has been developed for stress evaluation as shown in Figure 12. The model has included part of the mounting plate for simulating the loading conditions during the press-in operation.

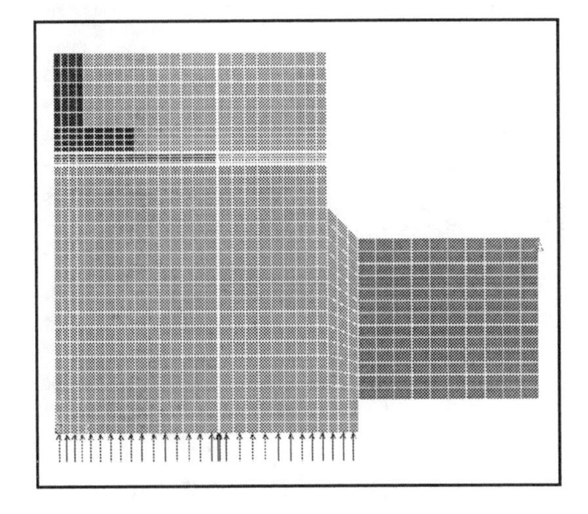

Figure 12. 2-D axisymmetry FEA stress model.

Loading

The model has included plasticity of the copper heat sink. Note that the initial thermal stress in the package induced during the manufacturing processes was not considered for modeling simplification without losing accuracy due to the following reasons:

- the thermally-induced die stress is predicted to be about 3,000 psi, which is relatively insignificant as compared with the minimum die strength of about 18,000 psi[10].
- some degree of solder creep will further lower the thermal stress in the die.

Therefore, only mechanical loads were input to the model, i.e., the radial load due to a radial interference of 0.001 inch and the axial pressure load of 1,000 lbf at the beginning of the press-in operation.

Current Design (Baseline)

The maximum die stress under the combined radial and axial loading is predicted to be about 22,122 psi, which exceeds the minimum die strength. A stress plot is as shown in Figure 13.

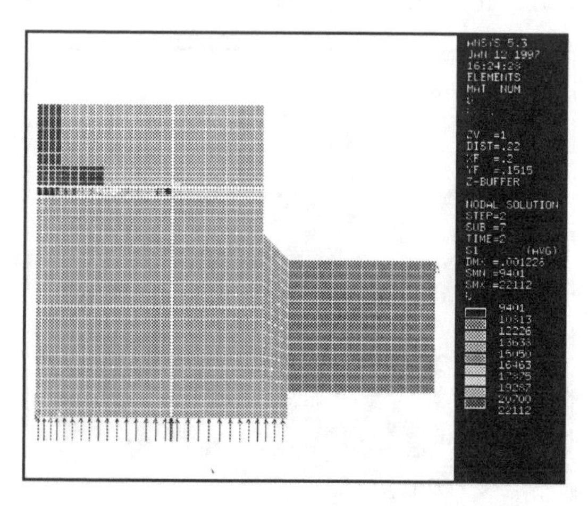

Figure 13. Stress plot for the baseline case.

From this plot, one can see:

(1) the die stress is mainly due to bending;

(2) there is a localized high stress along the edge of the lead head due to stress concentration.

Figure 14 is a deformation plot for the copper heat sink, which shows some serious surface deformation at the die attach area, resulting in die bending. Apparently, this

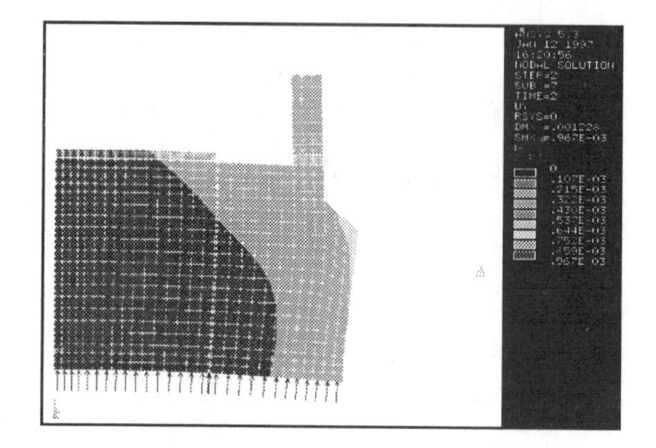

Figure 14. Plastic deformation plot (in axial direction) of the copper heat sink.

surface deformation is responsible for the die bending and high die stress. Note that the die attach solder fillet is not included in the model so that the results should represent the worst case. Apparently, this much die stress has caused some failures due to IR leakage though it may not be high enough to cause die to crack.

Design Improvements

Based on the simulation results, two major design improvements have been proposed, i.e.,

(1) cut a groove on the copper heat sink around the die attach area (i.e. a moat) to reduce the effect of press-fit loading to the die bending;

(2) increase the lead head diameter so that the lead head covers the whole die to minimize the stress concentration.

Again, the simulation was used to evaluate the above design improvements. The results are as summarized in Table 2.

Table 2. Simulation predicted maximum die stress (psi)

Design Configuration	Max. Die Stress, psi
Current (Base Line)	22,122
Big Lead Head (L/H)	12,785
Big L/H + .0335" Moat	9,765

According to the simulation, one can achieve a stress reduction factor of 56% as compared with the baseline design when both improvements are implemented. This should result in a significant reliability improvement.

The findings and results were presented to Ford EFHD and received very positively. As expected, Ford requested reliability test to demonstrate the simulation results. The required failure rate should be less than 10 ppm. How can we demonstrate a failure rate of less than 10 ppm cost effectively? The answer is: accelerated reliability test.

ACCELERATED RELIABILITY TEST

Methodology

The purpose of the accelerated reliability test is to conduct a reliability test on a small sample size to demonstrate reliability of a large sample size. The key to link the test results on a small sample size to the reliability of a big sample size is a failure function. Apparently, for this case, the die stress is considered to be the critical parameter for the IR leakage-induced failures of the parts. If the failure rate can be related to the die stress via a failure function, and provided stress values under test condition and assembling condition, one should be able to set up an accelerated reliability test procedure.

Failure Function

Assuming a failure function:

$$Q = A * (\sigma)^{\xi}$$

Where, Q: failure rate, ppm,

A: coefficient, constant,

σ: die stress, ksi (FEA predicted),

ξ: stress power, constant.

To determine A and ξ, one needs two data points. The current design provides one data point, i.e., $\sigma = 22.122$ ksi, Q = 101 ppm. Second data point was found from the previous design, i.e., $\sigma = 27.220$ ksi, Q = 2,800 ppm. Note that with the previous design, failure rate Q includes those due to both die fracture and IR leakage. Again the die stress is calculated from a stress FEA model. From these two data points, one can easily calculate the two constants, i.e.,

A = 2.885E-20,

ξ = 16.02.

Note that this failure function is based on some field data on the failure rate before and after 1995, and the simulation-predicted die stress values.

The Press-in Test Fixture

A test fixture was designed to have a mounting hole with smaller diameter than the those on the Ford mounting plates while it is big enough for the diode to be pressed in. The fixture is made of steel rather than aluminum for re-usability. With the proposed fixture, the average press-in force is measured to be about 1,604 lbf. This much press-in force is predicted to generate a high die stress of 40.360 ksi according to stress simulation. From the failure function, the failure rate is predicted to be 1.334E6 ppm, i.e., 100%.

Test Sample Size

Total of 10 groups of DOE parts were made for 10 different design configurations, i.e., standard and big lead head, with no moat, 0.010", 0.020", 0.030", or 0.040" moat. There are 12 samples in each group. Apparently, for the baseline parts, it is predicted from the failure function that 12 parts will all fail. If there is 1 out of 12 parts to fail, what is the equivalent failure rate in ppm for the parts on the Ford assemble lines? With the assumption that the die stress increases from the actual mounting plates to accelerated test fixture for a same scale factor for all design configurations, the equivalent failure rate can be easily calculated form the failure function, i.e., 5.45 ppm. Similarly, 2 failures out 12 parts will be equivalent to 10.05 ppm. 3 failures out 12 parts is equivalent to 16.39 ppm, etc. With this failure function, one can easily interpret the accelerated reliability test results to the failure rate under production conditions.

Test Results

For this test, the IR leakage was set to be the failure criterion. The test results for 10 groups of DOE parts (each group contains 12 parts) are as summarized in Table 3.

Table 3. Accelerated reliability test results in terms of parts failed per group.

Lead Head Configuration	Baseline (no moat)	0.010" Moat	0.020" Moat	0.030" Moat	0.040" Moat
Standard	12	4	5	1	0
Large L/H	7	0	2	0	0

FINDINGS & CONCLUSION

- 12 baseline parts all failed as predicted.
- Both large lead head and moat show significant improvement to the reliability of the parts.
- 0.030" moat & large L/H is found to be a robust design with zero failure.
- Integration of simulation and reliability test is proven to be an effective tool for achieving robust microelectronics package design.

Figure 15. SORF-8 package.

SORF-8 PACKAGE RELIABILITY IMPROVEMENT VIA DIE THINNING

DIE CRACK PROBLEM

The SORF-8 (with 6-cell LDMOS die) production line was experiencing a serious reliability problem due to die crack. This is a typical 8 lead SO (Small Outline) package as shown in Figure 15.

Since the 9 mil 6-cell LDMOS silicon die is attached to the 8 mil copper lead frame via Au-Si eutectic bonding, high thermal stress in the die is expected. The typical failure mode is as shown in Figure 16.

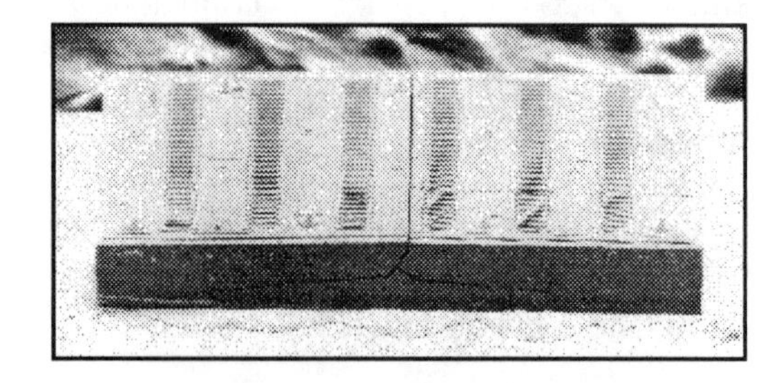

Figure 16. A typical cracked LDMOS die.

The pattern of the crack clearly indicates that the crack initiated from the middle of the top surface of the die. The questions are how high the die stress is, and if it is high enough to cause die to crack. Again, we started this investigation effort with stress simulation.

THERMAL STRESS EVALUATION

A 1/4-symmetry 3-D linear thermal stress FEA model of this package was constructed as shown in Figure 17. The maximum die stress due to eutectic die attach alone was predicted to be about 31,140 psi occurring at the middle of the top surface of the die, which correlated well with the typical failure mode observed in the field (Figure 16). Apparently, the package bending was responsible for the high thermal stress in the die. This bending can be also clearly depicted by an overlay displacement plot as

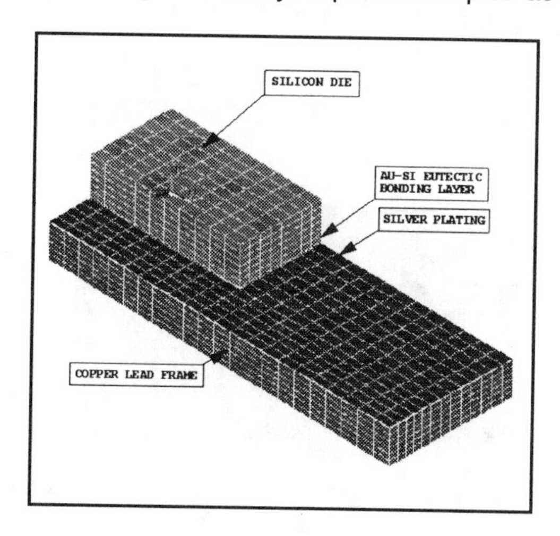

Figure 17. Stress FEA model for SORF-8.

shown in Figure 18. Obviously, the thin copper lead frame (8 mils) and the stiff Au-Si eutectic die bonding layer are the two major contributors for the die bending.

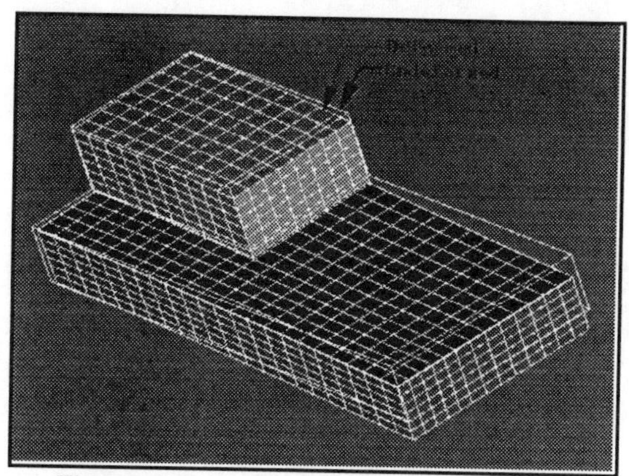

Figure 18. Overlay displacement plot.

DIE STRENGTH TEST

To assess the reliability of the package, a die strength test was conducted using the soft ball break apparatus developed by George Hawkins of Motorola[11]. A total 30 pieces were tested. The results are as summarized below:

- Average: 29,060 psi
- Std. Deviation: 6,150 psi
- Maximum: 38,550 psi
- Minimum: 19,757 psi

Comparing the FEA predicted thermal stress in the die with the die strength data, one concludes that die crack is expected right after die attach.

THE CHALLENGE

There are several obvious approaches to reduce die stress, i.e.,

1) increase thickness of the copper lead frame,

2) change copper lead frame to Alloy 42 lead frame,

3) solder die attach,

4) silver filled epoxy die attach.

All these approaches were evaluated via

simulation and found to be effective in reducing die stress. However, due to various limitations under production environment, none of them was feasible. This was a panic situation.

A NOVEL APPROACH: DIE THINNING

If thicker copper lead frame can reduce the package bending and die stress, what about a thinner silicon die? Quickly, a parametric thermal stress simulation was conducted. The results are as shown in Figure 19.

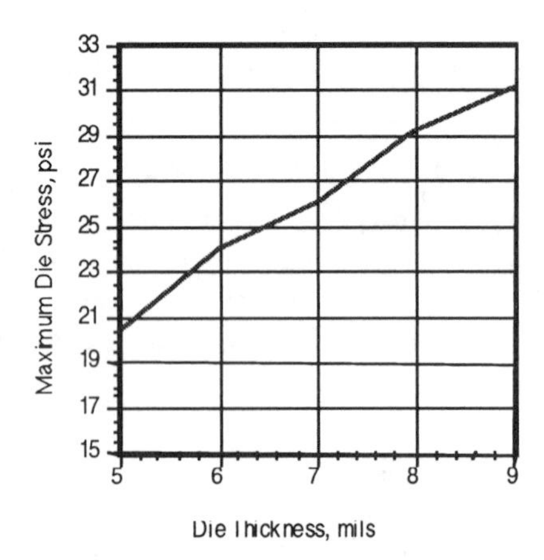

Figure 19. Maximum die stress vs. die thickness.

Practically, the minimum die thickness is about 6 mils due to handling difficulties. According the stress simulation, the die stress can be effectively reduced by about 23% by thinning the 9 mil die to 6 mils. This much stress reduction can make significant reliability improvement. Based on this novel idea, parts were made and tested.

RELIABILITY TEST

Thermal Cycle Test

Thermal cycle test results are as summarized in Table 4.

Table 4. Cracked die before and after temperature cycling (#die cracked/ Group)

Cycles	9 mil die	6 mil die
0	30/240	0/240
250	6/85	4/96
500	1/79	2/92
1,000	0/37	1/46
Total	37/240	7/240

Findings

I) Right after die bonding, Group #1 had 30 dice cracked out of 240 units whereas Group #2 had none.

ii) Group #1 had additional 7 dice cracked after 1,000 temperature cycles, which makes a total of 37 failures out of 240 units, i.e., a failure rate of 15.4%.

iii) All the cracked dice from Group #1 show the same pattern (Figures 20 & 21).

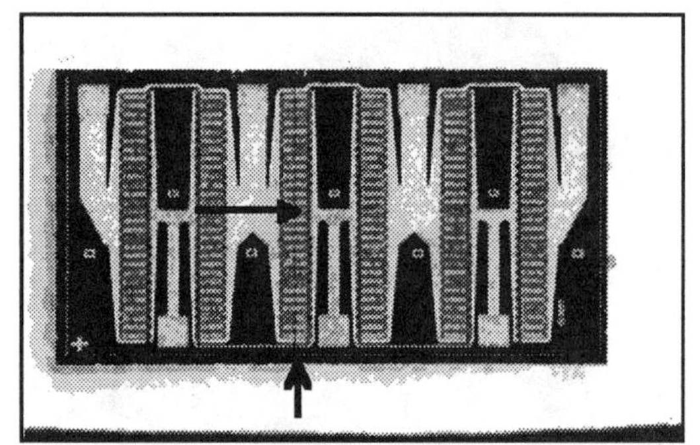

Vertical Crack

Figure 20. Typical 9 mil die crack pattern.

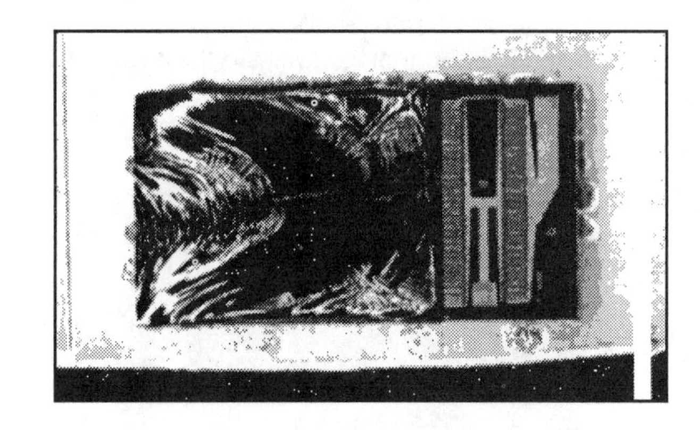

Figure 21. 9 mil die crack separated area.

iv) Group #2 also had seven units failed out of 240 units due to die crack. However, these are either edge cracks or cracks due to "non-wetting" as shown in Figure 22. Only one die from Group #2 showed the typical crack pattern observed from Group #1 as shown in Figure 22. The 6 mil die has in deed made

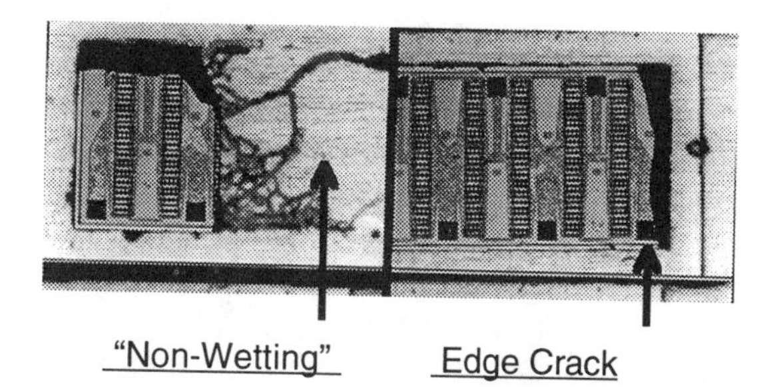

"Non-Wetting" Edge Crack

Figure 22. 6 mil die crack pattern.

significant improvement to the package reliability as compared with the current design with the 9 mil die, though it is not failure free.

HTSL (High Temperature Storage Life) Test

Both groups yielded no failures after 1,000 hours of HTSL test.

CONCLUSION AND DISCUSSION

- Combination of package simulation, die strength and reliability tests have proven to be a powerful tool not only for determining the root cause of die crack, but also for finding ways to improve package reliability.

- Die thinning has demonstrated to be effective in reducing thermal stress in the die while maintain the same thermal performance of the package.

CORRELATION OF FEA SIMULATION AND EXPERIMENTAL MEASUREMENT ON A WAFER LEVEL CHIP SCALE PACKAGE

BACKGROUND

A wafer level chip scale package (CSP) technology has been under development at Motorola SPS for several years. As a standard practice at Motorola SPS, the simulation has been extensively utilized for evaluating structural integrity, thermal performance, and RDS(on), through out the course of the project. In several occasions, some major design decisions had to be made based on the simulation results. Therefore, an assessment to the accuracy of the simulation results is needed.

THERMAL DEFORMATION MEASUREMENT FOR STRESS MODEL VERIFICATION.

The stress model of an MOSFET package was developed to simulate the structural behavior of the package during the complete manufacturing processes, reliability test, and surface mounting process (by the customer) as shown in Figure 23.

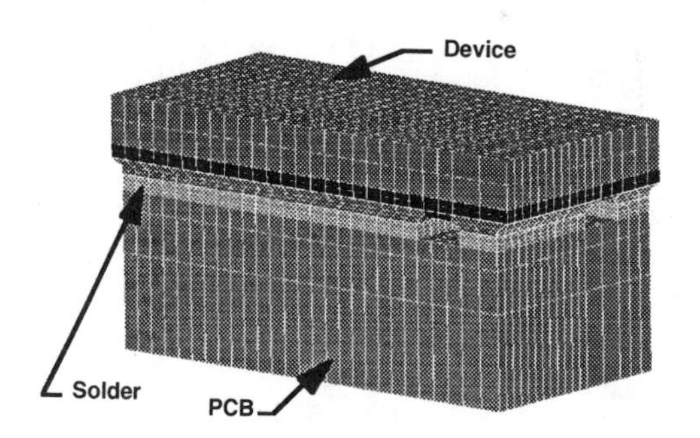

Figure 23. 3-D stress FEA model for MOSFET CSP package.

Note that only a small piece of PCB is included in the model when simulating the surface mounting process. This correlation study was conducted on the measured and FEA predicted deformation on the top surface of the package since deformation can be measured by several well developed optical techniques, e.g., moiré, laser moiré interferometry, laser holographic interferometry, laser speckle, digital correlation, etc. In this case, the deformation is very small and the initial deformation of the top surface can be neglected. In addition, the top of the package is a silicon piece, which has a smooth and reflective surface, so that the classical Michelson interferometer can be used to measure the out-of-plan displacement of that surface. Note that the Michelson interferometer has a sensitivity of half of the optical wave length. Therefore, it is an ideal technique for this application.

CORRELATION

FEA Simulation Results

A out-of-plan displacement plot is generated from the simulation results as shown in Figure 25.

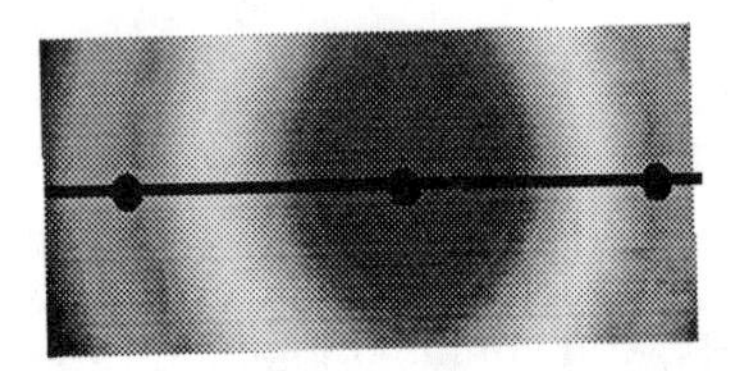

3a. $\Delta x = 0.224\ \mu m$ (FEA results)

Figure 25. Displacement plot from FEA simulation.

Note: the rigid body movement was excluded for generating this displacement plot.

Measurement Results

Six pieces were measured. A typical fringe pattern is as shown in Figure 24. The maximum surface deflection is about 0.32 µm, which is slightly more than half of the optical wave length of 0.633 µm for the He-Ne laser.

3b. $\Delta x = 0.32\ \mu m$ (measurement)

Figure 24. A typical interference fringe pattern for the MOSFET CSP package before surface mounting.

Vector Correlation

Instead of correlating the absolute maximum deflection value from FEA and measurement, the vector correlation was evaluated. The idea of vector correlation was proposed by Li and Malohn in 1988[5]. Since we do not have adequate number of fringes, this vector correlation study is rather primitive. however, the methodology can be clearly illustrated as follows,

- Select three points along the center line of the fringe pattern, read the fringe order, and calculate the deflection value at these points to obtain a measured displacement vector along the center line
- Pick up exactly the same three points from the FEA model and find the deflection values at these points to obtain an FEA predicted displacement vector.
- Do a vector correlation calculation as follows:

$$\xi = \mathbf{V}_{measured} \bullet \mathbf{V}_{FEA} / (||\mathbf{V}_{measured}|| * ||\mathbf{V}_{FEA}||)$$

where

$$V_{FEA} = \begin{Bmatrix} -0.139 \\ 0 \\ -0.128 \end{Bmatrix} \qquad V_{measured} = \begin{Bmatrix} -0.315 \\ 0 \\ -0.315 \end{Bmatrix}$$

Therefore, the correlation factor $\xi = 0.99$, which is considered to be excellent. In general, a correlation factor of greater than 0.90 is acceptable[5]. For this correlation study, more that three data points are needed. Further more, the above approach can easily be extended to 2-D or even 3-D correlation study. The calculation may be automated too.

THEN WHAT?

After the model was verified, it was used for a parametric design evaluation as to how the die size will effect the package reliability. The model has also been used for material evaluations. Since this is part of major ongoing development effort, we are continuously pursuing more progress in this correlation study.

CONCLUSION

- Integration of computer simulation and experimental measurement or reliability test have proven to be a powerful tool for achieving robust microelectronics package design as well as for reliability improvement of existing packages.

- Much work needs to be done in methodology development for correlation of computer simulation and experimental measurement/testing on microelectronics packaging.

ACKNOWLEDGMENTS

The authors wish to thank their colleagues at Motorola SPS, Mahesh Shah, Mike Schager, Hiep Le, Lisa Montez, Son Tran, and Yifan Guo for their help and assistance in making the above projects successful.

REFERENCES

1. Bocci, W.J., "Finite Engineering Analyses Applied to Microelectronics," Proc. 1986 National Aerospace and Electronics Conferences (NACECON '86), Vol. 4, May 1986, pp. 1150-1153.
2. John H. Lau, "Thermomechanics for Electronics Packaging," Thermal Stress and Strain in Microelectronics Packaging, Ch. 1, edited by John H. Lau, published by Van Nostand Reinhold, 1993.
3. David Dougherty, Quan Li, and Mahesh Shah, "Sensor Package Design Improvement Using Computer Simulation and Experimental Testing," ASME, 1995.
4. C.-P. Yeh, C. Ume, R.E. Fulton, K.W. Wyatt, and J.W. Stafford, Correlation of Analytical and Experimental Approaches to Determination of Thermally Induced Printed Wiring Board (PWB) Warpage," Thermal Stress and Strain in Microelectronics Packaging, Ch. 9, edited by John H. Lau, published by Van Nostand Reinhold, 1993.
5. Q.B. Li and D.A. Malohn, "Validation of FEM Modal Analysis Using Holographic Interferometry," IMAC-6, Florida, USA, Feb., 1988.
6. T. Hattori, S. Sakata, and T. Watanabe, "A Stress Singularity Parameter Approach for Evaluating Adhesive and Fretting Strength," Advances in Adhesively Bonded Joints, MD-Vol. 6, edited by S. Mall, K.M. Liechti, and J.R. Vinson, published by ASME, 1989.
7. ANSYS User's Manual, Vol. IV: Theory, Revision 5.1, Swanson Analysis Systems, Inc., 1994.
8. Charles M. Vest, Holographic Interferometry, Section 4.3, pp177, published by John Wiley & Sons, 1979.
9. Q.B. Li, "Robust Design of Pressfit Diode Package Via Simulation and Accelerated Reliability Testing," presented at Motorola Winter AMT (Advanced Manufacturing Technology) Symposium, 1/30/1997.
10. Quan Li, Motorola internal report: MSPS/CPSTG-IMO/TR-94/0003, 6/24/1994.
11. George Hawkins, Howard Berg, Mali Mahalingam, Gary Lewis, and Lynn Lofgran, "Measurement of Silicon Strength As Affected By Wafer Back Processing," Motorola Internal Report No. REPL-87-8, 4/15/1987.

MICROCONTROLLER COMMUNICATIONS

900696

Distributed Realtime Processing in Automotive Networks

Uwe Kiencke
Siemens Automotive
Regensburg, W.-Germany

Abstract

The formulation of software tasks as parallel processes allows their implementation within distributed microcontrollers. The requirements for Automotive Networks to support these applications are discussed. By introduction of a locality measure, a classification of networks can be made either into interactive distributed realtime processing or into classical communication.
Given a sufficiantly small locality, the physical network extension does not have an impact on the implementation. A concept i presented how to integrate process dispachting and synchronization. Based upon this concept, functions may be formulated independant of their location in a specific microcontroller.

1. Formulation of Software Tasks as Processes

In Automotive Control Systems, physical signals like temperature or pressure are measured. After processing them along given algorithms, output control signals drive the actuators. These procedures happen at least partially simultaneous and are interrelated. Microcontrollers for such applications must be cheap in order not to unduly increase the costs of a vehicle by its electronic content. Their processing and memory capacity are limited. On this basis control units have been developped, which were directed just to a limited subsystem. Most software programs therefore show little systematic structure up to now. Application problem oriented software and scheduling portions are usually mixed up together.

With the emergence of automotive networks and high-performence microcontrollers, functions become more complex and interrelated. Instead of a single control procedure, a variety of parallel tasks allow to optimize the overall driving perfomance of the car. New programming techniques, formulating tasks as parallel processes are required to handle the increasing complexity of the software.

1.1. Evolution from linear Programming to parallel Processes

In conventional linear programming different program sections are sequentially run through. In automotive realtime applications this is done cyclically, e.g. with the sample time of a control algorithm. An example is shown in Fig. 1 a where programs **A** and **B** are cyclically running through one after the other. Both programs represent the respective implementation of different control tasks.

It is an obvious suggestion to logically separate
- task solutions **A** and **B** from
- interrelationship between them,

as it is done in Fig. 1 b. Now programs **A** and **B** may be formulated totally independent from each other, while their interaction is handled by another mechanism. The result is better transparency in the program structure.

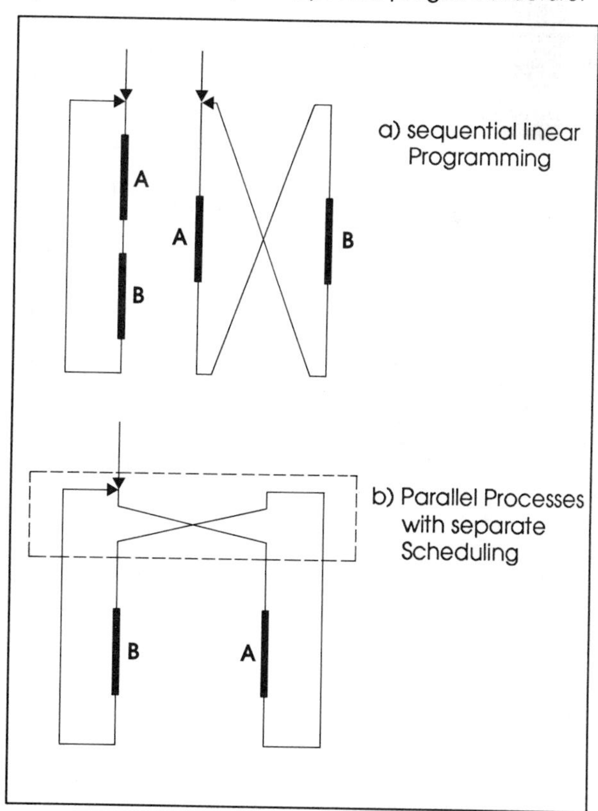

Fig.1 Evolution from linear Programming to parallel Processess

A simple example for scheduling is shown in Fig. 2 a. Processes **A** and **B** are only carried out, if preconditions **a** and **b** apply.

MA: if not **a** then goto MB;
A; goto MA;
MB : if not **b** then goto MA;
B; goto MB;

a. Unique Entry Points

a ?
b ?
A
B

b. Various History - dependent Entry Points

a1 ?
b ?
A1
B
a2 ?
A2

Fig. 2 Condition dependent Scheduling of parallel Processes

The conditions **a** and **b** may be thought being modified with in processes **A** and **B**. A further evolution step is shown in Fig. 2 b, where process **A** is partitioned into two parts $\mathbf{A_1}$ and $\mathbf{A_2}$. Process dispatching may now happen after partial execution of **A** already.

In order to restart execution at the right point, the entry point MA[i] into **A** must depend from the history before. It is set and stored before eventually jumping to MB.

MA1: MA[i]: = MA1;
if not $\mathbf{a_1}$ then goto MB;
A1;

MA2: MA[i]: = MA2;
if not $\mathbf{a_2}$ then goto MB;
A2; goto MA1;

MB: if not **b** then goto MA[i];
B; goto MB;

In general, more sophisticated scheduling strategies are applied, encorporating external interrupts, exchanging operating memory space, introducing priority schemes. The execution of processes is no longer bound to a single processor, but may be carried out on multiple processors.

1.2. Interaction between Processes

During execution parallel processes exchange information between themselves. There are two basic principle ways to perform this [1].

a. Cooperation

Data Memory Space of interacting processes is partiallly overlapping, allowing to read and write data from the different programs (Fig. 3). This mechanism is restricted to parallel processes running within a computer with one memory readily accessible from all instructions flows.

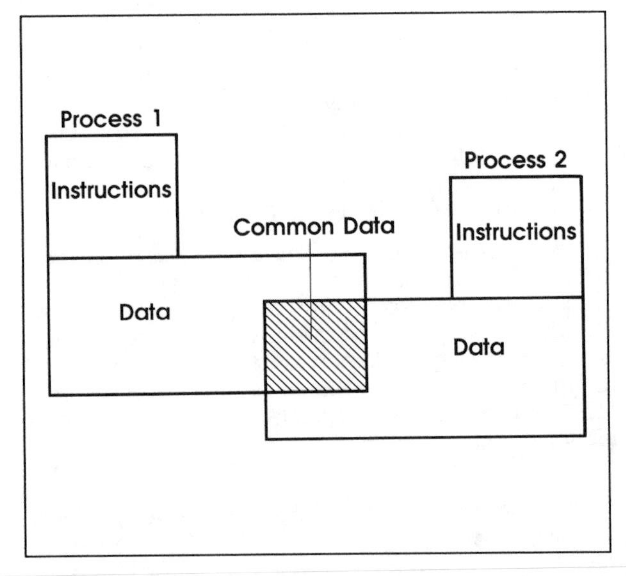

Fig.3 Interaction by Cooperation within Common Data Memory

b. Communication

Without overlapping memory space, data must be transferred from one data buffer to the other (Fig. 4). Communication has a direction, from transmitter to receiver.

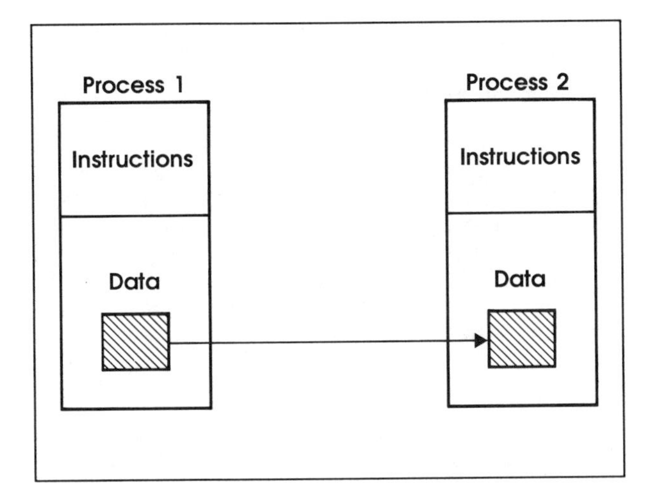

Fig. 4 Interaction by Communication between Data Buffers

The application of cooperation requires the access of different processes to eventually the same data. This access must be done under mutual exclusion in order to avoid inconsistencies of the data. For this purpose a module cooperation is introduced.

```
module cooperation;
export RESERVE, RELEASE;
type semaphore = (released, reserved) = released;

    Kernel procedure RESERVE (var B: semaphore);
    begin
    while B = reserved do;          {Wait}
    B: = reserved                   {Reserve}
    end;

    Kernel procedure RELEASE ( var B: semaphore);
    begin
    B: = released                   {Release}
    end;

    end cooperation.
```

The state of the semaphore varible B expresses

B = released: None of the processes is in the exclusive path, i. e. accessing the buffer

B = reserved: One of the processes is in the exclusive path.

The kernel procedures used in the approach must not be interrupted. While a process is waiting, it may be suspended and make room for another one to be resumed [2]. If each access to the communication buffer is framed between RESERVE and RELEASE procedures, inconsistencies can be avoided.

The second wayto interact between processes is by communication. In order to ensure correct operation, the sequence of events must be controlled (Fig. 5).

A process **A** can only transmit data after they became available in the source buffer SB. Process **B** can only dispose of the data after their arrival in it's target buffer TB. Installing a required sequence of events is called synchronisation. Adding data extends this to communication, which may be handled by the following module.

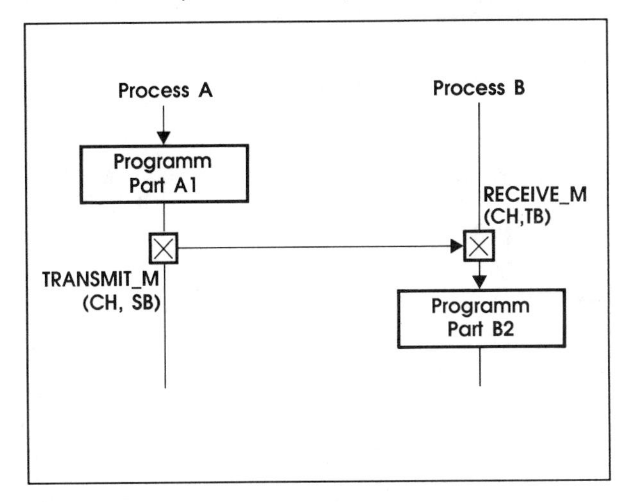

Fig.5 Synchronization of Processes

```
module communication;
export TRANSMIT_M, RECEIVE_M;
type channel    =    record
                     S : signal = (reset, set) = reset;
                     M: message
                     end;

    Kernel procedure TRANSMIT_M (vor CH: channel;
    SB: message);
    begin
    CH.M: = SB;        { Write message into channel }
    CH.S: = set        { Give synchronization signal }
    end;

    Kernel procedure RECEIVE_M (vor CH: channel;
    TB: message);
    begin
    while CH.S:
    = reset do;        { Wait for synchronization signal }
    CH.S: = reset;     { Reset synchonization signal }
    TB: = CH.M         { Read message from channel }
    end;

    end communication;
```

The actual communication is then performed by the operations

TRANSMIT_M (CH, SB)
RECEIVE_M (CH,TB).

The channel is storing the message between source and target buffer.

Taking avantage of the communication operation TRANSMIT and RECEIVE, a variety of enhanced synchronization structures may be built up. The interrelationship may extend over more than just two processes, requiring a respective number of channels CH (Fig. 6). A message may be transmitted to or received from a plurality of other processes (Fig. 7). For each communication **i**, a dedicated channel CH**i** must be created, comprising the messages M**i** and synchronization Signal S**i**.

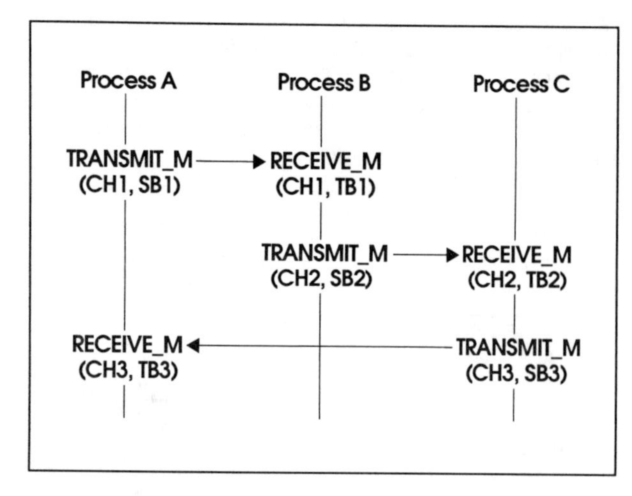

Fig. 6 Enhanced Synchronization and Communication Structures

2. Automotive Networks

In order to support the communication operations, described in the previous paragraph, automotive networks must meet a number of specific requirements.

2.1 Requirements for Distributed Realtime Processing

a. Latency Time

Several processors linked to the network may try to acess the transmission medium at the very same time. The next transferred message is selected by means of an arbitration method, introducing a delay for the non-transferred messages. The time duration between the request for and the actual start of transmission is called latency time. Depending on realtime constraints, latency must not exceed a certain value for highly prioritized messages. A reduction of latency time can be achieved by increasing the message transfer rate.

b. Deadlocks

In some arbitration schemes such as CSMA/CD all processors back off from the networks in case of a simultaneous attempt for access [3]. After different time delays, another trial is then started. Each such contention takes away a portion of the networks overall transfer capacity, thereby reducing the remaining effective transfer capacity. In case of traffic peaks, access attempts become so frequent, that the network will be totally blocked, stalling the information transfer completely. In order to prevent such situations, deadlock-free arbitration schemes must be applied.

c. Communication Buffer

The TRANSMIT and RECEIVE operations rely upon a channel CH, which takes over the message from a source buffer SB and transfers it to a target buffer TB. The processors must have got such buffers in their networks interface. In order to maintain the consistency of the stored data within, access from the networks as well from the processor must be mutually exclusive. Since the transmission over the network is carried out in realtime, the already introced Kernel access control procedures such as RESERVE and RELEASE must not last longer than a small fraction of the message duration. In order to avoid the introduction of wait states into the instruction flow within the processor, RESERVE and RELEASE procedures are furthermore contrained to a fraction of the shortest execution time of processor instructions.

d. Multicast Addressing

The enhanced process synchronization in Fig. 7a requires that messages can be received in multiple processors. In Controller Area Network CAN (4), this is supported by ACCEPTANCE FILTERING. Every message is broadcasted on the network, labeled with an IDENTIFIER. Within each of the receiving processors it is then locally decided whether to accept a mes sage or not. Owing to this feature messages arrive simultaneously in all the target buffers TB.

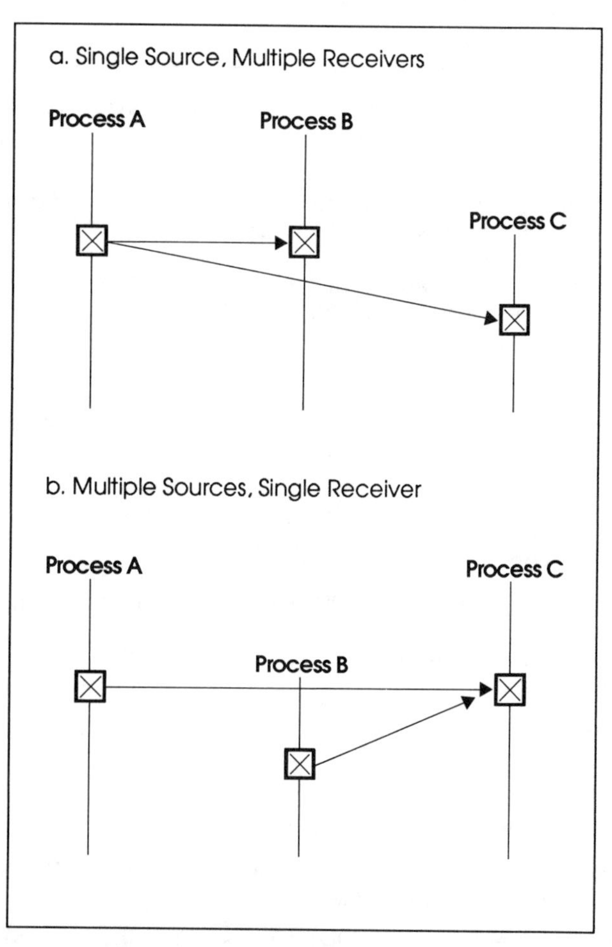

Fig. 7 Enhanced Synchronization and Communication Structures

e. Reception from multiple Sources

In Fig. 7b a communication structure is shown, where a process C gets synchronization signals from multiple source processes A and B. This feature has not yet been incorporated in any of today's automotive communication protocols. A proposal for implementation might be the concept of GENERIC INDENTIFIERs.

Every message on the network can be differentiated from others by it's unique IDENTIFIER. When messages from multiple sources shall pass through one single ACCEPTANCE window, not just one but a class of several IDENTIFIERs determine this window. A GENERIC IDENTIFIER therefore consists of a class of IDENTIFIERs, which should all be accepted. As an example, open doors shall generate respective warning messages which are transmitted over the network. The receiving process is charged to issue a warning signal to the driver in case of one or more of the doors being open. Thus the GENERIC IDENTIFIER in this example consists of all IDENTIFIERs assigned to the various "Door-open" - messages coming from all doors.

The procedure of ACCEPTANCE FILTERING can be generalized in order to support GENERIC IDENTIFIERs. Messages are now received when not only one but a generic class of IDENTIFIERs is matching that of the incoming message in the filtering process. A convenient simplification could be made by introducing some constraints upon the selection of IDENTIFIER codes. The codes of two IDENTIFIERs within a generic class may be restricted to form a sequence in which two adjacent codes have a Hamming distance of no more than 2, e.g. differ in only one bit. The differing bits can then be easily marked by "Don't care" tags, which are excluded from the comparison operation within ACCEPTANCE FILTERING (Fig. 8). Under these conditions, just a single comparison operation is still suffficient to cover also GENERIC IDENTIFIERs, easing realtime requirements within the processor interface to the network.

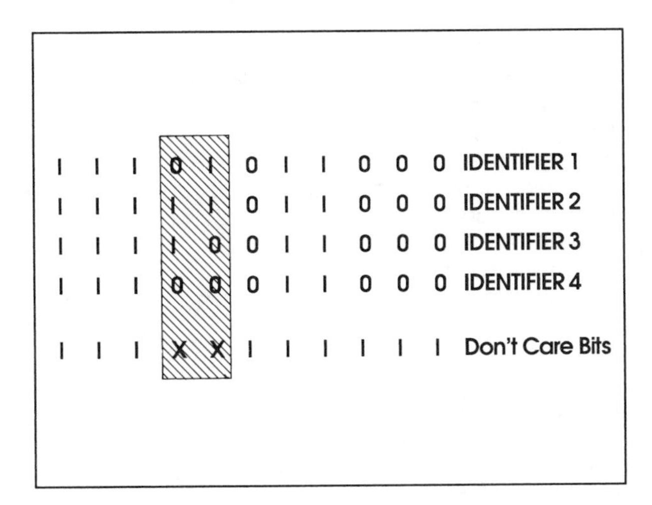

Fig. 8 Implementation of GENERIC IDENTIFIERs by code restriction and Don't Care Bits

f. Object Orientation

As shown in Fig. 6, each synchronization or communication procedure requires a separate logical channel CHi which comprises seperate data locations in source and target buffers SBi, TBi as well as separate synchronization variables Ch.Si. In CAN (4) all data and respective variables are integrated into COMMUINICATION OBJECTs together with their IDEN TIFIER in order to enhance transparency of the implementation as well as the application.

g. Error Handling

Multiple source and/or multiple destination of messages result into special requirements for error handling. The synchronization of processes remains only logically consistent, if the integrity of the respective transmitted messages can be guaranteed for all sources and receivers, even in case of local errors, which might hit at only one processor. Rather than releasing one part of the receiving processes upon reception of the synchronization signal, and keep waiting at the other part due to a local error, all processes must proceed alike by definition. In CAN[4], local errors generate an error message, assuring consistent reception in all processors of a network and error correction by retransmission.

2. 2. Determination of Message Transfer Rates

In distributed, interactive realtime systems, a large number of relatively short messages are exchanged in order to support synchronization and communication between processes. The transfer capacity of the network must be sufficiently high enough in order

a. to transfer at least all messages densely packed to each other within a given cycle time (Fig. 9)

$$\sum_{i=1}^{n} t_{Mi} < t_{Cycle} \quad (1)$$

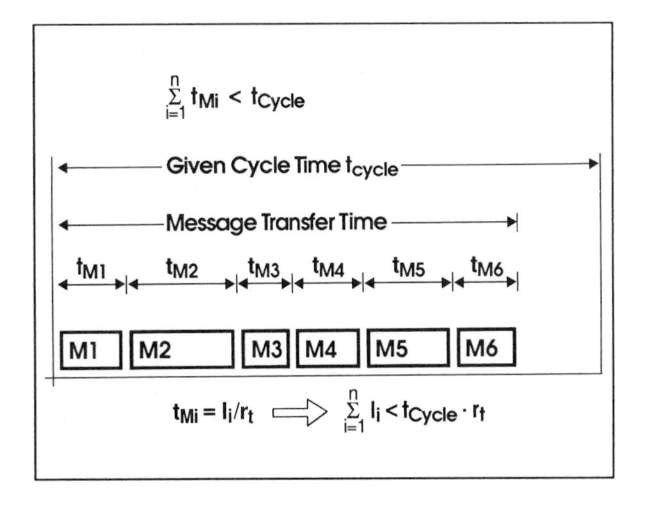

Fig. 9 Transfer Capacity of Automotiv Netwoks

b. and additionally to guarantee limited latency times for highly prioritized messages in case of contention (Fig. 10).

The transfer time t_{Mi} of a message M_i over the network depends upon the message length l_i and the tranfer rate r_t, according to

$$t_{Mi} = l_i/r_t \quad (2)$$

Condition **a.** can then be met by choosing a sufficiently large transfer rate.

$$r_t > (1/t_{Cycle}) \sum_{i=1}^{n} l_i \qquad (3)$$

A further increase of the transfer rate r_t is required to meet the additonal condition **b.** In case of a simultaneous request of several messages for transmission, only one can be actually transferred at a time over the network, whereas the others are delayed. In the example of Fig. 10 latency times are

$$t_{l2} = t_{M1} = l_1/r_t \qquad (4)$$

$$t_{l3} = t_{M1} + t_{M2} = (l_1 + l_2)/r_t \qquad (5)$$

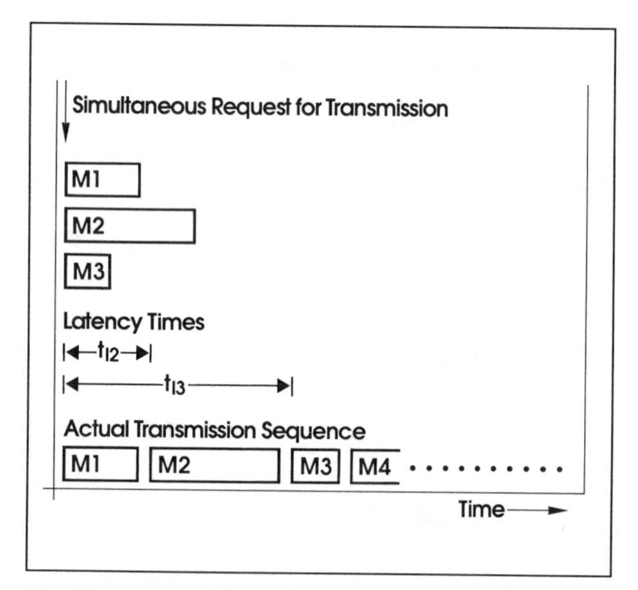

Fig. 10 Latency Time after simultaneous Request for Transmission

Requests for transmission come from different processors in different locations at arbitrary time instants, which are determined by the realtime processes generating the messages. Latency times are thus stochastic, depending on the actual contention situation on the network. A guaranteed value for latency introduces additional constraints on the tranfer rate, e.g.

$$r_t \geq l_1/t_{l2,max} \qquad (6)$$

$$r_t \geq (l_1 + l_2)/t_{l3,max} \qquad (7)$$

2.3. Physical Limits for Message Transfer Rates

Increasing the transfer rate r_t is done to reduce the message transfer time t_{Mi}, opening additional room for more messages or to meet tougher time constraints for latency times. There is however a physical limit in such a reduction, which comes from the maximum propagation speed C of information accross a network, which is approximately

$$C \approx 2 \cdot 10^8 \, m/s \qquad (8)$$

i.e. about two thirds the speed of light.
Introducing a maximum network extension d, the propagation time becomes

$$t_p = d/C \qquad (9)$$

One top of message length l_i and transfer rate r_t in equation (2), message transfer time is now determined by the propagation time t_p.

$$t_{Mi} = l_i/r_t + t_p \qquad (10)$$

The overall delay time of a message **i** from request for transmission to it's completion is then

$$t_d = t_{Mi} + t_{li} \qquad (11)$$

$$= l_i/r_t + t_p + t_{li} \qquad (12)$$

If the propagation t_p becomes dominant in equation (2), the increase of transfer rate r_t will no longer lead to a significantly enhanced transfer capacity (Fig.11). A major requirement for all applications must therefore be not to waist transfer capacity but to invest into systematic concepts for the definition of functions, the partitioning into processes as well as the formulation and normalization of the transferred signals [5]. Only messages required for the synchronization and communication between processes should be allowed for transfer. E.g. a simple cyclic copying of all source buffers SB into all target buffers TB without a direct functional justification would generate so many messages in a given cycle time, that transfer rates are immediately pushed to the physical limits in larger applications.

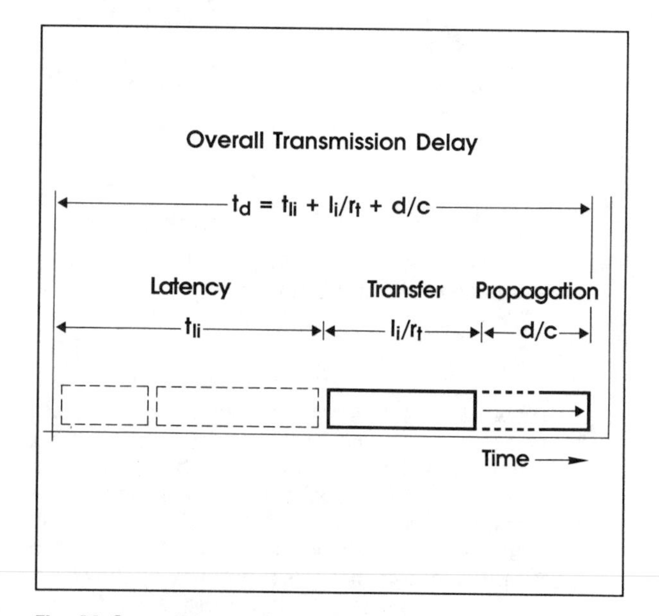

Fig. 11 Overall Transmission Delay Time

In order to compare application requirements to network capabilities, a LOCALITY measure is introduced. The LOCALITY L of a network is defined as the ratio of the propagation delay t_p to the transfer time l_i/r_t

$$L = (t_p \cdot r_t)/l_i \tag{13}$$

$$= (d \cdot r_t)/(c \cdot l_i) \tag{14}$$

The overall message delay time in equation (12) then becomes

$$t_d = (1 + 1/L)\, t_p + t_{li} \tag{15}$$

In the example of message M3, equation (7) may be formulated as

$$t_{l3,max} = (l_1 + l_2)/r_t \tag{16}$$

Introducing an average message length l_i and the number **m** of higher prioritized messages, this becomes

$$t_{l3,max} = m\, l_i/r_t \tag{17}$$

From (15) and (17) the delay time results as

$$t_d = (1 + (1 + m)/L)\, t_p \tag{18}$$

The LOCALITY L depends on the local extension of the network, the length of messages and the selected transfer rate. InFig. 12 the LOCALITY L of different networks is shown. As long as

$$L \ll 1 \tag{19}$$

Network	$\overline{l}_i$/[bit]	r_t/[bit/s]	d/[m]	L
CAN	10^2	10^6	40	$2 \cdot 10^{-3}$
Ethernet	10^3	10^7	10^3	$5 \cdot 10^{-2}$
Satellite Communication	10^4	10^8	10^6	50

Fig. 12 Locality of different Networks

distributed, interactive realtime processing is possible without any constraints. Since the limited propagation delay of messages is not yet felt, applications can be implemented in distributed processors without special precautions.
From

$$L \geq 1 \tag{20}$$

on, however, the network extension has an impact on the implementation. This is no longer the area of distributed realtime processing, but rather the field of classical communications. The main task here is no longer to interact by mutual time constrained exchange of information, but to transmit information from one location to the other.

In current automotive networks [4,6], the arbitration method of bit-by-bit contention is used for regulating the access of messages to the network. This procedure assumes bit information travelling over the entire network and back to the transmitter within each bit time, tightening even further the requirements for the LOCALITY measure. Maximum transfer rates in such networks are around 1 MBit/s at extensions of 40 meters.

3. Distributed Parallel Processing

The handling of parallel processes is part of operation systems and realtime control applications. The principle process scheduling procedures are already well known [1]. They shall be described as a basis for process dispachting in distributed networks.

3.1 Process scheduling

In each realtime processor a process scheduling scheme is installed, which selects one process to be active and get control over the processor (Fig. 13). The state "Non-active" contains all processes, which are not cooperating in a current application. From that predefined reservoir, processes may be created, i.e. put into the "Ready-state". Since the number of processors is usually far less than the number of ready processes, only one process at a time can be selected to resume operation, i.e. gain control over the processor. This process stays active, until it enters a critical portion of the program, where it is blocked by mutual exclusion (chapter 1.2, module cooperation) or by waiting to receive a signal (chapter 1.2, module communication). The process is then transferred into the "Wait-state". There it will no longer be considered in the contention to get access to the processor. When the expected event – removing the blocking – has arrived, e.g. RELEASE or RESERVE, the processor is released and transferred to the "Ready-state" again.

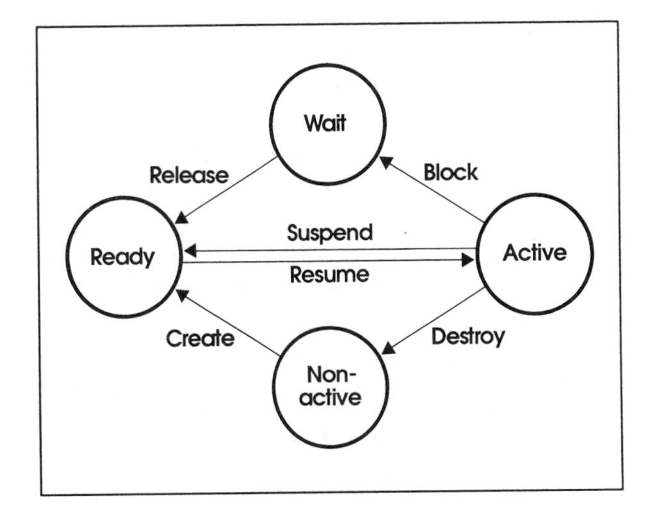

Fig. 13 State Diagram of Realtime Processes

Another way for a process to give up control over a processor is the momentary completion of a task or after a TRANSMIT operation. The process is then suspended and directly transferred into the "Ready-state".

In each of the process states, there is a set of processes. In case of applying a priority strategy, sequentially ordered files of processes are stored in respective queues

- file of non-active processes FN
- file of ready processes FR
- file of active processes FA
- file of waiting processes FW

The following "data object" may be suited to handle these files by a number of specific operations. It is presented just for the principle understanding and is not intended to be detailed. The files shall be of the type sequence.

```
module data object;
export INSERT, DELETE, FIRST, NEXT, LAST, EMPTY;
type sequence = {Definition of a file};

procedure INSERT (var F: sequence; E:↑ element);
{insert an element addressed via E into the file F}

procedure EXTRICATE (var F: sequence; E:↑ element);
{delete an element addressed via E from the file F}

function FIRST (var F: sequence); ↑ element;
{identify the first element within the file F}

function NEXT (var F: sequence; E: ↑ element): ↑ element;
{identify the element next to element E within the file F}

function LAST (var F: sequence): ↑ element;
{identify the last element within the file F}

function EMPTY (var F: sequence): boolean;
{check the cardinality of file F; EMPTY is true for cardinality
= 0, else EMPTY is false}

end data object.
```

Based upon these operations on the data files, process disatching may now be formulated. Two data types shall be declared, which type is state = (non-active, ready, active, waiting).

and

```
type process =  record
                STATE: state;
                REG: register
                end.
```

Assuming to have only one processor in one location, the variable PCURRENT contains the name of the currently active process, and variable PNEXT the name of the next one. All non-active processes shall already exist the file FN, from where they may be identified by a name P, extricated and transferred into the "Ready-state".

The procedure CONTEXT is called within the procedure RESUME, after the currently active process has been blocked or suspended and a next process has been assigned to resume it's operation. The task is to exchange all processor variables, commonly known as context switch. The specific implementation must take care of the hardware constraints of the actual processor.

The set of process dispatching operations upon the set of processes make up a modul "process dispatching".

```
module process dispatching;
export CREATE, DESTROY, BLOCK, RELEASE, SUSPEND,
        RESUME;
import INSERT, EXTRICATE, FIRST, NEXT, LAST, EMPTY;

var
process_set =  record
                FN, FR, FA: sequence of process = empty;
                PCURRENT, PNEXT: ↑ process;
                end;

procedure CREATE (P: ↑ process);
begin
EXTRICATE (FN, P); P ↑. STATE: = ready; INSERT (FR,P)
end;

procedure DESTROY (P: ↑ process);
begin
EXTRICATE (FA,P); P ↑. STATE: = non-active; INSERT (FN,P)
end;

procedure BLOCK (var FW: sequence of process; P: ↑
process);
begin
EXTRICATE (FA,P); P ↑. STATE: = ready; INSERT (FW,P)
end;

procedure RELEASE (var FW: sequence of process; P: ↑
process);
begin
EXTRICATE (FW,P); P ↑. STATE: = ready; INSERT (FR,P)
end;

procedure SUSPEND (P: ↑ process);
begin
EXTRICATE (FA, P); P ↑. STATE: = ready; INSERT (F,P)
end;

procedure RESUME;
begin
PNEXT: = FIRST (FR);
EXTRICATE (FR, PNEXT);
PNEXT ↑. STATE: = active;
INSERT (FA, PNEXT);
CONTEXT (PCURRENT, PNEXT);
PCURRENT: = PNEXT
end;

end process dispatching.
```

The assignment of values to a variable P ↑. STATE is not compelling, however still done in order to explicitly show the process state. The file of active processes FA has been used in spite of anticipating only one processor per location, in oder to maintain generality.

3.2. Integration of Communication across a Network into Process Dispatching

The dispatching scheme within each processor linked to an Automotive Network may be extended in order to cover also synchronization and communication. Provided that the network's LOCALITY is sufficiently small as stated in equ. (19), the process implementation itself is then independant of it's location in a specific processor. Only the separately formulated communication operations must be adapted to the process location.

To be sure, a process should always be located in the processors, where most of the synchronization and communication can be done within that processor itself, minimizing the amount of traffic across the network. The advantage is however, that in case of system evolution and eventual reassignment of process locations, modifications are restricted to inter-process-communication.

If a process cannot continue it's operation, because it is blocked at the entry of a critical program part of mutual exclusion or by waiting for a RECEIVE message, it is transferred into the file FW of waiting processors. The process is then released again, when a message is coming in or when the critical program part is released. The module comminication from chapter 1.2 may than be extended.

```
module communication;
export TRANSMIT_M, RECEIVE_M;
import BLOCK, RELEASE, RESUME, EMPTY, FIRST,
      PCURRENT;
type channel = record
               S: signal = (reset, set) = reset;
               M: message ;
               FW: sequence of process = empty;
               end;

    Kernel procedure TRANSMT_M (var CH: channel;
    SB: message);
    begin
    CH.M: = S.B; CH.S: = set;
    if not EMPTY (CH.FW) then RELEASE (CH.FW, FIRST
    (CH.FW))
    end;

    Kernel procedure RECEIVE_M (var CH: channel;
    TB: message);
    begin
    while CH.S = reset do
    begin
        BLOCK (CH.FW, PCURRENT);
        RESUME
    end;

    TB: = CH.M; CH.S: = reset
    end;

    end communication.
```

SUMMARY

The separate formulation of software tasks as parallel processes on one side and communication between those processes on the other allows to come to more transparent, systematic solutions. Synchronization and communication operations between processes are required within the individual processor as well as for message exchange across a network. Given a low LOCALITY of the network, the process formulation itself is independant of the implementation location witin a specific processor. Communication may be merged into the process dispatching scheme of each processor. The independantly formulated processes are thus integrated into a distributed, realtime control system for the overall functional optimization of automobiles.

Acknowledgement

The author has based the description of Process Scheduling upon the work published in [1]. He furthermore wishes to thank Prof. Horst Wettstein from the University of Karlsruhe for valuable contributions made in numerous mutual discussions.

REFERENCES

[1] Horst Wettstein, "Architektur von Betriebssystemen", 3. Edition, 1987, Carl Hanser Verlag, Munich, Vienna.

[2] IEEE Trial-Use Standard Specification for Microprocessor Operating Systems Interfaces, IEEE Standard 855, 1985.

[3] Andrew S. Tanenbaum, "Computer Networks", 1981, Prentice-Hall International, London, pp 292-295.

[4] Uwe Kiencke, Siegfried Dais, Martin Litschel, "Automotive Serial Controller Area Network", SAE-Paper 860391.

[5] Uwe Kiencke, "Zusammenwirken elektronischer Systeme im Kraftfahrzeug - verteilte Realisierung oder Integration - Schnittstellen, lokale Netzwerke", 2. Aachener Motorkolloquium 25.-26.10.89, S. 743-752.

[6] W. J. Johnson, J. R. Volk, "A Proposal for Vehicle Network Protocol Standard", SAE-Paper 860392.

930005

A Gateway for CAN Specification 2.0 Non-Passive Devices

Craig Szydlowski
Intel Corp.

ABSTRACT

The Controller Area Network (CAN) protocol, developed by ROBERT BOSCH GmbH, offers a comprehensive solution to managing communication between multiple CPUs. In September 1991, the CAN protocol was revised (CAN Specification 2.0) to add an extended message format that increases the number of permitted message identifiers. The CAN protocol now supports 11- and 29-bit message identifiers allowing standard and extended formats, respectively.

CAN Specification 2.0 implements a new message bit, the identifier extension bit (IDE bit) which allows CAN devices to differentiate standard and extended formats. However, most existing CAN implementations are based on the previous CAN protocol specification and will not recognize extended format messages and will respond with an error message. These chips are CAN Specification 2.0 non-passive.

Until semiconductor makers develop a variety of chips to implement 29-bit message identifiers, CAN users may require a gateway to interconnect networks using 29-bit message identifiers to existing networks that only use CAN 1.2 11-bit message identifiers. In this context, a gateway is a system that translates messages in one protocol to another and vice versa.

A gateway between two networks is typically designed to minimize transmission latency, to minimize lost messages (overruns) and to manage bus error issues. Following a brief description of the CAN message format, general gateway design considerations and performance estimates will be addressed. In addition, a method to send remote frames across the gateway is discussed.

GATEWAY EXAMPLE

The gateway considered in this paper is shown in figure 1. The microcontroller in the center, labeled 'GATEWAY', translates messages between the networks above and below. The network labelled 'standard' transmits CAN 1.2 standard messages (11-bit identifier) while the network labeled 'standard/extended' transmits CAN 2.0 both standard and extended messages (11- and 29-bit identifiers). The standard messages for the CAN 1.2 and CAN 2.0 networks are bit-wise and electrically identical.

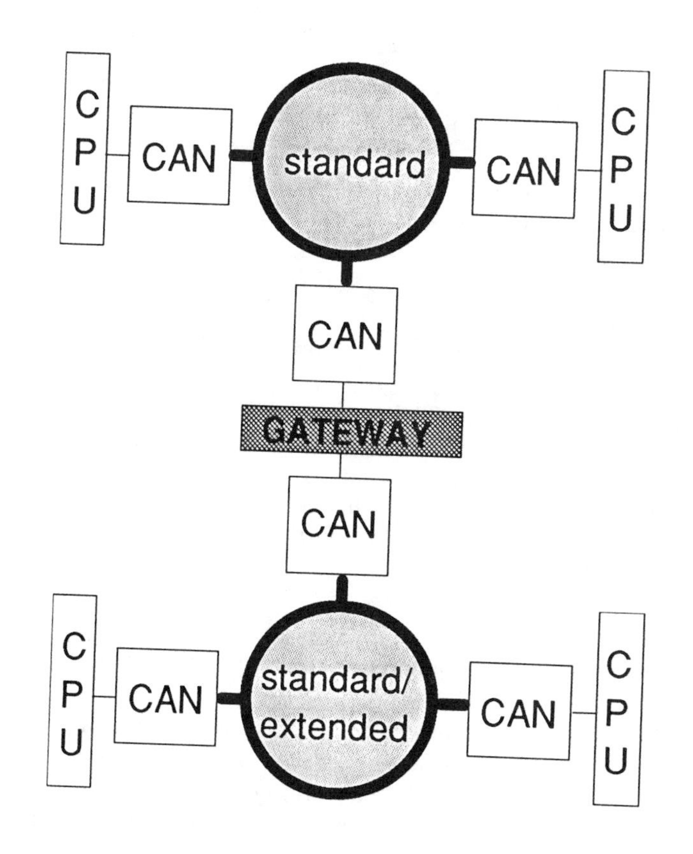

Figure 1: Gateway Example

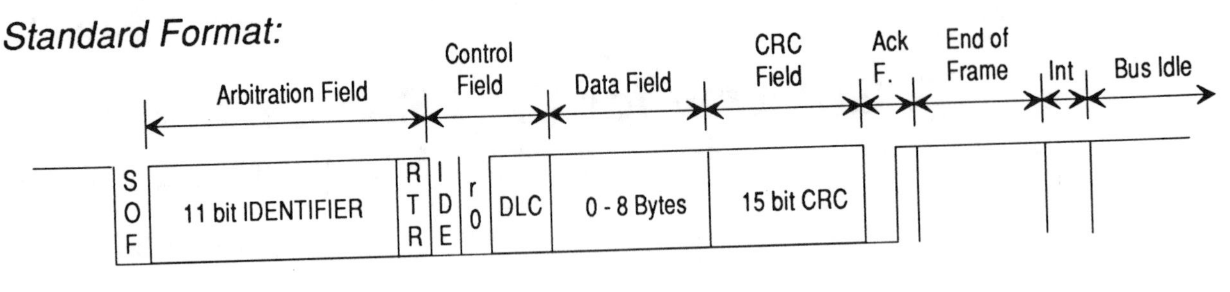

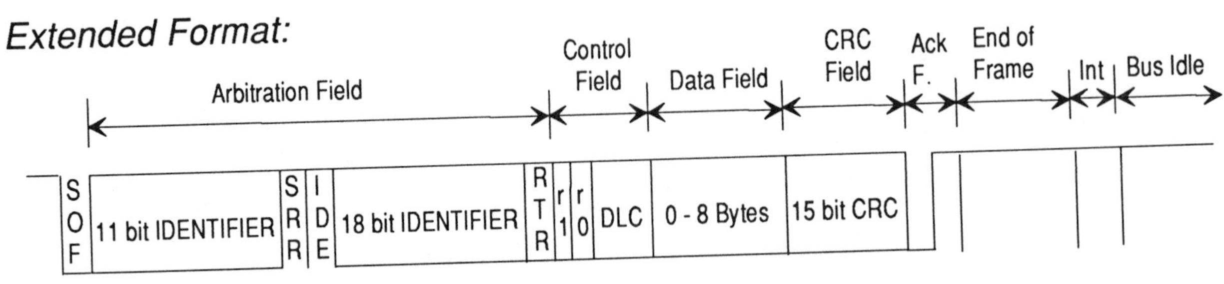

Figure 2: Standard and Extended Formats

The gateway microcontroller communicates with two CAN chips, one from each network. Examples of gateway functions are:

a) bridging standard messages without translation.
b) bridging standard and extended messages with message identifier translation.
c) managing re-transmission issues when errors occur.

CAN PROTOCOL MESSAGE FORMATS

The CAN protocol supports two message formats which differ in the length of the message identifier field. Messages using 11 message identifier bits are called standard and those using 29 message identifier bits are called extended. Both standard and extended message formats support four frame types:

1) Data - carries data
2) Remote - sent when a node requests data from another node
3) Error - sent when a node detects a message error
4) Overload - sent when a node requires extra delay

The following describes the standard and extended message formats for data and remote frames shown in figure 2.

SOF: Start Of Frame (dominant bit) marks the beginning of a data/remote frame.

Arbitration: One or two fields which contain the message identifier bits. The standard format has one 11-bit field and the extended format has two fields, 11- and 18-bits wide.

RTR: Remote Transmission Request bit is dominant for data frames and recessive for remote frames. This bit is in the arbitration field.

SRR: Substitute Remote Request bit is used in extended messages and is recessive. This bit is a substitute for the RTR bit in the standard format. This bit is in the arbitration field of the extended format.

IDE: Identifier Extension bit is dominant for standard format and recessive for extended format. This bit is in the arbitration field of the extended format and in the control field of the standard format.

Control Field: Reserved bits r0 and r1 are sent as dominant bits. The 4-bit Data Length Code (DLC) indicates the number of bytes in the data field.

Data Field: The data bytes are located in the data frame (0-8 bytes). A remote frame contains zero data bytes.

CRC Field: This field is composed of a 15-bit Cyclical Redundancy Code error code and a recessive CRC delimiter bit.

ACK Field: Acknowledge is a dominant bit sent by nodes receiving the data/remote frame and is followed by a recessive ACK delimiter bit.

End of Frame: Seven recessive bits ending the frame.

INT: Intermission is the three recessive bits which separate data and remote frames.

The minimum message lengths of standard and extended message formats are summarized below for data and remote frames. These bit counts will be used to assess gateway message transmission latency and overrun susceptibility. The actual lengths of these messages may differ because 'stuff' bits are added to the message. Stuff bits assist synchronization by adding transitions to the message. A stuff bit is inserted in the bit stream after five consecutive-equal value bits are transmitted; the stuff bit is the opposite polarity of the five consecutive bits. All message fields are stuffed except the CRC delimiter, the ACK field and the End of Frame.

Standard Format

Message Field		number of bits
SOF		1
Arbitration Field		12
identifier	11	
RTR	1	
Control Field		6
IDE	1	
r0	1	
DLC	4	
Data Field		0-64
CRC Field		16
ACK Field		2
End of Frame		7
Total		44-108 bits

Extended Format

Message Field		number of bits
SOF		1
Arbitration Field		32
identifier	29	
SRR	1	
IDE	1	
RTR	1	
Control Field		6
r1,r0	2	
DLC	4	
Data Field		0-64
CRC Field		16
ACK Field		2
End of Frame		7
Total		64-128 bits

HARDWARE CONFIGURATION

For this analysis, it is assumed both CAN implementations are stand-alone chips (figure 3). The gateway microcontroller reads and writes to the CAN chips as if they were smart RAMs. When a message is received, the gateway executes external read operations of the receiving CAN chip. Next, the gateway executes logical instructions to translate the message. Then the gateway executes external writes to the CAN device on the second network to program a transmission. Therefore, the microcontroller executes external read and write operations and translates message identifiers.

READ AND WRITE OPERATIONS - The time required to execute read and write operations is dependent upon the interface-timing characteristics of the gateway microcontroller and the CAN chips. Figure 3 shows a microcontroller interfaced to two CAN devices. This configuration uses a 16-bit address/data bus which doubles the communication throughput between the microcontroller and the CAN chips. The CAN chips drive independent interrupt lines to the microcontroller allowing the interrupting CAN device to be easily identified.

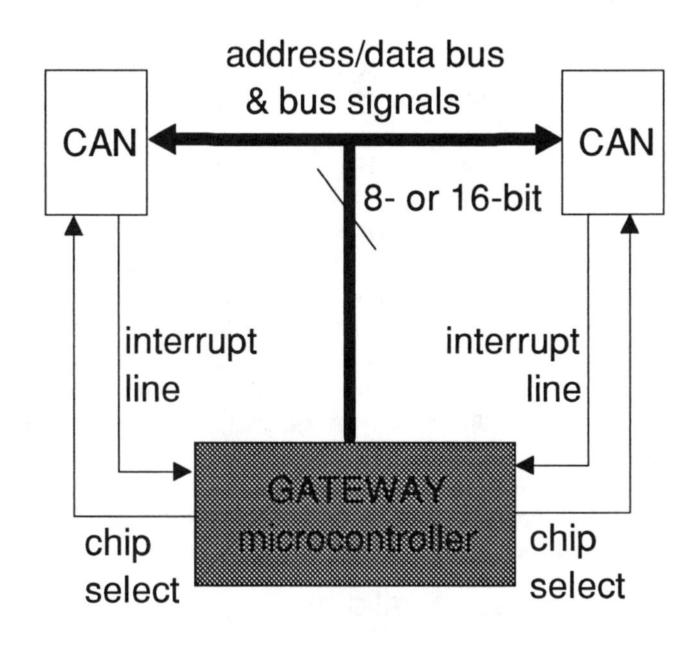

Figure 3: Hardware Example

CAN CHIP MESSAGE CONFIGURATION - The CAN chips must be configured to receive and transmit messages. Typically, CAN chips support 2 to 15 message objects that are configured to receive or transmit. Message objects consist of RAM bytes which store a message identifier, data bytes and control bytes. To receive messages more reliably, a CAN chip must be able to receive an incoming message while it is still processing a previously received message. This may be accomplished 1) by using a buffered receive message object or 2) by implementing a buffered receive using two receive message objects that are alternately validated and invalidated (figure 4). It is assumed that all receive message objects employ acceptance mask filtering which allows all relevant CAN bus messages to be accepted by a single buffered receive message object. Without acceptance filtering, a prohibitive number of receive message objects would be required to receive all necessary messages. A receive message object must be 'valid' before the last bit of the incoming message is received to capture the message.

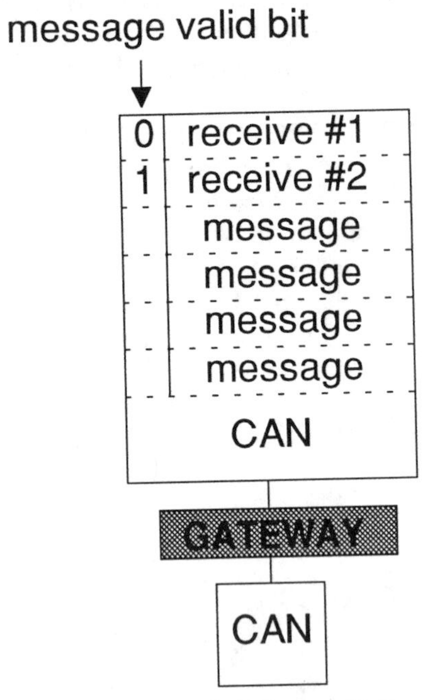

Figure 4: Receive Buffer

After using one or two message objects per CAN chip to receive, the remaining message objects should be used to handle remote frames and to transmit messages. The gateway must be prepared to store multiple transmit messages that may be waiting for CAN bus access (figure 5). The number of transmit message objects required is dependent on the gateway message traffic and the ability of these messages to get on the second network. Overall bus loading and the priority of the messages crossing the gateway strongly influence the number of required transmit message objects.

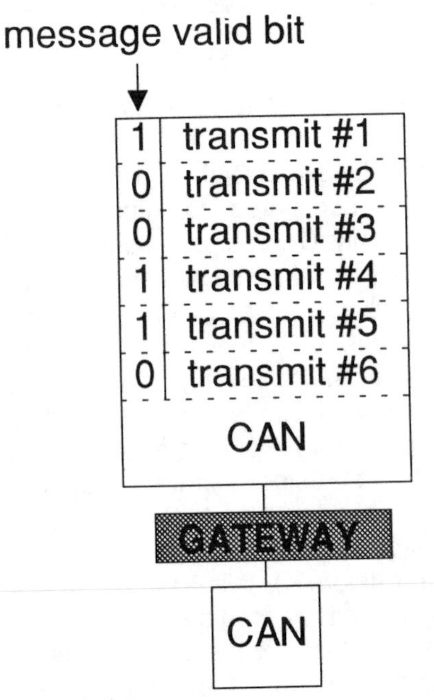

Figure 5: Transmit Message Objects

One method to manage transmit message objects is to implement a stack which stores unused message objects numbers. When a message transmission is needed, a message object number is "popped" from the stack. Similarly, when a message transmission is completed, its message object number is "pushed" on the stack. The gateway may need to manage the order of transmit message objects so low priority messages are not scheduled to transmit first. The Intel 82527, for example, uses the message that is stored in the lowest numbered message object for arbitration, regardless of whether the other message objects on the chip have higher arbitration priority. Therefore, a low priority message in message object #1 may delay the transmission of the other messages.

GATEWAY MESSAGE TRAFFIC

A gateway must be prepared to translate messages from one protocol to the other, and vice versa. A gateway for the CAN protocol must be concerned with two of the four CAN protocol frame types: data and remote. The other two frame types, overrun and error, are handled in hardware by the gateway's CAN chip on the respective network. Therefore, the gateway (microcontroller) has no added responsibilities. However, the gateway must be prepared to act when too many errors occur on either network as indicated by warning or busoff status bits. In this case, the gateway investigates the bus-status like other network nodes do.

DATA FRAMES

The gateway is typically designed to minimize transmission latency and to minimize lost messages (overruns). Analyzing latency and message overruns with respect to data frames must consider the two differences in data frames. First, data frames have between zero and eight data bytes. Second, the extended format has 20 more arbitration and control bits than the standard format.

Another issue to be considered is the effect of stuff bits. Stuff bits increase transmission latency by increasing the duration of a message transmission, a delay seen by messages waiting for access to the CAN bus. Stuff bits may actually help decrease the probability for message overruns by providing additional time for the gateway to manage message translation and message object programming. Since the number of stuff bits is dependent upon the bit patterns of message identifiers, data bytes and CRC codes, it is difficult to estimate their effect deterministically. Perhaps it is best to estimate an average number of stuff bits per message and the resulting transmission time.

TRANSMISSION LATENCY - Message transmission latency is the time between a completed message transmission on one network to its completed transmission on another network. The latency may be divided into three parts: gateway access, gateway processing and bus transmission delay. Gateway access is the time a message must wait to be serviced by the gateway microcontroller. The worst case occurs when two messages are received simultaneously on both networks and one message must wait until the other has been processed. Gateway processing refers to the time to read the receive message from the first network, translate the message, and configure the message for transmission on the second network as shown in figure 6. The bus transmission delay is the time required to gain access to the CAN bus and the following message transmission time. The transmission latency will vary based on the time required for the message to win arbitration on the second network. Low priority messages, will of course, have longer transmission latencies than higher priority messages.

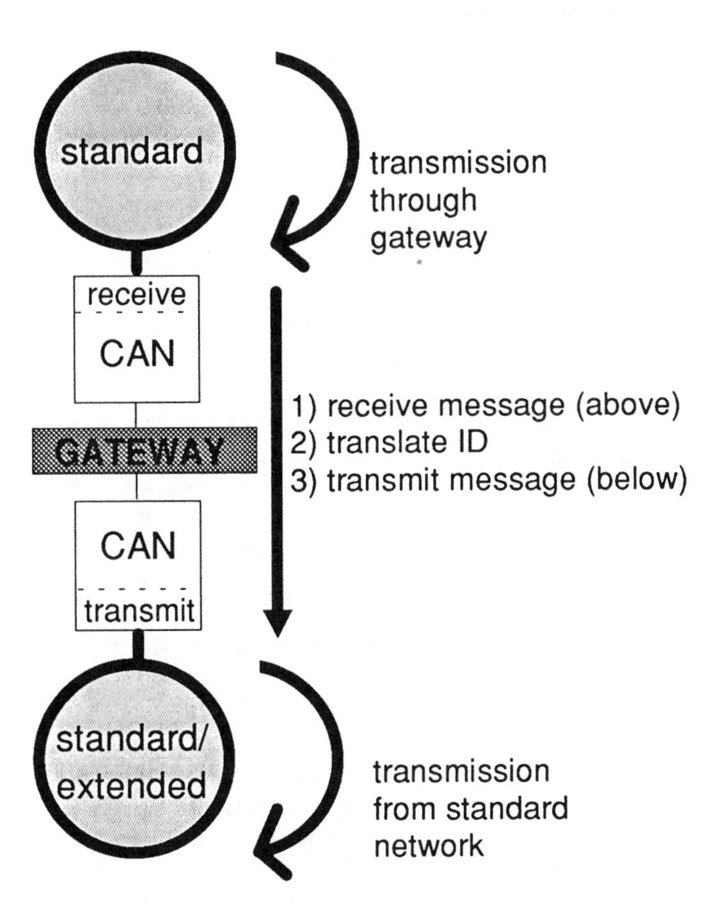

Figure 6: Gateway Processing

MESSAGE IDENTIFIER TRANSLATION - CAN 1.2 and CAN 2.0 network gateways can be used in two ways. The simplest implementation is to transfer standard messages between the two networks without message identifier translation. This performs the task on interconnecting CAN 1.2 and CAN 2.0 networks and allowing standard message to be shared. A second implementation may require message identifiers to be translated. Translation is required when messages have independent message identifiers on both networks. For example, an "engine temperature message" has a dedicated 11-bit standard message identifier on the CAN 1.2 network and a dedicated 29-bit message identifiers on the CAN 2.0 network. Translating between these two message identifiers may be implemented algorithmically or by using a look-up table. For this analysis, it is assumed the message identifier requires no translation and only the 11-bit standard identifier messages are transmitted across the gateway.

GATEWAY PROCESSING ANALYSIS - This section is an analysis of the minimum time required by the gateway to respond to a receive message from one network and to program a corresponding transmission on a second network. This latency analysis assumes the gateway microcontroller to be an Intel 16-bit 80C196 chip interfaced to an Intel 82527 operating at 16 MHz, using three wait-states and a 16-bit address/data bus. This assumption results in the execution of read and write operations in 1.875uS. The 80C196 responds to interrupts in 16 states following the execution of the current instruction or about 2.5uS.

The flowchart in figure 7 is an example of the minimum number of steps required for a gateway processing. The purpose of the following section is to estimate the time required to implement the flowchart program flow. This calculated time represents the bare minimum gateway execution time.

Flowchart 7 is referenced for the following gateway processing time estimate.

Flowchart block:

A) A message is received by one of the CAN chips. This CAN chip sends an interrupt to the gateway after the transmission of the last End of Frame bit (1.5uS).

B) The microcontroller responds to the interrupt request, enters the interrupt service routine (3.75uS).

C) The microcontroller reads the CAN interrupt register to get the interrupt pointer (1.875uS).

D) The microcontroller first determines whether the interrupt is from the error status register indicating a busoff or warning flag is set (1.375uS).

D1) If a busoff or error warning flag is set, the gateway should execute an error recovery sequence to determine the error source or to execute a software reset on get back on the bus.

E) Next, the microcontroller determines whether the interrupt was initiated by a reception or transmission.

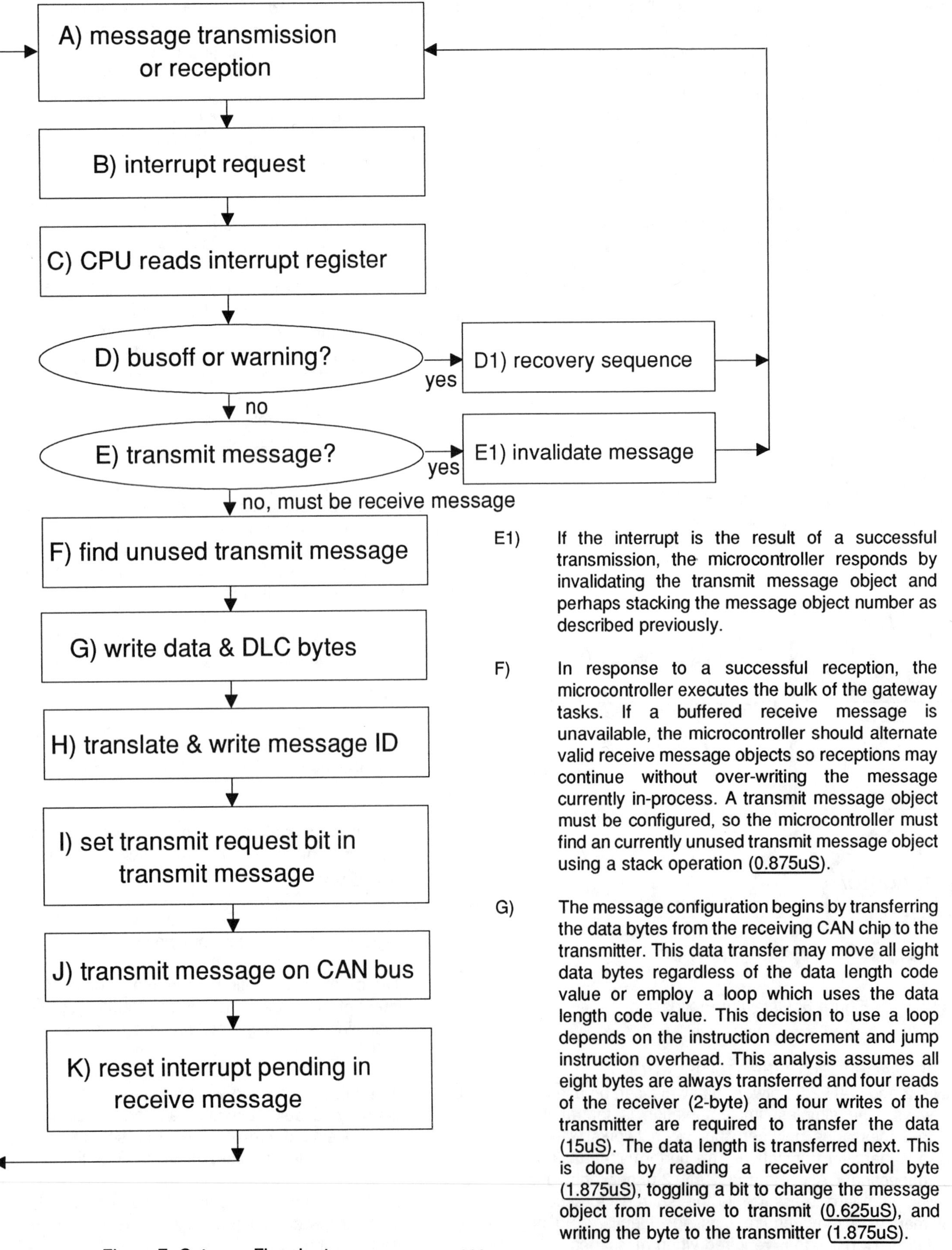

E1) If the interrupt is the result of a successful transmission, the microcontroller responds by invalidating the transmit message object and perhaps stacking the message object number as described previously.

F) In response to a successful reception, the microcontroller executes the bulk of the gateway tasks. If a buffered receive message is unavailable, the microcontroller should alternate valid receive message objects so receptions may continue without over-writing the message currently in-process. A transmit message object must be configured, so the microcontroller must find an currently unused transmit message object using a stack operation (0.875uS).

G) The message configuration begins by transferring the data bytes from the receiving CAN chip to the transmitter. This data transfer may move all eight data bytes regardless of the data length code value or employ a loop which uses the data length code value. This decision to use a loop depends on the instruction decrement and jump instruction overhead. This analysis assumes all eight bytes are always transferred and four reads of the receiver (2-byte) and four writes of the transmitter are required to transfer the data (15uS). The data length is transferred next. This is done by reading a receiver control byte (1.875uS), toggling a bit to change the message object from receive to transmit (0.625uS), and writing the byte to the transmitter (1.875uS).

Figure 7: Gateway Flowchart

H) The message identifier is transferred next assuming an 11-bit identifier and no translation. This requires a read of the receiver and a write of the transmitter (3.75uS).

I) The message object control registers are written to make the transmit message valid and to set the transmit request bit using a write (2-byte) instruction (1.875uS).

J) After setting the transmit request bit, the CAN chip is ready to arbitrate and transmit on the CAN bus (1.75uS).

The resulting gateway processing time from message received to message re-transmission request is approximately 40uS. This figure is used in the transmission latency calculation. The remaining operations (flowchart block K) of the interrupt service routine are reset the receive message interrupt pending bit and depart the interrupt service routine. The total interrupt service routine requires 45uS and figure 7 may be used to evaluate message overrun susceptibility.

TRANSMISSION LATENCY - A rigorous analysis of transmission latency through of a gateway should examine best and worst case scenarios. A few sample calculations are shown to approximate bare-minimum latencies. These latency calculations assume a gateway processing time of 40uS as previously demonstrated. This processing time is, perhaps, the very shortest realizable time and more time may be needed for genuine applications. Other assumptions are:

A) The number of stuff bits are estimated.
B) A message has "immediate" gateway access or is required to "wait" for 40uS another message to undergo full gateway processing.
C) CAN bus rates of 1MB/S and 250KB/S.
D) The second network is idle and the message transmission occurs immediately.
E) A 16-bit microcontroller/CAN chip interface. An 8-bit interface would add 18.75uS to the gateway processing time.

DATA FRAME TRANSMISSION LATENCY

CAN bus/S	data bytes	stuff bits	total bits	gateway access	latency time
1MB	1	3	55	immediate	95uS
1MB	1	3	55	wait	135uS
1MB	8	6	114	immediate	154uS
1MB	8	6	114	wait	194uS
250KB	1	3	55	immediate	260uS
250KB	1	3	55	wait	300uS
250KB	8	6	114	immediate	496uS
250KB	8	6	114	wait	536uS

MESSAGE OVERRUNS - A gateway should eliminate or minimize the occurrence of message overruns. A message overrun occurs when a CAN receive message object stores a second message before the first has been processed resulting in the corruption of the first message. The worst case for overruns is when both CAN chips receive messages at the same time. The gateway must make one receive message wait while it first processes the other receive message.

Message overruns are most likely to occur when both CAN networks require back-to-back messages to be transferred across the gateway. In this case, two messages must be transferred across the gateway in the time required for a single message to be transmitted on the CAN bus plus three intermission bit times. For standard messages with 1 data byte and zero stuff bits, this time is 55 bit-times. For CAN buses operating at 1MB/S, this means two messages must be processed in 55uS; based on the processing time 40uS per message from the prior discussed, this does not appear feasible. For a 250KB/S bus, two messages must be processed in 220uS and this is potentially realizable.

Theoretically, buffered receive message objects do not reduce the possibility of message overruns when messages are transferred across the gateway at a continuous maximum rate. Practically, however, a buffered receive message object may allow two back-to-back messages to be processed without an overrun; a message overrun is deferred until three back-to-back messages are transmitted on the CAN bus.

REMOTE FRAMES

A remote frame is used by a node to request another node to transmit its data. The remote frame itself contains no data. When transmit message objects receive remote frames, they will send data after they win bus arbitration.

For gateway applications, remote frames pose a very special problem. This problem is that most CAN implementation do not save the message identifier of the remote message. Unlike a receive message object which stores the incoming message identifier, the transmit message object responding to a remote frame does not store the remote message identifier. If acceptance filtering is used, the exact message identifier of the remote frame is unknown. Therefore, when the gateway receives a remote frame, it cannot readily configure a corresponding remote message on the second network.

A way to resolve this remote frame message identifier issue is to require all remote frames that cross the gateway to be implemented instead with data frame referenced here as a 'substitute remote frame'. This data frame should contain two bytes that store the correct remote frame message identifier. To simplify matters, this data frame should use a specially assigned message identifier which informs the gateway that the message is a substitute remote frame. If the gateway receives a message with this specially assigned message identifier, the gateway configures a remote frame on the second network using the message identifier designated by the data bytes of the substitute remote frame.

This substitute remote frame method is conducted as shown in figure 8. First, the message identifiers of incoming messages are checked against the specially assigned remote message identifier. When a match occurs, the gateway configures a remote message on the second network using the message identifier stored in the matched message. Next, a node on the second network should respond to the remote frame by sending a data frame. This data frame is handled no differently than any other data frame and is transferred to the originating network without special regard.

The drawback of this substitute remote frame method is the gateway processing tasks increase because the message identifier of data frames must be checked against the specially assigned remote message identifier; this increases latency time. A second drawback is the use of data bytes to store message identifiers is inconsistent with the CAN protocol philosophy.

TRANSMISSION LATENCY - The minimum transmission latency of a substitute remote message is two times the latency of a data frame plus the transmission time to send a remote frame on the bus. After the reception of a substitute remote frame, there is 1) gateway processing, 2) a remote frame transmission (zero data bytes), 3) a data frame transmission in response to the remote frame, 4) a second gateway processing, and 5) a data frame transmission on the originating bus.

Transmission latency examples are shown below. The "total bits" refers to the number of bits in the data frame which is requested by the remote frame. The gateway access time, in the worst case, could occur twice. This means that both data frame transfers across the gateway must wait for a full gateway access time. Therefore, the 'wait' gateway access assumption adds 80uS (2*40uS).

SUBSTITUTE REMOTE FRAME TRANSMISSION LATENCY

CAN bus/S	data bytes	stuff bits	total bits	gateway access	latency time
1MB	1	3	55	immediate	234uS
1MB	1	3	55	wait	314uS
1MB	8	6	114	immediate	352uS
1MB	8	6	114	wait	432uS
250KB	1	3	55	immediate	696uS
250KB	1	3	55	wait	776uS
250KB	8	6	114	immediate	1.17mS
250KB	8	6	114	wait	1.25uS

OVERRUN - The analysis of overruns is practically the same as for data frames except substitute remote frames only have two data bytes.

GATEWAY MESSAGE HANDLING

To minimize message overruns, message traffic across the gateway must be considered. There are two ways to manage gateway traffic. First, ensure both CAN buses have a transmission rate which allows the gateway to transfer two messages in the time required to transmit a single message on either bus (see the transmission latency section). A second way to manage message traffic is to use acceptance filtering to limit the rate that message are transferred across the gateway. The CAN chips connected to the gateway microcontroller are programmed to select a subset of the messages for transfer across the gateway using "acceptance masks". In this case, both CAN buses may run at 1MB/S as long as the message traffic across the gateway is contained to a manageable level. It was shown previously (transmission latency section) that transferring all messages across the gateway when both CAN buses are transmitting at 1MB/S is a difficult task.

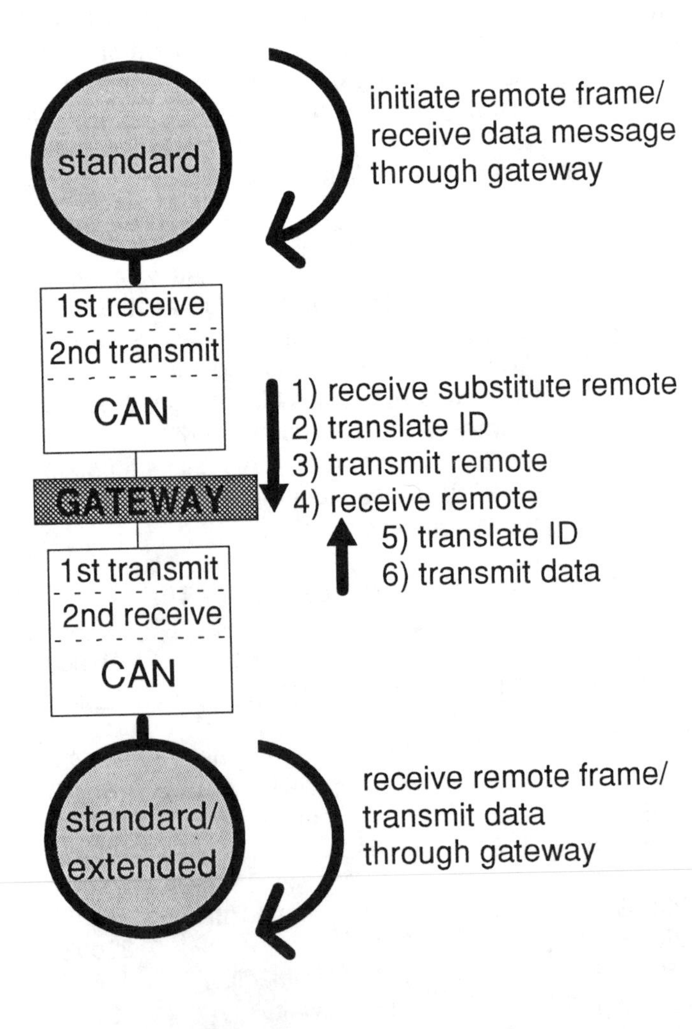

Figure 8: Substitute Remote Frame

CONCLUSION

Most existing CAN 1.2 chips will send error frames when CAN 2.0 chips extended frame messages are transmitted, because they are CAN 2.0 non-passive. To interconnect CAN 1.2 and CAN 2.0 networks together, a gateway may be employed[1]. A gateway should minimize transmission latency and lost messages (overruns). A simple gateway demonstrates the minimum latency of a CAN 1.2 to CAN 2.0 gateway ranges from 95uS to 1.25mS. To reduce the risk of message overruns, the message traffic across the gateway must be limited to twice the message transmission time of either CAN bus.

References:

1. "CAN Specification Version 2.0", Robert Bosch GmbH, Postfach 50, D-7000 Stuttgart, 1991.
2. Fred Phail, "Controller Area Network - An In-Vehicle Solution", SAE paper #880588, 1988.
3. Fred Phail, "In-Vehicle Networking - Serial Communications Requirements and Directions", SAE paper #860390, 1986.
4. Craig Szydlowski, "CAN Specification 2.0: Protocol and Implementations", SAE paper #921603, 1992.
5. 82527 CAN Literature Packet, Intel literature phone number (1-800-346-3028), order number 272304-001.

Endnote:

1. The Intel 82527 serial communications controller implements many of the key gateway features discussed in this paper. The 82527 is able to interface to both standard and dual standard/extended networks. The Intel 82527 supports the latest CAN 2.0 specification and manages both standard and extended format messages. The 82527 has 15 message objects including a buffered receive message object. Two acceptance filters provide added message handling flexibility. The 82527 interfaces to microcontrollers with 8- and 16-bit multiplexed address/data buses, non-multiplexed buses, or SPI compatible serial buses.

930006

Generic FMEA for Stand-Alone CAN Devices

Craig Szydlowski and Mukund Patel
Intel Corp.

ABSTRACT

Failure Mode and Effect Analysis (FMEA) is a useful tool to aid in the detection and prevention of product defects. FMEA's are used throughout the automotive industry to facilitate the delivery of failure-averse modules to customers. This paper applies FMEA techniques to generic stand-alone Controller Area Network (CAN) chips. The CAN protocol, developed by ROBERT BOSCH GmbH, offers a comprehensive solution to managing communication between multiple CPUs.

Device-level FMEA's investigate the system-level impact of electrical abnormalities on a pin-by-pin basis. Conditions such as pins shorted to power or to other pins are evaluated for their effect on the system. Severity levels are also estimated.

An FMEA for generic stand-alone CAN devices must consider failure modes for various pin types: power, clock, clockout, interrupt, mode selection, CAN bus, I/O ports, and address/data bus. In addition, this paper discusses how CAN devices detect and respond to bus errors and explores how the system may utilize this information in an FMEA analysis.

COMPONENTS OF A CAN BUS NODE

A CAN bus node has four components: host-CPU, CAN chip, CAN bus driver and the CAN bus. In this paper, these CAN node components are physically distinct, although it is clearly feasible to integrate two or more of these components onto a single chip. Figure 1 illustrates the CAN bus node.

The host-CPU is a microcontroller that performs an application function such as engine control or anti-lock braking. This host-CPU communicates with other modules to share sensor data and pertinent parameters and to implement diagnostic capabilities.

The host-CPU programs its dedicated CAN chip to transmit/receive messages from the CAN bus; the CAN chip responds with interrupts after relevant bus communications occur. The CAN chip relieves the host-CPU of the burden of monitoring all bus transmissions. The CAN chip interfaces to the CAN bus using a bus driver. This bus driver is capable of driving the bus while simultaneously sensing the bus value through an independent receive pin. This functionality is required since CAN is a "Carrier Sense, Multiple Access, with Collision Resolution" or CSMA/CR protocol(1). The CAN bus driver interfaces to a two-wire CAN bus with signals typically called CAN_H and CAN_L.

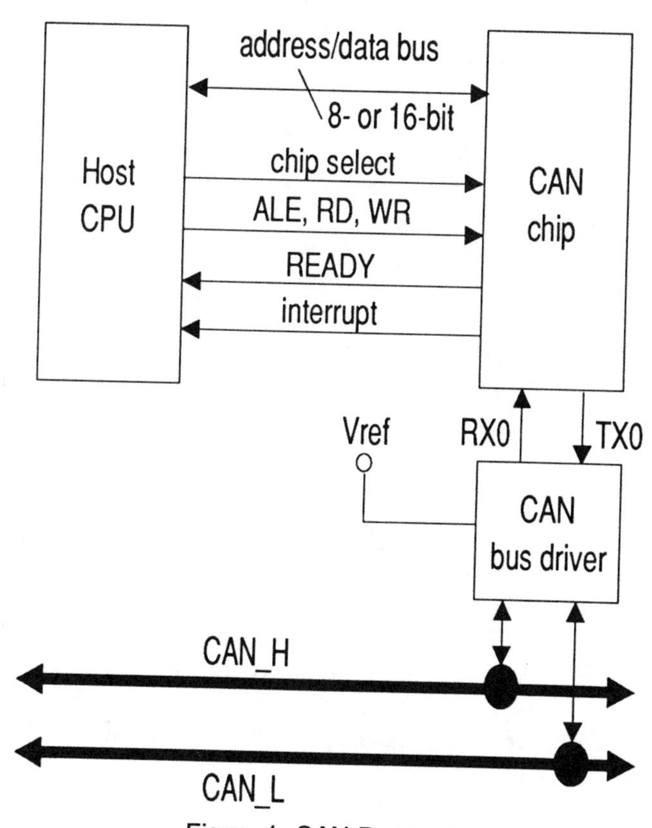

Figure 1: CAN Bus Node

FMEA: FAILURE MODE and EFFECT ANALYSIS

An FMEA document is used to identify situations in the module that jeopardize reliable operation. In particular, FMEAs describe possible outcomes when chip pins are open or are shorted to other pins. Unreliable chip operation is always possible anytime a short or open occurs within a module, however, the consequences to overall module function may differ depending upon the affected pin(s)(**2**).

CAN CHIP PINS - The pin functions of a stand-alone CAN chip may be loosely categorized as power, clocking, host-CPU interface, CAN bus interface and peripheral. Aside from peripheral pins, all (used) CAN chip pins should be considered severity level = 10 or 'critical' and proper/adequate connections are required to ensure basic chip functionality.

Another severity distinction with respect to CAN chip pins is whether improper connections result in the malfunction of the host-CPU. For example, two shorted CAN chip address lines may corrupt the host-CPU's address/data bus and may prohibit the host-CPU from accessing external program memory. Consequently, the host-CPU is unable to execute instructions and is therefore incapacitated. This outcome is usually more severe than an inoperative CAN chip.

POWER PINS - Proper VCC (power) and VSS (ground) connections are absolutely necessary for basic chip functionality.

CLOCKING PINS - External clocking from a crystal or ceramic resonator is typically required for chip functionality. Some CAN chips may allow clocking by CAN bus transmissions or a limp capacitor configuration. If the CAN chip CLOCKOUT signal is used to drive the host-CPU, it is a critical signal.

HOST-CPU INTERFACE PINS - CAN chips support various address/data bus configurations to interface to different host-CPUs.

1) 8- or 16-bit multiplexed address/data bus (chip select, address/data, ALE, WR, RD and READY)
2) 8-bit multiplexed address/data bus (chip select, address/data, AS, E and R/W#)
3) Asynchronous non-multiplexed address/data bus (chip select, address/data, R/W# and DSACK0#)
4) Synchronous non-multiplexed address/data bus (chip select, address/data, R/W# and E)
5) SPI compatible serial (chip select, SCLK, MISO and MOSI)

All of the pins associated with the address/data bus configuration are critical to ensure CAN chip functionality. As expressed earlier, these bus pins may affect other peripherals or memory which also reside on the host-CPU external bus.

CAN BUS INTERFACE PINS - The CAN chip interfaces to a driver using minimally TX0 and RX0 pins. A fully differential interface uses TX1 and RX1 pins as well. These pins are indirectly connected to the CAN bus and are therefore essential for proper CAN chip functionality. In addition, the CAN bus may be inoperative if TX0 is physically shorted to a dominant state which forces the CAN bus driver to drive the CAN bus dominant; this is the only CAN chip pin fault which may impair normal CAN bus operation. The CAN bus driver connects to CAN_L and CAN_H of the CAN bus and if either of these signals is shorted, the CAN bus will not function.

CAN PERIPHERAL PINS - Some CAN chips support I/O ports and other peripheral functions. Although peripherals may serve an important application function, these pins typically do not affect the functionality of the CAN chip, the host-CPU or the CAN bus.

An illustration of the previously described pins are shown in figure 2, the Intel 82527 CAN chip pinout (**3**).

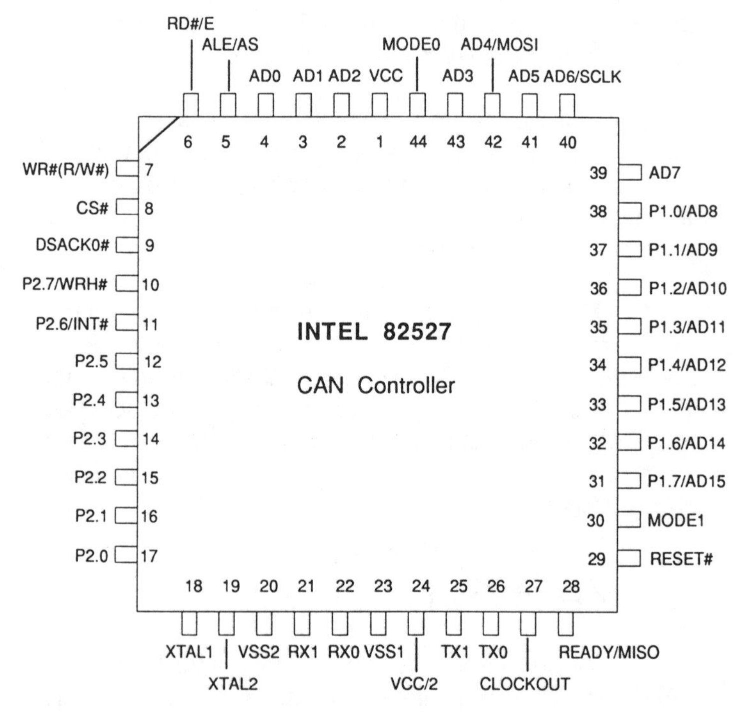

Figure 2: Pinout of the Intel 82527 CAN Chip

The FMEA analysis for these pin groups is shown in tables 1 through 9. Whenever a pin is considered shorted to another pin or supply, it is assumed that the original connection is still intact.

- Table 1: Power Supply Pins
- Table 2: Clocking and Chip Select Pins
- Table 3: Control Pins
- Table 4: CAN Bus and I/O Port Pins
- Tables 5,6: Intel Multiplexed Address/Data Bus Pins
- Table 7: Non-Intel Multiplexed Address/Data Bus Pins
- Table 8: Non-Multiplexed Address/Data Bus Pins
- Table 9: SPI Compatible Serial Pins

BASIC POINTS - There are some basic points relevant to failure modes and pin conditions:

1) Whenever two output pins are shorted together, the possibility of overcurrent exists. There is a strong potential for localized heating leading to signal line and/or chip destruction.

2) A failure may occur without system detection resulting in 2nd and 3rd order effects which corrupt the system. For example, the host-CPU may read corrupted data due to shorted data lines which leads to the software algorithm executing an inappropriate program flow.

3) Floating CMOS input pins often result in excessive ICC supply currents. An input voltage between 1 to 3.5 volts typically shorts VCC and VSS through n- and p-channel transistors. Therefore, it is assumed input pins are driven high or low and a short to an input pin is similar to a short to power or ground.

FMEA TABLES: THE MAIN POINTS

1) All CAN chip pins, besides peripheral pins, are critical to the functionality of the CAN node.

2) If the host-CPU uses the address/data bus to interface to chips besides the CAN chip, the failure of the CAN chip bus pins may be catastrophic.

3) Failure of the TX0/TX1 pins may be catastrophic to the CAN bus if the CAN bus driver responds by shorting the bus lines to power or ground.

CAN BUS DRIVER

The CAN bus driver is connected to the CAN_L and CAN_H signals of the bus. If either of these is shorted to power or ground, the CAN bus may not function. This conditional is detectable by the host-CPU using the CAN chip error, warning and busoff flags. To recover from this condition, the CAN bus driver may be designed to allow the host-CPU to float the bus driver outputs through the use of a control line. The simplest example is to connect the CAN bus driver to the CAN bus through a pass-gate composed of n- and p-type transistor which are controlled by the host-CPU (Figure 3). This method, however, will impact performance (speed) and electrical characteristics (voltage level reduction).

Physical layer implementations may be designed to sense that CAN_L or CAN_H is shorted and subsequently alter communication to utilize the functional CAN bus line.

Redundant bus drivers are only useful if the bus driver has significantly lower reliability than the other components of the module.

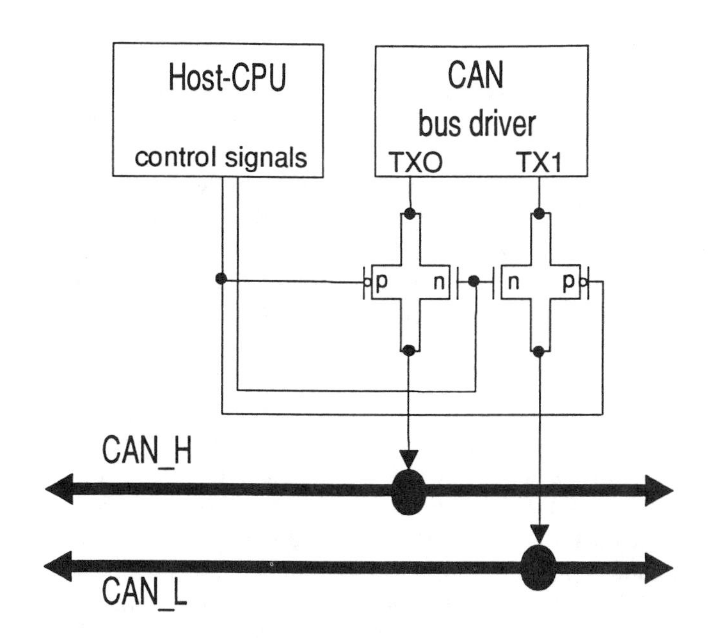

Figure 3: Pass-gate Between Bus Driver and CAN Bus

FAILURE MODE DETECTION

The host-CPU views the CAN chip as a smart RAM. The host-CPU may test its interface to the CAN chip by writing and reading diagnostic bytes to/from the CAN chip. For example, the CPU can write unique data to each data bytes, such as the data byte address (i.e. write the value AAH to location AAH, and so on). The CPU then reads back the data byte values. This diagnostic test verifies the address/data bus connections between the host-CPU and the CAN are functioning. Diagnostic tests also demonstrate the functionality of other pins such as XTAL1/XTAL2, chip select and mode pins.

The host CPU should interrogate CAN chip status/error registers which indicate the CAN bus status. Some CAN chips set flags when the CAN bus appears stuck at dominant or recessive levels.

The host-CPU may also program the CAN chip to request data from another node (remote frame) to verify its receive capability. Similarly, a diagnostic procedure may be implemented where test messages are transmitted and received on the network. The host-CPUs check the received messages to determine if any data corruption occurred within the CAN chip. For example, test messages using bytes such as AAH and 55H may check whether any of the bit-lines in the CAN RAM chip are shorted or open.

Through the use diagnostic bytes and programmed messages, a functioning host-CPU should verify that the CAN chip pins are wired without any shorts or opens. Peripheral functions may also be checked by routing parallel tracks to port pins of the host-CPU to test signal line states.

FAILURE MODE CORRECTIVE ACTION

Systems that detect catastrophic module failures and safely shut down or enter a limp-home mode limit the risk to the end-customer. One possible limp-home mode for a CAN node is redundant host-CPU interfaces to the CAN chip. If both parallel and serial interfaces are implemented, a host-CPU that identifies a parallel bus failure could resort to the serial interface, thereby restoring communication with the CAN chip.

CAN BUS ERRORS

The CAN protocol specifies error confinement procedures to assist the detection of CAN bus errors. Each CAN chip has two error counters: receive and transmits. These counters increment and decrement according to specific CAN protocol rules. Bus values that violate the CAN protocol typically increment these counters whereas orderly transmissions decrement these counters. When either of these counters reach a critical level, a warning flag is set. If these counters increment further, a catastrophic state is reached and the CAN bus will go "busoff" and float its CAN bus drivers.

These warning and busoff flags alert the host-CPU to take action. This action may be to run a diagnostic procedure to determine if its CAN chip is malfunctioning and requires a reset (hard or soft). If the host-CPU cannot correct the problem, it programs the CAN bus drivers to a float state. A functioning host-CPU must be able to identify a problem with its own CAN node and subsequently remove it from the CAN bus when a problem is detected. The only failure that a functioning host-CPU cannot respond to is when the CAN_L and CAN_H signals of the CAN bus driver are shorted.

CAN STUFF BITS

The format of CAN protocol messages are clearly defined in the CAN Protocol Specification(**4**). However, the actual lengths of these messages may differ because 'stuff' bits are added to the message. Stuff bits assist synchronization by adding transitions to the message. A stuff bit is inserted in the bit stream after five consecutive-equal value bits are transmitted; the stuff bit is the opposite polarity of the five consecutive bits. All message fields are stuffed except the CRC delimiter, the ACK field and the End of Frame.

The transmitting CAN node adds stuff bits to the message as needed while it send the message. All active nodes on the network must check to make sure stuff bits are added properly. Although stuff bits are not verified through the Cyclical Redundancy Code (CRC), they are checked by each non-transmitting active node on the bus. Consequently, no failure mode is particularly associated with stuff bits.

CONCLUSION

Failure Mode and Effect Analysis provides module designers with some insight into chip failure modes and the corresponding degree of severity. In this FMEA analysis, the TX0/TX1 and the address/data bus pins may be critical to the functioning of the host-CPU and the CAN bus, respectively. By understanding and anticipating failures, recovery schemes may be developed to further reduce the potential risk to the end-customer.

1. Carrier Sense, Multiple Access, with Collision Resolution is a type of serial data link that allows multiple nodes to be begin transmission at the same time. These simultaneous transmissions are arbitrated and one transmitter ends up retaining control of the link without corrupting the messages of the other transmitters.

2. This FMEA is only a guide and is NOT intended to guarantee device operation under any of the noted failure conditions.

3. The 82527 CAN Controller is a joint development of Intel and ROBERT BOSCH GmbH.

4. "CAN Specification Version 2.0", Robert Bosch GmbH, Postfach 50, D-7000 Stuttgart, 1991.

Generic FMEA For Stand-alone CAN Chip: Power Supply Pins

Pin Name	Pin Description	Failure Condition	Effect upon CAN chip	Effect upon host CPU	Effect upon CAN BUS
VCC	Main supply voltage (+5.0V).	Pin open (floating)	Loss of supply, chip does not function.	Inability to access CAN. A) Inoperative if clockout used.	Node will not function. CAN bus operation is normal.
		Pin shorted to VCC/VSS	VCC short is normal connect. VSS short - chip inoperative and excessive current.	Inability to access CAN A) Inoperative if clockout used.	Node will not function. B) Node may corrupt bus.
		Pin shorted to adjacent pin	Pin tied/driven 1) high, chip functions 2) low, chip inoperative and excessive current.	Inability to access CAN A) Inoperative if clockout used.	Node will not function. B) Node may corrupt bus.
VSS	Circuit ground input.	Pin open (floating)	Loss of supply ground, chip does not function.	Inability to access CAN A) Inoperative if clockout used.	Node will not function. CAN bus operation is normal.
		Pin shorted to VCC/VSS	VSS short is normal connect. VCC short - chip inoperative and excessive current.	Inability to access CAN A) Inoperative if clockout used.	Node will not function. B) Node may corrupt bus.
		Pin shorted to adjacent pin	Pin tied/driven 1) high, chip inoperative and excessive current 2) low, chip functions.	Inability to access CAN A) Inoperative if clockout used.	Node will not function. B) Node may corrupt bus.
VCC/2	Physical layer reference voltage output.	Pin open (floating)	Normal operation.	Normal operation.	Node may not function if physical layer interface malfunctions.
		Pin shorted to VCC/VSS	Chip may be inoperative, excessive current.	CPU may be inoperative if its VCC or VSS supply is affected.	Node may not function if physical layer interface malfunctions.
		Pin shorted to adjacent pin	Chip may be inoperative, excessive current.	Normal operation if shorted pin does not connect to CPU and VCC/VSS unaffected.	Node may not function if physical layer interface malfunctions.

A. The CPU will be inoperative if its XTAL1 input is driven by the CAN chip's clockout output.
B. If the supply pin shorts after successful initialization, the CAN chip may be left in a state that is driving the bus dominant.

Table 1

Generic FMEA For Stand-alone CAN Chip: Clocking and Chip Select Pins

Pin Name	Pin Description	Failure Condition	Effect upon CAN chip	Effect upon host CPU	Effect upon CAN BUS
XTAL1 and XTAL2	Input and output of the oscillator and inverter (external clock).	Pin open (floating)	Oscillator circuit corrupted, chip inoperative.	Inability to access CAN. A) Inoperative if clockout used.	Node will not function. B) Node may corrupt bus.
		Pin shorted to VCC/VSS	Oscillator circuit corrupted, chip inoperative.	Inability to access CAN. A) Inoperative if clockout used.	Node will not function. B) Node may corrupt bus.
		Pin shorted to adjacent pin	Oscillator circuit corrupted, chip inoperative.	Inability to access CAN. A) Inoperative if clockout used.	Node will not function. B) Node may corrupt bus.
Clockout	Clock output derived from XTAL1/XTAL2 used to drive the clock input of other chips.	Pin open (floating)	Normal operation.	A) Inoperative if clockout used.	CAN bus operation is normal.
		Pin shorted to VCC/VSS	Unused - normal operation. Chip may be inoperative, excessive current.	A) Inoperative if clockout used, or VCC/VSS affected and excessive current.	Node may not function. CAN bus operation is normal.
		Pin shorted to adjacent pin	Unused - normal operation. Chip may be inoperative, excessive current.	A) Inoperative if clockout used, or VCC/VSS affected and excessive current.	Node will not function. CAN bus operation is normal.
Chip Select	Input signal enables read and write access to the chip.	Pin open (floating)	CPU access to chip is unreliable.	CPU - Normal operation. Unreliable CAN access.	Node function is unreliable. CAN bus operation is normal.
		Pin shorted to VCC/VSS	VSS short - no CPU access. C) VCC short - corrupt data or possibly normal operation.	CPU may be inoperative if VCC/VSS affected and excessive current.	Node function is unreliable. CAN bus operation is normal.
		Pin shorted to adjacent pin	Chip may be inoperative, excessive current.	CPU may be inoperative if VCC/VSS affected and excessive current.	Node function is unreliable. CAN bus operation is normal.

A. The CPU will be inoperative if its XTAL1 input is driven by the CAN chip's clockout output.
B. If the oscillator connection floats/shorts after successful initialization, the CAN chip may be left in a state that is driving the bus dominant.
C. If the CPU external bus accesses other chips besides the CAN chip, the CAN data will be corrupted. If all external accesses are directed to the CAN chip, normal operation is possible.

Table 2

Generic FMEA For Stand-alone CAN Chip: Control Pins

Pin Name	Pin Description	Failure Condition	Effect upon CAN chip	Effect upon host CPU	Effect upon CAN BUS
RESET	Input signal used to initiate a hardware reset.	Pin open (floating)	Chip inoperative due to lack of start-up sequencing.	Inability to access CAN. A) Inoperative if clockout used.	Node will not function. CAN bus operation is normal.
		Pin shorted to VCC/VSS	Chip inoperative due to lack of start-up sequencing.	Inability to access CAN. A) Inoperative if clockout used.	Node will not function. CAN bus operation is normal.
		Pin shorted to adjacent pin	Chip inoperative due to lack of start-up sequencing.	Inability to access CAN. A) Inoperative if clockout used.	Node will not function. CAN bus operation is normal.
INT	The interrupt output pin informs the CPU that a CAN internal interrupt is pending.	Pin open (floating)	Normal operation if pin is unused, otherwise CPU fails to respond to CAN interrupts.	CPU operates normally, but it fails to respond to CAN interrupts.	Node functions improperly if interrupts used. CAN bus operation is normal.
		Pin shorted to VCC/VSS	Normal operation if pin is unused, otherwise B) excessive current.	CPU interrupt signal corrupted, CPU inoperative if VCC/VSS affected.	Node functions improperly if interrupts used. CAN bus operation is normal.
		Pin shorted to adjacent pin	Normal operation if pin is unused, otherwise B) excessive current.	CPU interrupt signal corrupted, CPU inoperative if VCC/VSS affected.	Node functions improperly if interrupts used. CAN bus operation is normal.
Mode	Input pins to select CPU-CAN interface modes (address/ data bus configuration)	Pin open (floating)	C) CPU access to CAN is questionable.	CPU - Normal operation. C) CPU access to CAN is questionable.	Node function is unreliable. CAN bus operation is normal.
		Pin shorted to VCC/VSS	C) CPU access to CAN is questionable.	CPU - Normal operation. C) CPU access to CAN is questionable.	Node function is unreliable. CAN bus operation is normal.
		Pin shorted to adjacent pin	C) CPU access to CAN is questionable.	CPU - Normal operation. C) CPU access to CAN is questionable.	Node function is unreliable. CAN bus operation is normal.

A. The CPU will be inoperative if its XTAL1 input is driven by the CAN chip's clockout output.
B. VSS short: pull-up always sources maximum current, VCC short: excessive current if interrupt active. Chip inoperative if VCC/VSS severely affect
C. The CAN chip may be programmed for the wrong address/data bus configuration and this depends upon the voltage on the mode pins upon RESET. Mode pins are usually wired directly to VCC and/or VSS.

Table 3

Generic FMEA For Stand-alone CAN Chip: CAN Bus and I/O Port Pins

Pin Name	Pin Description	Failure Condition	Effect upon CAN chip	Effect upon host CPU	Effect upon CAN BUS
RX0/RX1	Input signals from the CAN bus to the input comparator.	Pin open (floating)	Chip may go busoff after monitoring bit errors. No messages received.	Normal operation.	CAN bus operation is normal, but error frames transmitted by node until busoff.
	* RX1 may be unused with a twisted pair bus.	Pin shorted to VCC/VSS	Chip may go busoff after monitoring bit errors. No messages received.	Normal operation.	CAN bus operation is normal, but error frames transmitted by node until busoff.
		Pin shorted to adjacent pin	Chip may go busoff after monitoring bit errors. No messages received.	Normal operation.	CAN bus operation is normal, but error frames transmitted by node until busoff.
TX0/TX1	Output signals to the CAN bus.	Pin open (floating)	Chip goes busoff after monit-oring bit errors, messages not transmitted.	Normal operation.	CAN bus operation is normal, but error frames transmitted by node until busoff.
	* TX1 may be unused with a twisted pair bus.	Pin shorted to VCC/VSS	Chip goes busoff after monit-oring bit errors, no message transmitted A) over-current.	Normal operation if VCC/VSS unaffected.	CAN bus may be inoperative if the bus is driven dominant by the CAN bus driver.
		Pin shorted to adjacent pin	Chip goes busoff after monit-oring bit errors, no message transmitted A) over-current.	Normal operation if VCC/VSS unaffected.	CAN bus may be inoperative if the bus is driven dominant by the CAN bus driver.
I/O Ports	Bi-directional input/output port pins used for low speed applications.	Pin open (floating)	Unused, normal operation. Used, input/output data corrupted.	Normal operation.	CAN bus operation is normal.
		Pin shorted to VCC/VSS	Unused, normal operation. Used, possible input data corrupt, output overcurrent.	Normal operation if VCC/VSS unaffected.	CAN bus operation is normal.
		Pin shorted to adjacent pin	Unused, normal operation. Used, possible input data corrupt, output overcurrent.	Normal operation if VCC/VSS unaffected.	CAN bus operation is normal.

A. If the used TX pins are shorted to VSS or VCC, an excessive current condition may result and the device may become inoperative.

Table 4

Generic FMEA For Stand-alone CAN Chip: Intel Multiplexed Address/Data Bus

Pin Name	Pin Description	Failure Condition	Effect upon CAN chip	Effect upon host CPU	Effect upon CAN BUS
ALE	Address latch enable input used by the chip to latch the address driven by the CPU.	Pin open (floating)	No CPU access to CAN chip.	No CPU read access to CAN chip. A) CPU may be inoperative.	Node will not function. CAN bus operation is normal.
		Pin shorted to VCC/VSS	No CPU access to CAN chip.	CPU may be inoperative if VCC/VSS affected & excess current. A) CPU bus fails.	Node will not function. CAN bus operation is normal.
		Pin shorted to adjacent pin	No CPU access to CAN chip.	CPU may be inoperative if VCC/VSS affected & excess current. A) CPU bus fails.	Node will not function. CAN bus operation is normal.
RD	Read signal input used to indicate a read operation.	Pin open (floating)	No CPU read access to CAN chip.	No CPU read access to CAN chip.	Node will not function. CAN bus operation is normal.
		Pin shorted to VCC/VSS	No CPU read access to CAN chip.	CPU may be inoperative if VCC/VSS affected & excess current. A) CPU bus reads fail.	Node will not function. CAN bus operation is normal.
		Pin shorted to adjacent pin	No CPU read access to CAN chip.	CPU may be inoperative if VCC/VSS affected & excess current. A) CPU bus reads fail.	Node will not function. CAN bus operation is normal.
WR	Write signal input used to indicate a write operation.	Pin open (floating)	No CPU write access to CAN chip.	No CPU write access to CAN chip.	Node will not function. CAN bus operation is normal.
		Pin shorted to VCC/VSS	No CPU write access to CAN chip.	CPU may be inoperative if VCC/VSS affected & excess current. A) CPU bus writes fail.	Node will not function. CAN bus operation is normal.
		Pin shorted to adjacent pin	No CPU write access to CAN chip.	CPU may be inoperative if VCC/VSS affected & excess current. A) CPU bus writes fail.	Node will not function. CAN bus operation is normal.

A. The CPU may be inoperative if the external address/data bus is corrupted, especially if the CPU uses external memory.

Table 5

Generic FMEA For Stand-alone CAN Chip: Intel Multiplexed Address/Data Bus

Pin Name	Pin Description	Failure Condition	Effect upon CAN chip	Effect upon host CPU	Effect upon CAN BUS
AD0-AD7 or AD0-AD15	8- or 16-bit multiplexed address/data bus used for interfacing to the host-CPU. These pins are bi-directional.	Pin open (floating)	A) Unreliable read/write access to CAN chip.	A) Unreliable read/write access to CPU external bus. C) CPU may be inoperative.	Unreliable node function. CAN bus operation is normal.
		Pin shorted to VCC/VSS	A) Unreliable read/write access to CAN chip. B) Excessive current.	CPU may be inoperative if VCC/VSS affected & excess current. C) CPU bus fails.	Unreliable node function. CAN bus operation is normal.
		Pin shorted to adjacent pin	A) Unreliable read/write access to CAN chip. B) Excessive current.	CPU may be inoperative if VCC/VSS affected & excess current. C) CPU bus fails.	Unreliable node function. CAN bus operation is normal.
READY	Output to synchronize CPU accesses. Used to extend read/write operations.	Pin open (floating)	Normal operation if pin is unused, otherwise unreliable access to CAN chip.	Normal operation if pin is unused, otherwise C) unreliable CPU bus activity.	Unused - node functions. CAN bus operation is normal.
		Pin shorted to VCC/VSS	Normal operation if pin is unused, otherwise D) excessive current.	Normal operation if pin is unused, otherwise C) unreliable CPU bus activity.	Unused - node functions. CAN bus operation is normal.
		Pin shorted to adjacent pin	Normal operation if pin is unused, otherwise D) excessive current.	Normal operation if pin is unused, otherwise C) unreliable CPU bus activity.	Unused - node functions. CAN bus operation is normal.

A. The address and data are corrupted resulting in the CPU misconfiguring the CAN chip and reading corrupt data.
B. The address/data bus have dynamic states and a short will result in excessive current and possibly making the chip inoperative.
C. The CPU may be inoperative if the external address/data bus is corrupted, especially if the CPU uses external memory.
D. VSS short: pull-up always sources maximum current, VCC short: excessive current if READY active. Chip inoperative if VCC/VSS affected.

Table 6

Generic FMEA For Standalone CAN Chip: Non-Intel Multiplexed Address/Data Bus

Pin Name	Pin Description	Failure Condition	Effect upon CAN chip	Effect upon host CPU	Effect upon CAN BUS
AS	Address latch enable input used by the chip to latch the address driven by the CPU.	Pin open (floating)	No CPU access to CAN chip.	No CPU access to CAN chip. A) CPU may be inoperative.	Node will not function. Can bus operation is normal.
		Pin shorted to VCC/VSS	No CPU access to CAN chip.	CPU may be inoperative if VCC/VSS affected & excess current. A) CPU bus fails.	Node will not function. Can bus operation is normal.
		Pin shorted to adjacent pin	No CPU access to CAN chip.	CPU may be inoperative if VCC/VSS affected & excess current. A) CPU bus fails.	Node will not function. Can bus operation is normal.
E	Input signal to latch R/W# and to signal when data is read from the bus.	Pin open (floating)	No CPU read/write access to CAN chip.	No CPU read/write access to CAN chip.	Node will not function. Can bus operation is normal.
		Pin shorted to VCC/VSS	No CPU read/write access to CAN chip.	CPU may be inoperative if VCC/VSS affected & excess current. A) CPU bus reads fail.	Node will not function. Can bus operation is normal.
		Pin shorted to adjacent pin	No CPU read/write access to CAN chip.	CPU may be inoperative if VCC/VSS affected & excess current. A) CPU bus reads fail.	Node will not function. Can bus operation is normal.
R/WR#	Input signal to indicate a read or write signal.	Pin open (floating)	CPU may either read or write to CAN chip, but not both. CAN is inoperative.	No CPU read/write access to CAN chip.	Node will not function. Can bus operation is normal.
		Pin shorted to VCC/VSS	CPU may either read or write to CAN chip, but not both. CAN is inoperative.	CPU may be inoperative if VCC/VSS affected & excess current. A) CPU bus writes fail.	Node will not function. Can bus operation is normal.
		Pin shorted to adjacent pin	CPU may either read or write to CAN chip, but not both. CAN is inoperative.	CPU may be inoperative if VCC/VSS affected & excess current. A) CPU bus writes fail.	Node will not function. Can bus operation is normal.

A. The CPU may be inoperative if the external address/data bus is corrupted, especially if the CPU uses external memory.

Table 7

Generic FMEA For Stand-alone CAN Chip: Non-multiplexed Address/Data Bus

Pin Name	Pin Description	Failure Condition	Effect upon CAN chip	Effect upon host CPU	Effect upon CAN BUS
A0-A7	8-bit non-multiplexed address bus inputs used for interfacing to the host-CPU.	Pin open (floating)	A) Unreliable read/write access to CAN chip.	A) Unreliable read/write access to CPU external bus. C) CPU may be inoperative.	Unreliable node function CAN bus operation is normal.
		Pin shorted to VCC/VSS	A) Unreliable read/write access to CAN chip. B) Excessive current.	CPU may be inoperative if VCC/VSS affected & excess current. C) CPU bus fails.	Unreliable node function CAN bus operation is normal.
		Pin shorted to adjacent pin	A) Unreliable read/write access to CAN chip. B) Excessive current.	CPU may be inoperative if VCC/VSS affected & excess current. C) CPU bus fails.	Unreliable node function CAN bus operation is normal.
D0-D7	8-bit non-multiplexed data bus input/outputs used for interfacing to the host-CPU. These pins are bi-directional.	Pin open (floating)	A) Unreliable read/write access to CAN chip.	A) Unreliable read/write access to CPU external bus. C) CPU may be inoperative.	Unreliable node function CAN bus operation is normal.
		Pin shorted to VCC/VSS	A) Unreliable read/write access to CAN chip. B) Excessive current.	CPU may be inoperative if VCC/VSS affected & excess current. C) CPU bus fails.	Unreliable node function CAN bus operation is normal.
		Pin shorted to adjacent pin	A) Unreliable read/write access to CAN chip. B) Excessive current.	CPU may be inoperative if VCC/VSS affected & excess current. C) CPU bus fails.	Unreliable node function CAN bus operation is normal.
DSACK0#	Output to synchronize CPU accesses. Used to extend read/write operations.	Pin open (floating)	Normal operation if pin is unused, otherwise unreliable access to CAN chip.	Normal operation if pin is unused, otherwise C) unreliable CPU bus activity.	Unreliable node function CAN bus operation is normal.
		Pin shorted to VCC/VSS	Normal operation if pin is unused, otherwise D) excessive current.	Normal operation if pin is unused, otherwise C) unreliable CPU bus activity.	Unreliable node function CAN bus operation is normal.
		Pin shorted to adjacent pin	Normal operation if pin is unused, otherwise D) excessive current.	Normal operation if pin is unused, otherwise C) unreliable CPU bus activity.	Unreliable node function CAN bus operation is normal.

A. The address and data are corrupted resulting in the CPU misconfiguring the CAN chip and reading corrupt data.
B. The address/data bus changes states frequently resulting in shorting, possibly generating excessive current and making the chip inoperative.
C. The CPU may be inoperative if the external address/data bus is corrupted, especially if the CPU uses external memory.
D. VSS short: pull-up always sources maximum current, VCC short: excessive current when DSACK0# Chip inoperative if VCC/VSS affected.

Table 8

Generic FMEA For Stand-alone CAN Chip: SPI Serial Bus

Pin Name	Pin Description	Failure Condition	Effect upon CAN chip	Effect upon host CPU	Effect upon CAN BUS
SCLK	System clock input driven by the CPU.	Pin open (floating)	No CPU access to CAN chip.	No CPU access to CAN chip. A) CPU serial bus corrupt.	Node will not function. CAN bus operation is normal.
		Pin shorted to VCC/VSS	No CPU access to CAN chip.	A) CPU serial bus corrupt. CPU may be inoperative if VCC/VSS affected.	Node will not function. CAN bus operation is normal.
		Pin shorted to adjacent pin	No CPU access to CAN chip.	A) CPU serial bus corrupt. CPU may be inoperative if VCC/VSS affected.	Node will not function. CAN bus operation is normal.
MISO	Master (CPU) In Slave (CAN) Out - CAN chip output.	Pin open (floating)	No CPU access to CAN chip.	A) CPU serial bus corrupt. CPU may be inoperative if VCC/VSS affected.	Node will not function. CAN bus operation is normal.
		Pin shorted to VCC/VSS	No CPU access to CAN chip. B) Excessive current	A) CPU serial bus corrupt.	Node will not function. CAN bus operation is normal.
		Pin shorted to adjacent pin	No CPU access to CAN chip. B) Excessive current	A) CPU serial bus corrupt.	Node will not function. CAN bus operation is normal.
MOSI	Master (CPU) Out Slave (CAN) In - CAN chip input.	Pin open (floating)	No CPU access to CAN chip.	A) CPU serial bus corrupt.	Node will not function. CAN bus operation is normal.
		Pin shorted to VCC/VSS	No CPU access to CAN chip.	A) CPU serial bus corrupt. CPU may be inoperative if VCC/VSS affected.	Node will not function. CAN bus operation is normal.
		Pin shorted to adjacent pin	No CPU access to CAN chip.	A) CPU serial bus corrupt. CPU may be inoperative if VCC/VSS affected.	Node will not function. CAN bus operation is normal.

A. The CPU serial bus may be inoperative, thus affecting other bus peripherals.
B. The CAN chip output may short to other peripherals on the serial bus which may result in excessive current and possibly making the chip inoperative.

Table 9

941655

Tradeoffs Between Stand-alone and Integrated CAN Peripherals

Craig Szydlowski
Intel Corp.

The CAN Protocol[1] is currently implemented as on-chip peripherals integrated on microcontrollers and as stand-alone CAN chips. On-chip peripherals are available on several microcontroller architectures, including the MCS® 51 and the MCS® 96 microcontroller families. Likewise, there exists a variety of production-level stand-alone CAN chips such as the Philips PCA82C200 and the Intel 82527.

The decision to use an integrated CAN peripheral or a stand-alone CAN chip should consider the tradeoffs between both alternatives. These tradeoffs include implementation cost, design flexibility, level of CPU burden and system reliability. This paper discusses these tradeoffs from both qualitative and quantitative perspectives. The goal of this paper is to identify the key issues that differentiate these two alternatives for various design and production goals.

IMPLEMENTATION COST

The cost to implement a CAN peripheral in a system module may be divided into development and manufacturing costs. The development cost includes hardware and software engineering and design verification/qualification. The manufacturing costs include part procurement, product assembly and testing.

DEVELOPMENT COST - Figure 1 shows two hardware systems implementing CAN. System A requires three chips: a microcontroller or CPU, a stand-alone CAN chip and a CAN bus driver. The interface between the CPU and the CAN device is an address/data bus or a serial link such as the SPI protocol. A chip select signal is needed if other nodes are interfaced to the CPU's address/data bus. The CAN chip is driven by a low-tolerance input clock supplied by a crystal oscillator or the CPU's clockout. System A uses an interrupt line from the stand-alone CAN chip to the CPU to signal the reception of a message or the occurrence other CAN events.

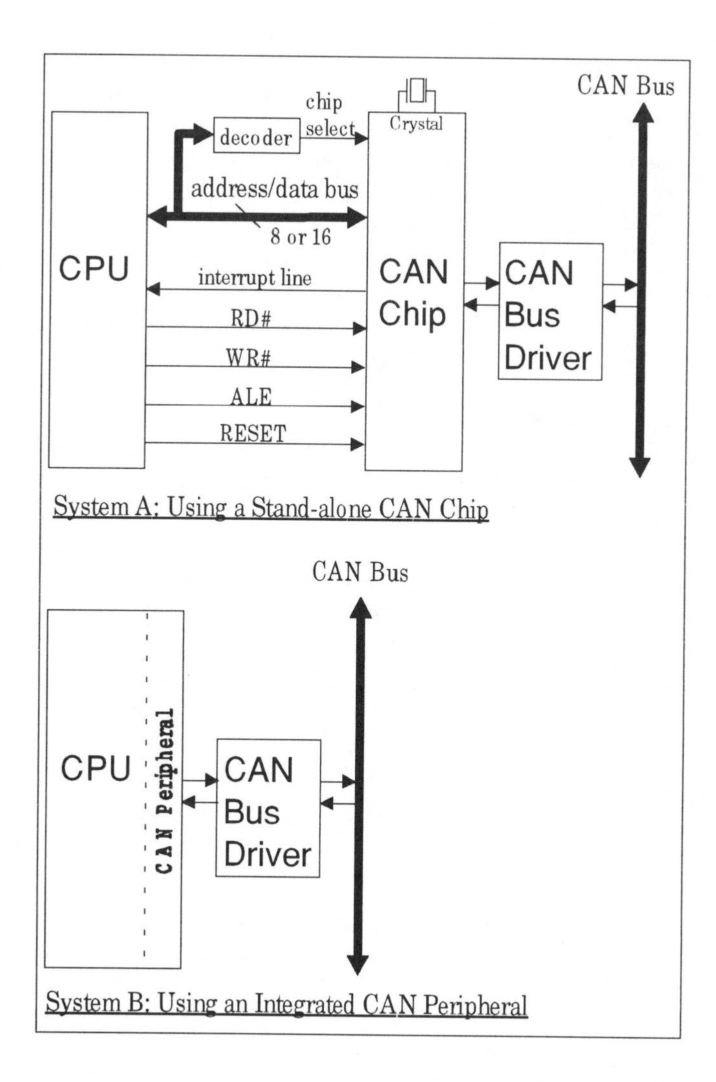

Figure 1: Hardware Implementations

System B implements a CPU with an on-chip CAN peripheral which clearly simplifies hardware design. In

[1] The Controller Area Network (CAN) protocol, developed by Robert Bosch GmbH, offers a comprehensive solution to managing communication between multiple CPUs. The CAN protocol is a serial communications specification for a multiplexed network to support control-oriented data. CAN implementations are prominent in automotive, agricultural, and industrial control applications.

addition, system B often uses less printed circuit (PC) board area and generates less board noise by eliminating the PC board traces used to interface the CPU and the CAN chip.

The software engineering cost is nearly the same for integrated or stand-alone CAN peripherals. In both cases, software must be developed for the CPU to read messages following reception and to write messages for transmission.

MANUFACTURING COST - The manufacturing cost for a system with an integrated CAN peripheral is lower than a system with a stand-alone CAN chip. The integrated CAN peripheral has many advantages over the stand-alone CAN chip since one less chip is required, Figure 1. First, production expense is less for a system with an integrated CAN peripheral since there are fewer items to order and to assemble. Second, systems using integrated CAN peripherals require less PC board area than systems with stand-alone CAN chips which reduces system form factor. Third, module testing and repair are easier with the integrated CAN peripheral because there are fewer components to verify.

The semiconductor manufacturing cost of CPUs with on-chip CAN can be less than the cost to produce separate CPU and CAN chips. The silicon area for an integrated peripheral is less than a stand-alone chip since the CPU interface circuitry can be optimized for its host-CPU and no additional area is needed to accommodate packaging (bond pads and Electro-Static Discharge (ESD) circuits). The integrated CAN peripheral has a significant die size savings since it is about half the silicon area of its stand-alone chip counterpart. A CPU with on-chip CAN also benefits from a packaging savings by requiring one less package and fewer testing steps compared to the two chip alternative.

In high volumes, semiconductor manufacturers can offer CPUs with integrated CAN peripherals for a lower price than separate CPUs and CAN chips. Today, however, there are few high-volume applications implementing CPUs with on-chip CAN. Consequently, more generic and higher-volume CPUs and stand-alone CAN devices enjoy production-economies of scale and possibly provide lower total system cost. As volumes for CPUs with integrated CAN increase, the manufacturing advantages through integration discussed previously will enable lower CAN node costs in the future.

Figure 2 shows a lower implementation cost for an integrated CAN peripheral. This trend considers the lower hardware development cost and the possible lower chip cost compared to the two chip alternative.

DESIGN FLEXIBILITY

Design flexibility allows engineers to upgrade their systems or to develop a similar system by applying existing hardware/software design experience. In either case, the flexibility to make changes often comes from exchanging the CPU to satisfy new requirements.

Although the software development is about the same for both stand-alone and integrated CAN implementations, software reusability may differ. Stand-alone CAN chips are designed to interface to different CPUs allowing the software developed for one system to be reused in another system, even if the CPU is different. Software developed for the integrated CAN peripheral of one CPU may not apply to a second CPU with on-chip CAN, especially if the CPUs are supplied by different vendors. This concept is shown in Figure 3 where the benefit from a stand-alone CAN chip configuration is more favorable when multiple designs are needed since the software and hardware development is reusable.

Therefore, exchanging the CPU to upgrade or to develop a new system may require some hardware modifications, but it is likely the same CAN chip may be used. By using the same CAN chip, the existing high-level language software may be reused, although the actual instructions differ among CPU architectures. The CAN chip has little impact on CPU interchangeability.

high
Utility/Benefit
integrated: low hardware development cost
stand-alone: more board space
low
less
more
Importance of Implementation Cost

Figure 2: Importance of Implementation Cost

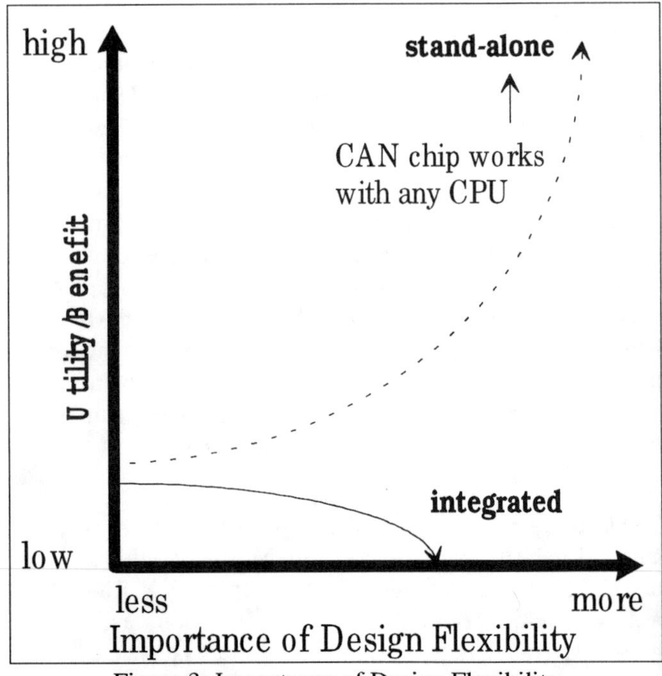

Figure 3: Importance of Design Flexibility

However an upgrade to a system that uses a CPU with on-chip CAN is significantly impacted by the CAN peripheral. The replacement CPU must also have on-chip CAN, otherwise a stand-alone CAN chip must be added to the system. Semiconductor makers are beginning to offer families of products with on-chip CAN providing users with CPU upgrade paths. Since the features of on-chip CAN peripherals are different among vendors, changing CPU architectures may require modifications to CAN software.

Some module makers must support a mix of networking protocols such as J1850, VAN and CAN. This flexibility is well supported by stand-alone protocol chips, through which only protocol chips vary among system module designs. In contrast, it is difficult to find a family of CPUs whose only difference is whether they have on-chip J1850, VAN or CAN.

LEVEL OF CPU BURDEN

Dedicated communications peripherals have been developed to support protocols such as CAN, VAN and J1850 to reduce the CPU burden to service a high-speed serial link. Even today's highest performance CPUs have difficulty supporting multiplexed communications above 100K bits per second using standard high-speed I/O pins instead of special peripherals. The level of CPU burden to maintain on-chip CAN is usually less than a stand-alone CAN chip since the CPU has faster and more efficient access to CAN registers. As a result, the CPU spends less time interacting with the on-chip CAN peripheral.

Figure 4 outlines the communications tasks at each CAN node with respect to the protocol, messaging and system/error response. The protocol tasks involve transmitting and receiving bits according to arbitration rules defined by the CAN protocol. Another protocol task is to calculate a 15-bit CRC code (cyclical redundancy error code) which is transmitted with each message and is verified at each CAN node. CAN peripherals complete all protocol tasks without CPU intervention.

With respect to the level of CPU burden, the messaging tasks must be serviced by the CPU. Messaging tasks require the CPU to write data to be transmitted, to read received data and to manage status/control registers in the CAN peripheral. Since the CPU "sees" the CAN peripheral as a smart RAM, messaging tasks are basically CPU read/write operations. A CPU with on-chip CAN will read/write to register locations using an internal bus. For a CPU interfaced to a stand-alone CAN chip, these read/write operations typically use the external address/data bus or a serial link. In addition to these read/write operations, the CPU may be required to manipulate message identifier bits and data fields as required by the messaging scheme. For example, the data byte may actually contain multiple parameters such as engine air flow and engine temperature. In this case, the host-CPU must execute shift bit and masking operations to prepare and interpret complicated byte configurations. The CPU burden required to manipulate message identifiers and data bytes is a function of the messaging scheme, and this burden is the same for on-chip and stand-alone CAN peripherals. The CPU burden differs for on-chip and stand-alone CAN only because the access time of CAN registers is different.

Protocol	Bitwise reception/transmission Bus arbitration Error code generation/checking
Messaging	Write data to be transmitted Read received data Manage control/status registers
System/Error Response	Node configuration System commands Local busoff

Figure 4: CAN Node Communications Tasks

System/error response is a general category for infrequent tasks initiated by the system or by an unusual number of bus errors. For example, the system may require nodes to dynamically allocate message identifiers during system initialization when similar nodes such as lamps are on the bus. The CPU also executes error recovery routines when the CAN peripheral is in "busoff state". Recovery from "bus-off" requires a hardware or software reset of the CAN peripheral.

The CPU-burden to communicate with the CAN peripheral is dependent on a few factors. The most critical factor is the amount of time required to read/write to the CAN peripheral. In the case of an on-chip CAN peripheral, the CAN registers are addressed using the internal address/data bus designed for high-speed access. In the case of a stand-alone CAN chip, the CPU uses its external address/data bus or a serial communications link. Figure 5 shows the level of CPU burden to receive CAN messages for three CAN bus transmission rates.[2] This analysis compared the CPU burden of an Intel 82527 stand-alone CAN chip to an Intel 87C196CA 16-bit microcontroller with on-chip CAN. The 82527 supports both 8-bit and 16-bit interfaces. The 87C196CA addresses the on-chip CAN as either scratch RAM or high-speed register RAM.

The level of CPU burden to receive messages for an integrated CAN peripheral ranges from 2.0% to 8.0% whereas a stand-alone CAN chip requires 4.2% to 16.7%. [3]The CPU burden for an on-chip CAN is approximately one-

[2]The level of CPU burden assumes a maximum bus loading, 100%, an average message length of 100 bits (8 data bytes) and the minimum number of CAN accesses and CPU operations necessary to receive a message (management of the interrupt pointer and control registers).

[3]This comparison assumes 16-bit word operations, three wait-states access of the 82527 and 4 µS for interrupt overhead.

Stand-alone CAN Chip	CPU Burden 250KBS	500KBS	1MBS
8-bit A/D Bus	**5.5%**	**11.0%**	**21.9%**
16-bit A/D Bus	**4.2%**	**8.4%**	**16.7%**

Integrat. CAN Peripheral	CPU Burden 250KBS	500KBS	1MBS
Scratch RAM	**3.6%**	**7.2%**	**14.4%**
Register RAM	**2.0%**	**4.0%**	**8.0%**

Figure 5: Level of CPU Burden For Various CAN Nodes

half the burden of a stand-alone CAN chip. The 87C196CA accesses 16-bits word in 400nS clocked at 20MHz whereas the 82527 16-bit word read access time is 1300 nS.

In some cases, message filtering techniques can be employed to reduce the number of messages a node is required to receive in a given period of time. This will reduce CPU-burden by limiting the number of interrupts to the CPU by "screening" unnecessary messages. During system development, it is necessary to model worst case message reception and the resulting CPU-burden so the system functions properly when the CAN bus is heavily loaded.

The acceptable level of CPU-burden to service a CAN peripheral is application dependent. In some cases, a system may only be required to receive a small range of messages.

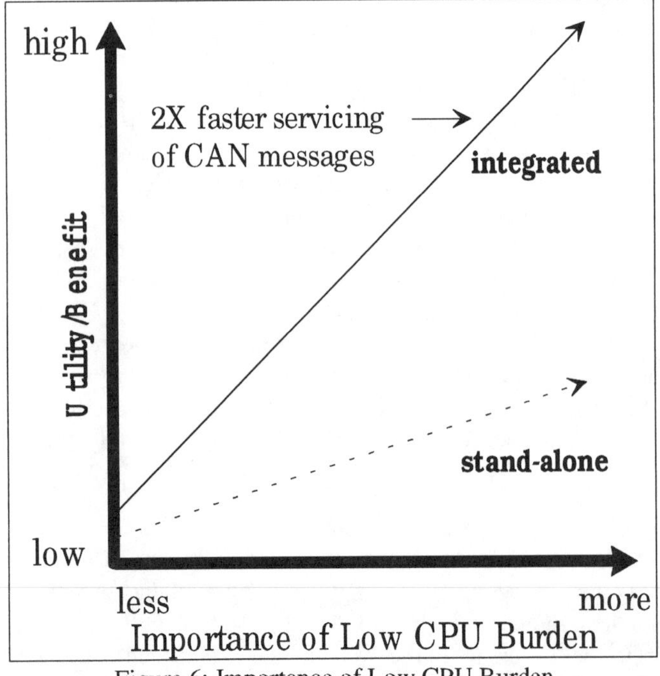

Figure 6: Importance of Low CPU Burden

An engine controller, on the other hand, may need to receive and transmit a large number of messages with high repetition rates.

Figure 6 indicates the lower CPU burden of on-chip CAN may provide a significant system advantage.

SYSTEM RELIABILITY

The reliability of an electronic system can be analyzed by considering the impact of the printed circuit (PC) board and the semiconductor components. A CAN node using an integrated CAN peripheral has a reliability advantage over a system using a stand-alone CAN chip because of its smaller form factor. As Figure 1 shows, the PC board implementing an integrated CAN peripheral does not have an interface between the CPU and the CAN peripheral and therefore requires fewer traces and plate through holes. An integrated CAN peripheral may provide improved chip-reliability over a stand-alone implementation since the integrated peripheral is burned-in and tested together with the CPU.

PC board reliability is a function of thermal stress. In particular, high use-temperature and temperature cycling aggravate the thermal expansion mismatch between the materials of the PC board by stressing the metal conductors and epoxy insulation. This additional stress increases the probability of cracks in the conductors leading to board opens. Temperature also accelerates insulator breakdown leading to shorts between board traces. Temperature cycling increases material fatigue caused by repeated expansion and contraction cycles.

"Plate through holes are Achilles' heel of the PC board, since they consist of thermally incompatible materials and are the basis of multi-layer interconnection. The second major cause of PC board failure is the loss of electrical insulation, termed insulation-resistance (IR) failure."[4] It is clear that PC board reliability decreases with additional traces, solder joints and plate through holes. Therefore, integrated CAN peripherals provide PC board reliability advantages.

The second aspect of CAN node reliability is the reliability of the individual semiconductor chips. CAN nodes implementing integrated CAN peripherals will typically have higher reliability than those implementing stand-alone CAN chips. First, the integrated CAN peripheral requires fewer circuits to interface to the CPU. The CAN peripheral is connected to an internal CPU bus eliminating the need for complicated interface circuitry and drivers. In contrast, the stand-alone CAN chip requires the integration of additional address/data bus circuitry, pin logic such as input/output circuits, Electro-Static Discharge (ESD) protection devices and mode select pins. As a result, an integrated CAN peripheral requires less interface circuitry and requires many fewer pins that could be susceptible to soldering issues and ESD exposure. The interface between the CPU and a stand-alone CAN chip requires as many as 21 traces or 42 solder joints for pin connections. In addition, the stand-alone CAN

[4] Clyde F. Coombs, Jr., Printed Circuits Handbook, Third Edition (New York: McGraw-Hill Book Company, 1988), p. 30.1.

chip may require a separate crystal oscillator which impacts system reliability as well.

Chip reliability is improved by thorough testing and burn-in. An integrated CAN peripheral is tested and burned-in with the CPU, exactly as the chip will be used in a system. The integrated CAN chip is from the same silicon processing as the CPU, so manufacturing variability is assessed during testing.

Although a stand-alone CAN chip is certainly tested and burned-in, the CAN product engineer must guarantee operation with a variety of CPUs which is a complex product engineering task. The stand-alone CAN may be interfaced to a CPU from a different manufacturing process and possibly from different vendors. Fortunately, most hardware designers consider worst case chip specifications to ensure designs operate across chip manufacturing variations.

A CPU with an integrated CAN peripheral will have a specified reliability based on the results of the product qualification and the historical trends of the manufacturing process. A typical defect per million (DPM) specification is perhaps 50-200 DPM. Measuring a lower DPM that is statistically significant requires a very large sample size and several additional months of evaluation.

A two-chip solution such as a CPU and a stand-alone CAN chip may actually have twice the DPM specification since the reliability of both chips must be summed. Even though the actual chip reliability of the stand-alone CAN system may be equal to the integrated CAN system, this is difficult to verify using reliability specifications derived on a chip-by-chip basis. Therefore, a CAN system with an integrated CAN peripheral may have better specified reliability than a system with a stand-alone CAN chip because of statistical limitations of comparing the reliability of single versus two-chip solutions.

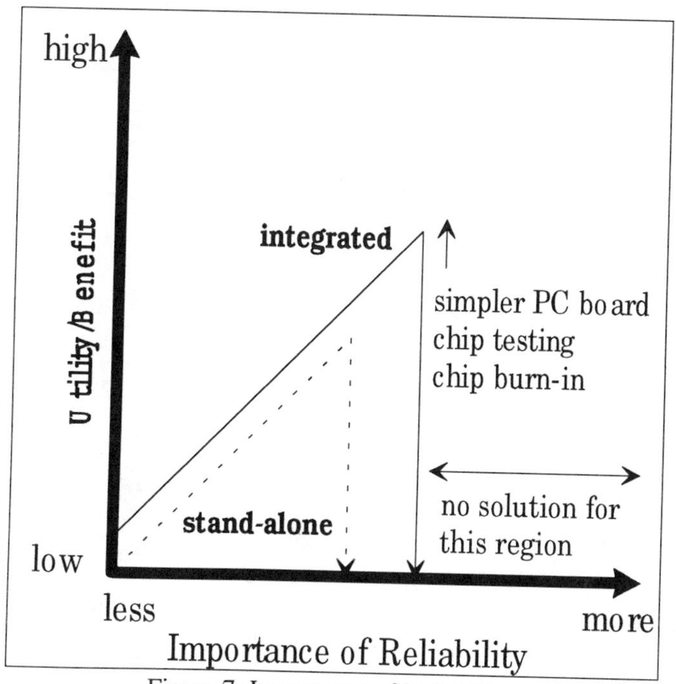

Figure 7: Importance of Reliability

Figure 7 shows the benefit of the additional reliability of a CPU with integrated CAN. The utility of both solutions falls when a very high level of reliability is required; this conveys the concept that an implementation which does not satisfy reliability targets is an unacceptable solution.

OVERALL ANALYSIS

The tradeoffs between using a stand-alone CAN chip and an integrated CAN peripheral may be viewed as the value of design flexibility versus implementation cost, level of CPU burden and system reliability. Yet today, many stand-alone CAN chips are shipping in high volumes, and the combined price of a CPU and a stand-alone CAN chip is very competitive. However, as CAN gains more market acceptance, on-chip CAN will be the solution of choice.

Figure 8 shows besides design flexibility, a CPU with on-chip CAN has many advantages compared to stand-alone CAN chips.

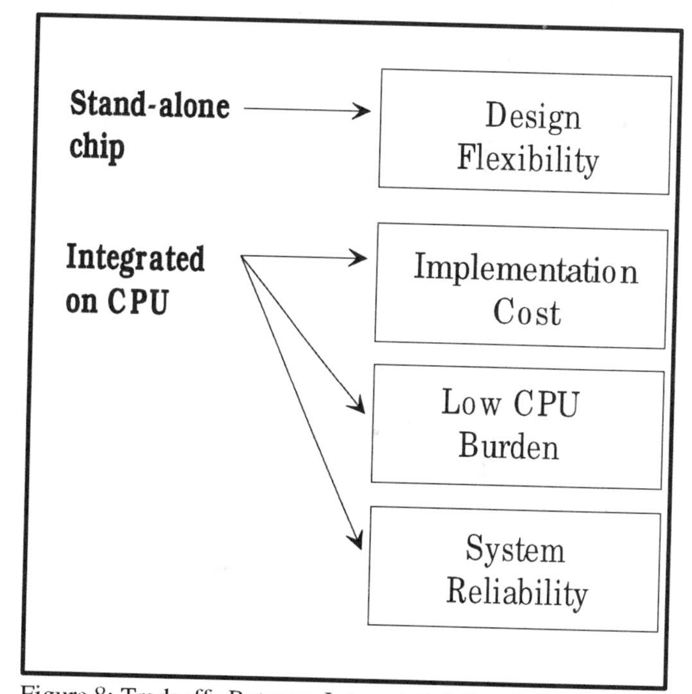

Figure 8: Tradeoffs Between Integrated & Stand-alone CAN

FOUR SCENARIOS

Four scenarios are presented to clarify design and production environments that weigh the tradeoffs between stand-alone and integrated CAN differently.

Scenario 1:

An engine control unit (ECU) manufacturer supplies a diverse set of car makers for 4, 6 and 8 cylinder engines. To meet various performance goals and control requirements, several CPUs from different architectures and vendors are used. Future emissions regulations will require system upgrading by implementing CPUs with more RAM and ROM needed for new diagnostic algorithms. A stand-alone CAN chip provides this ECU maker the flexibility to change CPUs to meet cost targets while satisfying a range of performance targets. By using the same CAN chip for all engine designs, the CAN hardware and software development is applied to all of the ECU designs which minimizes product line development costs. The reliability of the ECU is critical to ensure automotive safety, and in almost all cases, a stand-alone CAN design satisfies system targets.

Scenario 2:

Anti-lock braking systems (ABS) become standard equipment on most passenger cars and shipment volumes soar while module prices to car manufacturers drop. Most ABS designs are mature and developers work to strip out costs from their modules. PC board size must be reduced, assembly costs cut, and procurement costs minimized to ensure reliable high-volume manufacturing. Reliability is a key requirement because braking is a primary aspect of vehicle safety.

An integrated CAN peripheral on a high volume CPU provides the lowest CAN node cost. The CPU with on-chip CAN is lower cost than the two-chip alternative, PC board size is reduced, chip procurement cost is minimized and the added performance allows a less powerful CPU to run the system. The integrated CAN peripheral provides increased reliability as well. Since design flexibility is not needed for a mature high-volume product, a stand-alone CAN does not offer any advantages over an integrated solution.

Scenario 3:

A robot is designed in the event a nuclear accident occurs and unmanned cleanup is required. This robot is controlled by a dozen of networked microcontrollers providing a range of motion. Reliability is a critical concern given the potentially grave nature of the situation. The high temperatures associated with the nuclear waste increases the possibility of electronic malfunction, so reliability issues are carefully addressed. Few robots must be manufactured, so system cost is not a primary concern.

A high-performance CPU with integrated CAN is best to meet the reliability needs of the robot. The compact size of this CAN node contributes many reliability advantages.

Scenario 4

A maker of in-vehicle climate control is informed that a new car platform will control his module through commands transmitted on a CAN network. The current systems use one of three 8-bit CPUs from a common architecture, but with various memory configurations. About one-half million lines of software have been written to support thirty different vehicle types. Even though the 8-bit CPUs have satisfied all climate control applications to date, there is little CPU headroom to service a CAN node.

A new 8-bit CPU with on-chip and more program memory is the easiest design change. An integrated CAN peripheral will burden the CPU less than a stand-alone CAN chip because the on-chip CAN registers are accessed faster. With the added CAN functionality, the climate control maker should request additional program memory to store the CAN subroutine.

CONCLUSION

Today, a large variety of CAN peripherals are available as stand-alone chips and integrated on CPUs. Integrated CAN peripherals offer a number of advantages over stand-alone CAN chips such as lower implementation cost, lower CPU burden, and higher system reliability. The stand-alone CAN chip offers one compelling advantage over the integrated CAN peripheral: design flexibility. Until CAN becomes a standard peripheral on all new microcontrollers, stand-alone CAN chips will serve designers who use a variety of CPUs and regularly upgrade designs.

950039

TTP/A — A Time-Triggered Protocol for Body Electronics Using Standard UARTS

H. Kopetz
Technical University of Vienna

ABSTRACT

The TTP/A protocol is a character oriented time-triggered protocol for the design of low cost distributed real-time systems, particularly in the field of automotive body electronics. It is designed to operate with standard UART hardware interfaces to provide utmost compatibility with a large number of single chip microcontrollers. The distinguishing features of TTP/A over other protocols proposed for automotive body electronic applications are the support for composability in the temporal domain, the short error detection latency, and the protectiveness of the protocol.

INTRODUCTION

There is an increasing tendency to replace electromechanical control devices within an automobile by single-chip computer nodes. These single chip microcontrollers offer significant improvements in the cost/performance ratio and in the reliability over the devices they replace. The selection of the best interconnection technology between these nodes offers a new challenge to the design engineer. In addition to the paramount concern for minimal production cost and very high reliability of the interconnection structure, there is an increasing concern for desirable system properties, such as composability in the temporal domain, predictability, testability, and configuration flexibility. These system properties are determined to a large extent by the communication protocol that controls the information transfer between the different nodes.

The Society for Automotive Engineers has classified computer network applications within a vehicle into three classes [1]: class A and class B networks are for signal multiplexing and parameter sharing in the field of body electronics, such as power windows control or the transport of the information to the dashport, while class C networks are for safety critical control applications, such as the coordination of the operation of the engine, the transmission, and the intelligent brakes. There are substantial differences between body electronics applications and class C applications from the point of view of timeliness requirements and fault tolerance. In body electronics applications there is always a safe state--e.g., stop the current activity--that can be visited in case of a transient or permanent fault. In some class C applications the control system has to continue its operation, even during the occurrence of a fault.

In the last few years, a number of special communication protocols for automotive applications has been developed, such as, e.g., the Control Area Network CAN [2], and J1850 [3]. Most of these protocols are based on the Carrier Sense Multiple Access Collision Avoidance (CSMA/CA) media access method, where the bitwise arbitration of message identifiers allows the highest priority message to gain immediate control of the bus without the occurrence of collisions. These protocols are very responsive to the highest priority message, but cannot provide temporal composability: The dynamic media access strategy of these protocols makes it impossible to specify the node behavior to a level of detail that the timeliness properties of the system as a whole can be derived from the timeliness properties of the individual nodes (such that an analysis and validation of the timeliness of the nodes is adequate to verify the timeliness of the system). The composability property of the control system, sometimes called "plug and play", is of growing concern to the automotive industry: The assignment of the responsibility for a complete subsystem to a supplier necessitates a detailed and *composable* specification of all logical and temporal properties of the subsystem interfaces. Otherwise the system integrator is left with the entangling task of handling all the unintended and unspecified side effects caused by the integration: every sub supplier can demonstrate that its subsystem meets its specification, but the system as whole does not perform as intended.

Composability is a prime concern during the development of the TTP Protocol Family. The TTP/C protocol [4] achieves the temporal encapsulation of the subsystems by a static bandwidth assignment derived solely from the progression of a globally synchronized timebase. It requires special communication hardware for the fault tolerant clock synchronization. The support of active redundancy to mask any single fault makes TTP/C expensive for body electronics applications where such a level of fault-tolerance is not needed. Since, on the other side, composability is a very desirable property for body electronic systems as well, we designed a simpler version of TTP--the composable Fireworks protocol TTP/A --that operates on standard Universal Asynchronous Receiver/Transmitter (UART) interfaces and provides the short error detection latency needed for the implementation of 'fail-safe' body electronics applications. We call the protocol the "Fireworks" protocol, because every round of messages is started by a special control byte, the "Fireworks".

It is the objective of this paper to present the composable Fireworks protocol TTP/A.

This paper is organized into six main sections. After the introduction we analyze some of the requirements of body electronics systems and discuss the advantages of time-triggered protocols. Section three gives an overview of the TTP/A protocol. Section four describes the detailed operation of the protocol and presents an analysis of the influence of the quality of the crystal oscillator and the interrupt response time on the timing parameters of the protocol. Section five discusses implementation issues, such as the protocol data structures, the outlay of the physical layer, the data efficiency of the protocol, and the design of multicluster systems. The paper closes with a conclusion in section six.

WHY USE A TIME-TRIGGERED PROTOCOL FOR BODY ELECTRONICS?

In this section we investigate the arguments for the use of time-triggered protocols in body electronics systems. We start with an analysis of the requirements, analyze the properties of event-triggered versus time-triggered protocols and finally present some of the advantages of time-triggered protocols.

REQUIREMENTS OF BODY ELECTRONICS SYSTEMS -- Composability and Testability - Even within a single model range, many cars are equipped with different electronic subsystems according to customer demand. The network architecture must guarantee that all combinations of subsystems will operate as specified without any unintended side effects caused by the system integration. This property of an architecture is captured by the notion of *composability* in the value domain and in the temporal domain.

Constructive testability implies that the system test can be decomposed into a set of independent subsystem tests. It is thus necessary to specify the communication interfaces of the nodes to a level of detail such that the subsystems can be tested independently of each other in the value domain and in the temporal domain.

Protectiveness - In a conventionally wired car, a short circuit between two wires affects a single connection between two modules only. In a networked system such a fault may result in the complete loss of communication between all modules connected to the same bus. The consequence of a failure of a bus system are thus more serious than the consequences of a failure of a connection of a conventionally wired system. A critical (software) failure mode in a multiplexed system is the overload of the network by a "babbling" client--the *babbling idiot* failure. It is a desirable property of a communication protocol to be able to detect babbling clients and to protect the critical resource, the bus, from this kind of client failure. This property of a protocol is captured in the notion of *protocol protectiveness.*

Short Error Detection Latency - Whenever the remote execution of a body electronics control command is disturbed by the occurrence of a transient or permanent error, the remote node has to detect such an error within a short latency and has to visit a safe state autonomously without any delay caused by further, possibly unsuccessful, communication retries. In many cases, a safe state can be reached by just terminating the command execution immediately. Consider, e.g., the operation of a power window. In case the communication between the driver and the remote window controller is disturbed, no further movement of the window should take place. Another example is the operation of the stop lamp. In case the communication between the brake pedal and the stop lamp is interrupted, the stop lamp should be turned on autonomously, since this is the safe state. If the transient disturbance disappears after a few hundred milliseconds, the normal state can be reestablished.

In the area of body electronics, the occurrence of a transient fault is not disastrous if the error is detected within a short latency to avoid any unsafe state and an error recovery action within few hundred milliseconds is supported. A short error detection latency is thus a mandatory requirement for protocols for body electronic applications. The error detection latency should be in the same order of magnitude as the end-to-end response time of an application.

Timeliness - The timeliness requirements of body electronics systems are determined by the perceived reaction time of the human operator: In most cases an end-to-end reaction time between the actuation of a input device and the observation of the intended result of about 100 msec or less is perceived as an "immediate" reaction by a human operator. If, in a small number of situations, the reaction time is considerably longer--e.g., 200 msec, this is not considered a serious deficiency. In body electronics system the deadline of 100 msec is thus a soft deadline. This is in contrast to class C systems, e.g., fuel injection, where many deadline are hard. There is another important timeliness requirement on error detection: a receiver of a command, e.g., a power window controller, should detect a failure of the communication system and visit the safe state (motor stopped) within the same reaction time interval of 100 msec.

TIME-TRIGGERED VERSUS EVENT-TRIGGERED PROTOCOLS - In the real-time protocol community two fundamentally different temporal control strategies are known: time triggered--(TT} control versus event-triggered--(ET) control.

In a TT protocol the temporal control signals of the protocol are solely derived from the progression of the real-time. For example, a node has to send a particular message every 10 msec. TT systems require a global time base of specified precision. In a TT system the message rate is independent of the activity of the client. The preferred data model of a TT system is that of state data: an image of the current state of a control variable is carried in the message. State messages are not consumed on reading. A new version of a state message overwrites the previous version. The semantics of state messages is closely related to the semantics of a variable in a programming language. TT messages are idempotent: the receipt of duplicated messages has the same effect as the receipt of a single message. An example of an idempotent message is: "Activate the motor in clockwise direction".

In an ET protocol the temporal control signals are derived from events occurring in the environment of the protocol. For example, a new message is sent whenever the client issues a send command. In an ET system the message rate is determined by the activity of the client. The preferred data model of an ET system is that of event data: the difference between the new and the old state of a state variable, i.e., the event information, is carried in the message. Event messages are queued at the receiver and consumed on reading. Event messages are not idempotent: the receipt of duplicated messages has a different effect than the receipt of a single event message. An example of a non idempotent message is: "Reverse the motor to the other direction".

We call a protocol strictly event triggered if it contains no temporal parameters that limit the request rate of the client. In a strictly event triggered protocol it is impossible to exercise some form of protectiveness to avoid a babbling idiot behavior of the client. The protocol machine does not have any information that allows it to monitor the behavior of the client. Examples for strictly event triggered protocols are J1850 or CAN.

We call a protocol strictly time triggered if its progress is driven by a table containing the points in real-time when a particular message is to be sent. In this case it is impossible for the client to interfere with the protocol operation in the temporal domain, since the control of the protocol is autonomous. A TT protocol has full information about the future behavior of all clients and can therefore exercise excellent error detection and babbling idiot control. An example of a strictly time triggered protocol is TTP.

Some real time protocol take a mixed route between strict ET and strict TT control. Such a protocol contains some timing information about the expected behavior of the client to exercise a minimum amount of protectiveness. For example, the ARINC 629 [5] protocol allows a client to send one message only in an epoch of predefined length, or a token protocol, e.g. [6], allows the client to hold the token only for a maximum time, the token hold time THT.

Given the regular character of class C control applications within a car, it is evident that a TT protocol matches the application requirements of class C applications better than an ET protocol. The question is not as easy to answer for body electronics applications, since these applications are predominantly event triggered. It would therefore seem that an ET protocol is a natural choice for body electronics applications. However, there are some fundamental conflicts in protocol design between demand assignment on the one side and composability and error detection on the other side.

THE ADVANTAGES OF A TT-PROTOCOL - Composability in the time domain requires that the temporal properties of every node can be designed and tested in isolation and that the integration of a set of nodes into the complete system will not lead to any unintended and untested side effects. If all nodes compete for the single communication channel on a demand basis, then it is impossible to avoid side effects caused by the extra transmission delay resulting from conflicts about the access to the single channel, no matter how clever the media access protocol may be. If, on the other hand, the channel allocation is performed statically, implying that the transmission requests of each node are restricted a priori to subsections of the timeline, then any unintended interaction between the nodes can be avoided at design time. The price that has to be paid for the composability is the restricted flexibility and the lower channel utilization.

Another fundamental conflict exists between the requirement for flexibility and the requirement for protectiveness and error detection. Flexibility implies that the behavior of a node is not restricted a priori, i.e., any behavior is allowed. Error detection is only possible if the actual behavior of a node can be compared with some a priori knowledge about the "good" behaviors, and other forms of behavior are forbidden. Consider the example of an event triggered system with no regularity assumptions:

If there is no restriction in the rate of messages a node may send, it is impossible to avoid the monopolization of the network by a single node, i.e., the protocol has no protectiveness. If a node is not required to send a "life-sign message" at known regular intervals, it is impossible to detect a node failure within a bounded latency.

A time-triggered protocol trades flexibility and the capability of immediate transmission requests for composability, continuous error detection, and protocol protectiveness. Whether these are reasonable tradeoffs is for the system designer to decide.

OVERVIEW OF THE FIREWORKS PROTOCOL TTP/A

In this section we give an overview of the Fireworks protocol TTP/A. We start with a description of the design constraints and proceed with the presentation of the principles of operation of the protocol.

THE DESIGN CONSTRAINTS - After a detailed investigation into the characteristics and requirements of body electronics systems and an analysis of published protocols for this application domain [1,3] we came to the conclusion that any new protocol for body electronics must be designed with the following design constraints in mind:

- Composability and Testability
- Short Error Detection Latency
- Standard Serial Communication Interface
- High Data Efficiency

We have already discussed the first two topics in the previous paragraphs and focus here on the remaining two issues.

Standard Serial Communication Interface - Almost all single chip microcontrollers that are on the market support a standard UART interface. The hardware of the UART interface is optimized to the point that it will be very difficult for a new custom built controller to reach the same low cost level as that of a standard UART.

If a body electronics protocol is based on such a standard UART, the automotive design engineer can consider all the devices on the market and is free to select the microcontroller that is best suited for the application at hand. Provided the proper physical interoperability is provided at the physical layer, these single chip micro-controllers can be connected to form a low cost distributed system architecture. The constraints imposed by the requirement to employ standard serial communication hardware have a significant influence on the protocol design since special hardware mechanisms for the clock synchronization cannot be assumed.

After investigating the characteristics of the UART communication interfaces in a number of typical microcontrollers used in automotive applications [7], we decided to base the protocol on a character format of eight data bits and one parity bit, enclosed in two framing bits (a start bit and a stop bit). In addition to a standard UART channel, the protocol requires a timer.

High Data Efficiency - The advantages of a single wire operation in the field of body electronics are evident: it is possible to reduce the cabling effort while, at the same time, the reliability with respect to permanent failures is increased.

The single wire operation provides a number of constraints for the protocol designer: Firstly, the single wire only allows one sender at a particular interval of time. Secondly, the limited bandwidth of a single wire within a car (10 kbits) requires a high data efficiency of the protocol. Thirdly, a single wire operation will be subjected to a non-negligible number of transient transmission errors. Therefore the protocol must provide effective means to manage these transient errors.

PRINCIPLES OF OPERATION - TTP/A is based on one byte messages. Most of these messages are data messages, only one special message, the Fireworks message, is a control message.

In all TTP protocol versions the data messages are treated according to the state message semantics[8]: a new version of a data message overwrites the previous version and messages are not consumed when accessed by the application software. The state message semantics is closely related to the semantics of a variable in a programming language or a data cell in memory. From the point of view of the application software the communication network of a TTP system can be hidden behind the memory abstraction. The TTP communication interface is thus a strict data interface. There are no control signals crossing this interface, since the progression of the TTP protocol is controlled solely by the progression of time. Every protocol event occurs either at a predefined point of time (e.g. sending a message) or has to happen in a predefined time window (e.g., receipt of a message).

Round - In TTP/A all communication activities are organized into rounds (Fig. 1). A round is the transmission of a sequence of one character messages that is specified a priori in a Message Description List (MEDL). A round starts with a special control byte, the Fireworks, that is transmitted by the active master. The Fireworks byte serves two purposes: It is the global synchronization event for the start of a new round and it contains the name of the active MEDL for this round. The Fireworks is followed by a sequence of data bytes from the individual nodes as specified in the active MEDL. A round terminates when the end of the active MEDL is reached. Every round is independent of the previous round.

To be able to differentiate between a Fireworks byte and a data byte, the Fireworks has characteristic features in the value domain and in the time domain that differentiate the Fireworks byte from the data bytes: The Fireworks byte has an odd parity while all data bytes have even parity. The intermessage gap between the Fireworks byte and the first data byte is significantly longer than the intermessage gap between the succeeding data bytes. These characteristic features make it possible for all nodes to recognize a new Fireworks byte, even if some faults have disturbed the communication during the previous round. The characteristic features of the Fireworks byte simplify the reintegration of repaired nodes--a repaired node monitors the network until a correct Fireworks byte is detected.

Since the sequence of messages is determined a priori by the definition of the active MEDL, it is not necessary to carry the identifier of a message as part of the message. All eight data bits of a message are true data bits. This improves the data efficiency of the protocol, particularly for the short one byte messages that are typical for body electronics applications.

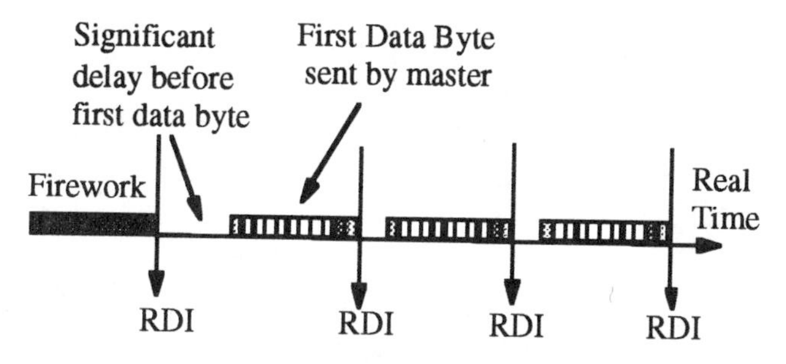

Fig. 1: Structure of a Round

Modes - From the point of view of the protocol operation, every round is independent of the previous round. In many applications, the termination of a round will cause the initiation of an identical next round by the active master. We call a sequence of identical rounds controlled by the same MEDL a *mode*. With the start of every new round a mode change can be initiated by the active master by packing the name of a new MEDL into the Fireworks byte.

To increase the flexibility without compromising the predictability, the protocol supports an operational mode that allows the on-line downloading of a new MEDL (in RAM based systems). The master can enter the download mode by sending the proper Fireworks byte. It then broadcasts the new MEDL in a specified format. Every slave node will construct its local MEDL from the information received by the master. After the downloading activity is completed, the master can activate the downloaded MEDL by a mode change. In a similar way, it is also possible to download programs dynamically.

Timeouts - The progression of the protocol through the active MEDL is controlled by a set of timeouts. The start of these timeouts is synchronized initially with the reception of the Fireworks byte and is again resynchronized with the reception of every new correct data message at every node, i.e., the "receive data interrupt " (RDI) of the UART controller is considered a global resynchronization event (Fig 2) for setting the timeouts for the *next* data byte.

To provide a high error detection coverage, the occurrence of this global resynchronization event RDI is itself monitored at every node during the reception of the *current* data byte. A correct RDI interrupt has to occur between two timeouts, the "Await Data Timeout" (ADT) and the "Data Processing Timeout" (DPT) (see Fig. 2). In case of a failure, we have to distinguish between two cases:

(i) If the sending node that is expected to send the current byte has failed to send a message or his message has been lost, no data byte will arrive at the receivers in the window <ADT,DPT>. In this situation, the DPT will continue the protocol operation with the proper error indication to its client.

(ii) If an RDI occurs outside the window <ADT,DPT>, a synchronization failure (either a faulty node sending a message at an incorrect point in time or a spurious interrupt caused by some transient disturbance) has been detected. In this situation the protocol will terminate the current round and will wait for a new Fireworks byte by the master.

In case the master does not send a new Fireworks within a specified time---the Multi-Master Timeout (MMT)-- a backup node takes up the role of the active master. The detailed calculation of all these timeouts is presented in the following section.

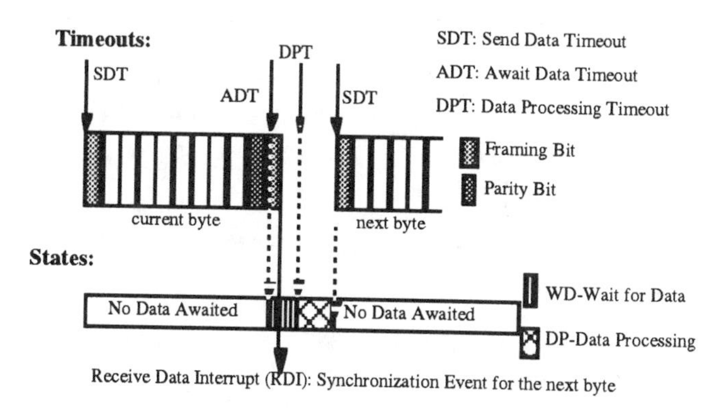

Fig. 2: Essential Timeouts of TTP/A

ERROR DETECTION AND ERROR HANDLING - The Fireworks Protocol TTP/A takes advantage of all error detection mechanism of the UART controller to detect value errors and provides a number of mechanisms to detect errors in the time domain with a short error detection latency. Note that in systems that support the "fail silent" abstraction, the error detection in the time domain is the primary error detection mechanism.

Error Detection in the Value Domain - The error detection in the value domain relies on the facilities of the particular UART controller and on data redundancy provided and checked by the application software. The TTP/A protocol requires that the controller supports odd and even parity. The Fireworks byte has odd parity, while all data bytes have even parity. In addition to the parity check, many UART controllers provide mechanism to detect various other kind of reception errors, such as noise errors detected by oversampling, framing errors, etc..

Whenever a data error is detected by a receiver and considered significant in the given application context, one of two alternatives for error handling can be selected under the control of the MEDL for this message: either leave the old version of the state variable unchanged or replace the corrupted data by a prespecified default value. The communication system of the receiver can also raise an interrupt to its host processor.

Error Detection in the Time Domain - The temporal control scheme of TTP/A is restrictive. After a new round has been initiated by the master, the temporal sequence of all correct send and receive events is specified in detail in the active MEDL and monitored by all nodes. If a "receive data interrupt" (RDI) is observed outside the specified window, a control error has occurred and the corresponding error flag is raised.

If an expected message is not received within the specified window, the missing data can be replaced by a default value or by the old version of the data. The very short error detection latency of TTP/A makes it possible to initiate fail-safe actions with minimal delay. A missing data message does not corrupt the control scheme. If a control error is detected--a message is received outside the expected window--then the present round is terminated immediately and the protocol is reinitialized to wait for a new Fireworks message by the master. If the master does not send such a Fireworks within a specified timeout (mulitmaster timeout MMT), then a backup master will take control of the network.

DETAILED DESCRIPTION OF THE BASIC PROTOCOL

STATE TRANSITIONS - The execution of the TTP/A protocol is controlled by a sequence of timeouts and by the "Receive Data Interrupt" from the UART controller. At any one time the protocol is in one of the following states:

SF	Send Fireworks--the master is allowed to send a new Fireworks message
WF	Wait for Fireworks--a node waits for a new Fireworks message
FR	Fireworks Received--a node has received a new Fireworks message
DP	Data Processing--a node is processing the received Fireworks or data
NA	No Data Awaited--no data message is expected
WD	Wait for Data--a node waits for a new data message
DR	Data Received--a node has received a new data message
ER	Error--an incorrect or unexpected event has occurred

After initialization, all slave nodes will be in the state "Wait for Fireworks" while the master is in the state "Send Fireworks". The following figures 3 and 4 depict the significant state transitions:

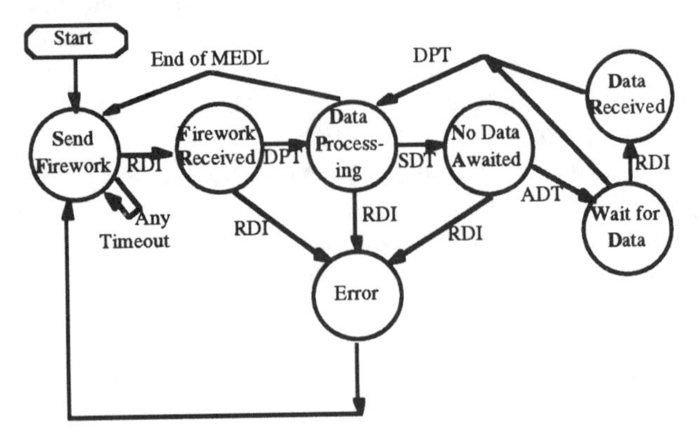

Fig. 3: State Transitions of the Master

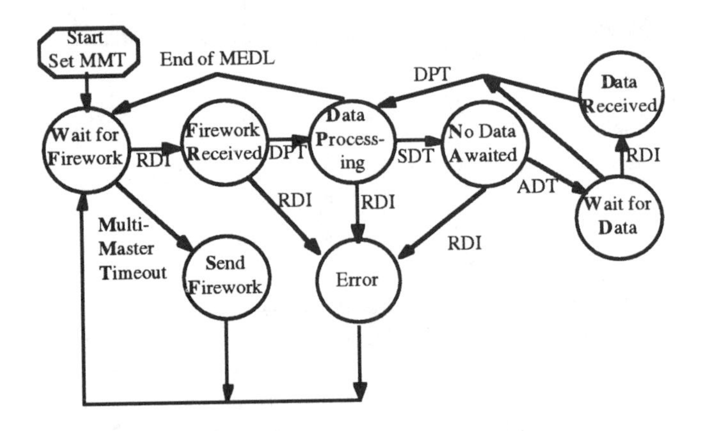

Fig.4: State Transitions of the Slave

CLOCK SYNCHRONIZATION - TTP/A synchronizes the local clocks of the nodes by the globally visible "Receive Data Interrupt" from the current sender (Fig. 5). The achievable precision Δ is determined by the following four parameters: the interrupt response time, the granularity of the local clock, the quality of the resonator, and the length of the resynchronization interval.

The interrupt response time is defined as the time interval between the point in time of arrival of the last bit (the stop bit) of the message in the controller and the point in time of reading the local free running clock. We call the difference over all nodes and all times between the maximum interrupt response time d_{max} and the minimum interrupt response time d_{min} the reading error ε. In a central synchronization scheme, the convergence function [9], i.e., the maximum difference between any two clocks immediately after synchronization, is $\varepsilon + g$, where g is the granularity of the local time. Between any two resynchronizations the clocks are free running. Their speed can differ by 2ρ, the drift rate from the nominal time, determined by the physical quality of the resonator. If a resynchronization is guaranteed to happen after Rsync, the achievable precision Δ of the ensemble is $\Delta = \varepsilon + g + 2\rho$ Rsynch.

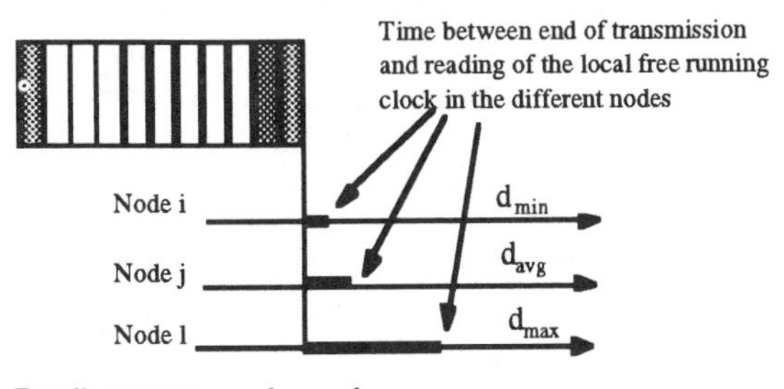

Reading error: $\varepsilon = d_{max} - d_{min}$

Precision $\Delta = \varepsilon + g + 2\rho$ Rsync

g: granularity of the local clock

ρ : maximum drift rate of the clocks

Rsync: Resynchronization Interval

Fig.5: Precision of the ensemble of clocks

SELECTION OF THE TIMEOUT PARAMETERS - In TTP/A we have to determine proper values for the following timeout parameters that activate state changes between the states indicated below:

ADT: Await Data Timeout (NA --> WD)
DPT: Data Processing Timeout (WD --> DP or DR -->DP or FR -->DP)
SDT: Send Data Timeout (DP --> NA or DP --> WF)
MMT: Multimaster Timeout (WF --> SF)

The following relations have to hold between these timeouts at all nodes at all times:

[RDI, ADT] < [RDI, RDI$^+$] < [RDI, DPT] < [RDI^{FW}, ADT^{FD}]

where
RDI$^+$: Arrival of next message;
RDI^{FW}: RDI of Fireworks byte;
ADT^{FD}: ADT of first data byte.

We base these timeouts on the synchronized local times with the above calculated precision Δ. If two nodes start the transmission of the next message at the same local time these times can differ, as seen by an omniscient outside observer, by Δ.

The RDI interrupt can happen at the node with the fastest resonator in the earliest case at dtrans$(1-\rho)$ after sending (where dtrans is the nominal transmission time of a single one byte message). Since ADT must occur before this earliest arrival, we select a value for ADT of dtrans$(1-\rho) - g$ (in relation to SDT). DPT must happen after the RDI interrupt of the node with the slowest resonator. Since this node can be one Δ behind and can have a delay of the interrupt by ε, a proper value for DPT in relation to SDT is dtrans$(1+\rho) + \Delta + \varepsilon + g$. The window when the next RDI may occur is thus 2dtrans $\rho + \Delta + \varepsilon + 2$ g.

Since we resynchronize on RDI, the interval [RDI,SDT] must be selected such that even the latest and the slowest node will have completed the processing of the previous message. This gives a value [RDI,SDT] = dproc^{max} + 2 dtrans $\rho + \Delta + 2\varepsilon + 2$ g, where dproc^{max} denotes the maximum processing time of a message.

The delay between the Fireworks byte and the first data byte must be significantly longer than the longest interdatebyte separation. It follows that [RDI^{FW}, ADT] must be at least dproc^{max} + dtrans$(1+3\rho) + 2\Delta + 3\varepsilon + 4$ g.

The maximum slot time between any two data messages is:
Slot-time^{max} =dproc^{max} + dtrans $(1+\rho) + 2\varepsilon + \Delta + 2$ g.

The slot time forms the basis for the calculation of the data efficiency of the protocol.

The Multimaster Timeout (MMT) is an application specific timeout parameter that is determined by the length of the longest MEDL. It has no influence on the data efficiency of the protocol in a fault-free environment.

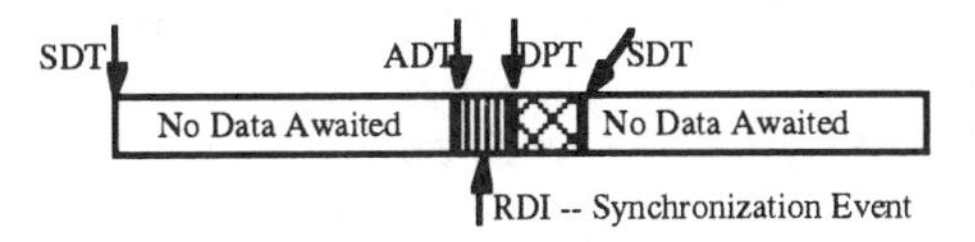

[SDT,ADT] = dtrans $(1-\rho) - g$ dtrans: transmission time
[SDT,DPT] = dtrans $(1+\rho) + \varepsilon + \Delta + g$ dproc^{max} : max. proc. time
[ADT,DPT]= 2 dtrans $.\rho + \varepsilon + \Delta + 2g$
[RDI, SDT] = dproc^{max} + 2 dtrans $.\rho + 2\varepsilon + \Delta + 2g$
[RDI, ADT] = dproc^{max} + dtrans $(1+\rho) + 2\varepsilon + \Delta + g$
[RDI, DPT$^+$] = dproc^{max} + dtrans $(1+3\rho) + 3\varepsilon + 2\Delta + 3g$
[RDI^{FW},ADT] = dproc^{max} + dtrans $(1+3\rho) + 3\varepsilon + 2\Delta + 4g$
slot-time^{max}= dproc^{max} + dtrans $(1+\rho) + 2\varepsilon + \Delta + 2g$

Fig.6: Time-out parameters of TTP/A

ADVANCED PROTOCOL FUNCTIONS

In this section we discuss advanced protocol functions that can be implemented with the TTP/A protocol. Many of these functions make use of the mode change mechanism that is supported by the basic protocol. It is not necessary that all nodes within a cluster support all modes. If a fireworks byte contains a mode that is not supported by a particular node, then this node will be passive during the execution of a round of the unsupported mode.

SPECIAL MODES

Sleep and Wakeup - In body electronics application some subsystems have to continue their operation even when the car is parked for long periods: e.g., the theft-avoidance system or the access control system. To minimize the load on the battery, these subsystem work autonomously in a low power mode without any traffic on the network.

In a TTP/A system a go-to-sleep mode informs all nodes of the network that the sleep state is to be entered. In the sleep state no messages are sent on the network, since any activity on the network will be interpreted as a wakeup event. Whenever a node wants to wakeup some other nodes, it sends any character (a wakeup byte) on the network as a wakeup event. The wakeup byte does not communicate any data, other then the fact that a new communication round is to be initiated. The corruption of the data field of the wakup byte by concurrent send operations is therefore of no concern. After the receipt of a wakeup event, every node will enter the "Wait for Fireworks" state. The master will then transmit a Fireworks byte to resynchronize all nodes and to initiate a new communication round.

Diagnostic Mode - In many scenarios a TTP/A node will be the only connection to a sensor or actuator. To be able to test the proper operation of the sensor/actuator, a special diagnostic mode has to be provided that can be activated by the master. The master is then in a position to test and calibrate the sensors and actuators connected to the TTP/A nodes.

MULTI-CLUSTER SYSTEMS - There are a number of single chip microcontrollers on the market that support two or more UART channels. These chips can be used as gateways in TTP/A systems. It is also possible to use a UART channel of a TTP/C chip as a gateway between a TTP/A and a TTP/C cluster. From the point of view of the data organization and the application software interface, TTP/A and TTP/C are compatible. This compatibility simplifies the design of the gateway software.

If more than one UART channel is connected to a single processor, then the reading error ε will increase, since not all channels can be on the highest priority level. This increase in ε has an effect on the time-out parameters of the protocol, as indicated in the previous sections.

The availability of single chip controllers with more than one UART interface makes it possible to implement multi-cluster body electronics systems economically. These multi-cluster systems provide an interesting alternative for the implementation of fault-tolerance at the system level. Critical nodes can be connected to more than one cluster such that the loss of a complete network will not cause a loss of the critical functionality.

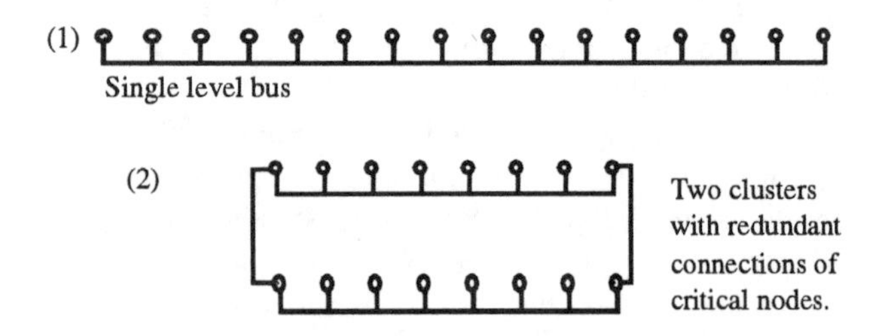

Fig.7: Multi-cluster body electronics system with redundant interconnections

In the above example (Fig. 7), a body electronics system with four critical nodes has been reconfigured from a linear single level bus configuration (1) to a two cluster configuration with redundant links between the four critical nodes (2). The addition of one more wire and four additional UART interfaces increases the reliability with respect to serious permanent cabling failures considerably.

IMPLEMENTATION ISSUES

The implementation of the Fireworks protocol will depend on the characteristics of the intended application and the capabilities of the given microcontroller. In the first part of this section we give some hints concerning the protocol data structures. We then analyze the protocol performance.

PROTOCOL DATA STRUCTURES - Message Descriptor List MEDL - The most important data structure in the TTP/A implementation is the Message Descriptor List MEDL. The MEDL determines the sequence of the data bytes within a round and specifies the attributes of the data. The best format of the MEDL implementation is determined by the characteristics of the given application.

In body electronics system, where a network will have up to 16 nodes and only a small number of nodes is interested in all data exchanged, the following format for the MEDL is proposed:

MEDL Header (2 bytes): Number of entries in MEDL (1 byte), MEDL control (1 byte)

MEDL Entry (3 bytes each): Entry number (1 byte), Data Control (1 byte), Data Address (1 byte)

The MEDL can be in ROM.

Mode Descriptor List MODL - For every mode a Mode Descriptor List (MODL) is required. The MODL will have one entry for each mode containing the mode name (1 byte), the mode control (1 byte) and the MEDL address (1 byte or longer). The MODL can be in ROM.

Smallest Implementation - The smallest implementation of TTP will be met in smart devices. These devices support the control of a single switch or the reading of a single sensor, have the software in ROM, receive a single byte of data and will send a single byte of data. The MEDL will thus have two entries with a length of eight bytes in ROM. The protocol software will require at most a few hundred bytes of ROM (depending on the instruction set of the microprocessor) and less than 10 bytes in RAM.

PHYSICAL LAYER - The physical integration of a number of UART devices into a bus system requires a proper physical layer. The physical layer can be based either on a two wire implementation or a single wire implementation.

It is not the intention of this paper to propose a new physical layer for body electronics systems. It is recommended to take one of the existing physical layer implementations for single wire bus systems, e.g., the J1850 physical interface. These implementations have been tested in practical vehicle operations.

DATA EFFICIENCY OF TTP/A - The data efficiency of TTP/A is the ratio of useful data transmitted in one message (eight bits of data and three bits of control information) divided by the sum of the slot time and the appropriate fraction of Fireworks byte. Let us assume a round with n data bytes. Then the data efficiency η is

$$\eta = \frac{dtrans\ 8/11}{slot\_time + (1/n)\ FW}$$

where dtrans is the transmission time of a one byte message, slot_time is the duration of a data slot, i.e., the duration between the start of two consecutive data messages, and FW is the duration needed to send the fireworks message including the intermessage gap between the fireworks and the first data message.

The slot time of a data slot is given by

$$slot\_time = dproc^{max} + dtrans(1+\rho) + 2\varepsilon + \Delta + 2g$$

The transmission time of the Fireworks byte, including the extended intermessage gap is

$$FW = dproc^{max} + dtrans(1+4\rho) + 2\Delta + 3\varepsilon$$

Let us assume that the Resynchronization period Rsynch = 5 dtrans, i.e. there may be up to two empty slots between any two data bytes. Then the precision of clock synchronization is

$$\Delta = \varepsilon + g + 10\rho\, dtrans$$

We can now transform the equation for the data efficiency to expose the contributions of the processing time the transmission time, the reading error, the granularity, and the drift rate to the data efficiency:

$$\eta = \frac{dtrans\ 8/11}{(dproc^{max} + dtrans)(1+1/n) + \varepsilon(3+5/n) + \rho\ dtrans\ (11+24/n) + g(3+7/n)}$$

To get some insight into the influence of the different parameters, we have calculated the data efficiency for a number of different scenarios (Fig. 9). In all cases we assume a transmission rate of 10 kbaud and a processing time of 50 μsec. The reading error ε is assumed to be 5 μsec if the protocol task is at the highest priority and 50 μsec if another task with a processing time of about 50 μsec can preempt the protocol task. The drift rate of the clock ρ is assumed to be 10^{-4} if the resonator is a quartz resonator, and 10^{-2} if the resonator is a ceramic resonator. The granularity of the local clock is assumed to be 4 μsec in all cases but the first one.

The first line in Fig. 9 describes a situation where 10 bytes are transmitted in a "perfect system" to get a reference point for the best achievable efficiency. Column η contains the data efficiency for the parameters listed in the previous columns. Some of the parameters are varied in the next four lines to get an insight in the contribution of the different mechanisms.

The last three lines refer to the benchmark example published in [6] where a loop with 32 total message bytes is considered. In this example the TTP/C protocol achieved a data efficiency of about 53 %, CAN about 23 % [4].

n	dtrans	dproc	ε	ρ	g	η %
10	1100	50	0	0	0	63.2
10	1100	50	5	10^{-4}	4	61.6
10	1100	50	5	10^{-2}	4	55.4
10	1100	50	50	10^{-4}	4	54.9
10	1100	50	50	10^{-2}	4	50.0
32	1100	50	5	10^{-4}	4	65.8
32	1100	50	5	10^{-2}	4	59.5
32	1100	50	50	10^{-2}	4	53.8

Fig.9: Data efficiency of TTP/A

CONCLUSION

In this paper we presented a new protocol for body electronics applications that differs from the presently available protocols, such as J1850, CAN, or the token protocols in three significant aspects:

Firstly, the TTP/A protocol supports composability in the value domain and in the temporal domain. It is possible to design and built large on board electronic systems out of well specified and tested components without the fear of unintended side effects caused by the system integration. Secondly, the TTP/A protocol provides a short error detection latency, since the a priori knowledge about the intended send and receive time of messages facilitates the immediate detection of a lost message or a failed node. A fail-safe body electronics application can thus visit the safe state autonomously with minimal delay. Thirdly, TTP/A systems can be implemented on state of the art UART interfaces that are standard on nearly all microcontroller chips. In addition to the UART channel the protocol implementation requires a standard timer. There is no need for special protocol hardware to support the clock synchronization or other protocol functions.

We have given a detailed description of the protocol and performed an analysis concerning the effects of the interrupt response time, the oscillator quality, and the granularity of the local real time clock on the data efficiency of the protocol. In most practical scenarios the data efficiency will be above 50 % . This high data efficiency for short messages compares favorably to other protocols proposed for body electronics applications.

REFERENCES

[1] SAE paper J2056/2 - Survey of Known Protocols, published in 1994 SAE Handbook, Vol., 2, pp.23.273 , Society of Automotive Engineers, Warrendale, PA, 1994

[2] Controller Area Network CAN, an IN-Vehicle Serial Communication Protocol—SAE J1583, March 1990, 1992 SAE Handbook, pp.20341-20355*

[4] Kopetz, H., Grünsteidl, G., TTP- A Protocol for Fault-Tolerant Real-Time Systems, IEEE Computer, January 1994, pp. 14-2

[5] Multi-Transmitter Data Bus—Par 1: Technical Description, ARINC Specification 629-2, Aeronautical Radio, Inc, 2551 Riva Road, Annapolis, Maryland, 21401, October 1991

[6] 1992 SAE Handbook, Vol. 2, Society of Automotive Engineers, 400 Commonwealth Drive, PA, USA, 1992 , p.20.301

[7] Ebner, Ch., Description and Comparison of Selected UARTS, MARS Bericht No. 22, Sept. 1994

[8] Kopetz, H., Damm, A., Koza, C., Mulazzani, M., Schwabl, W., Senft, C., Zainlinger, R., Distributed Fault-Tolerant Realtime Systems: The MARS Approach, IEEE Micro, Vol. 9, Nr. 1, pp., 25-40, Febr. 1989

[9] Kopetz, H., Ochsenreiter W., Clock Synchronization in Distributed Real-time Systems, IEEE Transactions on Computers, August 1987, pp.933-940

950291

Open Systems and Interfaces for Distributed Electronics in Cars (OSEK)

U. Kiencke and K. J. Neumann
IIIT, University of Karlsruhe

L. Frey, J. Graf, and T. Wollstadt
Adam Opel AG

J. Krammer, W. Kremer, F. Lersch, and E. Schmidt
BMW AG

J. Minuth, T. Raith, and T. Thurner
Daimler-Benz AG

H. Kuder and V. Wilhelmi
Mercedes-Benz AG

H.-J. Mathony and D. Schäfer-Siebert
Robert Bosch AG

R. John, M. Reinfrank, and K. Storjohann
Siemens AG

C. Hoffmann
Volkswagen AG

ABSTRACT

The individual development process for distributed, communicating electronic control units hinders the integration of Automotive systems and increases the overall costs. In order to facilitate such applications, services and protocols for Communication, Network Management, and Operating System must be standardized. The aim of the OSEK project is to work out a respective specification proposal in co-operation with several car manufacturers and suppliers. This will permit a cost-effective system integration and support the portation of system functions between different electronic control units.

1. INTRODUCTION

In a distributed control system, several controllers (stations) are connected via a communication link (Fig. 1), e.g. electronic control units within an automobile. Generally these control units are supplied by different companies, and they have different microcontroller architectures. For the connection of distributed system functions, stations exchange messages with standardized interfaces. A uniform network management guarantees the safe operation of safety-relevant, distributed systems [1].

The development costs for the communication and network management software may be significantly reduced, if the interfaces and procedures are standardized not just within one subsystem, but for the entire distributed system. The software should be implemented in a uniform operating system. In addition, operating systems with uniform interfaces offer different application programs available from various suppliers which co-exist in a single processor. In that way, the multitasking approach serves as an efficient means to cut costs.

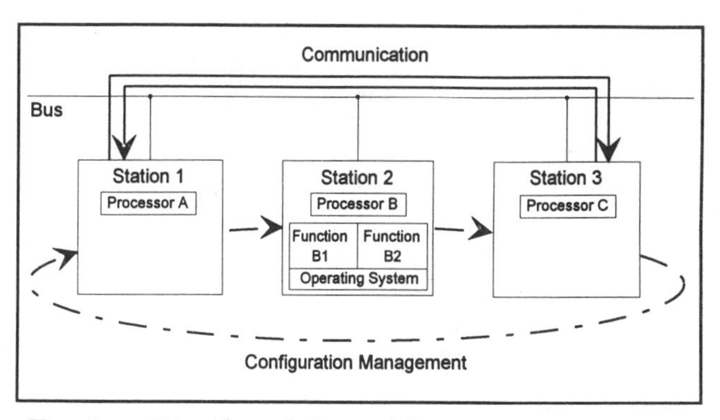

Fig. 1: Distributed Control System

The project "Open Systems and Interfaces for Electronics in Cars (OSEK)" thus aims to specify an open architecture for communicating vehicle systems [2]. This architecture comprises:

- **Communication** (Data exchange within and between Control Units);
- **Network Management** (Configuration determination and monitoring); and
- **Realtime Executive** (Operating system for Control Unit software).

With OSEK this expensive investment is only needed once, and it is possible to re-use it with minor modifications for various applications.

Particular targets of OSEK are:

- company-independent specification of interfaces, functions and protocols for Communication, Network Management and Realtime Executive;
- specification of a hardware- and software-independent user interface, which enhances portability and re-usability of application programs;

- efficient architectural adaption to respective applications by reconfiguration and scaling; and
- functional verification and implementation of prototypes in selected pilot projects.

It is not the aim of OSEK to engage into an implementation of products. These should be left open for e.g. software houses or microcontroller manufacturers.

Another multi company project in the German Automotive industry is MSR [3][4]. The aim here is an enhanced development efficiency by an improved information exchange between car manufacturers and their suppliers, on the basis of a common procedural model and – again – uniform interfaces and tools. There are contacts between OSEK and the MSR initiative. In the long run, OSEK could offer the software implementation platform for functions previously defined within the MSR framework.

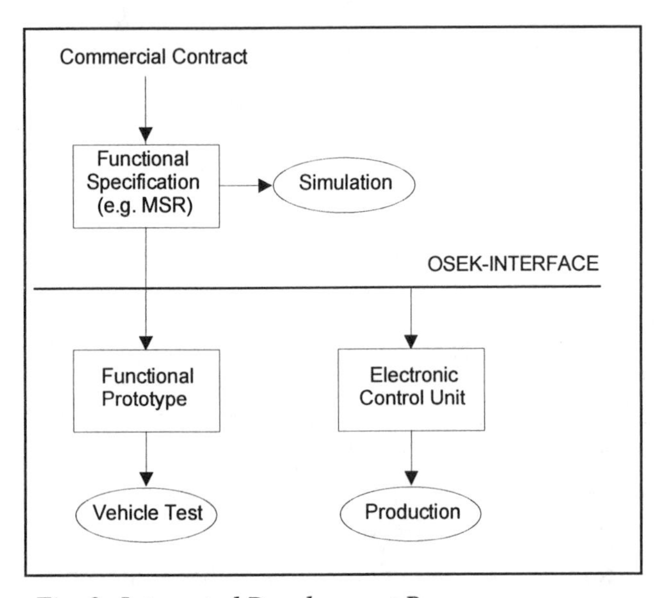

Fig. 2: Integrated Development Process

The aim of an integrated development process is shown in Fig. 2. A rough specification is supplied in a commercial contract between a car manufacturer and a supplier . This functional specification can be recursively more detailed, until it contains any specification requirement of the contracted subsystem. The MSR framework allows for the simulation of functions in advance, so that the subsequent development work necessitates significantly fewer modifications of target functions. The final MSR specification is done on the basis of the OSEK interface. From there, functional prototypes can be built and tested in vehicles. The suppliers use the same basis to develop their electronic control units for production purposes. Today's extremely costly and error-prone stage between prototypes and end products is thus overcome.

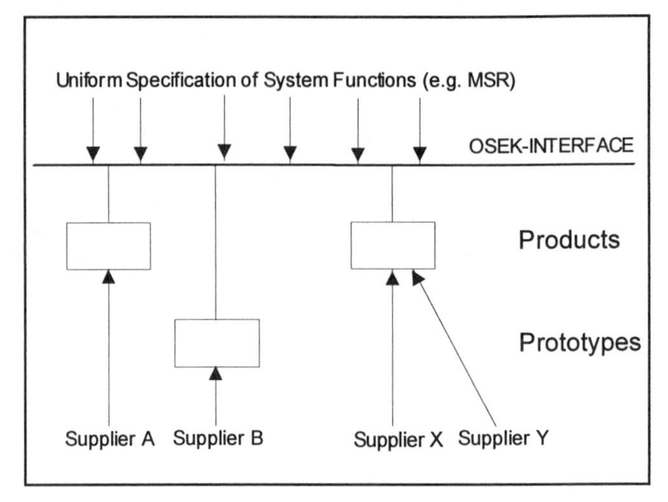

Fig. 3: Integration of Automotive Systems

When several different suppliers need to co-operate to incorporate their individual subsystems into a complete system, the MSR/OSEK approach is even more beneficial (Fig. 3). Since the OSEK interface can integrate end products together with prototypes, different timescales or eventual time delays at one supplier no longer impede vehicle tests of the complete system at the car manufacturer.

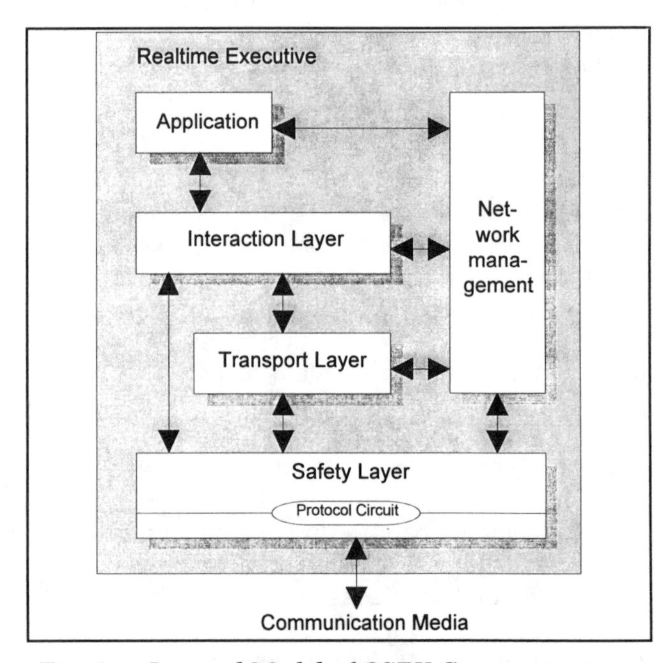

Fig. 4: Layered Model of OSEK Communication

In this paper, the OSEK project is presented. An overview is given of the rough specifications for communication, network management and realtime executive.

2. OSEK COMMUNICATION (OSEK-KOM)

The OSEK sub project communication (OSEK-KOM) specifies interfaces and protocols for in-vehicle communications. This comprises communication:

- between interlinked control units;
- within control units; and
- between control units and peripherals.

The control-unit software is thus simplified and can be re-used. The standardized interfaces keep the application software independent of special network protocols, such as CAN, ABUS, VAN, J1850, K-BUS, P-BUS, I-Bus etc.). OSEK communication can be provided for different protocols and microcontrollers.

As far as it is feasible in realtime systems, the communication is defined according to the ISO/OSI layer model [5], consisting of the *safety layer*, the *interaction layer* and optional in-between *transport layer* (Fig. 4). The physical layer below will be integrated in the interface circuitry. It is therefore not part of OSEK-KOM [6].

The *safety layer* provides services for the transmission of single messages via the network protocol used. The next link up is the optional *transport layer*. It provides services for the transmission of message packages of any length. There is a differentiation into transmission without acknowledge (1:n communication) and with acknowledge (1:1 communication). The *interaction layer* connects the interface to the application. Its services are fully independent of microcontroller architectures and network protocols. Service processes of the interaction layer are concurrent to application processes.

The interaction layer enables communication by:

- **Variables** for the transmission of system states, such as engine speed, engine temperature; and
- **Channels** for the transmission of various events, such as transgression of maximum engine temperature.

Intertask communication by such means

- offers a well-defined communication interface;
- is based upon a uniform, previously developed communication software;
- guarantees data consistency by organizing simultaneous access to data from concurrent processes; and
- supports portability of application software which can be assigned to different control units as long as time factors are not adversely affected.

2.1 Communication by Variables

The interaction layer contains a common data buffer for variables from distributed application tasks. A communication is accomplished, when one transmitter actualizes a variable by *write*, and when one or several receivers read that variable by *read* (1:n communication, Fig. 5). The value of a variable may be read many times. There is no time or scheduling information in such variables.

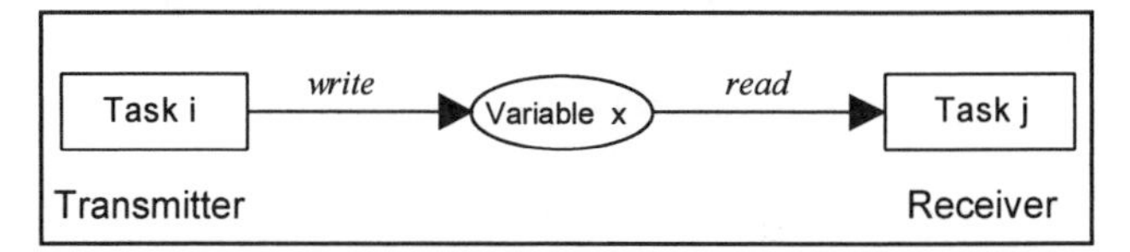

Fig. 5: Communication by Variables

The communicating tasks can be located in just one station or in several stations. That is why there is a distinction into station-global and network-global variables, allowing different implementations.

Implementation of Network-global Variables

The variable x is stored in the transmitter station. A copy x' of that variable is transferred to the receiver stations. A *write* access from the transmitter locally updates variable x. The interaction layer now takes care, that the copies x' are also updated by a message transmission. Subsequent *read* accesses are again local operations.

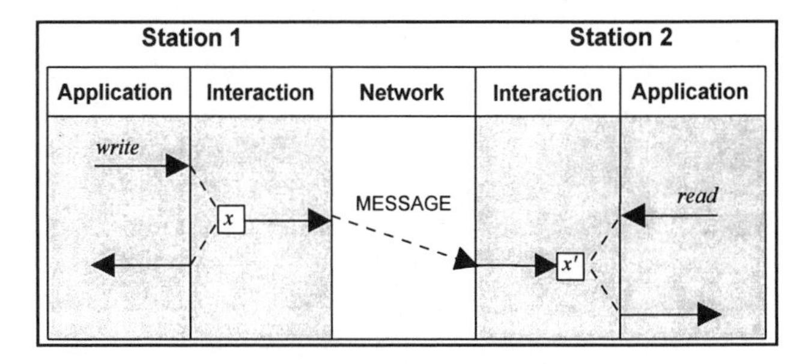

Fig. 6: Communication by Variables

Update of Copies of Variables

Network-global variables can be copied into receiving stations in different ways:

- event-driven: A *write* access triggers the transmission of respective variable data (transmission upon request); and
- time-controlled: Variable data are periodically transmitted, independently of any changes of their value. The time period is an attribute of the variable.

An eventual combination would transfer variable data either event-driven in case of changes, or time-controlled at the end of a given time period.

Monitoring the Actuality of Variables

Any variable can be monitored by giving it a particular timeout. A variable is considered to be no longer actual, if its value has not changed during the timeout period.

2.2 Communication by Channels

A channel is an object which comprises:

- a data structure for buffering messages;
- a data structure for storing task identifications; and
- operations *send* and *receive*.

In a communication by channel, data (messages) from one task are transferred to other tasks (1:n communication). The execution is done through a channel. The message will be stored into the channel, if the receiver is not yet ready for the communication. The operation *send* stores a message into the channel, *receive* reads that message from the channel and 'consumes' it, i.e. rendering its data in the channel obsolete (empty channel). The *receive* operation can only be performed once by a task, unless there is a new update.

Communication by channel implies the synchronization of the communication partners, i.e. the sequencing of participant tasks over time. A message can only be received if it has been previously stored into the channel. Transmission is however done asynchronously. There are two reception modes, either asynchronous (non-blocking) or synchronous (blocking). In the synchronous mode, the *receive* task is blocked until a new message arrives in the channel (Fig. 7). In order to prevent receivers from blocking continuously, the ready state of the receiver is limited by a timeout.

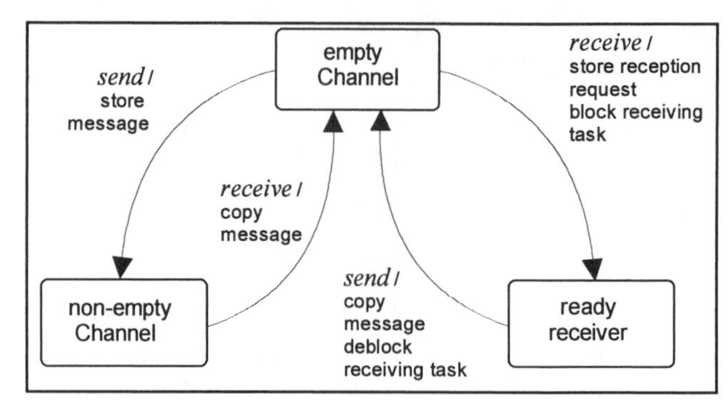

Fig. 7: State Diagram for 1:1 Channel with Synchronous Reception Mode

A fixed-capacity attribute is assigned to each channel, defining how many subsequent messages can be stored in the channel. Messages go through the channel according to the FIFO principle. They are received in the same sequence as they have been transmitted. In case of an overflow (capacity is exceeded), the oldest message is deleted in order to prevent transmitters from blocking.

Fig. 8 shows a 1:3 channel with a capacity of 2. Two messages can thus be stored in this channel. A list with 3 task identifiers is assigned to each of the message buffers. Whilst reading, the receiving task leaves its identifier at the buffer. A multiple reception of the same message can thus be prevented, separately for all tasks. A message is considered to be consumed, when all receiving tasks have performed their *receive* operation.

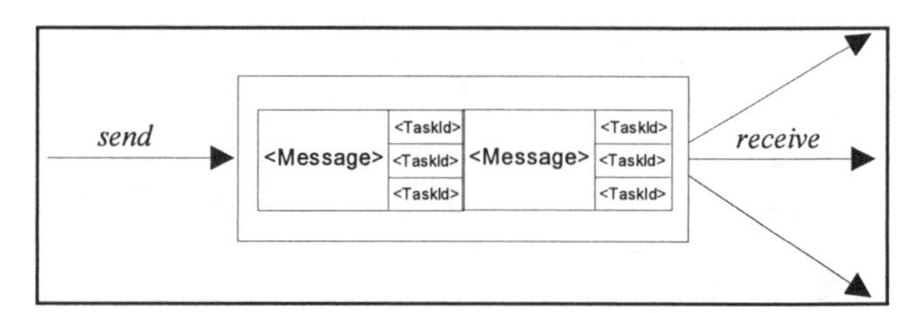

Fig. 8: 1:3 Channel with Capacity 2

As in the case of variables, channels are defined as station-global and network-global. A network-global channel is implemented by a channel buffer y in the transmitting station and channel buffers y' in the receiving stations. The message transfer is performed by the interaction layer.

3. OSEK NETWORK MANAGEMENT (OSEK-NM)

The reliability and availability of the communication link is guaranteed by an integrated network management (OSEK-NM). Its essential services are:

- defined and synchronized initialization of the network communication, determination of the configuration. Application tasks can only start their operation, after all data transmitting stations are fully operational.
- continuous configuration monitoring, detecting the addition or the eventual failure of stations. Changes in configuration are relayed to the application layer, where a functional reconfiguration must be decided (e.g. graceful degradation).
- defined and synchronized transfer of the network into the "sleep mode".

This configuration management makes up the core services of OSEK-NM. Further optional services are:

- initialization of operating resources and of objects defined by OSEK-KOM and OSEK-BS;
- control of operation modes;
- detection, management and messaging of failures;
- diagnostic support, e.g. error statistics, monitoring; and
- handling of network resources, e.g. temporary channels.

The network management relies upon station-address-oriented communication services which are provided by OSEK-KOM.

Configuration Management

A station transmits a status message over the network, when triggered by the previous station. By definition of predecessors and successors, a sequence develops, in which stations transmit their status messages. A logical ring is thus installed, where the status is forwarded from a station to its successor within the given configuration /7,8/. The additional traffic load on the network due to these status messages is relatively low.

The configuration is defined as the set of all stations, which are operational and accessible by communication. A station is regarded ready by the network management, first, when it transmits its status message cyclically or is triggered by its predecessor, and second, when it receives status messages of the other stations.

The configurations are as follows:

- The **Actual Configuration** is the set of all operational stations recorded by the network management at a time t. This set may change, e.g. differing among stations in case of failures, or stations being switched on or off by the driver.
- The **Reference Configuration** is given by the car manufacturer. It depends on the vehicle equipment and options available. This reference is used as a comparison to the actual configuration, which allows to eventually start operations of the application.
- The **Sum of actual Configurations** is the union of all actual configurations observed since the first start of the distributed system. It may serve as a substitute for the reference configuration.

All configurations are accessible through the application layer.

The state diagram of the network management within a single station is shown in Fig. 9.

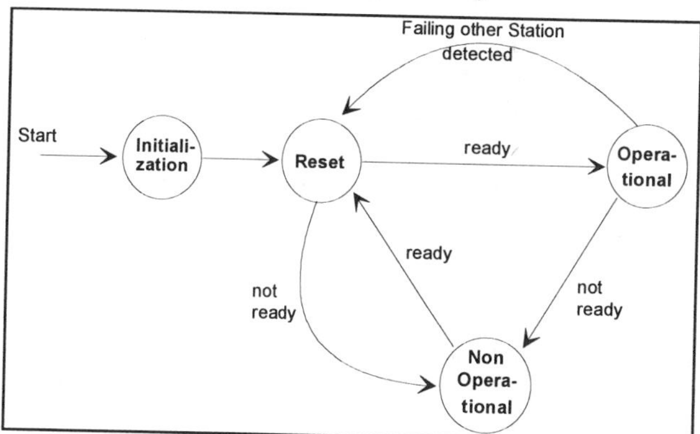

Fig. 9: State Diagram of Network Management in a single Station

Start and Configuration Monitoring

With the Start, stations are initialized one at a time. The duration of initialization depends mostly on the specific application. This is one reason, why some stations become operational sooner than others, and why the actual configuration may change at a later date.

Configuration management is done in parallel in all stations without any master function. As a first step, an eventually still stored previous actual configuration is reset to the reset configuration which consists just of its own station. Immediately thereafter, a status message is transmitted through the network, by which the station makes itself known to all other stations. Now OSEK-NM is transferred to the state Operational. All status messages on the network from other stations are then received, and the actual configuration is derived from that. Without any reception of status messages from other stations, the own station repeats its own status message after a timeout. It follows, that a logical ring can always be built up. With the reception of some other status messages, available stations are inserted into a logical ring according to their given addresses.

Status messages are triggered by a respective message from the predecessor station, and are directed to the successor station. Since all these data are accessible to the application layer, the availability of the reference configuration can be easily checked.

In case that a station is no longer operational, it reverts to the state Non-operational. From there, a respectively adapted status message is cyclically transmitted. A station failure can also be detected, when its status message is no longer transmitted within a given timeout. This is diagnosed in all other stations, which then go back to the Reset state, and determine the actual configuration from scratch.

Announcement of Stations

A new station is detected from its status message by the other stations within the actual configuration. It is incorporated into the logical ring.

Withdrawal of Stations

There is no defined specific procedure. A station just discontinues the transmission of its own status message.

Sleep Mode

The switch-down to sleep mode is initiated by one station. The corresponding request is entered into its status message, which is then transmitted from station to station. If all stations in the logical ring have acknowledged the request for sleep mode by forwarding it, and when this information transfer has completed a full loop, the entire network can be switched down.

4. OSEK OPERATING SYSTEM (OSEK-BS)

It will provide a uniform and efficient execution environment for all Automotive Electronic Control Units. Program Modules written e.g. in "C" language will be readily exchanged.

There seems to be an inherent contradiction between the targets for standardization and efficiency. This is due to the commercial realtime executives which burden applications with huge overheads. They:

- offer users a variety of services, functions, of which only a portion is incorporated into the actual application;
- guarantee 100% portability of tasks; and
- may be dynamically reconfigured and scaled.

Such a complexity appears to be an overkill in high-volume Automotive applications. OSEK-BS is therefore deliberately restricted in its capabilities. To overcome this, the basic targets are somewhat similar to those of commercial realtime executives. However, there are some significant differences. OSEK-BS:

- is configured and scaled just statically. The number of tasks, resources and services is specified in advance by a user;
- also aims at the portability of tasks. There is however no 100% portability;
- requires experienced users, since application errors are not always backed up by extensive routines; and
- operates from ROM code.

4.1 The OSEK-BS Concept

It is the basis for the independent development of different application programs. Their execution in realtime is controlled by events from the plant. The control-unit hardware is accessible through uniform interfaces. This enables the application software to have a certain degree of independence from the control-unit hardware.

Execution Layers

The overall application software is partitioned into software portions, which are concurrently processed according to their urgency in realtime. In order to cover the large variety of realtime constraints, OSEK-BS contains 4 different execution layers with different priority levels (Fig. 10). These priority levels determine the order of processing.

The **Interrupt level** holds the highest priority. It is dedicated for time-critical activities, which are characterized by short latency and execution times. In order to match these constraints, the Interrupt Service Routines (ISR) should not contain any data processing. The interrupt response and acknowledge can be very fast, if the real processing of data and other actions are shifted to the job level.

The next priority level is taken by the **internal operating-system functions**, which are defined by core and optional services.

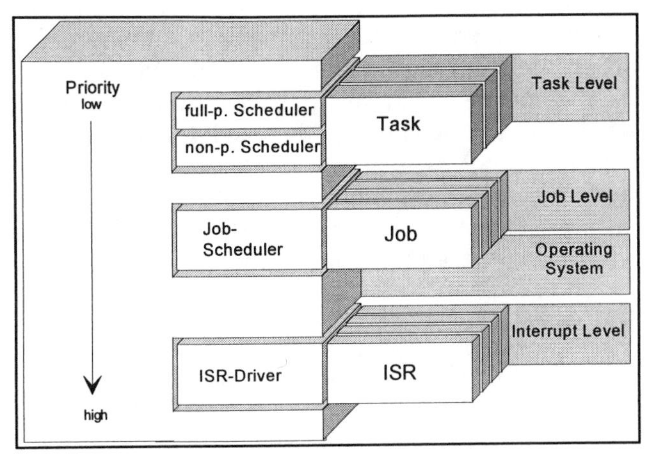

Fig. 10: Execution Layers of OSEK Operating System.

At the **Job level**, programs cover complete functions which are not interruptable (atomic) by other jobs or tasks. Within a job, there can be no status waiting for an event. Several attending jobs are sequentially executed. Therefore jobs are a special kind of tasks. Typical jobs are:

- Interrupt postprocessing;
- internal functions of the operating system such as timer manipulation, acceptance filtering of arriving messages; and
- execution of application routines reacting on asynchronous events.

Because of their atomic nature, such jobs increase interrupt latency times. They are however still short enough not to justify the installation of full tasks.

Task Level and Task Management

Tasks are organized by static priorities. They imply full context switch and the saving of variables into a stack. There are 4 task states.

running	The processor is assigned to the task, the respective instructions are executed.
ready	All preconditions for execution are met, the task is waiting for access to the processor.
waiting	At least one precondition for execution is still lacking.
passive	The task is not part of the execution.

Fig. 11 shows the state diagram

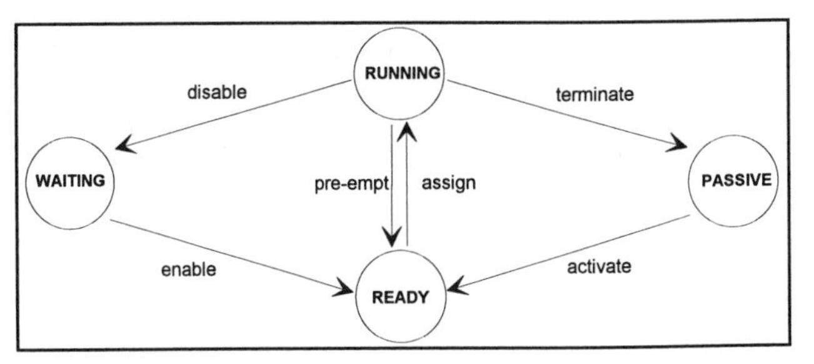

Fig. 11: State Diagram of Tasks

The essential elements of task management are the scheduler and the dispatcher. In case of an event, the scheduler decides according to priorities and to realtime constraints, which one of the tasks is going into the state *running*. During a task switch, the dispatcher saves the context of the task *running* up to that point, and installs the context of the next task.

Two principle scheduling procedures can be defined:

- **Full pre-emptive** scheduling allows the displacement of *running* tasks, if a higher-priority task becomes ready.
- **Non pre-emptive** scheduling allows the displacement of *running* tasks, only, if intended by the user or if terminated. This can only be done in case of short execution times.

OSEK-BS also offers:

- **Mixed pre-emptive** scheduling which combines the two alternatives depending on the task.

Non pre-emptive tasks are useful when the execution time comes close to the duration of the task switch, if RAM space can be saved by not storing a task context, or if a task may not be interrupted by another one. For a clean portability of non preemptive application software into a full preemptive environment, the user must have designed in the interruptability principle beforehand.

Services of OSEK-BS:

a) Core Services:

• Task Management	Management of task states, scheduling and dispatching.
• Interrupt Management	Procedure for fast eventually bufferd interrupt handling
• Job Management	Execution of short program files (jobs), without context and without interruption by other jobs and tasks.

• Event Control	Management of events for task synchronization.
• Time Control	Cyclical release of certain tasks into the *ready* state, which are then activated.
• Angle Control	Optional version similar to time control.

b) Optional Services:

• Time Management	Provision of absolute time, services to calculate relative times and to exchange time units.
• Angle Management	Similar to time control.
• Error Management	User support in case of errors.
• Intertask Communication	Station-global variables and channels.
• Semaphore Management	Exclusive access to atomic operations, commonly used resources and devices.

5. SCALING AND CONFIGURATION

The concepts and services of OSEK shall be applied to all stations of a distributed system, from the top-end to the bottom-end. They must support different bus protocols such as CAN, ABUS, VAN, J1850, K-BUS, P-BUS, I-BUS etc. An adaption to the required capabilities and the used resources is done by scaling and configuring.

The core services are mandatory basic functions. They are part of any implementation. The optional services are then used to extend the performance of OSEK for a specific application. The configuration is done by static parameters.

6. SUMMARY

The aims of the OSEK initiative are to specify uniform services, interfaces and protocols for the communication, network management and the operating system of a distributed realtime system. The cooperative effort of many car manufacturers and suppliers is not just in order to achieve a standardization. It guarantees, that practical application issues are considered on a broad basis within the OSEK specification, ensuring its acceptance by development engineers in their daily programs.

When this paper was filed for publication to the SAE in July 1994, a rough specification of OSEK had just been completed. Currently a detailed specification is being worked out. It should be noted, that the OSEK initiative is not restricted to a limited region of the world. All potential partners willing to be actively involved in a fruitful OSEK co-operation are most welcome onto the OSEK bandwagon.

REFERENCES

/1/ Kiencke U., "Distributed Realtime Processing in Automotive Networks", *SAE Technical Paper* No. 900696, 1990.

/2/ Mathony H.-J., Kaiser K.-H., Unruh J., Raith T., Thurner T., "Open Systems and their Interfaces for the electronic in Cars - OSEK", *13. Tagung 'Elektronik im Kraft-fahrzeug' im Haus der Technik*, Essen, 1993.

/3/ Besel, K.-G., Hirth, T. "Design systems for the MSR-Project", (German-language), *VDI-Berichte* Nr.: 1009, S. 503, 1992.

/4/ Leohold, J., "The MSR-Project: Tool support for new ways of cooperation between vehicle manufacturer and supplier", (German-language) *VDI-Berichte* Nr.: 1009, S. 491, 1992.

/5/ ISO "Information Processing Systems - Open Systems Interconnection - Basic Reference Model", *ISO* 7498, 1984.

/6/ Mathony H.-J., Kaiser K.-H., Unruh J., "Network Architecture for CAN", *SAE Technical Paper* No. 930004, 1993.

/7/ Raith T., Thurner T., Kocher H., "Netsoftware for databus systems in vehicles", (German-language), *VDI-Berichte* Nr.: 819, S. 171, 1990.

/8/ Kühner T., Häußler B., Thurner T., Müller K.-H., "Standardized netsoftware moduls for car ECU's - realized on a CAN-System", (German-language), *VDI-Berichte* Nr.: 1009, S. 653, 1992.

950295

Modeling Automotive Intercontroller Communication

James E. Beck
Carnegie Mellon Univ.

Christopher A. Lupini
Delco Electronics Corp.

ABSTRACT

Simple, analytic intercontroller communication (ICC) models are presented for four specific communication links used in automotive applications: Class 2, CAN, SPI and SCI. These models predict the transmission time and the worst case blocking duration for data transfers based on the bit rate of the medium, the clock rate of the host microcontroller and the amount of data being transmitted. The accuracy of the models is verified through comparison with empirical transmission time measurements for each link, and relative performance is discussed under normalized conditions.

INTRODUCTION

A method is needed to model intercontroller communication (ICC) in multiple processor automotive electronic systems. Such a model could be used when considering the problem of mapping cooperating software procedures to multicomputer products, i.e. sharing a specific computing task among several microprocessors. Specifically, a model is desired that will predict the transmission time and worst case blocking duration for communicated data based on the amount of data to be transferred and the characteristics of the physical communication medium and processing nodes. Candidates for the modeling activity were chosen from common automotive intercontroller and intracontroller communication methods.

This paper presents a generic model for ICC which is independent of the physical communication medium. The generic model is then extended to represent the characteristics of four specific communication links used in automotive applications. A summary of the communication links and their associated models is given at the end of the paper.

A GENERIC ICC MODEL

A generic ICC model developed by Sih [15] is summarized in Figure 1 . The $\square$ symbol is a placeholder for addition or multiplication and is dependent upon the particular transfer scheme. "Hops" refers to the number of intervening processing nodes the data has to pass through to get from the source to the destination.

Sih's model predicts transmission time for an ICC transfer in a generic multicomputer, based on the amount of data being transferred and the number of hops (if any) between the source and destination nodes. Incorporated into the model are three overhead terms and a binary operator which are chosen to represent the characteristics of the communication medium being considered. The model is general and can be tailored to characterize data transfers over a wide variety of communication networks with varying topologies and protocols.

$$C = (x + yD) \square zH$$

WHERE:

x = *data size independent overhead term*
y = *data size dependent overhead term*
z = *forwarding overhead term*
C = *transmission time*
D = *transmitted data size*
H = *numberof hops*
$\square \in \{+, \times\}$

Figure 1 Generic ICC Model

PHYSICAL COMMUNICATION LINKS

In this section, four specific communication links are introduced which are currently used or will be used in commercial automotive electronic systems. The characteristics of each link are described and incorporated into the generic ICC model, producing a set of link–specific modeling equations. The models will only characterize transfers over the medium; the effects of queuing will not be considered. Note that this is equivalent to the assumption that all link interfaces are managed by infinite–sized priority queues.

CLASS 2 – The Class 2 bus [1,2,3,5] is GM's implementation of SAE's J1850 communication standard [13,14]. It is a serial bus which implements the Carrier Sense Multiple Access / Collision Resolution (CSMA/CR) communication protocol. Each transmission consists of a bus arbitration phase followed by a data transfer phase. The bus is zero dominant, and collision resolution is achieved via a binary countdown [16] during the arbitration phase. Variable pulse width signaling is used, and the average bit rate achieved during standard operation with the current implementation is 10.4 kbps. The Class 2 bus interface is provided by dedicated, custom hardware which appears to the CPU as a memory–mapped peripheral [9].

Since a communication bus provides direct connectivity between all processing nodes, messages follow a direct path from source to destination (no hops). Hence, Sih's model reduces to EQ (1):

$$C = (x + yD) \qquad (1)$$

Each Class 2 message contains a 3 byte header used during the bus arbitration phase and a 1 byte footer which contains a Cyclic Redundancy Check (CRC) for error detection. Additionally, each message begins with a Start Of Frame (SOF) character and ends with an End Of Frame (EOF) character, which have a combined effective width of 5 bits. Therefore, each message contains 37 bits of overhead. Furthermore, during normal operation, the maximum message length is limited to 8 data bytes. This requires data transfers to be packetized, or "strip–mined" into sequences of 8 byte messages. Inserting this overhead figure into Sih's model and strip–mining yields the transmission time model for the Class 2 bus shown in Figure 2 .

$$C = \left\lceil \frac{D}{8} \right\rceil \left(\frac{37}{BR}\right) + \left(\frac{8D}{BR}\right)$$

WHERE:

C = transmission time (seconds)
D = transmitted data size (bytes)
BR = Class 2 bit rate (bps)

Figure 2 Transmission Time Model for Class 2 Bus

The first term in the equation represents the Class 2 message overhead accumulated over the number of messages obtained after strip–mining. The second term is a measure of the aggregate time needed to transfer the actual data. Note that there is no data size independent overhead term since strip–mining was required. Rather, the amount of overhead per data transfer is data size dependent and quantized, based on the maximum message length supported by the Class 2 bus.

The communication medium is assumed to support real–time message traffic, indicative of distributed control in an embedded (e.g. automotive electronic) system. Therefore, if a high priority data transfer arrives when a low priority transfer is active preemption should occur. However, since a transmitted message cannot be preempted without loss of data, preemption can only occur on message boundaries [8]. Priority inversion (aka "blocking") occurs when a high priority transfer is delayed by a lower priority one [8,17]. This corresponds to the period of time when a high priority transfer must wait until the next message boundary before preempting a low priority transfer. Since priority inversion must be considered when determining real–time schedulability, the worst case blocking duration for the Class 2 bus will be quantified. Since the maximum Class 2 message size is 8 data bytes, the maximum blocking duration is the time needed to transmit an 8 byte message, which is shown in Figure 3 .

$$B = \left(\frac{37 + 8 \times 8}{BR}\right) = \left(\frac{101}{BR}\right)$$

WHERE:

B = Max Blocking Duration (seconds)
BR = Class 2 bit rate (bps)

Figure 3 Maximum Blocking Duration for Class 2 Bus

CONTROLLER AREA NETWORK (CAN) – CAN is a protocol developed by Bosch in the mid 1980s which was intended for real–time control applications [6,7]. Like Class 2, CAN uses a CSMA/CR serial bus protocol. Unlike Class 2, however, CAN uses fixed pulse width NRZ signaling. The CAN protocol can be implemented on different media types, allowing galvanic or fiber optic implementations. Depending on the implementation, the bit rate can vary from between 5 kbps to a maximum of 1 Mbps. Many commercially available hardware devices exist that implement the CAN protocol.

Since CAN is a multicast bus, the reduced form of Sih's model given in EQ (1) again applies. Like Class 2, CAN messages contain a header used during bus arbitration which is 38 bits in length, and a footer which is 18 bits long. A 1 bit SOF and 7 bit EOF symbol are sent with each message, and an intermission frame is inserted between each message which is 3 bits wide. This gives a grand total of 67 bits of overhead per message. The NRZ signaling also specifies bit–stuffing, which causes an extra, inverted bit to be transmitted whenever a string of 5 consecutive 1s or 0s is encountered. Bit–stuffing applies to the entire header and data fields, and 15 bits of the footer which contains the CRC. Thus, bit–stuffing applies to 53 of the 67 overhead bits and the data field. The worst case situation will be modeled, resulting in an extra bit for every 5 bits encountered for which bit–stuffing applies. Like Class 2, the maximum length CAN message is limited to 8 data bytes. This again requires data transfers to be strip–mined into 8 byte messages. Beginning with Sih's model, adding the fixed message overhead, strip–mining and then bit–stuffing yields the transmission time model for CAN shown in Figure 4 .

$$C = \left\lceil \frac{D}{8} \right\rceil \left(\frac{67}{BR}\right) + \left(\frac{8D}{BR}\right) + \left\lfloor \frac{\left\lceil \frac{D}{8} \right\rceil 53 + 8D}{5} \right\rfloor \left(\frac{1}{BR}\right)$$

WHERE:

C = transmission time (seconds)
D = transmitted data size (bytes)
BR = bit rate (bps)

Figure 4 Transmission Time Model for CAN

The first term of this equation represents CAN message overhead accumulated over the number of messages obtained after strip–mining, the second term represents the time needed to transmit the actual data, and the third term represents the time needed to transmit the worst case number of bits added due to bit–stuffing. Again, the overhead is data size dependent and quantized due to strip–mining.

Consider the worst case blocking duration for CAN. This will be equal to the time needed to transmit a maximum length (i.e. 8 byte) message, as shown in Figure 5 .

$$B = \left(\frac{67 + 8 \times 8 + \left\lfloor \frac{(53 + 8 \times 8)}{5} \right\rfloor}{BR}\right) = \left(\frac{154}{BR}\right)$$

WHERE:

B = Max Blocking Duration (seconds)
BR = bit rate (bps)

Figure 5 Maximum Blocking Duration for CAN

SERIAL PERIPHERAL INTERFACE (SPI) – SPI is a full–duplex, serial, synchronous point–to–point communication link commonly available on microcontrollers [10,11]. As the name implies, SPI is typically used to interface a microcontroller to peripheral devices, however it is also capable of supporting ICC. Signaling is achieved via a simple, synchronous NRZ format. SPI is most often used in a master–slave configuration, but multiple–master implementations are feasible. The bit rate is a selectable derivative of the microcontroller clock frequency. The SPI interface is typically supported by dedicated hardware built–in to the microcontroller.

SPI links could conceivably be used to create a multicomputer of arbitrary topology, capable of supporting multihop communication. This has never been done at General Motors, thus no information is available concerning data routing characteristics for this type of implementation. Therefore, only the restricted case of a single SPI link connecting two processing nodes will be considered. Once again, the reduced form of Sih's model given in EQ (1) applies.

No presupposed message structure is mandated for the SPI link. In fact, SPI can conceivably support any arbitrary message structure. Therefore, message structure assumptions are required before an ICC model can be derived. Accordingly, the message structure depicted in Figure 6 will be assumed, which is adapted from a standard message format used in automotive applications [4].

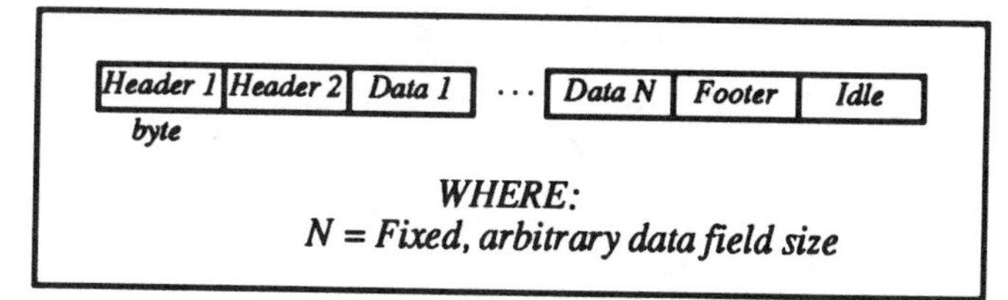

Figure 6 Assumed SPI Message Structure

The delay from an SPI message transfer request to the start of a transfer is programmable, with a minimum value of 0.5 bits needed for synchronization. The delay between successive transfers is also programmable, with a minimum value of 32 CPU clock cycles. The number of bits per transfer is programmable, but 1 byte transfers will be assumed. Based on the message structure shown in Figure 6 , there are 4 bytes of overhead per message, and a maximum of *N* data bytes per message, where *N* is arbitrary but fixed. Thus, there is a grand total of 0.5 bits and 32 CPU clock cycles of overhead per byte, and an additional 4 bytes of overhead per message. Inserting these overhead terms into Sih's model and strip–mining data transfers into *N*–byte messages yields the transmission time model for SPI shown in Figure 7 .

$$C = \left(\left\lceil \frac{D}{N} \right\rceil 4 + D \right) \left(\frac{8.5}{BR} + \frac{32}{CK} \right)$$

WHERE:

C = transmission time (seconds)
D = transmitted data size (bytes)
N = Max message data field size (bytes)
BR = bit rate (bps)
CK = CPU clock rate (Hz)

Figure 7 Transmission Time Model for SPI

The first term of the product represents the total number of bytes in the data transfer, including the SPI message overhead bytes accumulated over the number of messages obtained after strip–mining. The second term in the product represents the time required to transmit a single byte, including the SPI–specific transmission overhead.

Although SPI is not a multiple access link, if the processing nodes in question are multitasking then from the perspective of a processing task the SPI is a shared resource. As such, blocking between data transfers spawned by different tasks on the same node can occur. Once again, the worst case blocking duration is equal to the transfer time of a maximum length message, which is given in Figure 8 .

$$B = (4 + N) \left(\frac{8.5}{BR} + \frac{32}{CK} \right)$$

WHERE:

B = Max Blocking Duration (seconds)
BR = bit rate (bps)
CK = CPU clock rate (Hz)
N = Max message data field size (bytes)

Figure 8 Maximum Blocking Duration for SPI

SERIAL COMMUNICATION INTERFACE (SCI) – The last communication link to be considered is the SCI. Like the SPI, the SCI is commonly available on most microcontrollers. The SCI is a full–duplex, serial, asynchronous data link, (i.e. UART) [10,11]. SCI may be utilized to provide On Board Diagnostics–II (OBD–II) capability [12]. Again, the bit rate is a selectable derivative of the microcontroller clock frequency, but SCI bit rates are typically lower than SPI rates. Like the SPI, dedicated hardware is typically included within the microcontroller to support the SCI interface, however additional level–shifting circuitry is often needed to provide +/– 12V compatibility with the RS232 standard.

Once again, the restricted case of a single SCI link connecting two processing nodes will be considered and EQ (1) will be applied. When the SCI is idle, there is a synchronization delay from the time a transfer is requested to the time the transfer begins which is equal to one bit period. Back to back transfers can be accomplished, however, with no subsequent delay between transfers. The number of bits per transfer is programmable, but the most common mode is a single start bit followed by 8 data bits followed by a single stop bit. Like the SPI, the message format used with SCI is arbitrary. Therefore, the message structure shown in Figure 6 will once again be assumed. Thus, in the worst case, there is a total of 1 bit period of synchronization delay per data transfer, 4 overhead bytes per message and 2 overhead bits per byte. Data transfers will again be strip–mined into *N*–byte messages. Inserting this into the reduced form of Sih's model yields the SCI transmission time model shown in Figure 9 .

$$C = \frac{1}{BR} + \left(\left\lceil \frac{D}{N} \right\rceil 4 + D \right) \left(\frac{10}{BR} \right)$$

WHERE:

C = transmission time (seconds)
D = transmitted data size (bytes)
BR = bit rate (bps)
N = Max message data field size (bytes)

Figure 9 Transmission Time Model for SCI

The first term of the equation represents the synchronization delay. The second term represents the total number of bytes in the data transfer multiplied by the amount of time needed to transmit a single byte.

Lastly, the maximum blocking duration for the SCI link is shown in Figure 10 .

$$B = \frac{1}{BR} + (4 + N) \left(\frac{10}{BR} \right)$$

WHERE:

B = Max Blocking Duration (seconds)
BR = bit rate (bps)
N = Max message data field size (bytes)

Figure 10 Maximum Blocking Duration for SCI

CONCLUSION

The characteristics of the communication links and their associated ICC modeling equations are summarized in Table 1.

To verify the accuracy of the models, transmission time measurements were obtained in the laboratory for each communication link. Figure 11 plots transmission time versus total data size (not message size) for each link, showing both measured values and values predicted by the models. First, consider the Class 2 graph. Recall that the Class 2 bus specifies variable pulse width signaling. This causes each Class 2 transmission to contain a random sequence of short and long pulses. For the measured data shown in the Class 2 graph, short and long pulses are 64 us and 128 us in duration respectively, yielding an average bit rate of 10,416 bps and maximum and minimum bit rates of 15,625 bps and 7,812 bps respectively. The Class 2 graph plots the transmission times measured for data transfers with random bit streams, along with the transmission times predicted by the Class 2 model for the average, maximum and minimum bit rates. Note that the average bit rate model tracks the measured data, which is bounded by the maximum and minimum bit rate models. Next, consider the CAN graph, which plots measured transmission times against the CAN model with and without bit–stuffing. Again, the CAN models with and without bit–stuffing bound the measured data. Last, consider SPI and SCI. As the graphs indicate, these models are both quite accurate.

Figure 12 compares transmission time versus data size for all links, assuming a constant bit rate of 10 kbps and a common maximum individual message frame size of 8 data bytes. Each graph increases monotonically, indicating that transmission time increases steadily with the total number of data bytes to transfer (block size, or data size). With a normalized bit rate and maximum message size, data transfers take the longest with CAN and the shortest with Class 2 for all transfers greater than 2 data bytes in length. Note that Class 2, which is a multiple access link, outperforms both SPI and SCI, which are point–to–point channels. The relatively poor performance of SPI and SCI is attributable to two factors. The first is the overhead associated with the message structure assumed for these links. The second is the additional overhead per transmitted byte which is needed for synchronization in the case of SCI and is a result of the hardware implementation of the link interface for SPI. Likewise, the long transmission times predicted by the CAN model results from two factors. First, the CAN protocol specifies more message overhead than any other link. Second, the pessimistic assumption of worst–case bit–stuffing inflates predicted transmission times beyond what would typically be observed for this bus in practice. The discretization visible in the graphs is due to strip–mining, and the quantum of discretization corresponds to the maximum allowable message length, which was set to a constant value for all links. Lastly, note that for large data transfers the graphs begin to diverge. This is due to differences in overhead accumulated over a larger number of messages.

	Type	Data Width	Protocol	Data Rate (bps)	Transmission Time (seconds)	Max Blocking Duration (seconds)
Class 2	Bus	Serial	CSMA/CR	10.4 k	$\left\lceil \frac{D}{8} \right\rceil \left(\frac{37}{BR}\right) + \left(\frac{8D}{BR}\right)$	$\left(\frac{101}{BR}\right)$
CAN	Bus	Serial	CSMA/CR	[5k,1M]	$\left\lceil \frac{D}{8} \right\rceil \left(\frac{67}{BR}\right) + \left(\frac{8D}{BR}\right) + \left\lfloor \frac{\left\lceil \frac{D}{8} \right\rceil 53 + 8D}{5} \right\rfloor \left(\frac{1}{BR}\right)$	$\left(\frac{154}{BR}\right)$
SPI	Point to Point	Serial	Synchronous	$\left[\frac{CK}{510}, \frac{CK}{4}\right]$	$\left(\left\lceil \frac{D}{N} \right\rceil 4 + D\right)\left(\frac{8.5}{BR} + \frac{32}{CK}\right)$	$(4 + N)\left(\frac{8.5}{BR} + \frac{32}{CK}\right)$
SCI	Point to Point	Serial	UART	$\left[\frac{CK}{262112}, \frac{CK}{32}\right]$	$\frac{1}{BR} + \left(\left\lceil \frac{D}{N} \right\rceil 4 + D\right)\left(\frac{10}{BR}\right)$	$\frac{1}{BR} + (4 + N)\left(\frac{10}{BR}\right)$

D = Data Size (Bytes)
BR = Bit Rate (bps)
CK = CPU Clock Rate (Hz)
N = Max Message Data Field Size (Bytes)

Table 1 Communication Link Summary

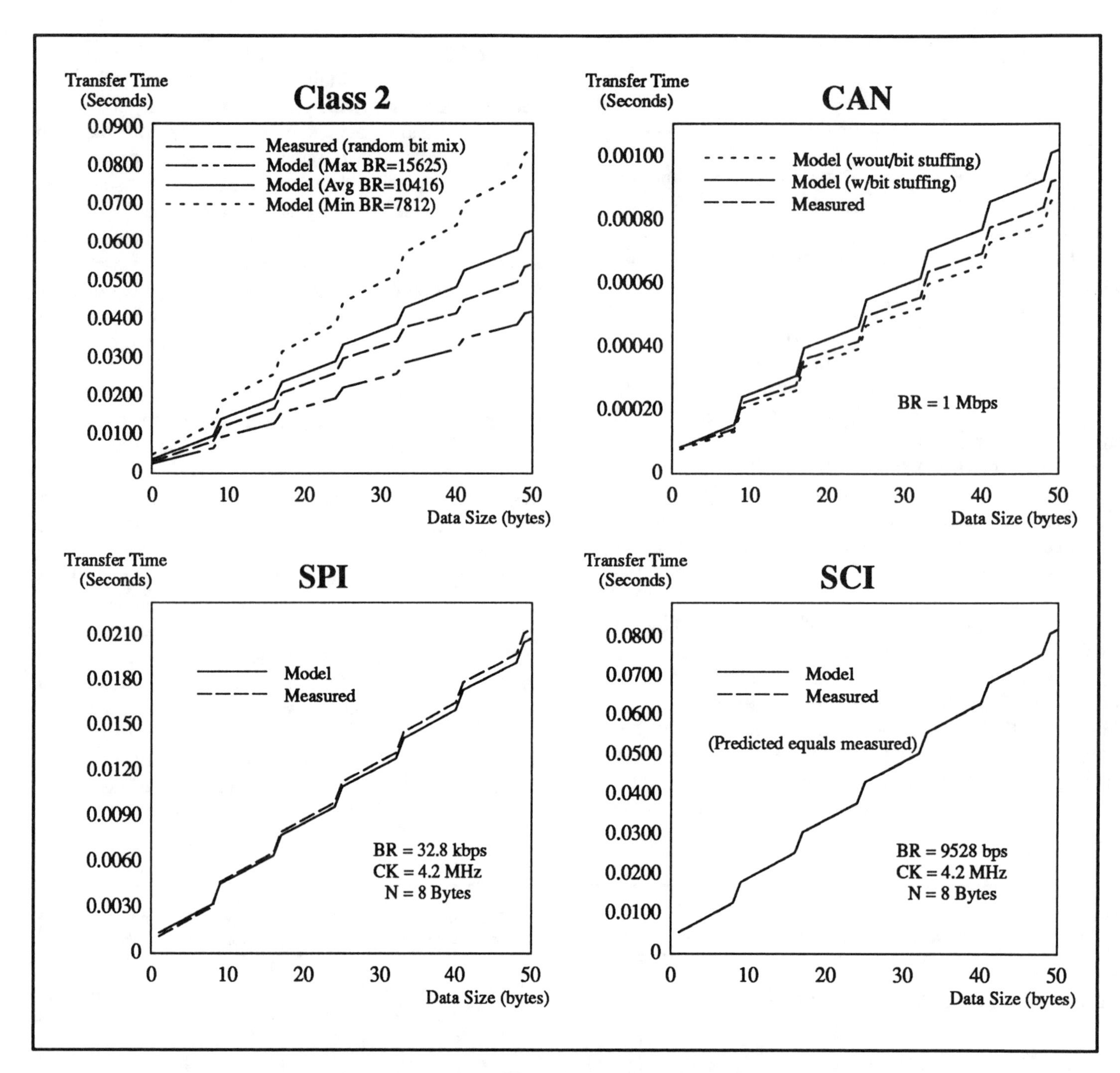

Figure 11 Transmission Time: Modeled vs. Measured

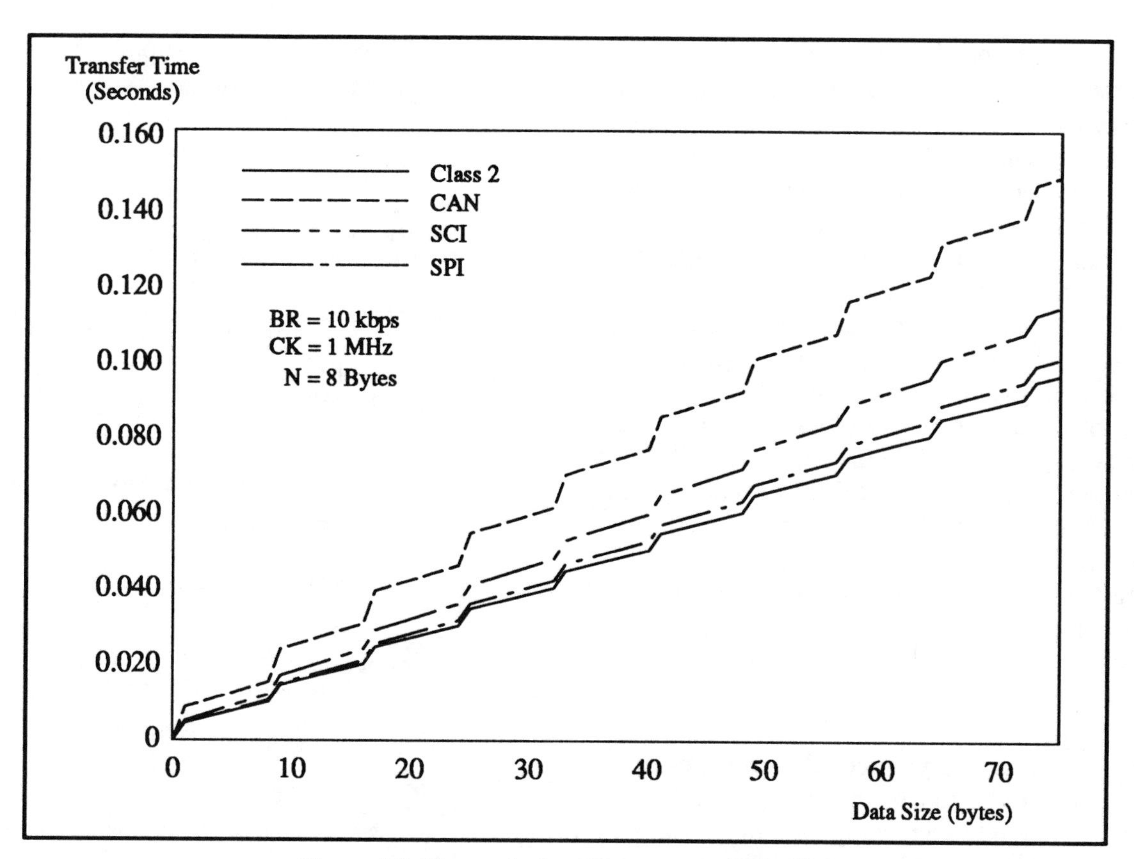

Figure 12 Transmission Time versus Data Size

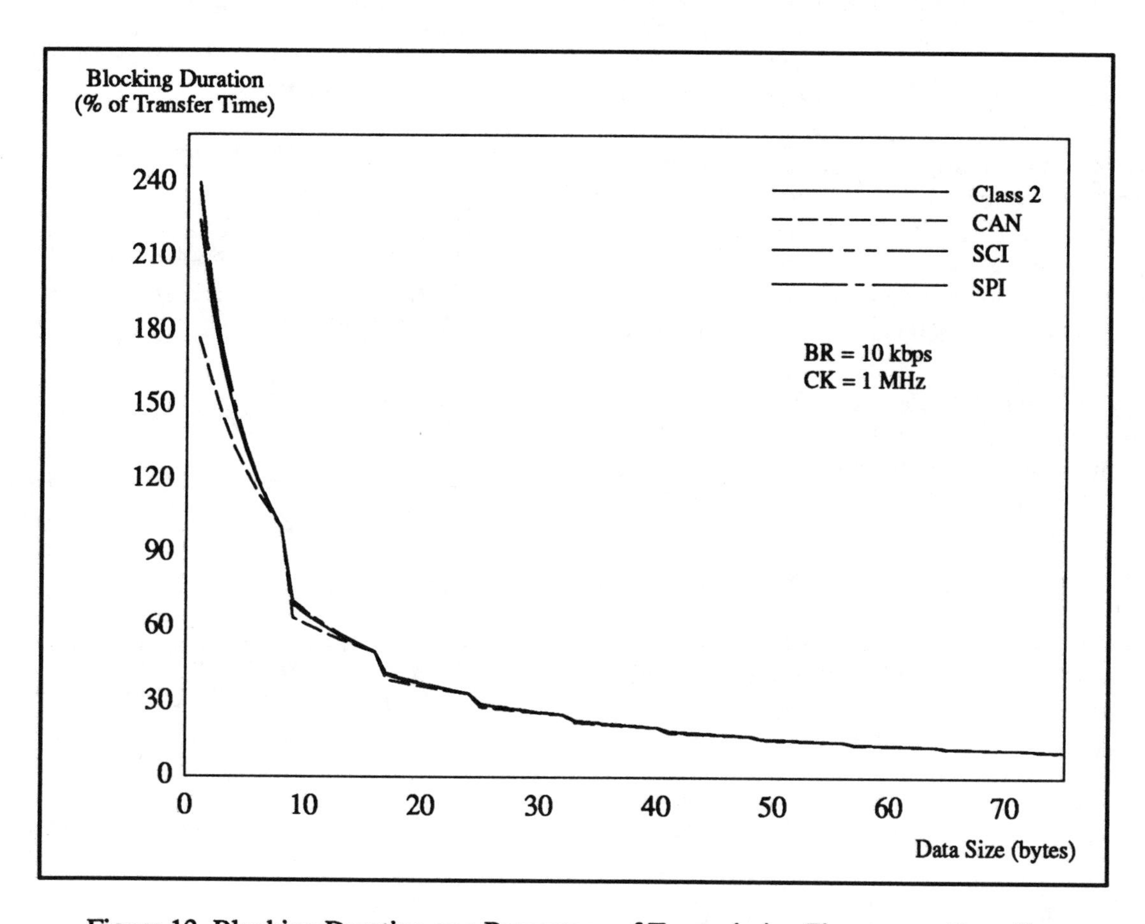

Figure 13 Blocking Duration as a Percentage of Transmission Time versus Data Size

Blocking as a percentage of transmission time versus data size is shown in Figure 13 , assuming a constant bit rate of 10 kbps and common maximum message size of 8 data bytes. First, note that blocking as a percentage of transmission time decreases with increasing data size for each link. This is because the worst case blocking duration is constant with respect to data size, and is determined solely by bit rate, message structure and maximum message size. Again, the graphs are discretized due to strip–mining, and the quantum of discretization corresponds to the maximum message size. Notice that for small data transfers the worst case blocking duration can be significant compared to transmission time, regardless of the communication link. If any of these communication links are used to carry periodic real–time message traffic, schedulable bandwidth utilization will be significantly degraded due to blocking effects, especially when usage is dominated by small (< 10 byte) transfers.

SUMMARY

A concise set of analytic models were presented which were shown to be capable of modeling ICC over four communication links commonly used in automotive applications: Class 2, CAN, SPI and SCI. Such models are useful when designing embedded systems that require multiple processing nodes. Specifically, ICC transmission time estimates must be considered when partitioning software procedures and data across processing nodes in a multicomputer. The reason for this is twofold. First, ICC transmission times contribute to the latencies of distributed real–time functions which are constrained. Second, the level of ICC specifies hardware communication requirements which have a direct bearing on cost. Additionally, transmission time and worst case blocking estimates can be used in conjunction with scheduling equations to analytically predict the real–time schedulability of communication links. This is intuitive, since transmission times and blocking durations directly contribute to the cumulative utilization level of each link which is, after all, a shared resource subject to scheduling.

ACKNOWLEDGMENTS

The authors wish to thank Jim Kinsey and Rich Woerner for sharing their expertise on the SPI and SCI communication links, and Mark Lowden and Dan Siewiorek for providing helpful feedback on early revisions of this manuscript.

REFERENCES

[1] J. Barnes. *General Motors Class 2 Messages.* Tech Report XDE–3104, Delco Electronics Corp., 1991.

[2] J. Barnes. *General Motors Class 2 Message Strategy and Application.* Tech Report XDE–3001, Delco Electronics Corp., 1993.

[3] C. Lupini and T. Braun. *Class 2: An Introduction to Medium Speed Multiplexing.* SAE paper 920222, February 1992.

[4] J. Barnes. *Serial Data Bus for Communications between Microcomputer Assemblies.* Tech Report XDE–5024, Delco Electronics Corp., 1994.

[5] J. Barnes and C. Lupini. *General Motors Class 2 Communications Requirements.* Tech Report XDE–3106, Delco Electronics Corp., 1993.

[6] C. Lupini. *Controller Area Network Protocol Application Document.* Tech Report XDE–4005, Delco Electronics Corp., 1994.

[7] *Controller Area Network Specification 2.0.* Robert Bosch GMBH, 1991.

[8] J. Lehoczky and L. Sha. *Performance of Real–Time Bus Scheduling Algorithms.* ACM Performance Evaluation Review, Special Issue, Vol. 14, No. 1, 1986.

[9] C. Lupini. *DLCS/P: Data Link Controller Serial/Parallel Application Document.* Tech Report XDE–3003, Delco Electronics Corp., 1994.

[10] M68HC11 Reference Manual. Motorola Inc., 1991.

[11] QSM: Queued Serial Module Reference Manual. Motorola Inc., 1991.

[12] *ISO International Standard 9141–2: CARB Requirements for Interchange of Digital Information.* International Organization for Standardization, 1991.

[13] *SAE Surface Vehicle Standard J1850: Class B Data Communication Network Interface.* Society of Automotive Engineers, February 1994.

[14] *ISO International Standard 11519–4: Class B Data Communication Network Interface.* International Organization for Standardization, 1994.

[15] G.C. Sih. *Multiprocessor Scheduling to Account for Interprocessor Communication.* Doctoral Thesis, University of California, Berkeley, 1991.

[16] A. Tanenbaum. Computer Networks. Prentice Hall, 1981

[17] H. Tokuda, C. Mercer, Y. Ishikawa and T. Marchok. *Priority Inversion in Real–Time Communications.* Proceedings of the IEEE Real–Time Systems Symposium, pp. 348–359, 1989.

960120

A Synchronization Strategy for a TTP/C Controller

Hermann Kopetz, René Hexel, Andreas Krüger,
Dietmar Millinger, and Anton Schedl
Technical University of Vienna

ABSTRACT

The provision of a system-wide global time base with good precision and accuracy is a fundamental prerequisite for the design of a time-triggered automotive real-time control system. In this paper we investigate the issues of clock synchronization in such a system. Our synchronization scheme allows every node to have a different oscillator. Based on typical parameters we derive the achievable precision and accuracy, which is in the microsecond range. The description of a prototype implementation shows the applicability of our approach.

INTRODUCTION

Within the automotive environment, two major application domains for future multiplexing systems can be distinguished. *Body electronics* (SAE classes A and B) comprise all functions that are not directly concerned with the movement of the car (e.g., lighting, dashboard displays, or power window control). *System electronics* (SAE class C), on the other hand, subsume functions directly controlling the vehicle's movement (e.g., engine control, vehicle dynamics control). These different applications impose contrasting demands on the underlying communication system. Therefore, automotive communication networks of the future will typically consist of a set of clusters connected to each other by gateway nodes. Such a multicluster system allows to execute different communication protocols that are tailored to the needs of the corresponding application.

Presently, the communication protocols proposed for the automotive environment (J1850 [10], CAN [11], VAN, ...) are used to implement body electronics systems. The SAE does not consider them to satisfy the requirements of safety critical class C applications [12, 13]. A promising solution explicitly considering these requirements are time-triggered communication protocols like TTP [7] that derive all control information (e.g., when to send a message) from the progression of time.

A necessary service of a time-triggered architecture is the provision of a system-wide fault-tolerant global time base of sufficient precision. In multicluster systems -- and sometimes even in a single cluster system -- it can not be assumed that all nodes will contain oscillators with the same nominal frequency. The design of a synchronization system within a set of clusters that will generate a uniform time base with a precision in the μsec range, despite the fact that each node may have an oscillator with a different nominal frequency, is an interesting research challenge.

The objectives of this paper are the presentation of a fault-tolerant synchronization strategy for a multicluster real-time system, where no assumptions are made about the base oscillator frequency in each node. Our strategy integrates internal and external clock synchronization into a single coherent time base. A discussion of a TTP/C controller prototype will demonstrate an implementation of the synchronization scheme.

The paper is organized as follows. In the next section we explain our architectural assumptions and introduce a uniform format for the representation of time in a multicluster real-time system. Section three focuses on the problem of internal synchronization and describes a macrotick generation logic that allows the generation of a global time base within the physical second standard from an arbitrary oscillator frequency. Section four is devoted to the topic of external synchronization and discusses the functions of a time-gateway. The TTP/C controller prototype implementation is described in section five. In section six we analyze the achievable precision and accuracy in a typical scenario from an automotive onboard system. The paper is concluded in section seven.

ARCHITECTURE

The assumptions about the base architecture have been influenced significantly by our experiences in the design of the MARS system [4].

SYSTEM STRUCTURE - Let us consider a time-triggered distributed system that consists of a set of clusters interconnected by gateways (Fig. 1).

Each cluster is composed of a number of nodes that are exchanging messages via a common communication channel. A node consists of a host CPU, a memory, I/O devices, and a communication controller for a time-triggered communication protocol (Fig. 2). The controller communicates with the host CPU via a dual ported random access memory (DPRAM) that contains periodically updated images of the state variables in the environment. We call this DPRAM interface between the communication controller and the client CPU the Message Base Interface (MBI).

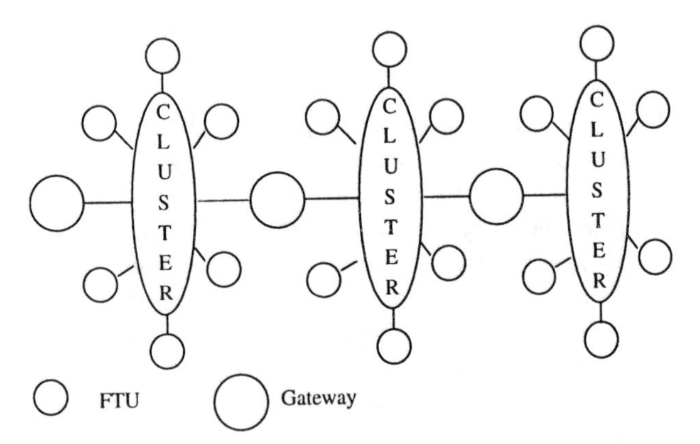

Fig. 1: Set of interconnected clusters

The communication between the nodes is controlled by the Time-Triggered Communication Protocol TTP [7]. The TTP/C version of this protocol is an integrated communication protocol that provides all services needed for the implementation of time-triggered fault-tolerant real-time systems, such as predictable message transmission, internal clock synchronization, membership service, prompt error detection and fault management.

In TTP the access to the communication channel is controlled by a time-division multiple access strategy (TDMA) that is based on the fault-tolerant common time base generated by TTP. Since in a time-triggered system the point in time when a message has to be sent, is known a priori to all nodes, the time difference between the expected and the observed arrival time of a message is a measure of the deviation of the sender's clock from that of the receiver. This information, which is continuously collected during system operation, is sufficient for the implementation of a fault-tolerant global clock synchronization algorithm, such as the Fault Tolerant Averaging Algorithm FTA [3].

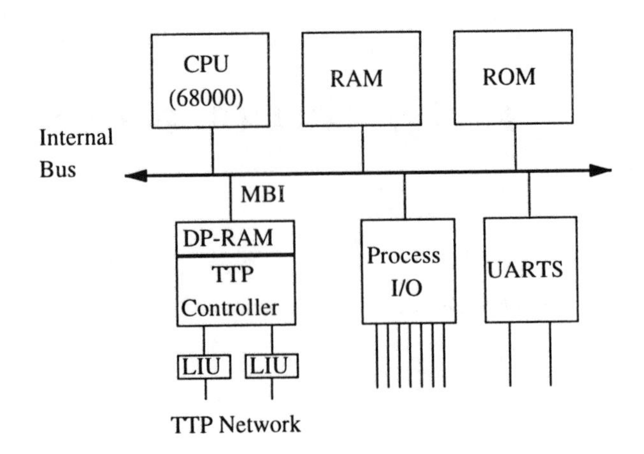

Fig. 2: Structure of a node

TIME REPRESENTATION - In a multicluster system, where a node may have an oscillator with a nominal frequency that is relative prime to the oscillator frequencies of other nodes, a common representation of time must be chosen that is independent of the characteristics of the individual node. This representation of time has to satisfy a number of criteria:

(i) it should be understandable to humans

(ii) it should be independent of detailed implementation decisions and the speed of the communication channel

(iii) it should be easy to manipulate by the computer

An accepted international standard of time is the chronoscopic International Atomic Time TAI. TAI has as a measure of time the physical second and has as the start of its epoch the midnight of January 1st 1958 [1]. TAI serves indirectly as the basis for all international time zones, such as UTC. In contrast to UTC, which requires the insertion of an occasional switching second, TAI is chronoscopic, i.e., uniform.

We propose a data structure for the representation of time as shown in Fig.3.

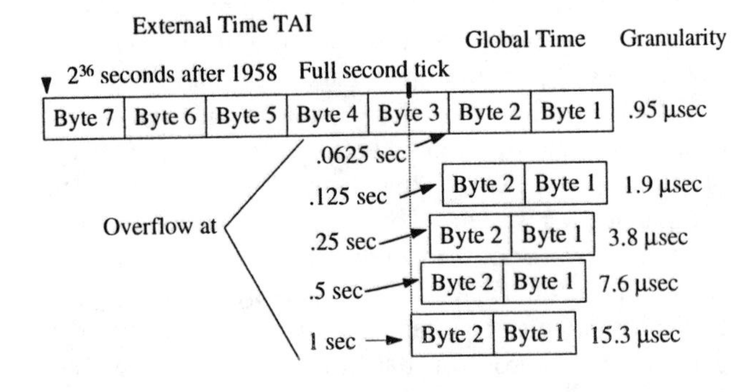

Fig. 3: Representation of the time

This seven byte data structure consists of two fields:

- a five byte (byte 3 to byte 7) external time field that contains in the first four and a half bytes the TAI representation of the current full second and in the last half byte the fractional part of a second (granularity 1/16 seconds, i.e., 62.5 msec). Considering that TAI starts at Jan.1st, 1958 this data format will not overflow until after the year 4000.

- a two byte (byte 1 and byte 2) global time field that contains the fractional part of the second as a fraction of the power of two, down to a smallest granularity of $1/(2^{20})$ second, i.e., exactly a granularity of 0.953 674 316 406 25 µsec.

Since this smallest granularity may be below the precision that can be achieved in a low bandwidth distributed system, we provide four further options for the base granularity, leading up to a granularity of 15.3 µsec. One of these options has to be selected during system configuration. In all cases we want to keep the global time field sixteen bits wide. Consequently, the last four bits of the external time field and the first four bits of the global time field may overlap.

The External Time TAI and the global time are part of the message base interface MBI. The communication controller updates the time fields autonomously. In a sixteen bit architecture the host CPU can read the global time without any integrity concerns. When reading the external time, the NBW access protocol [6] has to be used to guarantee the integrity of the time information.

The described representation of time is closely related to the representation of time proposed in the Network Time Protocol (NTP) [8]. The fractional part of the second is identical, however the full second count is different. NTP refers to the non uniform UTC, while we propose the chronoscopic TAI as the reference of the full second.

TIMESCALES - The local oscillator within a node has to serve the following purposes (Fig. 4)

(i) it has to generate the timing signals for all computational units of the node (oscillator tick);

(ii) it has to provide the time reference for node-local time measurements and for the generation of the bit encoding and sampling of the incoming bit stream at the communication link (microtick);

(iii) it has to generate the timing signals for the global and external time (macrotick).

In our view it is important to strictly separate these conceptually different functions during the system design.

Oscillator tick (g_o) - The oscillator tick is determined by the physical characteristics of the oscillator. In the general case it cannot be assumed that the selection of the nominal oscillator frequency is in the sphere of control of the computer system designer. This selection is often determined by electromagnetic interference concerns. Other sensitive electronic equipment within a vehicle dictates the use of an oscillator with a frequency within a limited frequency band. Furthermore it cannot be assumed that all nodes within a cluster will be driven by an oscillator with the same nominal frequency f^{nom}.

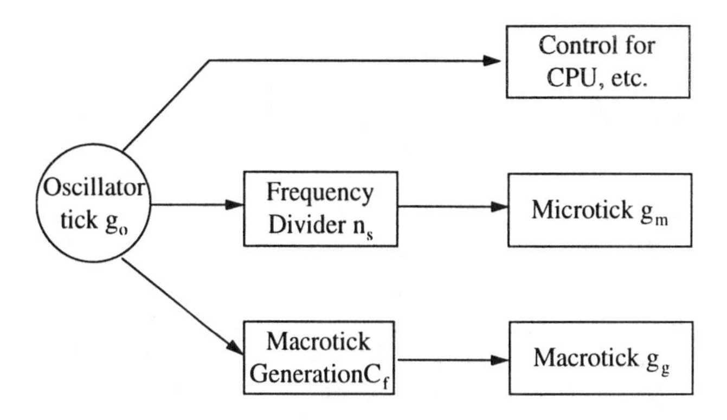

Fig. 4: Functions of the oscillator

Microtick (g_m) - The granularity of the microtick within a node is determined by the required quality of the time measurements within a node. The most important local time measurement in a TTP system is the measurement of the interval between the expected arrival time and the actual arrival time of a message. The quality of this time measurement depends on the reproducibility of the rise time measurement of the incoming bitstream signal on the communication channel. It is thus determined by the bit cell length, its shape, and the sampling granularity of the communication controller. A bit sampling rate of 16 samples/bitcell seems to be accepted practice. TTP systems that are to be deployed in the automotive environment have to support a transmission rate between 10 kbit/seconds and 1 Mbit/second, implying a bitcell length of between 100 µsec and one µsec. We propose a microtick granularity in the order of 1/16 of the chosen bitcell length. The microtick granularity is derived from the oscillator by a frequency divider n_s.

Macrotick (g_g) - The granularity of the macrotick g_g (the global time) has to be selected from one of the values of Fig. 3, so that the physical second is an integer power of two of the macrotick. According to [5] the granularity of the global time g_g must also satisfy the following relation

$$g_g > \Delta_{int}$$

where Δ_{int} is the synchronization precision of the ensemble of clocks. This precision Δ_{int} will be estimated in a later section. If no assumptions can be made about the frequency of the oscillator, then the conversion factor C_f from the oscillator tick to the macrotick cannot be assumed to be an integer. In general the conversion factor C_f will thus have an integer part I_f and a fraction F_f. To achieve the synchronization objectives, a macrotick generation logic has to generate macroticks of slightly different durations in order to synchronize the macroticks within a cluster. The maximum difference in the duration of the macroticks is one oscillator tick g_o.

INTERNAL SYNCHRONIZATION

The internal synchronization establishes a global time base within a cluster by mutual fault-tolerant synchronization of the local clocks. The global time within a given cluster is an abstract notion that is approximated by the local view of the global time in each node.

PRINCIPLE OF OPERATION - The heart of the global time is the local oscillator of each node that oscillates with its given frequency f_{osc}, determined by the physical shape of the crystal. The actual oscillator frequency f_{osc} can deviate from the nominal oscillator frequency f^{nom} because of a mechanical imprecision of the crystal or because of environmental effects (e.g. varying temperature). Since in our system the granularity of the global time, the macrotick, is a predetermined fraction of the full second, an adjustable oscillator-to-macrotick frequency conversion has to be put in place:

$$f_{macro} = f_{osc}/C_f$$

where C_f is the adjustable oscillator-to-macrotick conversion factor and f_{macro} is $1/g_g$, the frequency of the global time. This conversion factor C_f can be decomposed into two parts

$$C_f = C_n + C_c.$$

Hereby C_n is the conversion factor between the nominal oscillator frequency f^{nom} and the nominal macrotick frequency f_{macro}, whereas C_c is that part of the conversion factor that has to be periodically adapted in order to correct the systematic and stochastic variations of the oscillator and to bring the local view of the global time into synchronism with the ensemble of clocks.

The calculation of C_c is carried out periodically in each node by the Byzantine resilient Fault-Tolerant Average Algorithm [3]. This algorithm requires as its input the number of clocks, the maximum number of faulty clocks, and the measured deviations between the clock of the node from all other clocks. These deviations are measured continuously by the Time Difference Capture Logic described below.

MACROTICK GENERATION LOGIC - The Macrotick Generation Logic (shown in figure 5) performs a software-controlled frequency division by a fixed point value C_f. The correction factor C_f combines both, the nominal conversion factor from the nominal oscillator frequency f^{nom} to the macrotick frequency and the necessary (dynamically changing) correction factor to synchronize the local clock with the ensemble. C_f can be decomposed into an integer part I_f and a fraction F_f. These two values are assigned to Macrotick Generation Logic registers as shown in figure 6.

In the following the function of the Macrotick Generation Logic shown in figure 5 is explained:

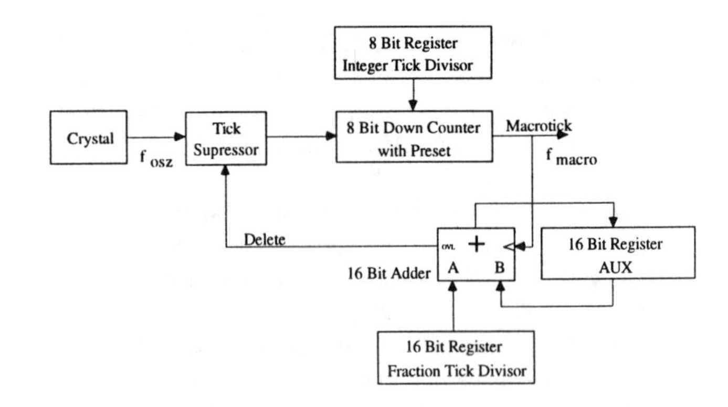

Fig. 5: Macrotick Generation Logic

The *8 Bit Down Counter* performs the division of the oscillator frequency f_{osc} by the value I_f. It is initialized with the value I_f stored in the Integer Tick Divisor register (ITD). It then decrements its contents whenever it receives a signal from the oscillator. If the counter reaches zero, it produces a macrotick and presets its counting register with the value stored in ITD.

The integer division of the oscillator frequency performed by the 8 Bit Down Counter neglects the fractional part of the divisor, F_f. Thus, after $1/F_f$ macroticks (i.e. $1/F_f$ integer divisions) the division error due to the neglected fraction exceeds 1 oscillator tick per macrotick. This error is compensated by suppressing the next oscillator tick, i.e., by delaying the 8 Bit Down Counter by one tick.

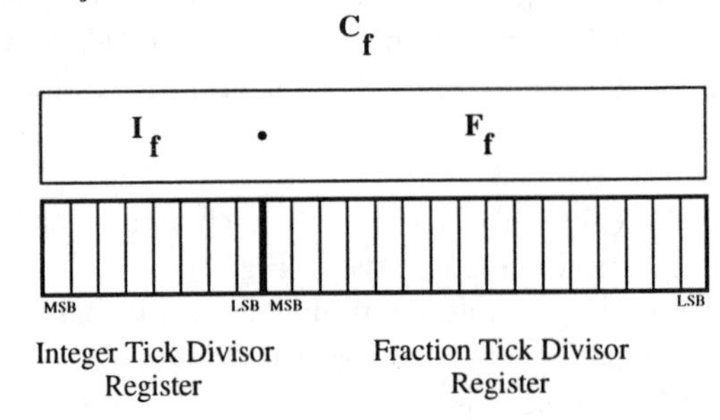

Fig. 6: Macrotick Generator Register Assignment

The actual quantity of the division error is computed by the *16 Bit Adder* unit. At every macrotick, the 16 Bit Adder adds the value F_f stored in the FTD (Fraction Tick Divisor) register to the binary value stored in the *AUX register* and writes the newly calculated value back to the AUX register. This means that at every macrotick the emerging division error is accumulated in the AUX register. If the accumulated error value exceeds 1.0 (oscillator ticks per macrotick), the signal *Delete* is set to high, telling the *Tick Suppresser* to suppress one tick of the oscillator signal. At the same time, the cumulated error value is modified to reflect the suppression of one oscillator tick. This is done by truncating the error value at 1.0 and storing the remaining fractional part in the AUX register.

TIME DIFFERENCE CAPTURE LOGIC - It is the objective of the *Time Difference Capture* (TDC) Logic to measure the time difference between the expected arrival time of a message and the actual arrival time of this message. The TDC measures these time differences with the granularity of the receiver microtick. The expected message arrival time is an event determined by the receiver's clock, whereas the actual message arrival time is an event determined by the sender's clock. The observed time difference is thus a measure for the deviation of the sender's clock from the receiver's clock.

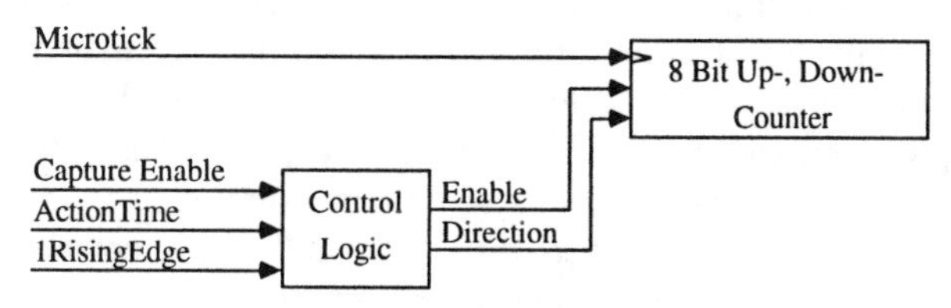

Fig. 7: Time Difference Capture Logic

The actual measurement is done by the TDC logic shown in Figure 7. The TDC unit consists of an *8 Bit Up- and Down Counter* which is clocked by the microtick signal. Three additional control signals control the counting direction and duration. The capturing operation is enabled by the *Capture Enable* signal, which is controlled by the protocol software. The signal *Rising_Edge* is generated by the receiver unit of the communication controller. Rising_Edge is set to high after the first rising edge of a new message frame is detected on the physical communication channel (i.e., this is the point in time of the actual message arrival). The signal *Action_Time* is derived from the local view of the global time. It is set to high as soon as the local view of the global time reaches the expected message arrival time. When either the Action_Time signal or the Rising_Edge signal is set to high, the counter starts to count the number of microticks until the other signal arrives. The accumulated number of ticks is the deviation between the expected and the actual message arrival time given in microticks. The sign of this number (i.e. the counting direction of the counter) is determined by the order of the Action_Time signal and the Rising_Edge signal. This order denotes the direction of the deviation of the local clock.

EXTERNAL SYNCHRONIZATION

External synchronization is concerned with linking the global time of a cluster to an external standard of time. For this purpose we need a *time server*: an external time source that periodically broadcasts the reference time in the form of a *time message*. This time message has to raise a synchronization event (such as the beep of a wrist watch) in a designated node of the cluster and has to identify this synchronization event on the agreed time scale. Such a time scale needs a constant measure of time, e.g., the physical second, and has to relate the synchronization event to a defined origin of time, the so called *epoch*. We call the node that interfaces to a time server a *time gateway*.

PRINCIPLE OF OPERATION - Let us assume that the time-gateway is connected to a GPS receiver. This GPS time server periodically broadcasts time messages containing a synchronization event and information allowing this synchronization event to be placed on the TAI scale. The time gateway has to synchronize the global time of its cluster with the time received from the time server. This synchronization is unidirectional and thus asymmetric, as shown in Fig.8.

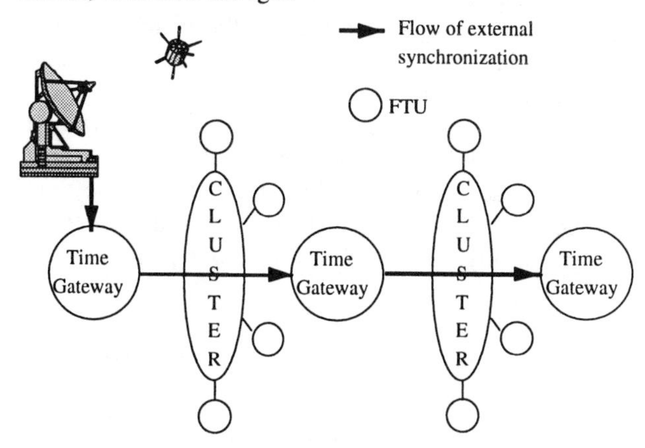

Fig. 8: Flow of External Synchronization

If another cluster is connected to this "primary cluster" by a "secondary time gateway" then the unidirectional synchronization functions in the same manner. The secondary time gateway considers the synchronized time of the primary cluster as the time reference and synchronizes the global time of the secondary cluster.

Whereas internal synchronization is a cooperative activity among all members of a cluster, external synchronization is an authoritarian process: the timeserver forces its view of external time on all its subordinates. From the point of view of fault-tolerance, such an authoritarian regime has one drawback: If the authority sends an incorrect message all its obedient subordinates will behave incorrectly. However, in the case of external clock synchronization the situation is under control because of the "inertia" of time. Once a cluster has been externally synchronized, the fault-tolerant global timebase within a cluster acts as a *monitor* of the time server. A time gateway will only accept an external synchronization message if its content is sufficiently close to its view of the external time. The time server has only a limited authority to correct the clock rate of a cluster. The enforcement of a maximum common mode correction rate--we propose less than 10^{-4} sec/sec--is required to keep the relative time-measurement error small. The maximum correction rate is checked by the software in each node of the cluster.

The implementation must guarantee that in no case it is possible for a faulty external synchronization to interfere with the proper operation, i.e., with the global time of a cluster. The worst possible failure scenario in case the external time server is failing maliciously is a common mode deviation of the global time from the external time with the maximum correction rate. The internal synchronization within a cluster will not be affected by this controlled drift from the external time.

TIME MESSAGE - The time message is a special message for external clock synchronization that is formed in the time gateway and processed in the communication controller of the receiving node. The format of the time message is shown in Fig. 9. The time message has a length of six bytes--the same length as the initialization (I-frame) message of the TTP protocol [7]. It is thus possible to use the same basic TDMA slot for the time message and the I-message. The time message has to be sent periodically (presumably with a long period) on the bus.

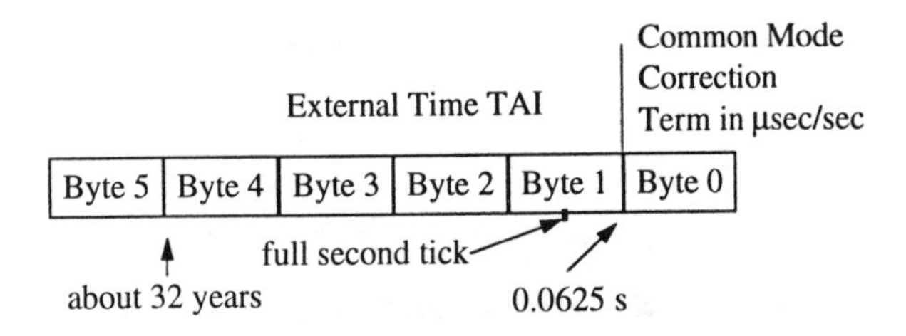

Fig. 9: Format of the time message

The time message contains two fields. Bytes one to five of the time message contain the external TAI time value of the current 1/16th second (see also Fig.3). Byte zero of the time message is the rate correction byte. It contains the requested common mode rate change in µsec/sec for all nodes in the cluster to bring the global time of the cluster in alignment with the current external time. The rate change byte is processed by the protocol software in each node and causes a modification of the C_C part of the conversion factor from the oscillator frequency to the macrotick granularity.

TIME GATEWAY - The time gateway has to control the timing system of its cluster in the following ways:

(i) it has to initialize the cluster with the current external time

(ii) it has to periodically adjust the rate of the global time in the cluster to bring it into agreement with the external time and the standard of time measurement, the second.

(iii) it has to periodically send the current external time in a time message to the nodes in the cluster in order that a reintegrating node can reinitialize its external time value.

The time gateway achieves this task by periodically sending a time message with a rate correction byte. This rate change byte is calculated in the time gateway software. First the difference between the occurrence of a significant event, e.g., the exact start of the full second in the time server, and the occurrence of the related significant event in the global time of the cluster is measured by using the local timebase (microticks) of the gateway node. This measurement is supported by the communication controller hardware. Then the required rate adjustment in ppm (parts per million) is calculated, considering that the rate adjustment is bounded by the agreed maximum rate correction in order to keep the maximum deviation of relative time measurements in the cluster below an agreed threshold and to protect the cluster from malicious faults of the server.

PROTOTYPE IMPLEMENTATION

Currently, a prototype of a TTP/C controller is being implemented. The following section shows how the previously introduced clock synchronization strategy is applied in this prototype.

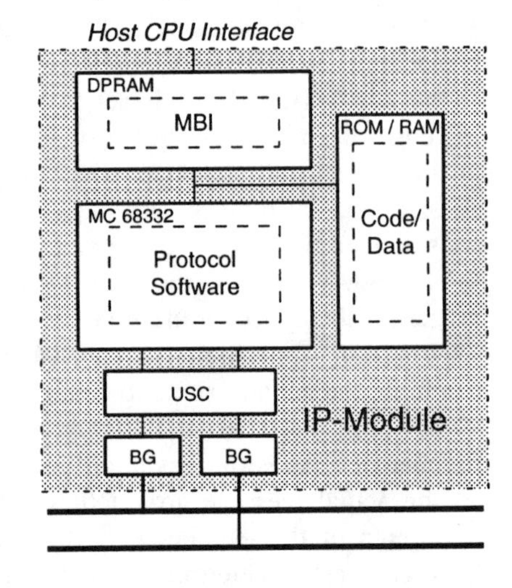

Fig. 10: IP Module Block Diagram

HARDWARE - The TTP/C controller is implemented on a standard industry pack (IP) module [2]. IP modules are supported by a wide variety of mainboards, allowing for utmost flexibility in system design. A multicluster gateway, for instance, can be implemented by connecting two TTP/C controllers to a single motherboard. Figure 10 shows the hardware devices mounted on the TTP/C IP module.

MC68332 Microcontroller (MCU) - The microcontroller comprises the main elements of the TTP/C controller. It consists of a CPU that executes the protocol software. It also contains a Time Processing Unit (TPU), an autonomous, micro-programmable processing unit designed for handling timer related functions. In the TTP/C system the TPU has been micro-programmed to handle the low-level parts of clock synchronization. Communication between TPU and CPU is handled through a small on-chip parameter RAM.

Dual Ported RAM (DPRAM) - The DPRAM accommodates the Message Base Interface (MBI), which is the top level interface between the TTP/C controller and the host system. Any data exchanged between the application program running on the host CPU and the controller is passed through the MBI.

Flash EPROM - The flash EPROM contains the TTP/C protocol software and all static data that are used by the controller.

Serial Communication Controller (USC) - The USC is the interface between the CPU and the communication subsystem. It converts CPU data to a modulated bit stream that is suitable for transmission over the communication media. Received bit streams are converted back to CPU readable data.

Bus Guardian (BG) - The Bus Guardian is the bottom level interface between the controller and the underlying media. It prohibits any untimely access to the communication channel, which could jam the whole cluster, possibly leading to catastrophic consequences. The Bus Guardian significantly increases the failure detection coverage of a node. It thus ensures that no single component failure can lead to a failure of the whole TTP/C cluster.

SOFTWARE - The main task of the system software is to execute the Time-Triggered Communication Protocol (TTP/C). One of the major tasks of the protocol is clock synchronization. In the prototype implementation, clock synchronization is split into two levels: low-level and high-level clock synchronization.

Low-level clock synchronization is performed by a set of micro-programs written for the TPU of the MC68332 microcontroller. The set of micro-programs consists of the following parts:

Macro Tick Generation - This function simulates the macrotick generation logic described earlier. It derives the macrotick clock of the system from the microticks derived from the crystal oscillator. The macrotick generator reads its Integer Tick Divisor (ITD) and Fraction Tick Divisor (FTD) values from the parameter RAM where they will be supplied by the CPU executing the high level clock synchronization routines. The function writes the current value of the synchronized global time into the parameter RAM.

Time Difference Capturing - The TDC mechanism simulates the time difference capture logic hardware; it measures and records time differences between the expected arrival time of a message and the time a message was actually received. The expected receive time (action time) is passed to the TPU by the CPU. For a certain period around this instant, a measuring window is opened, during which the time difference capturing function is active and waits for a frame. The begin of a received frame is signalled to the TPU by the USC. Once the TPU detects this signal, it measures the difference between the current time and the action time (in microticks) and records this difference in the parameter RAM.

Message Transmission Control Logic - This TPU function ensures that message transmissions are started very accurately at certain, pre-defined points in time, which essentially influences the achievable precision of the clock synchronization mechanism. Shortly before transmission commences, the USC is loaded with the transmit frame data and readied. It will, however, not start transmission before the Clear To Send (CTS) line has been asserted by the TPU. Driven by the macrotick generation function, the message transmission control function triggers the assertion of the CTS line at the action time predefined by the CPU. Together with the time difference capturing mechanism this allows the protocol software to measure clock deviations with a very low reading error.

Interrupt Generation - This routine is linked with the macrotick generation function and causes CPU timer interrupts at certain times specified by the CPU itself. This allows the corresponding interrupt service routine of the CPU to periodically update the time fields of the MBI.

The *high-level part of clock synchronization* is performed by the CPU. It is responsible for collecting the timing differences measured by the TPU routines and to periodically calculate clock state and clock rate correction factors using the fault-tolerant average (FTA) algorithm. The state correction factors are immediately applied to the macrotick generation logic and thus provide short-term synchronization for the cluster. The rate correction factors are averaged over multiple synchronization periods before they are applied to the macrotick generator. They compensate for the unavoidable drift of the crystal oscillators used in the various nodes.

For *external clock synchronization* a so-called time message has to be transferred from one cluster to another via the MBI. The time message contains the current TAI and a rate correction value for the global cluster time of the neighbouring cluster. This common mode correction term contained in the time message is calculated by gateway nodes from the deviation of the time bases of the neighbouring clusters. The gateway controller of the source cluster uses a TPU output channel that is triggered by its macrotick function and toggles a line once every second. The destination controller has a TPU input function that picks up the signal generated by the source controller and measures the difference to its own second generator. The next common mode correction term is calculated by the CPU on the destination side of the gateway from the timing differences that have been accumulated between two transmitted time messages.

PRECISION AND ACCURACY

The key quality parameter of the internal synchronization is the precision, i.e., the maximum time interval between respective ticks of any member of the cluster. In [3] a detailed analysis of the precision of the fault-tolerant average algorithm (FTA) for internal clock synchronization has been carried out. According to this analysis, the precision Δ_{int} is given by

$$\Delta_{int} = (\varepsilon + \xi)\,((N-2k)/(N-3k))$$

where ε is the reading error, i.e. the maximum random error in reading the state of the clock of one node by another node, ξ is the drift of the clocks, determined by the product $2\rho R_{int}$. The drift rate ρ is the maximum drift rate of the oscillator. R_{int} denotes the length of the resynchronization interval. The number N refers to the number of clocks in the ensemble and k refers to the maximum number of Byzantine faulty clocks.

Let us now investigate the parameters that can be achieved in a typical automotive system in order to get a good estimate for the achievable precision.

Reading error ε - The reading error ε is determined by the variability in the edge detection of the start of a new message (Fig.11) plus the variability in the message transmission times.

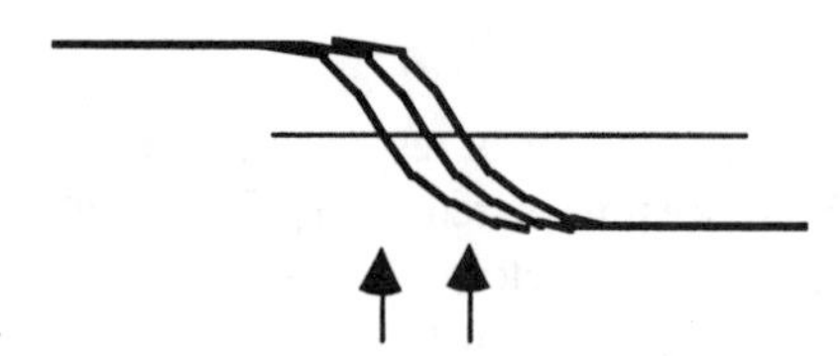

Fig. 11: Variability in the detection of the start of a new message

This variability is determined by the physical shape of the signal on the transmission channel and the granularity of the local time measurement (the microticks) in the node that records the edge detection. Considering that a bitcell consists of sixteen microticks, we can assume that the rising edge of the bitcell can be determined within three microticks. If we assume a bitrate of 100 kbit/sec and thus a bitcell of 10 μsec, then the reading error will be 3.(10/16) μsec, i.e., 1.875 μsec.

Drift ξ – The drift ξ is the product $2\rho R_{int}$ of the length of the resynchronization interval R_{int} and the drift rate ρ of the oscillator. In a typical automotive application, such as the automotive benchmark of [12], a cluster consists of 7 nodes with a TDMA cycle of 2.5 msec and a cluster cycle of about 10 msec. If we resynchronize once within a cluster cycle, then R_{int} is 10 msec. According to our experiments [14], the short term drift of a crystal oscillator of a typical computer within an interval of 10 msec is less than 10^{-6} sec/sec. The drift ξ within R_{int} is thus in the order of 10^{-8} second.

Byzantine error factor - If we assume an ensemble of seven nodes and at most one Byzantine error in a cluster cycle, then the Byzantine error factor is 5/4, i.e., 1.25.

Achievable Precision - In our example the achievable precision is given by

$$(1.875 + 0.01) * 1.25\ \mu sec = 2.35625\ \mu sec$$

In such a system the proper base granularity of the global time would be

3.8 μsec

as seen from figure 3, since the precision has to be better than the granularity.

The *accuracy* is the maximum deviation between the external time standard and the internal representation of the external time. It depends on the quality of the external time signal. If we assume that a good GPS receiver will have a time resolution of better than one μsec, then the time measurement within a time gateway will introduce an additional digitalization error of one microtick. It is smaller than the granularity error of the global time.

If we increase the speed of the communication system to 500 kbits/second, the dominating error term, the reading error, will be reduced to 0.35 μsec. In such a system, the smallest granularity of figure 3, 0.95 μsec, can be supported.

CONCLUSION

In this paper we described a synchronization strategy for a fault-tolerant distributed time-triggered real-time system that contains a number of autonomous clusters. We presented two hardware units for the support of clock synchronization, the macrotick generation logic and the time difference capture logic. We also described the simulation of these hardware units in a prototype implementation of a TTP/C controller. The topic of internal and external synchronization has been discussed.

The limiting term for the precision and accuracy of such a system is the performance of the time difference capture logic that measures the time difference between the clocks of a cluster by capturing the exact time of arrival of the leading edge of a bitstream arriving at a node. This term depends on the size of the bitcell, i.e. the bandwidth of the communication system.

We have shown that in a typical automotive system with a bandwidth of 100 kbits/second a timebase with a global precision of better than 4 μsec can be supported.

ACKNOWLEDGMENT

This work has been supported in part by a research grant from HP Labs in Palo Alto.

REFERENCES

[1] G. Becker, "Die Sekunde", *PTB Mitteilungen* vol. 85, Jan 1975, pp.14-28,
[2] Greenspring Computers, Inc., *IndustryPack Logic Interface Specification*, 1993
[3] H. Kopetz, W. Ochsenreiter, "Clock Synchronization in Distributed Real-Time Systems," *IEEE Transactions on Computers*, August 1987, pp.933-940
[4] Kopetz, H., Damm, A., Koza, C., Mulazzani, M., Schwabl, W., Senft, C., Zainlinger, R., Distributed Fault-Tolerant Realtime Systems: The MARS Approach, *IEEE Micro,* Vol. 9, No. 1, pp., 25-40, Febr. 1989
[5] H. Kopetz, "Sparse Time versus Dense Time in Distributed Real-Time Systems", *Proc. of the 14th Distributed Computing System Conference,* Yokohama, Japan, IEEE Press, June 1992,
[6] H. Kopetz, J. Reisinger, NBW: A Non-Blocking Write Protocol for Task Communication in Real-Time Systems, *Proc. of the IEEE Real-Time System Symposium*, IEEE Press, Dec. 1993
[7] H. Kopetz, G.Grünsteidl, TTP- A Protocol for Fault-Tolerant Real-Time Systems, *IEEE Computer*, January 1994, pp. 14-23
[8] D.L. Mills, Internet Time Synchronization: The Network Time Protocol, *IEEE Transactions on Communications,* Vol. 39, No. 10, Oct. 1991, pp. 1482-1493
[9] J. Rushby, Formal Methods and the Certification of Critical Systems, *SRI International SCL Technical Report SRI-CSL-93-07,* Menlo Park, Cal., USA, November 1993
[10] SAE J1850, July 90, Class B Data Communication Network Interface, published in *1994 SAE Handbook,* Vol, 2, Society of Automotive Engineers, Warrendale, PA, 1994
[11] SAE J1583, March 90, Controller Area Network (CAN), an In-Vehicle Serial Communication Protocol, published in *1994 SAE Handbook,* Vol, 2, Society of Automotive Engineers, Warrendale, PA, 1994
[12] SAE J2056/1 June 93, Class C Application Requirements, published in *1994 SAE Handbook,* Vol, 2, pp.23.366 - 23.272, Society of Automotive Engineers, Warrendale, PA, 1994
[13] SAE J2056/2 April 93, Survey of Known Protocols, published in *1994 SAE Handbook,* Vol, 2, pp.23.366 - 23.272, Society of Automotive Engineers, Warrendale, PA, 1994
[14] The short-term Stability of Crystal Oscillators: Experimental Results, Research Report No. 1/1995 , Institut für Technische Informatik, Technical University of Vienna, Austria, 1995

960123

Integrated J1850 Protocol Provides for a Low-Cost Networking Solution

David S. Boehmer
Intel Corp.

ABSTRACT

The J1850 specification was officially adopted by SAE as the standard protocol for Class B in-vehicle networks on February 1, 1994. J1850 has been a recommended practice for seven years and has gained wide acceptance throughout North America. Today J1850 is implemented in many production vehicles for data-sharing and diagnostic purposes. The widespread integration of J1850 in-vehicle networks is contingent upon low-cost implementation into applications such as ABS, engine, transmission and instrumentation. In order to address this need, Intel Automotive has developed a very low-cost J1850 protocol handler designed specifically to be integrated onto high-performance MCS®96 microcontrollers.

This paper introduces Intel's J1850 protocol handler which supports both GM and Chrysler versions of J1850. Integrating the J1850 protocol onto microcontrollers allows for faster access and therefore reduces service overhead. The described J1850 protocol handler will handle network protocol functions including access, arbitration, in-frame responses, error detection and delay compensation. A digital noise filter automatically rejects unwanted noise pulses. Three dedicated interrupts and byte-level message buffering provide for efficient interrupt handling. This implementation is consistent with the industry's goal to drive J1850 implementation below $2.00 per node.

INTRODUCTION

The use of in-vehicle networks is becoming increasingly common as government emissions regulations such as CARB and OBDII are requiring improved on-board diagnostics. Also contributing to this growth is the trend toward increased data sharing between ECUs. Data sharing can help to reduce "redundant" processing as well as minimize the number of sensors employed in the automobile. Both of these growth factors (diagnostics and data sharing) are ideally served by medium-speed Class B in-vehicle networks.

To classify various standards, SAE has defined 3 basic classes of in-vehicle networks with the primary differentiation being that of speed:

Class A:	<10Kb/s (low speed)
Class B:	10Kb/s to 125Kb/s (medium speed)
Class C:	125Kb/s to 1Mb/s or higher (high speed)

Two of the most predominate in-vehicle networking standards today are CAN and SAE J1850. CAN is a high-speed Class C protocol developed and maintained by Robert Bosch GmbH which was implemented in production vehicles beginning in the early 1990s. CAN's high-speed capability (up to 1Mb) makes it suitable for real-time vehicle control applications.

SAE's J1850 is a medium speed standard whose lower cost per node compared to Class C protocols makes it more suitable for diagnostics and non real-time data sharing functions, which occur at relatively slow transmission rates and network a greater number of nodes throughout the automobile. Lower speed protocols can often be implemented with lower cost physical layer and peripheral components compared to high-speed protocols. J1850 was officially adopted by SAE as the standard protocol for Class B in-vehicle networks on February 1, 1994. J1850 has been a recommended practice for seven years and has gained wide acceptance throughout North America. There are two basic versions of J1850. The first is a 10.4Kbp/s Variable-Pulse Width (VPW) type which uses a single wire for the bus. The other is a 41.6Kbp/s Pulse-Width Modulation (PWM) which uses a 2-wire differential bus. The J1850 protocol handler discussed in this paper supports the 10.4Kbp/s VPW type.

As with in-vehicle networking in general, J1850's incorporation into production vehicle designs is strongly dependent upon the implementation cost per node. To address this issue and help the industry drive down the cost of implementing a Class B network, the described protocol handler was developed to provide J1850 10.4Kbp/s VPW support at a low cost, without sacrificing J1850 functionality.

The primary goals in developing this J1850 protocol handler were as follows:

1. Provide a simple, low-cost J1850 protocol handler targeted for integration onto 16-bit MCS®96 microcontrollers.
2. Minimize CPU burdening for reception and transmission of J1850 messages.
3. Minimize external circuitry required to implement a node.
4. Support both GM (Class II) and Chrysler 10.4Kbp/s VPW requirements (In-Frame Responses (IFRs), single and 3-byte headers as well as other differences).

A benefit of integrating the J1850 protocol onto the microcontroller is that it frees up a communication port (often a synchronous serial or parallel port) for other purposes. Having a direct interface to the core on the internal peripheral data bus also provides for much faster accesses without having the overhead of transferring data over a serial bus external to the microcontroller. In other words, the user can write data to be transmitted faster and read received data faster. This in turn reduces processor overhead.

A J1850 node has two components. One component is the digital circuitry which processes messages according to protocol rules. The other is the bus driver which is an analog circuit that interfaces the waveshaped bus to the digital circuitry. It was concluded that the most cost effective J1850 implementation integrates the digital portion onto the microcontroller and implements the analog circuitry as a stand-alone bus transceiver. This allows the logic to be manufactured on a high-density CMOS process and the bus transceiver to be manufactured on a process optimized for analog and power devices. Although this approach requires a two-chip solution, the smallness of a transceiver when compared with an entire stand-alone protocol chip is very significant and offers a significant PCB area savings.

87C196LB: 16-BIT MICROCONTROLLER WITH INTEGRATED J1850 PROTOCOL HANDLER

The 87C196LB is the first MCS96 microcontroller to integrate the J1850 Network Protocol. The 87C196LB is a

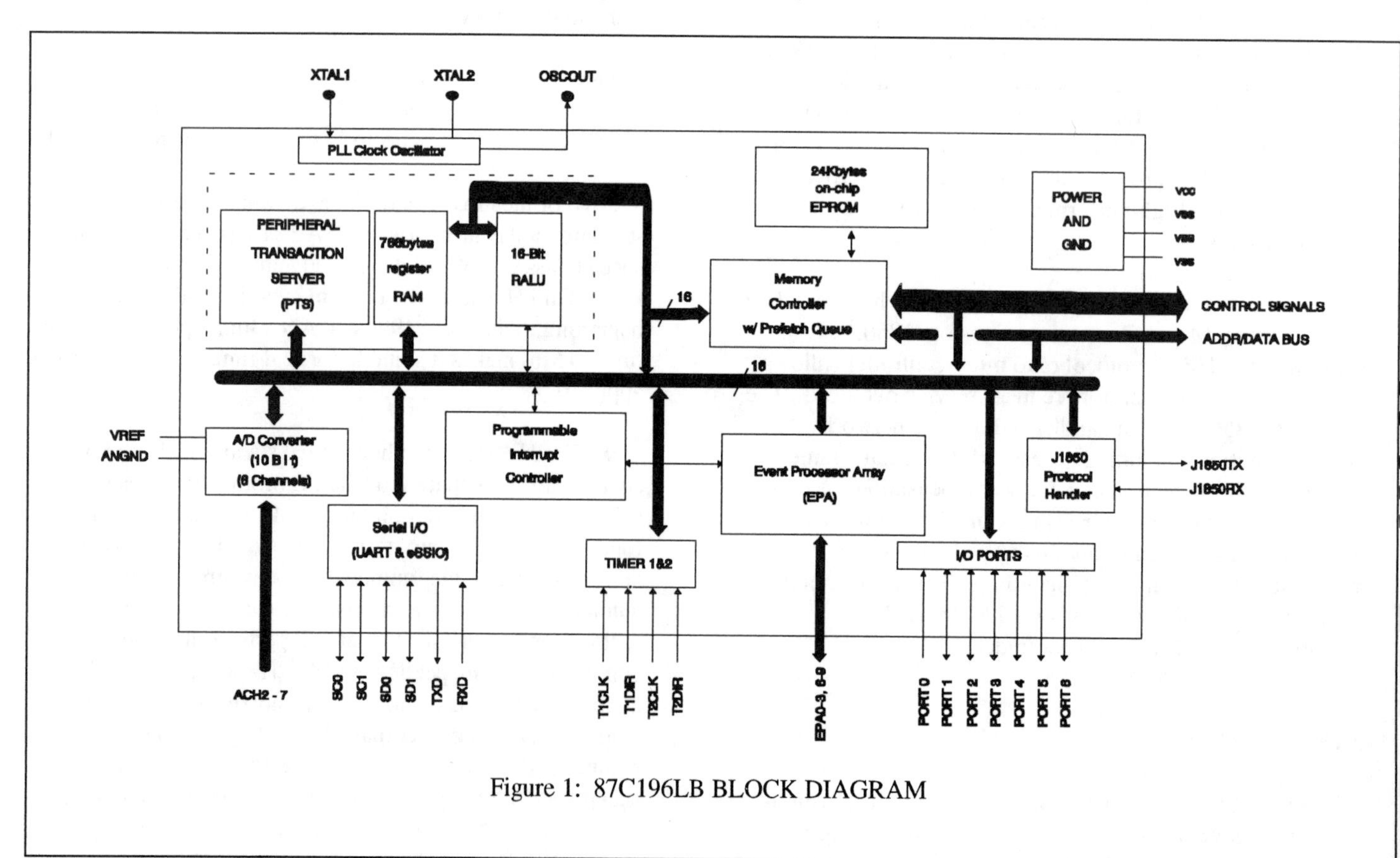

Figure 1: 87C196LB BLOCK DIAGRAM

highly-integrated, high-performance 16-bit microcontroller based upon the 8xC196Kx architecture. A block diagram is shown in figure 1. In addition to J1850 network protocol support, the 87C196LB integrates other peripheral modules that make it attractive for Automotive applications such as anti-lock braking control (ABS). 24 Kbytes of EPROM memory is integrated on-chip for code storage. Also included is 768 bytes of directly-accessible register RAM. High-Speed Input and Output is accomplished through 10 Event Processor Array (EPA) channels, six of which can be used for input capture, the other four are dedicated for compare-only functions such as software timers or periodic A/D conversions. To complement the 10 EPA channels, two 16-bit Timer/Counters are integrated to provide a fully configurable time-base. Both a Serial I/O port (UART) and an enhanced Synchronous Serial I/O port (SSIO) are included on the LB. The SSIO was enhanced to provide for programmable clock phase and polarity. This feature allows for flexibility when interfacing with other synchronous serial protocols such as SPI. The 87C196LB also includes a 10-bit, 6-channel A/D converter.

J1850 PROTOCOL HANDLER ARCHITECTURE

The J1850 network protocol handler integrated onto the 87C196LB (shown in figure 2) consists of four primary building blocks; Control State Machine (CSM), Symbol Synchronization and Timing circuitry (SST), Cyclic Redundancy Check circuitry (CRC) and Special Function Registers (SFRs).

CONTROL STATE MACHINE - The control state machine (CSM) handles all sequencing of framing elements, IFR support, bus contention detection, bit-by-bit arbitration, CPU data transfers, and trapping of various error conditions.

SYMBOL SYNCHRONIZATION AND TIMING CIRCUITRY - The Symbol Synchronization and Timing circuitry consists of a clock prescaler, digital filter, synchronization and bit timing circuitry and delay compensation circuitry.

The clock prescaler circuitry must comprehend various microcontroller operating frequencies to provide a single reference frequency to the J1850 timer/counters. This is accomplished by providing user programmable clock prescaler bits within the J_CFG SFR (bits 0 and 1). The user code programs these bits once during initialization of the J1850 protocol handler. The prescale bits support 8, 12, 16 and 20MHz operating frequencies.

A digital filter is implemented to automatically reject noise pulses of 8us or less in duration.

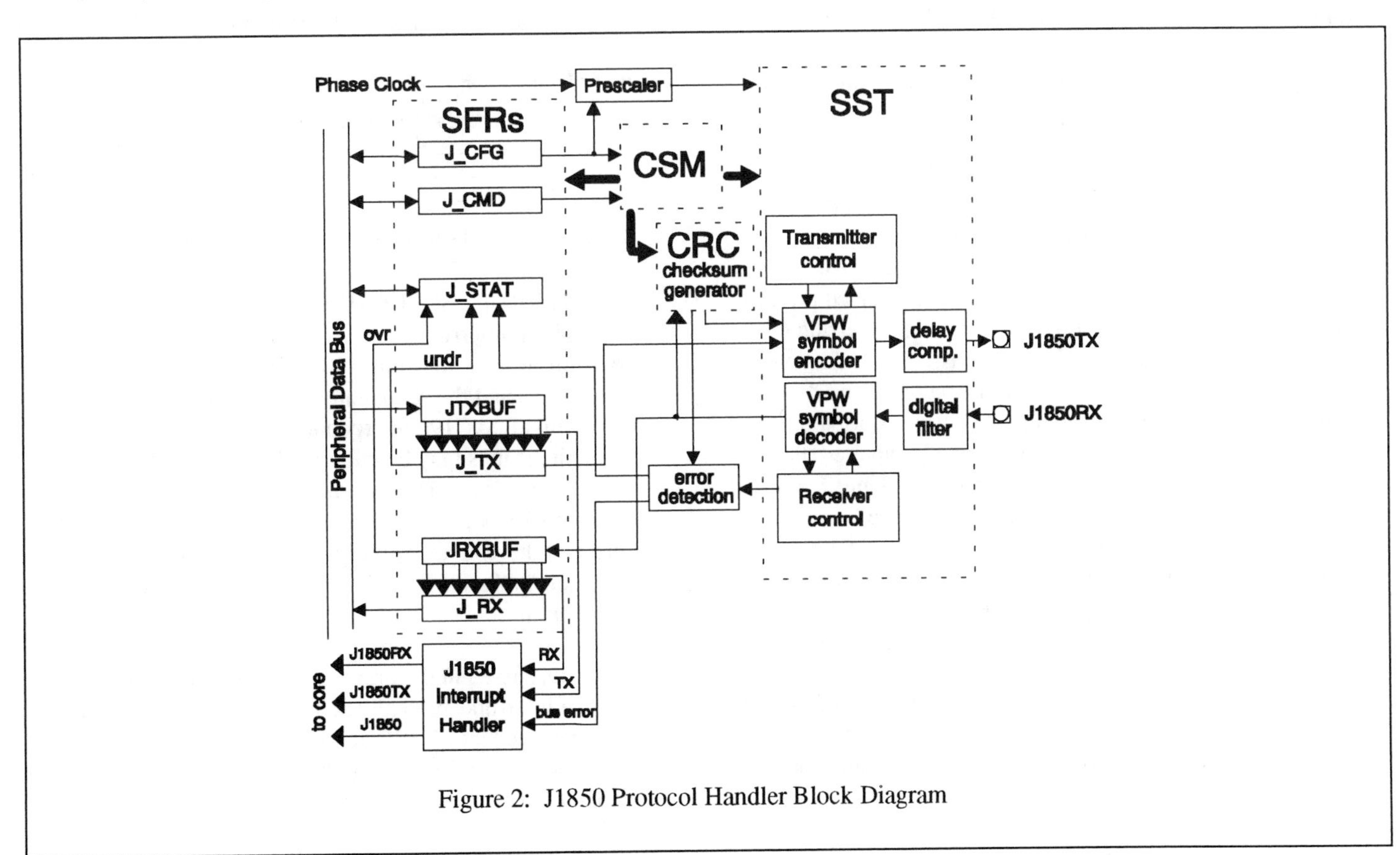

Figure 2: J1850 Protocol Handler Block Diagram

The synchronization and bit timing circuitry ensures that processed bus symbols are properly timed and synchronized with the J1850 bus, as defined by SAE J1850, Revised MAY94.

As J1850 is used on more car platforms, more silicon vendors will be developing J1850 transceivers for the growing market. One inherent issue for a J1850 protocol handler is that it must compensate for variations in propagation delays from manufacturer-to-manufacturer. To address this issue, the J1850 protocol handler described in this paper implements delay compensation circuitry to provide the flexibility to adjust symbol times. This results in valid bus symbol times on the J1850 bus. Transceiver manufacturers specify both a TX delay (typically in the range of 14uS) and a RX delay (typically less than 1us). The user can tailor the delay time required to meet bus symbol timing requirements by programming the proper value into the J_DLY register.

CRC CIRCUITRY - The CRC circuitry handles CRC calculations for both transmitted and received messages as specified by SAE J1850 Revised MAY94, section 5.4.1. This circuitry uses a special polynomial equation to calculate a CRC for transmitted messages. The 1's complement of the calculated CRC is appended as the last byte of the message. At the receiving node, the CRC circuitry calculates a CRC for incoming messages. A CRC result of C4h is always obtained for error-free message receptions. If a value other than C4h is obtained, the bus error flag will be set in the J_STAT register. If enabled, a J_STAT interrupt to the core will be triggered to flag the error.

TRANSMIT / RECEIVE BUFFERS - To provide for efficient data transfers, the J1850 protocol handler described in this paper implements double-buffered transmit and receive registers. The double-buffered receive register (J_RX) allows reception of a second data byte to be received prior to the previous byte being read-out by the user. Referring to figure 3, the receive register is labeled as J_RX and its buffer is labeled as JRXBUF. As symbols are received and decoded from the J1850 bus, the resultant data bits are shifted one-by-one into JRXBUF. After a complete byte is captured, the complete data byte is automatically transferred into J_RX and a receive interrupt (J1850RX) is triggered to the microcontroller core. The J1850 interrupt signals the user code that a complete byte has been received and is ready to be read from J_RX. If the received byte is not read-out prior to the next byte being received, an overrun error is flagged via the J1850STAT interrupt as well as the OVR/UNDR flag in the JSTAT register. Any bus or timing errors flagged during reception are also flagged in the JSTAT register and can alternatively trigger the J1850STAT interrupt to the core.

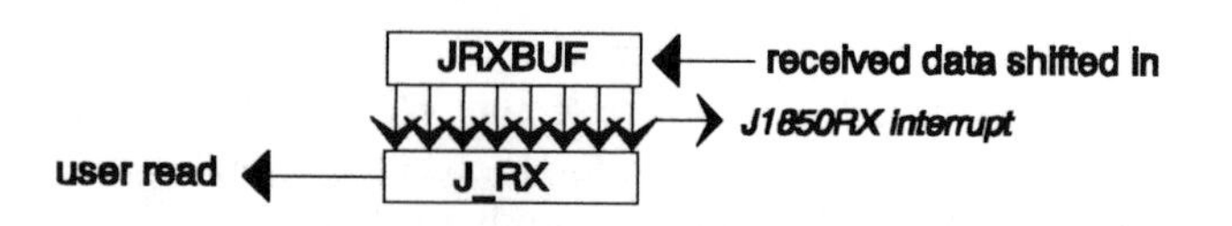

Figure 3: Double-Buffered Receive Register

Similarly, the double-buffered transmit register (J_TX) allows the user to write a second byte to the transmit buffer while the first byte is still being shifted out. To transmit a message, the user first initializes the transmitter by writing the command register (J_CMD). The architecture of this protocol handler offers very low overhead as the J_CMD register only has to be written once for each message transmission. After writing J_CMD, the user writes the first two bytes to be transmitted to J_TX. Since both the transmit register and buffer are cleared after initializing J_CMD, the first byte is automatically transferred through JTXBUF into the J_TX register (refer to figure 4). The second of these bytes will be loaded into JTXBUF since J_TX is occupied by the first byte. After the first byte is transmitted to the J1850 bus, the contents of JTXBUF will be automatically transferred into J_TX. This transfer will trigger the J1850TX interrupt to the core, which signals the user that the third byte to be transmitted should be written to the transmit SFR (J_TX). This interrupt/write process continues until the entire message has been transmitted. At this time the J_STAT interrupt and the MSGTX bit within the J_STAT register will indicate successful message transmission.

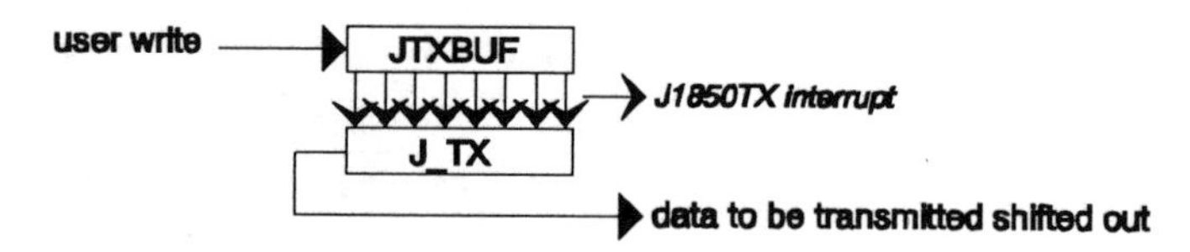

Figure 4: Double-Buffered Transmit Registers

INTERRUPT STRUCTURE - The 87C196LB supports three J1850 interrupts directly to the core. These include the transmit receive interrupt (J1850TX), the receive interrupt (J1850RX) and the status interrupt (J1850STAT). As is the case with all 8xC196 microcontrollers, each of the three J1850 interrupts may be individually enabled/disabled through software. This provides the user with the capability to optionally poll for change of status as opposed to being interrupt driven. This gives the user flexibility to configure the interrupt handler to best serve the needs of their particular application.

As discussed in the previous section, the transmit and receive interrupts basically alert the user that a byte has either been received into J_RX and is ready to be read-out (J1850RX interrupt), or that the transmit buffer is ready for another byte to be written to it for transmission (J1850TX interrupt).

The J1850STAT interrupt signals a change of status for the J1850 protocol handler. This interrupt goes active when one of the following conditions is detected:

1. A bus error; due to a CRC error, a bus symbol timing error, an incomplete byte received or a "no echo" condition being detected.
2. A complete message has been received and no error has been detected.
3. A complete message has been transmitted and no error has been detected.
4. An overrun has occurred for the receive buffer or an overflow has occurred for the transmit buffer.
5. A break symbol has been detected on the bus.

Each of the above conditions are differentiated by individual status bits in the status register (JSTAT). The status bits are updated even if the J1850STAT interrupt is not enabled. This gives the user the flexibility to optionally poll for change of status.

The MCS 96 architecture also provides the Peripheral Transaction Server (PTS), which is a microcoded hardware interrupt processor which provides for high-speed, low overhead interrupt handling. The PTS is ideal for automating the transfer of data to and from transmit and receive buffers. When used with the J1850 interrupt handler, the user can configure the PTS to automatically transfer incoming messages to RAM (for received data). For transmitted data, the PTS can automatically transfer data from a RAM table to the transmit buffer. The user utilizes the PTS by specifying a PTS "cycle" to occur in place of a standard interrupt. The user specifies the number of transfers to take place prior to an "end-of-PTS" interrupt being triggered to the core. This signals that an entire message has either been received or transmitted.

IFR SUPPORT - The J1850 protocol handler supports In-Frame Responses (IFRs) through a combination of software and automatic Normalization Bit (NB) generation and detection in hardware. For IFR receptions, the J1850 protocol handler will automatically detect a valid NB and alert the user code (via the J1850STAT register's IFR RCV'd status bit) that an IFR is being received into the receive register (J_RX). For IFR transmissions, the user code will determine in software if an IFR response is required for a given received message. If so, the user must respond by writing a "1" to the IFR bit in the J1850 command (J_CMD) register. This bit being set will signal the transmitter that the next byte written to the transmit buffer will be an IFR. It also signals the CSM that a NB symbol is to be automatically sent, followed by IFR data, after a valid EOF symbol is detected on the bus.

SPECIAL FUNCTION REGISTERS - Six special function registers (SFRs) provide an interface to the microcontroller's core. These SFRs provide the user with the necessary control to handle communication over the J1850 bus. The SFRs are located in standard SFR memory space on the 87C196LB (1F00h to 1FDFh) and are directly accessible. The SFRs include a configuration register, command register, status register, transmit and receive buffers as well as a delay register.

Configuration Register - The configuration register (J_CFG) allows the user to configure the integrated J1850 protocol controller for use within a particular application. J_CFG typically is written only once during initialization. The first two bits, PRE0 and PRE1 are prescaler bits that select the proper time base for the J1850 module clocks. This prescaler is required to accommodate the various frequencies at which the 87C196LB microcontroller can be clocked.

The NBF bit (Normalization Bit Format) is required to select the proper normalization bit format to signal whether the IFR includes a CRC or not. The format used is dependent upon the automobile manufacturer's specification as described below. Default out of reset is NBF = 0.

	GM (NBF=1)	Chrysler (NBF=0)
IFR does not include CRC	active long NB	active short NB
IFR includes CRC (not supported)	active short NB	active long NB

Figure 5: Normalization Bit Formats

The 4X MODE bit is implemented to allow reception and transmission of symbols in 4x mode (41.6Kbp/s vs. 10.4Kbp/s). 4x mode is typically not used for normal communications, its use is primarily for in-module programming and diagnostics. 4X mode is entered/indicated when the 4X MODE bit is set to a "1".

7	6	5	4	3	2	1	0
NBF	rsv	4x mode	rsv	rsv	rsv	PRE 1	PRE 0

Figure 6: J1850 Configuration Register (J_CFG)

Command Register - The command register (J_CMD) provides the user control of the state machine and transmit/receive buffers. This register is written once for each message transmitted over the network. The first four bits (MSG0:MSG3) specify the number of bytes to be transmitted in the next message. This number includes the header byte(s) but not the CRC byte. As bytes are transmitted, these bits are decremented allowing the user to determine the status of message transmission in progress. The MSG bits signal the CSM when J1850TX (buffer ready) interrupts should no longer be triggered, i.e., the last message byte has been written to J_TX and the next byte transfer should not trigger an interrupt. This signal is also used to signal the CSM that the next byte to be transmitted should be the CRC.

Writing a "1" to the ABORT bit will abort any transmission in progress and flush the transmit buffer.

The IGNORE bit was implemented to allow a simple method of message filtering via software. If the user code should detect a header byte that indicates the message in progress is not "addressed" for its particular node, the user may write the ignore bit to disable receive interrupts until a valid EOF is detected. This prevents the unnecessary servicing of receive interrupts.

When set, the IFR bit indicates that the next byte written to the transmit buffer will be an IFR and that after EOD is detected, a NB followed by the IFR should be sent. The AUTO bit enables auto retry when arbitration is lost on the first byte. This feature is especially useful for Type 2 IFRs where multiple responders will arbitrate for the bus until all IFR response bytes are successfully transmitted.

7	6	5	4	3	2	1	0
Auto	IFR	ignor e	abort	MSG 3	MSG 2	MSG 1	MSG 0

Figure 7: J1850 Command Register (J_CMD)

Status Register - When read, the status register (J_STAT) provides the user the current status of the receive buffer as well as interrupt status. Bits 0-4 going active will trigger the J_STAT interrupt to the microcontroller core. Bits 5-6 are status bits only and will not trigger an interrupt.

The BUS ERR (bus error) flag is set for any of the following four conditions: CRC error, bus symbol timing error, incomplete byte received and no echo. The MSG RX bit signals that a complete message has been successfully received. Likewise, the MSG TX bit goes active after a message is successfully transmitted as indicated by the EOD symbol being detected on the bus after the last message byte is transmitted.

The OVR/UNDR (receive buffer overrun/transmit buffer underflow) bit goes active when either the J_RX SFR overruns or the J_TX SFR underflows. A J_RX overrun is defined as a symbol being received when both the receive register (J_RX) and receive buffer (JRXBUF) contain received bytes. This is basically the case where the user has not read-out the receive buffer fast enough. Likewise, a J_TX underflow is defined as the case where the last bit of a message byte being transmitted is sent, the MSGx count has not expired and the transmit register does not contain another message byte to be transferred into the transmit buffer (J1850BUF). This is basically the case where the user has not written the next byte to be transmitted in time to prevent disruption of data flow onto the J1850 bus.

The BREAK bit indicates that a break symbol has been detected on the bus. The J1850 protocol handler will respond to the detection of a break symbol by automatically switching to normal mode of operation.

The BUS STAT bit indicates whether the bus is idle or busy. The BUS CONT bit indicates when contention is detected on the J1850 bus. The IFR RCV bit indicates that an IFR byte has been received and is ready to be read from the J_RX receive register.

7	6	5	4	3	2	1	0
IFR rcv'd	BUS Cont	BUS Stat	BRK rcv'd	Ovr/ Undr	MSG TX	MSG RX	BUS ERR

Figure 8: J1850 Status Register (J_STAT)

Transmit Register - The transmit (J_TX) register is a byte register used to transfer data to the J1850 bus. J_TX is buffered to allow sufficient time to write the next byte to be transmitted. Writing to J_TX automatically initiates a transmission.

7	6	5	4	3	2	1	0
DB7	DB6	DB5	DB4	DB3	DB2	DB1	DB0

Figure 9: J1850 Transmit Register (J_TX)

Receive Register - The receive (J_RX) register is a byte register used to transfer received data from the bus to the CPU. Like the transmit register, the receive register is buffered to allow sufficient time for user code to respond to the receive interrupt and read-out a received byte.

7	6	5	4	3	2	1	0
DB7	DB6	DB5	DB4	DB3	DB2	DB1	DB0

Figure 10: J1850 Receive Register (J_RX)

Delay Register - The delay register (J_DLY) allows the user to adjust the timing of symbols to compensate for inherent variations in propagation delays which may exist between different transceiver manufacturers. The first five bits of this register specify the amount of desired delay compensation in micro-seconds (us). The upper three bits are reserved. For a given delay time written to this register, that delay time is subtracted from all subsequent symbol transmission times as measured at the output of the 87C196LB's J1850TX pin. This allows the actual symbol time detected at the bus to be controlled and thus provide optimal symbol times on the bus.

7	6	5	4	3	2	1	0
rsv	rsv	rsv	DLY 4	DLY 3	DLY 2	DLY 1	DLY 0

Figure 11: J1850 Delay Register (J_DLY)

TRANSCEIVER INTERFACE

The 87C196LB's J1850 protocol handler interfaces to an off-chip transceiver IC through a two-pin interface. For purposes of this paper, we choose to refer to the Harris Semiconductor HIP7020 bus transceiver. The HIP7020 comes packaged in an eight-pin SO package.

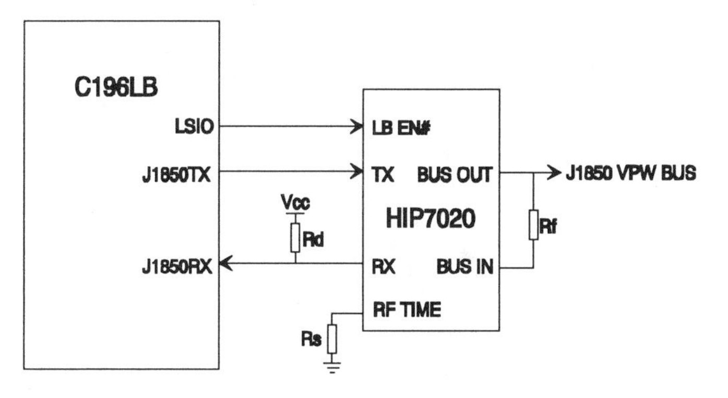

Figure 12: C196LB to Transceiver Interface

The above diagram shows the 87C196LB microcontroller interface to an HIP7020 transceiver IC. The interface is fairly simple and consists of a TX and a RX connection where the J1850 protocol handler transfers properly timed bus symbols to and from the transceiver in a digital fashion. For transmissions, the transceiver performs necessary waveshaping and places the data on the J1850 bus. For receptions, the transceiver converts the waveshaped waveforms on the bus to a CMOS digital signal which is transferred via the RX connection.

Also shown is an additional loop-back enable function. When active, loop-back enable internally ties the TX input to the RX output and isolates both signals from the J1850 bus.

SUMMARY

Intel has designed a low-cost J1850 protocol controller to fully support 10.4Kbp/s VPW communications. The protocol controller was specifically designed to be integrated onto a high-performance, 16-bit MCS 96 microcontroller referred to as the 87C196LB. This controller provides support for all four types of IFRs, NB insertion, CRC generation, and 4x mode. The J1850 protocol handler presented offers benefits that include minimal CPU burdening, PCB real-estate savings, and freeing-up of external communications to an off-chip controller. Three dedicated interrupts to the microcontroller core along with double buffered transmit and receive registers, allow for efficient data handling. These features make this microcontroller attractive for applications such as ABS that require a low-cost Class B network for purposes of improved diagnostics and data sharing.

REFERENCES

[1] SAE Vehicle Network for Multiplexing and Data Communications Standards Committee, SAE J1850 Standard, "Class B Data Communications Network Interface", Rev MAY94

[2] Intel Corporation, 8XC196LB Target Specification, October 18, 1995

[3] Intel Corporation, 8XC196Kx, 8XC196Jx, 87C196CA Microcontroller User's Manual, June 1995

[4] Harris Semiconductor, HIP7020 Data Sheet, February 1994

TYPICAL MICROCONTROLLER APPLICATIONS

880095

Core-Oriented Microcontrollers Provide the Flexibility and Performance Required in Today's ETR Designs

Massood Aghai-Yazdy
National Semiconductor Corp.
Microcontroller Group
Santa Clara, CA

Abstract :

Todays radio designer faces many important decisions when designing an Electronically Tuned Radio (ETR) of which selecting the microcontroller is one of the most critical. Certainly ,the selection of the right microcontroller not only increases system performance,but also reduces board space,power consumption and cost. The 8-bit core oriented microcontrollers from National Semiconductor are designed with integrated features which will provide system flexibility and higher integration while meeting the above criteria. These microcontrollers represent an excellent complement to the other standard devices in forming a complete radio design.

Introduction :

Microcontrollers first found their way into the front end of radios to provide automated tuning ,time/frequency display,desired station store and recall,and time-of-day clock functions . As ETRs have evolved to become more complex , microcontrollers are reqnired to process,compute,and store more data while keeping the cost , power consumption,and CPU time to a minimun. Radio manufacturers have also insisted upon higher system integration to decrease their board real estate ,coil adjustments,and manufacturing cost without sacrificing system flexibility and performance.These stringent demands certainly require a powerful core-oriented microcontroller with the right blend of features,RAM and ROM .

IC vendors have provided many alternatives to meet the demands of an ETR design . This article will focus on the advantages of the standard core-oriented microcontroller and how it meets the demand of a mid-range ETR.

Section I will address the microcontroller requirements in ETRs while demonstrating the advantages of a standard core-oriented microcontroller and its features relative to board real estate ,power consumption,code efficiency ,and cost savings. In addition , system flexibility, reliability,and cost trade-offs ,with regard to ETR front end partitioning ,will be discussed. Section II will review National's COP888 core-oriented microcontrollers which have been designed with midline ETRs in mind .

SECTION I

SYSTEM DESCRIPTION :

This section will review mid range ETR demands on microcontroller hardware and software.These demands include the I/O count ,on board features,data throughput,serial communication, code efficiency,power saving features,and interface to other standard products. A block diagram of a mid range ETR is shown in fig 2.The microcontroller,in conjunction with the phase locked loop and display driver, performs station display and automatic tuning.In the audio section, the microcontroller performs fader,balance, treble, base,and volume level adjustments via a digitally controlled tone device .In addition ,it interfaces to the other blocks in the ETR to select DNR and to mute the radio between stations.It is essential for the radio designer not only to understand the microcontroller requirements ,but also the level of integration needed to achieve the highest flexibility and performance with the lowest tracking adjustments,external components,and cost.This article will address all of the above issues.

A-HARDWARE REQUIREMENTS :

I-power savings :

The ETR designer always searches for ways to reduce power consumption in his design , particularly when the car ignition is switched off. Selecting a microcontroller with the proper features and the ability to utilize these features effectively can reduce the power consumption drastically.First CMOS process for inherent low power consumption is required.Second , features such as HALT with multi-input wakeup and IDLE are neccessary to further reduce power consumption. Needless to say ,any microcontroller output connected to an unpowered device,must be forced low or tri- state before entering the HALT or IDLE mode.Such practices assure lower power consumption.

II-I/O :

Digital I/O :

ETRs today have many types of switches on the front panel or in remote locations for controlling radio functions .Some of these switches are simply a standard keyboard matrix : memory station switches,scan, seek,and time / frequency select . Manual tune up/down keys are usually rotary switches with indentation ;these switches generate gray code . Five microcontroller outputs and 4 inputs are required for 5X4 keyboard interface .

A/D :

The majority of ETRs utilize analog knobs for adjusting volume ,fader,balance,treble,and base levels. In order to improve distortion existing within the analog tone and to increase cost savings associated with using 2 wires instead of 6 for remote control , digitally-controlled tone devices can be used . Five analog inputs are required for these functions. Another analog input may be neccessary for sensing the dimming switch. Therefore,a total of six channel A/D inputs are required to digitize the control information before it is sent to the tone device via the microcontroller.

A microcontroller with an A/D on-board reduces the board real estate ,and cost while increasing the reliability

UART :

Some ETRs locate the volume,tune,and other control keys in the steering column (remote from the receiver), where the driver can easily access them . This function requires a UART for minimun wire connections.Generally ,a software UART meets the demands of low and mid-range ETRs,in spite of the increased timing constraints and software overhead.The selection of software over hardware can reduce the system cost significantly.

Display Driver :

The display functions in ETRs are threefold,station frequency,time-of-day,and annunciators.The display and its companion display driver are usually changed from one car radio to another .Consequently , the overall radio partitioning must keep the display and its driver separate from other devices and boards. This naturally increases system flexibility while it decreases manufacturing cost. A microcontroller must incorporate a serial interface for communication with external VFD or LCD display drivers.

Manual and automatic tuning with accurate station detect and mute:

Manual and automatic tuning function is performed in conjunction with a Phase Locked Loop. Automatic tuning is based on the invocation of the two switch inputs, SCAN and SEEK.

Anti Theft :

Car radio theft has dramatically increased over the years. A radio with a security code programmed during manufacturing is now on the market which causes ETR to become inoperable when removed from its compartment.This function requires EERAM for storing the security code. An external EERAM with serial interface is still the most economical solution . However ,the integration of the microcontroller and EERAM may be desirable for reducing the board space,and software overhead.

Desired station store and recall with last station recall after power up and mute :

All ETRs provide memory station recall for desired AM/FM channels and recall of the last station selected after power up.These functions require EERAM for non volatile storage.

III-Timers

Idle timer for time-of-day clock :

Some ETRs are designed to provide a time-of-day clock in conjunction with radio functions.A clock defeat provides the option to keep the display blank if there is another clock in the car.There are usually two methods to implement a real time clock.The microcontroller can use the external 50 HZ signal provided by a PLL or an on board free running timer (idle timer) to perform the time keeping .The advantage of using a free running timer in the idle mode is that it is desirable to shut off the PLL,RF,and IF blocks when the radio is not in use ; this reduces power consumption dramatically.

PWM timers for display dimming :

Some VF display drivers ,to keep a high quality display, utilize duty cycle for brightness control. This feature requires a PWM timer.

IV-Watchdog for system reliability:

Watchdog capability is a requirement for any automotive ETR where overvoltage or electromagnetic interference might cause the micro to get stuck in infinite loops .

B- SOFTWARE REQUIREMENTS :

The ETR designer,after determining hardware feature requirements, must evaluate software trade-offs.A microcontroller with multi- function and a majority of single byte/single cycle instructions makes efficient software realizable.This ,in effect, increases the system data throughput ,while decreasing the cost associated with utilizing more RAM and ROM . Software development cost will also decrease significantly.The following are some common ETR routines.

1- keyboard scan/decode/debounce
2- channel frequency computation and conversion to a proper format for both display and PLL.
3- time keeping routine
4- store and recall of the desired stations
5- station detect and mute
6- information transmission to the PLL,display driver,and tone
7- frequency conversion between the U.S. and European radio mode
8- display algorithm
9- BCD to seven segment display conversion

A desirable microcontroller for an ETR must support ROM table look up, bit and byte manipulation ,BCD arithmetic, and efficient block moves . For a microcontroller with a Harward architecture (the RAM and ROM are separately addressed), a specific instruction is required to perform an efficient ROM table look up for BCD to seven segment display conversion. Such instructions allow large tables to be located in ROM instead of RAM ,resulting in more efficient RAM utilization . Another useful instruction for ETR programming design is a JUMP INDIRECT instruction for keyboard decode algorithms .This instruction allows a jump to an address specified in ROM based on keyboard status input in the accumulator.This instruction increases the system data throughput while increasing code efficiency significantly for a 8X4 or 8X8 keyboard interface.

SECTION II

The COP888 family is a new generation of microcontrollers from National Semiconductor .This microcontroller family , having integrated features into a single chip,provide the best flexibility,and price/performance trade-offs for ETR designs .These products are fabricated utilizing NSC's MMCMOS process ,CORE methodology and are available in numerous packaging options including condensed TAPE PAK (tm) .The COP888CL,COP888CF,and COP888CG are all implemented utilizing the same core, the same high efficiency instruction set,and the same development system support .Consequently ,the cost associated with new program development is minimized . It is also conceivable that the same code can be utilized ,regardless of the system partitioning . Fig 3 illustrates the COP888 core in conjunction with the on board features for each family member .

CORE architecture :

Core ,as the word implies, is the section of the microcontroller which remains unchanged across the product line.National Semiconductor COP888 core consists of the CPU ,MICROWIRE/PLUS ,1 processor independent PWM timer,hooks for 16 vector interrupts,two power saving features ,the necessary I/O for interface to on board peripherals and hooks for increasing the on board RAM and ROM up to 32K .The COP888 core incorporates a memory mapped architecture which provides many advantages for ETR program development .First, the number of instructions have been minimized without the loss of performance . This in effect makes the software programming easier with reduced program development time . Second ,the majority of the instructions have more addressing modes which increases program code efficiency.Finally ,the PLA size is minimized which reduces the cost. COP 888 Core is currently being fabricated utilizing the 2 micron MMCMOS process.The core runs at a maximun speed 1 usec per instruction clock cycle.

MMCMOS process :

The MMCMOS is a double metal,single poly process with the following advantages :

1- low power
2- high noise immunity
3- scalable to 1 micron
4- high density
5- high speed
6- capacity to integrate A/D capacitor,EE,and bipolar modules.
7-use of epitaxial layer for latch up immunity

COP888CF :

The COP888CF with 8 channel A/D ,two processor independent PWM timers ,watchdog,MICROWIRE/PLUS serial communication port,and two power saving features HALT and IDLE ,is a good choice for ETRs today.The COP888CF block diagram is shown in fig 3.The HALT and IDLE modes with multi-input wakeup provide the power saving features required in ETRs.The watchdog timer with programmable window selection protects the micro from getting stuck in infinite loops while providing programming flexibility for different timing loops. The two powerfull 16-bit timers with two auto-reload registers generate PWM signals without processor intervention.This meets the display dimming and D/A function requirements.The MICROWIRE/PLUS ,three wire serial communication bus, provides the interface to display driver, PLL,and tone IC's. The COP888CF has 4K of ROM and 128 bytes of RAM and it is available in a 28,40,or 44 pin packages.

COP888CL :

For a radio partitioning scheme ,where the A/D must be separate from the the microcontroller,the COP888CL is an excellent choice.The COP888CL has the same features as COP888CF except the A/D module has been removed.

COP888CG :

The COP888CG incorporates a full duplex UART,three PWM timers,watchdog timer,and two power saving features ,HALT and IDLE with multi-input wake up.The COP888CG with on board UART meets the demand of an ETR patitioning with remote control and remote functionality testing. The RAM is 192 bytes and the ROM size is 4 K.

Conclusion :

Selecting the right microcontroller increases the system performance and can reduce the board space,as well as reduce the system cost.The standard 8-bit core oriented microcontrollers from National Semiconductor supports the ETR requirements with higher system flexibility and integration. Above all ,these microcontrollers are fabricated utilizing the hand-packed layout design for the lowest cost solutions.The variety of RAM/ROM sizes and on board peripherals make this family flexible for different ETR partitioning ,preserving investments in software. Finally these products will have their core in the National's standard cell library for ETR custom designs.

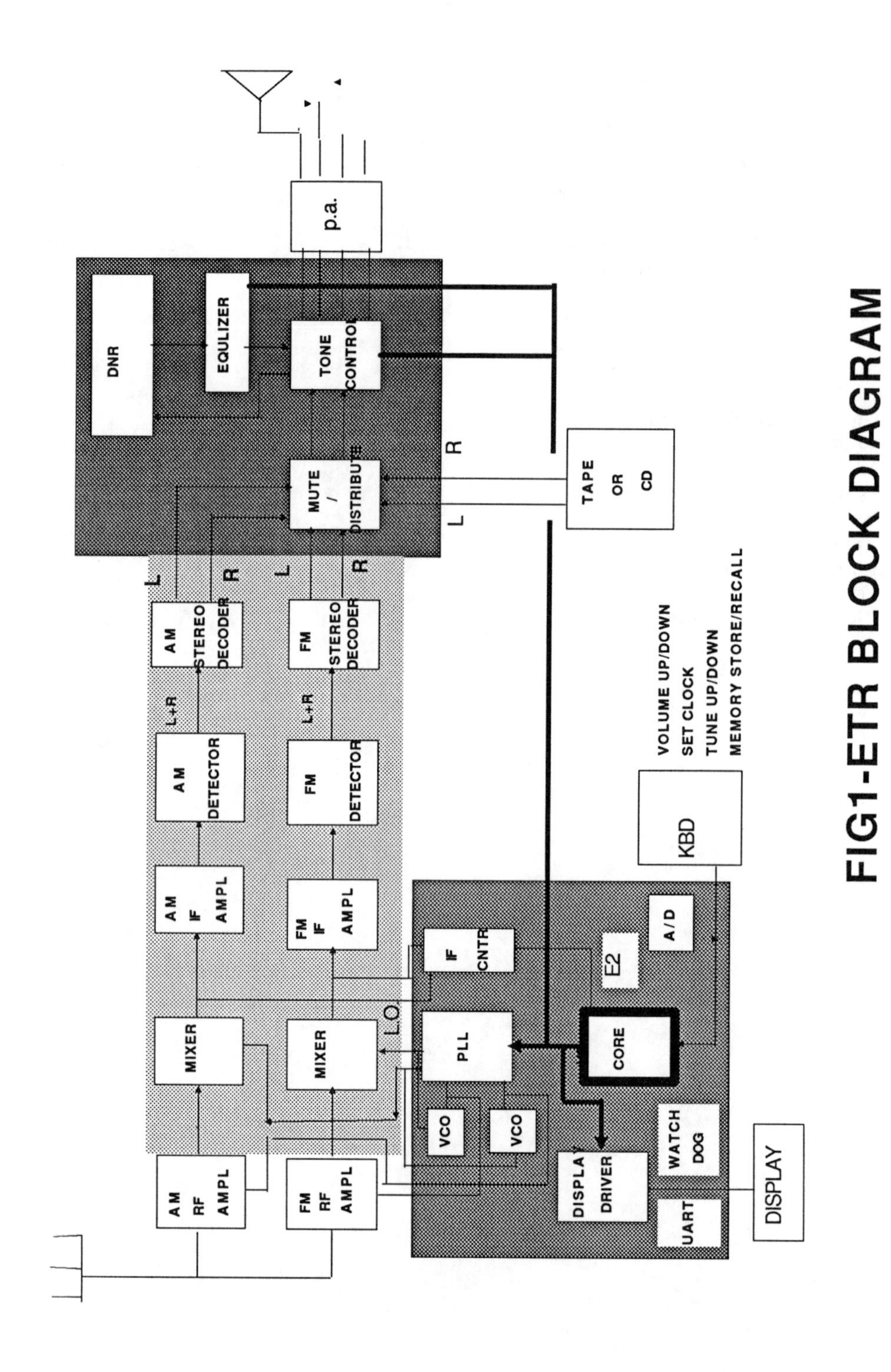

FIG1-ETR BLOCK DIAGRAM

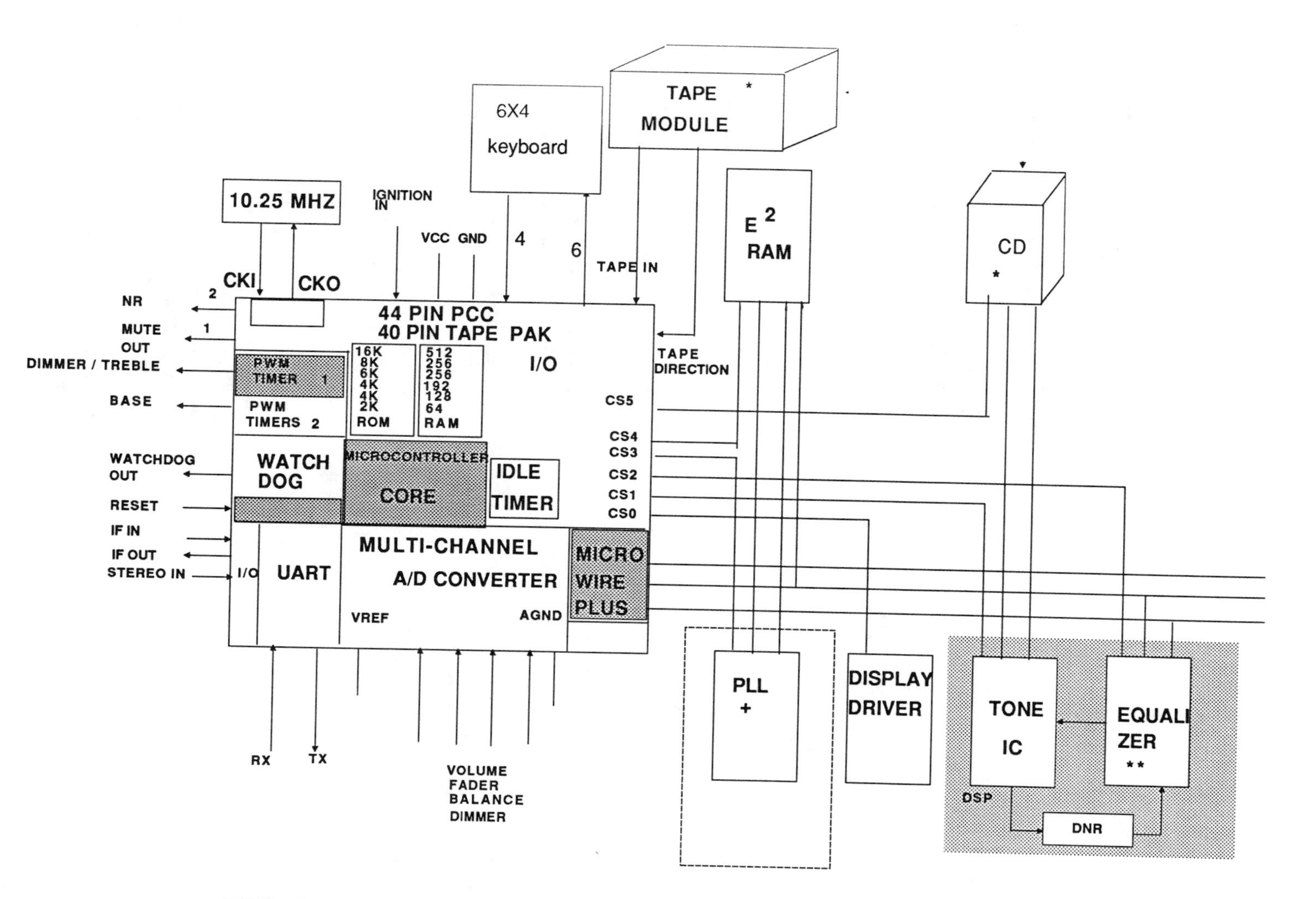

10.25 MHZ
IGNITION IN
6X4 keyboard
TAPE MODULE *
E 2 RAM
CD *
VCC GND
4
6
TAPE IN
CKI
CKO
2
1
NR
MUTE OUT
DIMMER / TREBLE
BASE
WATCHDOG OUT
RESET
IF IN
IF OUT
STEREO IN
44 PIN PCC
40 PIN TAPE PAK
I/O
PWM TIMER 1
PWM TIMERS 2
16K 8K 6K 4K 4K 2K ROM
512 256 256 192 128 64 RAM
TAPE DIRECTION
CS5
CS4
CS3
CS2
CS1
CS0
WATCH DOG
MICROCONTROLLER
CORE
IDLE TIMER
I/O
UART
MULTI-CHANNEL
A/D CONVERTER
VREF
AGND
MICRO WIRE PLUS
RX
TX
VOLUME
FADER
BALANCE
DIMMER
PLL +
DISPLAY DRIVER
TONE IC
EQUALI ZER **
DSP
DNR

FIG 2

COP888CL

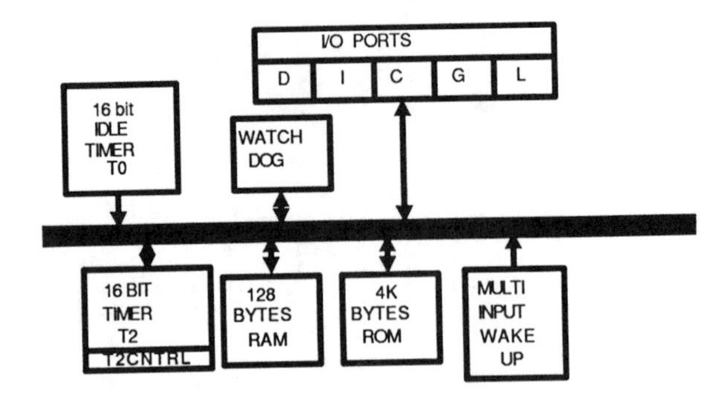

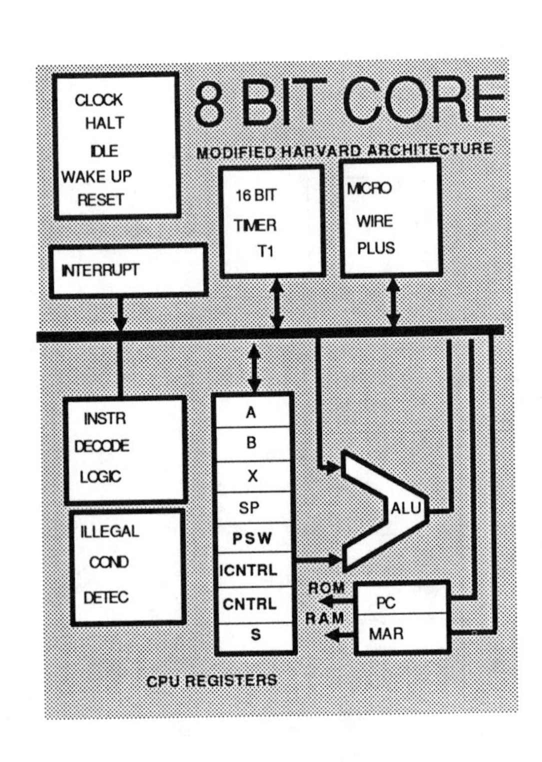

COP888CF

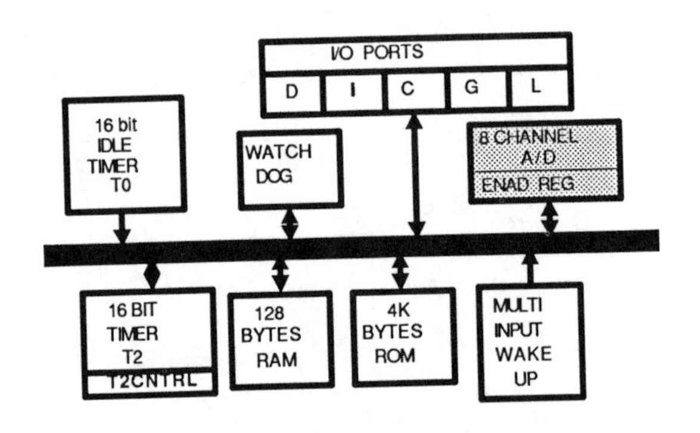

COP888CG

FIG 3

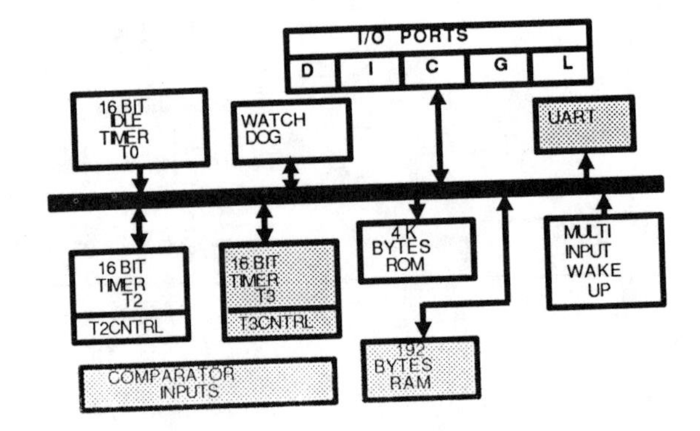

880136

Trends in Electronic Engine Control and Development of Optimum Microcomputers

Naoki Tomisawa and Shuichi Toki
Research & Development Dept.
Japan Electronic Control Systems Co., Ltd.

ABSTRACT

Electronic fuel injection systems for automotive engine control are now capable of greater control precision and increased function through the progress in microcomputer performance and cost efficiency. The dependence of engine control systems on microcomputer performance is expected to remain significant for the future as well, in such areas as improved learning control techniques and new control theories.

This paper first refers to sequential fuel injection systems under cylinder-by-cylinder control and adaptive learning control techniques for which a new control theory has been employed, and then presents a new microcomputer integrated with a multifunctional timer module for extremely high real-time control capabilities by means of its 16-bit core CPU. A single chip ASIC memory unit integrated with EPROM and RAM devices serves as a peripheral LSI for the microcomputer.

THIS PAPER PRESENTS a 16-bit microcomputer with an ASIC memory unit developed for engine control applications. The microcomputer incorporates a hardware timer module with extremely high real-time control capabilities, and has successfully eliminated cumbersome software interrupt routines to permit dedicating the CPU block to logical operations.

In addition, by employing a dual configuration for the microcomputer and memory blocks, highly efficient chip design has been achieved and contributed to extremely low production costs despite the large 32K-byte ROM and 2K-byte RAM.

1. ENGINE CONTROLLER TRENDS

Historical controller trends that we have experienced are shown in Fig. 1. Their future trends are forecast as presented below.

(1) MULTIFUNCTIONAL CONTROL – Unlike the single function fuel injection control of early days, the integrated control of ignition, knocking, and idling speed is rapidly penetrating the market today. This trend towards multifunctional controls is sure to expand to cover the turbocharger, distributor-less ignition, self-diagnostics, automatic transmission (AT), and other areas.

(2) CONTROL SOPHISTICATION – Individual control functions themselves are also expected to become more sophisticated. In the area of fuel injection, cylinder-by-cylinder sequential injection will become commonplace, and increasingly more systems will correct air-fuel mix ratio fluctuations on an individual cylinder basis by self-adaptive (learning) control.

Knocking control will also evolve into either cylinder-by-cylinder or intra-cylinder pressure sensor-activated combustion pressure control, integrated with idling speed control by employing a modern control theory.

Not only have the definition and precision of individual control functions been increased, but a need has also a risen to achieve higher control speeds for the accurate control of transient engine modes.

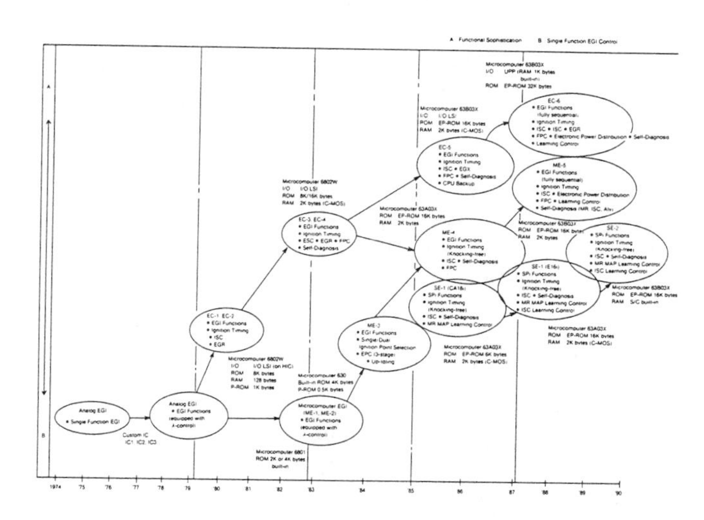

Fig. 1 Historical Engine Controller Trends

(3) COMPACT SIZE – Controller mounting space has been steadily decreasing due to shrinking vehicle sizes, larger interiors, and greater freedom for overall

vehicle design.

In addition, weight and cost reductions by shortening engine harness cables will impose increasingly more stringent limitations on controller mounting space, calling for still smaller controllers in the future.

A forecast is given in Fig 2 of shrinking controller size and associated mounting density trends.

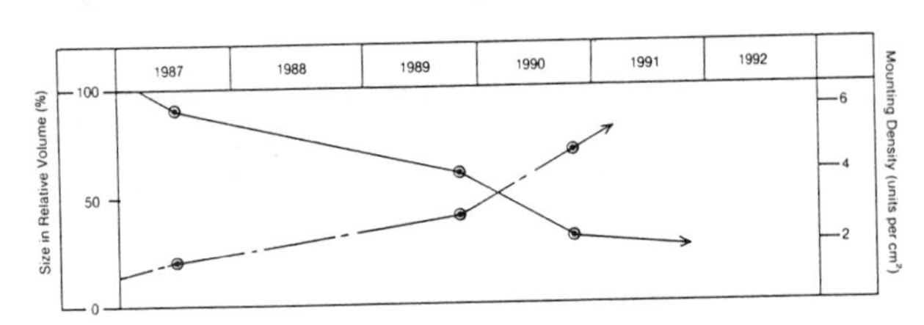

Fig. 2 Controller Size Trends (in relative size of our own products)

(4) LOWER COSTS – Along with the remarkably advanced processing capability of microcomputers, LSI costs have also come down.

With engine controllers, enhanced microcomputer capability will induce further transitions to software of the functions formerly handled by hardware, increasing the LSI share in total costs. Thus LSI cost savings are envisioned to lower total controller costs still further with the passage of time.

In view of the current progress in microcomputer technology, we forecast that controller costs of the 1990s will come down to about 70 to 60% of their current costs.

2. PROGRESS IN CONTROL FUNCTIONS

Some of the new control systems presented below are currently in production by virtue of the enhanced performance of microcomputers.

(1) SEQUENTIALLY CONTROLLED FUEL INJECTION SYSTEM – A sequential fuel injection system provides cylinder-by-cylinder control over fuel injection at individual cylinders and is synchronized with the intake cycle of each cylinder. Such systems are rapidly coming into extensive use to achieve optimum combustion within individual cylinders of the engine.

Initially, this system merely injected fuel to each cylinder independent of the other cylinders just before the intake cycle, but has since advanced to precisely control injection timing and volume, as presented below.

(i) Injection Timing Control – The best approach for injecting fuel into individual cylinders at high efficiency is to have the fuel ride the intake air flow in each intake cycle of the cylinders.

For this purpose, fuel injection needs to be concluded before the latter half of each intake cycle arrives when the intake air flow velocity drops down (see Fig.3 FUEL IN LIMIT).

Thus, the conclusion of fuel injection points to FUEL IN LIMIT timing. The fuel injection timing must precede the FUEL IN LIMIT timing by a fuel injection pulse width (Ti) that fluctuates broadly with the engine operating mode. To exercise this control over each cylinder individually, extremely high-speed operations are required of the CPU.

(ii) Injection Volume Control – As may be seen in Fig 3, the intake air flow into an injected cylinder has not built up sufficiently at the fuel injection starting point. The volume of fuel injected into that cylinder, however, is determined by the integrated volume of the entire cycle of the intake air flow.

A contradiction therefore develops: while the computation of a fuel injection pulse width needs to be concluded before the pulse-controlled injection cycle begins, the integrated volume of intake air flow must have been measured.

In consequence, the need arises to project the entire intake volume out of an intake flow build-up in the cylinder scheduled for an upcoming fuel injection cycle.

Specifically, the boost pressure near a cylinder intake valve is projected out of the throttle valve opening area variation rate and engine speed, and the aggregate intake volume then inferred from the projected boost pressure and the actual air intake volume build-up.

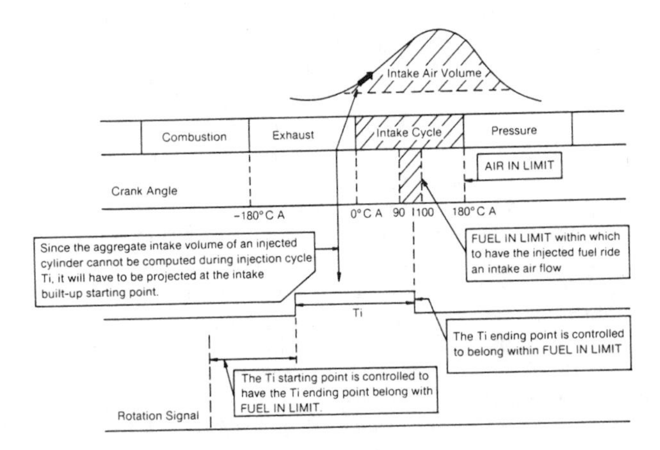

Fig. 3 Sequential Injection Timing Chart

In addition, computations of the injection pulse width based on an actual intake volume must also be kept even after the fuel injection has begun, and when the actual volume deviates from the earlier projection, the pulse width for an ongoing injection must be revised midway through the cycle.

It then follows that the microcomputer must be equipped with mutually independent fuel injection pulse generating timers, each covering one of the cylinders. They must also function flexibly enough to permit the revision of a generated pulse width midway through the generation of that pulse.

(2) OPTIMUM LEARNING CONTROL – The learning control system is employed to preserve the initial performance of an engine over the long term, by absorbing any deviation that may have been built by a production process into the electronically controlled fuel injection system components and compensating for any aging variation of the cylinder efficiency and other engine parameters, as well as any operating environment changes in atmospheric pressure, temperature, and humidity.

Almost all conventional learning systems, however, are the "repetitive learning" type that relies on data maps.

This is disadvantageous in that if minute learning domains are set for higher learning correction accuracy, the frequency of learning exercises are reduced, in turn lowering the updating speed.

Being studied of late is an approach that, instead of making corrections based on a data map, analyzes the factors for any feedback control deviation generated and makes corrections with theoretical correction formulas best suited for the error factors.

The latest control techniques are outlined below taking up learning control for the air-fuel mix ratio as an example.

(i) Factor Analysis – When air-fuel ratio feedback control computations are based on an O_2 sensor-triggered λ control, the current engine operating mode, control deviation, its deviating direction, speed, and other parameters are stored instantaneously. By inference from the databases thus formed, the error factors are analyzed. For the inference operation, the influence of component deterioration, environmental variation, and other factors is stored in advance, and factor analysis made from the databases mentioned earlier.

For such inference operations, a large volume of databases must be handled, very high capacity ROM and RAM devices are required. High operating speed is also required, necessitating a high-speed CPU for the analysis.

(ii) Inferential Learning – The influence of individual component deterioration, environmental variation, and other factors stored to implement the factor analysis is subjected to self-revision in the event that the steady monitoring of satisfaction degrees for the air-fuel ratio learning achievement (that is, whether air-fuel ratio deviations have actually been reduced by the correction) reveals any deviation from the optimum degree.

(iii) Cylinder-by-Cylinder Correction – Since air-fuel ratio deviations originate from the manufacturing differences in individual components of the engine, these deviations can be expected to differ from one cylinder to the next. Consequently, a system will probably emerge in the near future that, in conjunction with the sequential injection system mentioned earlier, detects the O_2 sensor output variation of each individual cylinder separately for cylinder-by-cylinder learning.

All the above control techniques rely on the high processing capability of the CPUs, so that the emergence of a truly high performance CPU is eagerly anticipated.

3. FUNCTIONS REQUIRED OF ENGINE CONTROL MICROCOMPUTERS

Performance requirements for tomorrow's engine control microcomputers are discussed in the following.

(1) FUNCTIONS AND OPERATING CAPACITY – Along with increasingly diversified control functions and higher sophistication of individual control parameters, the need arises for processing a large number of control parameters at high speed. Furthermore, as discussed in the sequentially controlled fuel injection system example, highly sophisticated real-time processing is required.

In the past, the extensive pulse I/O processing needed for engine control was achieved by software-dependent interrupt requests. The CPU was also heavily burdened to maintain sophisticated real-time processing at the same time by preassigned engine control computations. Limitations, however, will be imposed on such software-based processing, by the increased multifunctionality and higher sophistication of engine control computations needed for the real-time processing requirements of tomorrow.

Hence, we propose the provision of engine control-dedicated hardware timers which would hardly impose hardly any software burden and yet flexibly accommodate a variety of pulse I/O processing. This approach will reduce housekeeping tasks of the software and permit the CPU to concentrate on its originally intended engine control computations.

Higher operating speed will be required of the core CPU, but since, as discussed in the paragraph on "optimum learning control", the use of an inferential processing concept is seen likely for the future handling of databases, we prefer to retain a full 16-bit CPU. A 32-bit CPU will not be required in the near future, in view of the control accuracy possible with existing engine control sensors and actuators. Rather, the disposing of unrequired bits would necessitate redundant processing.

For an engine control system of which both real-time control and computation capabilities of high sophistication are required, the use of a multi-CPU configuration is also conceivable, where pulse I/O processing would be handled by one of the CPUs and computational processing by the other.

However, as control becomes increasingly more complex and further sophisticated, the inter-CPU transfer of data would have to be achieved in greater volume and at a higher speed, increasing the overhead processing and eventually defeating the originally intended purposes.

For this reason, we have ignored the concept of a multi-CPU system at this time and decided on a single CPU system in which a core CPU with high computational capabilities and multifunctional hardware timers will be integrated.

Figure 4 shows an outline of the required functions.

	Required Function	Purposes Served
1	Built-in Real-Time Control Hardware Timers, each of which: • Accommodates fully sequential control (over an individual cylinder independently). • Accommodates ignition and energization angle controls. • Incorporates numerous PWM (ISC, EGR etc.) • Incorporates numerous pulsive input measuring timers (for the engine speed, vehicular speed input, etc.).	• Reduces the software overhead processing. • Enhances real-time control capabilities.
2	• Full 16-bit Processing Core CPU	• Higher speed • Enhanced computational accuracy • Accommodates a high level language.
3	• Built-in AD Converter (10-bit)	• Reduces handling overhead. • Consolidation of chips (for a compact design)

Fig. 4 Functions Required of Engine Control Microcomputer

(2) COSTS – A chip configuration must be sought that meets the functional requirements and is also the most economical.

The cost of the engine control microcomputer has been minimized by reviewing its chip area, design rule, reliability, etc. and by configuring it with two chips, one containing the CPU, timers, and A/D converter, and the other the ROM and RAM units, expanded I/O ports, and timers.

(3) SOFTWARE DEVELOPMENT – The programs of conventional system that relies on software interrupt requests for pulse I/O processing have become complex because multiplexed interrupt requests, unique computational processes for shortening processing times, and a large number of interrupt requests are involved to assure highly sophisticated real-time control. Such programs requiring special software techniques have had to disregard the general-purpose applicability and maintainability of software, and raise the problems listed below.

[1] No program can be maintained except by the party who has generated it.

[2] Program bugs are liable to occur under the conditions overlooked when generating the program.

[3] Virtually no compiler replacement in a high level language is possible.

These problems disappear when the pulse I/O processing system is configured with dedicated hardware timers.

Provision of the hardware timers also leads to the generation of a group of easily maintained programs, so that software with fewer bugs can be developed in a shorter time. This further facilitates the accommodation of a high-level language, but where the CPU is concerned, its RAM capacity should be increased and stack operating commands and further improved to achieve an even better response.

Regarding the ROM and RAM capacities, the larger the better of course, to cope with increased multifunctions and higher control sophistication, and to better organize the group of programs for easier maintenance. But in view of current capacities and projected future functions, ROM and RAM configurations have been assigned for our purposes at the following levels:

ROM: 32K bytes x 8 bits
RAM: 2K bytes x 8 bits

4. LSI TRENDS FOR MICROCOMPUTER AND PERIPHERALS

The integration level of ICs in general has doubled annually for over a decade in conformance with Moore's law. As an example to illustrate the situation, the DRAM integration enhancement curve is shown in Fig. 5.

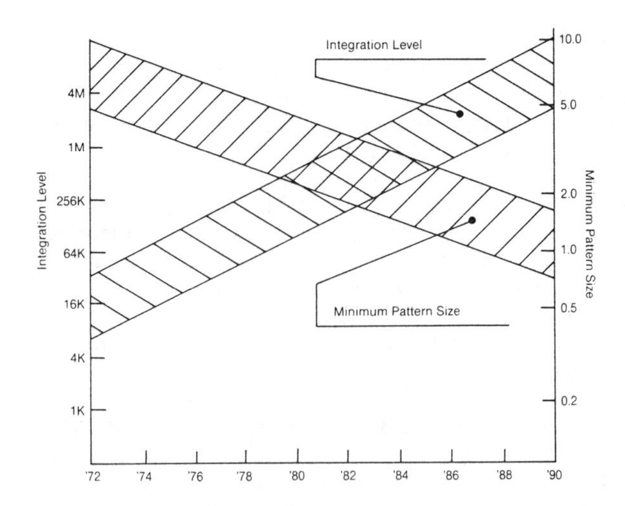

Fig. 5 Integration Enhancement of Memory LSIs

As of 1987, the minimum pattern size, or design rule, of volume produced memory devices and microcomputers has been micro-miniaturized down to the 1.6 to 2.0 μm level.

Figure 6 shows the microcomputer chip cost versus software cost interrelations for the fuel injection control system incorporating a microcomputer and rated at a given control functionality level. Its horizontal coordinate represents the chip functions and its vertical coordinate, the total cost of the system.

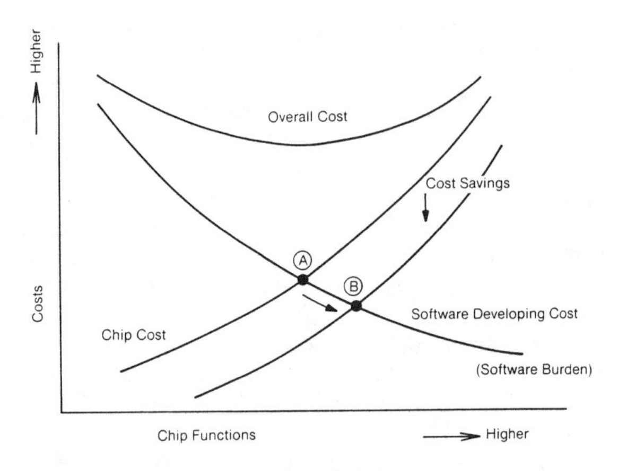

Fig. 6 Cost Impacts of Software and Chip Functions

In the figure, [A] is the point at which the total software developing cost and chip cost is minimized. As the integration level is raised and the relative chip cost lowered, the minimal point will be shifted to [B] of higher chip functions. In other words, with an elevated chip integration level, not only may costs be held unaltered by converting to hardware some of the areas formerly processed with software, but the software burden may also be alleviated.

To closely approximate the ideal image, real-time microprocessors have been configured of the functional blocks as shown in Fig. 7. The low speed processor block configured of CPU + ROM + RAM + I/O is basically engaged in software-based processing. Emphasis for this block is placed on the ease of its software development and its high reliability rather than its operating speed, and requirements for the CPU serving the block are beginning to include its readiness for the support by a high level language. For this reason, the CPU capacity is being phased out of 8 bits into 16 bits or higher.

Its high speed processor block on the other hand, is engaged in high-speed processing of a μs order, so that its hardware scale is tending to be enlarged for the minimization of any software intervention.

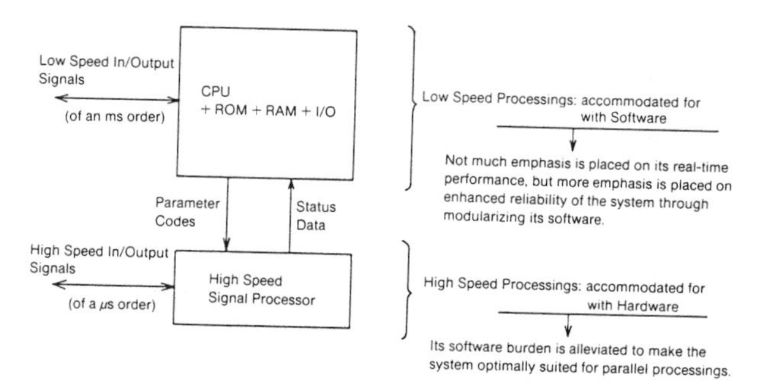

Fig. 7 Functional Blocks of Real-Time Control Microcomputer

This paper reports on, and details a 16-bit single chip microcomputer, the M37790STJ, incorporating extremely powerful real-time hardware functions and developed by taking full advantage of the high integration level achieved with a 1.3 μm design rule that rivals that of 1 M-bit DRAMs.

5. MICROCOMPUTER SPECIFICATIONS

The M37790STJ is an automotive applications-dedicated single chip microcomputer, and its functions may be taken maximum advantage of by employing it together with the M5M72561J ASIC memory device,a chip in the same family.

An outline of its functions is presented in Fig. 8, and its functional block diagram in Fig. 9.

- Storage Capacity
  - Built-in Memory Units .... No RAM or ROM
  - Memory Space .... 18m bytes
- Command Execution Time .... 0.5 μs (with shortest commands at 8MHz)
- Interrupts .... 22 factors (including Reset) Priority sequenced at 7 levels
- Timers (18-bit timers)
  - Real-Time Output Timers 8
  - Cascade Timers 4
  - Free-running Timer 1
  - Input Signal Measuring Timers 3
- PWM .... 3 units
- Serial I/O (switchable between Clock Sync and Asynchronous .... 1 ch
- A-D Converter (at 10-bit resolution) ...... 8 ch
- Watch-dog Timer .... 1 unit
- Wait Function (by Software One-Wait or RDY input)
- Hold Function (by HOLDREQ input)
- General-Purpose Ports (including In/Output, Output-dedicated and Input-dedicated Ports) .... 33 ports
- Package .... 84-pin PLCC
- Process .... CMOS
- Single Power Supply .... 5V ± 10%
- Operating Temperature Range .... -40 to 85°C

Fig. 8 Outline Functions

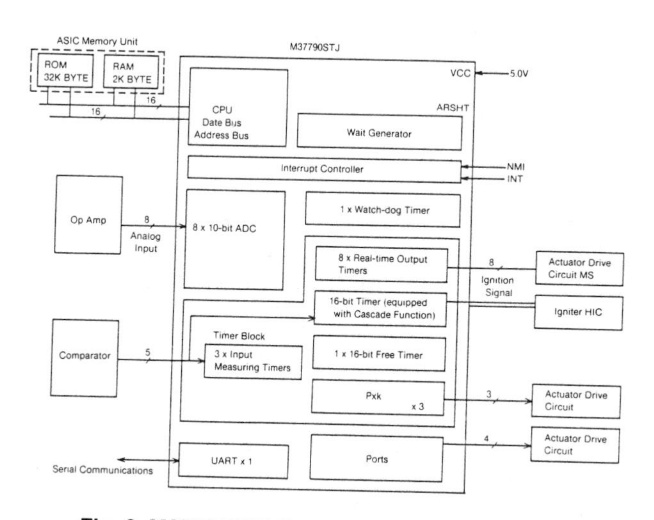

Fig. 9 M37790STJ Functional Block Diagram

The M37790STJ microcomputer and M5M72561J ASIC memory device employed together may serve up to 8-cylinder MPI fully sequential fuel injection system engine control purposes, in view of the number of real-time output ports available.

(1) REAL-TIME CONTROL BLOCK – Thanks to the enhanced integration level of the M37790STJ that permits many circuits to be integrated on a single chip, sixteen 16-bit timers have been built in for high-speed parallel processing of multiple signals.

Except for the PWM block, these timers have been grouped into three blocks. Eight timers TA1 through TA8 serve fuel injection purposes, and four timers TB1 through TB4 that synchronize with engine operation serve ignition timing, knocking detection timing, and other signal-generating purposes. Timers TC1 through TC3 that measure input cycle times serve either the vehicle or engine run speed measuring purposes.

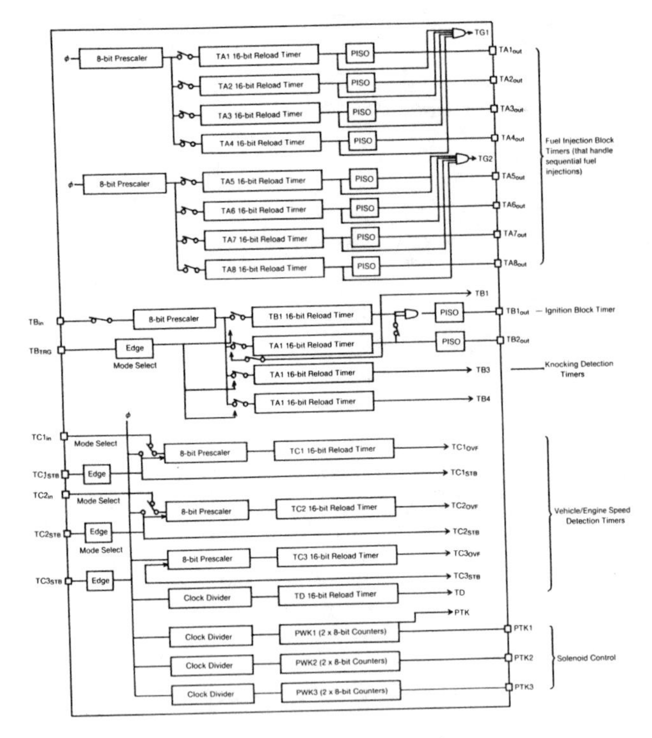

Fig. 10 Timer Layout

(2) REAL-TIME OUTPUT BLOCK TIMERS – The real-time output block timers TA1 through TA8, have each been configured as shown in Fig. 11. Both timer counters and timer buffers enable read and write operations, permitting their counts to be altered midway through the run of a timer counter. This capability makes the revision of injection timings in the middle of any ongoing fuel injection cycle simple to achieve. The underflow output of a timer drives its real-time output port, and every time an underflow occurs, the timer shifts its PISO (parallel-in, serial-out) register data for output from the port.

An example of the operation of these real-time output block timers is shown in Fig.12. At each PISO write, timer start, and timer underflow, the PISO data is shifted and the port output updated. There are four modes of these timer operations, among which the normal one-shot mode may be optimally employed for fuel injection purposes. Specifically, T0 corresponds to the waiting time before an injection, and T1 to the duration of injection. (See Fig. 11.)

During operation of the engine at its rated speed and under the fuel injection control, for which the above mode has been employed, a software-free mode may be phased in after writing the individual T0, T1, and PISO data and starting the timers up, to alleviate the control-related software burden.

In addition, PWM waveforms accurate to 16 bits can also be generated by phasing in a complementary continue mode, for application to hydraulic pressure control, for example.

Incidentally, symbol $\overline{T}$ in Fig. 12 denotes a 2's complement of T.

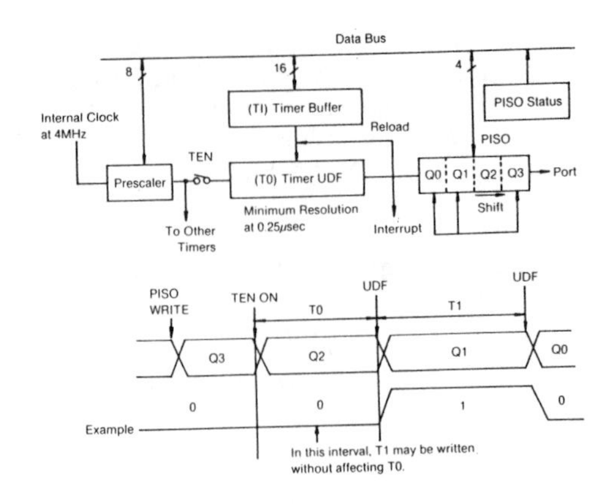

Fig. 11 Configuration and Operational Outline of Real-Time Output Block Timers

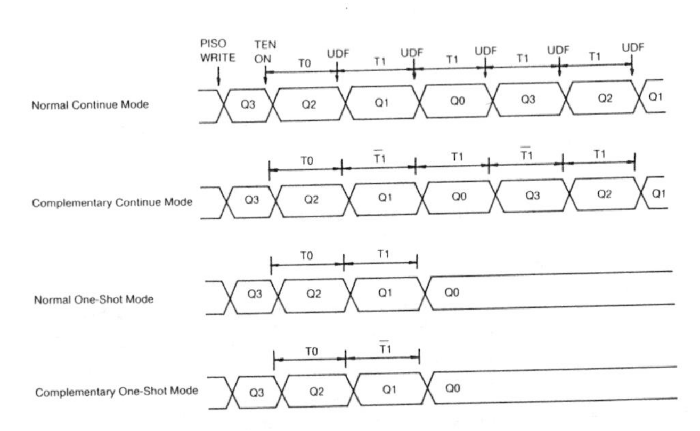

Fig. 12 Operating Modes of Real-Time Output Block Timers

(3) INPUT MEASURING BLOCK TIMERS – The configuration of input measuring timers TC1 through TC3 is shown in Fig. 13. The counters may be written and the measurement registers read. In a normal operating mode when a TC_{STB} input is given, not only will the count at that point in time be read into a measurement register, but the counter will also be cleared. Accordingly, at every TC_{STB} input, the time elapsed since the last TC_{STB} input remains intact in the measurement register. In other words, no software burden will be levied for cycle time measurements.

Figure 14 shows an example application where TC2 and TC3 have been employed to discriminate engine cylinders and detect the engine speed. Signal 1° CA from the crank angle sensor indicating a rotational engine angle is impressed on $TC2_{IN}$ and signal 180° CA (in the case of a 4-cylinder engine) from the crank angle sensor similarly impressed on $TC2_{STB}$ and $TC3_{STB}$. By designing the pulse width of signal 180° CA to differ among individual cylinders, any of the cylinders may be identified merely by reading the TC2 measurement register data, and the synchronization with engine operation assured by reading the TC3 measurement register data.

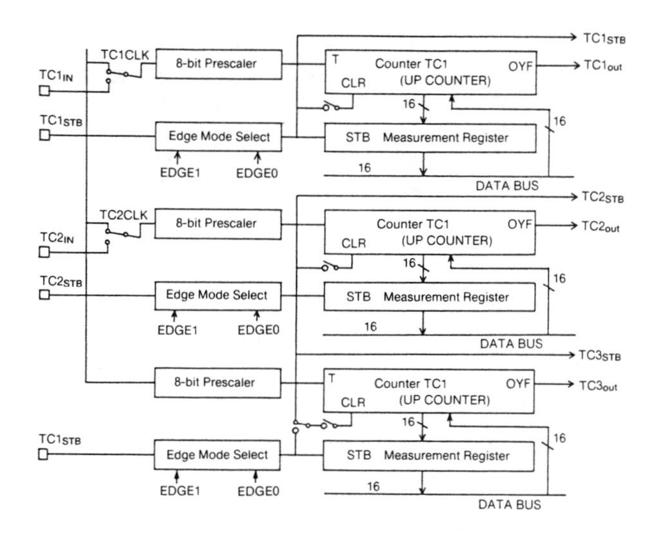

Fig. 13 Configuration of Input Cycle Time Measuring Timers

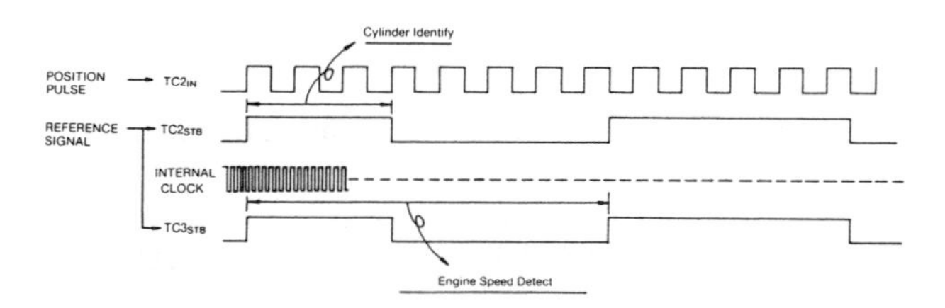

Fig. 14 Example Application of Input Cycle Time Measuring Timers

6. ASIC MEMORY UNIT

The M37790STJ microcomputer incorporates no memory units, and requires ROM and RAM devices to be connected externally. If standard EPROM and RAM devices are employed for this purpose, the system parts count including the decoder parts will increase that much higher. ASIC memory chip M6M72561J contain integrated on a single chip all such externally connected ports, timers, and other parts, and resolves this problem of an increasing number of parts mentioned above.

An outline of the M6M72561J functions is shown in Fig. 15, and its functional block diagram in Fig. 16.

The M6M72561J has been equipped with a BYTE pin which specifies the data bus usage in either 16 or 8 bit increments to enable its connection to either a 16-bit or 8-bit CPU as preferred. When writing with an EPROM writer, a BYTE=H (V_{pp}=12.5V) mode should be phased in for the usual 27C256 mode write operations.

- Storage Capacity
  - EPROM .... 32K bytes
  - RAM .... 2K bytes
  - Decoded Outputs for External Memory 2
- Timers
  - 8-bit Timers 2
- I/O Ports
  - In/Output Ports 8
  - Output-dedicated Ports 8
  - Input-dedicated Ports 14
- Standby Function: engaged by $\overline{CE}$ Input
- Package .... 68-pin PLCC
- Process .... C-MOS
- Operating Temperature Range .... -40 to 85°C

Fig. 15 Outline of M6M72561J (ASIC Memory) Functions

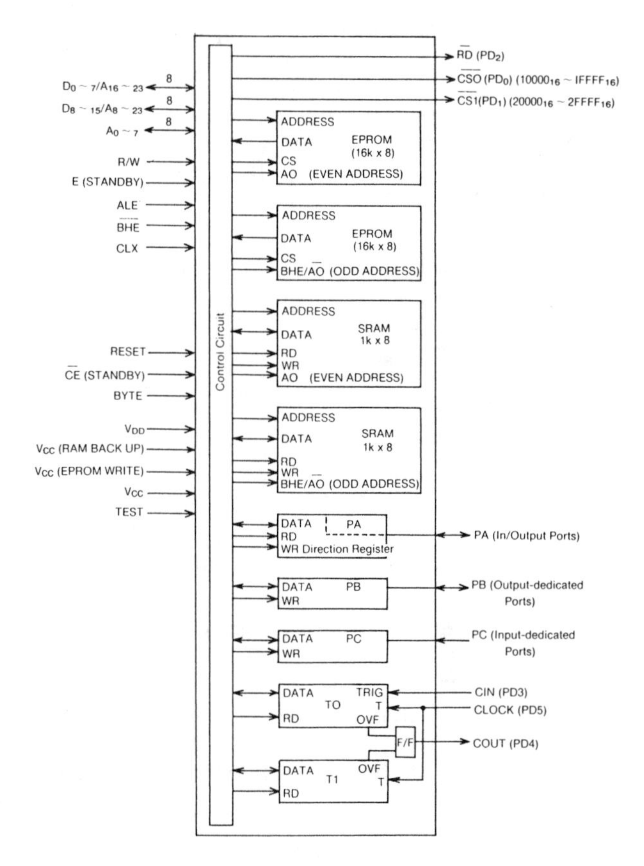

Fig. 16 M6M72561J (ASIC Memory) Functional Block Diagram

7. BENCHMARK TEST RESULTS

Benchmark test results of this microcomputer are shown below in contrast with those of a 68 series single chip 8-bit microcomputer.

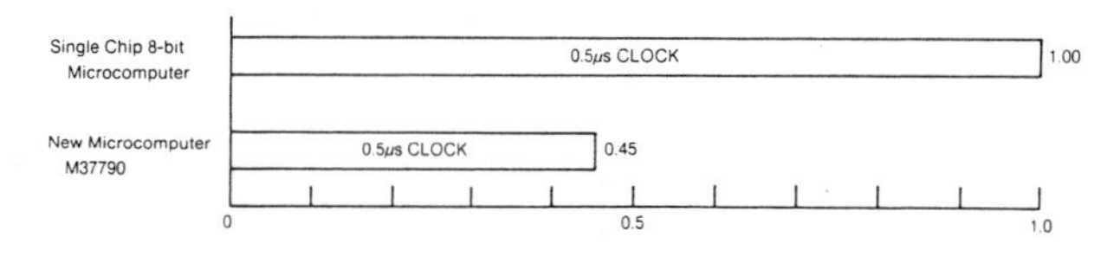

Fig. 17 Execution Time

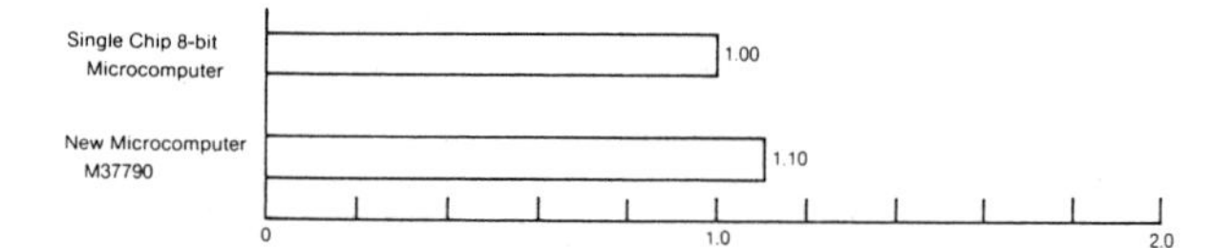

Fig. 18 Object Code Size Efficiency

8. OUTLINE OF EVALUATION SYSTEM

Presented below is an outline of the system with which control logics have been developed and evaluated for developing this microcomputer and assessing its performance.

(1) MICROCOMPUTER DEVELOPMENT – Figure 19 shows the block diagram of an engine controller for which this microcomputer has been employed. The engine to which the controller has been applied is a 4-cylinder, full sequential control, multi-point injection type for which a hot wire air flowmeter (L-Jetronic) has been employed.

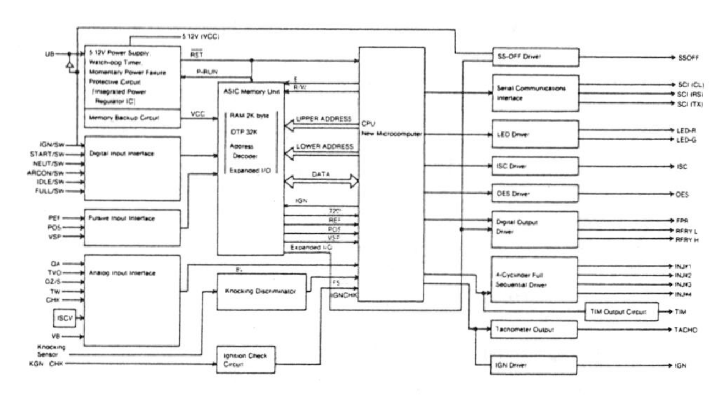

Fig. 19 Engine Controller Block Diagram

The computational processing capabilities of this controller have been measured and its control logics developed, both with the equipment shown in Fig. 20.

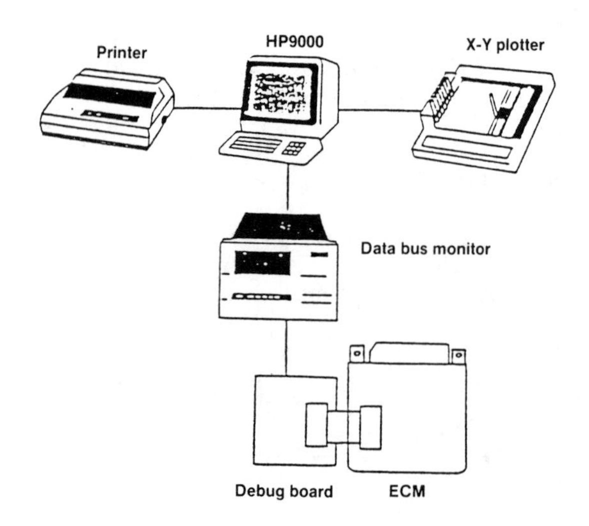

Fig. 20 Program Evaluation System

(2) EVALUATION OF NEW CONTROL LOGICS – Performance parameters of this microcomputer have been verified by designing a system that incorporates enhanced sequential control logics achieved with the controller presented in the previous section. The assessment system used for this purpose is shown in Fig. 21.

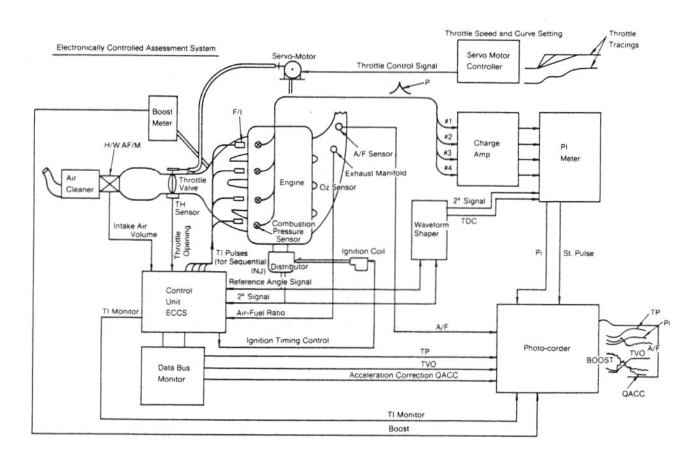

Fig. 21 Control Logics Assessment System

9. ADVANTAGES OFFERED BY THIS MICROCOMPUTER

The results of evaluating this microcomputer are presented below, as assessed with the evaluation/assessment systems discussed earlier.

(1) MICROCOMPUTER PERFORMANCE – A comparative study has been made as shown in Fig. 22 and 23. engine controller (in Fig. 19) against a nearly identical system built with another conventional microcomputer (68 series single chip CMOS 8-bit microcomputer).

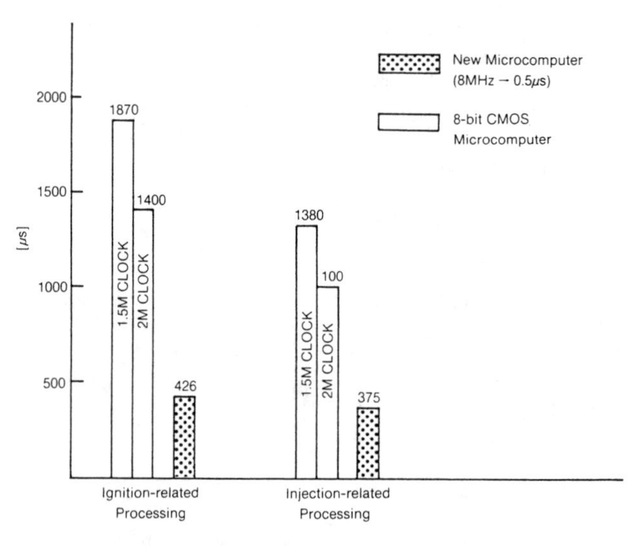

Fig. 22 Comparative Execution Processing Times

The above performance reveals that the job processing efficiency of the new microcomputer has been enhanced 2- to 5-fold over that of the 68 series single chip CMOS microcomputer.

This has been brought about by improvements in the core CPU itself (with a higher operating speed and further enriched command organization), and a greatly reduced software interrupt processing overhead, just to cite salient factors. The enhancement achieved has made it possible to introduce optimum learning and other new control techniques, and the results have been worthy of a next generation microcomputer, fulfilling our original expectations.

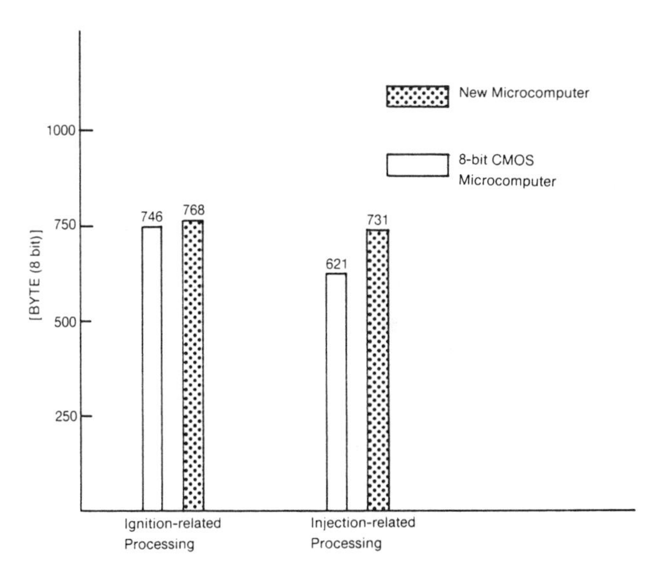

Fig. 23 Comparative Memory (ROM) Byte Count Requirements

A comparative study of byte counts required for the memory when generating nearly identical programs indicates that the number of bytes required for the new microcomputer has been slightly higher than that for the 68 series single chip 8-bit CMOS microcomputer. This is because more processing is required of the full 16-bit data for surplus bit disposal where 8-bit data suffices. Further enhancements to the program structure are expected to resolve this problem in the future.

(2) EVALUATION OF NEW CONTROL LOGICS – To take advantage of the enhanced program processing efficiency and to evaluate real-time control capabilities of the multifunctional hardware timer module, engine performance has been assessed with the sequential fuel injection system discussed earlier and combined with intake air volume projecting control.

The assessment has been made relative to rapid acceleration patterns out of a normal operation mode, on the basis of the combustion pressure pickup behavior in individual cylinders, air-fuel ratio fluctuations, and other criteria.

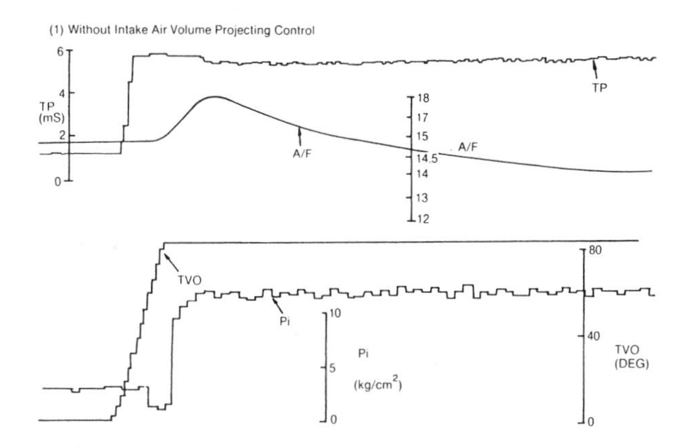

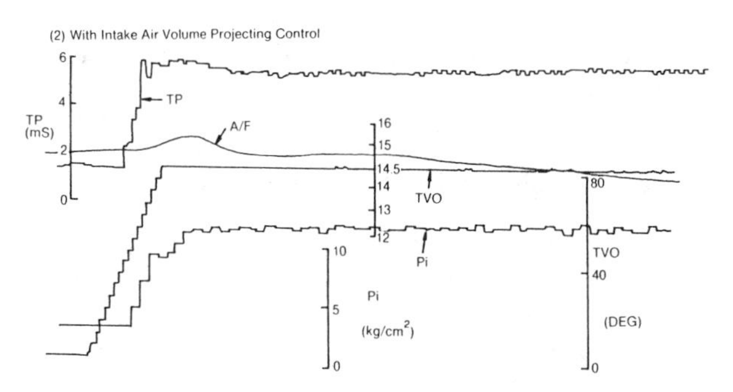

Fig. 24 Advantages Offered by Improved Sequential Control System

(at W.O.T. accelerations out of an N = 1,600 rpm mode by stepping on the throttle for 150ms)

(1) Without Intake Air Volume Projecting Control
(2) With Intake Air Volume Projecting Control

During rapid accelerations under conventional sequential control, the combustion pressure (PI) has dropped down causing flameouts due to the generation of response delays in the early stages of the acceleration in the control system as a whole. These response delays have been brought about by delays in the programmed computation and sequential timer software interrupt overhead processing, errors through failure to detect the aggregate intake air volume of injected cylinders, and many other factors. The response delays have also aggravated air-fuel ratio fluctuations, and have invited greater vehicular acceleration pickup shocks and degraded emissions, through amplified combustion pressure level differences among the cylinders.

Under the control by this microcomputer, real-time control responses have been vastly improved by the hardware sequential timers, and the aspect, in conjunction with introduction of intake air volume projecting control made possible by the enhanced computational efficiency, has permitted the combustion pressure (PI) to climb smoothly along with the intake air volume pickup in individual cylinders, without any flameouts in the process as revealed in Fig. 24.

Further enhanced control techniques, the optimum learning control introduction, and other future improvements are expected to hold air-fuel ratio fluctuations at an extremely low level, and will make cost savings of the system as a whole possible, such as by the reduced use of precious metals for catalytic purposes, for example.

10. SUMMARY AND CONCLUSIONS

In this paper, future trends of the electronic engine control system have been reviewed, and a presentation has been made of the microcomputer system developed to serve as a next generation automotive engine controller.

By employing the developed engine control microcomputer, we were able to achieve the following results:

[1] By developing a multifunctional hardware timer module dedicated to the engine control I/O processing, the software interrupt processing overhead burden has been successfully and substantially reduced, and highly sophisticated real-time control achieved at the same time.

[2] By virtue of the thus earned feasibility to have the CPU concentrate on computations, its computational processing capabilities have been enhanced, while for the software development, the generation has been made possible of a group of easily maintained programs.

[3] Reviews have been made for the optimum sharing of functions between the CPU and memory devices, and minimal costs successfully achieved with a dual-chip configuration for the highest chip area utilizing efficiency.

[4] The enhancement achieved in computational processing has permitted the latest control theories to be introduced, to further improve engine control capabilities.

The engine control microcomputer that we have developed has made it possible to build a dual-chip system that easily accommodates even 8-cylinder or other large-scale engines.

With this microcomputer as a stepping stone, we intend to continue our efforts towards a single-chip ASIC microcomputer in which the microcomputer and ASIC memory devices are consolidated.

BIBLIOGRAPHY

(1) Yoshitaka Hata: SAE Paper 860592,
New Trends in Electronic Engine Control – To the Next Stage.
(2) Naoki Tomisawa: SAE Paper 860594,
Development of a High-Speed, High-Precision Learning Control System for Engine Control.
(3) Masaharu Hayakawa:"Interface" Magazine,
Software Techniques to Fully Exploit the Functions of a Single-Chip Microcomputer.

881138

Two Functions, One Microcontroller Four-Wheel ABS and Ride Control Using 80C196KB

Steven McIntyre
Intel Corporation

INTRODUCTION

Since the beginning of the 1980's, electronics have played a crucial role in the advancement of the automobile. Innovations such as electronic sequential fuel injection, distributor-less ignition, electronic cruise control, four-wheel steering, active suspension control, four-wheel anti-lock brakes, traction control, transmission control, and even electronic air foil control are becoming more and more popular automotive features.

With these introductions comes a basic module integration problem - the CPU horsepower to combine such things as engine and transmission control, or suspension and anti-lock braking control in the same module. Combining of module functions provides numerous advantages; it would cut the number of wires to and from vehicle resources, reduce the need for inter-module communications, reduce duplication of processing functions, and improve overall performance of the vehicle as a whole.

This paper considers the combination of four-wheel anti-lock brakes and dual rate damping shock ride control. Intel's 12 MHz 80C196KB easily supports multiple module functions. It has the CPU power to perform the calculations for anti-lock brakes (ABS), the feature set required for measuring wheel speeds, and the I/O to support suspension systems.

Higher performance CPUs are needed mostly for single-chip centralized processing. Advocating of centralized processing or parallel processing is not the intent of this paper. If centralized processing is the preference, however, Intel's 12 MHz 80C196KB has the CPU power to handle the job.

ABS SYSTEMS

Anti-lock braking systems (ABS) are designed to prevent wheel lock-up. ABS, regardless of road conditions, improves vehicle response in the following ways:

1) increased directional stability helps prevent skidding
2) improved avoidance maneuvering during braking (3/4 wheel systems only)
3) stopping distance is shortened when wheel lock up is eliminated.

Electronically controlled anti-lock braking is not new. A primitive version was available in 1969 on the Lincoln Continental Mark III. It was not until 1978, when ABS was introduced into passenger cars, that ABS became a household word. Since that time, many automotive module manufacturers have produced ABS modules as standard and optional features on passenger cars and trucks. ABS has received wide acceptance because of the safety they provide. West German insurance companies have shown that it is possible to avoid roughly 7% of all traffic accidents involving passenger cars and lessen the consequences of another 10 to 15% by the means of ABS.[1]

ABS also extends the life of a car's tires and braking system. Because of all these benefits, the penetration of electronic anti-lock controls in U.S. automobiles will increase from 7.2% in 1991 to 50% by 1995 according to Automotive News, November 24, 1986.

Basic Components of ABS Systems

ABS is composed of three main components:

1) Wheel speed sensor(s)
2) Brake modulator(s)
3) An electronic control unit (ECU)

Wheel speed sensors gather information about the vehicle's wheel angular velocity. The ECU processes this information and sends control information to the brake modulators. The brake modulators regulate the amount of braking applied to each wheel individually. Figure 1 shows an ABS control circuit.

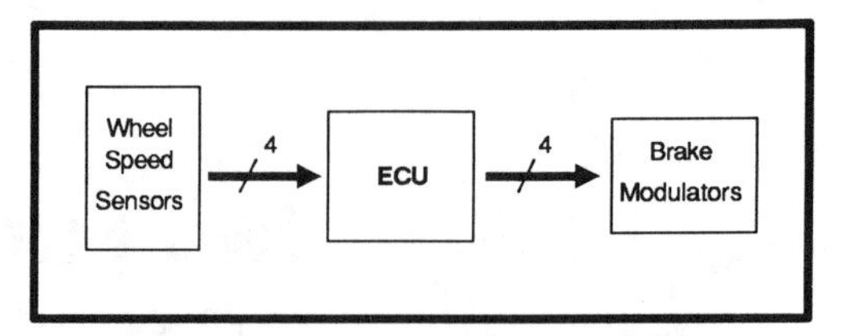

Figure 1. Four Wheel ABS Control Cicuit

The most distinguishing characteristic of an ABS is the number of channels and sensors in the system. The three most common configurations are four channel/four sensors, three channel/three sensors, and one channel/one sensor. This paper will focus on four channel/four sensor ABS. Figure 2 shows a typical ABS 4 channel/4 sensor system.

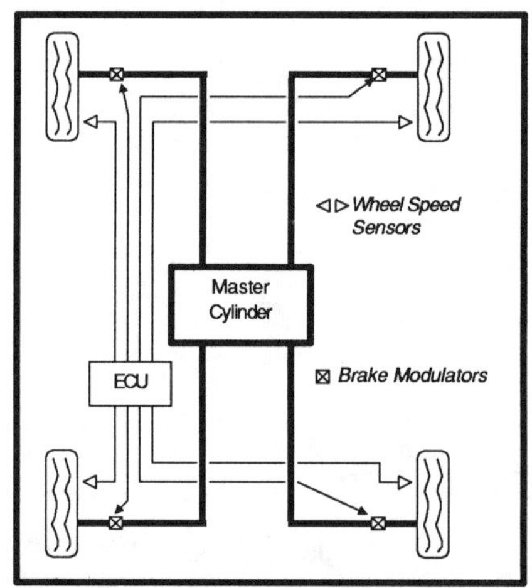

Figure 2. Four Channel/Sensor System

Close Examination of the ECU

The ECU houses all the electronics which monitor and control the ABS system. It is comprised of input and output filtering circuitry, a microcontroller (CPU), fault memory, analog-to-digital converter and safety monitoring circuitry. A block diagram is shown in Figure 3.

Input Stage The inputs to the ECU are signals from the wheel speed sensors. Systems can use from one to four wheel speed sensors.

A typical wheel speed sensor consists of a pulse wheel and a wheel speed pickup. The pulse wheel has teeth around the perimeter and is fastened to the wheel hub. When the tire rotates, the inductive wheel speed pick ups produce a sinusoidal waveform with a frequency proportional to the wheel speed. Pulse wheels typically have between 40-100 teeth and

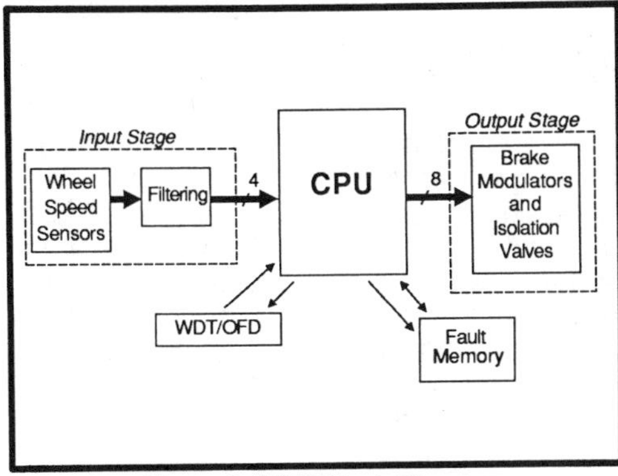

Figure 3. ECU Block Diagram

supply signals between 30-6,000 Hz.

When a vehicle is started, the ECU checks all sensors for opens and shorts. Any malfunctions are reported, and depending on the error severity, the system either continues or shuts down.

The input signals are noisy (as most signals in the vehicle are), so some form of filtering is required. When the CPU is powerful enough, it performs digital filtering. Typically additional analog filtering must be added to the ECU. Filtering techniques are fairly unique to each manufacturer, but it typically involves zero crossing detectors with some hysteresis (ranging between 100-500 millivolts). Some wheel speed sensors do the filtering and conversion in the sensor pick up. The output from the sensor is a clean 0-5 volt digital pulse waveform.

CPU Processing The CPU must perform pulse counting, or pulse width measurements on the 30-6,000 Hz to determine individual wheel speeds. The CPU uses these values to determine wheel speed, wheel retardation, reference velocities, wheel slip, and wheel acceleration and deceleration. Based on these values, outputs to the brake modulators are calculated.

All these calculations must take place in a certain period of time, referred to as "loop time". The faster the loop time, the faster the module can detect individual wheel lock ups. A loop time for a four-channel system is typically five-seven milliseconds. The number of teeth on the pulse wheel and the number of sensors directly influence the loop time and processor speed.

Each tooth produces an interrupt that the CPU must process. An estimate of the number of interrupts per second is roughly 6,000 for each sensor using a 90-100 tooth sensor, or 24,000 interrupts per second for a four channel/sensor system. Systems that perform four-wheel ABS, using a 90-100 tooth sensor, with a loop time of five milliseconds will use close to all of the five millisecond loop time for CPU processing, during extreme input conditions.

Output Stage A brake modulator controls the amount of braking pressure applied to the wheels, which is controlled by solenoid valves with a response time of three to eight milliseconds. There are different types of solenoid valves: a build pressure valve, a release pressure valve, and an isolation valve.

The output section of the ECU contains circuitry to check the brake valves for opens and shorts, as well as control outputs for brake modulators. There is also feedback from these devices which is used to monitor proper operation. Current regulators or high powered operational amplifiers are used to operate the brake valves. Pulse width modulated outputs are used to operate these valves. The output frequencies of these outputs are around 500 Hz.

Watchdog Timer (WDT) and Oscillator Failure Detect (OFD) The safety of an ABS system is critical. The systems require constant monitoring to ensure proper operation. While the CPU monitors the mechanical systems, an external watchdog timer (WDT) and oscillator failure detector (OFD) monitors the CPU and ensures its proper operation. Software drives the external WDT. If the WDT is not periodically modified, the processor will either reset, or shut down. The same is true of the OFD circuit. If the clock frequency is above/below a set frequency the ECU will shut down to conventional braking, and alert the driver of the failure. These functions could be on the CPU chip or off, but most manufacturer's prefer it to be off-chip.

Fault Memory Fault memory is used to store parameters when an error in the system occurs. This memory can later be accessed for debug, safety, and diagnostic purposes. Typically, the fault memory consists of EEPROM. This could be on the CPU chip or off. The number of read/write cycles is about 1,000, depending on the application and most systems require between 64 and 128 bytes.

Analog-to-Digital Converter An A/D converter is not an essential requirement for ABS, but it is used by manufacturers for various reasons. These include monitoring brake pressure, calibrating the system, monitoring battery voltage, and testing solenoids. No more than 8-bit resolution is required for any of these functions.

ABS Performance and System Requirements

The performance of an ABS involves two factors: 1) The actual performance of the system, including the stopping distance, steerability and the range of speed over which the system is operational; 2) The users perceived performance which includes how the brake pedal feels and the overall smoothness of the braking system. The number of channels, the type of valves, number of speed sensors, and type of controller play an important part in the overall performance of the ABS system.

ECU System Requirements The selection of the number of channels and sensors in the ABS system is an indication of the complexity of the operating algorithms and CPU power required. Additional sensors and channels mean additional calculations and longer software algorithms. For a one-channel/sensor system, an estimated 4K of ROM and 128 bytes of RAM is required. For a three- or four-channel/sensor system, 8K of ROM and 256 bytes of RAM is needed. This memory includes the diagnostic and monitoring software for checking the module operation. Some systems need 12K bytes of ROM and 320 bytes of RAM to get the job done.

The additional memory is typically used for added diagnostics and safety routines. In addition, the extra memory can be used for additional ABS calculations for a smoother brake, or even traction control. Traction control (ASR - from the german Antriebsschulpfregelung) can be done in several ways. The inverse of ABS is traction control. In a passive traction control system the engine ECU controls the throttle rather than the ABS ECU controlling the brake. The traction control software would output a digital signal to the engine ECU indicating wheel slip during acceleration. The engine control ECU will take care of the throttle control (a passive design), yielding a total system memory size of 12K bytes of ROM and 320 bytes of RAM.

To date, ABS modules contain either a 16 or 8 bit controller with the adequate processing speed to handle ABS calculations. In order to detect wheel lock-ups faster, a 16-bit processor, or an 8-bit processor with 16-bit math and timers is required. To calculate the wheel acceleration/deceleration the following equation is used.

$$W_{a/d} = (V_w - V_{w1}) / T_s \qquad \text{Eq 1}$$

The main loop time (T_s) is a major component in the calculation. The smaller the loop time, the faster wheel lock-up can be detected. V_w and V_{w1} are wheel speed velocities at two different loop intervals.

A typical loop time (T_s) for ABS/ASR is five milliseconds. This produces an adequate lock-up response and allows enough time for the processor to do all four wheel speed calculations. In fact under worst case conditions, all five milliseconds could be used for CPU processing of collected data.

An 8 MHz 80C196KB is appropriate for this five millisecond loop time requirement.

Interrupt Processing Software is set up to interrupt the processor each time a tooth on a wheel sensor is detected. The 8096 family of products has an input staging FIFO (first-in-first-out) RAM connected to a 16-bit timer that can store a history of input events (Figure 4). No other processor family (8- or 16-bit) can store a history of high speed events without impacting CPU performance and processor RAM resources. The FIFO design enables the processor to do other calculations while inputs are being captured. Up to eight events can be logged in the FIFO from four high speed inputs before an interrupt is required to dump the history of logged events. This impacts the interrupt overhead of the ABS system performance by 1/8th of systems not using the 8096 products. Later it will be shown that by using this FIFO, critical CPU processing time will be returned to the 80C196KB device.

For a four sensor system all four high speed inputs are required. The 80C196KB has a total of four high speed capture inputs, one for each wheel speed indicator.

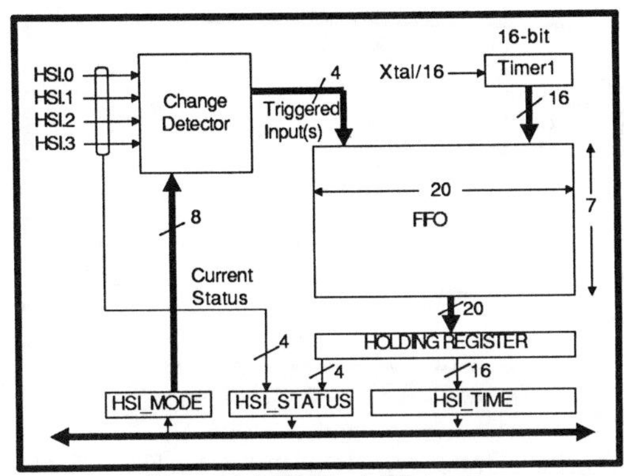

Figure 4. HSI Block Diagram

Other requirements of the CPU are 16x16 multiply instructions, 24/16 divides, as well as additions and subtractions. This math is used for all wheel/vehicle velocity/adhesion calculations. The faster the math, the more time the CPU has left to do other things.

Output requirements are also needed, but are not as critical as the inputs. Brake modulator control is done one of two ways depending on the type of brake valves used. Some require one output control, which is able only to release pressure. Other systems require two control signals; one for inlet and one for outlet.

There are some valves on the market that require only one control that do both build and release pressure. The single valves offer better performance than the two valve systems. With two-valve systems, undefined switching conditions can occur because of varying switching times between the inlet and outlet valve.

The feel of the brake pedal also varies between systems. In some systems, the driver can feel a pulsing in the brake pedal when the ABS is activated. In others, the brake pedal is isolated from the brake circuit and the driver will not feel the pulsing brake. The only effect that these mechanical differences have on the microcontroller is a change in the number of outputs required; one output per channel to perform the isolation and one or two outputs required for the brake modulation. The frequency of the output signals to the brake modulators and isolation solenoid valves are in the 500 Hz range.

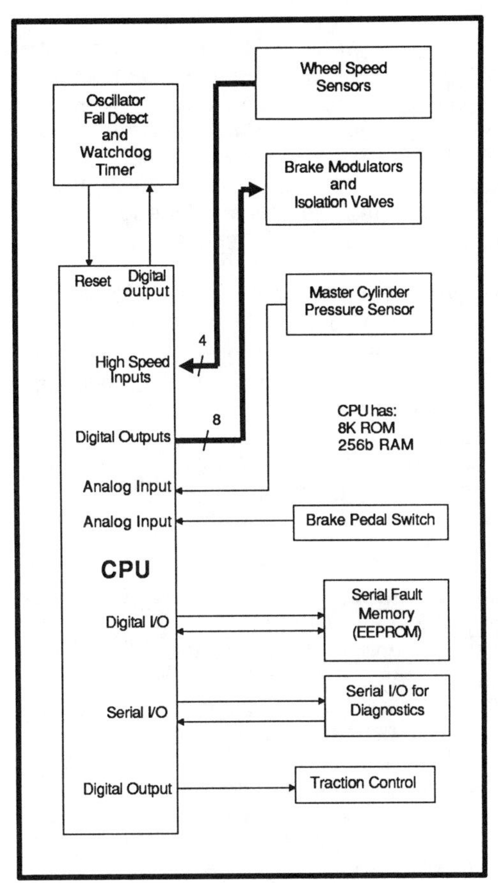

Figure 5. The ABS, ECU

Figure 5 illustrates a typical ABS system ECU block diagram. The CPU needs to have analog-to-digital converter capability, and serial I/O for diagnostic communications. Additional digital I/O (or serial I/O) is also required to communicate with the external fault (EEPROM) memory.

RIDE CONTROL SYSTEMS

Ride control systems are suspension systems which optimize both maneuverability and ride comfort without compromising either one. Suspension systems range from dual rate damping and variable rate damping to a completely active (hydraulic) system. Dual rate damping systems a fast acting rotary solenoid on the shock absorber which modifies the damping rate from either a "soft" comfortable rate or a "firm" rate for improved handling. Variable rate systems modify the shocks damping rate variably from "soft" to "firm" depending on road conditions and vehicle body fluctuations.

Some ride control modules are "manual dial" and others are automatic. Manual suspension systems are driver invoked. The driver simply "dials" in the ride desired, while more sophisticated systems are fully automatic. Each type of system has its place depending on cost, safety, and overall response.

The suspension system in the following example is a dual rate damping shock absorber with load leveling. The damping settings are selected by the CPU module, based on wheel turning position and rate, braking, acceleration, and road undulations. Load leveling is added to the system and requires little CPU power, but the additional functions require more I/O.

Basic Components of an Adjustable Ride Control System (ARCs)

The components that make up the ARC system are adjustable shocks/struts, rotary actuators, position sensors, a steering sensor, and an electronic control module (ECU). Figure 6 shows a conceptual diagram of the system.

Adjustable shock/struts The rear adjustable shock absorbers and front adjustable shock struts are a special design. Firm damping is tuned by a conventional piston assembly, while soft damping is obtained by opening a rotary valve and allowing oil to bypass the piston valving. The ratio of something which determines firm to soft rides varies, but is nominally a 3:1 ratio. Each shock also contains an air-bag assembly for load leveling. The air-bags replaces the need of coils or leaf springs. The air-bags require a compressor and vent input for air pressure control.

Actuators The actuators are rotary solenoids traveling about 60°. They have a response time of 30 mS against a 200 g-cm load and a 6 Amp stall current. Two signals from the ECU are required. One for the actuator setting and one to monitor the setting.

Position sensors The position sensors, or height sensors are two output sensors indicating a "too high", "too low", or "trim" height adjustment. The processor senses the digital signal and determines load leveling adjustments. If the sensors are read very rapidly, road conditions (undulations) can be determined.

Adjustments to the damping during such undulations can aid in the control and comfort of the vehicle.

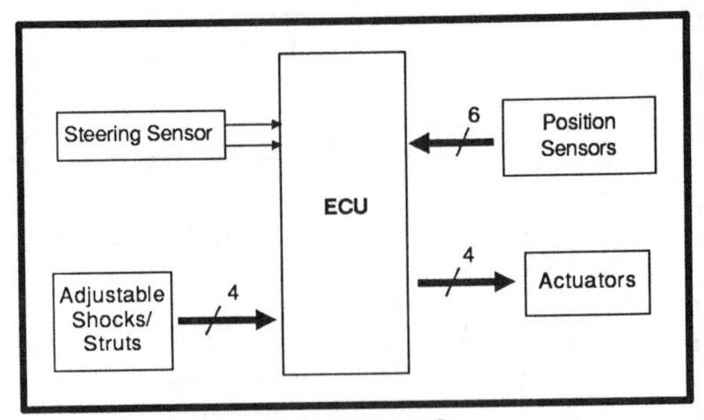

Figure 6. ARC System Components

Steering Sensors The steering sensor is mounted to the steering column. It consists of two LED/photo-diode sensing pairs that pass over a perforated disc. The disc has 20 holes with spacing every 18° of the circumference. Two outputs contain information about the position and direction of the wheel. The output signal is shown in figure 7.

ECU and other ECU inputs The ECU also monitors master cylinder brake pressure, brake pedal pressure switch, as well as accelerations. These inputs are analog inputs with a 0-5 volt range (after scaling). Vehicle speed is required for the system to compare to steering. The ECU also has outputs for WDT and OFD similar to the ABS module.

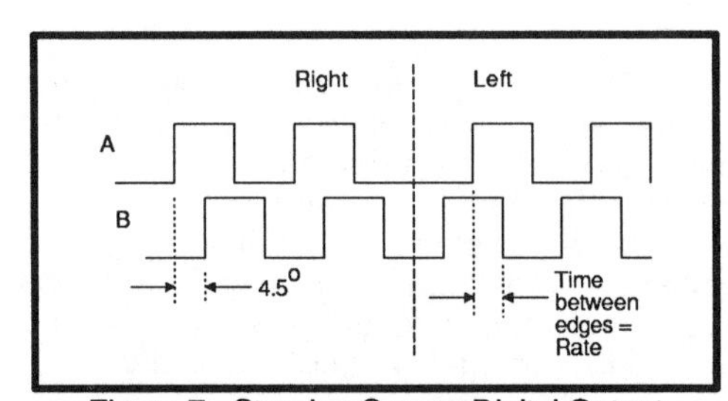

Figure 7. Steering Sensor Digital Output

A Closer Look at the ECU Module for ARC

Monitoring such things as steering, brake switch, vehicle accelerations, position sensors, feedback error detection, and vehicle speed are simple for even an 8-bit microcontroller. All that is basically required is low speed I/O. A conceptual diagram of the system being described is shown in figure 8.

Vehicle Speed and Acceleration Vehicle speed comes from the engine controller module. It's frequency ranges from 0 to 300 Hz, roughly corresponding to two times the vehicle speed in miles per hour (300 Hz = 150 mph). The acceleration input comes from the engine control module in a signal that indicates throttle position. It is digital in nature, but some applications use an analog sensor direct from the throttle. The microcontroller can use either an analog input used or a digital input.

Steering Sensor As shown in figure 6, the steering sensor is a dual signal output, referred to as a quadrature waveform. If signal A's rising edge precedes signal B's rising edge, then the direction is right. If signal B's rising edge precedes A, the direction is left. Directional changes occur when one signal has two edges before the other signal has one. Rate is determined by the time between edges.

Absolute steering position is relative. The center of steering is found mathematically, through binary searching over time. Each edge is 4.5^{o} apart. Within 80 seconds of vehicle operation (operation is indicated by speed >18 mph) the center of steering is determined. Starting at $\pm 72^{o}$ and ending at 4.5^{o} a binary search is performed. If the steering passes through this region the assumed center is updated and searching continues using the new assumed center. After the center is found, it is constantly updated based on time at a given point on the disc.

Shocks and Actuators Signals to the actuators and from the shocks are digital. The actuator signals are a logical "one" when activated (firm) and a logical "zero" when off (soft). The default condition is off (soft). The shock signals are monitored to indicate the current position of the actuators. This is needed to maintain control during critical maneuvers.

Brake Pressure Switch The brake pressure switch is an analog signal. This means that either discrete external hardware be placed in the ECU module to determine an analog trip voltage, or the CPU must contain an A/D converter.

Error Detection Safety is always the highest concern of any vehicular electronic module. An ARC system is no exception. It requires constant monitoring to ensure proper operation. An external WDT and OFD is used to monitor the CPU in the system. Its functions are similar to that of ABS error detection.

Position (Height) Sensors The height sensors output a 2-bit code indicating the condition of the sensor on two digital output lines. A digital "11" translates to a "trim" condition, "01" means "too high",

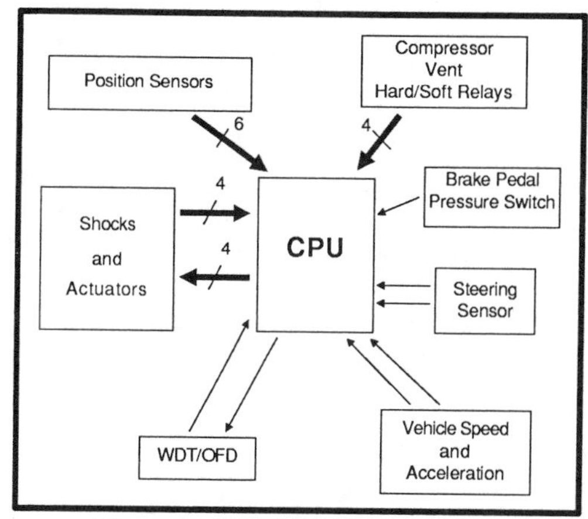

Figure 8. ARC-ECU Block Diagram

while "10" means "too low". The "00" code is reserved for the illegal condition and will indicate that monitoring of the height sensors is no longer valid. The module will not be shut down, but undulation and load leveling will be disabled.

CPU Performance Requirements for ARC

Performance of the ARC system is not as critical as an ABS system. In fact, ABS would be of higher priority if the two systems were combined. The overall loop time of such a function could be about one half the response time of the actuators, or about 15 mS. With the CPU power of the 80C196KB, a 5 MHz input clock can be used to perform the suspension control function. Under worst case conditions about six milliseconds are used for CPU processing (using a 5 MHz clock).

The software for an ARC function is designed to be a FOREGROUND/BACKGROUND program. Although the actuators are updated every 15 mS in the BACKGROUND, computations and system monitors are continuously performed in the FOREGROUND. A two loop flow chart of the software for this ARC design is shown in figure 9.

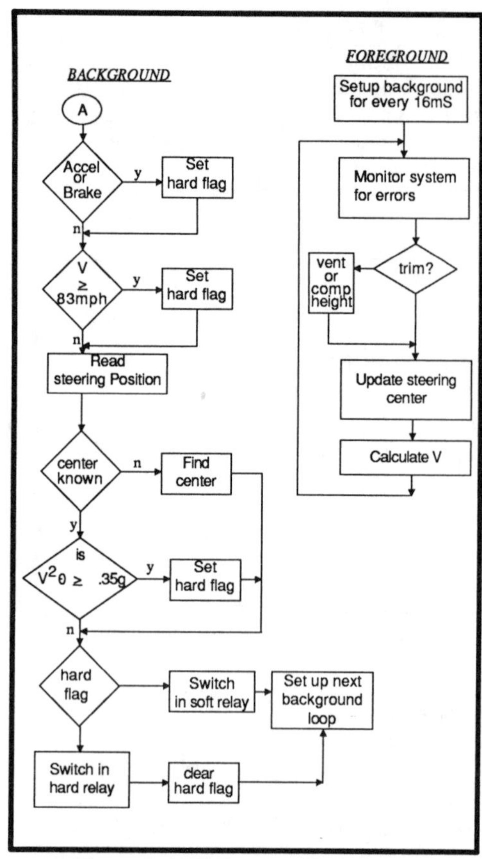

Figure 9. ARC Program Flow Chart

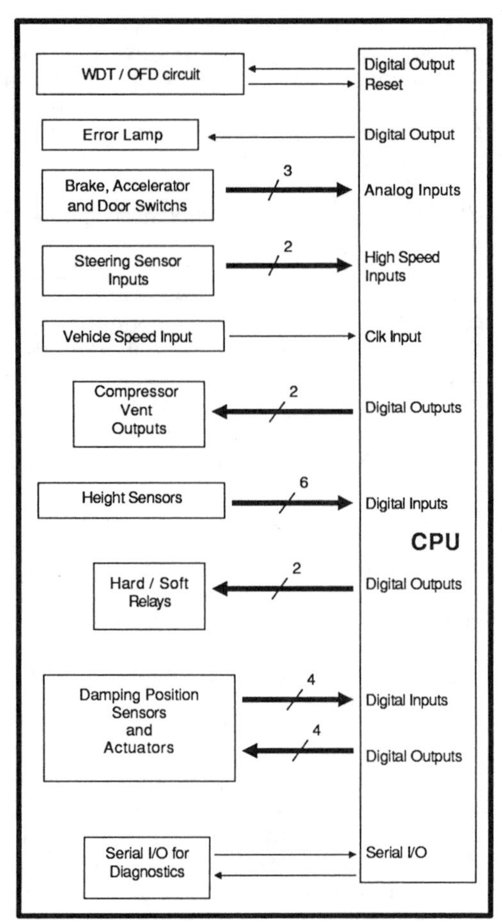

Figure 10. The ARC w/ Load Leveling

ECU System Configurations An 8- or 16-bit microcontroller can be used. The CPU performance is not critical, but I/O is. A total of 24 I/O lines (both digital and analog) are required. Four to 8K bytes of ROM with 128 to 256 bytes of RAM are required depending on the complexity of the diagnostic and calibration code.

If an 80C196KB device is used to perform the ARC function, an input frequency as low as 5 MHz can be used. As this frequency indicates, the CPU performance requirement is very low.

I/O requirements are different from CPU requirements, however. In fact, three analog inputs are required for accelerator pressure sensing, brake pressure sensing, and an analog door sensor (used in a load leveling system to indicate a possible change in load). Three high speed inputs, with time capture capabilities are used for vehicle speed and steering sensing, ten digital inputs are used for height sensing and damping position sensing, and eight digital outputs are used for actuators, firm relay, soft relay, compressor, and vent switches. Figure 10 shows the complete ARC system with load leveling capabilities.

COMBINING ABS AND ARC

Keeping in mind that the combined system should be able to accomplish both four-wheel ABS with traction control and adjustable ride control (ARC) with load leveling, several things can be noted. For example, several functions of each system can be shared, if the two functions are combined. ARC requires a reference vehicle speed. This reference speed is already found in the ABS/ASR portion. The reference vehicle speed is found by first determining if the vehicle is in acceleration or deceleration. Then either the lowest or highest individual wheel speed value is used for the reference vehicle velocity. Since this value is used in the ABS calculations the need for one high speed input and ROM/RAM memory to manipulate the vehicle speed for ARC can be eliminated.

The brake pressure switch is also required for both systems, as well as an external error detection circuits and an error lamp. These can be combined in one module.

The ARC portion of the module produces such things as steering angle and rate. Although the steering angle is not used in ABS systems, it would be used as extra data in the anti-lock calculations.

A Closer Look at the ECU

The actual module schematic for the combined ABS/ASR plus ARC with load leveling is shown in figure 11.

ABS The ABS system in the combined module is identical to the one given in the standalone system. The I/O required to do ABS is four high speed inputs for individual wheel speed detection, four outputs for modulator isolation valves, four outputs for decay/build brake modulation valves, a brake pressure switch input (analog in nature and also used for ARC), and an analog input for monitoring master cylinder brake pressure.

Traction control uses a single output for throttle control. Although traction control can be performed in several ways, this design is a passive one. It simply sends a signal indicating wheel slip during accelerations to the engine ECU. In turn the engine ECU will take the appropriate action.

An EEPROM is used for error detection and two digital I/O lines are required for this communication. The serial port of the 80C196KB is used for module diagnostic support.

Total memory used for the entire ABS/ASR portion of the module is about 8K of program code ROM and 256 bytes of register RAM. This also includes code to monitor the systems performance and code for module diagnostics via the serial port.

The diagnostic code was modified for ROM and RAM savings. Diagnostics is done via the serial port. The diagnostic intelligence was placed outside of the CPU. By moving the intelligence out of the CPU chip, about 4K of program ROM and 128 bytes of register RAM was no longer required.

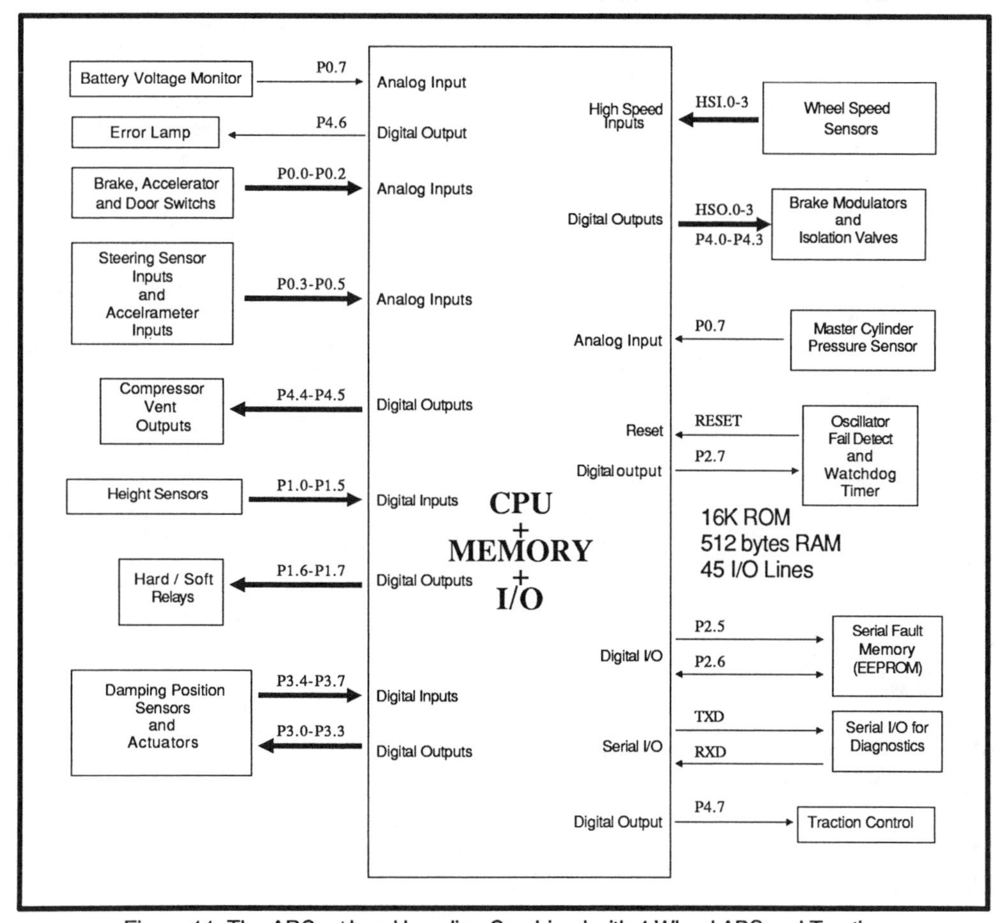

Figure 11. The ARC w/ Load Leveling Combined with 4 Wheel ABS and Traction

ARC The adjustable ride control design used in the combined system has been modified slightly. Most of the modifications were in the steering sensors. Rather than using the quadrature approach to steering sense, an analog sensor is mounted to the rack. The sensor sends a voltage corresponding to the actual wheel angle instead of an angle from the steering wheel. Using this method will produce tighter steering angles. The smallest increment of steering angle using the slotted disc approach is 4.5^o. The analog sensor resolution varies with sensors, but with the 10-bit analog-to-digital (A/D) converter on the 80C196KB, and a $0\text{-}180^o$ rack swing potential, angle increments of 0.2^o can be detected. The sensor requires only the use of one of the device's analog inputs.

Other features include hard and soft relay control, four outputs for shock actuator control, four inputs for damping switch monitoring, six inputs for height sensing, a compressor switch control, a vent control, a door sensor switch, an analog brake pedal pressure switch (shared with ABS control), an analog accelerator pedal pressure switch monitor, and two analog inputs for an accelerometer.

The accelerometers are added to the system in two positions. One is mounted to determine body roll, and the other is mounted to determine vehicle pitch. The two measurements are used in the calculations for shock damping.

Combined Functions Several resources can be shared, when the two functions are combined. The brake pedal pressure switch, wheel velocities, wheel angles, diagnostic routines, and failure detect circuitry.

The memory resources are also shared. The routines that calculate vehicle speed, diagnostics, failure detection, and brake pressure sensing all take memory (ROM/RAM) resources.

The combined systems save about 20% of the memory (both register RAM and program ROM) used by the systems individually. The total memory requirements for the ARC control is 6K of program ROM and 128 bytes of register RAM.

Another shared feature includes external EEPROM communication. The external EEPROM was used for logging errors in the system. It only requires two low speed I/O lines for communication. One for the synchronous clock output and one for data I/O. Most serial EEPROMs tri-state the data lines and only one I/O line can be used.

Performance Requirements

The ABS/ASR portion of the combined module demands the most performance. When calculating wheel acceleration/deceleration (eq 1), the "main loop" time (T_S) is a major factor in the calculations. The tighter the loop time, the faster a lock-up can be determined. In the combined system example a loop time of five milliseconds is used. V_W is the wheel velocity at present, while V_{W1} is the wheel velocity during the previous loop. These calculations are done for each wheel being sensed prior to taking any channel action for anti-lock output commands.

The equations for V_W and V_{W1} are:

$$V_W = TC / (FTIME\text{-}ITIME) * MPH1 \qquad \text{Eq 2}$$

The wheel speed velocity (V_W) is a function of the number of negative transitions (TC) in a given time base (FTIME-ITIME). MPH1 is a constant value representing the output frequency generated by the sensor at 1 mph. The output frequency is a function the rolling tire's radius and the number of teeth per pulse wheel.

Wheel Speed Sensing Wheel speed was another area of performance savings using the 80C196KB. Each negative edge is time stamped and recorded by the CPU. In most CPU architectures, each time a negative transition edge occurs an interrupt is generated by the CPU. The time it takes to process that interrupt may be very small (about 60-70 uS using an 8 MHz clock) but more than makes up for in the quantity of interrupts per second. With a wheel speed of 150 mph, using between 90 and 100 tooth pulse wheel, and generating a 6,000 Hz signal, a four-wheel speed inputs would generate 24,000 interrupts per second, 120 every 5 millisecond loop. Figure 12 shows the wheel speed sensing interrupt service routine.

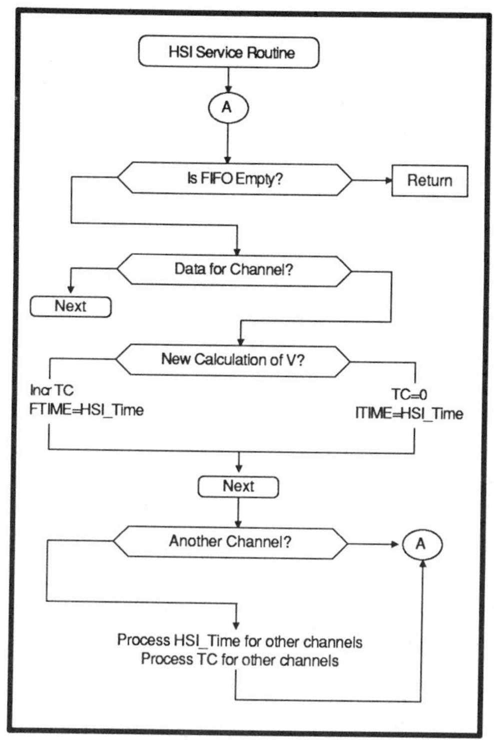

Figure 12. High Speed Input Interrupt Service Routine

The 80C196KB has the capability to store eight high speed event times in a first-in first-out (FIFO) buffer. The interrupt structure associated with the FIFO has the ability to interrupt on every event, every four events or when full. Most code written using high speed inputs interrupt when any event is logged in the FIFO. This is a waste of 80C196KB resources. When the FIFO is completely full, the processing time required to empty its contents takes 70 uS (using 8 MHz clock). The frequency of the inputs are considerably less than once every 70 uS per input. Therefore, a maximum of four FIFO entries are possible at any one time. Interrupting the processor when the FIFO is full would not leave any room for storing any further. If the processor is programmed to interrupt every four events the interrupt service routine would occur four times less. The interrupt overhead (this is the time required to enter and exit the service routine) saved would be substantial. The FIFO would have room to data log four more events.

Six microseconds, of the 70 uS used to empty the FIFO, is used for interrupt service routine overhead. Of the five milliseconds loop time, 720 uS are used for interrupt overhead using the "interrupt every event" technique. If the "interrupting every 4th event" technique is used, only 180 uS is for interrupt overhead. 540 uS is saved for other processing. Increasing the clock frequency to 12 MHz would reduce the overhead to 120 uS. For this reason, the FIFO feature on the 8096 family of products is perfect for ABS/ASR applications.

Wheel position sensing Wheel turning position is read directly from an analog sensor and converted to a digital number by the 80C196KB's A/D. The converted analog value corresponds to the wheel angle (0-180°). If the A/D conversion is performed with respect to time it is possible to find the rate of change of the wheel. The center position of the wheel is calculated using the binary search method and updated constantly over time. The only difference is the data collection method (analog versus quadrature input).

The Working Code Example

The main loop of ABS/ASR is the time required to do all slip/adhesion calculations, analyze the system, and output action to valves. In a standalone ABS/ASR system an eight MHz 80C196KB would be adequate. In the combined ABS/ASR and ARC system a 12 MHz 80C196KB is used.

With a 12 MHz 80C196KB and using the interrupt techniques described above, the 5 mS loop time for code processing is reduced to 2.75 milliseconds. The saved 2.25 milliseconds are spared for other CPU processing.

In the ARC system background program the loop time is 16 mS. The clock frequency is five MHz. But only six milliseconds (at 5 MHz) are used for CPU processing under worst case conditions. Increasing the processor input clock frequency from five to 12 MHz reduces the CPU processing time from six to 2.5 milliseconds. The old method of steering angle detection took a considerable amount of time for decoding of high speed input channels and processing rates. Wheel angle measurements are more direct with the analog sensor approach. The analog method saved about 10% of the overall loop time, thus reducing the loop time to 2.25 milliseconds.

The ARC system flow chart shown in figure 9 still holds true. Some slight modifications (mostly time saving techniques used in reading and maintaining steering positions) are made to meet the 2.25 mS time requirement.

A 12 MHz version of the 80C196KB with expanded memory capability would meet all the system requirements of the combined system in a single chip solution. The total system memory requirements were 14K of ROM and 384 bytes of RAM. This includes diagnostic code and code required to monitor module performance.

CONCLUSIONS

Combining four-wheel ABS and traction control with dual rate damping shock suspension and load leveling can be accomplished with a single module microcontroller, Intel's 80C196KB. A 12 MHz clock provides the extra horsepower for centralized processing of these functions.

The high speed input FIFO is a unique feature of the 8096 family of products. It proves itself as a definite advantage over systems that do not use the 8096 products. Its interrupt structure yields a CPU process savings of 10.8% over conventional high speed capture structures. This feature makes the 8096 family a perfect match for ABS/ASR. It provides the high speed input capture with low CPU overhead in interrupt processing.

With the integration of features on automotive modules, such as ABS/ASR and ARC with load leveling, the need for more program and data memory became apparent. This lends itself to a proliferation path of the 8096 family into higher amounts of on-chip memory such as RAM and ROM.

REFERENCES

Bosch Automotive Handbook; 2nd Edition; "Suspensions"; "Braking Equipment"; SAE Isbn 0-89883-518-6

Reinecke, E.; "An Anti-lock System with Extended Safety and Control System Functions"; Int. J. of Vehicle Design, vol6, nos 4/5, pp 561-566; 1985

Petersen, E. and Quicke, K.; "New Anti-lock Systems for Commercial Vehicles Realized with Single-Chip Microcomputers"; Wabco Westinghouse GmbH, Hanover; IMechE Paper #C205/81

Maisch, W. and Schramm, H.; "Further Development of the Anti-lock Braking Systems for Commercial Vehicles with Compressed Air Brakes"; IMechE Paper #C191/85

Klein, Hans-Christof; "Anti-lock Brake Systems for Passenger Cars, State of the Art 1985"; SAE Paper #865139

Gerstenmeier, Jurgen; "ABS Electronics, Current Status and Future Prospects"; IMechE Paper #C239/85; "Electronic Control Unit for Passenger Car Anti-skid (ABS)"; IMechE Paper #C186/81; "Traction Control (ASR)- An Extension of the Anti-lock Braking Systems (ABS)"; SAE Paper #861033

Soltis, Micheal W.; "Programmed Suspension Provides Ride Control"; Automotive Engineering; February 1987

Hirose, Maanori; Matsushige, Seiichi; Buma, Shuichi; Kamiya, Kohji; "Toyota Electronic Modulated Air Suspension System for the 1986 Soarer"; IEEE 1986

Kerastis, Micheal W.; Lizell, Magnus B.; Guy, Yoram; "Advances in Electronic Suspensions"; SAE paper 1986

Hussain, Seyd; "The Application of High Speed Integrated Digital Microcontrollers in Modern Anti-Lock Systems"; SAE paper 1985

885106

Anti-Spin Control

P. R. Crossley

ABSTRACT

The increasing number of individual control microprocessors in modern vehicles provides a new opportunity to automotive engineers to develop new control systems relying on interaction between these individual systems. Anti-spin control is one such new control system which relies on interaction between the powertrain and anti-lock braking microprocessors.
Anti spin control (or traction control) is aimed at achieving improvements in vehicle stability by the control of wheel spin on low or split coefficient of friction surfaces whilst accelerating. The system is analagous to the operation of anti-lock brakes during braking.
As the system controls dynamic powertrain torque a system computer model has been produced and used to develop the control system. This paper describes the analysis and development of this control system

THE WIDESPREAD INTRODUCTION of electronic control systems into road vehicles has generally been to provide an individual feature or meet an individual need. Electronic ignition has provided a more accurate, repeatable and reliable source of ignition spark timing, electronic ABS has enabled close to maximum braking forces with reliable stability regardless of driver ability. These systems, when grouped together in a vehicle can provide greater than their sum of individual features if they are allowed to interact. This is well understood for engine and transmission electronics where interactions allow the control to work as a powertrain system and so realize benefits in fuel economy, driveability and emissions (1), (2)*. A logical extension of this approach is to enable interaction between powertrain and chassis electronics to allow new control features and take the first step towards total vehicle control. Traction control is one of these first steps.

* Numbers in parentheses designate references at end of paper.

OBJECTIVES

Traction control can be considered analagous to antilock braking systems (ABS) but operating in the opposite sense i.e. the ideal system will provide maximum traction regardless of conditions and always maintain vehicle stability and driver control.

Figure 1 shows two typical longitudinal force versus percentage slip relationships for

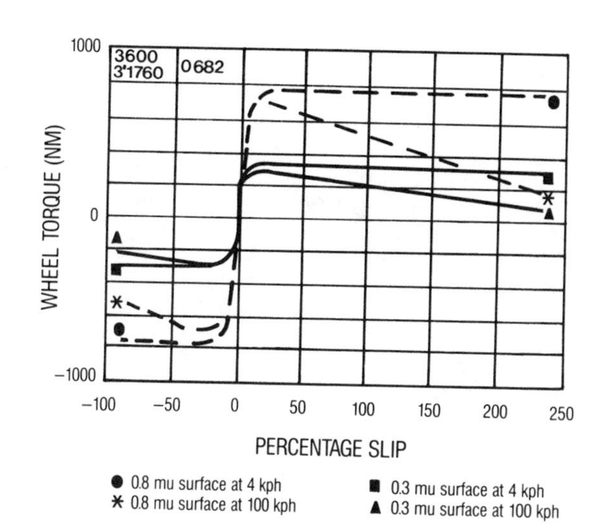

Figure 1 Longitudinal force (expressed as the torque on the wheel) as a function of tyre slip for various operating points (5).

a modern road tyre at two road speeds. The force generated increases rapidly with tyre slip until a maximum is reached, typically between 8-20%. The force then decreases with percentage slip, the final point yielding a reduced force. In practice without electronic control, the latter part of the curve is not usable because as the slip relating to the maximum force is exceeded the wheel rapidly becomes locked (in the case of braking) due to the negative slope.

Figure 2 shows a typical relationship between lateral and longitudinal force for a modern road tyre. This indicates that at maximum longitudinal force the lateral force available for steering and vehicle stability is already severely reduced. By the point of lock up or rapid spin shown by the extreme ends of the curve, the lateral force is so low that loss of steering and vehicle instability can easily occur.

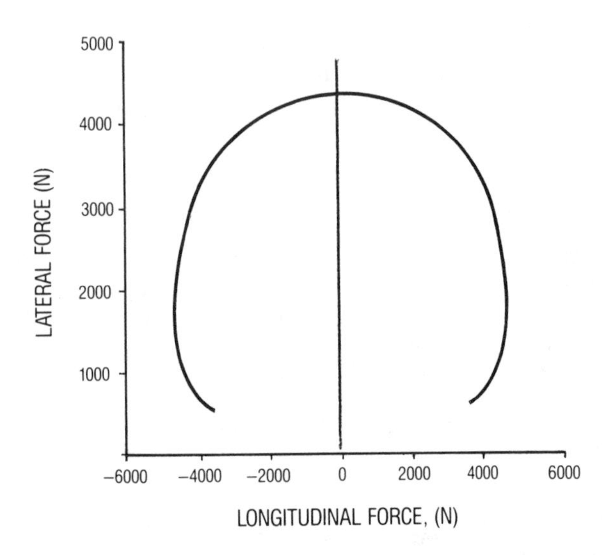

Figure 2 A typical longitudinal vs lateral force curve for an automobile tyre.

Traction control, like ABS, is aimed at maintaining near to peak longitudinal force while ensuring the lateral force available is not reduced to critical levels.

The characteristics shown are typical of certain surfaces, but during development of traction control systems it is essential to allow for other surfaces where the relationship between traction and stability are quite different.

This paper is specifically concerned with the development of a possible traction control system. Various approaches to the problem are feasible. The reduction of the driven wheel torque can be achieved by:-

(1) Throttle valve control for SI engines
(2) Fuel flow control
(3) Spark advance control for SI engines
(4) Brake actuation or
(5) a combination of the above.

In addition the distribution of drive torque to the driven wheels can be achieved by either:-

(1) Brake actuation to individual wheels or
(2) The uses of a limited slip differential.

The scheme addressed in this particular work uses a combination of throttle and brake intervention on a rear wheel drive vehicle. An electronically controlled throttle valve is used for engine torque modulation. The brake intervention scheme uses a similar electro-hydraulic system to that currently used in ABS. The system is outlined in Figure 3. In principle the brakes are required to provide slip control on a split coefficient of friction surface and for a very rapid response to controlling wheel spin on uniform, low coefficient of friction surfaces. The throttle control (which is slower) is used following the initial control of wheel spin by the brakes.

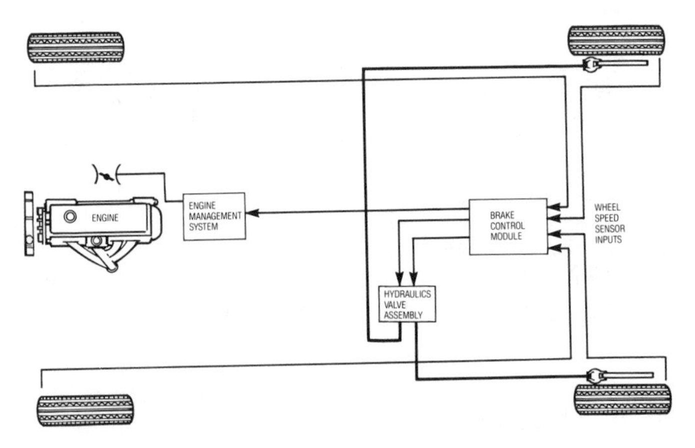

Figure 3 System outline of the proposed traction control scheme.

APPROACH.

The control system proposed above is multi-variable. In addition the control system is required to operate over the entire vehicle operating range, for example low and high engine and vehicle speeds. The powertrain system behaves differently at these different extremes making the system non-linear. The tyre/slip characteristic, the torque converter characteristic are also known to be non-linear.

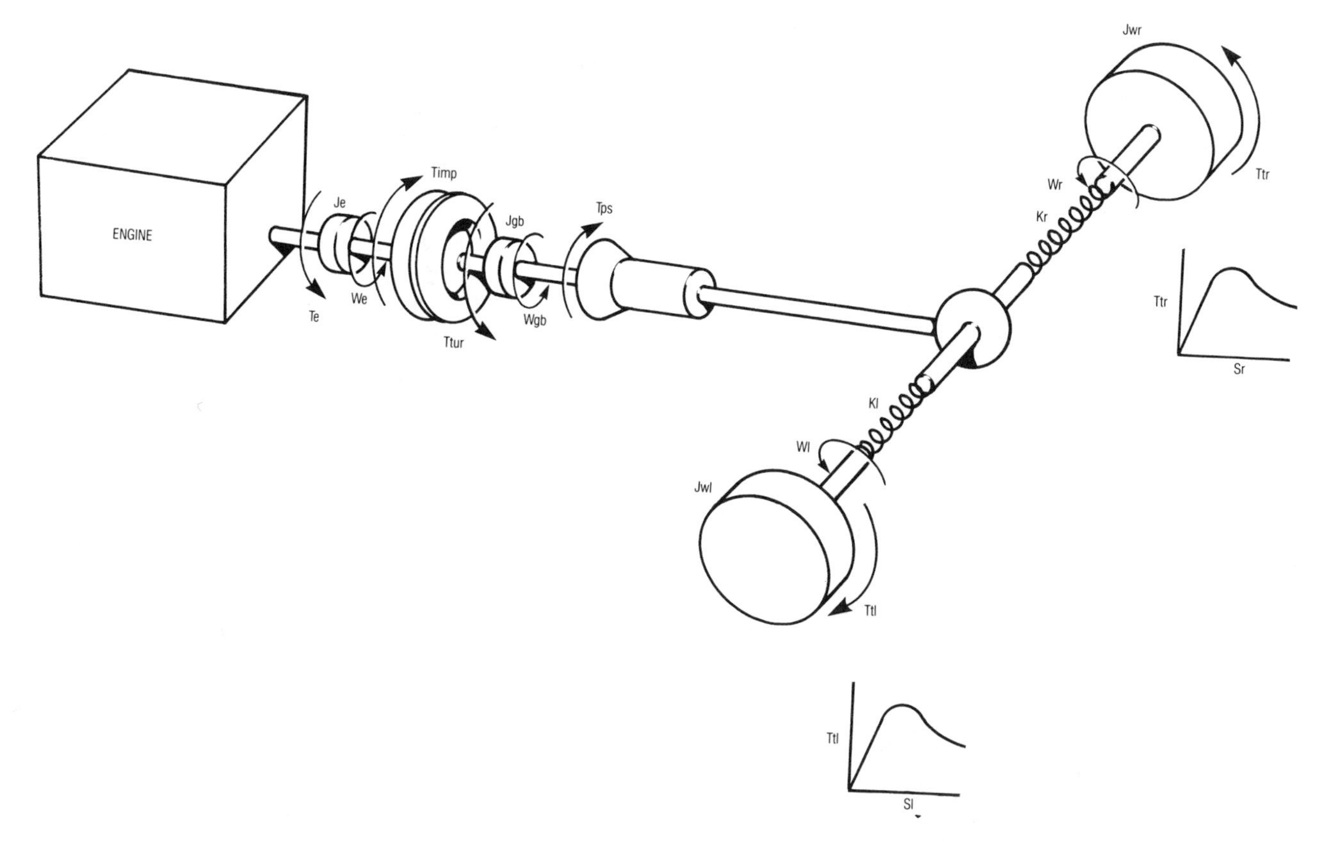

Figure 4 Schematic overview of the mathematical model.

A further complication of the scheme shown in Figure 3 is that it involves the interaction of two computers with the vehicle drivetrain. The communication process itself poses potential difficulties as this will introduce delays into the control loops.

In view of the inherent complexity of the powertrain control system there is a clear need for a systematic approach to the control algorithm design. In order to assist in the process of controller development a low order mathematical model of the system has been developed which represents the dominant characteristics. Based on engineering judgement the model has been restricted to include phenomenon in the 0-20 Hz frequency range. This restriction allows the model to be simplified by lumping distinct inertias and distinct compliances into aggregated inertias and compliances. It is suggested that a simplified model of this nature can provide great insight into the control problems and is more amenable to control analysis than a relatively complex model covering broader frequency bands and too many degrees of freedom. A schematic overview of the model is given in Figure 4. This paper is partially concerned with the development of the computer simulation model for use in control studies related to the development of the electronic control system. In addition the paper outlines the approach used in controller design with reference to a series of simulation results.

POWERTRAIN MODEL

The intention of the powertrain model was to represent the dominant dynamics of the driveline between the throttle and the driven wheels including brake actuation dynamics. In addition a torque/slip tyre-road interface characteristic is included for each driven wheel. The whole powertrain model, shown schematically in Figure 4, was developed in an incremental modular manner. Distinct sub-models, such as the engine or tyre/road interface, were developed as seperate modules whose function could be tested independantly prior to assembly into the entire model. This process was facilitated by usage of the modelling and control analysis package MATRIX_X (9). A further benefit of the graphically oriented modular structure of $MATRIX_X$ is that it permitted the incorporation of sub-models developed for other studies to be used. These distinct powertrain sub-models will now be described.

ENGINE MODEL -- The engine model is a simplified version of the model developed by Powell et al (3). The model is physically based and includes the following phenomena:-

(1) the throttle actuator characteristic.
(2) intake manifold dynamics
(3) induction-to-power stroke delay.
(4) engine pumping losses.
(5) engine rotational dynamics.

The throttle actuator dynamics are often very fast in comparison with the remainder of the engine dynamics and can be ignored for some studies. However for more realistic simulations, at the expense of simulation speed, a throttle actuator model can be included. The model is based on the response of the throttle plate position to a step input in demand. The representation was derived using a recursive maximum likelihood routine in the system identification routes of $MATRIX_X$.

The intake manifold dynamics are represented as a first order lag. The time constant for this lag varies with the engine operating point.

A representation of the delay associated with the discrete nature of the internal combustion engine is included. This delay is related to the firing frequency of the engine, and is engine speed dependant.

Engine pumping losses are physically based and the basis for this model is succinctly described by Powell (3).

The engine rotational dynamics are represented by a single lumped inertia combined with the torque converter impellor inertia. The engine torque developed by the engine is opposed by the torque converter impellor torque and a constant load representing the auxiliary loads. We therefore have

$$J_e W_e = T_e - T_{imp} - T_l \quad (1)$$

TRANSMISSION/DRIVESHAFT MODEL -- The gearset inertias and gearbox shaft compliances are associated with high frequency phenomena outside the range of interest for this particular study. For this reason the gearset inertias can be represented as a lumped inertia reflected at the gearbox input shaft and combined with the torque converter turbine inertia. The compliance of the gearbox shafts are ignored. We therefore have for the gearbox dynamics

$$J_{gb} W_{gb} = (T_{tur} - rT_{ps}) \quad (2)$$

The detailed operation of the torque converter is not required for this study and hence a black box approach was adopted. This essentially consists of representing the torque converter by its static torque characteristic. The model is based on work by Hrovat et al (6) and is fully described by Davey (5).

The propshaft compliance is high in comparison with the compliance of the axle shafts. The final drive ratio of the differential unit also tends to make the axle compliances more significant. Therefore the propshaft is assumed rigid and the compliance associated with the whole driveline is lumped into the axle shafts. The compliance values used were determined by experiment. Figure 4 illustrates this assumption.

Analysis of the differential using a classical Newtons Law type approach is difficult and it is helpful to consider energy flows though the differential. Hrovat et al (7) have applied bond graph modelling techniques to an entire vehicle drivetrain including the differential. From this analysis the equations of motion for the differential can be readily deduced.

TYRE MODEL -- The dynamics of interaction of a pneumatic tyre with the road have been subjected to much analysis to give insight into the tyre/road interface (8). What is required for the present study is an input/output model which generates the longitudinal force as a function of tyre slip ratio. A tyre/road characteristic based on experimental data and represented by a set of parameterised equations has been used. A typical characteristic is shown in Figure 4 for each wheel. This yields a coefficient of friction (Tractive force/Normal force) as a function of normalised slip defined by,

$$S_l = (r_w w_l / v) - 1 \quad (3)$$

$$S_r = (r_w w_r / v) - 1 \quad (4)$$

In order to simulate a more realistic road surface representation the coefficient of friction used can be made to take on a constant value onto which a small amplitude noise signal is superimposed. The lateral force provided by the tyre has not been modelled.

BRAKE MODEL -- A high order input/output model using classical system identification techniques has been used to represent the dynamics of

(1) the electro hydraulic actuation valves
(2) the hydraulics of the brake circuit
(3) the hydraulics of the brake actuator.

This yields a linear transfer function of brake actuator pressure in terms of the discrete signals applied to the valves. A piecewise linear brake torque/brake pressure characteristic based on data available has been used. Brake torque is assumed to be dependant only upon brake actuation pressure. Temperature effects have not been included in the model.

VEHICLE DYNAMICS -- The vehicle model was intended to represent the longitudinal movement of the vehicle. Not included in the model are the dynamics of the suspension or weight transfer effects. A constant and equal normal force on each drive wheel permits the application of a propulsive force on the vehicle through the tyre/road interface.

Opposing motion is a representation of road load:-

$$F_1 = a + bv + cv^2 \tag{5}$$

The parameters a, b and c, define the vehicle rolling resistance and aerodynamic drag and were determined by experiment.

TRACTION CONTROL

The control of wheel spin during vehicle acceleration is a multi-variable non-linear problem. The following control problems exist:-

(1) the compliance associated with the driveline and the inertias of the engine, transmission and driven wheels can result in various oscillatory modes of the system if the combined brake and throttle control system is not well co-ordinated. This has important implications where, because of the non-linear nature of the system, a controller designed for one operating point may well be unstable at another operating point.

(2) the engine and brake intervention need to be well-coordinated in order to avoid engine-brake fight on uniform low mu surfaces. For split mu conditions the action of brake intervention should be as an intelligent limited slip differential in which some engine-brake interaction is desirable. For the uniform low mu situation, brake intervention should be used to control the initial wheel speed flare and the slower acting engine used to control the steady state wheel slip.

The control system designed should satisfy the above broad performance specification. The control study reported here is concerned with the design of the throttle strategy for the system shown in Figure 3. This approach involves the communication of a brake control module with the engine management system. This can lead to a significant delay in the throttle control loop. This delay is made up of:-

1 signal processing within the brake control module.

2 transmission of information from the brake control module to the engine management system and

3 calculation delays within the engine management system. This delay is also variable.

In view of the communication and measurement delays present, a simulation study has been performed in order to examine the scheme's feasibility. Both the measurement delay and engine management loop time have been varied and the consequent effect on controller performance examined.

Figure 5 illustrates the response of the throttle control system. In this simulation the vehicle is accelerating on a good road surface when after 1.2 seconds it encounters a

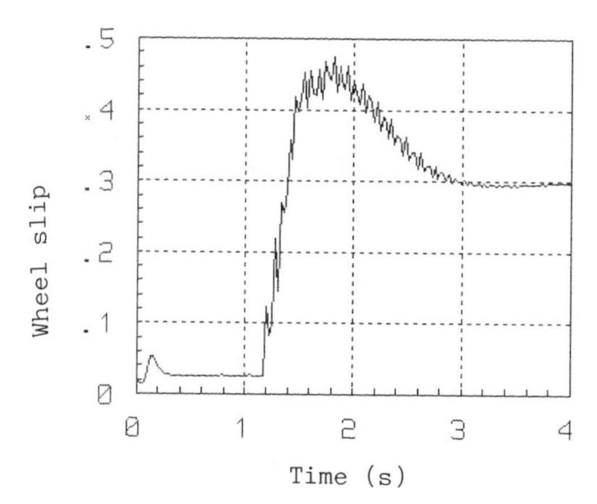

Figure 5 Response of the controller to a step change in coeffecient of friction. Nominal loop time.

low mu surface representing packed snow. The target normalised wheel spin (as defined in Eq.(3) and Eq.(4) above) is 0.3. The control action has not been optimised but does reduce wheel spin within approximately 1.5 seconds with a small amount of undershoot between 3 and 4 seconds into the simulation.

Figures 6 and 7 illustrate how the response varies when the loop time of the engine management system is increased and decreased by 50%. The response time of the controller is similar for each case. In the case of an increasing loop time, Figure 6, a small ripple in wheel spin has been introduced. This illustrates a marginally stable situation being approached which the simulation indicates occurs when the loop time is doubled. In the case of a decreasing loop time, Figure 7, the response remains stable.

The results illustrate the robustness of the control algorithm used and permit an examination of the bounds of variability in the loop time of the engine management system. A similar study has been performed in which the transmission delay was also varied in order to determine how large a delay is feasible.

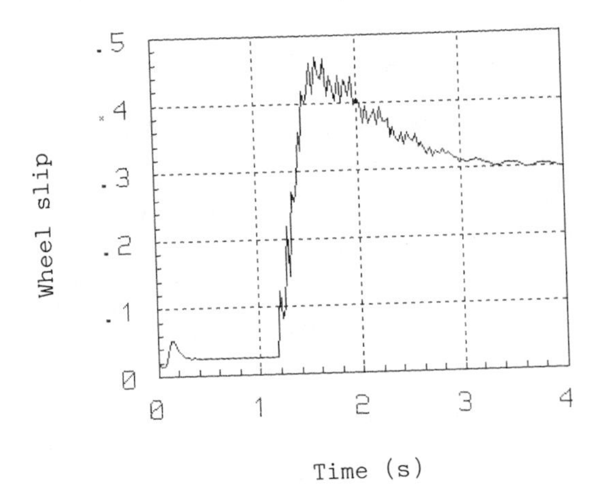

Figure 6 Response of the controller to a step change in coeffecient of friction. Loop time increased by 50% over the nominal value.

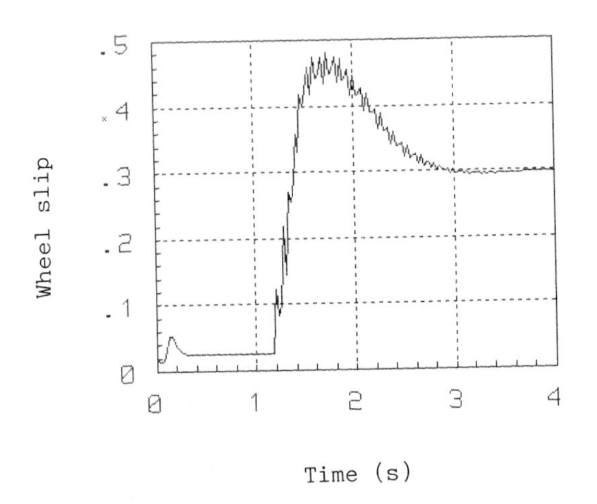

Figure 7 Response of the controller to a step change in coeffecient of friction. Loop time decreased by 50% below the nominal value.

SYSTEM VARIATIONS -- A number of vehicle parameters are known to affect the dynamics of the system, such as engine speed. At low engine speed, for example, the dynamic lag associated with the manifold is long compared with higher engine speeds. The controller needs to be re-tuned to account for these system variations. Computer simulation provides an ideal framework in which these variabilities can be examined since simulation experiments are repeatable. Thus using exactly the same input conditions (for example the coefficient of friction of the tyre/road interface and driver demand) for different vehicle operating regimes, the sensitivity of the controller can be examined and re-tuned to give the best response. An illustrative example of this procedure can be seen by referring to Figures 8-10.

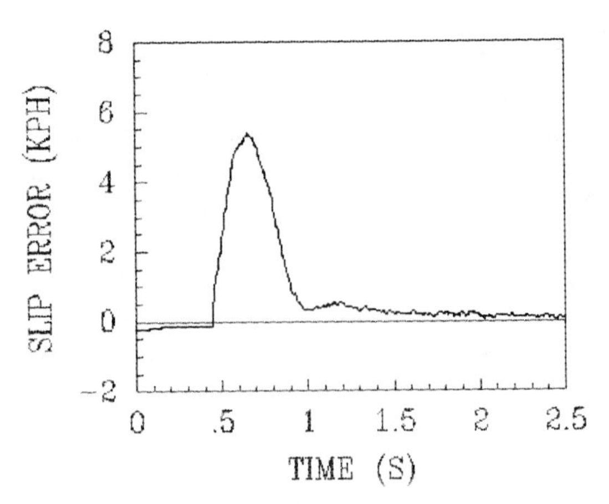

Figure 8 Response of the controller to a step change in coeffecient of friction when correctly tuned.

Figure 8 shows the response of the control system to a step change in coefficient of friction whilst the vehicle is accelerating. This simulates the vehicle being driven from a good road surface onto packed snow. The parameter shown is slip error where zero slip error represents the optimum value for traction. The wheel spin is contained within approximately 0.6 seconds.

Figure 9 shows the response of the same controller, this time at a different operating point. The wheel spin is now contained within approximately 0.7 seconds, but with significant 'overshoot' illustrating a deterioration in the controller performance. If the controller is re-tuned for these new conditions, see Figure 10, the response of the controller now contains the wheel spin again within approximately 0.6 seconds and without any significant 'overshoot'.

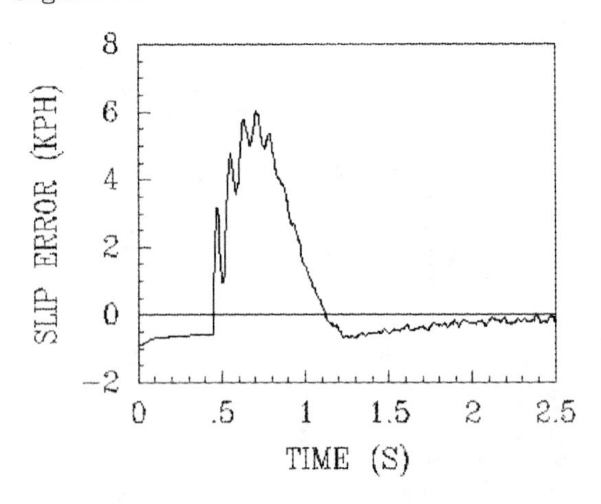

Figure 9 Response of the controller to a step change in coeffecient of friction when the controller is incorrectly tuned.

The simulation results illustrate:-

1) that the controller used is robust for this particular system variation and
2) that the simulation permits the controller to be readily re-tuned to account for system variations.

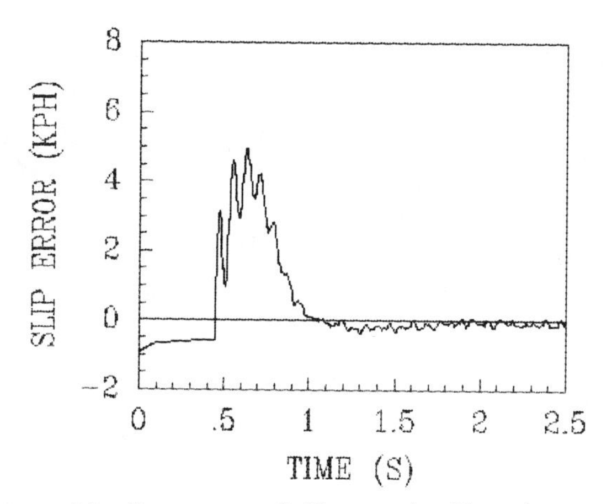

Figure 10 Response of the controller to a step change in coeffecient of friction when the controller is re-tuned.

CONCLUSION.

This paper has described a simulation model used for addressing traction control issues. The modular environment used ($MATRIX_X$) has permitted the incorporation of powertrain sub-models developed elsewhere to be quickly and efficiently incorporated into the whole powertrain model.

A scheme for the control of wheel spin has been proposed and has been shown to be feasible through the use of simulation studies. In particular the control strategy developed has been shown to be insensitive to engine management system loop time variations of approximately 50%. This simulation can be seen to have a significant input into electronic hardware design. Furthermore a robust control algorithm has been designed which is tolerant of system variations requiring alteration only to restore the speed of response as the operating point changes. These simulation results are encouraging and provide a solid basis for future experimental developments.

NOMENCLATURE

F_l road load.

K_l stiffness of left axle shaft

k_r stiffness of right axle shaft

J_e effective polar moment of inertia of the engine flywheel and the torque converter impellor.

J_{gb} effective polar moment of inertia of the torque converter turbine and gearbox

J_{we} effective polar moment of inertia of the left driven wheel

J_{wr} effective polar moment of inertia of the right driven wheel.

S_l normalised slip ratio of left driven wheel

S_r normalised slip ratio of right driven wheel.

r gearbox ratio (output shaft speed/input shaft speed)

r_{fd} final drive ratio (output shaft speed/input shaft speed).

r_w radius of the driven wheels.

T_e Engine torque.

T_{imp} torque converter impellor torque

T_{ps} drive shaft torque

T_l auxiliary loads torque

T_{tur} torque converter turbine torque

T_{tl} tyre/road interface torque to left driven wheel

T_{tr} tyre road interfaced torque to right driven wheel

V vehicle speed

W_e engine flywheel angular speed

W_{gb} gearbox input shaft angular speed

W_l left driven wheel angular speed

W_r right driven wheel angular speed.

ACKNOWLEDGEMENTS.

I would like to acknowledge the contributions of J. Main who initiated this work, D Hrovat and L. Roller each of Ford Motor Company. I also thank the Directors of Ford Motor Company for permission to publish this work.

REFERENCES

(1) MAIN, J.J. "Ford ELTEC Integrated Powertrain Control', SAE Paper 860652

(2) 'Electronic Engine Management and Driveline Control Systems', SAE Publication SP-481, 1981.
(3) POWELL, B.K. & COOK, J.A. 'Discrete simplified external linearisation and analytical comparison of I.C. engine families ', ASME Journal of Dynamic Systems, Measurement & Control, Vol 109, December, 1987.
(4) POWELL, B.K. & POWERS,W.F. 'Linear Quadratic Control Design for Non-linear IC Engine Systems', Proc. Int.Soc.of Automotive Technology and Automation Stockholm, Sweden, October, 1981.
(5) DAVEY, C.K. 'Computer Modelling of an Anti-Spin Control System'. Sixth Int. Conf. on Automotive Electronics Oct 1987 London.
(6) HROVAT, D & TOBLER, D. Graph Modelling and Computer Simulation of Automotive Torque Converters' The Journal of the Franklin Institute.
(7) HROVAT, D TOBLER , W.E & TSANGARIDES, M.C. 'Bond graph Modelling of Dominant Dynamics of Automotive Powertrains', Proc of ASME Winter Annual Meeting 1985.
(8) CLARK, S.K, Ed, Mechanics of Pneumatic Tires, Monograph 122, National Bureau of Standards, Washington D.C, 1971.
(9) Integrated Systems Inc. Santa Clara, Calif, 'MATRIXX user guide version 6'. 1986.

890286

VFD Systems Directly Driven by Single Chip Microcomputer for Automotive Applications

Soichi Shinya, Masami Suzuki, and Hiroshi Yamaguchi
Futaba Corp.
Gary Wires
Futaba Corp. of America

ABSTRACT

Vacuum Fluorescent Displays (VFD's) are being used widely as the informational display panel in the automobile because they are a self-emissive, high luminance display device that feature crisp and easy-to-read images. This paper reports on the development of a VFD that can be directly driven by the outputs of a single chip microcomputer operating from a 12 volt supply. This system is most ideal for automotive applications.

THE BASIC STRUCTURE of the VFD, illustrated in Figure 1, is a triode consisting of a cathode, grid, and anode. The cathode, a tungsten wire coated with Ba, Sr, and Ca, is directly heated to a temperature of about 600°C. Thermoelectrons generated from the cathode are accelerated towards the anode by the voltage applied to the grid, which is stretched over the anode. The electrons collide with the phosphor layer, which is formed on the anode with the use of thick film technology, to emit light.

In general, the anode is a graphite layer that is connected to the outer leads either by a silver pattern formed by thick film technology, or an aluminum pattern formed by thin film technology. To maintain a vacuum inside the package, the base plate, containing the patterns, and the front glass are sealed together using frit glass. Phosphors with low excitation voltages, such as ZnO:Zn, are typically used where the drive voltage ranges from about 12V up to 100V.

OBJECTIVE

Since the development of the multidigit flat VFD in 1972, the application of VFD's has expanded from use in desk top calculators to audio equipment, VCR's, appliances, and automotive applications.

The first use of VFD technology in an automotive application began with the clock, and has been followed by use in the audio system, message indicator, and instrument panel. Multicolor capability, multifunctions, and large-scale analog features have been achieved through technological improvements.

The static drive system is driven from the 12V battery voltage and is now most commonly used in the electronically tuned radio (ETR) in the automobile. However, the increasing number (33 to 58) of display segments make it difficult to drive the display directly from the CPU I/O ports. Thus, the use of driver integrated circuits, with a serial-in, parallel-out architecture, are required for the CPU to VFD interface. One disadvantage of this approach is that a relatively large PCB area is consumed due to the number of terminals.

Recently, the duplex drive VFD (1/2 duty cycle) system was developed. The display is driven directly from the CPU I/O ports using a 12V battery source, and has sufficient luminance for the automobile, plus high reliability.

SIMPLIFIED SYSTEM

The developmental process and system construction are next described. The trend of VFD's and driving methods by generation are shown in Figure 2 for consumer products, column A, and for automotive ETR applications, column B.

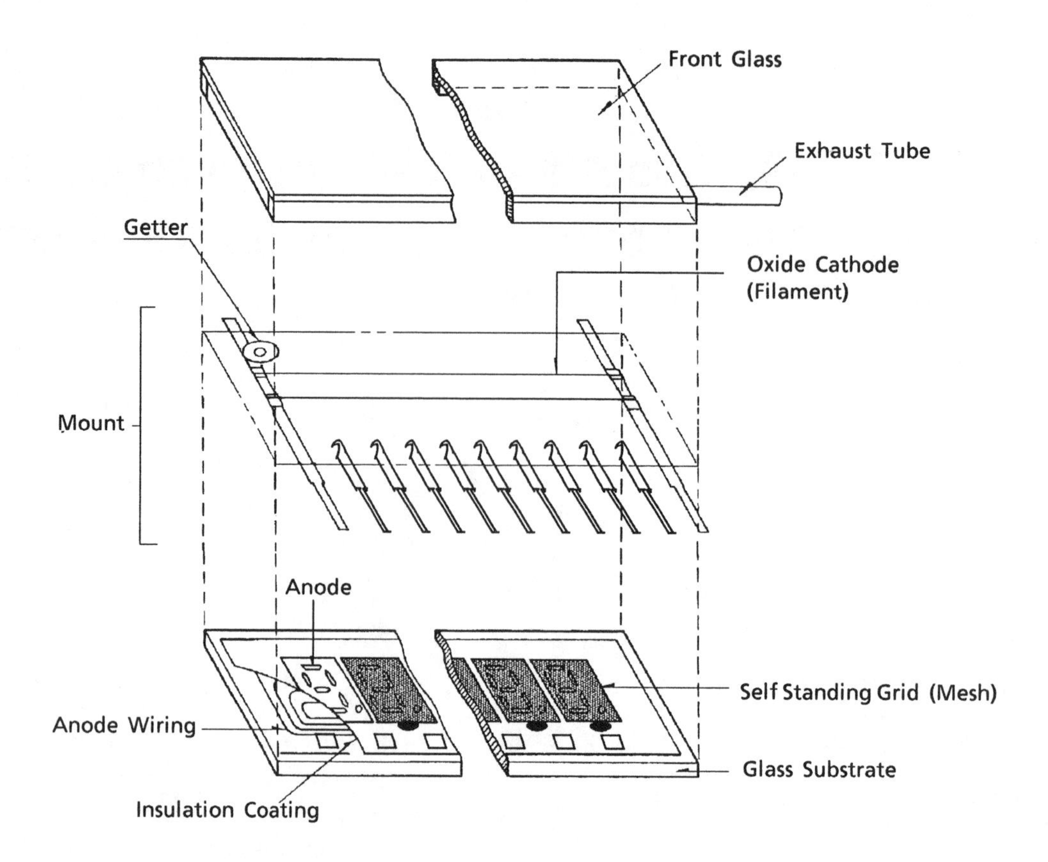

Fig.1 Basic Structure of VFD

The major components in the ETR include the PLL, controller, drivers, VFD, and power source.

Since it was hard to obtain sufficient luminance at 12V with the first generation VFD, a DC/DC converter was needed to provide a higher drive voltage. In the second generation, the desired luminance could be obtained at 12V, thus eliminating the need for a DC/DC converter.

The third generation VFD drive system was designed with a simpler structure, due to advancements made in semiconductor process technology, using one microcomputer chip that included the VFD driver, PLL, and controller. Due to the lower luminance requirement of consumer products, the direct drive system using one microcomputer chip had been realized earlier because dynamic drive was possible.

This paper discusses the fourth generation VFD. It enables operation with duplex drive (1/2 duty cycle) and replaces static drive while maintaining the required high luminance level for automotive applications. A single in-line lead (SIL) package is achievable which simplifies the connection to the PCB when compared to the dual in-line (DIL). Also, as the information content increases in the VFD, the number of terminal leads on the VFD and drivers is reduced by utilizing the duplex drive system.

Figure 3 shows the improvements made on the luminous efficiency of the phosphor material. The development of the self-standing grid has allowed the VFD to have a minimum luminance level of 1400 cd/m^2 when driven using a duplex drive system and 12V battery source. Developmental efforts will be continued in order to realize the 5th generation VFD capable of being driven at a 1/4 duty cycle.

CHARACTERISTICS OF THE DUPLEX VFD

As previously described, the improvement in the luminous efficiency and the development of the self standing grid have largely contributed to the realization of achieving a higher luminance level at 12V drive voltage and 1/2 duty cycle.

As shown in Figure 3, the luminous efficiency improved from 14.4 lm/W to 16.1 lm/W (111%) in a period from 1987 to 1988. In addition, the gap between the filaments and grid electrodes has been reduced from 0.65mm to 0.5mm with the new self-standing grid process,

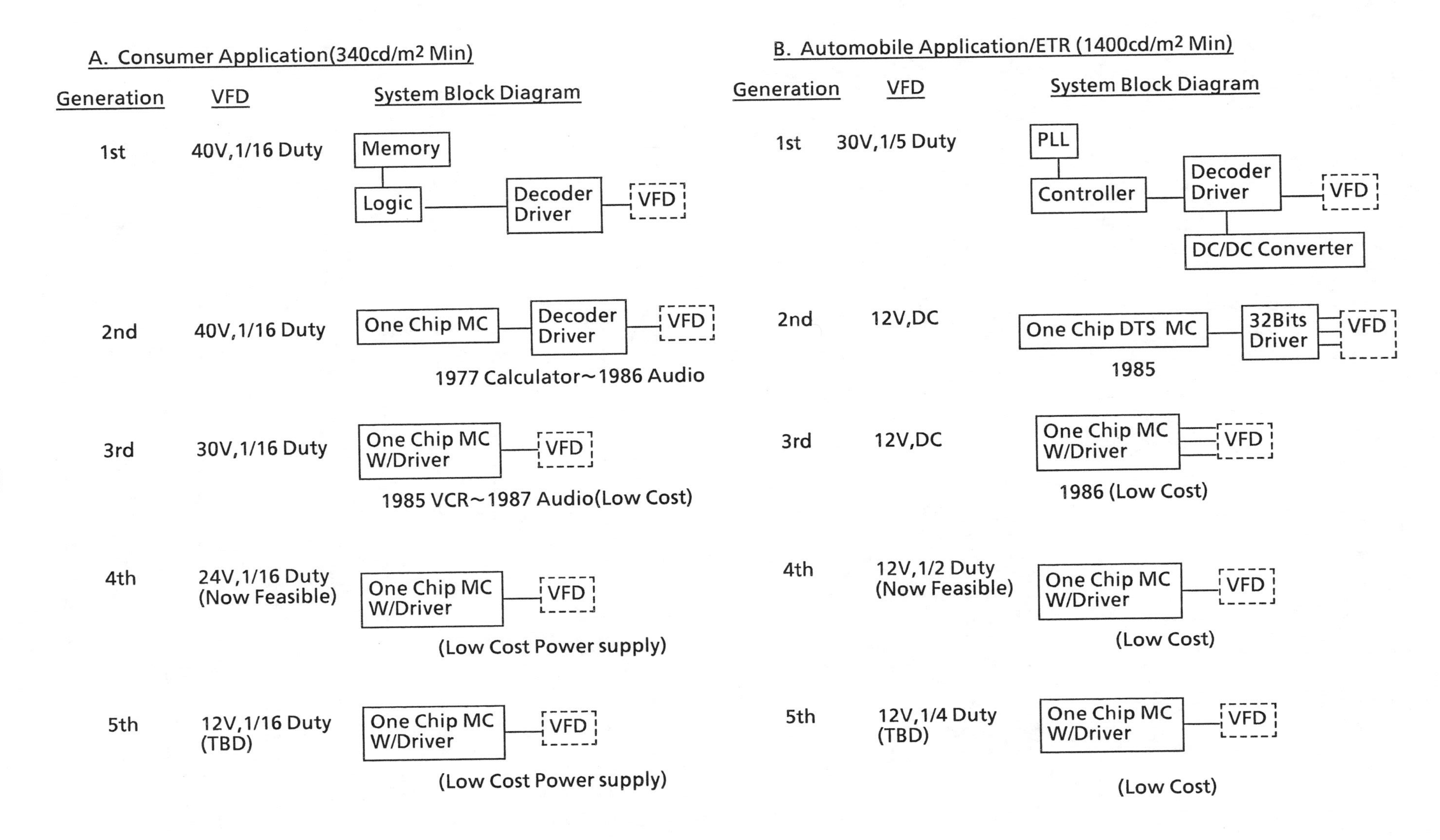

Fig. 2 Trends of VFD Technology and Driving Method

which results in an increase in luminance of $(0.65/0.5)^2 \times 100 = 169\%$.

The relative luminance of a VFD driven by the duplex drive system, when compared to that of a conventional static drive system, can be calculated as shown in Eq. (1).

$$\underbrace{1.11}_{\text{Luminous efficiency}} \times \underbrace{1.69}_{\text{Perveance}} \times \underbrace{1/2}_{\text{Duty cycle}} = 0.94 \qquad \text{Eq. (1)}$$

When driven by the duplex system, the luminance obtained is almost the same (94%) as the static type VFD. Figure 4 shows the measured luminance versus voltage characteristic curve of a VFD with a 1/2 duty cycle. The reliability tests performed, and confirmed, are shown in Table 1, and the lifetime data is shown in Figure 5.

Photographs of conventional static type VFD's are shown in Photo 1 and Photo 2, while their counterpart duplex types are shown in Photo 3 and Photo 4, respectively. Table 2 shows a comparison of the specifications of the 4 product types.

Photo 5 shows a module consisting of a static type VFD and driver IC on a PCB, while Photo 6 shows a chip-in-glass VFD, a special type VFD that contains a driver IC chip within the glass package.

A comparison of the relative system cost for each system is shown in Table 3. As can be seen from the table, the 1/2 duty cycle system is the most cost effective because an external driver IC is no longer necessary. Other advantages to consider include a simplification in the PCB layout, a reduction in the required interconnect traces, and an increase in the PCB space available for other components.

CONCLUSION

For the car audio system, a duplex drive VFD capable of being driven directly from the CPU was developed. The VFD has the required luminance and reliability levels needed by a display device in an automotive application. This new VFD drive system can also be used in the odometer, speedometer, and other gauges of the automobile due to the space reduction, simpler PCB layout, ease of assembly, and cost reduction of the system.

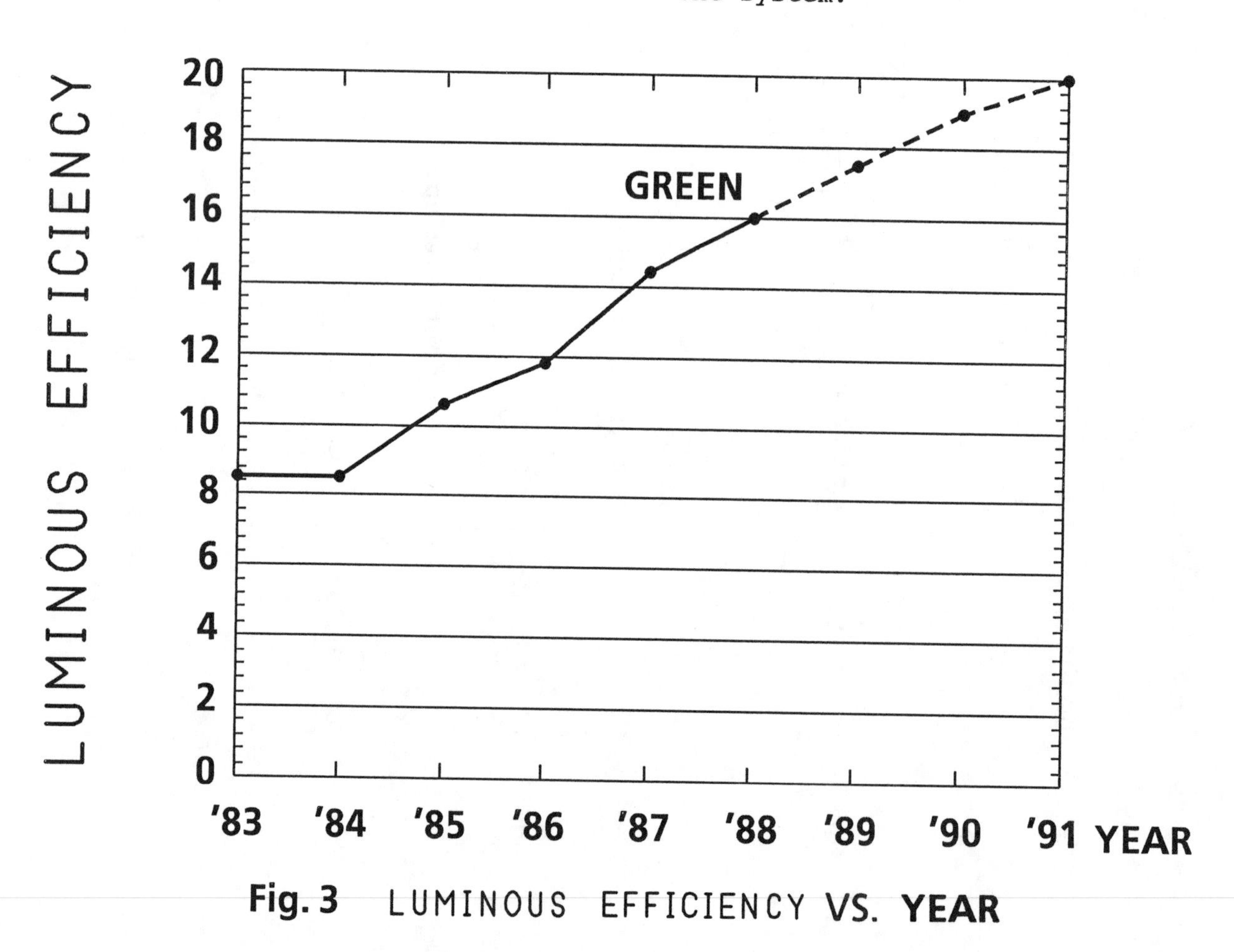

Fig. 3 LUMINOUS EFFICIENCY VS. YEAR

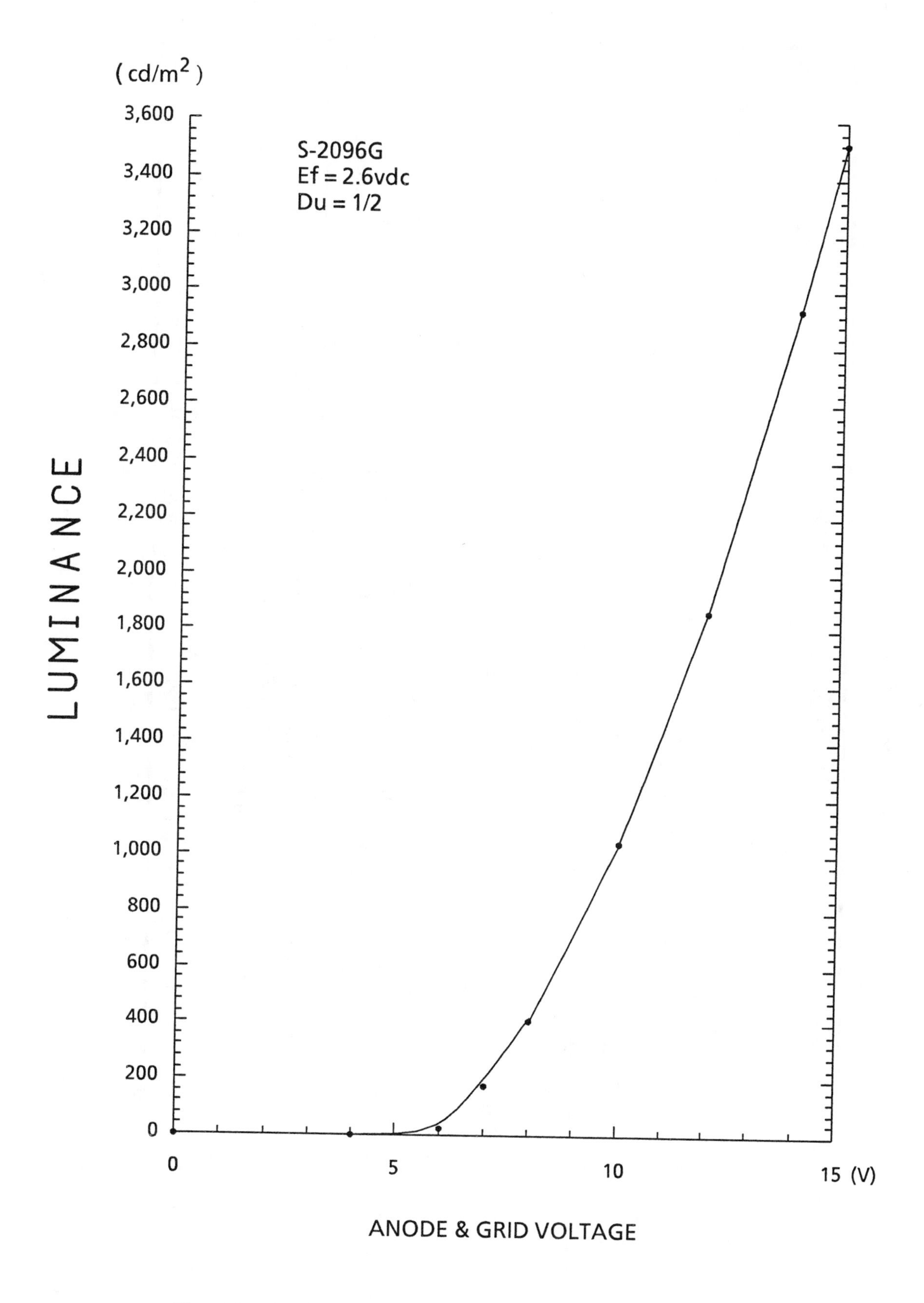

Fig. 4 Luminance VS. Anode & Grid Voltage

Table 1 RELIABILITY TESTS PERFORMED

ITEM	TESTING CONDITION	METHOD
THERMAL SHOCK	30min.; $+85^{+0}_{-3}$ °C; -55^{+0}_{-3} °C; 30min. Non operation 100 cycles	MIL-STD-202F method 107D
POWER TEMP OPERATION	At 85±2°C operation for 1000 hours	MIL-STD-202F method 108A
TEMP CYCLING	15min; $+85^{+3}_{-0}$ °C; $+25^{+10}_{-5}$ °C; -40^{+0}_{-3} °C; 30min. Operation 1000 cycles	MIL-STD-202F method 102A
VIBRATION (With holder)	1.5mm total excursion, 10~55Hz frequency, sweep time cycle 1 minute, vibration applied for 2 hours in each of X, Y, and Z directions under non operating condition, 6 hours total	MIL-STD-202F method 201A
SHOCK	100G maximum acceleration, 6ms duration time, half sine wave applied 3 times in each of X, Y, and Z directions (18 times total, unlighted)	MIL-STD-202F method 213B
LIFE	18 cycles total At room temperature, for 1000 hours under normal operating condition	JIS C5036

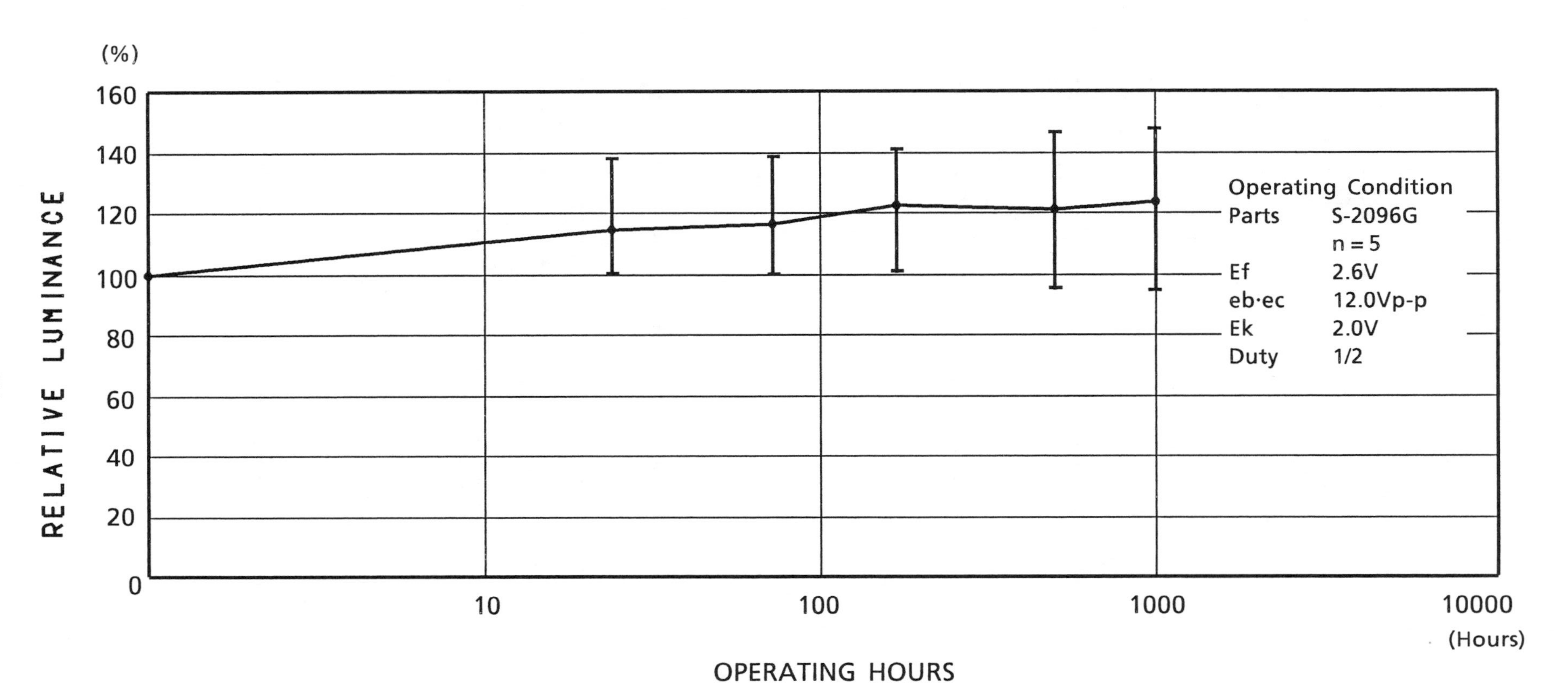

Fig. 5 Life Characteristics

Tabe 2 Specifications of VFD's

	4-BT-87Z	S-2096G	6-MT-90N	S-2119G
Duty	Static	1 / 2	Static	1 / 2
Package(mm)	28.5×76.5	28.5×76.5	20.5×58.4	20.5×60.0
Filament Voltage(Vdc)	2.6	2.2	1.9	2.2
Anode & Grid Voltage(V)	13.8	12	13.8	12
Luminance(cd/m^2)	2060	2060	2060	2060
Total Power Consumption (W)	0.95	1.32	0.43	0.52
Number of Lead Terminal	64	38	50	27

Table 3 The Cost Comparison Assumption of Relative System Cost

Generation	4th	2nd	2nd
VFD Models	S-2096G	AH009E (Module)	S-2042NI(CIG)
System	Duty 1/2	Static	Static,Chip-in-Glass
VFD	100	100	120
Driver	Not Required	80	60
PCB	Not Required	20	Not Required
Assembly+Others	Not Required	20	60
Total	100	200	240

Photo 1 Conventional Static Type VFD (Futaba Part No. 4-BT-87Z)

Photo 2 Conventional Static Type VFD (Futaba Part No. 6-MT-90N)

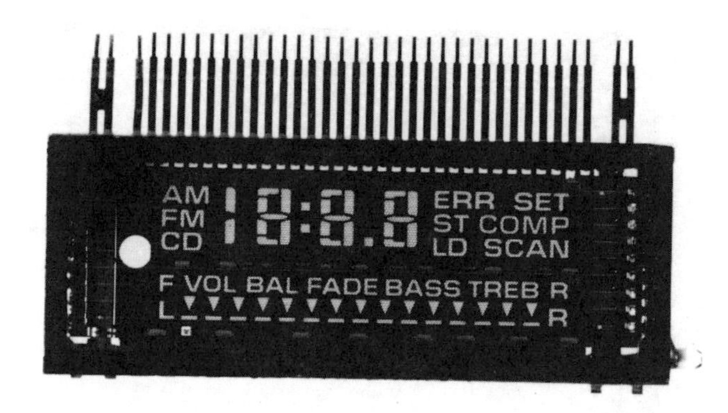

Photo 3 Duplex Type VFD (Futaba Part No. S-2096G)

Photo 4 Duplex Type VFD (Futaba Part No. S-2119G)

Photo 5 Module with Static Type VFD and Driver (Futaba Part No. AH009E)

Photo 6 Chip-in-Glass VFD

890303

The Application of Microcontrollers to Automotive Smart Sensors

Manuel Alba
National Semiconductor Corp.

ABSTRACT.

Today's sophisticated automotive electronics controllers require higher quality information about vehicle parameters. This information is provided to them by means of various kinds of sensors, using different types of electric signals. Although many of these sensors have been quite adequate in the past, new precision and stability requirements have emerged that dictate the use of new approaches <1>.

An approach that offers higher performance, as well as flexibility, utilizes a microcontroller as an integral part of the sensor element. Until very recently, the cost of this approach was too high to be considered for automotive applications. Fortunately, new microcontroller offerings -and associated circuitry- offer high performance at low cost, making it feasible to design microcontroller-based sensors for the vehicles of the early 1990's.

Key techniques of this microcontroller-based approach are reviewed, including examples of some of the most important applications.

INTRODUCTION.

Accuracy requirements for many sensor elements in early 90's automotive applications are in the sub-1% area over the whole temperature range (-40 to +125 degrees centigrade).

This is mainly driven by two forces: a) the ever present need for higher performance and b) the advent of sophisticated systems where safety is critical.

On the performance side, we have seen a rapid evolution in the capabilities of the microcontrollers used in automotive systems. The migration to 16-bit machines is quite evident in systems like Engine Control. These microcontrollers are fast, have efficient architectures and instruction sets, and usually contain a number of attractive peripherals on-board. In order to achieve higher performance though, it is not sufficient to increase processing throughput: sensor information must also be upgraded (many of the sensors used today are still the same that fed information to previous generations of microcontrollers).

In many ocassions, you will find these 'central' controllers allocating some of their precious processing time and memory space to tasks related to the collection and correction of sensor signals (eg. linearization, averaging, etc.). It is a desirable goal to provide these central controllers with high quality signals that do not require any further correcting. This not only relieves the central controller, but better signals are also provided to it, since the best signal conditioning and correction is performed where the signal originates.

On the safety side, we have recently seen an unprecedented avalanche of new automotive systems that are controlled electronically. Proper operation of many of these systems is essential for passenger safety: ABS, Active Suspension, Electronic Power Steering, Air Bags, Passive Restraints, etc. Consequently, with very little room for error, accuracy requirements for sensors associated with these systems have increased.

Smart Sensors that incorporate a microcontroller in the same housing have been proposed as the solution for higher quality signals <2>. However, the cost of this approach has traditionally been an obstacle. Fortunately, the cost of microcontrollers that integrate the necessary functions to implement a Smart Sensor has been consistently decreasing tahnks to several factors: smaller die sizes due to the utilization of finer geometries (below 2 microns), utilization of larger size wafers (6" in diameter), more efficient manufacturing, etc. Therefore, it is now feasible to design Smart Sensors with microcontrollers on-board.

A microcontroller-based Smart Sensor offers many advantages: the ability to distribute processing to the nodes where physical variables are being sensed, thus off-loading the central processor; the ability to provide environmental compensation and signal correction where the variables are being sensed, thus resulting in a higher quality signal being fed to the central processor; the ability to collect and store diagnostics information, distributing it later on demand; in many cases, the ability to save in manufacturing costs; etc.

SOLID STATE SENSORS.

The Smart Sensor advent is tightly coupled to the increased use of solid state sensors in vehicles, in the form of pressure sensors, accelerometers, air-flow sensors, etc <3>.

Since these sensors produce outputs that vary considerably with temperature, a need for effective temperature compensation arises. For example, the bridge resistance of pressure sensors can vary as much as 8% or 9% over the -40 to +125C range.

The use of a microcontroller as a companion part to a solid state sensor allows for very effective temperature compensation, linearization and calibration of the end product.

THE MICROCONTROLLER-BASED ALTERNATIVE.

We will concentrate in the most obvious case for microcontroller-based Smart Sensors, that in which the analog signal of a solid state sensor is being processed (eg. pressure) <4,5,6>. The overall methodology can be understood by looking at Figure 1.

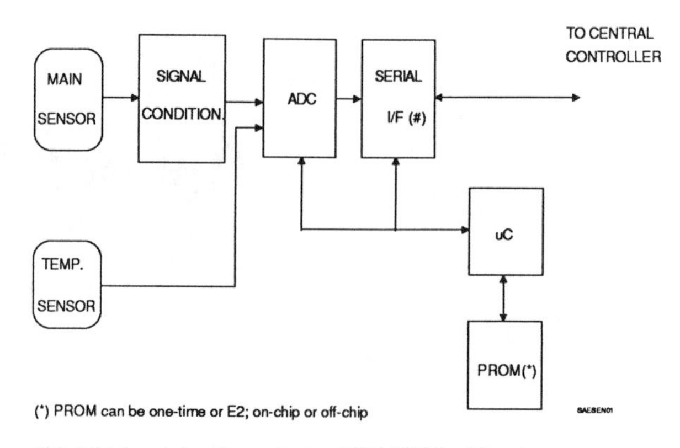

Figure 1. Smart Sensor for Linear Signals

Since the signal derived from the sensor is typically small, a signal conditioning stage is necessary to provide proper signal amplification. Next, the analog signal is converted to digital form by the Analog-to-Digital Converter (ADC). The signal derived from the temperature sensor is also converted to digital. Using these two pieces of information, the microcontroller is able to compensate for any temperature effects on the main sensor by means of using the compensation data stored in the Non Volatile Memory (NVM). The corrected data is then communicated to a central controller over the serial interface.

The most typical compensation scheme generates the correction data at the time of manufacturing. Using a pressure sensor as an example, the process is as follows: the Smart Sensor is placed in a chamber where precise temperatures can be generated; calibrating pressures are supplied to the Smart Sensor and its voltage output is registered for each of them; an nth order polynomial is used to fit a curve to the data obtained for the n+1 calibrating pressures; the resulting coefficients are then stored in NVM and subsequently utilized to compute an actual pressure when the sensor is used in a vehicle <7>. (Alternatively, if a large size EPROM is chosen as NVM, a large number of points can be stored and the correction can be done by means of a simple table lookup).

This process is performed with different temperatures in the chamber and various sets of coefficients are generated.

In other words, the equation utilized is <4,8>:

$$P = C_0 + C_1V + C_2V^2 + ... + C_nV^n$$

where:

P = corrected version of measured pressure X

V = voltage derived from pressure sensor while measuring pressure X

C0..Cn = correcting coefficients stored in NVM

The way to compute the correction coefficients necessitates that n+1 calibrating pressures (P0 through Pn) be applied to the sensor, in order to obtain n+1 output voltages (V0 through Vn). Once this data is obtained, the following system of equations can be formed:

$$\begin{aligned} P_0 &= C_0 + C_1V_0 + C_2V_0^2 + ... + C_nV_0^n \\ P_1 &= C_0 + C_1V_1 + C_2V_1^2 + ... + C_nV_1^n \\ P_2 &= C_0 + C_1V_2 + C_2V_2^2 + ... + C_nV_2^n \\ &\vdots \\ P_n &= C_0 + C_1V_n + C_2V_n^2 + ... + C_nV_n^n \end{aligned}$$

This can be be translated into the following matrix, from which the correction coefficients will be computed:

$$\begin{bmatrix} P_0 & 1 & V_0 & V_0^2 & ..... & V_0^n \\ P_0 & 1 & V_1 & V_1^2 & ..... & V_1^n \\ P_0 & 1 & V_2 & V_2^2 & ..... & V_2^n \\ \vdots & \vdots & \vdots & \vdots & & \vdots \\ P_0 & 1 & V_n & V_n^2 & ..... & V_n^n \end{bmatrix}$$

Recursive techniques can be then utilized to obtain the values of C0 through Cn. For instance, using the Gauss Jordan method we arrive at the results after n+1 iterations, with a final matrix of the form:

$$\begin{bmatrix} C_0 & 1 & 0 & 0 & ..... & 0 \\ C_1 & 0 & 1 & 0 & ..... & 0 \\ C_2 & 0 & 0 & 1 & ..... & 0 \\ \vdots & \vdots & \vdots & \vdots & & \vdots \\ C_n & 0 & 0 & 0 & ..... & 1 \end{bmatrix}$$

The non-linearity of the sensor utilized determines the order of the polynomial required. In several cases, a third order polynomial is adequate; this means that 4 calibration values are needed and 4 coefficients are derived (C0, C1, C2, C3). In theory, the larger the number of coefficients the better, but this is subject to practical constraints: NVM space limitations, the ability of the microcontroller to perform multiplications at a fast speed, etc.

An example is shown in Figure 2, where a 2nd and a 3rd degree polynomials are used for comparison. Both fitting curves converge on the calibration pressure extremes of 0 and 30 psi, and in a common intermediate point. By adding an extra calibration point in the case of the 3rd degree polynomial, we notice that we obtain a better fit, which differs from the 2nd degree one by as much as 4%.

The magnitude of the measured voltages and the linearity (or lack of it) of the sensor will play an important role in the magnitude of the correcting coefficients. That will determine if 8-bit or larger coefficients need to be stored in NVM, and therefore the amount of memory needed. Notice that the computed coefficients can be positive or negative.

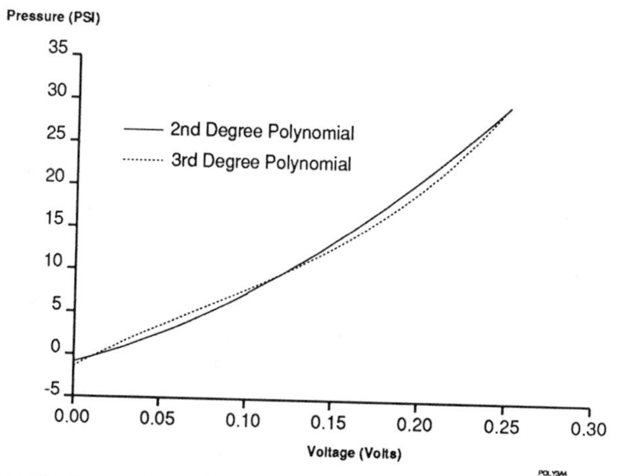

Figure 2. Comparison of 2nd. and 3rd. Degree Correction Polynomials

Once this process is carried out at ambient temperature, the chamber is adjusted for some other temperatures of interest (eg. the extremes of -40 and +125C and a couple of points in between), and is repeated for each. Therefore, a set of nth order curves is generated and table lookup for the different temperature segments can be performed, using the set of coefficients that best fits the operating temperature at the time the measurement is made during vehicle operation. Also, an interpolation scheme can be implemented, using the two nearest curves.

It is important to mention that this system can be cost effective, since a large batch of sensors can be calibrated at the same time, thus avoiding the expense of the one-by-one laser trimming often used in analog implementations. Also, it is important to notice that all the intensive calculations required to compute the coefficients are better handled by an external computer dedicated to that task, not by the sensor microcontrollers. However, in the case of future self-calibrating sensors, these computations could be done by the microcontroller itself <4,9>.

SIGNAL CONDITIONING - The signals derived from a solid state sensor can be quite small, especially if a capacitive scheme is utilized. Therefore, it is critical to provide adequate signal amplification as close to the sensor as possible.

This approach requires operational amplifiers with well above average features for low offset voltage, low offset voltage drift with temperature, high gain, low noise, low current consumption, etc.

Low offset voltage is important to avoid introducing errors into the system at the first stage (if an offset is present anyway, it can be compensated for with software). Obviously, it is highly desirable that, whatever the offset voltage value be, it remain stable over the temperature range of interest. Thus, a low offset voltage drift value is also quite important.

High gain is also critical in error reduction. For example, it can be demonstrated that an operational amplifier open loop gain of 10,000 (80 dB) can introduce a gain error of 1% in a closed loop system designed to provide a gain of 100 (eg. a 50 mV signal span that needs to be amplified to 5 volts). This results in a loss of 3 LSB's in an 8-bit system and 10 LSB's in a 10-bit system (Figure 3). This is an unacceptable error contribution that must be avoided.

As we can see in the figure, in order to avoid error contributions at this stage, the open loop gain of an operational amplifier must exceed 50,000 (94 dB) in an 8-bit system and 200,000 (106 dB) in a 10-bit system.

Low current consumption is also important and can be best achieved by utilizing CMOS operational amplifiers.

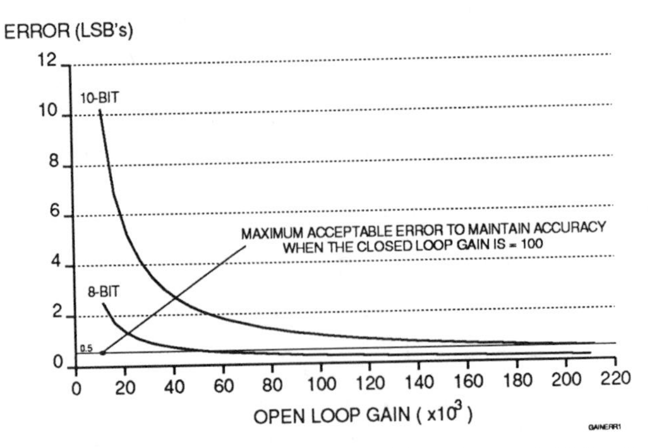

Figure 3. Errors Due to Finite Open Loop Gain in 8- and 10-bit Systems

An example of an operational amplifier that meets the requirements of Smart Sensor applications is found in National's LMC660 (quad version) and LMC662 (dual version), built with CMOS technology <10>. This op amp has low offset voltage (3 mV maximum), low offset voltage drift (1.3uV/degree C), high gain (126 dB) and low distortion (0.01% at 10 khz). On top of this, it features single-supply operation (+5 to +15 volts) as well as rail-to-rail swing.

Using the example we mentioned before, it is clear that the high open loop gain of this op amp (126 dB) allows it to participate in an 8-bit or 10-bit data acquisition system without a loss of LSB's due to gain error.

Also, the quad version (LMC660) can be used in an instrumentation amplifier configuration for improved performance.

MICROCONTROLLERS FOR SENSORS- As we said before, not long ago it was difficult to justify the use of microcontrollers for Smart Sensor implementations. The problem was not only the cost of the microcontroller itself, but also the fact that many other building blocks were needed (timers, watchdog logic, analog-to-digital converter, UARTs, etc.) and that microcontrollers were physically too big to fit in many sensor enclosures.

Fortunately, the cost of microcontrollers has been decreasing gradually, their level of integration increasing rapidly and their packaging technology improving greatly. This makes possible the utilization of microcontrollers in Smart Sensors.

An example of this is National's

COP888CF. This is an 8-bit microcontroller which contains several key functions on-board: 8-bit analog-to-digital converter (ADC) with 8-channel multiplexer, 16-bit timers, watchdog logic, Microwire serial interface for peripherals, etc <11> (see Figure 4).

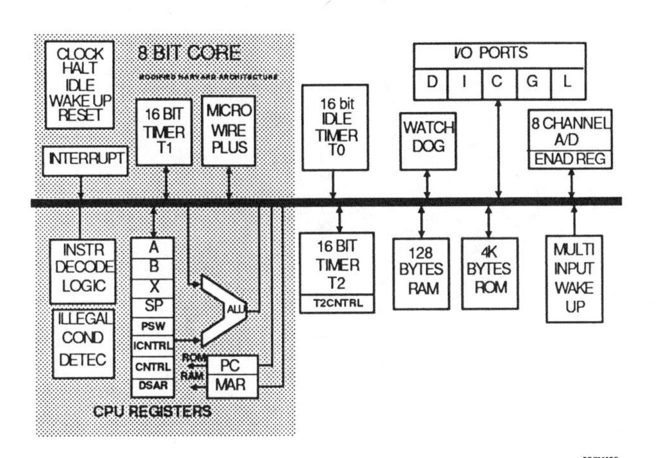

Figure 4. COP888CF Block Diagram

A key feature in the case of this microcontroller is the on-board ADC, which is implemented using the Successive Approximations Register technique and that supports ratiometric conversions (Figure 5). This results in very fast conversion times - as fast as 7.2 usec with a 10 Mhz clock, which is more than adequate for automotive sensor applications -. The extra speed is useful in obtaining more accurate results, since multiple readings can be taken sequentially (oversampling) and pre-processed (eg. averaged) to yield a more reliable result.

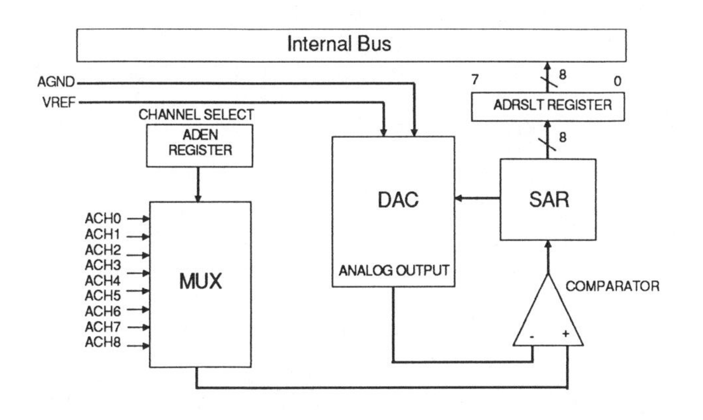

Figure 5. Block Diagram of ADC for the COP888CF

Eight single-ended or four differential channels can be served with this ADC. An example of channel usage is a pressure Smart Sensor with a differential output: in this case, a differential channel is dedicated to it and 1 channel is dedicated to a single-ended temperature sensing device. The temperature information is used to perform pressure signal correction inside the microcontroller. The ADC can be programmed to perform single or continuous conversions. Pins are also provided for analog groung (Agnd) and an external reference (Vref).

Another important feature of the COP888CF are the three 16-bit timers, T0, T1 and T2 (Figure 6). Timer T0 - the IDLE Timer - is a countdown timer that supports the IDLE mode of operation, in which power consumption is reduced to around 30% of the normal one. It also supports the Watchdog Logic, a very critical function that helps in the prevention of catastrophic failures by detecting infinite loops in a program.

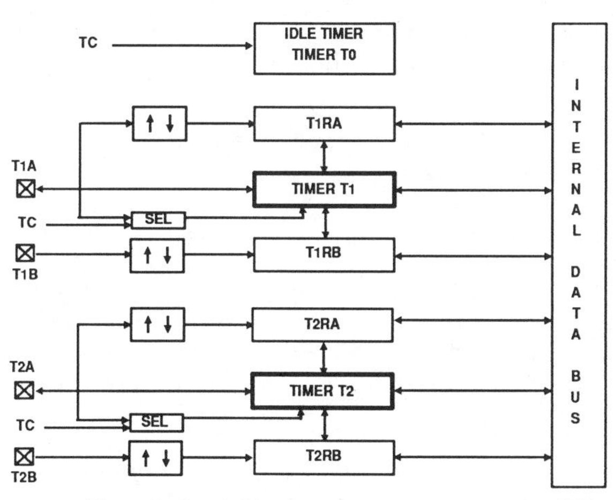

Figure 6. Block Diagram of COP888CF Timers

Timers T1 & T2 have each two 16-bit auto-reload/capture registers (T1RA and T1RB in the case of T1 and T2RA and T2RB in the case of T2). They can generate Pulse Width Modulated (PWM) signals independently of the processor, as well as count external events and perform input capture. As denoted in Figure 6, the counter and capture modes can operate on the leading or falling edges of signals. In the figure, tc denotes the system clock.

In PWM mode, the user loads the ON time into the timer and its RB register, and the OFF time into its RA register. In this form, the PWM signal is generated independently of the processor. The importance of the PWM feature is that it can be utilized to generate 65536 different duty cycles that can represent the digitally corrected version of the analog signal being sensed. If this PWM signal is fed to a simple RC filter, we essentially have an analog interface that is compatible with today's systems. If the central controller accepts digital sensor signals, the PWM could be used directly.

If we used one of the timers in the Input Capture mode, it runs at the fixed tc rate and its associated RA and RB registers operate as capture registers. In this mode, we can sense periodic signals like the ones produced by a wheel sensor or the frequency generated by a capacitive micromachined sensor, after they pass through a signal conditioning stage. For instance, the captured data representing wheel RPM's can be transformed into vehicle speed, acceleration, jerk, etc. by the microcontroller. That is, raw data can be adjusted, corrected and finally processed into valuable information by the Smart Sensor and then fed to a central controller which is relieved from performing those computations (for example, an ABS controller).

An alternative use for one of the two timers is for the generation of the timing required to implement a software UART. Speeds of 9600 baud are easily achieved, providing the user with yet another alternative to communicate with the software controller. This serial interface can also be used to communicate with the external computer used for calibration at the time of manufacturing.

On the computational side, the COP888CF is capable of performing a 16 by 16 multiply in 544 usec (clock= 10 Mhz), with the code occupying only 36 bytes of ROM <12> (Figure 7). In the case of a third order polynomial - continuing with our pressure sensor example - and assuming 16-bit values, 6 multiplications and 3 additions are required, resulting in a total pressure calculation time of less than 2.8 msec (you save a multiplication by storing the result of the squared voltage and using it to compute the cube value).

On the packaging side, this device is available in Plastic Leadless Chip Carrier and, in the near future, will be available in Tapepak. These high density packaging technologies allow the microcontroller to fit in small sensor housings.

NON VOLATILE MEMORY - As we mentioned before, one of the greatest advantages of Smart Sensor implementations in the case of solid state sensor elements, is the ability to provide sophisticated

```
16 x 16 MULTIPLY

; NSC COP800C    (36 Bytes)
; MULTIPLICAND IN [0,1], MULTIPLIER IN [2,3]
; PRODUCT IN [2,3,4,5]

                     CNTR = OFF
3  2   BEGIN:   LD      CNTR,#17   ; CNTR <- 17
1  1            LD      B,#4
2  2            LD      [B+],#0
2  2            LD      [B],#0
2  2            LD      X,#0
1  1            RC
1  1   LOOP:    LD      A,[B]
1  1            RRC     A
2  1            X       A,[B-]
1  1            LD      A,[B]
1  1            RRC     A
2  1            X       A,[B-]
1  1            LD      A,[B]
1  1            RRC     A
2  1            X       A,[B-]
1  1            LD      A,[B]
1  1            RRC     A
1  1            X       A,[B]
1  1            LD      B,#5
1  1            IFNC
3  1            JP      TEST
1  1            RC
1  1            LD      B,#4
3  1            LD      A,[X+]
1  1            ADC     A,[B]
2  1            X       A,[B+]
3  1            LD      A,[X-]
1  1            ADC     A,[B]
1  1            X       A,[B]
3  1   TEST:    DRSZ    CNTR
3  1            JP      LOOP
5  1            RET
```

Figure 7. Mathematical Capabilities of the COP888CF

compensation for environmental variables. In orther to do this digitally, Non Volatile Memory (NVM) is necessary for the storage of the compensation coefficients.

This memory can take the form of ROM, EPROM, EEPROM or Battery Backed-Up RAM. Due to the cost constraints of automotive applications, the latter seems to be impractical. Masked ROM is also impractical, since the coefficient storage must be done at the time the sensor is manufactured, not at the time the microcontroller is. Hence, let us talk about the other two.

EPROM technology is a very stable and reliable technology that can be easily used in these type of applications <13>. Due to obvious constraints of Smart Sensor applications, it is not practical to utilize an IC package with a window for ultraviolet erasing of the EPROM. Therefore, this type of memory has to be used in the OTP mode (one time programmable).

This limits the flexibility of this approach since, if at the moment of sensor manufacturing anything goes wrong, the data cannot be reprogrammed. This problem can be partially solved by using a larger memory size than the one needed, although the slack portion would be wasted in most cases.

Another key consideration is that EPROM for this type of application must be on the same die as the microcontroller, since small size EPROMs of the kind needed for coefficient storage in Smart Sensors are not available as stand-alone units. However, the larger size of an EPROM could be used to store a large number of already computed data points, implementing the corrections by table lookup.

A more flexible approach requires the use of small Electrically Erasable PROMs or EEPROMs. Data can be stored and erased multiple times in this case, typically more than 10,000 times over the life of the product. This not only allows for more flexibility at the time of manufacturing, but also allows for recalibration over time (either on- or off-vehicle).

Small size EEPROMs of the type needed for Smart Sensors are readily available as stand-alone units at very low cost. This gives the user the flexibility of using a microcontroller with EEPROM on-board or off-board, depending on cost and reliability considerations at the time of doing the design. An added potential benefit of using EEPROM technology is that diagnostics information can be stored in real-time.

EEPROM technology has progressed substantially, so it can be used reliably in applications subject to the underhood temperature range (-40 to +125C).

Examples of parts suitable for Smart Sensors are National's NMC93C06, NMC9326 and NMC93C46, which are implemented with CMOS technology (also produced in NMOS versions) <14>. They are available in Small Outline (SO) 8-pin surface mount packages (ideal for this case) and contain 256, 512 and 1,024 bits of memory, respectively. Alternatively, they are all available in a 14-pin SO package, in order to provide users with a footprint that will allow them to upgrade to larger memory sizes.

As we can see in Figure 8, the key to making these EEPROMs fit in a small IC package is the use of the 3-pin serial interface Microwire (SK clock, DI data input, DO data output) that can easily interface to microcontrollers like the COP888CF.

The NMC93C06 can store 32 coefficients of 8-bits each or 16 of them if 16-bit ones are being used (in the latter case, an up to third order polynomial for 4 different temperatures can be used). The NMC93C26 and the NMC93C46 can store 2 and 4 times as many coeficients, respectively.

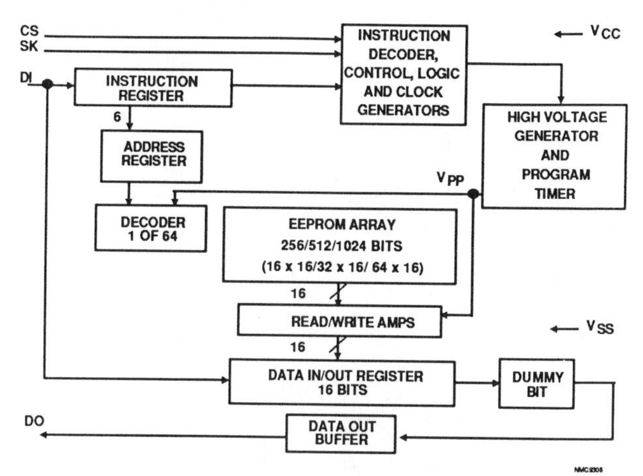

Figure 8. Block Diagram of NMC93C06/26/46 EEPROM's

PRACTICAL IMPLEMENTATION.

Figure 9 illustrates the implementation of a microcontroller-based Smart Sensor using some of the components mentioned throughout the paper. This implementation features a single +5V supply, CMOS technology, -40 to +125C temperature range and high density packaging for all the main components.

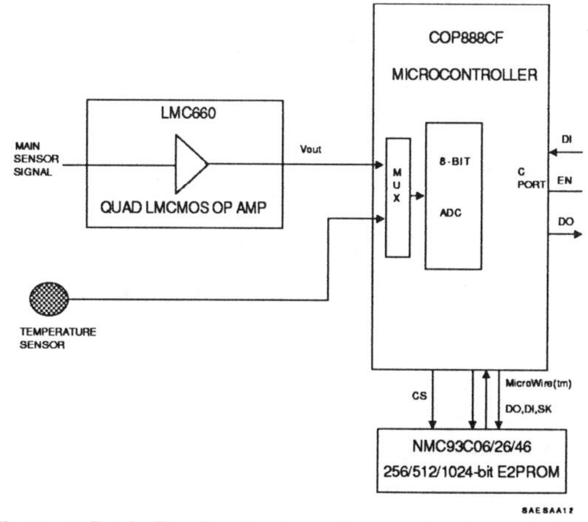

Figure 9. Basic Blocks of a Smart Sensor Implementation

The main sensor signal is amplified and conditioned by the LMC660 Quad Operational Amplifier and then it is converted to digital by the ADC that resides inside the COP888CF Microcontroller. This in turn communicates with the off-board NMC93CX6 EEPROM via the Microwire serial interface, which stores the correction coefficients. Communication to the central controller can be implemented via a software UART, a PWM signal or an analog signal derived by RC filtering the PWM signal (a timer would be needed in all cases). In the future, a serial interface like the recently adopted J1850 SAE Multiplexing Standard could be used.

Figure 10 shows that the aforementioned components, in SO package and Tapepak versions, occupy an area smaller than a U.S. 5 cents coin (21 mm in diameter). This is of critical importance for Smart Sensor implementations, where space is a premium.

Figure 10. Smart Sensor Components in High Density Packages

CONCLUSION.

Microcontroller-based Smart Sensors that were not feasible a few years ago are now a sensible alternative where high accuracy over a broad temperature range is desired. This is especially important in the case of solid state sensors, which are making major inroads into automotive applications.

All the components necessary to achieve these implementations are now available at a cost that will allow them to penetrate the high-end applications first and, eventually, the mainstream applications.

BIBLIOGRAPHY.

1. J. Paulsen, "Powertrain Sensors and Actuators: Driving Toward Optimized Vehicle Performance", Convergence '88 Proceedings, SAE/IEEE Publications.

2. M. Alba, "A System Approach to Smart-Sensors and Smart-Actuators Design", SAE Congress 1988, Sensors & Actuators Proceedings, SAE Paper 880554.

3. R. Knockeart, "Integrated Micromachined Silicon: Vehicle Sensors of the 1990's?", Convergence '88 Proceedings, SAE/IEEE Publications.

4. C. Gross, T. Worst, "Intelligent Interface for Electronic Pressure Scanners", ISA Proceedings.

5. P. Frere, S. Prosser, "Temperature Compensation of Silicon Pressure Sensors for Automotive Applications", Sixth International Conference on Automotive Electronics Proceedings - 1987, Institution of Electrical Engineers.

6. J. Brignell, A. Dorey, "Sensors for Microprocessor-Based Applications", Journal of Phys. E; Sci. Instrum., Vol. 16-1983.

7. W. Wolber, Notes from the "Future Sensors and Measurements: Intelligent Sensors" Seminar, SAE Continuing Education.

8. P.N. Mahana, F.N. Trofimenkoff, "Transducer Output Signal Processing Using an 8-bit Microcomputer", IEEE Transactions on Instrumentation and Measurement, June 1986.

9. M. Arington, M. Railey, "Principles of Automatic Alignment and Calibration Using Embedded Microcomputers", IEEE Transactions on Instrumentation and Measurement, March 1985.

10. National Semiconductor, "Linear Databook 1", 1988 Edition.

11. National Semiconductor, "Microcontrollers Databook", 1988 Edition.

12. National Semiconductor, "Benchmark Comparisons - COP800", December 3, 1986.

13. M. Kraska, "Digital Linearization and Display of Non-Linear Analog (Sensor) Signals", 1988 IEEE Workshop on Automotive Applications of Electronics, IEEE Publications.

14. National Semiconductor, "Memory Databook", 1988 Edition.

890762

An Integration Approach on Powertrain Control System

Yoshinori Ohno, Kazuhiko Funato, and Kazuo Kajita
Toyota Motor Corp.

1. Abstract

Engine control systems were the precursor of scale automotive electronics systems using microcomputers.
Toyota Motor Corporation introduced high−level, total control of the power train by applying system integration through introducing a multi−CPU system to the 1988 MY Toyota Camry.
Integration in the ECU has been promoted in parallel with system integration. By adopting single−chip microcomputers, monolithic ICs, and hybrid ICs all designed and developed for car electronics, and semiconductor barometric pressure sensors for incorporation into ECU's, etc., ever−expandable functions can be provided in a smaller and more lightweight ECU package with higher reliability.

THE USE OF ELECTRONICS FOR AUTOMOBILE CONTROL in Toyota began with fuel injection control. An introduction of microcomputers as a solution to stringent emission regulations and improvement of fuel economy requirements, gave a trigger to enhanced that engine control system has advanced to the precision digital control as the total engine control system through the incorporation of various control functions including spark advance and idle speed control.
As the proliferation of automatic transmission, adoption of electronics to the automatic transmission is promoted. This has enabled improvements in electronic control of shifting and lockup schedule, along with improved fuel economy and shift quality (Fig. 1).

Function	'74	'75	'76	'77	'78	'79	'80	'81	'82	'83	'84	'85	'86	'87	'88
E F I															
A/F feedback															
E S A															
I S C															
Diagnostic															
ECT interface															
Back up circuit															
E G R															
Knock control															
E C T															
Torque control															

Fig. 1 Integration of Powertrain Electronics

High level integration and improved processing speed achieved by electronics promoted integration of different control systems to provide advanced performance which cannot be obtained in individual control systems. Variety of integrated control systems is now studied and high−level, integrated systems are marketed.
There are three ways to go about achieving this integration.

1) Unify all control functions into a single integrated ECU which carries out complex control. (small scale)

2) Arrange for communication between a several number of control systems for complex control. (medium scale)

3) Link each control system through network communications, supply data to each control system as it is needed and thus achieve overall control. (large scale)

Integration of circuits in an ECU itself is also a key to provide expanding functions without increasing the ECU size or number of devices. To meet such a demand, Toyota developed integrated electronic parts for automobiles; these parts are optimally installed inside an ECU. High density packaging technology could design reliable systems without increasing the number of parts and ECU size (Fig. 2).

As an example of this integration, this paper discusses the integration of the engine control and transmission control ECU's into a single ECU to make a system which carries out overall control as well as the integration within the ECU itself.

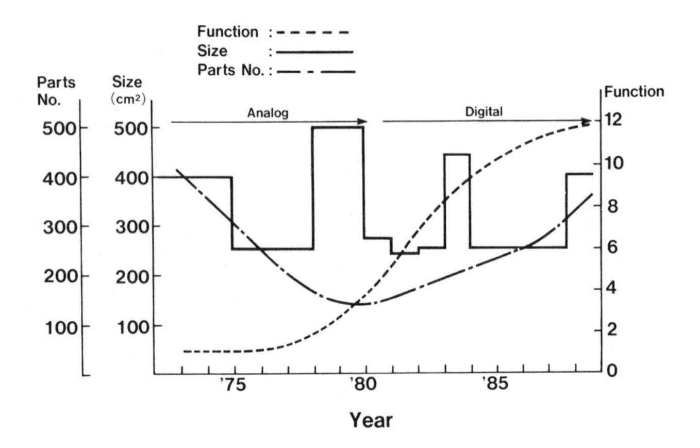

Fig. 2 Progress of ECU Functions and Size

2. Outline of the '88 MY Camry 2VZ−FE System

A multiple CPU system which is an example of system integration is adopted in the 1988 MY Toyota Camry with 2.5 L V6 2VZ−FE engine. This system combines the fuel injection, spark timing, Idle Speed Control (ISC), automatic transmission control and the on−board diagnostic functions. A system diagram is shown in Fig. 3.

Fuel injection control calculates the optimum fuel injection duration for the engine conditions and carries out air−fuel ratio feedback control. This system also includes a learning function which compensate the base air−fuel ratio automatically for changes over time to achieve an intelligent control system.

Spark timing control is achieved by means of the ECU determining the advance angle from an ignition map stored in memory based on engine speed and loard.

ISC system controls the engine's fast idle speed during warmup and the idle up speed during air conditioner compressor operation. A stepping motor is used as the actuator which drives the ISC control valve.

Transmission control operates shift control solenoids in accordance with a shift pattern map stored in the ECU based on the throttle angle, vehicle speed and engine speed. To further reduce shock during shifting, the spark timing is changed to control the output torque.

The on−board diagnostic system detects system malfunctions and turns on the check engine lamp to warn the driver that repairs are needed. In addition, this system stores the trouble code, making it possible for service personnel to determine the nature of malfunctions rapidly and accurately.

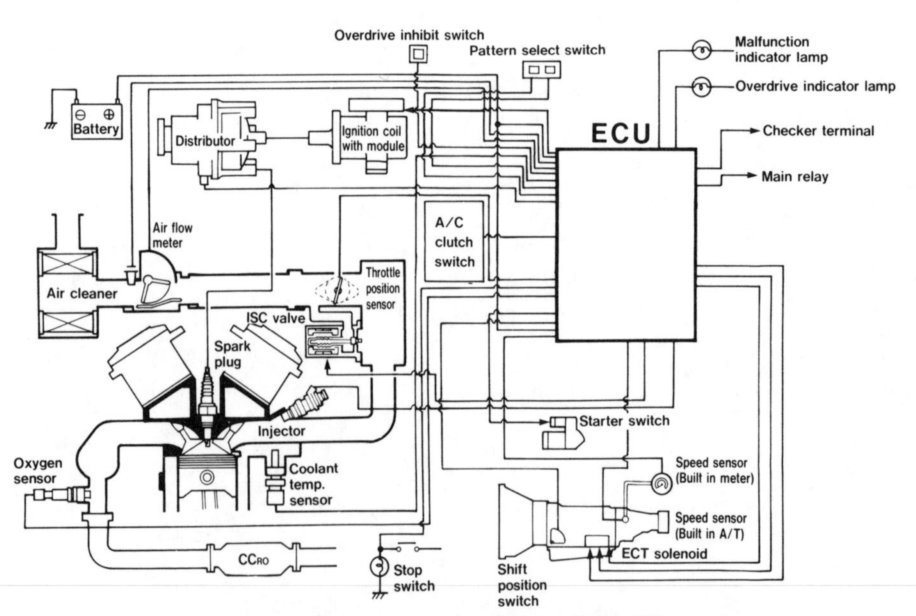

Fig. 3 '88 MY Toyota Camry 2VZ−FE

3. Powertrain Integration

(1) System Integration

In the past, engine control and transmission control were performed by different ECU's, but there are many input signals which can be shared in these two systems. Therefore, in order to achieve overall control of the powertrain, the integration of these systems into a single ECU was undertaken.

Two single—chip microcomputers with a high level custom design were used in this ECU to comprise a multi—CPU system, with one microcomputer chip controlling the engine and the other controlling the transmission, with mutual communication between the two making overall control possible. Fig.4 shows TCCS internal block diagram.

In this multi—CPU system, output torque is controlled by recognizing shifting progress conditions on the basis of engine rpm and vehicle speed during shifting and immediately before shifting is completed—during which time torsional vibration of the drivetrain becomes a problem—and ignition timing is retarded to reduce the amount of shock during shifting. The retard angle is determined on the basis of the shift type and throttle position. This torque control is achieved by communication between two CPUs. The ignition angle retard request and amount of retard angle signals are output from the transmission control CPU and the throttle angle signal and ignition retard angle control signal are output from the engine control CPU.

The transmission control CPU not only sends shifting instructions to the shift control solenoid determined from the vehicle speed and throttle position but also monitors the engine speed. When the right speed conditions are established, the transmission control CPU sends an ignition angle retard request to the engine control CPU, instructing that the ignition timing be retarded in accordance with the throttle position and the shift pattern. The engine control CPU judges if the ignition timing is within a range where it can be retarded in accordance with the ignition retard angle request; if it is, it retards the ignition accordingly. The CPU controls the rise in exhaust temperatures by compensating the air—fuel ratio Simultaneously, ignition retard angle control is limited by the coolant temperature, battery voltage and duration of retard, etc. In some cases the ignition retard angle request from the transmission control CPU cannot be complied with. In such a case, the engine control CPU outputs a inhibit signal to the transmission control CPU.

Fig. 5 shows the timing chart for signal transmission between the two CPU's.

The signals between the two CPU's are transmitted along 5 signal lines and the amount of ignition retard angle is expressed by the information in the three bits ESA1, ESA2 and ESA3. The Hi or Lo status of the ignition angle retard request signal (command request) indicates whether a command is requested or not.

The ignition angle retard control status signal (status) indicates one of three statuses—inhibit, enable and run—with Hi, Lo or pulse output.

The data for ignition angle retard and air—fuel ratio compensation are processed at high speed using the signals described above and the appropriate adjustments made within 150ms when a gear shift is made.

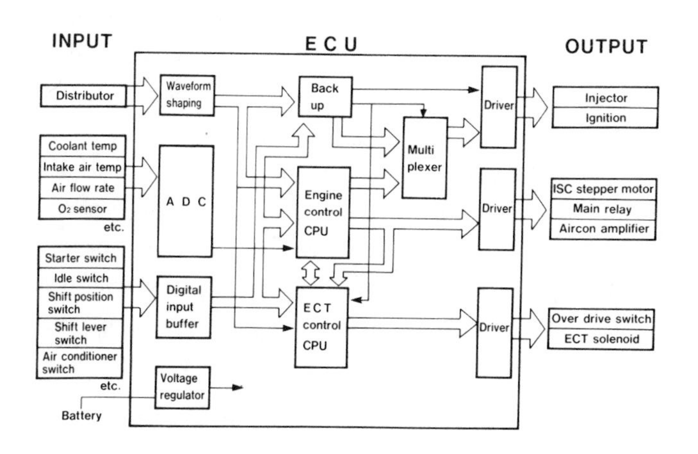

Fig. 4 Block Diagram of the ECU

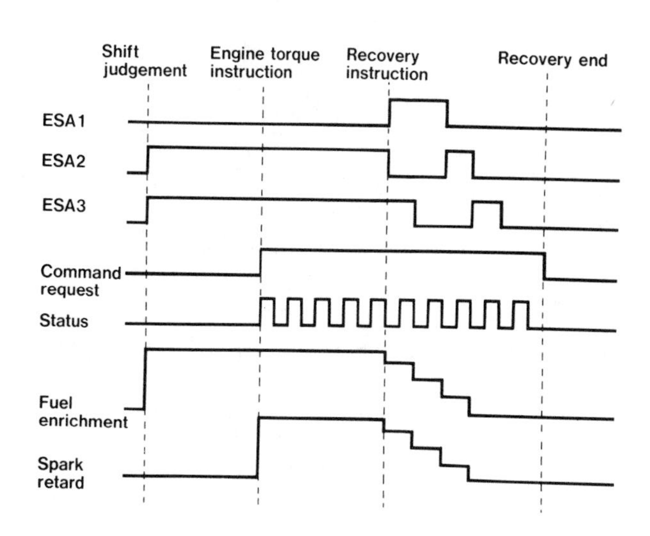

Fig. 5 Communications Between 2CPUs

In fail safe logic, the engine control CPU and the transmission control CPU have a master—slave relationship. The engine control CPU is the master. It includes a watchdog timer in the voltage regulator IC and a backup IC, which monitor for abnormal states. The transmission control CPU, which is the slave, is monitored by the master CPU. Thus, fail safe measures are always put into effect when an abnormal state occurs in either CPU.

If trouble occurs in the master CPU, the watchdog timer in the voltage regulator IC is activated immediately, sending initialize signals to the master CPU and carrying out recovery operations. If the trouble continues, it is detected by the backup IC. This causes the backup IC to take over computation for ignition and injection from the CPU mode, switching to predetrmined ignition and injection signals; this system enables the engine to continue operating and prevents breakdown. At this time, the slave CPU monitors the failure signals from the backup IC and changes the shift pattern to a pattern which is safe for the current running state. In addition, the engine malfunction indicator lamp lights up to warn the driver. If trouble occurs in the slave CPU, a watchdog timer created from software in the master CPU immediately starts to work, supplying initialize signals to the slave CPU in order to carry out a recovery operation. The ignition angle retard control normally carried out during shifting through communications between the two CPU's is forbidden during this time. Furthermore, the trouble code for the slave CPU malfunction is recorded for the purpose of diagnosis. (Fig. 6)

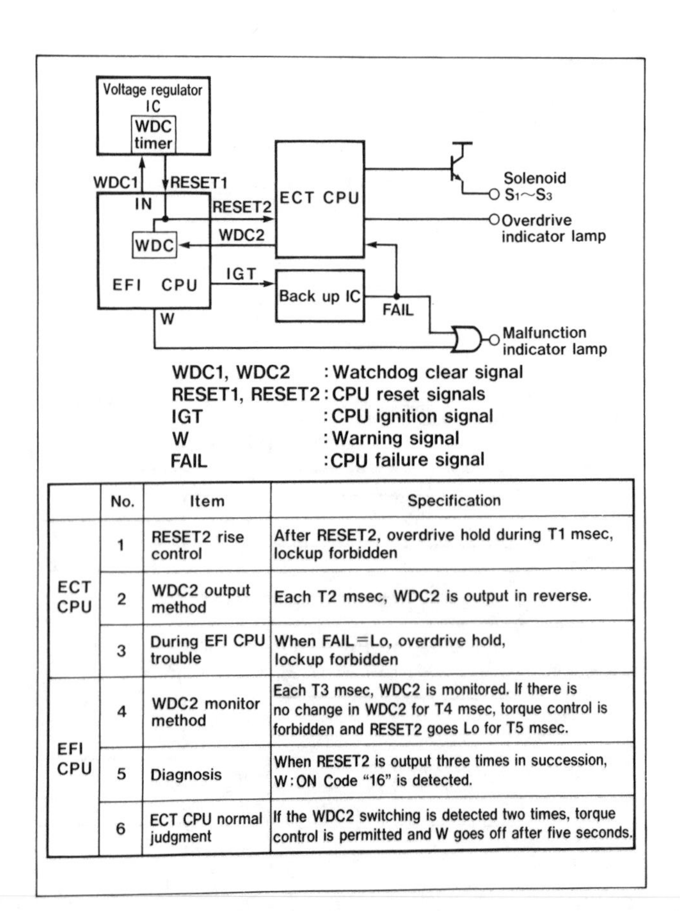

	No.	Item	Specification
ECT CPU	1	RESET2 rise control	After RESET2, overdrive hold during T1 msec, lockup forbidden
	2	WDC2 output method	Each T2 msec, WDC2 is output in reverse.
	3	During EFI CPU trouble	When FAIL=Lo, overdrive hold, lockup forbidden
EFI CPU	4	WDC2 monitor method	Each T3 msec, WDC2 is monitored. If there is no change in WDC2 for T4 msec, torque control is forbidden and RESET2 goes Lo for T5 msec.
	5	Diagnosis	When RESET2 is output three times in succession, W: ON Code "16" is detected.
	6	ECT CPU normal judgment	If the WDC2 switching is detected two times, torque control is permitted and W goes off after five seconds.

Fig. 6 2CPUs Fail Safe Logic

(2) ECU Integration

The ECU is configured from a microcomputer, custom IC, transistors, diodes, resistors, condensers and other components. Particularly in regard to the ECU, a lot of effort is going into integration through the adoption of custom designed single—chip microcomputers with high level functions, custom designed monolithic IC's, hybrid IC's with high density packaging and semiconductor barometic pressure sensors built into the ECU. As these functions increase year by year, it becomes more and more possible to reduce the size and weight of the ECU.

Single—chip Microcomputer:

Toyota began to use CMOS design 8bit microcomputers with advanced functions in 1984, and the family of such devices has grown since that time.

This 8bit microcomputer is a custom designed single—chip type microcomputer develpoed jointly by Toyota, Nippondenso and Toshiba.

A single—chip microcomputer uses a single LSI integrated package which inclueds the CPU, Input/Output parts, Memory and other peripherals, with the effect that the size and cost of the ECU are reduced while its reliability is improved.

This single chip microcomputer uses compact command codes to make effective use of the on—chip ROM.

For example, use of bit manuplation/bit condition branch commands is considered to be an extremely effective means of reducing the memory capacity required for programming and consolidates bit operation commands.

The frequency of use of bit operation commands in powertrain control is about 15—20% of all commands.

Chips with memory capacities of 12K, 8K, 6K, and 4Kbytes are made, making it possible to select a chip which matches the scale of the system for which it is adopted. On larger scale systems, multiple chips can be used, but high level functions in high density packages make real time processing possible without the necessity of increasing the package area. (Fig. 7, 8).

Function		T5897	T7616	T6312	T7433	T7616 +T7433
Technology		C-MOS	C-MOS	C-MOS	C-MOS	C-MOS
Memory	ROM	4KB	6KB	8KB	12KB	18KB
	RAM	192B	224B	256B	384B	608B
Instruction set	Multiply	8×8	8×8	8×8	8×8	8×8
	Divide	16/8	16/8	16/8	16/8	16/8
	Bit manupulation	Yes	Yes	Yes	Yes	Yes
High speed I/O	Capture	4	4	4	4	8
	Compare	4	4	4	8	12
Interruption	Factor	15	15	16	16	15+16
	Level	8	8	8	8	8+8
Package		DIP42P	DIP42P	DIP42P	DIP64P	DIP42P +DIP64P
Application		Small scale	Medium scale		Large scale	

Fig. 7 Custom Microcomputer for Toyota Moter Corporation

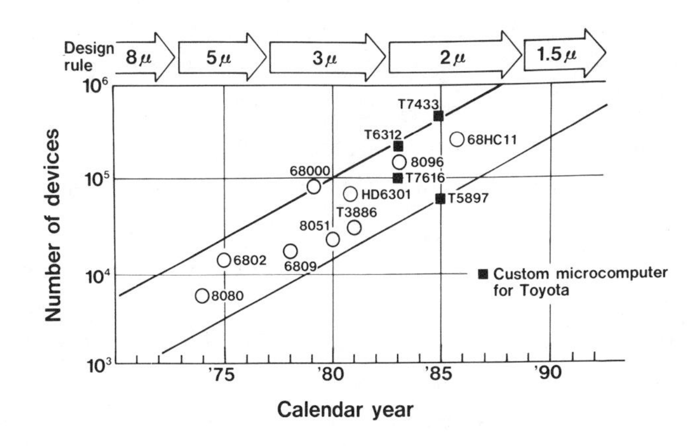

Fig. 8 Trends in Microcomputers

Monolithic IC's:

A large number of custom designed monolithic IC's with sophisticated functions are used in ECU's for powertrains. One example of monolithic IC's is a 11bit analog to digital (AD) con-

between analog inputs and digital inputs. The AD converter and digital buffer can be integrated into a single IC, making it possible to reduce the size and cost of the ECU while improving its reliability.

This AD converter includes a high resolution 11bit analog to digital conversion circuit and 16 channel input circuit and carries out high speed serial communications, receiving data for the microcomputer. With a CMOS design, this high speed device has a conversion time as fast as 246μs and is contained in a 28 pin DIP package. (Fig. 9)

verter with 16 chanel input port. This device can be switched

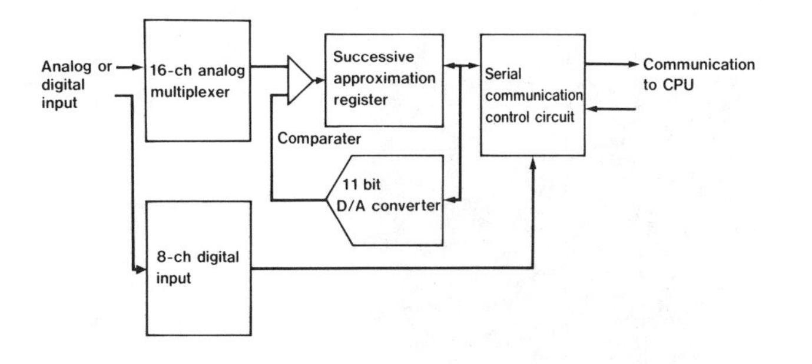

Fig. 9 16-ch Input 11 bit A/D Converter Block Diagram

Hybrid IC's:

Hybrid IC's are used to make highly efficient circuit blocks for powertrain control, and aid the development of small, lightweight ECU's with high level functions.

Hybrid IC's consist of custom monolithic IC's, transistors, diodes, capacitors and resistors surface mounted on ceramic substrates and contained in single in-line packages (SIP) with good packaging efficiency.

Some examples of hybrid IC's are input buffers, output buffers, voltage regulator circuits and parallel/serial conversion circuits. (Fig. 10)

Hybrid IC's have a high mounting efficicency and many variations can be produced in a short development time. However, they lose out to monolithic IC's from the standpoint of cost.

The method of first achieving functional integration and size reduction in a monolithic IC and then incorporating this monolithic IC into a hybrid IC allows the realization of ultra-compact IC's.

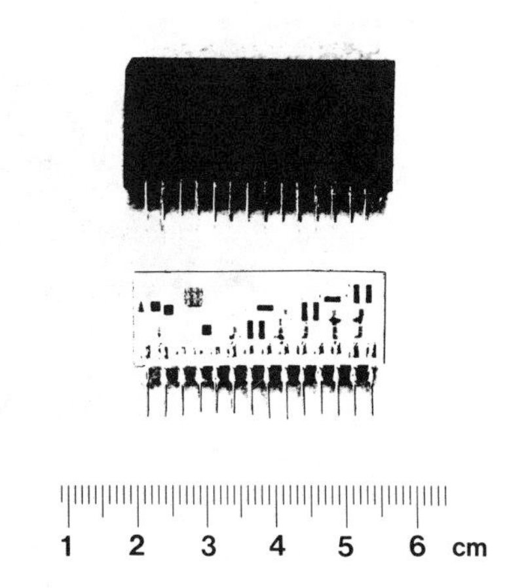

Fig. 10 External View of Hybrid IC

Semiconductor Barometric Pressure Sensors Built into the ECU:

One example of the integration of sensors with the ECU is the semiconductor barometric pressure sensor built into the Electronic Control Unit for the 1988 MY Toyota 4 L 6 cylinder engine, which made it possible to achieve a small, lightweight and less costly ECU.

Based on out accumulated experience with stand alone intake air, turbo and barometric pressure sensors, a high precision sensor are achieved with a measuring range of 460−800 mmHg with an accuracy of ±10 mmHg.

This sensor is constructed by mounting a semiconductor pressure gauge, a monolithic IC for amplification, resistors and condensers on a ceramic substrate, with fine adjustments carried out by laser trimming. A single in−line package which is the same size as a hybrid IC is used. (Fig. 11)

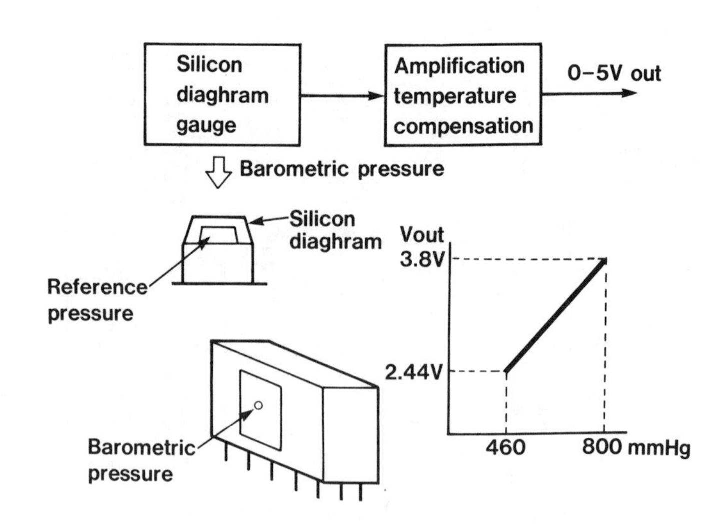

Fig. 11 Semiconductor Barometric Pressure Sensor Built into the ECU

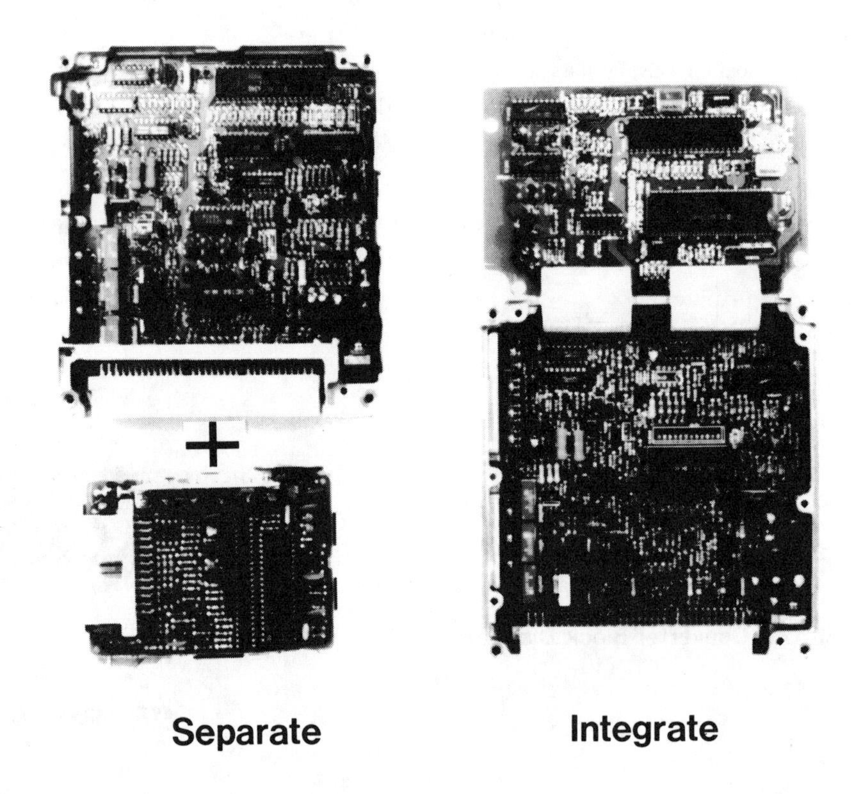

Fig. 12 Comparison of ECU's

4. Summary

Through integration with high level functions and organic links between systems, more precise and accurate control capabilities have been achieved. This system delivers better drivability, fuel economy and safety.

Along with the integration which resulted in this system, there is a tendency for the size of the ECU to increase, but through the use of fruits of integration technology such as hybrid IC's, monolithic IC's, surface mounted devices (SMD's) and custom microcomputers, it has been possible to keep the size and weight of the ECU down, to reduce the number of components and achieve higher reliability.

System integration, discussed in this paper, will realize more advanced control. In parallel with the integration of systems, integration in ECU will also make further advances.

5. References

(1) T. Kawamura, et al
"Toyota's new single−chip microcomputer based engine and transmission control system" SAE850289

(2) S. Mizutani, et al
"Automotive Electronics in Japan−The Next Five Years" SAE851654

(3) H. Ono
"Electronic Engine & Driveline Management"
1988. 10. 18. SAE Australia Seminar Proceeding

890765

Six Cylinder EFI Control Using a Low-Cost Microcontroller and the Universal Pulse Processor IC

Prabhas Kejriwal
Hitachi Microsystems International, Inc.
San Jose, CA

Robin Blanton
Hitachi America, Ltd.
Dearborn, MI

ABSTRACT

This paper describes a solution to control applications which require a large number of counters/timers with a maximum resolution of 5 microseconds. The solution comes from integrated circuits using a Universal Pulse Processing (UPP) core. The UPP core provides up to 24 channels of very versatile counter/timer/shift register functions with a built-in Arithmetic Logic Unit (ALU) and associated control functions. The concept of the UPP is explained and an application example is shown. The proposed application illustrates the use of a microcontroller and a UPP to control a conceptual six cylinder fuel injected engine.

INTRODUCTION

Control systems normally require the generation of waveforms with controllable timings and measurement of level change timings of certain signals. Often the number of timers available in a microcontroller is limited and is insufficient for the task. When a time resolution of 50 microseconds or more is adequate, software routines driven by some periodic interrupt may provide sufficient additional timing functions, while reducing CPU throughput. When system performance requires a higher level of throughput, the use of external counter/timer chips is necessary. In systems where additional timers are required, some typical solutions have been to add a custom counter/timer chip, to use multiple microcontrollers, or to add several standard counter/timer chips. **Figure 1** shows a typical block diagram of such a system using multiple HD63B40 Programmable Timer Modules.

Universal Pulse Processor (UPP) chips provide a potential solution to this need for additional hardware with the benefit of reducing total system cost and complexity. One UPP chip, such as the HD63140, provides up to 8 internal and 16 external channels of very versatile counter/timer/shift register functions. Any unused I/O channels can be used as parallel input/output. In addition to the UPP timer core, other on-chip functions include an 8 channel, 10 bit analog to digital converter, 1K bytes of SRAM, and a watchdog timer. **Figure 2** shows a block diagram of such a system using a UPP peripheral chip.

Figure 1. Current Engine Control System

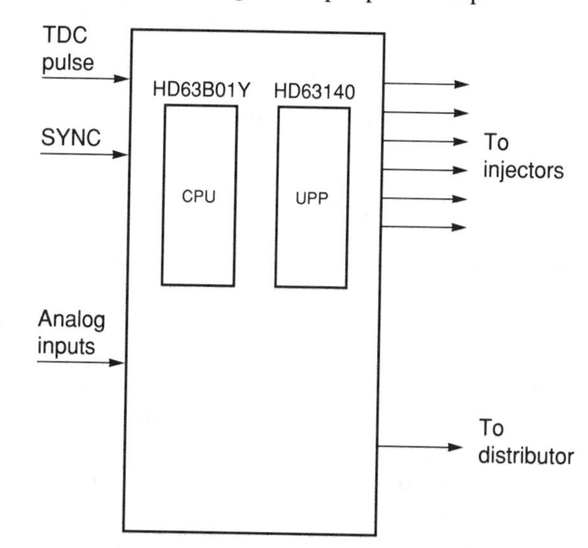

Figure 2. New Engine Control System Using UPP

The total system cost is reduced through the reduction of the part count from 6 LSIs to 2. The cost of an HD63140 UPP chip is less than the total cost of the programmable timer modules, and the A/D converter that it replaces. System cost is further cut by the reduction of the circuit board area used, and decreased part count lowers assembly costs. The lower chip count, with its associated lower number of interconnections, should also lead to a more reliable system. Another advantage of a solution using a UPP peripheral chip is that it reduces the software required by the microcontroller for the generation of routine timing outputs and measurement of input pulses. This will allow the microcontroller to use its CPU power for more sophisticated engine control algorithms and to control higher RPM engines. **Table 1** shows a comparison of the software overhead associated with the two systems.

Table 1. Comparison of software overhead

Item	Current System	New System
Bus Speed	2 MHz (A/D)	2 MHz (CPU) 4 MHz (UPP)
A/D Converter	10 bit × 16 ch 100 μs ± 1 Lsb (+25°C)	10 bit × 10 ch 42 μs ± 2 Lsb (–40~85°C)
Number of ICs	7	3

A brief description of the UPP concept and a description of its potential use in a six cylinder Sequential Multi-Point EFI control application follows.

UNIVERSAL PULSE PROCESSOR CONCEPT

A **UPP** is a programmable I/O module with a 16-bit **A**rithmetic **L**ogic **U**nit. **Figure 3** shows the block diagram of a UPP core. Its features include:

1. 16 bit Arithmetic Logic Unit (ALU)
2. 24 I/O nodes, 16 brought out to I/O pins, 8 internal only
3. 24 16-bit data registers which can be used as counter/timer/compare/capture/shift registers
4. 16 step program space (within the function table)
5. 16 available functions (op codes)

The function table in a UPP specifies up to 16 user definable operations that are performed during execution. There are an additional 4 pre-programmed operations in the function table, for a total of 20 program steps. During execution, the control passes through the table entries which execute the specified operations. The number of steps in the table that are sequenced through can vary between 1 and 20. The programmable contents of the register called **M**aximum **F**unction **N**umber **R**egister (**MFNR**) determine the number of program steps to be executed.

With some restrictions, registers and I/O pins can be shared among functions. The following functions are available.

Functions which sample a counter/timer register value at the occurrence of external trigger:

1. FRS—Free running counter/timer with sampling
2. INS—Interval counter/timer with sampling
3. UDS—Up/Down counter/timer with sampling
4. GTS—Gated Counter/timer with sampling

Functions which cause an event (level change in an output node) when a Counter/Timer register matches with a compare register:

5. FRC—Free running counter/timer with compare
6. INC—Interval counter/timer with compare
7. PWC—Pulse width counter/timer with compare
8. OSC—One shot counter/timer with compare
9. FFC—Fifty-fifty duty counter/timer with compare
10. GTC—Gated counter/timer with compare

Other Counter/Timer functions:

11. TPC—Two phase up-down counter
12. CTO—Combination trigger one shot counter/timer

Shift Functions:

13. SIT—Shift input
14. SOT—Shift output
15. SPO—Shift parallel output

Miscellaneous:

16. NOP—No operation

There is a choice between a counter and a timer in many of these functions. The difference between the two is that while a counter counts external pulses, a timer counts internal clock pulses. A counter can be up or down, whereas a timer is always up.

Figure 4 shows the block diagram of an HD63140 chip with a UPP core.

MICROPROCESSOR–UPP INTERACTION — The UPP is recognized by a microprocessor as an I/O device. As part of the initialization, the microprocessor will initialize the registers, set-up the function table and enable the UPP. The data transfer between the microprocessor and the UPP occurs through the UPP data registers on an eight bit data bus. The UPP can also be set-up to interrupt the processor when a specified event occurs.

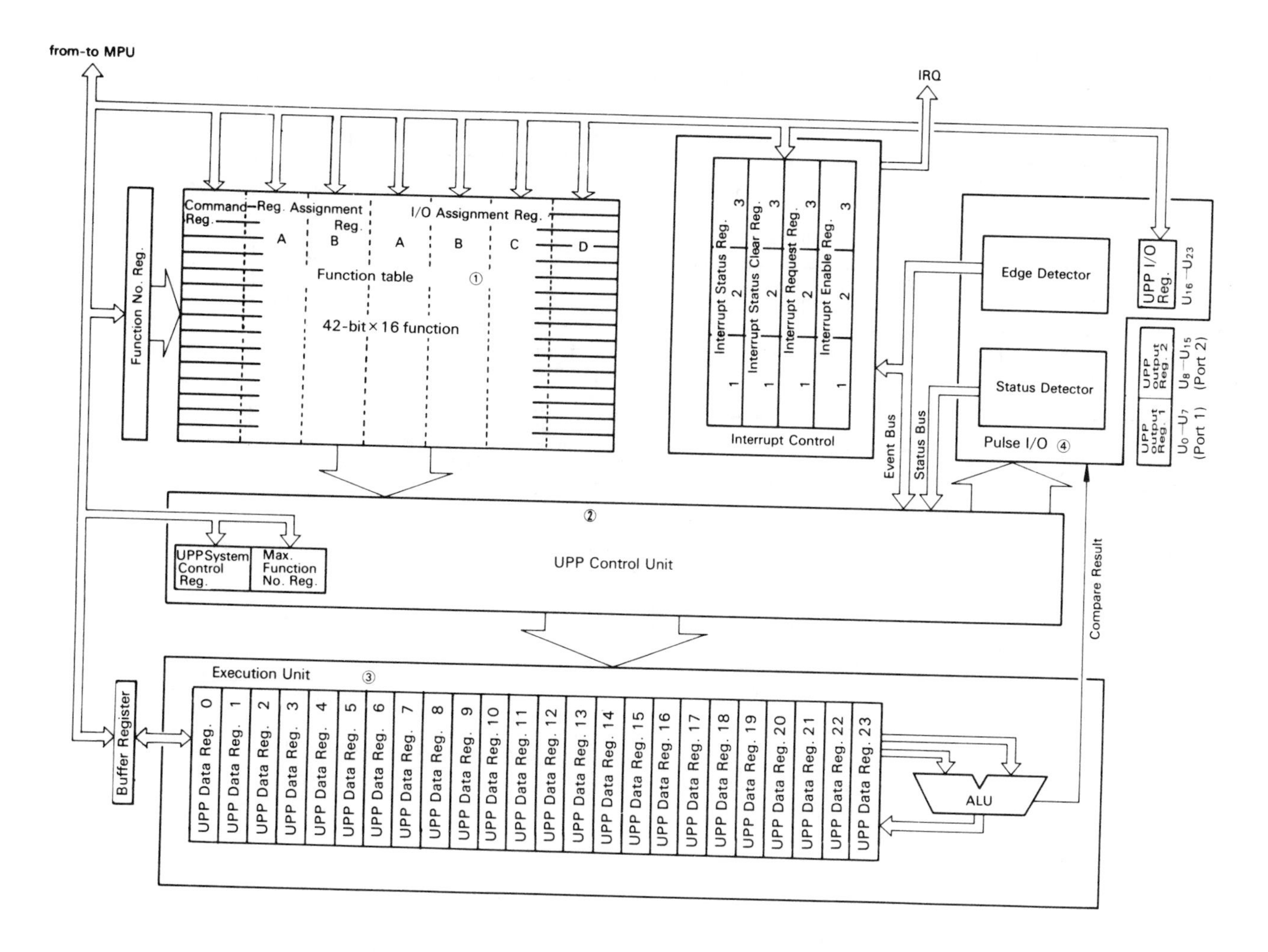

Figure 3. UPP Core Block Diagram

COUNTER/TIMER INTERACTION—There are three levels at which the different functions in an UPP can interact with each other:

1. Common I/O Pins—The output signal of one function can be the input signal of another function by using a common I/O pin.

2. Common Data Register—Data can be transferred from one function to another by using a register that is common to the two functions. For example, an up-counter may be used as a common reference register for multiple pulse-width-modulation outputs. **Figure 5** shows an example of this.

3. Software control—The microprocessor software can read the output of one function, and change the register of another function based on the initial reading.

In a system using multiple counter/timer ICs, the common data register mechanism is not available. The common I/O mechanism is available to a limited extent, but since the number of channels in a counter/timer IC is small, the interaction which is possible in this way is limited.

SIX CYLINDER FUEL INJECTION ENGINE CONTROL

A simplified six cylinder sequential fuel injection engine controller is illustrated. This controller uses the HD6301Y, which is a low cost 8 bit microcontroller and an HD63140 (UPP). The HD6301Y has 16 K bytes of ROM, 256 bytes RAM, two timers, a serial port and many I/O pins.

In this implementation, the UPP generates all of the signals to control the fuel injectors and spark output signal. The timing for these outputs are based on two assumed inputs; a sync input pulse that slightly precedes the intake Top Dead Center (TDC) of cylinder 1, and a common TDC input for all 6 cylinders that indicates the fuel intake cycle. The CPU sets the values of fuel injection delay, injection duration, spark delay and spark duration. If the engine is running at a constant load and speed, no CPU intervention is required. The UPP also provides the CPU with a measure of the engine cycle time.

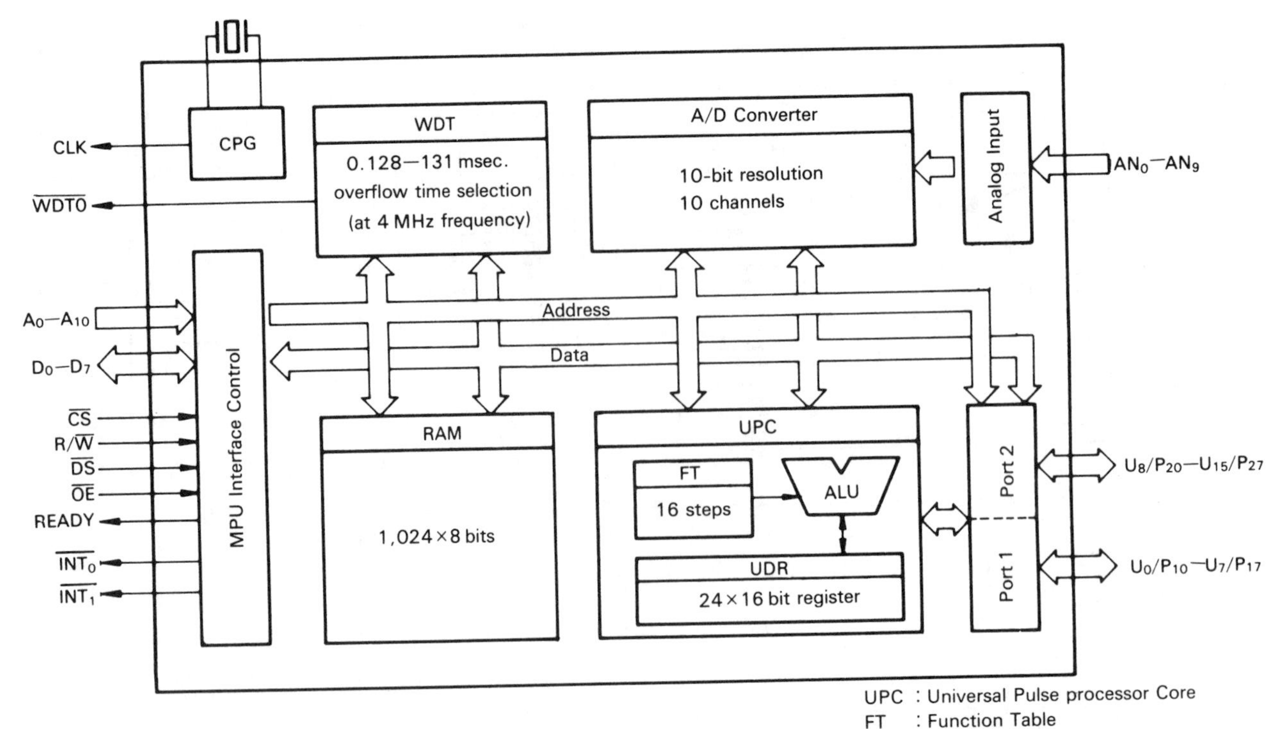

Figure 4. HD63140 Block Diagram

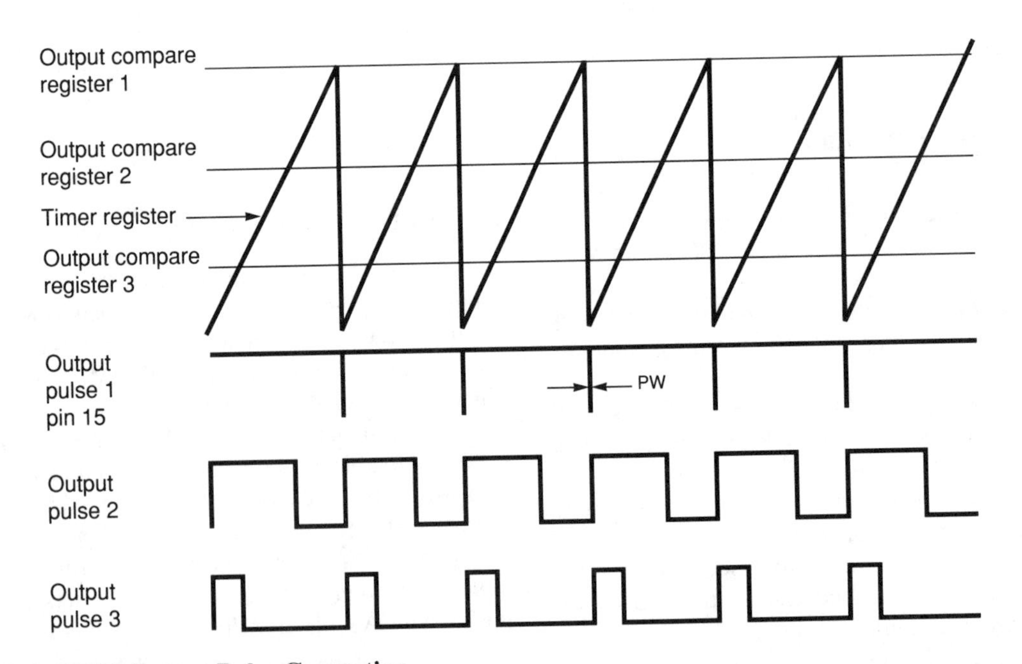

Figure 5. Multiple PWM Output Pulse Generation

HARDWARE

Figure 6 shows the interconnections between the HD6301Y microcontroller and the UPP. The HD6301Y is used in the expanded mode so that the data and address buses are externally available. The UPP operates at 4 MHZ and requires a 16 MHZ crystal. The UPP provides a clock on its CLK output pin with a frequency that is half that of the crystal. This 8 Mhz clock is connected to the EXTAL input of the HD6301Y. The CPU further divides the EXTAL input and runs at 2 MHZ. The data bus (D0-D7) and address lines (A0-A10) are directly connected to the UPP. The upper 5 address lines are decoded to generate the

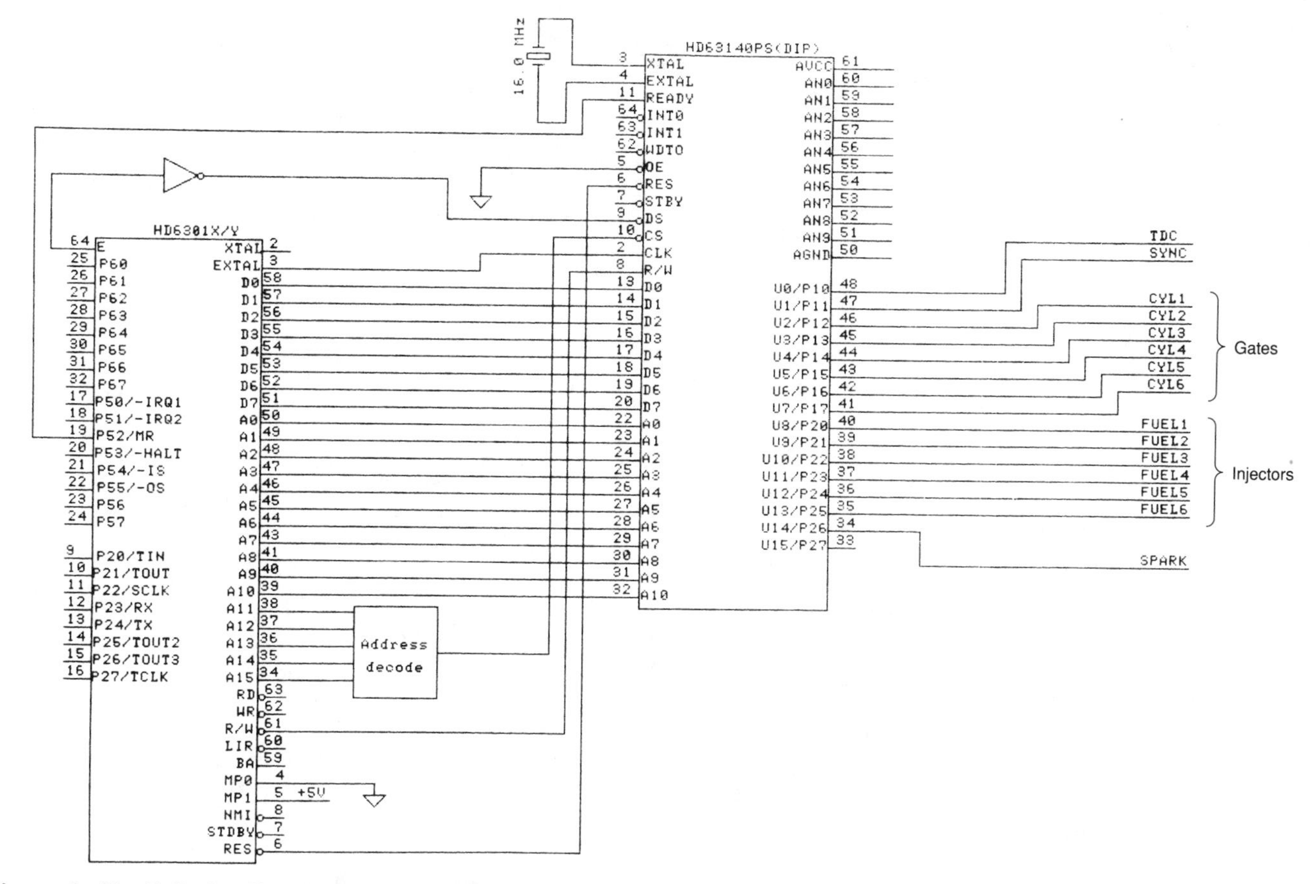

Figure 6. Six Cylinder Fuel Injection Engine Control Schematic

chip select for the UPP. The READY line of the UPP is connected to the MR (memory ready) input of the HD6301Y so that the CPU can synchronize the UPP for register accesses (which take longer than normal memory cycles).

UPP I/O

The UPP generates fuel injection and spark control signals. **Figure 7** shows the signal timings while the I/O pin assignment is shown in **Figure 6.** Fifteen out of 16 available I/O pins are used. The first two of these (TDC and SYNC) are inputs and the rest are outputs. Below is a brief description of the signals:

TDC – TOP DEAD CENTER (I) — A LOW to HIGH transition takes place in this signal every time one of the 6 cylinders is in the TDC position at the beginning of the intake stroke.

SYNC (I)—This is a synchronizing input. The rising edge of this signal corresponds to cylinder 1 being at the TDC position at the beginning of the intake stroke.

FUEL 1 – FUEL 6 (O)—These signals are positive pulses which control the fuel injection timing for each of the 6 cylinders. The CPU adjusts the start time (with respect to the corresponding TDC) and duration of these pulses.

SPARK (O)—This is a positive pulse train which controls the spark. Spark is assumed to occur at the rising edge of the spark output signal. The CPU adjusts the start time (with respect to the rising edge of TDC) and duration of these pulses. Note that there is only one signal for the 6 cylinders.

CYL 1 – CYL 6 (O)—These are the gating signals for the 6 cylinders. They are active from the TDC edge of a particular cylinder to the TDC edge of the subsequent cylinder. These output signals are not used in this schematic, but rather these signals are brought to output pins in order to be made available as inputs to the fuel duration functions.

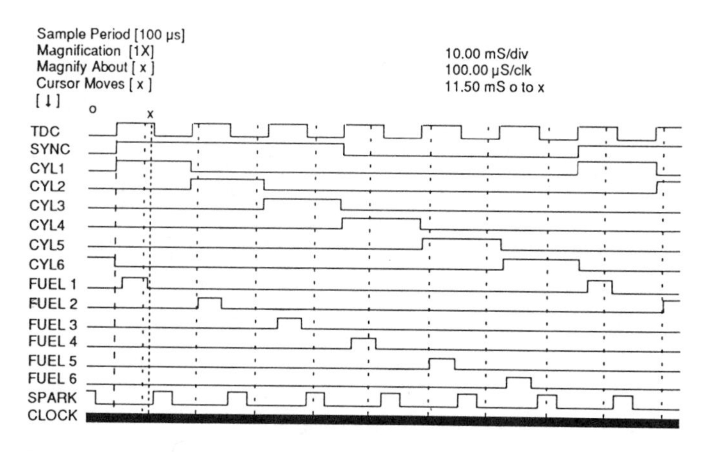

Figure 7. Signal Waveforms

UPP PROGRAM

Figure 8 shows a schematic representation of the UPP program which implements the engine control function. Each rectangular block represents a step of the program. The large letters in the top center of a block is the mnemonic of an instruction. The sequence number of the step is next to it on the right, followed by '#'. The inputs are on the left side of a block. A "+" sign next to an input denotes a rising active edge while a "–" sign denotes a falling active edge. The outputs are on the right side of a block. A HI or LO indicator next to an output connection specifies the polarity of the output pulse or gate signal. The text in the smaller rectangles inside the function block specifies the registers used and their functions. **Figure 9** shows the data that is stored in the UPP function table to achieve this.

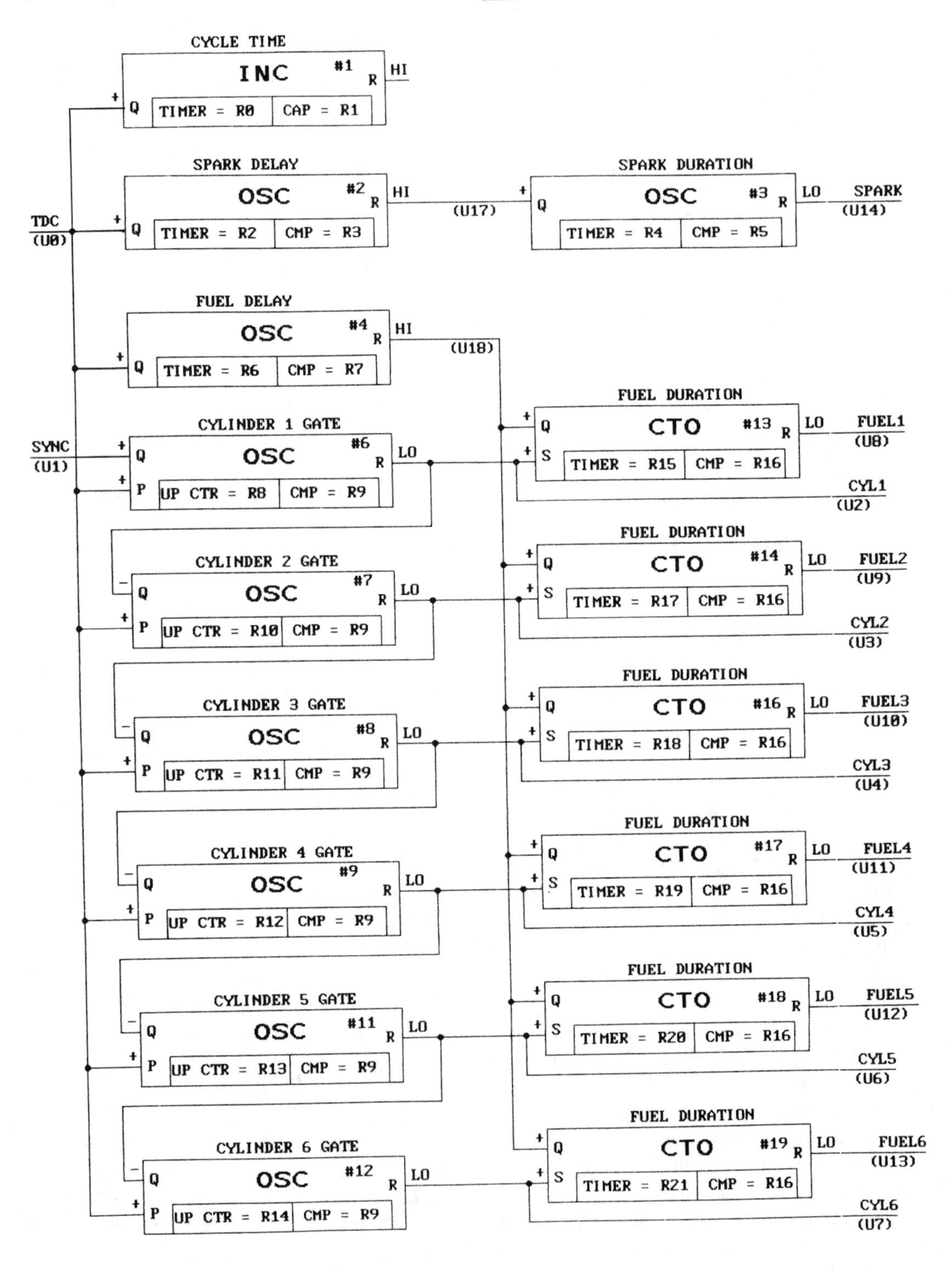

Figure 8. Six Cylinder Fuel Injection Engine Control UPP Program

Enter address of desired table 7E00

number	CMR	RASRA	RASRB	IOARA	IOARB	IOARC	IOARD
1	80	0	1	0	0	0	0
2	88	2	3	20	0	1	0
3	f0	0	0	0	0	0	0
4	f0	0	0	0	0	0	0
6	f0	0	0	0	0	0	0
7	f0	0	0	0	0	0	0
8	f0	0	0	0	0	0	0
9	f0	0	0	0	0	0	0
11	f0	0	0	0	0	0	0
12	f0	0	0	0	0	0	0
13	f0	0	0	0	0	0	0
14	f0	0	0	0	0	0	0
16	f0	0	0	0	0	0	0
17	f0	0	0	0	0	0	0
18	f0	0	0	0	0	0	0
19	f0	0	0	0	0	0	0

```
Data Registers initialization values:
REG  0: ffff 8000 ffff 0003 ffff ffff ffff ffff
REG  8: ffff ffff ffff ffff ffff ffff ffff ffff
REG 16: ffff ffff ffff ffff ffff ffff ffff ffff
```

Enter address of desired table 7ECC

number	CMR	RASRA	RASRB	IOARA	IOARB	IOARC	IOARD
1	10	0	1	0	20	10	0
2	71	2	3	0	20	11	0
3	70	4	5	0	31	e	0
4	71	6	7	0	20	12	0
6	78	8	9	20	21	2	0
7	78	a	9	20	42	3	0
8	78	b	9	20	43	4	0
9	78	c	9	20	44	5	0
11	78	d	9	20	45	6	0
12	78	e	9	20	46	7	0
13	b0	f	10	0	32	8	2
14	b0	11	10	0	32	9	3
16	b0	12	10	0	32	a	4
17	b0	13	10	0	32	b	5
18	b0	14	10	0	32	c	6
19	b0	15	10	0	32	d	7

```
Data Registers initialization values:
REG  0: 0000 ffff 0000 1000 0000 2000 0000 0500
REG  8: 0000 0001 0000 0000 0000 0000 0000 0000
REG 16: 2000 0000 0000 0000 0000 0000 ffff ffff
```

Figure 9. UPP Function Tables

This program uses all 16 of the available program steps. Fifteen of the 16 available external I/O pins, 2 of 8 available buried I/Os and 22 of the available 24 registers are used. The pulse width resolution depends on what is stored in the **M**aximum **F**unction **N**umber **R**egister (MFNR). Since the program runs through a total of 20 steps, with a 4 MHZ clock (250 nanoseconds per step) the output pulse resolution will be 5 microseconds. A brief description of the program logic follows:

CYCLE TIME — An INS (**IN**terval timer/counter with **S**ampling) function (step #1) is used to measure the speed of the engine. R0 is used as the timer. At the rising edge of the TDC pulse the contents of R0 are captured into R1, and R0 is cleared. Therefore, the contents of R1 multiplied by the pulse resolution T (5 microseconds) gives a value of time equal to one-third of an engine revolution. If it was desired, an interrupt to the micro-controller could be generated at each capture.

SPARK CONTROL—Two OSC (**O**ne-**S**hot timer/counter with **C**ompare) functions (steps # 2 and 3) are used to generate the SPARK control signal. Step #2 controls the delay of the start of SPARK from the TDC edge (determined by the contents of R3) and step #3 controls the SPARK duration (determined by the contents of R5).

The rising edge of the TDC starts the R2 timer of step #2 and causes a HI to LO transition at the output U17. A LO to HI transition on U17 takes place at the end of the spark delay period when R2 reaches R3. The U17 output is the trigger input to the function at step #3. A resulting LO to HI transition takes place at the output pin SPARK (U14) with the LO to HI transition on U17, at the end of the spark delay period. This also starts timer R4. A HI to LO transition takes place at U14 after the end of the spark duration period, when R4 reaches R5.

CYLINDERS 1-6 GATES — Six OSC functions (steps # 6,7,8,9,11 and 12) are used to generate the gating signals CYL 1 – CYL 6. These signals can be used to gate the spark control output to generate control signals for individual spark plugs. R9 is used as a compare register in all 6 steps and is initialized to 1. The signal TDC (rising edge active) is used as the clock source for these steps.

The counter R8 in step #6 is cleared and starts counting at the rising edge of the SYNC pulse. At this time the output of #6 (CYL 1) also goes LO to HI. The next rising edge of TDC increments R8 to 1. This results in a successful comparison and the CYL1 output then changes from HI to LO. This repeats every SYNC pulse yielding the gate signal for cylinder 1. The falling edge of CYL 1 in turn acts as a trigger to step #7. The output CYL 2 goes LO to HI at the falling edge of CYL 1 and one TDC pulse later returns LO. It feeds the gate output of one cylinder to the trigger input of the next step in the same way, and generates all the gate outputs.

FUEL DELAY — Another OSC function (step # 4) generates a delay from the TDC edge equal to the fuel delay period. This logic is the same as the logic of step #2. The rising edge of U18 is delayed from the rising edge of TDC by a period equal to the contents of R7 multiplied by the pulse resolution T.

FUEL CONTROL OUTPUTS—Six CTO (**C**ombination **T**rigger **O**ne-shot counter/timer) functions (steps # 13,14,16,17, 18 and 19) generate fuel control signals (FUEL 1 – FUEL 6) from the fuel delay signal and the 6 cylinder gating signals (CYL 1 – CYL 6). The CTO function is similar to OSC except that it uses an additional trigger enable input. The fuel delay output (U18) is used as trigger input in these steps. The timer of a given cylinder is started by the rising edge of the fuel delay signal occurring during the corresponding gate signal. This also produces a LO to HI transition on the appropriate fuel control output. When the timer reaches a count equal to the contents of the compare register (R16), a HI to LO transition on the output takes place.

All 6 fuel control outputs use a common compare register to establish the pulse width of the fuel injection signal. In a dynamically changing system it is possible that these pulse widths may need to be changed from cylinder to cylinder.

In the typical microcontroller system shown in Figure 1 it would take a fair amount of software overhead to keep track of this output compare function and to implement changes rapidly. One of the powerful features of the UPP is the fact that all comparisons take place on a 'greater-than-or-equal-to' basis. This means that if the new calculated pulse width is less than the pulse width presently being generated, a new value can be written into the compare register and the comparison of 'greater-than' will cause the pulse output to be immediately terminated. **Figure 10a** shows an example of how a typical microcontroller would generate a PWM output. **Figure 10b** illustrates the ease with which the UPP can accurately generate changing PWM ouputs.

GROUP INJECTION — The versatility of the UPP also allows such techniques as simultaneous or group injection. Each of the outputs U00 through U15 of the UPP can be accessed as parallel I/O. These parallel ouputs can be set or reset either individually or as a group by the microcontroller. If a group injection sequence is desired, the microcontroller needs to issue a stop command to the pulse processor and set all fuel control ouputs high for a predetermined amount of time. Reset the fuel control outputs low and re-start the programmed UPP timing functions to allow the sequential portion of the injection pattern to begin.

TIMING CONTROL REGISTERS —

Spark delay = (R3) * T
Spark duration = (R5) * T
Fuel delay = (R7) * T
Fuel duration = (R16) * T
T = Pulse resolution = 5 microseconds

SUMMARY

An engine control system constructed with the UPP chip offers several advantages:

COST — The UPP provides a system that has many powerful timing features at a lower overall system cost. To provide the same number of counter/timer functions would require several counter/timer chips or multiple microcontrollers with on-board counter/timer channels.

HARDWARE INTEGRATION — The UPP is a high integration device. It combines many functions that are needed for engine control systems onto one chip. Some functions, such as the 8 channel 10 bit A/D converter, exceed the abilities of similar functions integrated into existing microcontrollers such as the 68HC11.

RELIABILITY — The UPP allows the construction of a system with a lower part count. Decreasing the system part count will lead to an increase in the overall system reliability due to fewer components and fewer external connections.

SOFTWARE — The UPP provides a system that can lower the software burden of the microcontroller. In a complex system where a large number of counter/timer functions are needed, a microcontroller may spend the largest part of its time

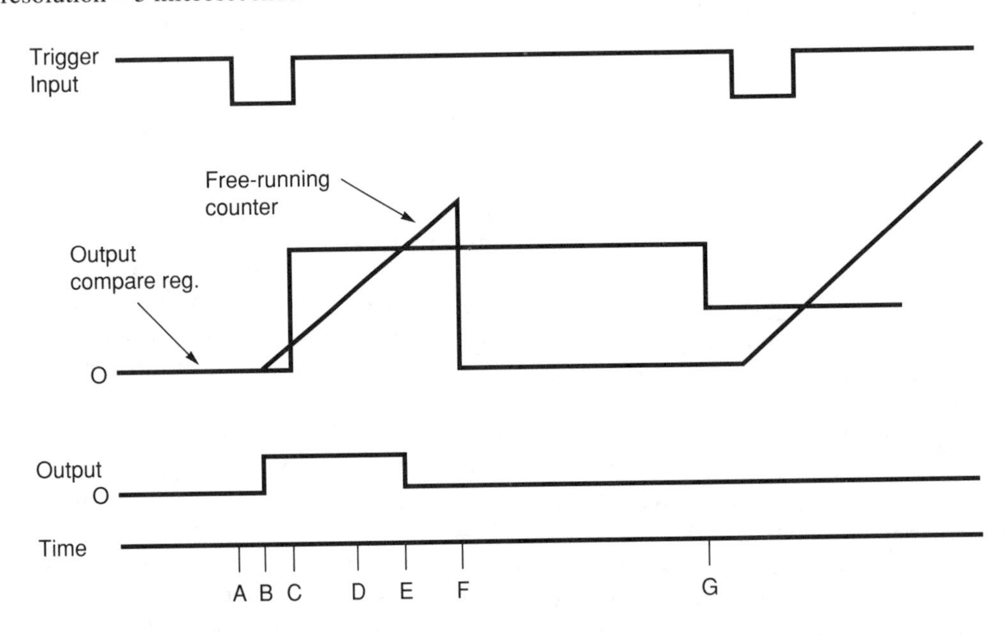

A. CPU must monitor trigger input, and then start free-running counter.

B. Compare register value of 0000 causes low to high output transition.

C. CPU reprograms compare register value to value of desired pulse width.

D. CPU reprograms output level for next compare.

E. Next compare causes high to low output transition.

F. CPU resets the FRCs to zeros.

G. New compare register value is written if needed.

Figure 10a. PWM Output Using the Microcontroller's On-Board Counter/Timer

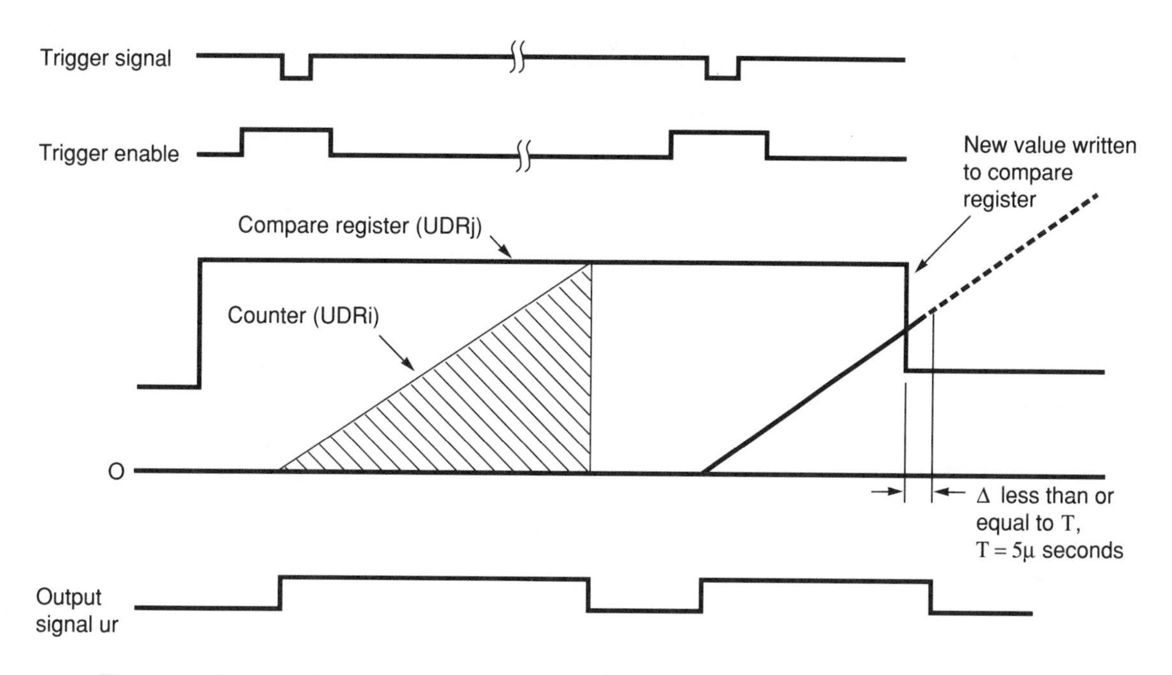

Figure 10b. PWM Output Using the UPP

maintaining counters and timers in software and/or continually resetting and reloading hardware timers and compare registers. With a UPP, once the system has been initialized the microcontroller can dedicate itself to other tasks without having to constantly tend the UPP. If the engine is running at a constant load and constant speed then no further CPU intervention is needed. The CPU can spend its time monitoring other system functions and inputs.

CONCLUSION

The concept of a Universal Pulse Processor has been demonstrated by the example of an engine control unit. The system demonstrated meets the needs of applications requiring a large number of counter/timers with a pulse resolution as small as 5 microseconds. The system demonstrated makes use of the large number of counters, timers and registers on the UPP to implement a controller that could be constructed for a lower overall system cost than a typical system using several external components.

REFERENCES

1. Hitachi America Ltd.: HD63140 and HD6301Y data sheets.
2. N. Tomisawa and S. Toki, "Trends in Electronic Engine Control and Development of Optimum Microcomputers", Society of Automotive Engineering: Paper Number 880136, February 1988.

891646

Engine Control–What Does It Take?

Steven M. McIntyre
Intel Corporation
Chandler, AZ

ABSTRACT

The amount of LSI logic to control and monitor engine performance is constantly being integrated. Injection control modules range in functionality from simple fuel metering to sequential fuel injection with distributorless ignition. This paper will review the type of I/O processing power required for engine control today and look at what will be required of I/O processing of tomorrow.

INTRODUCTION

Controlling engines electronically is becoming more and more common due to stiffer regulation of emissions. Controlling the fuel into individual cylinders is becoming a *required* function. In today's engine control units (ECUs) fuel metering is a standard function. This task is either throttle body (TBI), multiport (MPI), or sequential fuel injection (SFI). Each of these require CPU controlled outputs based on engine information such as camshaft position, oxygen content in the exhaust gases, throttle position, air density, etc. The placement of these outputs and the ability to move edges with high resolution may make the difference between a poor-running engine and a high-output power machine.

Although ignition control is not a standard function in today's ECUs, it will become more and more popular as processor integration and cost/performance curve dictates. In today's engines this function can be in stand-alone modules (one for ignition and one for injection), handled by a distributor based on an ECU-multiplexed output, or integrated within the ECU but handled through two CPUs.

The placement of the ignition control signals is also based on camshaft position and engine RPMs. The key to both engine control output signals is the camshaft position signal.

CAMSHAFT POSITION SIGNALS

There are many ways to get engine position and engine RPMs. One is to use a two-signal input to the ECU, one from the crankshaft and one from the camshaft. Since the crankshaft rotates twice for each cylinder cycle (4 strokes), obtaining exact stroke information from this engine signal output would be difficult. The camshaft only rotates once per cylinder cycle. A reference signal (usually a single "tooth") tells the processor which stroke is the power stroke for cylinder one.

0148-7191/89/0807-891646$02.50

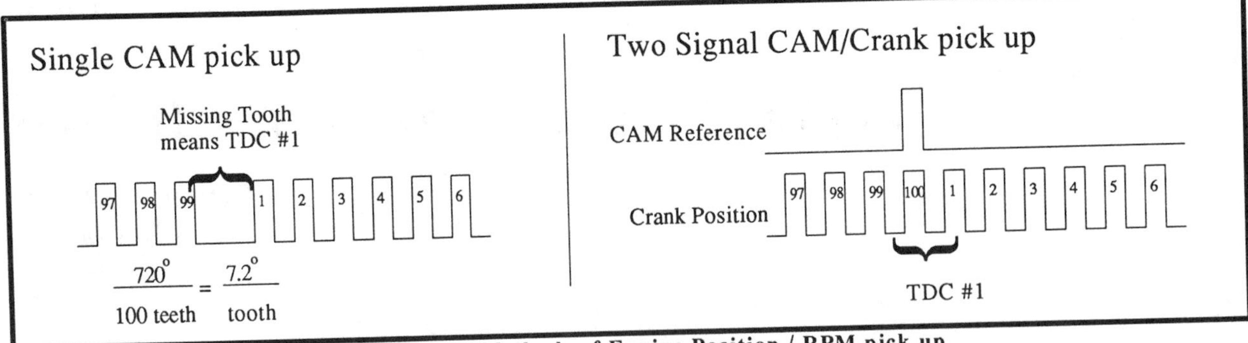

Figure 1. Two Methods of Engine Position / RPM pick up

Through these two signals engine RPMs and position can easily be obtained, but may be costly due to more external hardware used to gather the information.

An alternate method is a single camshaft signal with more "teeth" and represent the first cylinder power stroke position as a missing tooth. An example of both waveforms this would produce is shown in figure 1.

Since the camshaft (CAM) rotates only once for each 720 engine cycle degrees, the CAM is a better place to pick up both engine RPMs and engine position. The more teeth the CAM wheel has, the better the resolution the ECU has to detect exact engine position.

In the Figure 1 example (a), a one hundred tooth wheel is used. Each tooth represents 7.2° of the 720 total degrees.

Most microcontrollers have devised a method to detect when a rising or falling edge occurs on a processor input, typically called *input capture.* Input capture records the "real time" (in processor units) when an edge is detected. From this information the microcontroller programmer can determine engine speeds or engine position in degrees. Usually, though, there are some resolution errors because of the microcontroller's ability to detect that incoming edge.

TODAYS PROCESSOR INPUT and OUTPUT

As it was stated, input captures would determine the time when a CAM signal edge is found. Based on this information, an output signal is generated to drive the fuel injectors or spark plug coils. Figure 2 shows how an output signal is generated.

Notice that when the *nth* CAM signal is found, the fuel injector or spark output is turned on. This positive edge output triggers a interrupt to the processor as well as turn on the injector or spark. Next the

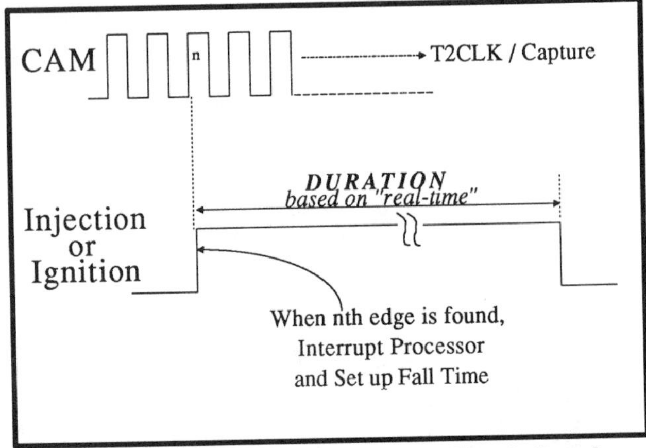

Figure 2. Spark and Injection Output Signals Based on CAM Input Signal using 8096

microcontroller interrupt service routine sets up the duration of the output pulse based on other collected information. The microcontroller output capable of such a task is called *compare outputs* or *high-speed outputs.* This feature can transition outputs, trigger interrupt events, or in some cases reset timers and start analog-to-digital conversions. All based on a pre-programmed timer/counter value. However, the outputs are only based on one timer/counter and are not based on both the CAM position and internal processor's "real time". The interrupt-driven compare outputs take away valuable CPU time in order to perform their function.

A good example of *compare outputs* in silicon is the 8096 HSO units. It can accomplish all of the above mentioned functions including toggle more than one pin using only one memory location, but each edge is still based on one timer/counter base.

Figure 2 shows an 8096 compare outputs connected to control an engines spark ind injectors. The outputs rising edges are placed at fixed CAM position intervals (7.2° resolution) while the falling edge is based on the processors "real-time" placement (2uS resolution).

OTHER ECU INPUTS and OUTPUTS

The ECU not only determines the position and speed of the engine. It must know critical engine parameters like engine temperature, throttle position, manifold air density, battery voltage, and percentage of oxygen present in the exhaust gases.

All these inputs to the ECU are used to better control the fuel injectors and ignition. Having a processor with an analog-to-digital (A/D) converter is required to analyze these signals.

Most of the microcontrollers today have methods of converting an analog voltage into digital information for further processing. The A/D resolution varies with silicon manufacturers.

A good example of an analog-to-digital converter would again be on the 8096. It has eight analog inputs that can be preprogrammed to convert at specific time intervals with high resolution (10-bit). In addition, it's conversion times are below 30 microseconds.

A co-processing loop would determine the correct OFFSET and DURATION values for each cylinder and placed in the RAM while determining the exact CAM speed and position. A look-up table approach to these OFFSET/DURATION values could be performed, but if the I/O control is "automatic" it might save enough time to perform the mathematical calculations to accomplish better engine control.

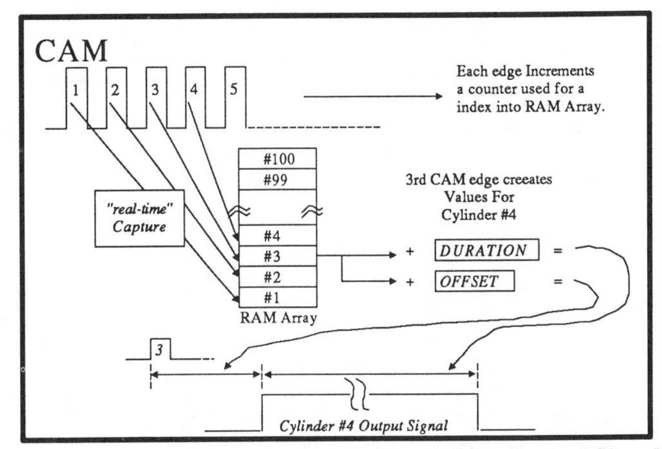

Figure 3. "Automatic Input and Output Engine Control Signals

TOMORROWs INPUT and OUTPUT IMPROVEMENTs

The following improvements could be made to tomorrows I/O on microcontrollers:

1. The analog inputs could have better resolution than 10-bit or at least have better error adjusting capabilities. The A/D converters on microcontrollers currently have no way of adjusting for offset and gain errors. This adds to the overall error in the A/D and reduces the integration advantage over discrete A/Ds.

2. The same input signal from the CAM can be *captured* and pulse accumulated (count pulses) in order to put the engine position (in increments of 7.2 degrees) information into processor "real time" terms. This will better rising edge placement as well as allow for easier processor access.

3. The output signal to the injector and ignition could be "automatic". A RAM location is set up with the OFFSET (from the CAM position in "real-time" units) and an additional RAM location is set up for the DURATION. This would allow maximum pulse placement flexibility with low or no CPU overhead. Resolution of this signal needs to be in the one microsecond range, therefore rising-edge placement (OFFSET) needs to be to equal zero.

A WORKING EXAMPLE : CAM

Suppose the CAM position sensor produced a signal such as the one in Figure 1, and that the output signals for the spark and injectors are based on this signal as shown in Figure 2. If every rising edge of the CAM signal were "time stamped" into a RAM location and indexed via the edge number for easy access, an array of 100 numbers would be generated (one for each CAM rising edge). The array pointer would be reset when the "missing tooth" was found.

This table of "real-time" values could be used to generate the output signals. OFFSET and DURATION values would be added to these numbers and placed in their respective control registers. Figure 3 illustrates such a task.

Notice that the output signal placement is totally flexible based on "real-time" (internal microcontroller time) resolution. If this internal timer had the ability to be scaled, resolution would be up the individual programmer.

These input and output tasks would happen without microcontroller intervention and CPU processing would not interfere with or be interfered by I/O processing. A background loop would simply constantly update OFFSET and DURATION for each cylinder.

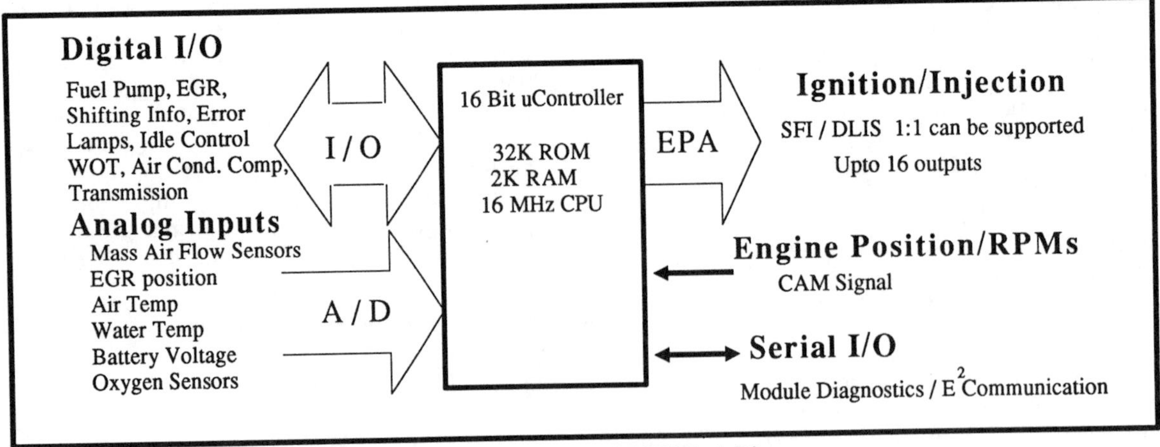

Figure 4. Tomorrows Typical ECU Block Diagram

Adjustments of OFFSET and DURATION could be made at any time, including during the output event.

One implementation of such I/O in silicon is that of *Event Processor Array (EPA)* and *Peripheral Transaction Server (PTS)*. These two feature combine to form the previously mentioned "automatic input and output processing" ideal for next generation engine controlling.

WORKING EXAMPLE : A/D

Because a microcontroller has no external pins dedicated to trimming the zero offset and full scale errors from the actual A/D performance, the user must use the absolute error to build his system.

This limitation takes away valuable information from the system and could impact overall performance.

Suppose the analog-to-digital converter were able to convert on its own reference voltage and analog ground. If this were possible the processor could determine through software the zero offset and full scale errors.

After determining these errors a single register is loaded with the corresponding error adjustments. This number is automatically added to or subtracted from the A/D result. The user reads the adjusted A/D result value.

This feature would give back to the user what he lost through integration (error adjustments).

CONCLUSION

The main input and output signals in ECUs are based on generic microcontroller I/O features (Capture/Compare and A/D). The I/O used to control tomorrows engines will need to be "fine tuned" to meet a new set of unique requirements. Figure 4 illustrated what a typical microcontroller would look like in tomorrows ECU.

This paper reviewed only the I/O processing: a dedicated I/O processor with the ability to manipulate injection and ignition signals with the resolution needed for tomorrows engines, and a self adjusting analog-to-digital converter that compensates for losses due to integration.

Ideally this I/O processor to control injection and ignition would be integrated into the microcontroller, better utilizing the single ECUs' overall size, high-temperature requirement, and integrated function.

Having this I/O processor behind a powerful 16-bit CPU would make engine control easier to integrate, and in turn more cost effective.

REFERENCES

"Trends in Electronic Engine Control and Development of Optimum Microcontrollers", Naoki Tomisawa and Shuichi Toki, SAE Paper #880136.

891647

Automotive Real-Time Control Systems Engine Control Using a MC 68332

Anthony G. Lobaza
GMC

ABSTRACT

With the increasing sophistication of today's automobiles, the electronics required for control are also becoming more complex. The evolution of the microprocessor has now advanced to the 16/32 bit plateau. In addition, several integrated circuits have been combined into one. This paper will discuss microprocessor controlled systems in automobiles, particularly, how a Motorola MC68332 could be used to control an engine.

INTRODUCTION

Computer controlled engine systems in automobiles became popular in the seventies as a means of meeting Federal Emissions Standards. At that time, the functions they controlled were trivial compared to their current day uses. Now they are being used extensively throughout several systems in a vehicle.

Besides engine control, several other automotive uses of microprocessors are becoming popular (see Figure 1). An example of this is anti-lock braking systems. These systems are designed to allow the driver to retain the ability to steer and/or maintain control during sudden stops. Another fairly common use of a microprocessor is to control the heating ventilation and air conditioning system. The driver is required only to select a temperature for the vehicle and the setting is reached automatically. A digital instrument panel also requires a microprocessor. The Buick Riviera and Reatta take the concept of a digital instrument panel a step further by implementing several functions--gauge display, climate control, radio control--from one display screen. Other microprocessor controlled systems on a vehicle which are not as obvious to the driver include: transmission control, suspension control, and lighting control.

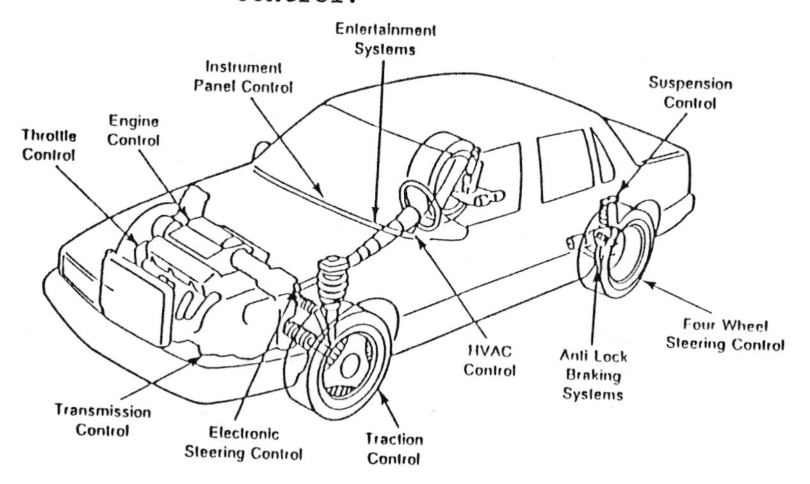

Figure 1 -

Automotive Control Systems

Computer controlled systems are also under development by several companies, or, in some cases, are in low volume production. These systems include: traction control, throttle control, and steering control. Seemingly high tech by today's standards, they will probably be standard equipment in the not too distant future.

Still the most common use of a computer in an automobile is for engine control. The number of signals to be monitored and generated are numerous (see Figure 2). Therefore, both the hardware and software requirements for an engine controller are very complex. Because of this complexity, as well as the integration of more than one system control task into a single computer, the need for a more powerful microprocessor has become a reality. Because of this need, the MC68332 was designed by Delco Electronics and Motorola. The following will discuss how the advanced features of the MC68332, specifically the Time Processor Unit (TPU), can be used to control an engine.

MC68332 OVERVIEW

The MC68332 is divided into the five following modules:

- Central Processing Unit
- System Integration Module
- Time Processor Unit
- Queued Serial Communication Module
- RAM Module

All of these modules communicate with each other on the Intermodule Bus (see Figure 3).

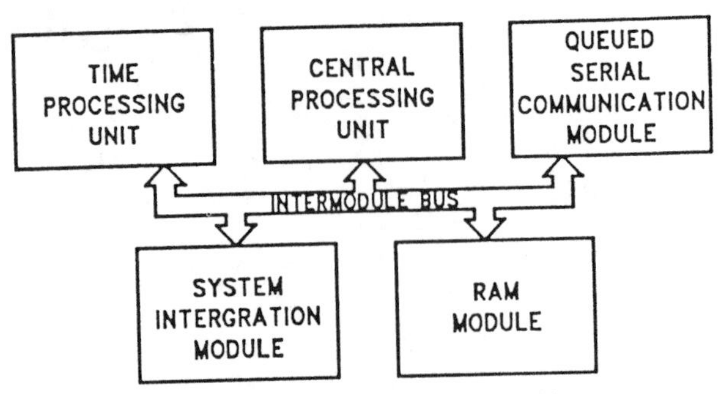

Figure 3 - MC68332 Block Diagram

The Central Processing Unit is a MC68000 based system which incorporates certain features from both the MC68010 and MC68020. It also has several new features specific to it such as a table lookup instruction and a low power stop mode. The System Integration Module has several responsiblities, including clock generation, system protection, chip selecting, external bus interfacing, and system testing. The Time Processor Unit is a high resolution timer which is capable of handling both simple and complex timing tasks. It can be programmed to perform one of several predefined functions on any of its sixteen channels. For communications there is the Queued Serial Communication Module. This module is split into two independent submodules, the Queued Serial Peripheral Interface and the Serial Communications Interface. Finally, there is 2K of RAM in the device which is powered by standby power.

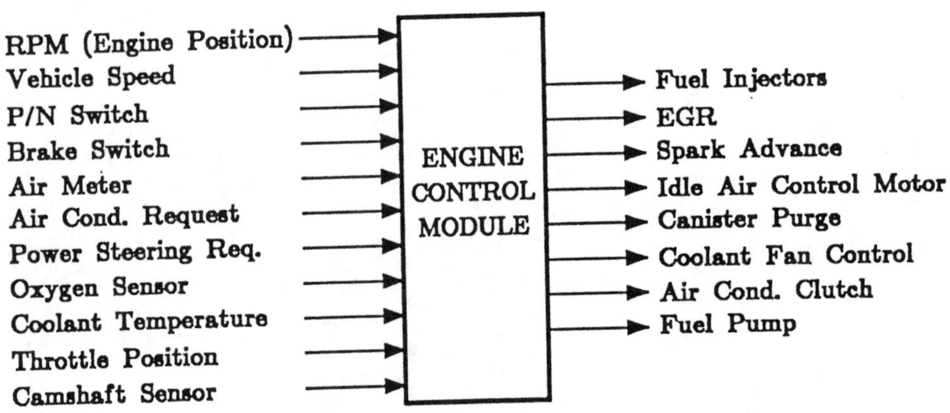

Figure 2 - Engine Control Requirements

ENGINE CONTROL REQUIREMENTS

The input and output requirements for an engine vary greatly based upon the type of control scheme used and the options on the engine. For this discussion, the system will be kept simple: a six cylinder engine with sequential multiport fuel injection, electronically controlled spark, and a fuel control scheme based upon air flow derived from a mass air flow sensor. The basic inputs and outputs necessary for this type of control are shown in Figure 2.

Input Requirements for the MC68332

For both the RPM signal and vehicle speed signal, a square wave will be received and directed to a respective channel of the TPU. If the TPU channel is set up for input capture, the frequency of the signal will be automatically computed by the TPU. When a programmable number of pulses have been received, the TPU will interrupt the microprocessor and record in one of its registers a count proportional to the RPM or vehicle speed. Based upon the frequency of each square wave, the RPM and speed parameters can be calculated, as well as the relative position of the engine.

From the air meter there are two possible output signals--an analog signal or a frequency signal--depending upon the type of meter utilized. With an analog signal, an analog to digital converter must be used. Due to noise and resolution problems, the frequency meter is preferred over the analog. Just as with the RPM and vehicle speed signals, an input capture can be programmed into the TPU. By reading one of the registers associated with the particular channel, the mass of air flowing into the engine can be calculated based upon the frequency from the air meter.

Other input signals which must be considered are the discrete inputs. These signals include the brake switch, the park/neutral switch, the air conditioner request, the power steering request, and the camshaft sensor. To handle these signals there are two possibilities. The first is to use a TPU channel and configure it as a discrete input. This, however, is not the most efficient use of the TPU since it is capable of handling much more complex signals. Therefore, using a discrete I/O device would make more sense and leave the TPU channels open for other options. Furthermore, because of the internal architecture of the MC68332, connecting any external device to it is simple. If the device is to go on the parallel bus, the chip select, wait states, and all other control signals can be programmed with software. For serial communications, the device can be connected to the Queued Serial Peripheral Interface of the MC68332. Similar to the parallel bus, several programmable options are associated with the serial bus, including the ability to queue up to sixteen outgoing messages.

The final category of input signals is analog voltages. These signals include voltages from the oxygen sensor, the coolant temperature sensor, and the throttle position sensor. Obviously, for these signals, an analog to digital converter is necessary. As with the discrete I/O device, either a parallel or serial interface can be used to take advantage of the above mentioned features.

Output Requirements for the MC68332

One complicated output signal which needs to be generated is the electronic spark timing signal. Because of the delayed ratioing function of the TPU, however, its generation is made simple. By measuring the period of a position input signal, an output signal can be generated as a function of the input signal (see Figure 4). When designating a channel of the TPU for delayed ratioing the two parameters registers, RATIO1 and RATIO2, determine the high and low time of the output signal. The equations are as follows:

High Time = Input Period * RATIO1

Low Time = Input Period * RATIO2

For more spark advance, the ratio parameters of the output signals need to be reduced. Increasing the ratio parameters will result in spark retard.

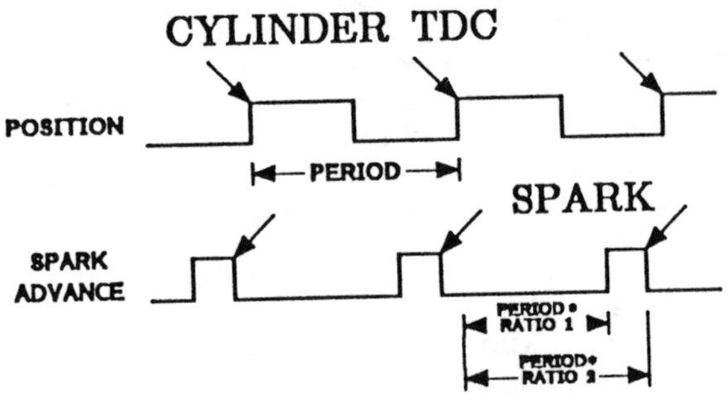

Figure 4 - Electronic Spark Timing

The most complicated output signals to produce are the fuel injector pulses. To generate these signals, the proper relationship must be established between the camshaft sensor signal, engine position sensor, and fuel injector pulse width. The camshaft sensor will determine which of the two banks of the engine is ready for fuel, while the engine position sensor will determine which of the three cylinders of the selected bank should be fueled. All that remains for the microprocessor to accomplish is to determine where to start and stop the fuel pulse so that the correct amount of fuel is injected at the proper time (see Figure 5). By using the delayed ratioing function of the TPU, an injector pulse can be generated for each of the six cylinders. Software which loads the proper TPU

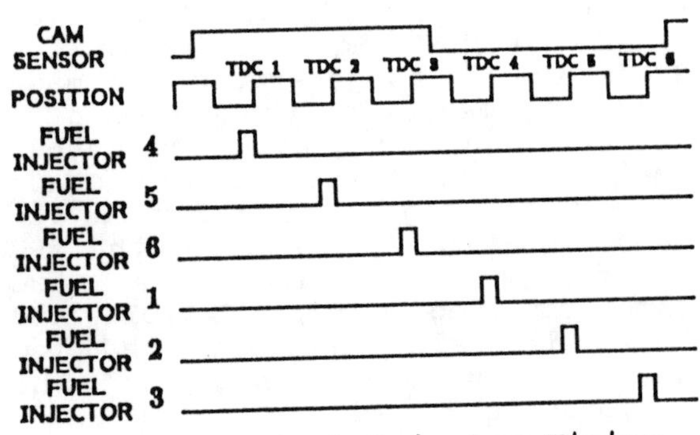

Figure 5 - Fuel Injector Timing

parameter registers can be used to fuel the desired cylinder.

Since a stepper motor is used for idle air control, output signals must be generated to control it. To accomplish this, the stepper motor function of the TPU can be utilized. With this function, the control is simplified both in the hardware and software areas. All that needs to be done by the microprocessor is the writing of the desired motor position to the position register and the motor acceleration to the acceleration register. This feature simplifies the software by not requiring it to track each step individually and also eliminates the need for external devices which generate stepper motor control signals.

The canister purge system can be implemented with the same simplicity as the stepper motor control. For this system, a pulse width modulation (PWM) signal is needed. Consequently, a TPU channel would be programmed to generate a PWM. By simply providing a pulse width period and a pulse width high time, the TPU will generate the proper control signal.

The exhaust gas recirculation system has a variety of control signals which must be generated depending upon the control mechanism used. Pulse width modulation, stepper motor control, or digital output signals may be required. However, no matter what the control requirements are, the TPU can be programmed to satisfy them.

As with discrete inputs, discrete outputs such as the coolant fan control, air conditioning clutch control, and fuel pump control can be provided by the TPU. For the coolant fan control, a channel of the TPU may be desirable since a PWM control signal may also be a requirement. For the others, a discrete I/O device would be more practical.

Software Requirements

Although all of the above features of the MC68332 have stressed I/O functionality, especially with the TPU, this microprocessor has several other advanced features which enhance its real-time control operation.

The most important feature is that the microprocessor is able to run at 16.7 MHz which makes it capable of satisfying very high throughput requirements. This becomes crucial when several systems are controlled, or very complex algorithms need to be processed. Since the use of high level languages is also becoming more common, as well as the writing of modular code, the throughput overhead associated with each must be tolerated by the system. Along with using high level languages, doing calculations with variables that are in engineering units also makes writing software easier and more understandable, but more CPU intensive. Therefore, the ability to operate at high frequencies has become a necessity.

As algorithms become more complex and actuators more precise, control systems must advance in the same direction. For this reason, 8/16 bit microprocessors are being replaced by 16/32 bit systems. The MC68332 offers high resolution along with several high precision instructions to take full advantage of its architecture.

Another advantage of the system, with respect to software, is the addition of a table lookup instruction. Because of the nonlinearity of certain engine parameters, table lookup conversions may be more effective than the solving of complex equations. For example, the mass airflow sensor outputs a frequency based upon air flow. Because the frequency is not directly related to the mass airflow through an equation, a table associating the airflow to the frequency is necessary. Due to memory constraints, the number of elements in each table has to be limited. Therefore, if a frequency from the air meter is not exactly replicated in the table, an interpolation is performed by the instruction based upon the two closest values.

Finally, several aspects of control are often based upon time intervals. For this reason, a programmable timer is internal to the MC68332 which can be programmed by the user to cause interrupts.

SUMMARY

This paper has merely touched on some of the unique capabilities of the MC68332, such as its complex timing ability, increased throughput capacity, improved instruction set, and internal timer. Although this document only addressed engine control, more than one automotive system could be simultaneously controlled by a microprocessor as powerful as the MC68332.

CREDITS

This manuscript was originally prepared for, and presented at Electro/89.

891648

A New MOTRONIC System with 16 Bit Micro Controller

Martin Zechnall and Gunther Kaiser
Robert Bosch GmbH

ABSTRACT

The functionality of engine management systems has grown rapidly over the last few years. The paper presents a new Motronic concept, the engine management control M3. The Motronic family M3 is a modular design destined to control engines with up to eight cylinders individually. The main features of this system and the ECU's concept are discussed.

1 THE SYSTEM

The first digital engine management system at Bosch, the Motronic, was introduced in 1979 in volume production [1, 2, 3, 4, 5, 6]. Since then, in response to market demands, the system's performance has been continuously improved. Today's standard will be illustrated in the following with the newly designed modular Motronic system M3.

The Motronic family M3 is a modular design engine management system destined to control engines with up to eight cylinders individually. The now realized type M3.1 for six cylinder engines, whose main features are shown in Figure 1, will be presented in this paper.

- Sequential fuel injection
- Distributorless ignition
- Hot wire air-mass sensor
- Adaptive closed loop Lambda control
- Adaptive canister purge control
- Adaptive idle speed control
- Adaptive knock control, cylinder-individually
- Diagnostics
- Others

Fig. 1: **Motronic M3.1, main features**

Basic input signals for the calculation of injection time and ignition angle are engine speed and air mass flow. As shown in Figure 2, the detection of engine speed and reference mark needs only one inductive sensor in conjunction with a toothed disk with 60 teeth, with 2 teeth missing for TDC reference purposes, mounted onto the crank shaft.

By measuring and comparing time intervals between consecutive teeth the position of the TDC reference mark is determined by the ECU's controller. The engine speed is calculated by measuring the time interval between a predetermined number of teeth, usually several times per crank shaft revolution.

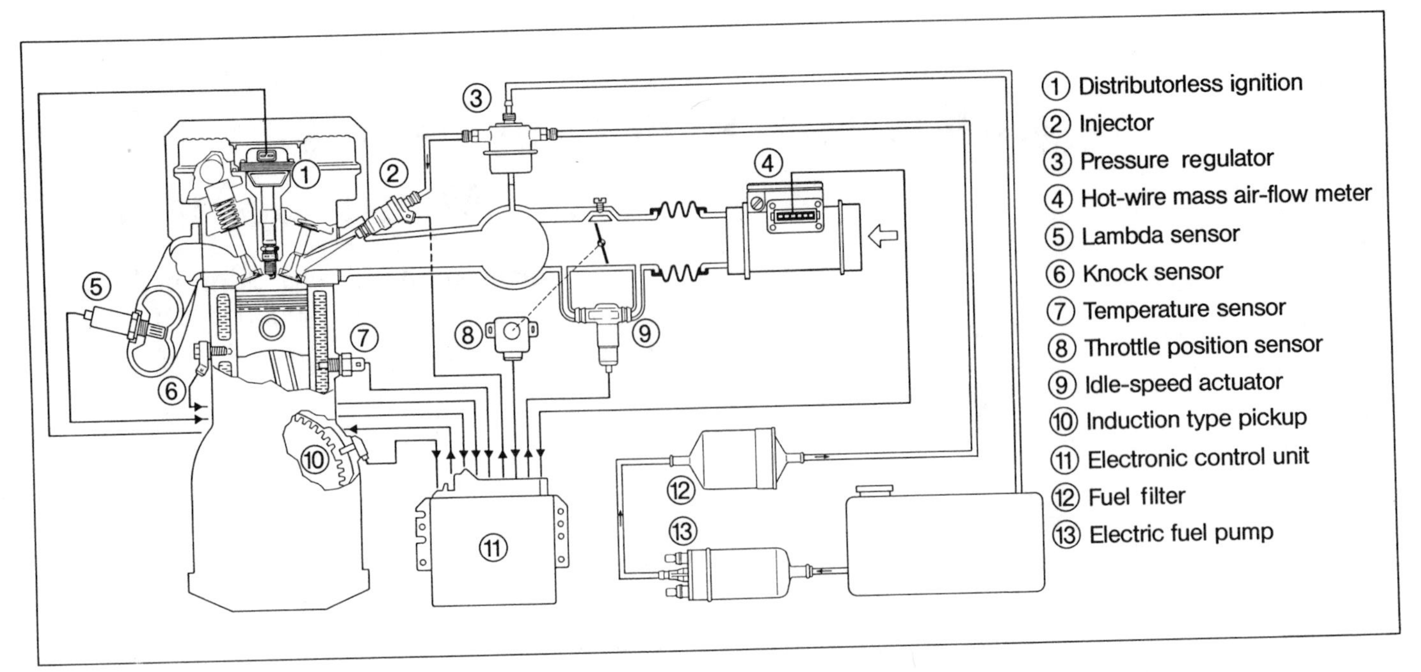

Fig. 2: **Motronic M3.1, system diagram**

The correct adjustment of required injection time and ignition advance is based on the amount of air mass per stroke. It is calculated by measuring the air mass flow together with the engine speed. There are different sensor types to measure the air mass flow directly or indirectly such as the vane type air flow meter, manifold pressure sensor or hot film air mass sensor and others. To obtain best overall performance the hot wire air mass sensor is the preferred solution. The hot wire principle should be well known [7, 8]; the sensor itself is shown in Figure 3. The hot wire as the main element is cooled by the intake air; the necessary heating current to keep the temperature of the sensing element constant is nonlinearly proportional to the air mass flow. By scanning the sensor's output voltage via the ECU's A/D converter in short time intervals, a high resolution is achieved. A sophisticated algorithm detects and eliminates the influence of momentary reverse air flow which exists in certain engine operating conditions.

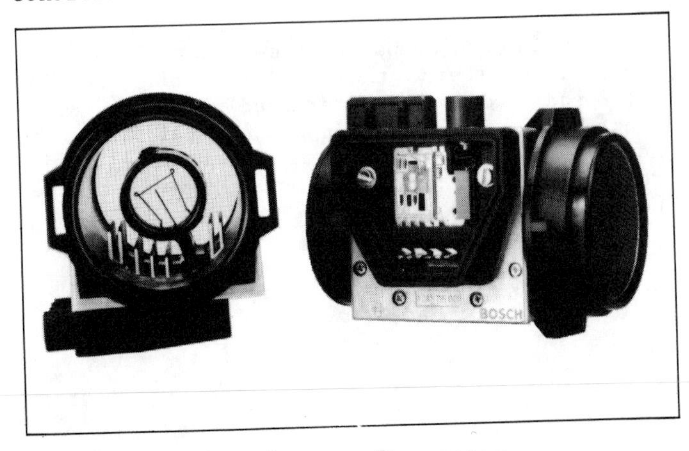

Fig. 3: **Hot wire air mass flow meter**

To optimize engine performance, a sophisticated sequential fuel-injection system has been adopted (Figure 4). Normally idle roughness and exhaust-emissions are reduced by avoiding injection of fuel directly into an open intake valve.

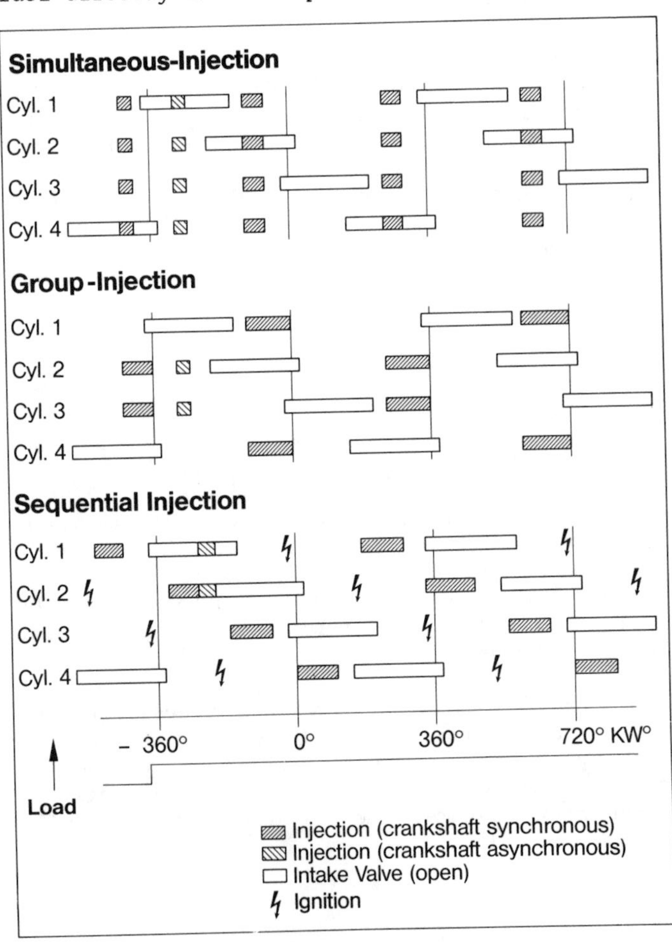

Fig. 4: **Comparison of the different injection modes**

Therefore, during steady state conditions, the end of the injection period of each cylinder's injector is adjusted in relation to the opening period of the intake valve of the respective cylinder. During dynamic engine conditions, the calculated injection time has to be adjusted to the actual value of air mass per stroke in order to obtain the correct air/fuel ratio and thus avoid poor driveability and increased exhaust emissions. The injection time of those injectors, which are in operation, is actualized. If the injector has just finished its injection time, the adjustment for the actually required amount of fuel is achieved by additional injector pulses up to the finished intake cycle of the respective cylinder.

The correct timing of the sequential fuel injection needs a synchronization to the camshaft position. For this purpose a second inductive sensor is used, detecting one reference mark each camshaft revolution.

Waiting for synchronization of camshaft and crankshaft reference mark during cranking results in lengthening of the starting time. To overcome this, the sequential fuel injection mode is switched to a simultaneous one during the first cycles of cranking. After synchronization is achieved, the simultaneous mode is converted to sequential injection timing. A sophisticated algorithm provides for correct air fuel ratio of each cylinder during this phase.

Distributorless ignition is a basic feature of the Motronic M3. The system M3.1 controls up to six single ended or three double ended coils (Figure 2). The power stages for running the ignition coils are integrated into the Motronic-ECU. The ignition control circuitry was designed with consideration of a possible overlap of the dwell angles of the different coils.

2 THE ECU

The development of the Motronic ECU M3 was highly influenced by customer demands for

- increased functionality and
- small ECU housing.

To ensure the required high computing performance, a two-controller design is the basis of the digital system (Figure 5). The miniaturization of the ECU is achieved through intensive use of today's bipolar and CMOS-integration technologies and the utilization of hybrid assembly technologies for some of the power stages (Figure 6). Additionally, SMD-components are extensively used on both sides of the PCB.

- 2-Controller-system: Intel 8397BH + 80C51
- External RAM, EPROM
- CMOS-ASIC
- Bipolar-ASIC interfaces
- Bipolar-ASIC power stages

- 2 Printed circuit boards
- SMD on both sides
- Hybrid power stages
- 88 Pin-connector
- Overall dimension: 185 × 136 × 33 mm

Fig. 5: **Motronic M3.1, electronic control units, main features**

Fig. 6: **Motronic M3.1, electronic control unit**

A reduction of the size of the required 88-pin ECU connector is attained through the introduction of low diameter signal pins in addition to the normal high current pins. Despite the increased functionality of the system, a reduction of the size of the ECU housing of about 30 % is the result of the procedure described.

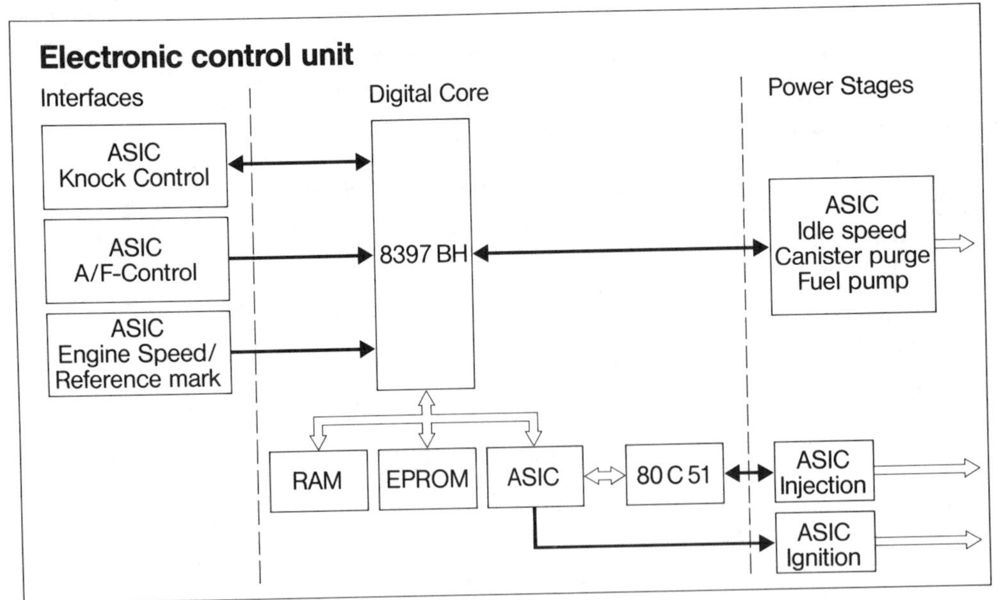

Fig. 7: **Motronic M3.1, electronic control unit, block diagram**

Core of the digital system (Figure 7) is INTEL's 8397BH 16 Bit-controller [9]. With its on-board 10 bit/8 channel A/D-converter, 232 byte RAM, 8K ROM and a high speed I/O-structure, this controller seems to be a good compromise between the demands for high computing power, high integration density of peripheral components and today's given possibilities of the semi-conductor integration technology in view of maximum die size and package pin count of a controller for the rough environment of automotive applications. For external program memory extention, typically an EPROM is used. Beside part of controller's program it stores all of the car-specific data. The use of this concept allows a change of ECU parameters during volume production within a short time. As we have experienced, this flexibility is needed mainly in the introduction phase of a new vehicle model. If data changes are unlikely, a ROM instead of an EPROM can be used for cost saving reasons.

The ROM on-board CPU contains no application-specific data to avoid the long time interval needed for new ROM masks at possible changes of the cars data set. It holds only standard routines suitable for several projects.

For extension of controllers on-board RAM a 8 k x 8 CMOS SRAM is used. A standby current supply enables the storage of adaptive data and all of the information needed for diagnosis of the system. Despite the possible loss of data after disconnection of the car's battery, this is the solution preferred to todays available large storage sized and expensive E^2PROM's. After a loss of the data set due to a battery disconnetion all of the adaptive and diagnostic data are re-astablished during few driving cycles.

The basic system is extended by a second controller for sequential fuel injection. The Slave-CPU 80C51 used handles all of the required I/O-operations and control algorithms for sequential fuel injection. Without car-specific data, the needed control program fits completely within the on-board ROM of the 80C51. All of the required control data are calculated by the Master-CPU 8397BH and transmitted to the Slave-CPU via a parallel interface.

The interface circuitry is realized with an ASIC made in CMOS technology. Beside the modularity, the advantage of this concept is the unloading of the Master-CPU from time consuming calculation tasks needed for controlling the complex sequential fuel injection scheme.

To simplify the control of the power stages for distributorless ignition, a double buffered output register is integrated into the CMOS ASIC which interfaces the Slave-CPU with the Master CPU. Preloaded data in the first stage of the output register will be shifted into the second stage by a trigger signal generated with a high speed output channel of the Master-CPU. The advantage of the concept is the control of up to eight ignition coils with only one high speed output channel of the Master-CPU.

To reduce the number of parts required, and to save space on the PCB, extensive integration in bipolar technology for most input and output interface circuitry is employed. The power drivers for injection valves, idle speed actuator, fuel pump, etc., are multiple power stages with integrated diagnostic functions. To save space, the IC and all of the needed interface circuitry is mounted onto a ceramic substrate (Figure 8).

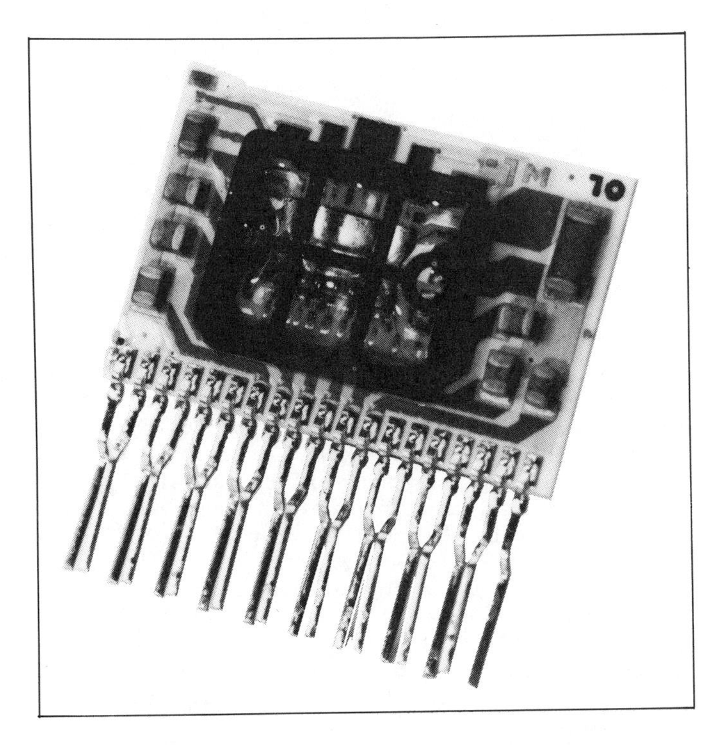

Fig. 8: **Motronic M3.1, hybrid power stages**

3 DIAGNOSIS

Diagnosis in engine management systems is one of the most widely discussed items. The main aim of diagnosis is to get information about failures in the ECU-box, peripheral components and in the wiring harness to support engine inspection. Any change in behavior that could cause a lack of performance concerning exhaust emission, or a failure which could risk the engines life, should be detected. Using the micro computer's ability to perform internal testing and memorization facilitated the introduction of self diagnosis test methods (Figure 9).

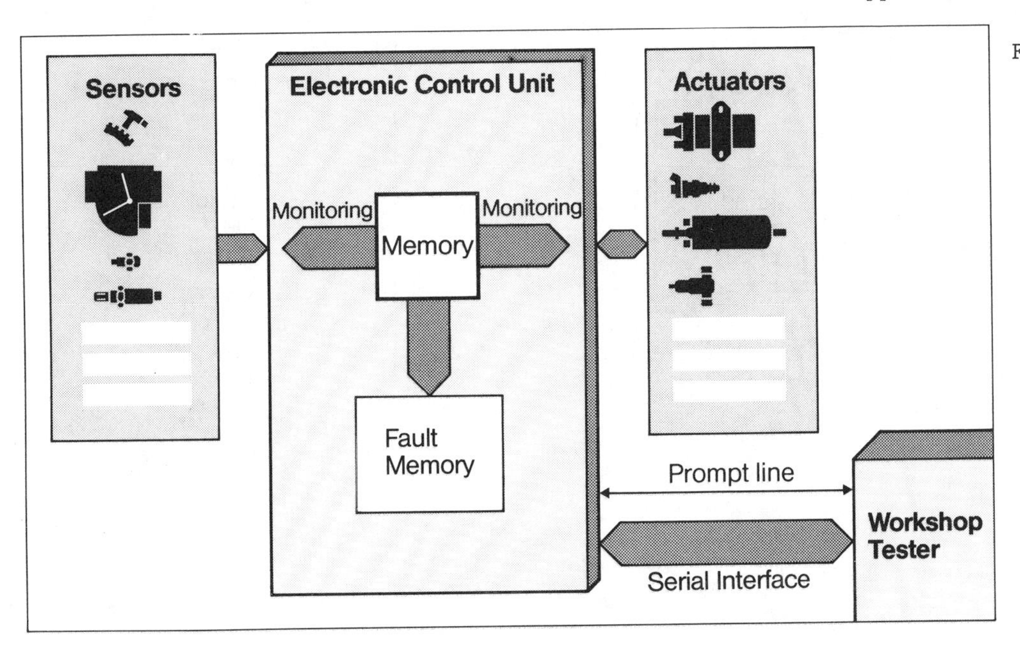

Fig. 9: **Engine control with self-diagnosis, block diagram**

Sensor signals are tested for normal operating range and output stages may be triggered for service purposes while the engine is not running. Detected failures are stored in a non volatile RAM to be read out in the service station and to support engine inspection. The complex processing of self diagnosis during real time operation of the engine management control affects CPU speed and increases significantly the RAM and ROM size. Today's diagnosis systems need more than 30 % of the memory capacity.

4 APPLICATION TOOLS [10]

To fit the car-specific data, the application tools VS20 and VS100 are used. The application tool VS20 (Figure 10) displays input parameters of the Motronic ECU such as engine speed, engine load, engine temperature, etc., and allows modification of output control signals such as injection quantity and ignition timing.

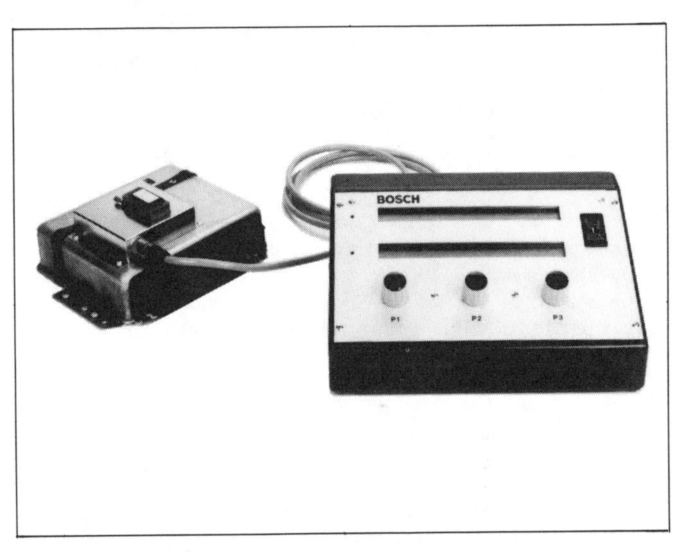

Fig. 10: **Motronic M3.1, electronic control unit, application tool VS20**

For data transfer, a serial link is used between the ECU and application tool VS20.

The application specific data, maps and tables, held in the EPROM of the ECU, are modified by the application tool VS100 (Figure 11).

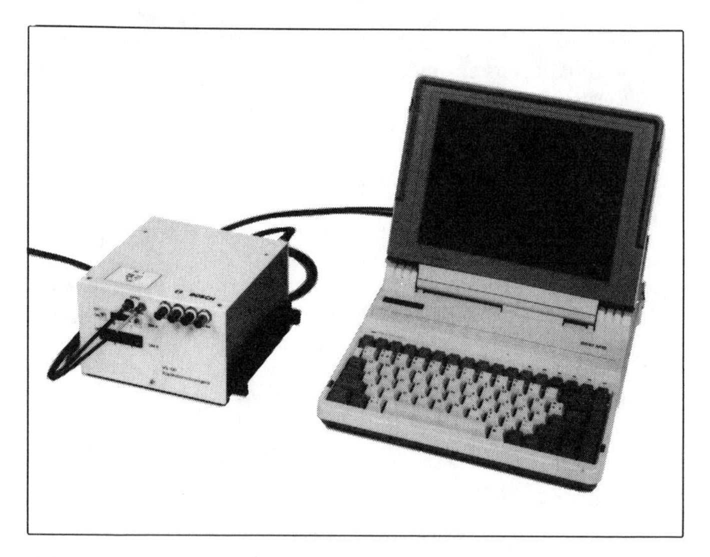

Fig. 11: **Motronic M3.1, electronic control unit, application tool VS100**

Connected to the digital address/data bus of the ECU the system VS100 emulates the program and data memory. An effective online modification of car-specific data is possible with this tool.

5 DEVELOPMENT TRENDS OF THE ECU

The electronic content of automobiles has significantly increased in the past years. Systems like engine control, drive by wire, transmission control and traction control considerably improve the performance of the car. But the required interaction between these systems created a communication problem.

Up to now, the different communication links have been realized by dedicated interfaces and connection lines (Figure 12). The resulting wiring harness is difficult to assemble and maintain and complex to diagnose. To solve this problem, a high speed serial communication link, the Controller Area Network (CAN) [11, 12, 13] has been developed for interconnecting electronic control units (Figure 12). The CAN interface is optimized specifically for automotive requirements. The performance data of the CAN bus are shown in Figure 13. The CAN bus allows efficient coordination of the drive train and engine management systems, which results in a significant performance improvement.

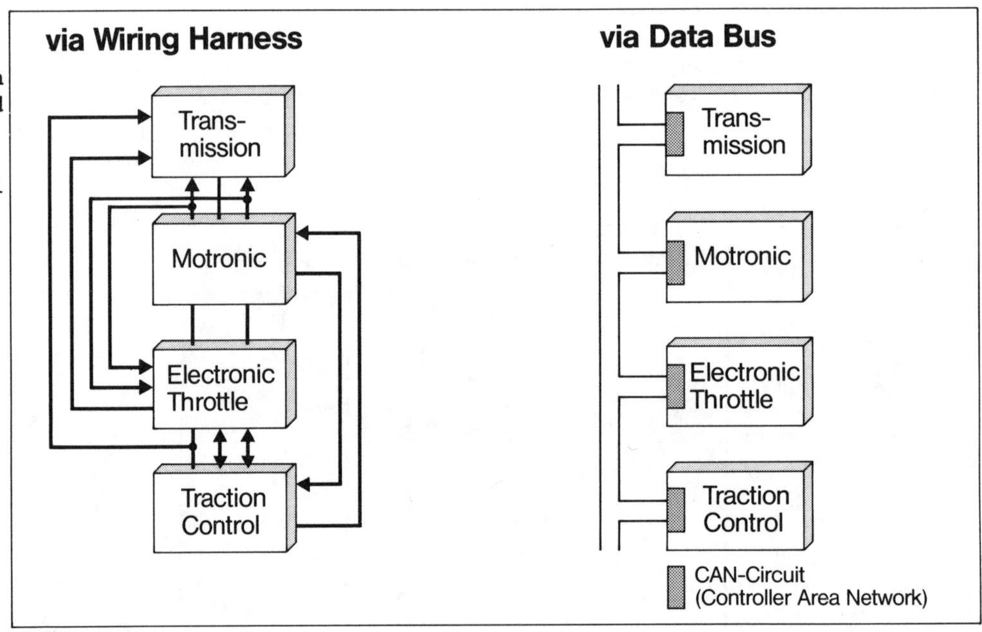

Fig. 12: **Communication between engine management and drive train systems, conventional method and high speed serial interface CAN**

Topology: Bus Configuration, One Logical Bus Line Only
Transfer Medium: Screened Copper Wire, In the Future Optical Fibre
Geometrical Extension: Maximum 40 m (At 1 MBit/s)
Transfer Rate: 10 KBit/s to 1 MBit/s
Data Capacity: 0 to 8 Bytes
Number of Identifiers: Up to 2032
Message Length: Maximum 130 Bits
Recovery from Errors: Maximum 29 Bit Times, Typically 17–23 Bit Times

Fig. 13: **Performance of the high speed serial interface**

The next member of the M3 family contains the CAN interface as well as the other drive train systems. The substitution of the convential links and interfaces leads to further enhancement of the ECU's packaging density.

For a smaller ECU housing, a higher integration of digital components is a must in future. Improvements in semiconductor integration technology will provide the possibilities for a dramatic reduction of the number of parts needed for the digital system in the near future.

REFERENCES

[1] I. Gorille, "Digital Engine Control for European Cars", SAE-Paper 1980 No. 800165

[2] H. Schwarz, H. Denz, M. Zechnall, "Steuerung der Einspritzung und Zündung von Ottomotoren mit Hilfe der digitalen Motorelektronik MOTRONIC", Bosch Technische Berichte 7 (1981) 3

[3] H. Decker, I. Gorille, S. Rohde, M. Zechnall, "New Digital Engine Control Systems", IEE London 1985

[4] G. Felger, G. Plapp, "A new single point fuel injection system with adaptive memory control to meet most stringent emission standards", C221/85 IMechE 1985

[5] D. E. Bergfried, U. Mayer, R. Schleupen, P. Werner, "Engine Management Systems in Hybrid Technology", SAE-Paper 1986 No. 860593

[6] G. Kaiser, M. Zechnall, G. Plapp, "Closed Loop Control at Engine Management System MOTRONIC", SAE-Paper 1988 No. 880135

[7] J. S. Sumal, R. Sauer, "Bosch Mass Air Flow Meter: Status and Further Aspects", SAE-Paper 1984 No. 840137

[8] R. Sauer, "Hot-Film Air Mass Meter - A Low-Cost Approach to Intake Air Measurement", SAE-Paper 1988 No. 880560

[9] Intel Automotive Product, Handbook 1989

[10] W. Borst, G. Coza, M. Henn, M. Zechnall, "Development Aids For Digital Engine Control Systems", ISATA Conference publication, Sept. 1980

[11] U. Kiencke, S. Dais, M. Litschel, "Automotive Serial Controller Area Network", SAE-Paper 1986 No. 860391

[12] Intel report "First High Speed In-Vehicle Serial Communication Device", Intel Europe 1988, Order Number E00104-001

[13] S. Dais, M. Chapman, "Impact of Bit Representation on Transport Capacity and Clock Accuracy in Serial Data Streams", SAE-Paper 1989 No. 890532

900247

Microcontrollers for Modularity in Current ETRs

Ramesh Sivakolundu and Abdul Aleaf

National Semiconductor Corp.

Embedded Control Group

Abstract

The current automotive ETR market trend indicates that modularization of the various radio features both in terms of hardware and software would result in a highly integrated and flexible design that meets the price/performance of the low,mid and high end ETR's. This document deals with the various modular features and partitioning that would be required to implement them. Core oriented microcontrollers provide the required flexibility and performance.

The intent of this paper is to describe a cost effective implementation of a modular ETR by the Microcontroller Applications Group at National Semiconductor Corporation. The COP888xx family of full feature, fully static microcontrollers provide the required core architecture and efficient system solutions.

INTRODUCTION

The development of electronic products has been taking place in the context of increasingly global market for products, and the segmentation of these markets into different user lifestyle groups. Changes in the usage pattern through lifestyle developments have created fundamental shifts in product needs just as dramatic as technological advances. In practice these two factors are interactive, and reinforce each other to encourage the generation of new product types. A split in the design of products between young users wanting lively novelty in their environment, and the more mature user who wants sophistication made easily accessible.

This segmentation of customers leads to a situation where products are no longer differentiated only by features, but also by their pattern of use. Intelligent, adaptable products with the same hardware, although visible differentiation between range models, is a strong expectation at the moment. As products become more intelligent, the interactive interface becomes more important.

Car audio equipment can be considered as a semi-integrated component of the car interior environment. The modular treatment of car interior equipment is likely to increase as the number of options for customers proliferates in the same manner as domestic audio systems, while the available space around the driver is still restricted. Standardization of dimensions will rationalize production and installation costs, but sets limits on miniaturization. The effects of these marketing pressures will therefore consolidate the identity of car audio equipment.

Developments such as these emphasize the importance of product planning phase, to ensure engineering and design advances are focused in a direction relevant to the customer. General purpose microcontrollers provide the flexibility in ever changing customer oriented market. With a proper selection of the microcontroller, on can design a product with sufficient emphasis on hardware and software. In order to reduce the turn around time of products, hardware and software integration is very important. Higher system integration to decrease IC count and manufacturing cost, without sacrificing system flexibility and performance, demands a powerful core-oriented microcontroller with the right blend of features. This article will focus on the advantages of the standard general purpose controller and how it meets the demands of the ETR market.

Section I will address the requirements of the microcontrollers in ETRs while demonstrating the advantages if a standard controller and its features. Section II will review National Semiconductor's COP888 family of microcontrollers, which has the required modularity and features in today's ETRs. Section III will discuss the EMI/Noise considerations associated with digital circuitry.

SYSTEM CONSIDERATIONS

This section reviews the various factors that influence the design of current ETR's along with the demands on microcontroller hardware and software. The microcontroller's on board features and their interface to other standard products used in the design would also be addressed.

MODULARITY/PARTITIONING - The current automotive ETR design trend indicates that the car radio should be fundamentally different from the receivers used at home, for in the moving vehicle it is impossible to divert your attention towards the radio. Most of the ETR's have numerous features such as Preset Stations, Fader, Automatic Preset Programming, Dynamic Noise Reduction, Remote Control, Automatic Volume Control, Anti-Theft Coding, to name a few. This makes it necessary to modularize the ETR design in such a way that features can be added to the basic low end design. Modularization also lets the automotive industry to satisfy the varied taste and interest of the consumer without much of an impact on the design turn around time.

The technical advances in car audio equipment also makes it possible to design ETR's based on how consumer would wish to use them, rather than on the dictates of mechanical or electrical devices. The user interface is of particular concern as car audio equipment user looks for visually more expressive products. This necessitates the partitioning of the user interface from the main audio unit.

Keeping the primary objective to design a high feature content/low manufacturing cost radio the modular system approach would tailor the radios to individual car models and driver preferences. Thus can be achieved by utilizing separate microcontrollers on the main radio board and and on the user interface board. The interface board can incorporate all the functions which change from one ETR model to another. The main radio board includes features common to all ETR models. Modularity would virtually allow any combination of features and functionality to be integrated by selecting the optimal microcontroller combination for the two boards.

HARDWARE CONSIDERATIONS - A desirable microcontroller for both the main and interface boards should have a minimum I/O count for cost effectiveness, and certain on board features. Other demands on the microcontroller are code efficiency, reduced power consumption, data throughput and standard interface to other peripheral products. Selecting microcontrollers with the proper features and the ability to utilize those features effectively can reduce the chip count and result in higher reliability.

The main board controller would, basically, be involved in interfacing with the audio and RF section of the ETR and should have sufficient I/O lines for the interface. The microcontroller should have serial communication with programmable master/slave configurability, for on-board communication with the peripherals and also for inter-board communication. Fig 1 gives the block diagram of a typical main board. Most of the advantages of the modular ETR design lies in the interface board as it has the critical features which make the ETRs different from one another.

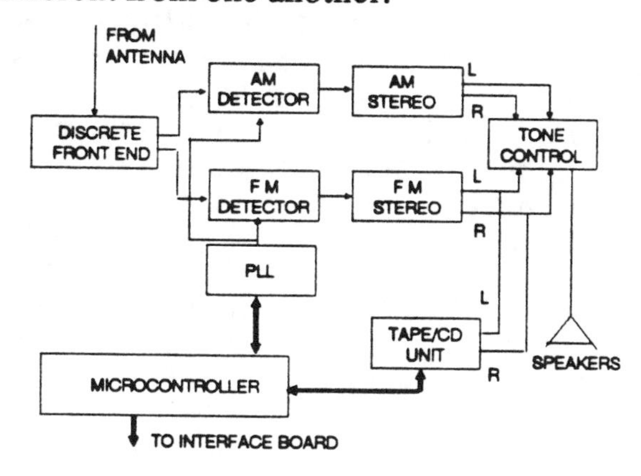

FIG. 1 MAIN BOARD BLOCK DIAGRAM

The interface board microcontroller can be selected based on the additional features and functions required for a certain radio. Fig.2 gives the block diagram of a typical interface board. The interface board is required to communicate with the display drivers, keyboard matrix for user input, and with the main board. Besides, these each additional feature adds certain requirements. The following sub-sections review the various factors involved in selecting the microcontroller.

Keyboard Interface - Appearance and ergonomic design of automotive audio equipment is an important aspect in todays competitive market.Most of the ETRs today have various switches on the front panel to get user feedback. The microcontroller input/output lines have to be efficiently utilized for this purpose. With the current technology the restraints on the form of the

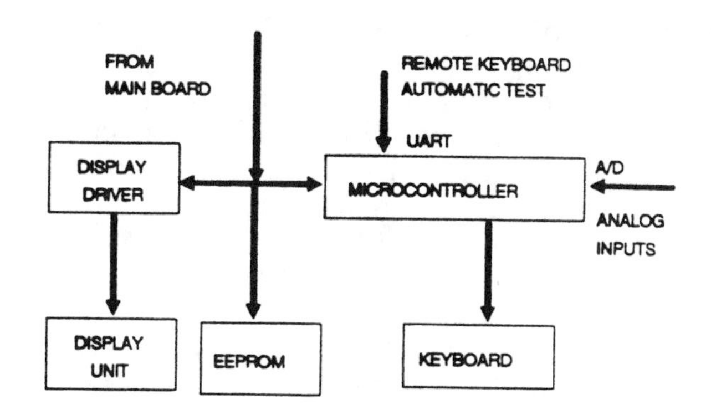

FIG. 2 INTERFACE BOARD BLOCK DIAGRAM

front panel are less and less internal components, but effectively psychological and anthropometric factors such as finger sizes, viewing sight lines, ease of understanding, feel of controls and safety standards. Keeping these in mind the effective use of the microcontroller input/output lines to read the various function selection becomes critical. Optimization of the number of I/O lines used up in this process would effectively decide the microcontroller to be used. Multiplexing the input lines between two or three features could reduce the pin out of the controller.

Analog switches for adjusting volume, fader, balance, treble and bass levels could dictate the requirement of an analog to digital converter (A/D). In order to improve distortion and to increase cost savings digitally controlled devices could be used, however, this increases the overheads on the software. A microcontroller with on board A/D could prove beneficial in reducing the chip count and increasing the reliability.

Display Driver - The display functions are threefold: station frequency,time of the day, and annunciators. The number if annunciators is dependent on the features in the radio and is usually changes from one model to another. Therefore, the overall partitioning must keep the display unit independent of the other devices and boards. This increases the system flexibility and decreases the system cost. The microcontroller selected should be capable of interfacing with most of the display drivers. The optimal design should handle the changes in the display with software modification and the hardware should remain unaltered. Serial interface for communication with VFD or LCD display drivers provide this requirement.

Station Preset- A station preset feature would require the microcontroller to interface to non-volatile memory, if power consumption is of certain importance, or could store the station information in the RAM and HALT the microcontroller for low power consumption. Interfacing a non-volatile memory would also enable the ETR to support anti-theft feature. Another feature that could be easily added is the automated station presets. The microcontroller could scan the entire frequency range and pick the stronger station and automatically store it as preset frequencies. This would require the microcontroller to interface with some device that would differentiate the strength of the signal being received.

Remote Access- Remote control from the steering column as a convenience feature would require a UART for the communication link. An UART would also facilitate the communication with an external tester for automated testing. The test equipment should be have access to address all the functions and features on the ETR. The automated testing delivers accuracy, repeatability and increased productivity when compared to manual testing, which could introduce human error and inaccurate data acquisition. The microcontroller should be capable of addressing each feature under test and communicate the test information accurately to the testing equipment. Such a feature would definitely reduce the ETR evaluation time and also provide means of storing the data acquired during test process.

System Reliability- This is a requirement for any system where electromagnetic interference and voltage surge could cause the microcontroller to execute from undefined program memory space or be in an indefinite loop. Watchdog capability in the selected microcontroller would give an added edge in detecting these conditions and resetting the microcontroller.

Besides, these features the ETRs have various requirements for timers. The time of the day would invariably require a idle timer if power consumption is considered. The dimming feature of the display requires a pulse width modulated (PWM) signal.

SOFTWARE CONSIDERATIONS- Bearing in mind the hardware features selected, the next step is to evaluate the software requirements and trade-offs. Selecting the optimal microcontroller would, in effect, increase the system data throughput and decrease the cost associated with utilizing RAM and ROM. Software development cost will also decrease significantly. Some of the commonly used software routines for an ETR system are:

1. Keyboard scan/decode/debounce routine.

2. Channel frequency computation - both for PLL and Display Driver routines.

3. Station detect and mute routines

4. Station preset store and recall.

5. Data transmission to PLL, Display Driver and other peripheral devices.

6. Interband spacing and top and bottom of the band values for various countries.

7. Time keeping routines.

8. Diagnostic Routines.

A desirable general purpose controller for a low-end ETR should have at least features that support ROM table lookup, bit and byte manipulation, BCD arithmetic. Considering the limited memory space, an ideal controller should also have a powerful single-byte, multiple function instruction set. Ease of decoding the keyboard entry and efficient method of accessing large lookup tables in ROM would effectively make software development a lot easier. Most of the information about channel frequency is normalized and the normalization factor depends on the interband spacing, bottom and top of the band information. The normalization helps in computation of information for PLL and Display Drivers. Significant code is needed for serial data communication with various peripheral devices and the choice of an optimal controller can save run time during program execution, if the controller has an interrupt structure suitable for the same.

The interrupt structure of the controller chosen has a tremendous impact on the flow structure of the software. Ideally every peripheral device and keyboard should be capable of interrupting or waking up the controller. This would result in reduced ROM usage and also decrease the reaction time of the controller to the event that it is supposed to respond.

In order to reduce the complexity of the keyboard on the front panel, the input/output lines of the microcontroller are multiplexed. Some of the switches has a different meaning when the cassette player is selected. The software has to do the necessary context switching and decode the switches accordingly. In most of the current ETRs the preset station switches are multiplexed to correspond to Dolby Noise Reduction (DNR), Automatic Music Sensing and other features of the cassette player. This becomes an overhead on the software, however, it reduces the hardware requirements. Another advantage of multiplexing the input/output lines is evident from the fact that by changing the software the cassette player can be added as an optional feature to the ETR.

Diagnostics is very important from any systems point of view and providing hooks in software for detecting possible failures would reduce service time and increase productivity. Testability of the software itself is another task that improves the efficiency of the system and results in a compressed product development schedules. This also establishes a standard for test methods and data presentations. Routines that perform these tasks should acquire the necessary information from the various devices and present them to the testing equipment. With on-board features on the controller, the test software becomes more complicated and the importance of the selection of the microcontroller becomes more evident. However, the test time and accuracy of the testing overrides the complexity of the software.

SECTION II

The COP888 family of full-feature, cost-effective microcontrollers uses a 8-bit single-chip core architecture fabricated with National Semiconductor's M^2CMOS processor technology. This family contains many advanced features, including low-power HALT and IDLE modes, MICROWIRE/PLUS serial communication, multiple general purpose multi-mode timers, multi-input wakeup/interrupt, watchdog, and clock monitor. The COP888 family members contain a full complement of maskable vectored interrupts. These high performance microcontrollers run at an instruction clock cycle of one microsecond for the majority of single byte instructions. Several addressing modes and a rich instruction set further enhance throughput efficiency and minimal program size.

The COP888CF, COP888CG, and COP888CL microcontrollers are the first members of this expandable 8-bit core processor family. They contain all of the COP888 family features previously mentioned. In addition, the COP888CF also contains an eight-channel, successive approximation A/D converter, while the COP888CG microcontroller contains a full-duplex, double buffered UART and two differential comparators.

ARCHITECTURE- The COP888 family of microcontrollers represents a modified Harvard-type architecture in that the indirect addressing allows ROM data tables to be accessed (LAID Load Accumulator Indirect instruction).

The branch structure of the COP888 architecture contains both absolute and relative jump instructions as well as indirect jump instruction (JID). This instruction is analogous fo the "case" statement in high level languages, and can be efficiently used to decode the key closure patterns and branch to individual key functions. This is a unique single byte instruction and replaces several compare and branch instructions thus improving the throughput and reducing the program size. Conditional branching is accomplished by a number of different test instructions and conditional skip instructions.

The COP888 microcontroller is totally memory mapped, with the exception of the Accumulator, Program Counter, and Idle timer. A block diagram for the COP888 family is shown in Figure 3.

FUNCTIONAL OVERVIEW- The COP888 microcontroller supports several features that provide the flexibility and price/performance trade-offs for ETR designs. The IDLE MODE with low standby power and ability to maintain real-time with associated IDLE timer can be effectively used in time of the day feature of the ETR. Flexible and reconfigurable input/output structure provides an efficient interface the the peripherals and keyboard matrix. Multi-input Wake Up (MIWU) from HALT and IDLE modes, along with the associated interrupts, could effectively perform the requirement of waking the processor. This feature could be exploited in the partitioned, two processor design approach of current ETRs. The interface controller could be in HALT or IDLE mode and the various switches on the front panel could wake the

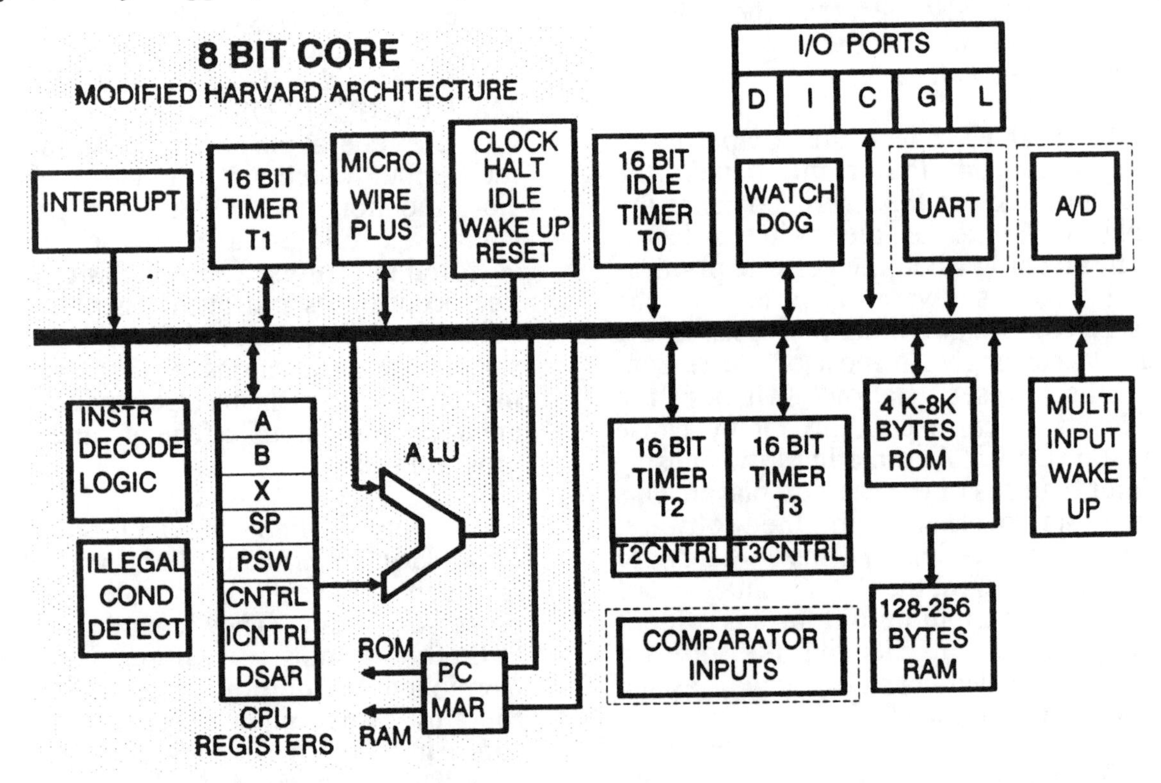

FIG. 3 COP888XX BLOCK DIAGRAM

controller. The COP888 family has eight I/O lines that can be configured as MIWU lines.

MICROWIRE/PLUS, a three wire Serial Data Communication System, allows the microcontroller to be programmed for either master or slave mode configuration. This provides the flexibility in interfacing with devices such as, non-volatile memory, display drivers, PLLs. Versatile 16-bit timers, each with two 16-bit autoreload/capture registers could serve the purpose of generating Processor Independent PWM with alternate reloading of "on" time and "off" time. The software overheads are minimal and the interrupt capability of the timers allow real/time dynamic duty cycle update by generating a vectored interrupt on each half of duty cycle. The flexibility in generating a PWM signal of the desired frequency and duty cycle could effectively serve the purpose of controlling the cassette motor.

WatchDog and Clock Monitor features furnish the system reliability requirements. The programmable WatchDog service window can be used to monitor the number of instructions between WatchDog services in order to avoid runaway programs or infinite program loops. The WatchDog consists of two independent logic blocks. The Upper logic block establishes the upper limit on the service window, while the Lower logic block defines the lower limit of the service window. The implication of the WatchDog logic is that WatchDog must be serviced at least once before the upper limit expires. Tho WatchDog may be serviced multiple times before the upper-limit expiration, as long as the WatchDog services occur no more often than the lower limit of 2048 instruction clock cycles. This is an unique feature in the COP888 family of microcontrollers that provide protection against situations where the software is stuck in loop doing WatchDog service only. The Clock Monitor is used to detect the absence of a clock or a very slow clock below a specified rate.

Fourteen maskable vectored interrupts allows a significantly efficient flow structure of the software, thereby reducing the ROM usage and the reaction time of the controller. The COP888 family supports interrupts for IDLE Timer, MICROWIRE/PLUS, Timers, Multi-input Wake Up, and UARTs.

The COP888CG/COP888EG incorporate a full-duplex UART, which meets the demand of an ETR partitioning with remote control and remote functionality testing. The UART baud rate is software selectable in conjunction with both a prescaler and baud select register. The UART contains a full set of error detection circuitry and a diagnostic test capability. The COP888CF with 8 channel A/D provides the interface requirements with analog switches for adjusting volume,fader,balance, treble and bass levels. The A/D converter supports both single ended and differential modes of operation, and contains a prescaler option that allows seven different clock selections. The minimum A/D clock cycle for the COP888CF microcontroller is 600 nanoseconds and each conversion takes 12 A/D clock cycles.

EMI CONSIDERATIONS

The Electromagnetic Interference (EMI) problems are primarily the interference generated during fast risetime transitions. These fast transitions result in adverse circuit effects such as crosstalk, common impedance coupling, and also cause harmonics of high order from periodic waves which result in emissions. Audio system designs are especially vulnerable to this type of interference.

Although most emissions conduct out of circuit board trace and radiate from there, the emissions radiated directly from the silicon chip cannot be ignored. With extensive use of digital logic in today's microcontrollers, certain levels of EMI will occur, The microcontroller radiation problems are primarily on the pins.

Interference can be propagated out of the signal lines or the power lines, so steps must be taken to control both. This can be done by filtering or rise time control, and is preferably done on the microcontroller, but can also be done externally.

The principle culprits are the periodic waves, such as internal CPU signals. Frequent internal switching as a result of many bits switching simultaneously increases the emission levels.

At National Semiconductor Corporation, design efforts continue to minimize or suppress the interference caused by the microcontroller utilizing proper control techniques.

CONCLUSIONS

The current electronic technologies has gradually removed the limiting parameters for product designers, so that product development can start with a very blank sheet of paper, and many design options are available.

Ironically microcontrollers has focused attention away from the technology itself, towards subjective, and undefinable human values in design. The function of control becomes so interchangeable, evidently because of the use of microcontrollers, the decision about the front panel for ETRs becomes a complex human factors problem. This new interface problem create the need for new design skills - a combination of software and hardware design, mechanical and industrial design, ergonomics and graphic design, information design and psychology.

Given these new design skills and design freedom created by microcontroller electronics, new generation of car audio products is now possible, which will give customers more features and capabilities, and can be truly usable, friendly, and enjoyable.

At National Semiconductor Corporation, design and development efforts, of the Embedded Control Group, continue to provide system designers with the flexibility and price performance ratio required in today's global market.

900238

Trends in Powertrain Integrated Control and Development of Optimum Microcomputers

Naoki Tomisawa
Japan Electronic Control Systems Co., Ltd.
Derek L. Davis
Intel Corp.

ABSTRACT

Through the rapid evolution of microcomputer capability in the field of automotive electronic fuel injection, a system has recently been developed to integrate powertrain control, unifying engine control and automatic transmission control.

Implementation of advanced control theory is one of the major issues in this field and will benefit greatly from microcomputer performance improvements.

This paper commences with a discussion of the trend toward higher performance requirements in integrated powertrain control, driven by a new concept in control: variable transmission shiftpoint control based on vehicle operating conditions.

The latter portion of the paper describes a new 16-bit dual CPU system for integrated powertrain control. The system achieves an improvement in processing through high-density communication. Software development is also enhanced by separating the functions of the engine-control microcomputer from those of the transmission-control microcomputer and by using mutually non-interfering software.

1. TRENDS IN INTEGRATED POWERTRAIN CONTROL

Engine controllers, starting with the analog fuel injection single-function control types that were predominant in the 1970s, have now centralized almost all engine control functions by unifying ignition control, idle speed control and other controls.

With optimum control of the powertrain system as a goal, the next step is expected to be an integration of the heretofore separated engine control and automatic transmission (AT) modules.

(1) TRENDS IN CONTROL INTEGRATION [1],[2] - First, let us consider trends in control of the vehicle as a whole. Powertrain control is the control of "running" among the basic vehicle functions of "running," "stopping," and "turning." Most current control systems utilize separate stand-alone control modules for engine control and AT control, communicating as required.

To achieve more sophisticated control (both finer and more intelligent) of the engine and transmission, higher density information exchange between the two systems is required. This leads to either development of high-speed inter-module communication links or integration of the modules into a single unit which performs centralized control of the powertrain system.

A further step will be to apply communication links to functional units such as the brake control system, which are in charge of the "stopping" and "turning" operations.

Thereafter, even more sophisticated system control will be pursued.

In this trend, powertrain control is central to the overall vehicle control system and is expected to serve as the host in communication.

The unification of the engine and AT control modules to achieve overall control of the powertrain system will reduce costs and is expected to be developed even further.

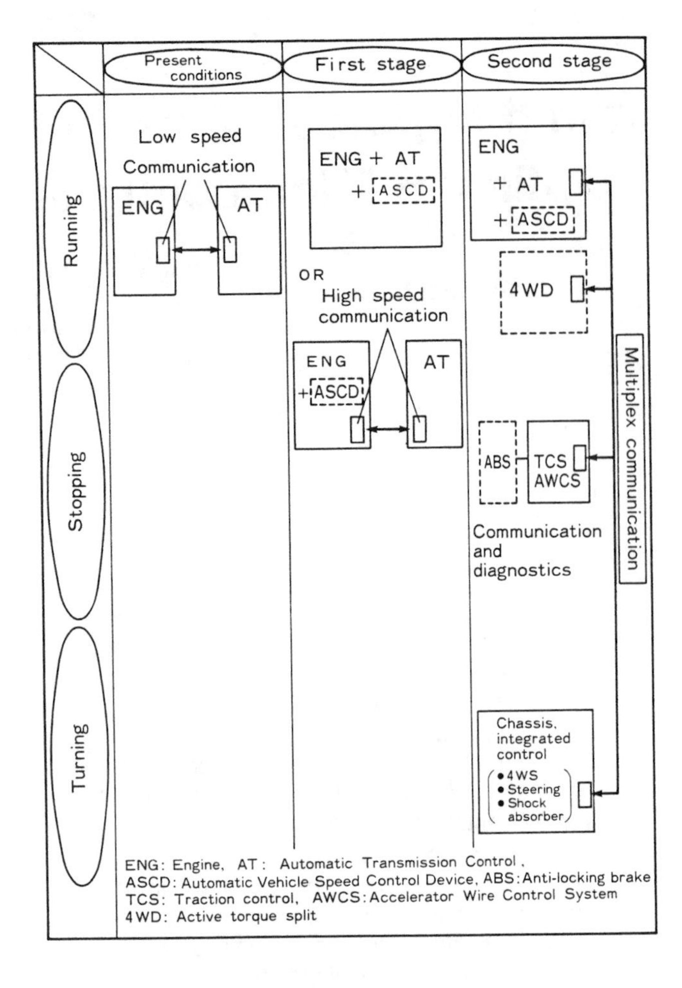

Fig. 1 Trends in Integrated Control

(2) TRENDS IN CONTROL TECHNOLOGY- As outlined above, vehicle control systems will migrate toward integration of control functions such as the powertrain system and the chassis system through communication links. Eventually, vehicular control will be a single integrated system. However, such an integrated system will allow not only finer functional control by sharing information between control modules, but also the use of more advanced control theory to obtain a performance impossible with conventional stand-alone modules.

One example is a new approach being studied for powertrain systems. This would incorporate control of the engine shaft torque at the AT shiftpoint according to information received from the AT modules, and an algorithm to optimize AT shiftpoint timing by recognizing and judging the vehicle operating conditions based on engine control module information. This will allow adjustment of the driving characteristics to conform to the desires of the driver.

2. PROGRESS OF POWERTRAIN CONTROL TECHNOLOGY

The following are descriptions of new control theory implementation in powertrain control.

(1) ENGINE TORQUE CONTROL - Engine torque suppression-based AT control information can be cited as an example of control integration in a combined engine/transmission system.

In practice, to reduce the shock of AT shift, an example is reported of suppressing engine torque at specific time points, calculated by the AT controller, near the start and finish of up-shift or down-shift. Generally, ignition timing is retarded and combustion pressure is controlled by reducing the fuel supply.[3]

Another example is engine idle speed control based on the delay in response of AT hydraulic pressure when shifting between Neutral and Drive. This improves idling stability and resistance to engine stall. Also reported is engine torque suppression during high standing-start acceleration and immediately before down-shifting. This limits stress on the automatic transmission body, allowing a reduction in the size and weight of the transmission and resulting in improved fuel consumption.

Realtime communication between the engine control system and the AT control system is required to support implementation of advanced theory, as described in the example above.

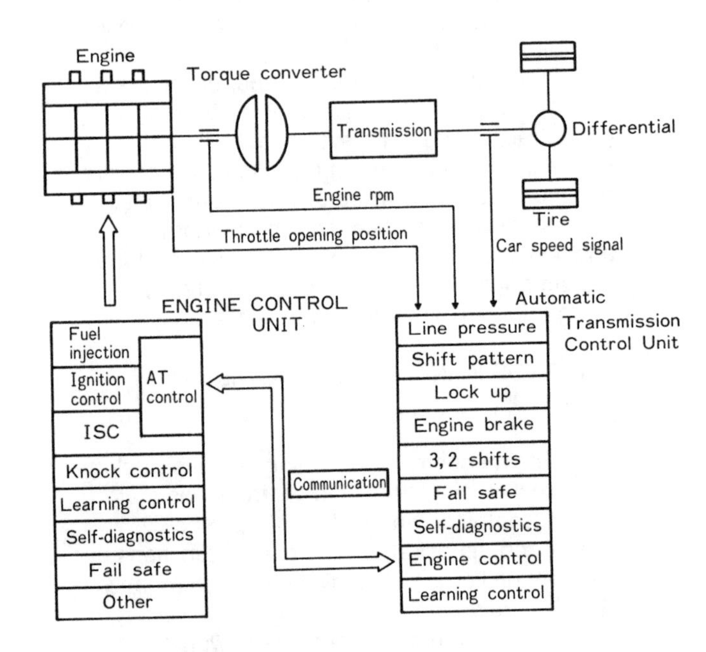

Fig. 2 Integrated Control of Engine Torque

(2) ADAPTIVE CONTROL BASED ON OPERATING ENVIRONMENT [4] - Adaptive control based on the operating environment is an intelligent control approach that will be introduced via powertrain control integration.

In current engine and AT control systems, each control parameter is determined adaptively, based on the target market, exhaust gas characteristics, and operability, resulting in constant characteristics for a specific vehicle or vehicle type.

For instance, shift timing and other AT system parameters are preset according to overhead and engine rpm.

Consequently, AT parameters best suited to one driver/environment combination (city streets, high speed, congested traffic) may be unacceptable for another driver/environment.

To remedy this, built-in intelligent control with a large dynamic range of control parameters is being studied. This system recognizes and evaluates the driver's intentions and driving environment, adjusting vehicle characteristics to conform to those conditions.

The system features an intelligent control block that uses inputs from the engine controller and AT controller as well as powertrain output data to perform intent analysis and environment recognition. On the basis on this analysis, optimum system control characteristics are selected from the built-in system characteristic targets (wide range).

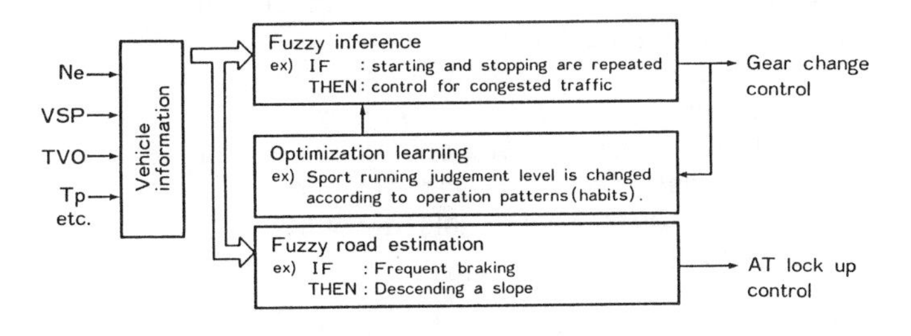

Fig. 3 Intelligent Control Block

Information that must be interpreted by the intelligent control block is classified as follows:

(i) analysis of driver intent; aggressive/passive behavior

(ii) recognition of operating environment; congested traffic, city streets, freeway etc.

To evaluate this information, the mode of operation and status of environment are quantified (standardized) with the built-in fuzzy membership function. Then an estimation is made, by fuzzy inference, to what extent the engine rpm, vehicle speed, throttle position, and overhead parameters (or differential values of each) comply with the standardized values.

Also, the adaptation of these parameters must incorporate the history of driver habits and operating patterns (adaptive learning).

Based on the interpretation performed by the intelligent control block, the systems are optimized by varying the AT shiftpoint timing and the hydraulic line pressure for slip-rate control. Intelligent control should be carried out and the AT system controlled according to the resulting evaluation. Most of the vehicle data required for operating environment recognition is contained within the engine system. This must be shared with the AT system in order to achieve these intelligent control objectives. In addition, the communication method chosen must not significantly increase CPU overhead in either the engine system or the AT system, particularly if inference computations must be performed.

3. PERFORMANCE REQUIRED FOR POWERTRAIN CONTROL INTEGRATION MICROCOMPUTERS

As discussed previously, integral-type control units are the most promising for integrated powertrain control systems that must reconcile trade-offs between performance enhancement and cost reduction. Let us consider the functions and performance required for microcomputers in integral-type control units.

(1) CONSTRUCTION OF MICROCOMPUTER CHIPS - In a control unit integrating engine control and AT control, software design, experiments, and compatibility with the vehicle should be considered first. The microcomputer is part of the plant unit in the engine system and the AT system, coordinating its operation to contribute to the overall vehicle performance and emissions. Each system is usually developed independently and in parallel with the development of the mechanics of each plant unit. Thus, if one control system under development is affected by modification of the other, it requires review. This applies to all stages of design, experiment, and compatibility of vehicle control.

To avoid this inconvenience, the system architecture should allow independent software development for engine control and AT control such that they will not interfere with each other. To control both the engine system and the AT system with one CPU, very high speed CPU operation would be required to assure that the processing requirements for one task (e.g. AT control) could not impinge on the processing resources available for the other task (e.g. engine control).

We think it difficult to realize this mutual non-interference using a single 16-bit CPU system.

Consequently, we have chosen a dual processor approach, specifying microcomputers with optimum functions for the engine system and the AT system.

Second, let us consider the cost of VLSI devices. Current engine control and AT control systems are very sophisticated. In particular, engine systems with cylinder sequential fuel injection and ignition control are already in commercial application. Thus, full 16-bit CPU microcomputers combined with a sophisticated realtime control system are already required.(5)

The costs of such VLSI devices are dominated conceptually by die size as shown in Figure 4, although architecture, design rules, etc. will have an effect. With the increase in die size, the cost curve rises sharply. This is because the yield decreases abruptly beyond a certain size.

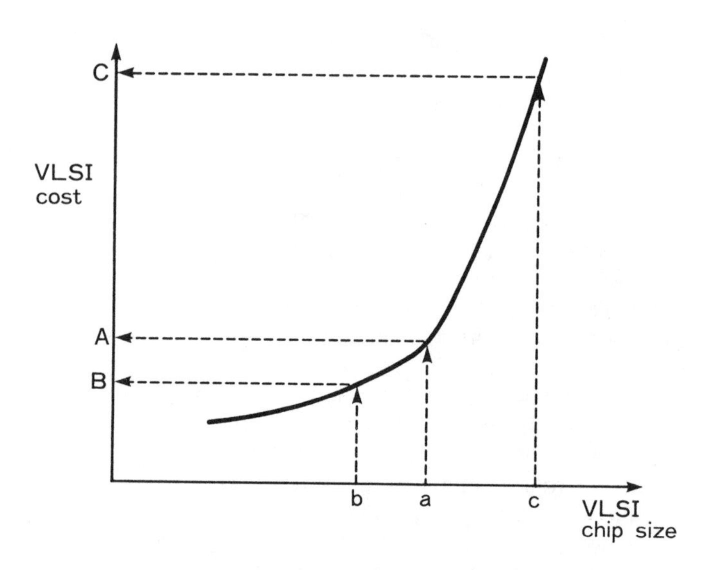

Fig. 4 VLSI Cost Curve

Engine control units must incorporate LSI devices of a fairly large die size as explained above. Assuming that the powertrain control module contains one microcontroller dedicated to engine control (die size "a") and one microcontroller to AT control (die size "b"), the LSI device cost becomes (A+B). Single CPU control (of both engine and AT) would require a die size of about (a+b=c), resulting in an LSI device cost of "C". Considering the sharp rise in the cost curve, (A+B) would be smaller than "C", leading to the higher cost of the single CPU configuration.

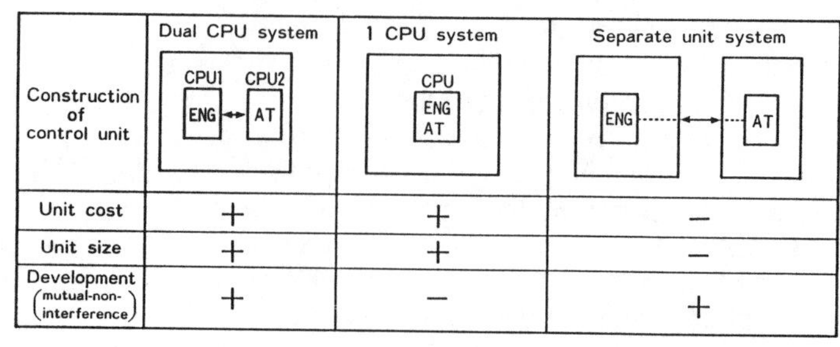

	Dual CPU system	1 CPU system	Separate unit system
Construction of control unit	CPU1 CPU2 ENG ↔ AT	CPU ENG AT	ENG ↔ AT
Unit cost	+	+	−
Unit size	+	+	−
Development (mutual-non-interference)	+	−	+

+ : Advantage, − : Disadvantage

Fig. 5 Characteristics of each Systems

Figure 5 summarizes the characteristics of each systems discussed above.

(2) INTER-CPU COMMUNICATION - In the dual CPU system for integrated powertrain control, the most significant issue is the communication of data between the two microcontrollers.

As explained in ENGINE TORQUE CONTROL, realtime communication is most important when the engine control parameters vary according to AT control information. Also, as discussed in ADAPTIVE CONTROL BASED ON OPERATING ENVIRONMENT, to facilitate intelligent control, communication between two or more functional units must be achieved without significantly increasing the overhead for the microcontrollers. Their performance is required for inference and other operations. To meet these requirements, dual-port RAM could be used for inter-CPU communication. However, the cost would be comparatively high. Alternatively, a new "slave port" implementation allows emulation of a dual-port RAM, providing external access to the internal RAM of the microcontroller.

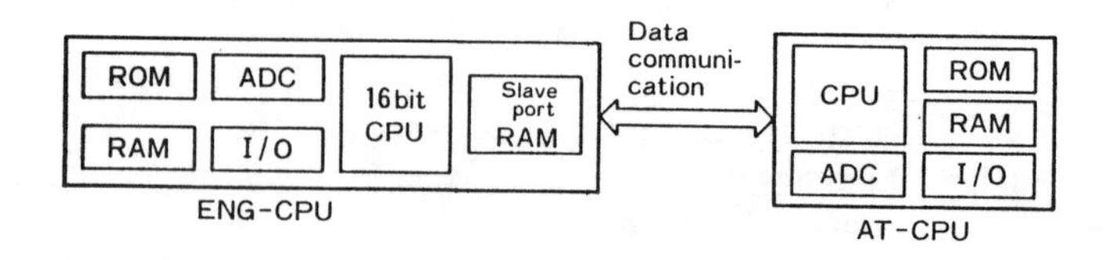

(3) REALTIME I/O OF EACH CPU (TIMER SYSTEM) - Increasing microcomputer software development efficiency is an important task. Software for the integrated powertrain control system, in particular, will be extremely large.

Conventional microcomputers that perform pulse I/O processing by software interrupts have needed complicated programs to handle multiple interrupt servicing and a special operating method to shorten processing time, to ensure a high degree of realtime control. Programs requiring these special software techniques must sacrifice software versatility and ease of maintenance.

This has become a significant problem in software development for integrated powertrain control and affects inference operations.

To solve these problems, we have developed a realtime processing system without increasing CPU overhead for the engine microcontroller or the AT microcontroller. This allows the computing resources of the microcontrollers to be used for the implementation of more advanced integrated control. This realtime processing system is an evolutionary extension of the Event Processor Array (EPA).

Figure 7 shows a fuel injection pulse output timer for the engine system.

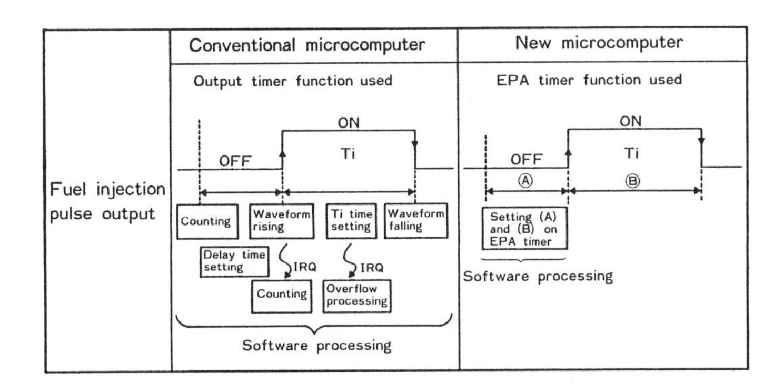

Fig. 7 Fuel Injection Pulse Output Timer

(4) DELAYED PWM INTERRUPTS - A number of hydraulic actuators are used for the AT control system. We prefer to provide a fail-safe system to detect abnormal operation of these actuators and prevent damage to the mechanics of the transmission. Abnormal operation of an actuator is generally detected by monitoring the ports after the solenoid is turned on or off using the circuit shown in Figure 8.

To avoid the effects of noise due to reverse EMF, the port monitoring is delayed by a software interrupt for a short time after the on/off switch point. However, as AT control sophistication increases in the integrated powertrain control system and more solenoids are utilized, this interrupt processing method would increase the overhead on the CPU. To cope with this, special delay timers automatically generate interrupt requests at specified times synchronized with the solenoid control output (PWM).

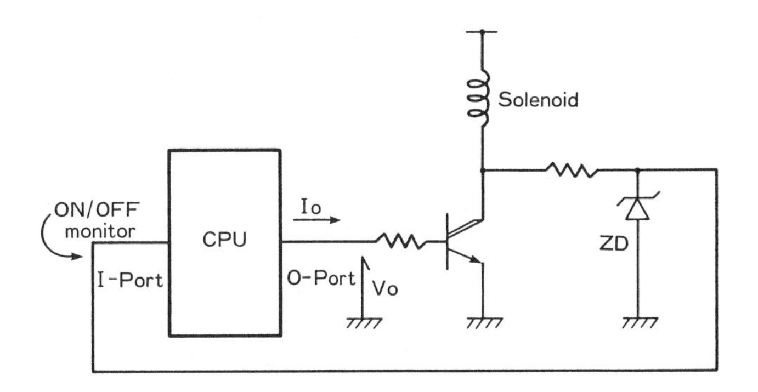

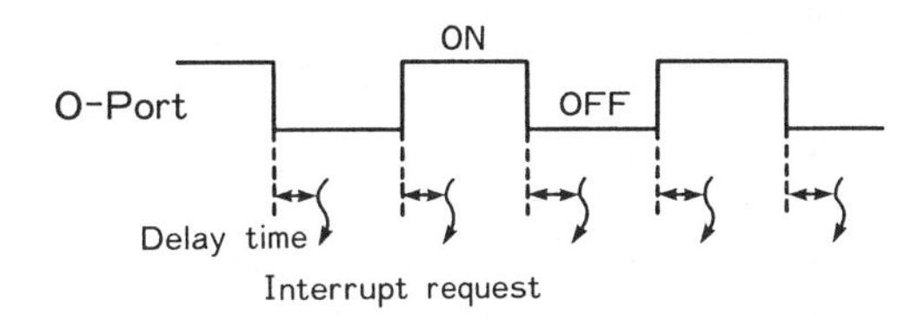

Fig. 8 AT Control Solenoid Drive Timer

(5) MEMORY - To avoid a system with high LSI device count and to reduce the physical size of the control module, RAM and ROM are incorporated on both the engine microcontroller and the AT microcontroller. To quickly address changes in module specification, yet retain a cost effective package, a One Time Programmable (OTP) implementation of ROM was selected.

On the basis of current capacity and future performance predictions, we have used the following ROM and RAM, although higher capacities would be better for implementing multi-function, sophisticated control, arranging programming systems, and making maintenance easier.

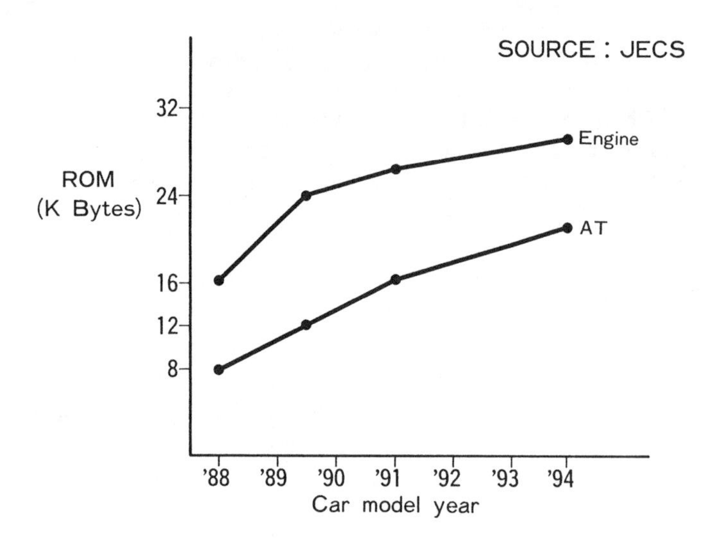

Fig. 9 Engine-AT Control Memory (ROM) Size

4. APPLICATION SPECIFIC DESIGN MODULARITY

(1) MICROCONTROLLER PROLIFERATION- In many automotive application areas, the level of capability of the electronic control module varies greatly from low-end to high-end. Thus, a microcomputer satisfactory for the high-end will be too expensive for the low-end. Conversely, an economical microcomputer for the low-end cannot meet the performance requirements for the high-end.

The result may be two or more entirely different computer architectures used throughout the module product line. This requires significant resources to support different:

- Development environments (both hardware and software tools)
- Application algorithms
- Application software
- Architectural knowledge and experience
- Production/test knowledge and experience

One effective approach to the problem is to generate several proliferations of a relatively high performance core architecture, targeted to address the memory and I/O requirements of specific applications. These application specific devices are tailored to provide the appropriate capabilities for the application, while at the same time enjoying the high-volume production required to drive costs down.

(2) MODULAR DESIGN METHODOLOGY - Maximizing design productivity is an important factor in implementing a proliferation strategy. Modular design techniques, when used at all levels of the design process (e.g. architecture, modeling, schematics, and layout), can significantly improve productivity.

EXAMPLES

Architecture - Limiting changes to specific modular portions of an existing peripheral design simplifies overall definition effort.

Design - Well isolated functional changes mean reduced time for RTL model and schematic development and simulation.

Product Test/Qualification - Modular testing allows much of the work invested on a previous proliferation to be reusable.

Layout - Reusing modular layout blocks and modular planning for memory array scaling (RAM, EPROM, etc.) can dramatically affect the overall effort required for a new proliferation.

But modular design not only improves resource efficacy for the integrated circuit supplier, it has the same effect for the electronic module supplier in areas of:

- Electronic control module hardware design
- Algorithm development
- Software development and maintenance
- Module qualification and testing

The balance of this paper discusses the attributes of two high performance 16-bit microcomputer proliferations implemented with the modular design methodology described above.

5. MICROCOMPUTER ATTRIBUTES FOR ENGINE AND TRANSMISSION CONTROL

The following sections detail specifications of two new application-specific microcomputers defined for automotive engine (Engine Microcomputer) and transmission control (Transmission Microcomputer). They are implemented on a high density 1 micron dual layer metal EPROM process, integrating all program memory, data memory, and peripheral functions as typically required by the target applications.

In addition, by appropriate system partitioning, the Transmission Microcomputer (TMC) is a strict subset of the Engine Microcontroller (EMC). This allows full system development for the TMC to be done using EMC silicon.

(1) HIGH PERFORMANCE 16-BIT CPU - The performance requirements of engine control and transmission control dictate that a high performance 16-bit realtime processing core be the basis for proliferations addressing these applications. In addition, the core must be able to handle the multitude of interrupts that are possible in such systems.

The core chosen for this family of products implements a 16-bit register-to-register architecture, operates at 16 MHz (4 native MIPS peak), and is capable of supporting 112 distinct interrupt sources.

The core is combined with EPROM and RAM memory arrays (which may be scaled) to create a so-called "macrocore."

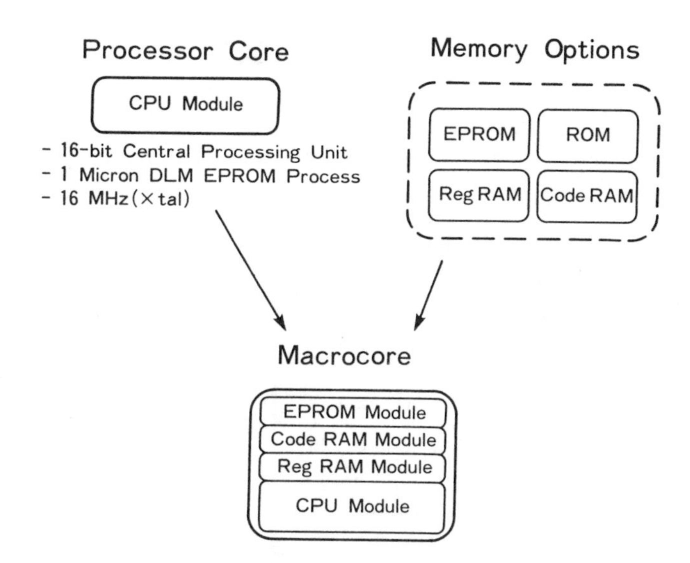

Fig. 10 Macrocore Development

Peripheral blocks are then added to this macrocore to generate specific proliferations. In this case, these are the Engine Microcomputer (EMC) and Transmission Microcomputer (TMC).

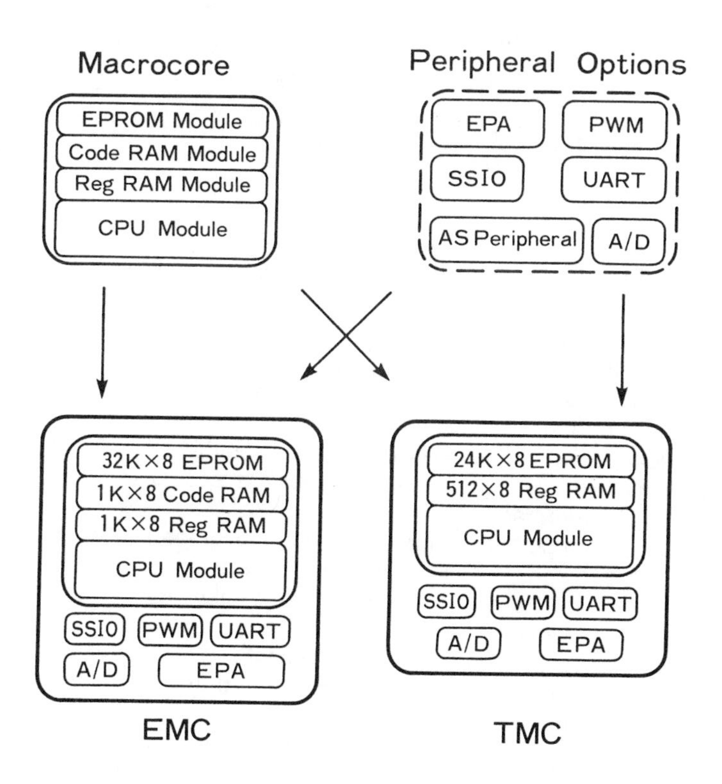

Fig. 11 Proliferation Development

(2) EMC/TMC MEMORY

Program Memory - Program memory is implemented on the EMC and TMC devices as Erasable Programmable Read-only Memory (EPROM). This EPROM array is organized to extend the full width of the chip such that the size may be increased or decreased without introducing unusable "holes" in the silicon. The EMC contains 32,768 bytes of EPROM while the TMC contains 24,576 bytes.

Data Memory - Data memory is available as Register RAM and as Code RAM. Both types of RAM are implemented in the same fashion as the EPROM, allowing array scaling. Register RAM is used extensively by the instruction set for register-to-register operations. Accesses to the Register RAM occur in parallel with accesses to the EPROM or Code RAM, resulting in higher data rates than are possible with a single bus architecture.

Code RAM is conceptually identical to Program Memory and program code may be executed from it. Code RAM is particularly useful for downloading (via UART) and executing diagnostic routines that would normally exceed the module's memory capacity.

The EMC contains 2048 bytes of both Register RAM and Code RAM. The TMC contains 512 bytes of RAM.

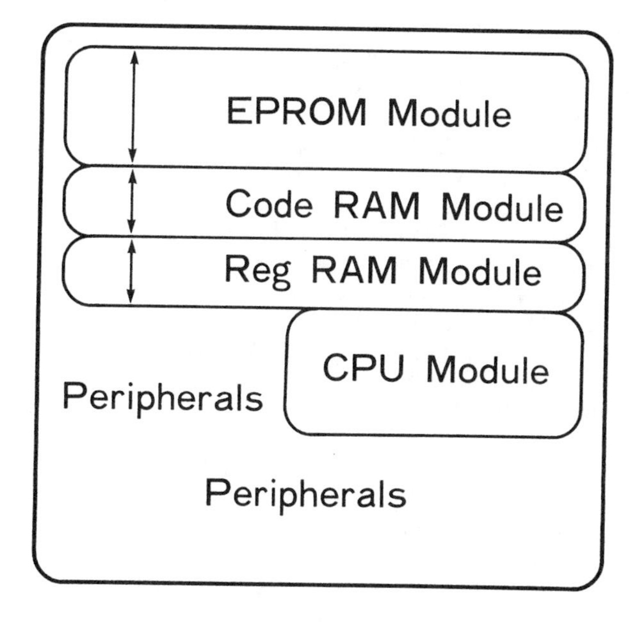

Fig. 12 Memory Array Organization

(3) EMC/TMC PERIPHERALS

Analog/Digital Converter - The 10-bit Analog to Digital Converter is architecturally capable of handling 16 analog inputs with a conversion time of less than 20 µsec. The EMC implements 11 of the channels; the TMC implements 8.

UART - The UART is a standard asynchronous serial communications channel, additionally capable of operating in a synchronous (clocked) mode. This is implemented on both the EMC and the TMC.

SSIO - The SSIO (Synchronous Serial Input/Output) unit is a synchronous-only communications channel, capable of operating from 15 KBaud to 2 MBaud. Both data and clock are bidirectional. The EMC and TMC each contain a single SSIO channel.

Pulse Width Modulator - The EMC and TMC each have 6 fully independent PWM outputs, frequency programmable from 4 Hz to 32 KHz.

Event Processor Array - The Event Processor Array (EPA) is a high speed event control peripheral that controls output event generation, input event capture, software interrupt generation, and other special events, all with a resolution of 250 nsec.

All events are referenced to one of two 16-bit time bases. Typically, for engine control one of the time bases will represent realtime and the other engine position. Dedicated capture/compare silicon at each "EPA pin" controls the functions for that pin (channel).

The TMC incorporates 8 of these "channels" for general purpose input/output/event control. The EMC contains the 8 general purpose channels as well as 8 dedicated output-only channels for injection control.

Because the EPA is modular at the timer and channel level, additional channels may be added or removed as required for a particular application specific device. This has no effect on the resolution of event control, but is only limited by the overall processing capabilities of the core processor.

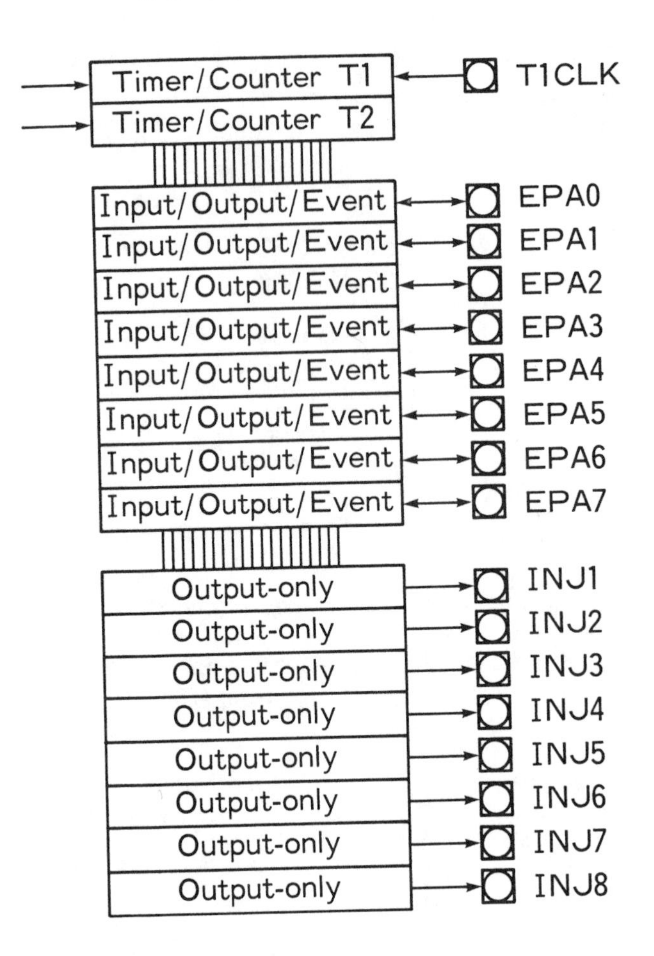

Fig. 13 Event Processor Array Organization

(4) REALTIME INTERRUPT SERVICING - In realtime management applications such as engine control and transmission control, required CPU performance correlates directly to interrupt service time. Thus, reducing the time spent in servicing interrupts results in higher effective CPU performance.

One potential source of interrupt overhead is due to software sorting of interrupts when several interrupts are mapped into a single interrupt vector. The problem is to design an interrupt system that doesn't carry a lot of hardware overhead for a proliferation with few interrupt sources, yet can be expanded to support another proliferation with many interrupt sources (without software sorting).

The core used for the EMC and TMC accomplishes this by allowing many interrupts (e.g. 16 from the EPA) to be mapped into a single interrupt line in the core, while still utilizing distinct interrupt vectors for each possible interrupt source. Thus the core interrupt logic remains small and identical from proliferation to proliferation, yet can efficiently support anywhere from 16 up to 112 interrupt sources.

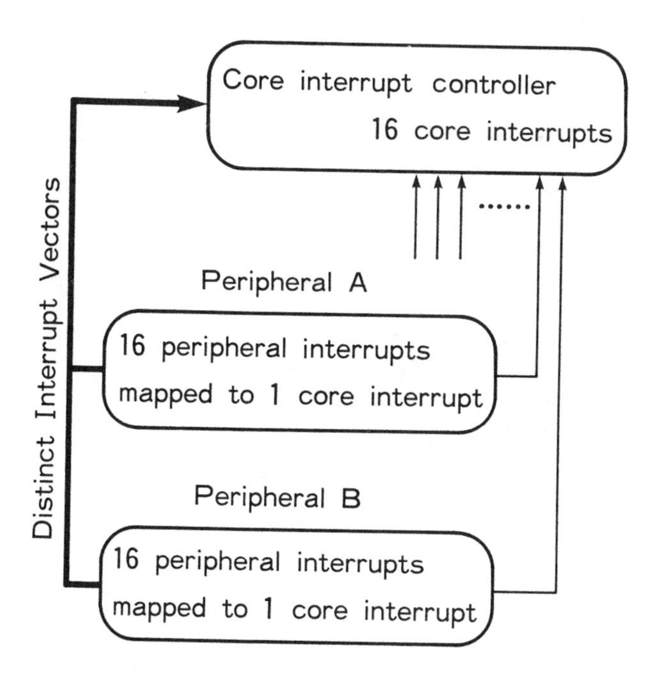

Fig. 14 Interrupt Expansion

Another source of overhead in servicing of simple interrupts is the time required to save the machine state before the interrupt is served and restore it thereafter. The processor must first save the program counter and the status flags. Then, if the interrupt is time-critical, the user must disable interrupts and push the interrupt mask. Only then can the true interrupt service routine start. At the end of the service, all of this information must be restored for the processor to continue from where it was interrupted.

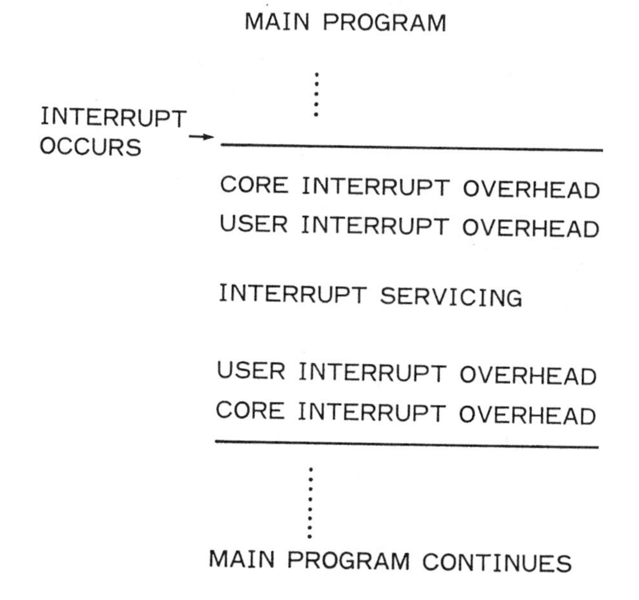

Fig. 15 Software-based Interrupt Servicing

By creating dedicated microcode for certain simple application-specific interrupt serving operations, this overhead can be completely eliminated. An example would be the interrupt service routine used when receiving a message on the serial port (UART). This executes at the receipt of each new byte of the message, transferring the data to consecutive locations in memory until the message is complete. But each time the routine executes, the overhead described above is incurred.

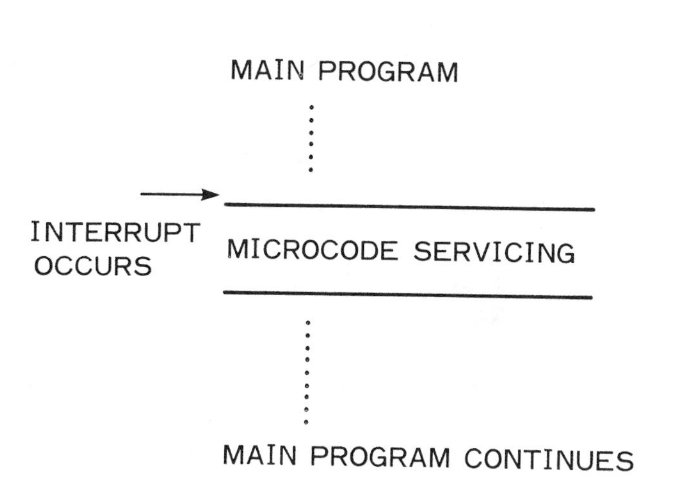

Fig. 16 Microcode-based Interrupt Servicing

A special microcode routine that automatically transfers that contents of the serial port buffer to consecutive memory locations executes very quickly because it does not have to store the state of the machine. In essence, it appears like a single instruction is injected into the normal program flow to serve that interrupt.

But because these microcode routines are non-interruptible and have complete control over the machine state, they do not have to be limited to simple operations. Several such application-specific routines are implemented on the EMC for special control of injection outputs, parallel communication with the TMC, etc. And since microcode is easily modified for a particular proliferation, the TMC has its own set of special interrupt servicing microcode for transmission control functions.

(5) INTER-MICROCOMPUTER COMMUNICATION - In an application where two microcomputers must share various realtime parameters (e.g. an integrated engine/transmission control module), a dual-port RAM is often used to provide a shared memory. Normally this requires a third device in the module, or implementation of a true dual-port RAM within one of the microcomputers.

This is not required for the EMC/TMC processor pair because the bus controller of each device contains a Slave Port. The Slave Port allows an external bus master (microcomputer, processor, etc.) to address the EMC or TMC as a peripheral located in the bus master's memory space.

With this approach, the TMC can directly access the EMC over the system bus and gain access to the internal memory of the TMC. This is treated as an interrupt by the EMC, but served with a special microcode sequence as described above. For limited data rates, this technique eliminates the need for an expensive internal or external dual-port RAM.

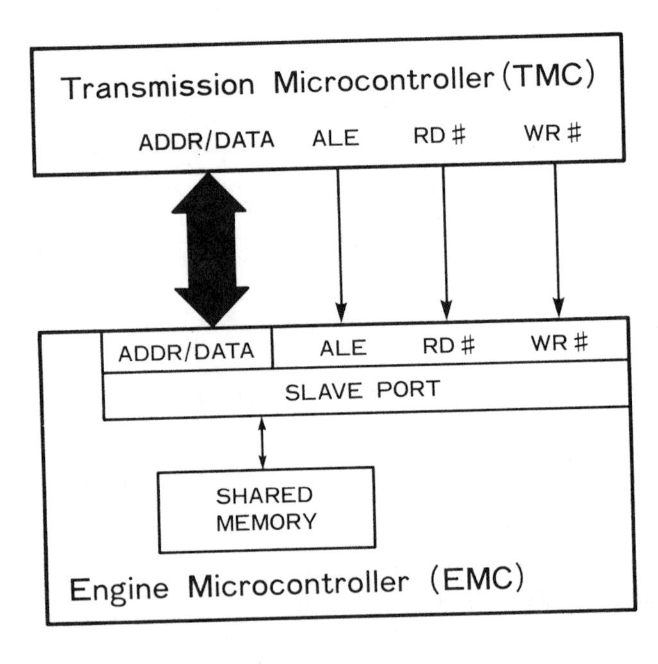

Fig. 17 DP-RAM Emulation via Slave Port

(6) SUMMARY OF EMC AND TMC ATTRIBUTES

	EMC	TMC
Memory		
Program (EPROM)	32K x 8	24K x 8
Data (RAM)	2K x 8	512 x 8
Communication Channels		
UART (Async/Sync)	1	1
Synchronous (serial)	1	1
Slave Port (parallel)	1	1
High Speed I/O Channels		
Output-only	8	-
General Purpose	8	8
Time Bases	2	2
PWM Channels	6	6
A/D Channels	11	8
Interrupt Sources	30	22

6. OUTLINE OF EVALUATION SYSTEM

The following is an outline of the system we used to develop and evaluate this dual CPU system.

(1) INTEGRATED POWERTRAIN CONTROLLER - Figure 18 shows the configuration of this controller. Engine control is completely separated from AT control with the dual CPU configuration. Thus, non-interference of software had been achieved. Both microcontrollers communicate information via the Slave Port, facilitating independent and mutually non-interfering programs for the engine and AT systems, although those systems are integrated and housed together in a case measuring 160 x 170 x 35 [mm].

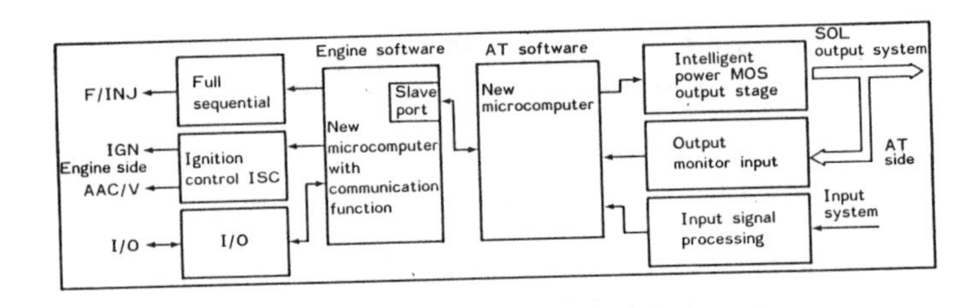

Fig. 18 Controller Block Diagram

(2) CONTROLLER EVALUATION SYSTEM- Figure 19 shows an outline of the system we used for developing and evaluating the controller programs.

This system can simultaneously operate the controller under development (integrated powertrain control unit) and the previous generation controller by entering vehicle and engine simulation information. If a difference occurs between the outputs of two control units, the flow of the software at that time will be captured and displayed. This system enabled us to verify the characteristics of the integrated control unit versus the individual control units on which the former generation of engine and AT control was based. Mutual non-interference of engine and AT control software was also verified.

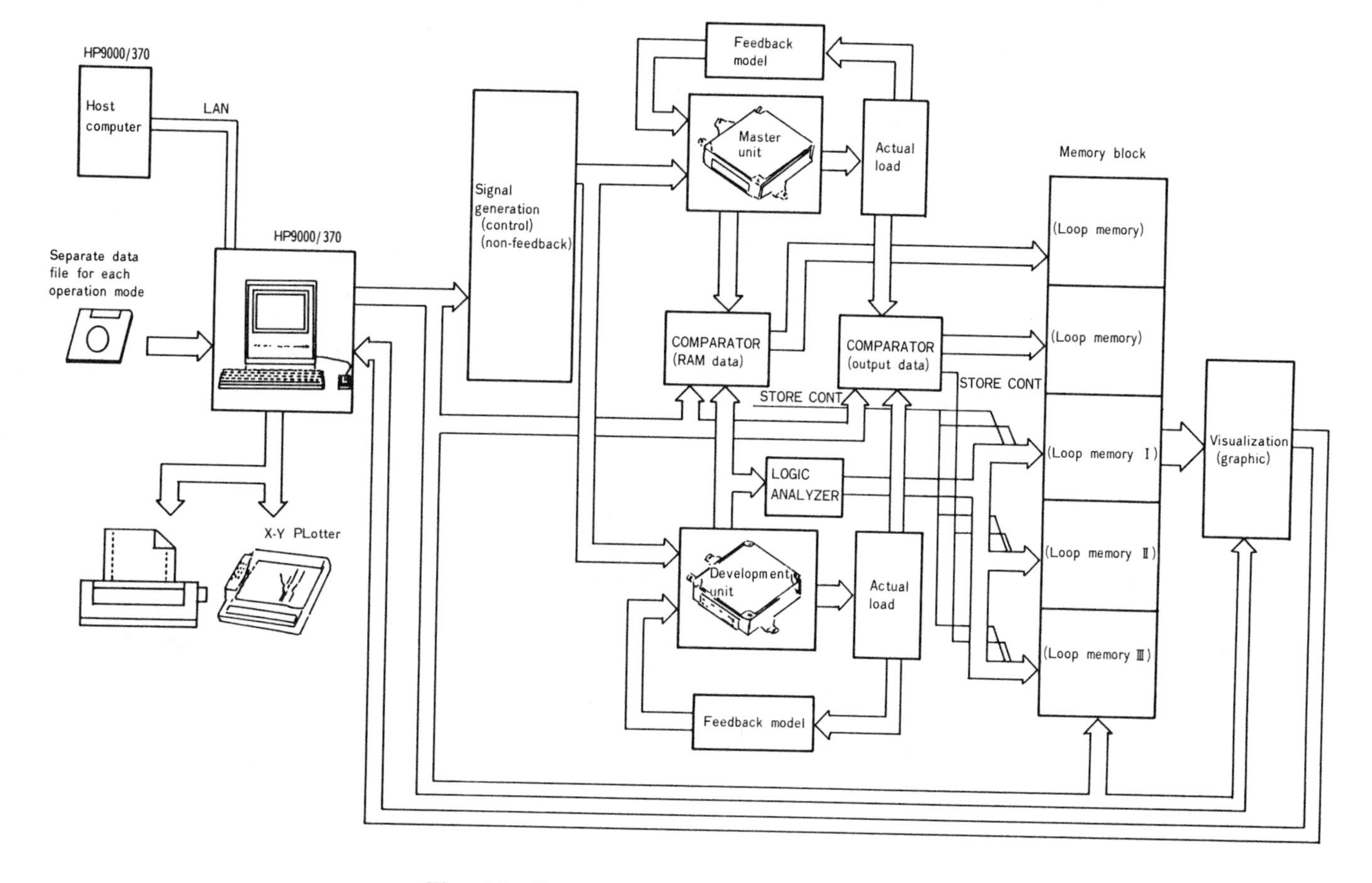

Fig. 19 Program Evaluation System

7. SUMMARY AND CONCLUSIONS

This paper discusses the development of optimum microcomputers for integrated vehicle control and integrated powertrain control.

With this system architecture, we have achieved the following:

[1] Separation of engine control system and AT control system functions through the development of a dual CPU system that permits mutually non-interfering control software.

[2] Development of the Slave Port communication mechanism to facilitate the high density communication required by advanced control algorithms.

[3] Improvement in software efficiency through use of the realtime processing capabilities of the Event Processor Array (EPA).

[4] Low-cost and compact control units through the use of highly integrated single-chip microcontrollers of high memory capacity.

REFERENCES

(1) GM: Electronic Future Concept, Convergence '88

(2) Nippon Denso and Future CAR Electronics Internal Combustion Engines, January 1989

(3) M. Schwab: Electronic Control of 4 Speed Automatic Transmission with Lockup Clutch, SAE paper 840448

(4) GM: Adaptive Vehicle, US-PAT. 4,829,434

(5) Naoki Tomisawa: Trends in Electronic Engine Control and Development of Optimum Microcomputers, SAE paper 880136

900780

On-Chip Realtime Operating System for the Engine Control System

Shoji Matsubara
NEC Corporation
Tokyo, Japan
Takashi Kuwahara
NEC IC Microcomputer Systems, Ltd.
Kawasaki, Japan
F. Bruce Gerhard, Jr.
NEC Electronics Inc.
Natick, MA

Abstract

The ongoing advances being made in electronics technology has made it possible to expand the range of microcomputer applications in automotive engine control systems. Progress in this field of application has given rise to very sophisticated control systems called total engine control systems, which are designed to provide interacting control with other subsystems (e.g., antiskid braking system, traction control system, transmission control system), instead of just controlling the engine itself. Improved engine control system performance and more versatile functions have made the application software controlling the system larger, with more complex control processing.

Until now, control processing in engine control systems could be implemented by the microcomputer's interrupt processing program alone. However, because engine control systems have grown in both size and complexity, it has become difficult to determine the sequence in which processing programs (tasks) are to be executed. Consequently, the basic software (realtime operating system) that efficiently makes this determination has become necessary.

This paper describes the use of microcomputers in engine control systems by using our newly developed μPD78602 16-Bit Single-Chip Microcomputer that incorporates a realtime operating system as firmware (alias Realtime Task Manager: RTM) as an example.

1. Introduction

The use of microcomputers in engine control systems (beginning in the latter half of the 1970s) has grown rapidly with the progress made in electronics technology over the years. Today, microcomputers are used in most electronically controlled gasoline engine systems.

Now that electronic gasoline engine control systems have achieved the initial objectives of purifying exhaust gas and reducing fuel costs, efforts are being made to use these systems to improve engine performance, the primary objective of engine control. This includes increasing output power per cubic volume of displacement, improving torque characteristics in the low-to- medium speed ranges, and obtaining quick response. However, improvements cannot be made by simply using newly developed mechanisms, no matter how innovative they may be, unless such mechanisms are combined with electronic control technology.

Advances in electronic control technology are largely due to the development of high-performance, high-function microcomputers. In fact, more and more 16-bit single-chip microcomputers are being used in today's modern engine control systems. Microcomputers are also used in what is called the total engine control system of luxury cars. In addition to enhancing engine function and performance, emphasis is being placed on developing more advanced control technology to interact with transmission control and other subsystems. This trend is expected to continue.

2. Trends in engine control system and software development

We have overviewed the current situation and trends regarding engine control systems. When examining the features built into medium and large-size engine control systems, we immediately notice that the number of control items are steadily increasing, and control is becoming more complex. For example, the engine control systems being used in the latest high - performance engines control many items, such as 1) fuel injection control, 2) ignition timing control, 3) speed control during idling, 4) EGR control, 5) knocking control, 6) supercharging pressure control, and 7) variable-mechanism control. In addition, learning control and self-diagnostic control are also included. Apart from these individual controls, there are some more advanced systems designed to control all engine-related operations. In addition to controlling the engine itself, such advanced systems provide interacting control with transmission control, traction control, and other control subsystems. These total engine control systems are being vigorously developed or put to practical use. Such sophistication in engine control systems is reflected in the increasing number of development man-hours.

Conversely, the development cycles of engine control systems are becoming shorter every year due to growing market demands and the rapid progress being made in electronic technology. Since the development periods are now shorter, the compatibility of electronic components used in these systems with future components must be ensured and future expansions must be accommodated. For these reasons, emphasis is now placed on selecting hardware and software offering greater flexibility.

Given the increasing sophistication of engine control systems and the limited development periods, success depends on how efficiently systems can be built to process the growing number of control items and to handle the increasingly complex control requirements. Significant advances in electronic technology are indispensable to achieving success. Whether we can attain it depends on improvements made to the functions and performance of microcomputers, which form the nucleus of engine control systems.

Thus, we now turn our attention to the functions and performance of microcomputers. The hardware used in microcomputers is undergoing higher sophistication to meet the growing demands for engine control systems. Some newly developed or commercialized products, for example, have powerful timer functions to cope with increased input/output processing loads, and 16-bit or 32-bit CPU cores to enable high-speed arithmetic/logic operations.

However, such development presents a problem in that the increased number of control items and diversified contents of control have resulted in greater software development man-hours. In such realtime-sensitive control systems as engine control systems, software configurations are closely related to microcomputer hardware. In addition, the microcomputer's interrupt handler is used to determine the program to be executed and in which sequence. Therefore, more complex control not only increases control program size,but also requires more complex interrupt processing, a problem that cannot be solved by simply enhancing the microcomputer hardware.

When microcomputers were first introduced into engine control systems, control programs were about several kilobytes in size. Today, control programs are as large as 20 to 40 kilobytes, sizes that can no longer be dealt with by one programmer alone. When considering that control programs are likely to continue increasing in size, it is important to devise methods to improve software development efficiency.

Several methods may be considered for this purpose. One such method to reduce the software development period and ensure the continued use of software resources is introducing a realtime operating system, something that has been studied in the machine control field in recent years. Consequently, we developed a 16-bit single-chip microcomputer called the μPD78602, which incorporates a realtime operating system on- chip.

3. What kind of realtime operating system should be incorporated in the microcomputer?

(1) Some facts about a realtime operating system

Before going into the main subject, let us outline a realtime operating system.

Most control programs used in engine control and similar systems are structured so that run states are changed by generating interrupts as a trigger. Such control programs are normally divided into multiple processing units so that appropriate units can be executed depending on the system's control status. These processing units are called tasks, and this method of programming is known as multitasking or multiprogramming.

Multitask execution is always done under operating system control, and the run state of each task is determined by the operating system. Various processing requests are issued from each task to the operating system. The status of each task changes as a result of executing the requested processing, and the next task to be executed is determined by the operating system according to task status. These processing requests are called supervisor calls, macro calls, or system calls. Therefore, an operating system may be considered a program consisting of multiple procedures to process system calls.

Because an operating system is normally configured with a set of macro instructions, it occupies part of the microcomputer's memory space. Note that system calls also use such instructions as software interrupt, branch, or subroutine call instructions to issue requests to the operating system.

Although there are many kinds of operating systems, those used for realtime control are called realtime operating systems. Realtime control means reading momentarily changing signals from sensors at high speed (as in engine control systems), immediately executing arithmetic operations on signal contents, and controlling output according to the results.

(2) Functions of a realtime operating system required for electrical automotive equipment

Realtime operating systems used for electrical automotive equipment generally require the following functions:

1) Tasks can be switched over in case of such events as interrupts.
2) After a system call is issued, it can be processed by the operating system and control is returned to the task in the shortest time possible (time known as overhead)
3) The duration in which interrupts are disabled while the operating system is processing each system call must be minimized.
4) A task can be activated from other tasks.
5) An intertask synchronizing mechanism must be built into the system.
6 Data can be exchanged between tasks.

Obviously, these functions are also important requirements for electrical automotive equipment control.

Realtime operating systems available today are configured with macro instruction, or software. Consequently, there are certain limits on reducing system-call overhead or improving interrupt response, regardless of the schemes employed. In fact, the system-call overhead or interrupt-inhibit time required for today's realtime operating systems is in the range of several ten microseconds to several hundred microseconds. With such a large overhead and slow response, today's operating systems are apparently insufficient for use in engine control systems where fast, realtime response is essential. Therefore, the applicability of newly developed operating systems for engine control systems depends on how fast they can process jobs.

4. Functions of Realtime Task Manager (RTM) built in the μPD78602

This section describes the features of the μPD78602 with an on-chip realtime operating system (developed by considering the functions and performance required for the control system applications described above), and the functions and performance of the realtime task manager (RTM) built into the device.

(1) Features of the μPD78602

The μPD78602 is a 78K/VI family 16-bit single-chip microcomputer that ranks top among NEC's original 78K-series microcomputers.

The μPD78602 comes with a powerful instruction set (184 instructions) offering excellent symmetricity and orthgonality. By using a "4-stage pipeline method" (capable of parallel instruction processing in each stage and branch address prefetch) and a "dual bus structure," high-speed arithmetic/logic operations with a minimum instruction execution time of 125ns (with 32MHz X'tal at Ta = 70°C) are made possible. Moreover, the μPD78602 incorporates a Realtime Task Manager (RTM) to enable the realtime operating system to switch between tasks at high speed by using dedicated multitask control hardware built into the device.

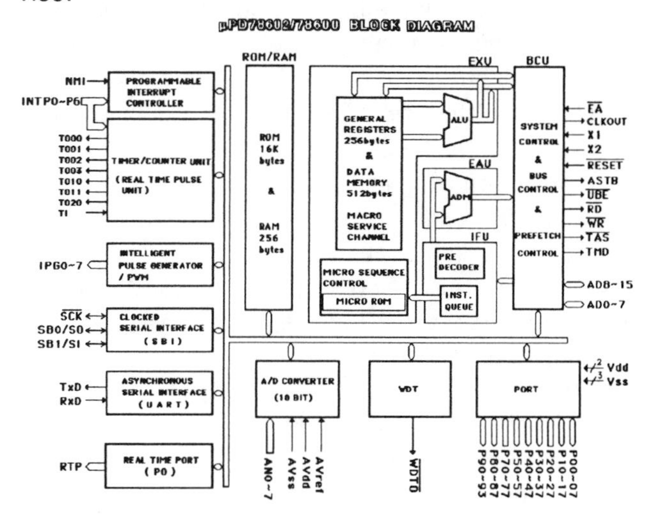

Figure 1 Block Diagram of the μPD78602

In addition, the μPD78602 has a multipurpose Realtime Pulse Unit (RPU), an Intelligent Pulse Generator (IPG), counter/timer, 8-channel 10-bit A/D converter, and a powerful interrupt controller to provide the macro service functions featured in the conventional 78K-series microcomputers.

Figure 1 shows a block diagram of the μPD78602. Figure 2 shows a block diagram of the timer system in the device.

1) RPU BLOCK DIAGRAM

TM0
φ/4, φ/8 → 16 / 18 - Bit Freerunning Timer
16 Bit Compare Register x 4
External INT 4 → 18 Bit Capture/Compare Register x 4
INT x 8
Timer Out x 4

TM1
TI, φ/8, φ/16 → 16 Bit Timer / Event Counter
16 Bit Compare Register x 2
External INT 2 → 16 Bit Capture / Compare Register x2
INT x 2
Timer Out x 2
to RTP (set)

TM2
TI, φ/8, φ/16 → 16 Bit Timer / Event Counter
16 Bit Compare Register x 2
External INT 2 → 16 Bit Capture Register x 2
INT x 2
Timer Out x 2
to IPG
to RTP (reset)

2) IPG BLOCK DIAGRAM

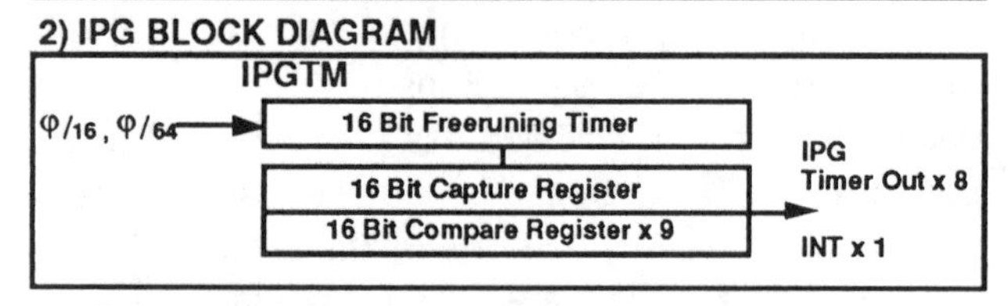

Figure 2 Block Diagram of the On-Chip Timer

Features:

- 16-bit CPU
- Powerful instruction set (184 instructions)
  - 32-bit transfer and arithmetic/logic operations
  - Signed/unsigned multiplication and division(16bits x 16 bits, 32 bits ÷ 16 bits)
  - Bit manipulate instructions (transfer, Boolean operation, set, reset, test)
  - Context switching instruction
  - String instruction
  - Sum-of-product calculate instruction
- Minimum instruction cycle: 125 ns with 32MHz X'tal
- General-purpose register: 16 banks each consisting of eight 16-bit registers
- Memory
  - On-chip ROM: 16K bytes
  - On-chip RAM: 1K byte
  - Memory space: 64K bytes
- Multipurpose pulse input/output unit (Realtime pulse unit: RPU)
  - 16/18-bit free-running timer x 1
  - 18-bit capture/compare register x 4
  - 16-bit compare register x 4
  - Pulse output x 4
  - 16-bit timer/event counter x 2
  - 16-bit capture/compare register x 2
  - 16-bit capture register x 2
  - 16-bit compare register x 4
  - Pulse output x 3
- Intelligent pulse generator (IPG)
  - 16-bit free-running timer x 1
  - 16-bit capture register x 1
  - 16-bit compare register x 9
  - Pulse output x 8
- High-precision 10-bit A/D converter (8-channel)
- Realtime output port
- Serial interface (with baud rate generator)
  - Asynchronous serial interface (UART)
  - Serial bus interface (SBI)
- Interrupt controller
  - Vectored interrupt function
  - Context switching function
  - Macro service function
- Realtime task manager (RTM) built in
- Turbo access manager control signal output function
- Watchdog timer
- Standby function (STOP/HALT)
- CMOS technology
- 84-pin PLCC package

(2) Functions of the realtime task manager (RTM) built into the μPD78602

For applications where response characteristics are very strict requirements as in engine control systems, software- implemented realtime operating systems are practically useless because of their slow response. To solve this problem, we designed the μPD78602 to handle a system call as one instruction and process it using hardware and microprograms. As a result, we managed to greatly reduce delays in system call processing time and interrupt response time.

Functionally speaking, we created the μPD78602 similarly based on μITRON specifications (Micro-Industrial-The Realtime Operating System Nucleus), a built-in type OS, to maintain compatibility with future expansion. We basically implemented μITRON by carefully selecting functions suited for applications in electrical automotive equipment, and by discarding all unnecessary parameters to increase processing speed. In addition, the memory space required for RTM construction for built-in use was minimized so that the on-chip user ROM area is not used for it.

The following lists the features of the RTM in the µPD78602.

1) All system calls (9 kinds, 14 instructions) are microprogram-implemented.
   As a result, the system call processing time is 7µs (typical) with a 32MHz quartz. The interrupt inhibit time is also 7µs (typical).

2) Task switching using register banks
   The processing time required to save data and restore it from general-purpose registers is reduced by using context switching.

3) Register bank sharing by multiple tasks
   This feature was made possible by having multiple tasks share the same register bank to enable the execution of more tasks than otherwise possible with a limited register bank capacity.

4) Static object generation
   All objects (i.e., tasks and semaphores) are generated at system generation. Thus, by doing away with dynamic generation, excess processing time during execution is eliminated.

5) Binary semaphores (Up to 16 semaphores can be set.)
   These semaphores operate at turn-on or off of a single bit. To increase processing speed, we have omitted calculating functions, wait conditions, and all other unnecessary functions.

6) Number of tasks controlled: 32 tasks
   Priority levels: 4 (for each task)

(3) Task status

Figure 3 shows the task status transition diagram of the RTM. The RTM used in the µPD78602 has the five task states described below.

RUN ---------This state means that a task is being executed after obtaining the right to use the CPU. Under no circumstances are two or more tasks in the RUN state.

READY -----This state means that though ready to go, the task cannot enter the RUN state because some other task is being executed.

WAIT --------This state means that the task is waiting to obtain the right to access a resource because its request was not met by a system call requesting the right to access the resource.

SUSPEND ----This state means that the task has been forcibly "interrupted" by a system call.

WAIT SUSPEND --- This state means that the task in WAIT state has been forcibly interrupted. This has a "double wait" effect.

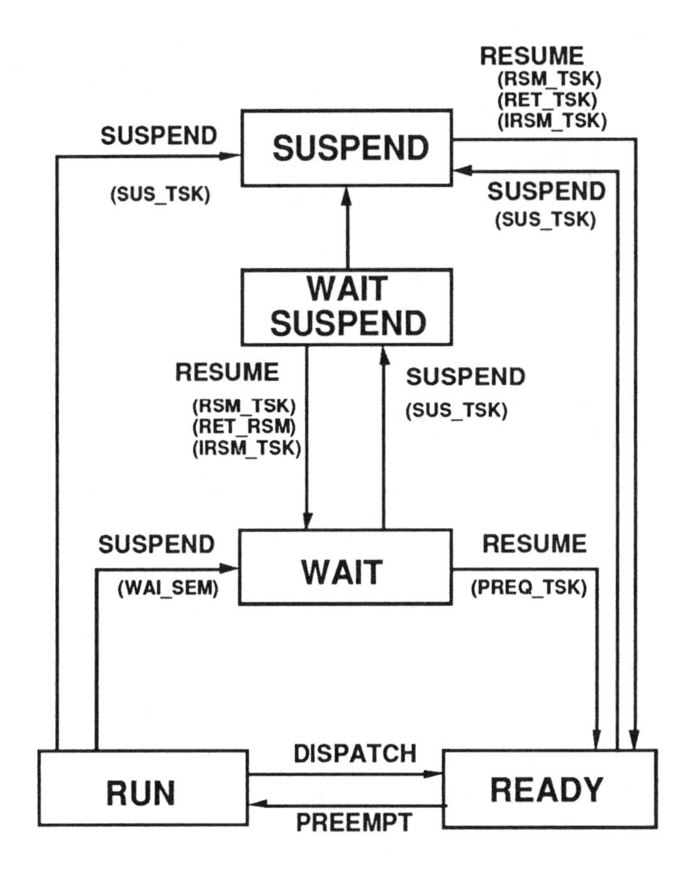

Figure 3 Task Status Transition Diagram of the RTM

Basically, three task states are always provided: the run state, ready state, and wait state. The task states defined for the RTM are, however, those considered to be absolutely necessary for it.

(4) Task management

Task management refers to interrupting and restarting of each task. There are 32 tasks that can be controlled, and it is possible to set up to four priority levels for each task.

Figure 4 shows system calls for task management, the contents of processing, and execution time. Figure 5 shows an example of the task management data structure in the READY state (ready queue).

The smaller the priority number, the higher the

priority level. (Priority 0 is the highest priority.) Within the same priority, the task that is first placed in the READY state among other tasks enters the RUN state.

TASK MANAGEMENT

SYSTEM CALL	CONTENTS OF PROCESSING	EX. TIME	CONDITION
SUS_TSK	SUSPEND TASK	1μs	WAIT → SUSPEND
		7μs	READY → SUSPEND
RSM_TSK	RESUME TASK	1μs	WAIT SUSPEND → WAIT
		6μs	SUSPEND → READY
IRSM_TSK	RESUME TASK WITH NO DISPATCHING	1μs	WAIT SUSPEND → WAIT
		2μs	SUSPEND → READY

Figure 4 System Calls for Task Management

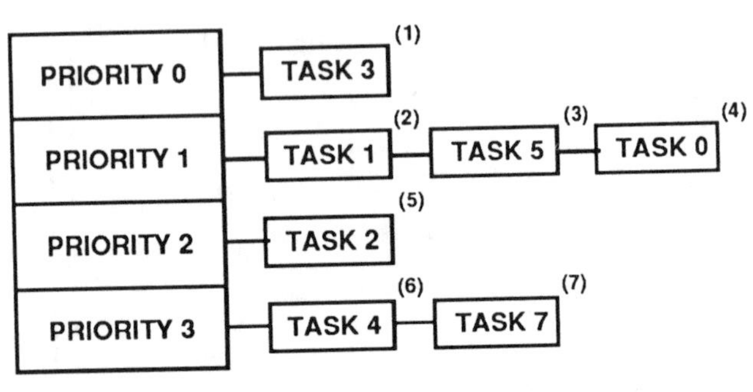

Figure 5 An Example of the Task Management Data Structure in the READY Queue

(5) Synchronization management

When multiple tasks operate in parallel contending for the access right to shared resources (i.e., data, peripheral hardware), data integrity is lost and erroneous operation results. To prevent this, intertask synchronization must be established by mutual exclusion or some other means. For this purpose, the RTM uses semaphore-based mutual exclusion.

A semaphore in this scheme functions as a flag to indicate the status of access to shared resources. When a task accesses any particular shared resource, the corresponding semaphore bit is set to '0' to inhibit access to the shared resource by other tasks. When the semaphore bit is '1,' access from the tasks is allowed. The RTM incorporates 16 semaphore bits.

Figure 6 shows the system calls for synchronization management (semaphore control), the contents of processing, and execution time.

(6) Interrupt management

Figure 7 shows the system calls for interrupt management, the contents of processing, and execution time. Interrupt management is used to control return from interrupt processing to a task, and to restart a task during interrupt processing.

(7) Initialization

Figure 8 shows the system calls for initialization and the contents of processing. The RTM is initialized by an initialization system call, and the program is executed under OS control. Consequently, the microcomputer required for the control system must be initialized (i.e., by setting the special function register (SFR), RTM task information) before initialization can be done.

(8) Data structure

Figure 9 shows the data structure of the RTM.

SYNCHRONIZATION MANAGEMENT

SYSTEM CALL	CONTENTS OF PROCESSING	EX. TIME	CONDITION
WAI_SEM	WAIT SEMAPHORE	1μs	SEMAPHORE : 1→ 0
		7μs	RUN → WAIT
PREQ_SEM	POLL and REQUEST SEMAPHORE	1μs	SEMAPHORE : 0
		1μs	SEMAPHORE : 1→ 0
SIG_SEM	SIGNAL SEMAPHORE	1.5μs	SEMAPHORE : 0→ 1
		7.5μs	WAIT → READY

Figure 6 System Calls for Synchronization Management

INTERRUPT MANAGEMENT

SYSTEM CALL	CONTENTS OF PROCESSING	EX. TIME	CONDITION
RET_INT	RETURN from INTERRUPT HANDLER	5.5μs	———
RET RSM	RETURN from INTERRUPT HANDLER and GET UP TASK	2μs	WITHOUT GET UP TASK
		7μs	SUSPEND → READY

Figure 7 System Calls for Interrupt Management

INITIALIZATION

SYSTEM CALL	FUNCTION	CONTENTS OF PROCESSING
INIT	INITIALIZATION	QUEUING UP THE TASK and DISPATCHING

Figure 8 System Calls for Initialization

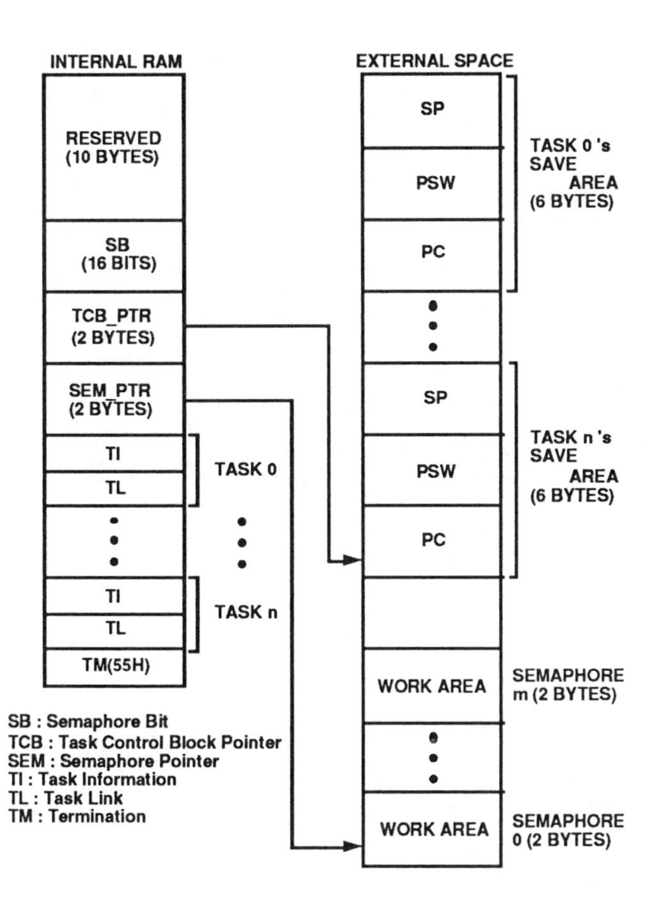

Figure 9 Data Structure of the RTM

5. Application example of RTM in μPD78602 (engine control system)

This section discusses how the RTM is used in engine control. For this discussion, let's assume a simple model of an engine control system whose specifications are as follows.

(1) Control specifications

Control in this model is applied to control fuel injection and ignition in a four-cylinder engine. For this, the pulses shown in Figure 10 are output according to the values obtained from the equations below.

① Fuel injection time (Ti) is calculated by using the following equation.

$$Ti = Tp \times Kt \times Ko \times Ts$$

where

TP =basic injection quantity calculated from suction quantity and engine speed

Kt = water temperature correction factor obtained from the table of water temperature correction factors by referencing it with the water temperature

Ko =oxygen feedback correction factor used to determine the correction factor necessary for proportional-plus- integral control based on the oxygen sensor status

Ts = battery voltage correction time obtained from the table of battery voltage correction times by referencing it with the battery voltage

② Ignition timing (Ia) is obtained from the map of spark advance by referencing it with the engine speed and basic injection quantity.

③ Dwell angle (Da) is calculated by using the following equation.

$$Da = Da1 \times Kb$$

where

Da1= Dwell angles obtained from the table of dwell angles by referencing it with the engine speed

Kb = battery voltage correction factor obtained from the table of battery voltage correction factors by referencing it with the battery voltage

④ Measurements

Measurements are made for five items: water temperature and other A/D conversions (four items), and engine speed. The measurement timing is such that water temperature A/D conversion is measured every 500ms, battery voltage A/D conversion is measured every 100ms, suction quantity A/D conversion and TDC pulse period (engine speed) are measured at every TDC pulse, and oxygen concentration A/D conversion is measured every 20ms.

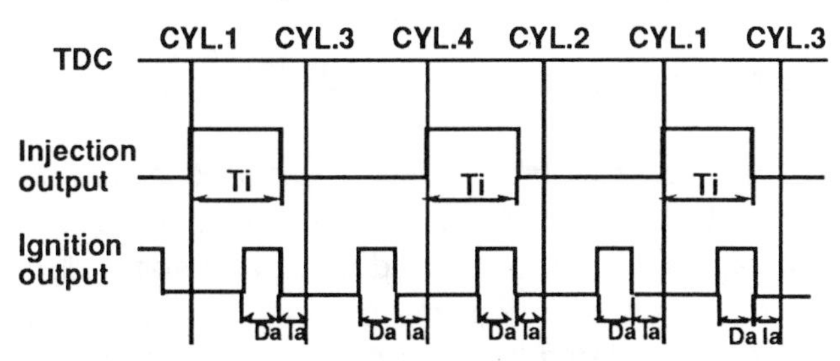

Figure 10 Basic Timing (TDC,Ignition Injection)

(2) Sample program based on conventional method

Figure 11 shows how a sample program for this system may be configured when conventionally created. In this example, TDC interrupt processing starts at the TDC timing and fuel injection is simultaneously initiated at a rate of once per every two TDC pulses. The fuel injection termination time is set, the ignition timing is set, and suction quantity A/D conversion is done during this interrupt processing.

In addition, TMR0 interrupt processing is executed to process interrupts at certain time intervals. A/D conversions for water temperature, battery voltage, and oxygen concentration are also done during this processing.

In the MAIN processing, each parameter is calculated based on the measurement results of each input, and the injection time and ignition timing are determined from the calculated values.

In program configurations where run states are changed by interrupt generated as triggers as in this example, the program size is increased as the number of control items or contents of correction is increased. This causes the following problems:

① Because the programming method necessary to share data or built-in hardware resources is not clearly defined, various different methods may be used depending on the programs. This makes it impossible to develop programs on a project basis (development by a team of programmers).

② Various approaches are used for dividing programs (i.e., by the contents of processing, execution time, or type of resource). Therefore, it is difficult to maintain software resources compatible with future expansion.

③ Because the programs only handle interrupt processing and main processing, tasks whose processing requires fast response and those which do not coexist in the processing. Therefore, when necessary to execute a task that requires fast response over the others, task processing must wait until the task currently being executed (which need not be fast) is terminated. As a result, the practical response speed is reduced.

④ To create a program in which a procedure having the highest priority can be executed by determinations made based on various states during main processing, the program may become very complicated because it is necessary to make such priority determinations for all processing or to determine whether the operating conditions are met.

Thus, for the engine control systems that are becoming increasingly complex, software development efficiency cannot be improved as long as conventional methods are employed in program development. This is because conventional methods have drawbacks that make them difficult to maintain software compatibility or to develop programs on a project basis. The end result is that the greater sophistication simply makes the programs more complex.

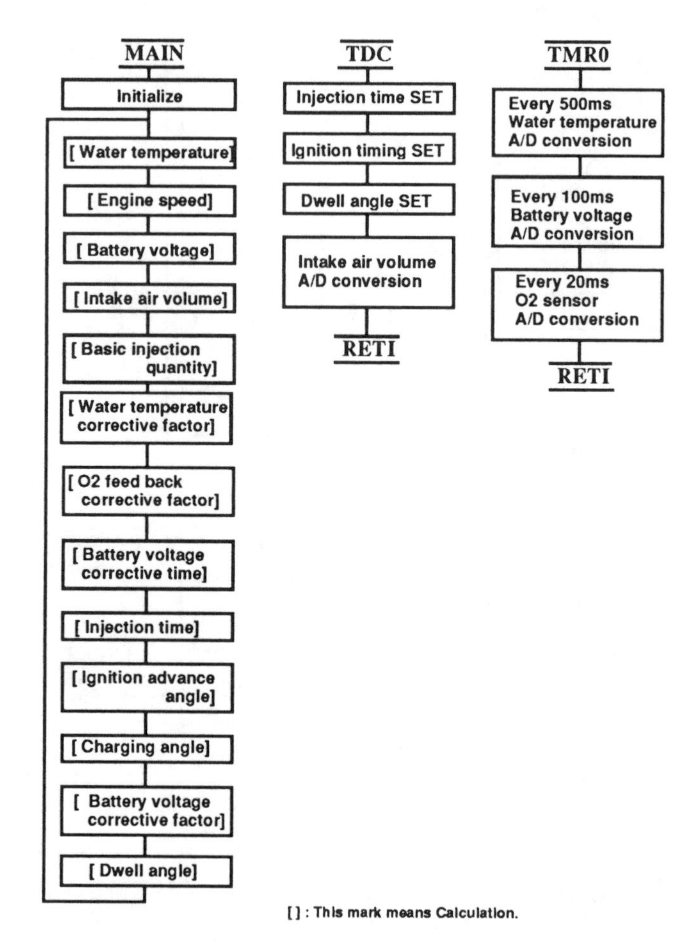

Figure 11 Sample Program Based on Conventional Method

(3) Sample program created by using RTM

Next, Figure 12 shows a sample program created by using the RTM. In this case, the conventional program is divided into seven tasks; each task is assigned a priority for optimized control. In the MAIN processing, only the injection quantity and ignition timing are calculated, and each parameter is calculated in TASK1 to TASK6. Parameter calculations are done when the data necessary to calculate a given parameter is entered. The start- up timing and contents of each task's processing are as follows:

TASK1 Started at TDC timing, this task calculates the engine speed and dwell angle from the calculated engine speed.

TASK2 Started at TDC timing, this task measures and calculates the suction quantity, then activates TASK6.

TASK3 Started every 500ms, this task measures and calculates the water temperature, then calculates the water temperature correction factor.

TASK4 Started every 100ms, this task measures and calculates the battery voltage, then calculates the battery voltage correction time and battery voltage correction factor.

TASK5 Started every 20ms, this task measures the oxygen concentration, then calculates the oxygen feedback correction factor.

TASK6 Started from TASK2, this task calculates the basic injection quantity from the engine speed calculated in TASK1 and the suction quantity calculated in TASK2, then calculates the spark advance of ignition from the resulting basic injection quantity and engine speed.

MAIN Processed in the background, this task calculates the injection time and ignition timing based on the parameters obtained in TASK1 to TASK6.

The priority levels of these tasks are such that TASK1, 2, and 6 are assigned priority 0 (highest), TASK3, 4, and 5 are assigned priority 1, and MAIN is assigned priority 3 (lowest). These defined priority levels enable urgent processing or processing to obtain important parameters to be executed preferentially over the others.

Interrupt request (TDC, TMR0) generation timing is the same as when the RTM is not used. The timing of built-in hardware that must respond at high speed is set during this interrupt processing. Other processing (i.e., A/D conversions, calculations) is executed in the appropriate TASK after being started within interrupt processing, thereby reducing the time required for interrupt processing.

In addition, semaphore '1' is used to indicate whether use of the A/D converter is granted. In this way, duplicated use of the converter can be avoided by having the task enter the WAIT state depending on the semaphore status.

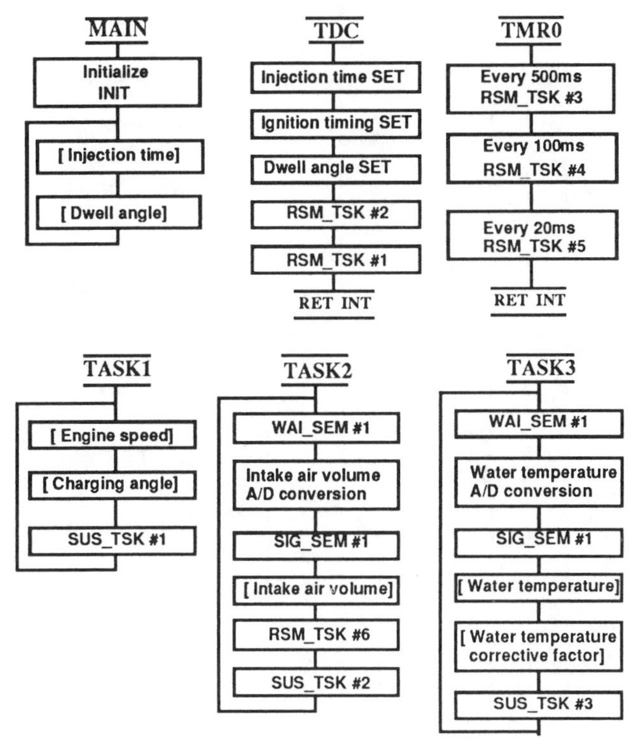

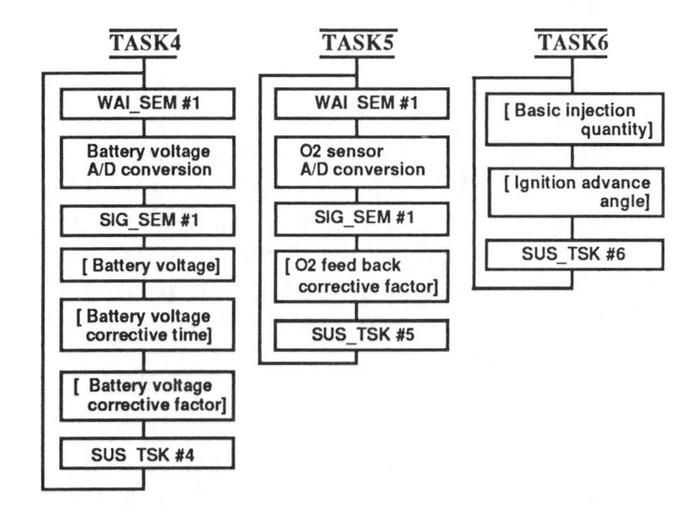

Figure 12 Sample Program Created by the RTM

The timing chart in Figure 13 schematically shows the program flow of the above operations.

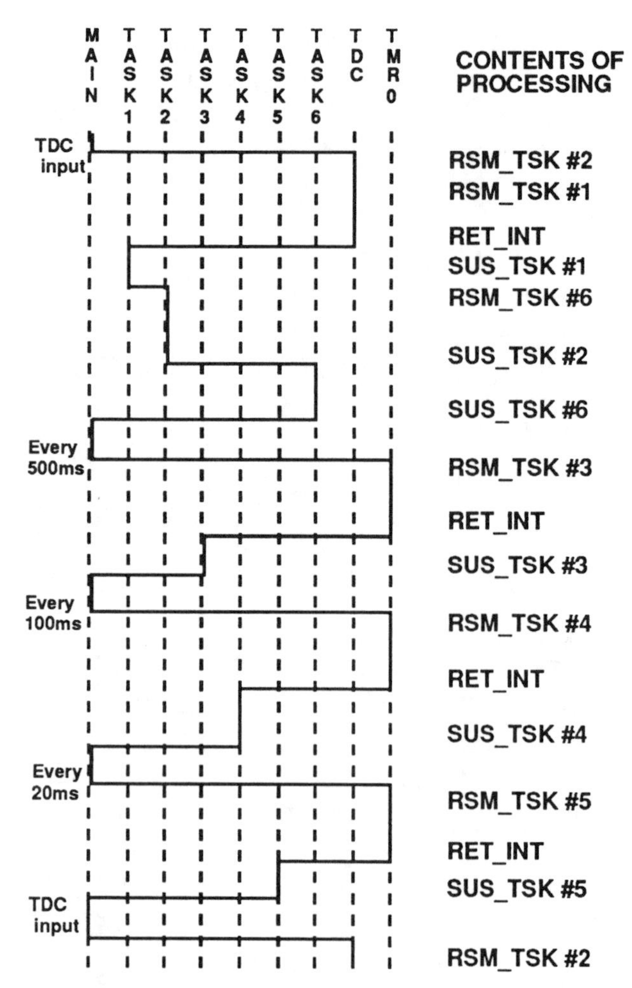

Figure 13 Timing Chart of the Program Based on the RTM

As discussed above, the use of the RTM provides advantages over conventional methods as follows:

① Because semaphores and dedicated instructions are provided for sharing data or built-in hardware resources, the entire program can be standardized according to a single method. This facilitates program development on a project basis (by a team of programmers) and reduces the development period.

② Because the program is divided by execution timing (i.e., by task), the compatibility of software resources can be maintained. In addition, program additions or alterations can also be done.

③ Because priority levels can be assigned to each processing (task), provided that operating conditions are met for high- priority processing during low-priority processing, it is possible to execute high-priority processing by temporarily stopping low-priority processing. For this reason, the practical response speed is increased.

④ Because the RTM manages the operating conditions and priority levels of processing, it is possible to execute processing of the highest priority among current tasks by simply executing a dispatching operation. Consequently, simple program creation is made possible.

Thus, given the increasing complexity of engine control systems, the use of the RTM makes it possible to reduce the software development period and maintain software compatibility. Moreover, the RTM facilitates construction of a fast-responding, realtime operating system without sacrificing overall system performance.

6. Conclusion

This paper has described the functions and performance of the realtime operating system (alias realtime task manager: RTM) incorporated as firmware in our newly developed μPD78602 16-bit single-chip microcomputer. This paper also discussed how the RTM is used in engine control systems.

Because the engine control system performance has becomes higher with more versatile functions, there are more control items and more complex control. This will no doubt impose a large burden on future microcomputer software development. Therefore, it is important to increase the efficiency of software development. This can be done by developing the processing programs (tasks) under OS control using the RTM, instead of tasks are conventionally executed after determining the sequence using only the microcomputer's interrupt processing program.

To implement RTM system calls in the μPD78602, we carefully selected only those functions needed for realtime-sensitive applications. Although this is the current state of implementation, we consider it necessary to optimize such implementation in the future by further examining its usefulness for the intended applications.

For the ROM and RAM built into the Single-chip microcomputer with RTM, we also plan to develop products of greater capacities to make them suitable for the growing size of future control programs.

910082

An Engine & Transmission Control System with New 16-bit Single Chip Microcomputer

Shigeru Kuroyanagi
Toyota Motor Corp.

Takayuki Ono
Nippondenso Co., Ltd.

Tetsuro Wada
Toshiba Corp.

Brad Cohen
Motorola Inc.

1. Abstract

The microcomputer is the most powerful component available at this time for application to highly functional and high-precision electronic control systems in automobiles.

Toyota Motor Corporation recently improved an 8-bit microcomputer and this has now been followed by the development of a new 16-bit microcomputer to permit major expansion of the functions available in such systems. This is a single-chip VLSI which is flexible and sophisticated, quite suitable for real-time control systems in automobiles. It provides large memory, both ROM and RAM, a poweful instruction set appropriate for use in real-time control,high-speed and intelligent input/output (I/O) functions, and higher speed of data communication functions for intercommunication between microcomputers.

This microcomputer has realized the development of a high-speed and high-precision combined control system centered upon engine and transmission control.

2. Introduction

Engine control systems using microcomputers have grown more complex in order to enhance engine performance. The functions of electronic control units (ECU) have also been enhanced in accordance with this trend.(Fig.1)

As of the 1985 model year, in parallel with the introduction of the knock control system, Toyota Motor Corporation developed a single-chip custom designed 8-bit microcomputer and started its applications. This microcomputer permitted highly functional and high-precision real-time processing, and is now a major part of the all engine control ECUs in Toyota models.

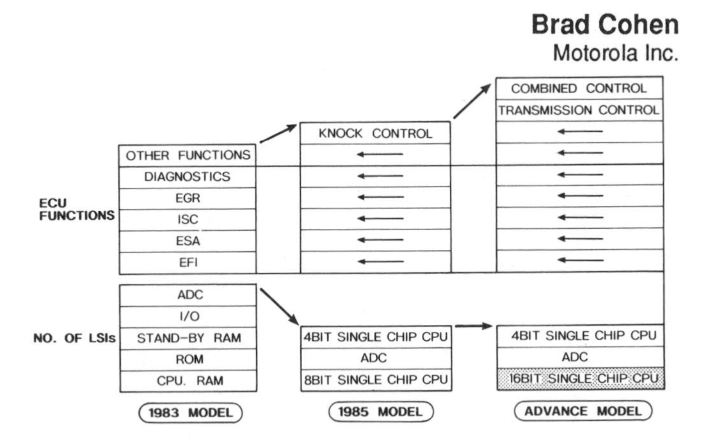

Fig. 1 Function Growth of Engine Control

Since that time, there has followed a continuous series of technological developments related to engine control functions, with the purpose of realizing lower emission levels, higher fuel economy and better drivability. A number of new functions have also become necessary for transmission control systems, primarily for the purpose of reducing the shock during gear shifts. A further requirement that has developed over the course of time is for interrelated functions in both engine and transmission control system, in order to achieve futher improvement of fuel economy and further smoothness during gear shifts.

In response to these various requirements, a large-scale ECU has been designed in order to unify the engine control system and the transmission control system into a single system. Naturally, the microcomputer which is the heart of the ECU must have higher capabilities and increased flexibility. In fact, these requirements are so complex that the 8-bit microcomputers in current use will soon become inadequate for high-end systems.

This paper will describe outstanding characteristics of a new 16-bit microcomputer, jointly developed by the Toyota, Nippondenso, Toshiba and Motorola, followed by the description of an advanced engine and transmission control system utilizing this microcomputer.

3. Development Goals

Toyota has modified and improved the custom designed 8-bit microcomputer series in current use and begun manufacturing a 2nd generation microcomputer,which was first used in the 1991 model year.

The 2nd generation microcomputers have been enhanced in a number of significant ways, as follows;

- Increase internal clock (12MHz to 16MHz)
- Increase of internal memory capacity up to 16 kilobytes (KB) ROM
- Addition of a serial I/O with direct memory access function to facilitate high-speed data communication between single-chip microcomputers
- Addition of a built-in analog-to-digital (A/D) converter
- Applying 1.5μ design rules

The 2nd generation microcomputers fully satisfy all functional requirements for the combined engine and transmission control systems that are currently in mass production.

However, requirements are expected to increase sharply, with respect to the variety of functions to be provided by the combined engine and transmission control system, and this will create the need for a new microcomputer, functioning at an even higher level than the 2nd generation 8-bit microcomputers. In order for this new microcomputer to accomplish both current and future requirements, a number of functions must be enhanced,including significant increases in operational processing speed, reduction of load on the central processing unit (CPU) for interrupt processing, high-speed and high-precision I/O ports, improved interface functions, and a large memory.

Thorough consideration has been given to design and manufacturing changes that would be required to guarantee the high reliability established by the previous 8-bit microcomputer series. The market record of the 1989 model vehicles has been outstanding ---- absolutely no defects have emerged, and the highest possible quality record has been achieved. Any new microcomputer must provide reliability comparable with this record.

On the basis of the requirements described above, the following eight targets were set for the development of the new microcomputer;

- 16-bit data length
- Increased CPU speed
- Development of intelligent functions for high-speed I/O ports
- Improvement of data communication between microcomputers
- large memory on chip
- Wide operational temperature and voltage range
- To adopt single-chip architecture
- High reliability

It was assumed that this microcomputer, satisfying the above requirements, would continue to represent state-of-the-art technology for even 3 to 5 years after its development in real-time control systems in automobiles.

Table 1 compares the 8-bit microcomputer in current use with the new 16-bit microcomputer.

Table 1 Comparison between 8bit CPU and New 16bit CPU

Items		8bit		New 16bit
		1st Generation (T7433)	2nd Generation (T5A41)	
Word Length(bit)		8	8	16/32
Min. Inst. Exec. Time(nsec)		500[12MHz]	375[16MHz]	250[16MHz]
16*16 Exec. Time(μsec)		55(*)	41(*)	1.5
32/16 Exec. Time(μsec)		100~200(*)	75~150(*)	1.6
ROM		12KB	16KB	48KB
RAM		384 B	768 B	2KB
Bit Operation		○	○	○
Bit Test & Branch		○	○	○
Serial I/O		8bit 2ch	8bit 2ch	16/8bit 2ch
SPI		——	8bit 1ch	16/8bit 1ch
Fast Pulse Input		4ch	5ch	8ch
Fast Pulse Output		8ch	8ch	16ch
DMAC		——	2ch	3ch
Interrupt	Requests	16	18	31
	Context Switch	——	——	○
Voltage(v)		4.5~5.5	4.5~5.5	4.0~5.5
Temperature(Deg. C)		-40~105	-40~105	-40~125

*: subroutine

4. Features of the New 16-bit Microcomputer

A block diagram of the new 16-bit microcomputer is shown in Fig.2.

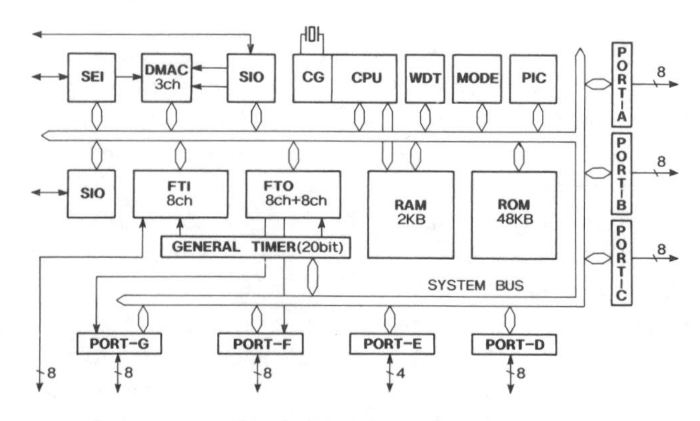

Fig. 2 Block Diagram of New 16bit Microcomputer

CPU

The CPU core of the new microcomputer is based on the MC68000, which is widely used in 16-bit control applications. A number of revisions have been effected to adapt this product for use in a real-time control systems in automobiles.

- A clock generator has been added to permit use as a single-chip microcomputer.
- With the use of a 16 MHz crystal oscillator, minimum instruction execution time is as fast as 250 nsec.
- A 32-bit internal data bus has been applied between the register set and internal RAM, doubling register stacking speed.
- A context switch function based on addition of another register bank has been added to accomplish high-speed interrupt processing.
- Dedicated hardware for multiplication and division has been added for table interpolation processing frequently used in engine control. With this hardware, multiply and divide processing speed has become approximately three times as fast as a MC68000 operating at the same clock frequency.

Fast-Timed Input Port (FTI)

FTI is a highly functional and intelligent input port, designed to reduce the load on the CPU during high-speed pulse input. All that is necessary with FTI is to write a command into the command memory; a search is then made for a leading or trailing edge on the designated channel, and the time is measured. It is further possible to measure pulse width, and frequency or phase difference. A block diagram and examples of input are shown in Fig.3.

FTI is a function that is indispensable for the signal processing of such parameters as engine speed and vehicle speed. It reduces CPU interrupt processing and this enables an increase in precision.

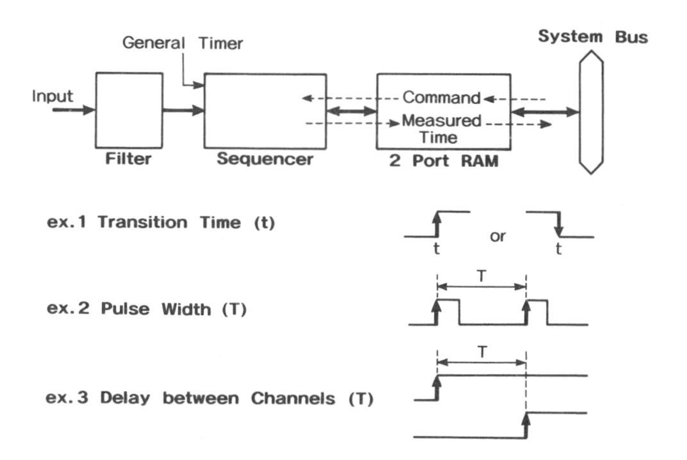

Fig. 3 Block Diagram of FTI and Examples of Measured Time

Fast-Timed Output Port (FTO)

FTO is a highly functional and intelligent output port, designed to reduce the load on the CPU during high-speed pulse output. Similar to FTI, when the FTO commands are written into the command memory, a designated pattern can be output at a designated time in response to a designated channel.
Furthermore, output can be inverted at designated time intervals, and it is simple to set the frequency and generate outputs that will effect duty ratio changes. A block diagram and examples of output are shown in Fig.4.

FTO permits the generation of high-precision reliable output signals for sequential injection and ignition control. Furthermore, duty ratio output for linear solenoids and the like can be obtained without imposing additional loads on the CPU.

Direct Memory Access Controller (DMAC)

DMAC is used for high-speed, high-volume transmission of data between single-chip microcomputers without any additional load on the CPU. The DMAC has three independent channels, and conducts direct memory access transfer between the serial I/O area and the RAM. A block diagram is shown in Fig.5.

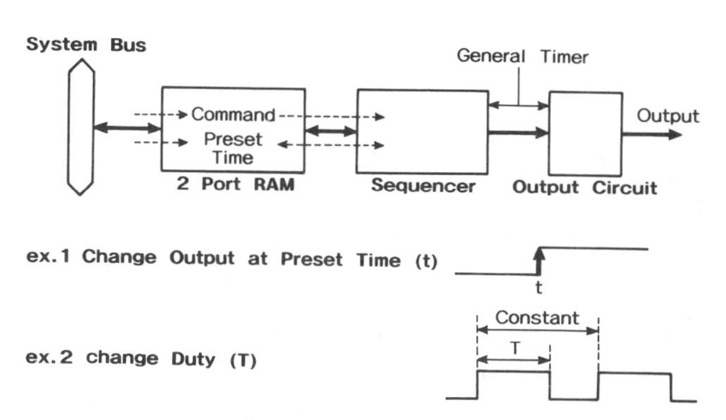

Fig. 4 Block Diagram of FTO and Examples of Output Pulse

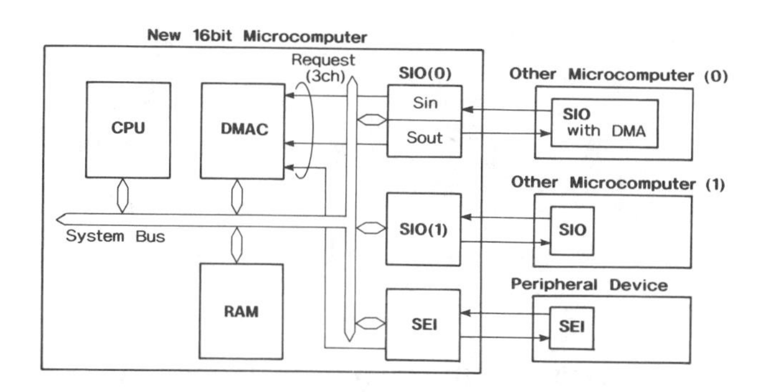

Fig. 5 Block Diagram of DMAC

For example, in systems which use multiple single-chip microcomputers for engine and transmission control, a good deal of data is used in common by the two systems, such as engine speed data and throttle opening data. Use of the DMAC in these cases enables one microcomputer to utilize the data collected by another microcomputer by simply reading its own memory, which vastly facilitates high-speed processing.

Serial Input/Output Port (SIO)

SIO is used for the serial transmission of data to and from other microcomputers or to and from peripheral devices. Individual buffer registers have been provided for transmission and reception, so that the transmission and reception areas can function independently. Two full duplex transmission modes are provided, i. e., an asynchronous mode and a synchronous mode. Two check functions have been added to increase the reliability of received data and to ensure that no malfunctions have occurred due to noise. Furthermore, data can be transmitted with the SIO reception or transmission area using the DMA function described above.

Serial Extension Interface (SEI)

SEI is used for serial transmission between independent peripheral devices. Fig.6 shows an example of the interface used with an external A/D converter. When the SEI is informed via the end of conversion (EOC) line that the A/D conversion has been completed, data is moved from the buffer register to the shift register, and the shift is commenced. The shift clock is output in 16 bits, the same length as data, from SCLK, and data are transmitted in a synchronized manner. This is issued as a command to the A/D converter, and at the same time, the last converted data is transferred to the shift register. When the 16-bit data transmission has been completed,

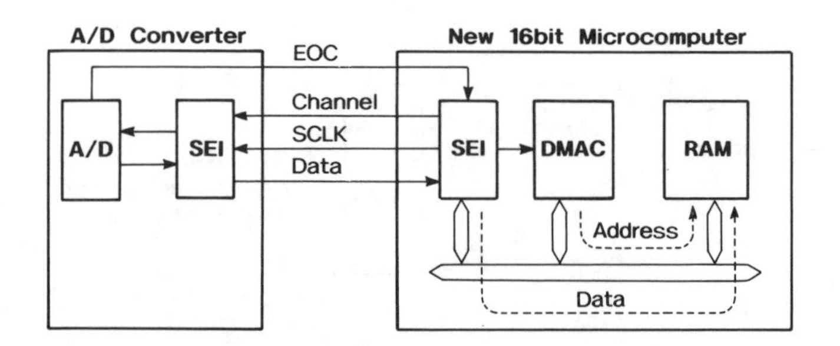

Fig. 6 Communication with A/D Converter

an interrupt request and a DMA request are issued simultaneously, and received data is transmitted to the internal RAM. In other words, even if the CPU does not activate the A/D converter at each individual operation, A/D conversion data can be received through interrupt processing.

Programmable Interrupt Controller (PIC)

The PIC is used to establish priority control in cases where interrupt requests overlap, or where an interrupt request overlap with the processing of another interrupt. There are 31 independent causes of interrupts, including FTI, SIO, non-maskable interrupts (NMI) and software interrupts. The priority order for interrupts is programmable, and seven different levels can be set. The addition of the context switch function has reduced interrupt overhead time to a mere 2.5 μsec.

Instruction Set

The instruction set for this microcomputer is a superset of the MC68000 instruction set.

Six new instructions have been added to specifically address the requirements of this automotive application. Four of these instructions (SETB, CLRB, BRSET, BRCLR) are bit operation instructions and have been widely used in the previous and current 8-bit microcomputer solutions. In the current engine control applications, these instructions often represent 15 to 20 percent of the total number of instructions in a given program. Without these instructions, use of a similar programming approach could increase the size of program memory by as much as 10 to 15 percent. The instructions were included such that programming techniques and application code proven with 8-bit solutions could continue to be applied with minimal execution time and efficient use of program memory. The other two instructions (MOVEC and MOVEMX) support access to the context switch control register (used in managing the two register sets) and enable execution of 32-bit transfers between CPU registers and onboard RAM, respectively.

Memory

This microcomputer contains 48 KB user ROM and 2 KB user RAM. The majority of the engine control systems currently in mass production are configured with 12 KB of ROM, and that of the transmission control systems are configured with 6 KB of ROM. Thus this microcomputer leaves a considerable amount of space to add new functions. Up to 0.5 KB of RAM can be used as standby RAM.

Other Features

This microcomputer has other features especially for automotive use, which will be briefly described here.

The process is a CMOS device in technology 1.2μ double layer metal, and approximately 740,000 transistors have been integrated on the single chip. The packaging is QFP with 120pins, in conformity with EIAJ specifications, and provides high density mounting while still maintaining resistance to humidity. A photograph of the chip and its package are shown in Fig.7.

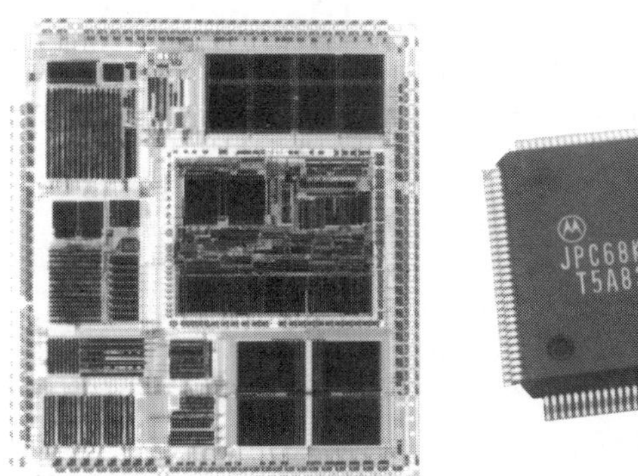

Fig. 7 Chip and Package

The operating temperature range is -40 to 125℃, and the operating voltage is between 4 and 5.5 volts, thus allowing a broad operating environment.

A 2 to 4μsec filter function has been installed in the timer input port to counteract possible degradation due to noise.

A power-down mode is provided.

5. System Overview

The configuration of the system utilizing this newly developed 16-bit microcomputer is shown in Fig.8.

This system has been designed as an advanced model for combined control centered upon a 3-liter six-cylinder DOHC engine and a four-speed automatic transmission.

The engine control system provides sequential injection control for the six cylinders, spark advance control with a knock control function, idle speed control, a variable intake control function and other controls. The transmission control system provides gears shifting control, hydraulic control within the transmission and other functions. Combined control functions include spark retardation and fuel cutting to reduce the engine torque (This reduces the shock when gears are shifted). Further, the system also provides other integrated controls through communication with the traction control ECU, cooling fan control ECU and air conditioner control ECU. It also has an on-board diagnostics function, and a backup function for use if the microcomputer should fail.

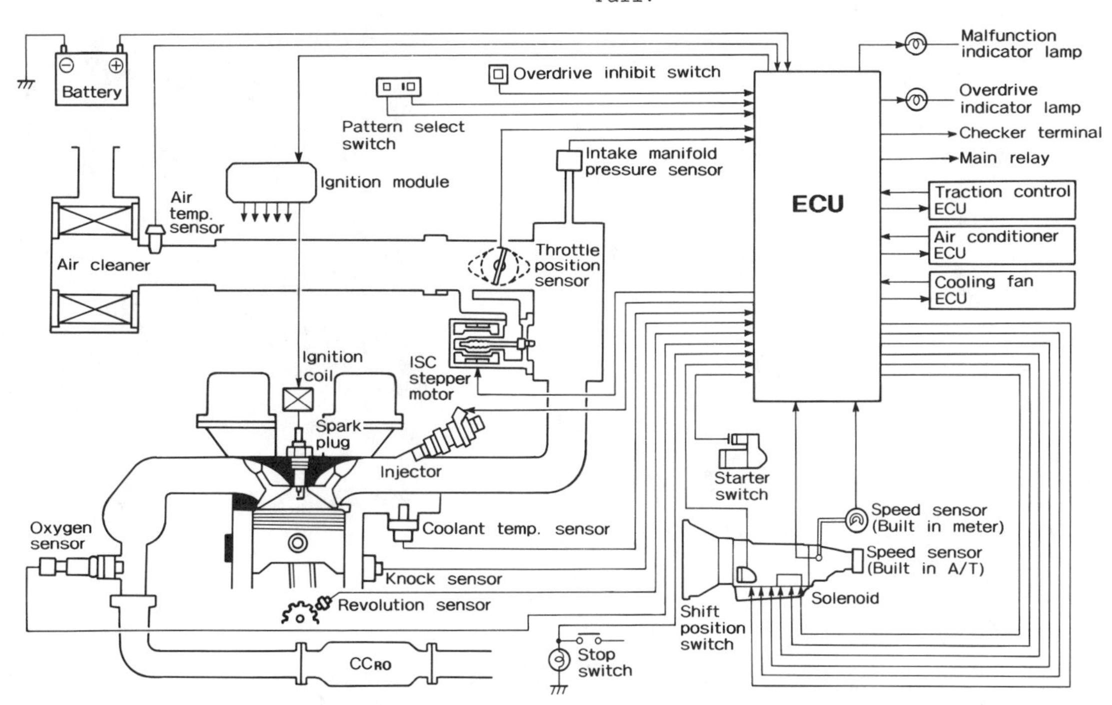

Fig. 8 Toyota Computer Controlled System

6. ECU

Expansion of memory, expansion of available I/O ports and improvement of high-speed processing capabilities are all required by addition of control system functions. This advanced system requires approximately 80 percent more ROM and approximately 30 percent more I/O signals than are provided in average-scale combined engine and transmission control systems currently mass producted. The I/O signals include input capture and output compare signals. Input capture signals are made necessary by the addition of a clutch revolution speed sensor in the transmission. Output compare signals are made necessary by the addition of linear solenoids used to control hydraulic pressure in the transmission and the adoption of sequential injection. Thus, this greatly increases the load on the microcomputer.

An ECU block diagram of this system functioning with the 8-bit microcomputer in current use is shown in Fig.9. In order to implement this system, it was necessary to use a T5A41 microcomputer for engine control, and another T5A41 for transmission control. T5A41 is the top-of-the-line product in the series. This configuration necessitates the use of serial I/O with direct memory access function for the transmission of data between the two microcomputers. The data communicated in this manner includes information concerning engine speed, intake manifold pressure and torque control-related data. Furthermore, fail-safe logic must be utilized between the two microcomputers, and therefore the two microcomputers operate in accordance with a master-slave relationship. The engine control microcomputer functions as the master, and is provided with a backup IC and a watchdog timer on its power supply IC to monitor abnormal operating conditions. The transmission control microcomputer is the slave, and is monitored by the master; this architecture permits fail-safe operation in the event of malfunctions in either of the two microcomputers.

Fig.10 shows the system under discussion configured with the new 16-bit microcomputer, and a photograph of this ECU is shown in Fig.11. The two T5A41 microcomputers that were used in the system shown in Fig.9 have been replaced the new 16-bit microcomputer.

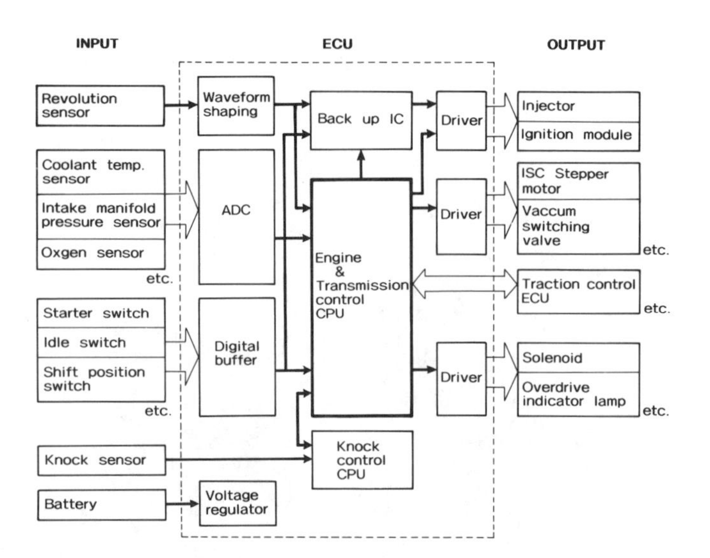

Fig. 10 Block Diagram of ECU with 16bit Microcomputer

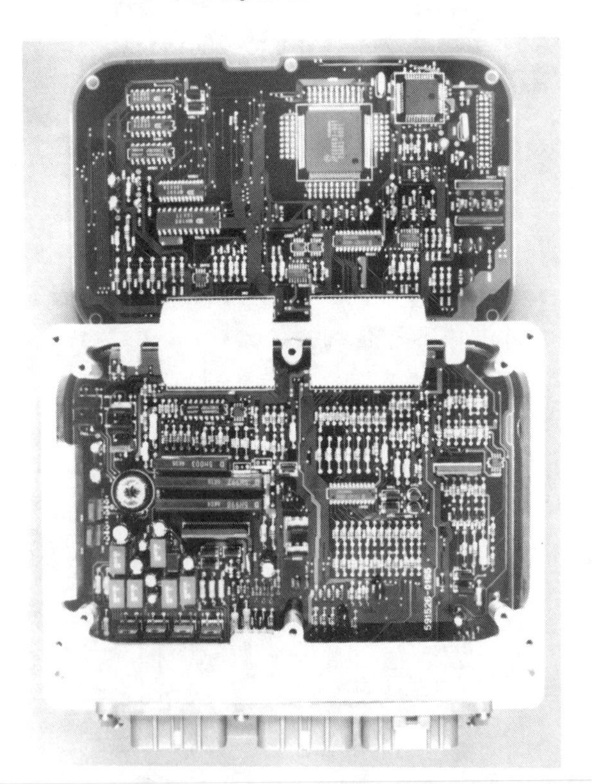

Fig. 11 Photograph of ECU with New 16bit Microcomputer

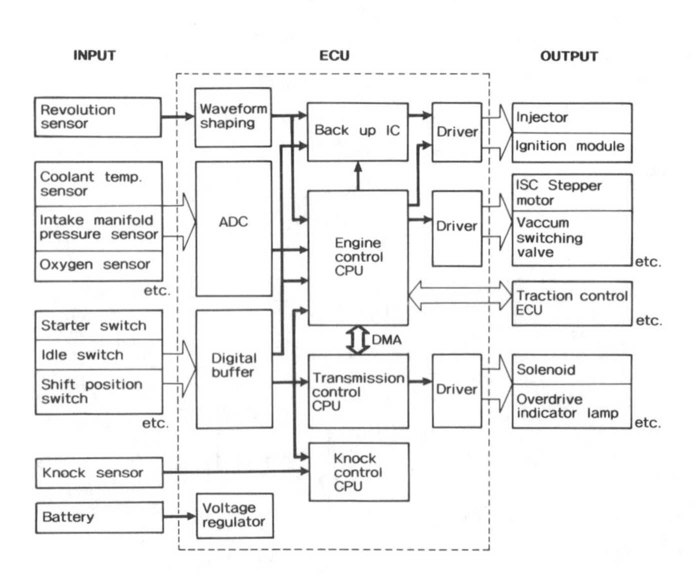

Fig. 9 Block Diagram of ECU with 8bit Microcomputer

This replacement was possible because it is approximately five times faster in processing speed, and approximately two times larger memory of the T5A41 when considering with the difference of bit length. Fig.12 shows a comparison of the processing time of the two microcomputers. This 16-bit application also eliminates the need for mutual data transmission and the master-slave monitoring function, a simplified program can provide more capacity for the addition of new controls in the future. Furthermore, this architecture greatly increases control precision. We illustrate this by two examples.

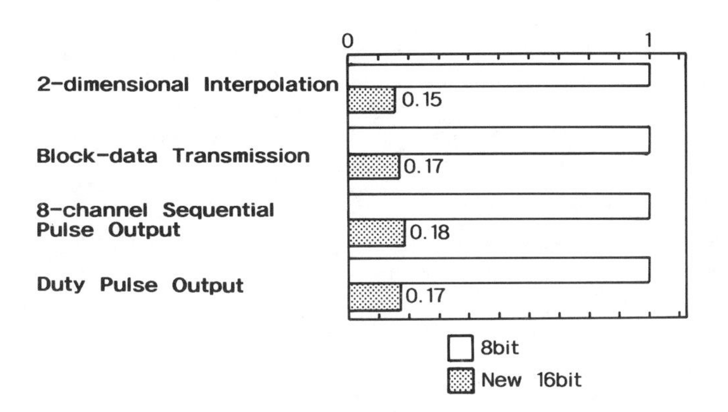

Fig. 12 Comparison of Processing Time

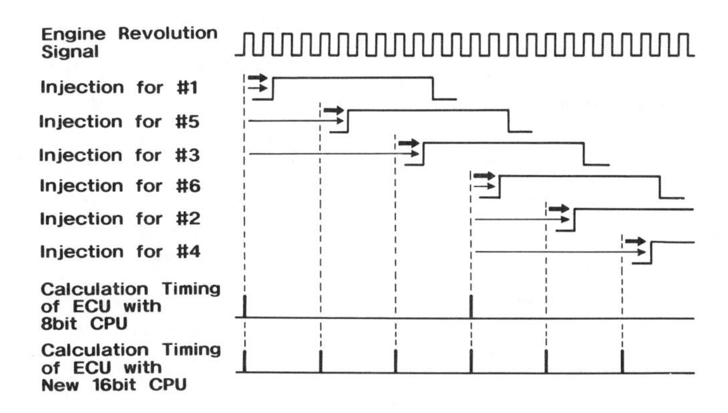

Fig. 13 Comparison of Calculation Timing between 2CPUs

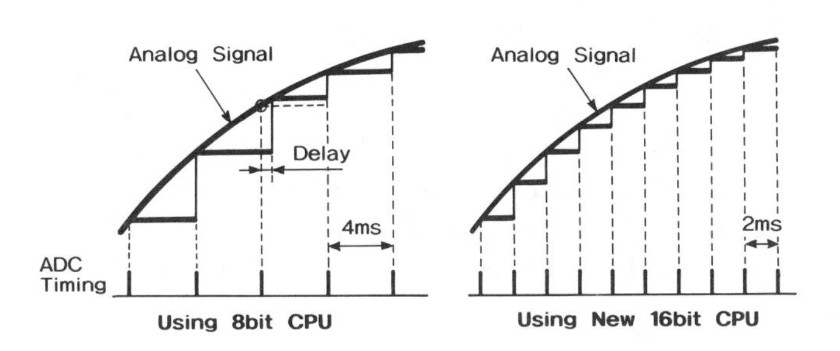

Fig. 14 Comparison of ADC Timing between 2CPUs

The first concerns fuel injection control. The new 16-bit microcomputer can perform more frequent calculations of the amount of fuel to be injected and thus achieve a higher degree of precision. Fig.13 shows a time chart for an engine speed of 6,000 rpm. The ECU using the 8-bit microcomputer calculated fuel injection quantities at intervals of approximately 10 msec, for it must maintain a balance with other controls. This resulted in injection to three cylinders on the basis of the same calculation. On the other hand, an ECU using the new 16-bit microcomputer can calculate fuel injection quantities at intervals of approximately 3 msec, hence, even at an engine speed of 6,000rpm, the amount of fuel appropriate for instanteous engine conditions can be injected into each cylinder, which best brings out the advantage of sequential injection.

Next, A/D conversion is performed more frequently, and this has increased the responsiveness to changes of analog data. Fig.14 shows A/D conversion when intake manifold pressure is in a state of transition. An ECU using the 8-bit microcomputer performs A/D conversion on a 4 msec timer interrupt processing basis in order to maintain a balance with other controls, but this can create a 1.1 msec lag at worst during periods of overlapping interrupts. On the other hand, an ECU using the 16-bit microcomputer can perform A/D conversion every 2 msec, which reduces the lag to a mere 0.3 msec at worst. This increases the responsiveness to changes of analog data concerning intake manifold pressure, thus increasing the quantitative precision of fuel injection amount and spark timing calculated by the control unit.

Even if expansion of functions is necessary in the future, multi-microcomputer configurations (various combinations of 8-bit or 16-bit microcomputers) will allow an ECU to be adaptable to the scale of systems by using the DMAC and the SIO function.

7. Summary

A new 16-bit microcomputer which is flexible and sophisticated, quite suitable for real-time control systems in automobiles has been developed. Use of this microcomputer has permitted the development of a high-speed, high-precision combined control system centered upon engine and transmission control, while reduction of the number of LSIs used and increased operating range have improved the reliability of the ECU. Furthermore, this microcomputer possesses capabilities for a number of advanced functions, and is adaptable to various kinds of automotive systems.

Development in this field has been facilitated by close cooperation among automobile, ECU and semiconductor manufacturers. Future efforts to enhance this type of cooperative relationship will be conducive to the development of even more sophisticated and technologically advanced products.

8. References

(1) T. Kawamura et al.
"Toyota's New Single-Chip Microcomputer Based Engine and Transmission Control System" SAE 850289

(2) Y. Ohno et al.
"An Integration Approach To Powertrain Control Systems" SAE 890762

(3) T. Inoue et al.
"Future Engine Control" SAE 901152

912782

The Power of a Personal Computer for Car Information and Communications Systems

Fred Phail
Intel Corp.
Chandler, AZ

ABSTRACT

Autonomous car navigation systems have been on the market since the mid-1980s. To date these systems have seen limited market acceptability due to high cost and lack of adequate map data bases and roadway infrastructure to make navigation systems more useful to the consumer.

Today, Japan, the US, and Europe all have programs which are beginning to address making our highways more intelligent, specifying the infrastructure necessary for successful implementation. Many geographic areas are now being digitized and stored on CD ROM. Automakers are also now beginning to address the utility of these autonomous systems. One way to increase utility is to take advantage of the power of the personal computer to make a cost-effective car information and communication system which shares the navigation "computer", display and CD player.

Some of the functions of the car information and communication system could include: operation and/or display of the automatic temperature control, audio and video entertainment center, trip computer, navigation system, cellular telephone, and car diagnostics. Certain features could be incorporated by taking advantage of the "computer" in the car. These features might include: autodialing, maintenance logs, "yellow pages" directory, paging, route determination and electronic mail.

This paper will discuss a conceptual car information and communication system that is based on a highly-integrated, two chip "PC", the Intel386(TM)SL, which could utilize a PC-like Windows(TM) operating system, notebook PC memory cards and CD ROM based maps.

INTRODUCTION

Autonomous in-vehicle navigation systems utilizing map matching and/or dead reckoning techniques have been on the market since the mid-1980s. Estimates indicate that approximately 50 navigation systems were in various stages of development and testing during the 1980s around the world (1). The first commercially available system which used map matching to enhance the accuracy of dead-reckoning was introduced by ETAK in 1985. Typical of many of the systems available today, the ETAK Navigator used a flux-gate magnetic compass and differential odometer for dead reckoning. The calculated path was then corrected via map matching from the stored information on a magnetic tape cassette (Figure 1). Once the car's initial position was identified by the driver and the destination selected, the arrowhead symbol directed the driver to the final destination.

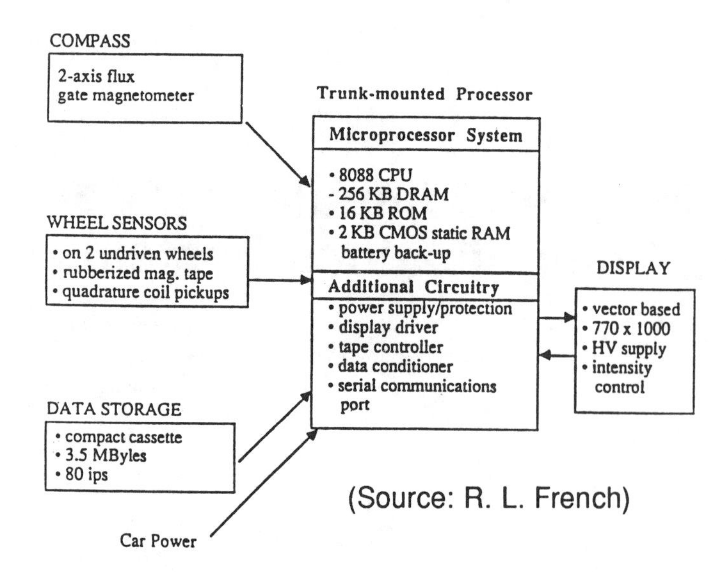

(Source: R. L. French)

FIGURE 1: ETAK NAVIGATOR SYSTEM DIAGRAM

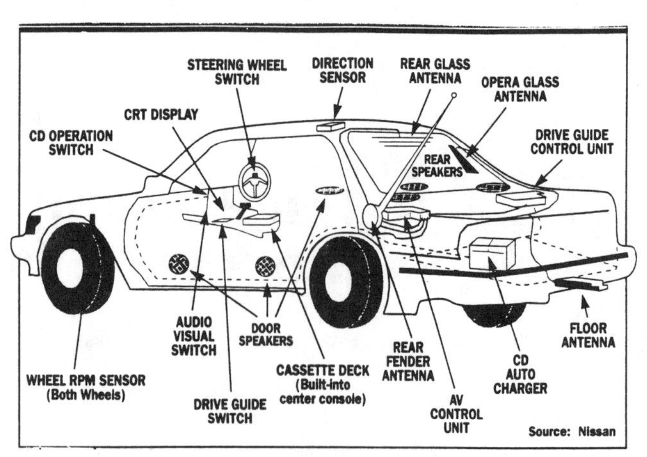

FIGURE 2: NISSAN MULTI-AV SYSTEM

A pioneer in the expansion of the functionality of a navigation system was Toyota with its Electro-Multivision in 1987 as introduced on the Toyota Crown for model year 1988. This system used a color map display which doubled as a color TV monitor. The system also included a radio with cassette player and one CD player which was multiplexed for music and reading the CD ROM based maps.

Today an updated version of the ETAK system using CD ROM maps available from Bosch - Blaupunct is called the Travel Pilot. GM and Clarion are also licensed to use the ETAK system. Pioneer and Mazda use the global positioning satellite system (GPS) for their navigation systems and the Honda Electro-Gyrocator calculates position inertially with a gyrcscopic device. Nissan's Multi AV system (Figure 2) is a map matching navigation system, and like the Toyota system, includes entertainment features. Many of these newer systems are improving driver utility through the additions of "yellow pages" information overlaid on the map and tied into the car cellular telephone. Route determination algorithms are also being pursued. These are the first steps in moving the autonomous navigation system toward an interactive car information and communication system.

EVOLUTION OF CAR INFORMATION AND COMMUNICATION SYSTEMS

James Rillings and Robert Betsold, in their Convergence 1990 paper entitled "An Advanced Driver Information System for North America" (2), discussed the three stages for the development of these systems:

o	Information Stage	1990 - 1995
o	Advisory Stage	1995 - 2000
o	Coordination Stage	2000 - 2010

The Information Stage will have primary emphasis on providing drivers with information to improve individual planning and decision making. There is no reliance on an infrastructure. Dynamic traffic information collected and transmitted by an infrastructure will be added in the Advisory Stage. Optimum route determination will be calculated and relayed to the driver. Many of the experimental systems of today are focused here. Automatic exchange of information from the vehicle to and from the infrastructure will highlight the Coordination Stage. The infrastructure will use combined information to provide coordinated routing and traffic signal control. The driver will be able to summon required services which can be automatically routed to the location of the vehicle.

Next-generation systems being developed for the Advisory Stage in the later part of this decade will evolve into car information and communication systems. The utility of the car information and communication system will increase over the autonomous navigation system as the driver is now able to "look ahead" to minimize travel time. Society will also benefit as a whole by virtue of fewer pollutants being released into the environment and less of our precious natural fossil fuels being used. In addition, potentially there may be fewer accidents and fewer frustrated, stressed-out drivers. The ability to process and display data easily, quickly and accurately during the Advisory Stage will be greatly enhanced through personal computer technology. In the later half of this decade, the PC will become as commonplace as the calculator is today and range from inexpensive handheld types to complicated supercomputers performing trillions of operations per second. There will likewise be a vast array of user reconfigurable displays and software which makes using the PC much more user friendly. All of this technology can be applied to the car information and communication system.

CAR INFORMATION AND COMMUNICATION SYSTEM FUNCTIONALITY

The car information and communication system can be much more than just the point of integration of navigation and telecommunications. It can serve as the focal point for car performance and power mode control, electric power management, diagnostics, security and trip computer.(3) As shown in Figure 3, the car information and communication system can be linked to other major electronic subsystems such as chassis control, powertrain control, low-speed multiplexed subsystems, LCD or CRT display, cellular telephone, entertainment center, and climate control via a high speed data bus. The high-speed link allows the car information and communication system to pass and receive information from the various subsystems without severely impacting the real-time data communication.

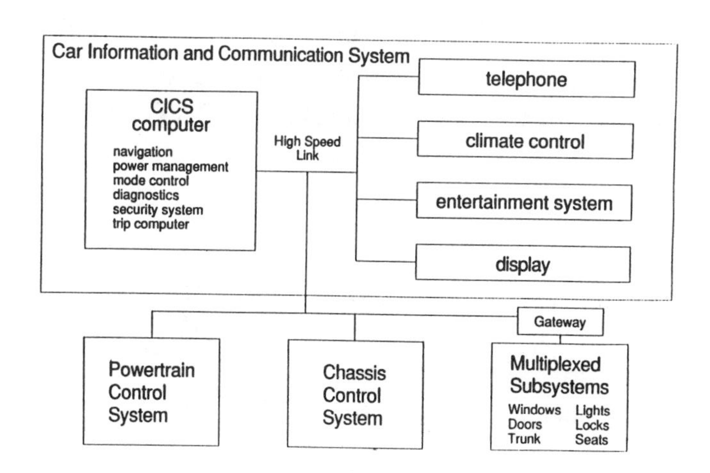

FIGURE 3: CAR INFORMATION AND COMMUNICATION SYSTEM INTERCONNECTION

The display is controlled by the car information and communication system much in the same way as the personal computer controls its monitor. An assortment of screens (as shown in Table I) can be called-up upon driver request. Once the appropriate screen is displayed, the driver can control the functionality of the subsystem or manipulate the data by using touch screen controls.

TABLE I
POSSIBLE SCREENS for a CAR INFORMATION and COMMUNICATION SYSTEM

Entertainment
Radio Stations
CD Selection
On/Off, Volume
Graphic Equalizer

Car Telephone
Phone List and Autodial
Yellow Page Information
Paged Message
Faxed Message

Main Menu
Screen Selection
Restricted Access
Log-on (keyboard)
PC Mode

Maintenance
Oil and Filter, Lube,
Fluids, Plugs, Filters,
Fuel Economy Log

TABLE I (cont.)
POSSIBLE SCREENS for a CAR INFORMATION and COMMUNICATION SYSTEM

Navigation	**Climate Control**
Maps	On/Off, Temperature
Yellow Pages	Set and Fan Control,
Route Selection	Zone Control
Electronic Mail	**Trip Computer**
Create	Ave. Fuel Economy
Send, Receive,	Instantaneous F. E.
File, Etc.	Elapsed Time
Personal Calendar	Miles to Empty, Etc.
Car Performance and Power Mode	**Security**
Sport, Economy,	Car Alarm Controls
Luxury, Secure,	
Secure Emergency,	
Occupied, Inactive,	
Start and Run	

Just as personal computer programs can be stored on a floppy-disk, car-specific programs can be input to the car information and communication system computer by using a FLASH memory card. The FLASH memory card consists of FLASH memory devices mounted to a printed circuit board and encapsulated (Figure 4). The card, as used on many notebook personal computers today, is small (about 2" X 3.5" X.125") and fits into a slot in the front panel of the car information and communication system computer.

Today there are thousands of programs written for the personal computer, and likewise many new programs could be written for specific commercial and non-commercial uses for the car information and communication system and sold as aftermarket devices. This will happen because there are many software vendors who understand the PC and can easily write the new programs. Table II shows just a few of these.

FIGURE 4: FLASH MEMORY CARD

TABLE II
POSSIBLE AFTERMARKET SOFTWARE

Non-Commercial	Commercial
Trip Planning	Real Estate
Car Maintenance and Expense Log	Delivery Van Route Plan and Log
Electronic Mail	Emergency Alarm/ Facility Locator
Shop Manuals	Taxi Cab Route Log
Tour Guides	
Day Planners	

Because FLASH memory has the capability to be electrically writable as well as readable, it can be used to transfer data from the car to the home or office. Such information as trip planning, trip records, car service log, diagnostics fault codes, electronic mail messages and fax information could be transmitted through the cellular telephone via modem to the computer.

An optional keyboard can be added to the car information and communication system to further enhance the useability of the personal computer functionality. The keyboard could be used in conjunction with a diagnostic computer at car service centers to step through the rigorous diagnostics routines in search of faults. The keyboard could also be used to reprogram the car information and communication system computer operating system much in the same way that DOS (TM) is updated today in personal computers.

A single CD player would be used and shared between the car entertainment system and the car information and communication system. There should be little degradation of the navigation functionality with one CD player if GPS is part of the system. The GPS information is used to compute the path of the vehicle along the map route displayed in the car while the CD player is being used for music. In urban canyons where GPS does not work well, a localized map of the city may be stored on a FLASH memory card so that when the CD player is used for music, navigation is displayed only from this simple map.

CAR INFORMATION and COMMUNICATION SYSTEM COMPUTER ARCHITECTURE

Daniel Frank and Oliver McCarter in their Convergence 1990 paper entitled "Evolution of the Office on Wheels Toward the Intelligent Vehicle/Highway System" (4) noted several key enablers for widespread use of car information and communication systems. These are:

1. Digital communication vs analog for cellular telephone
2. Reconfigurable displays that meet severe automotive requirements
3. User-friendly functions and features
4. Practical information infrastructure
5. An open system architecture

The remainder of this paper will focus on the final key enabler, open system architecture and discuss the necessary computer configuration which meets this intent. As stated previously, a personal computer platform can take advantage of the expertise that exists in hardware and software design and development when implemented as a car information and communication system. The majority of personal computers in use in the world today are based upon Intel Corporation's highly successful 8086(TM) family of microprocessor products. One current family members is called the Intel80386(TM)SL, microprocessor which was designed as a highly integrated chip set for small form factor personal computers such as notebook PCs. As shown in Figure 5, the chip set consists of a CPU and an ISA I/O subsystem. It offers the world standard Intel386(TM) microprocessor performance with low-power consumption and less board space. This is ideal for notebook PCs and car information and communication systems where space is at a premium.

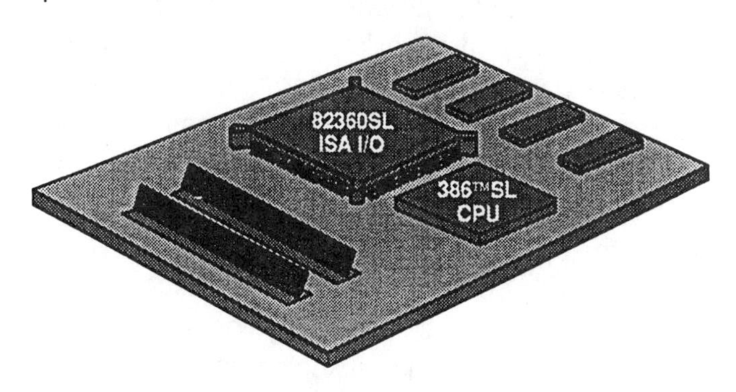

FIGURE 5: INTEL386(TM)SL

As a member of the Intel 8086(TM) microprocessor family, it can execute programs written for the original Intel 8086(TM) microprocessor based personal computer. Thousands of application programs have been written using various operating systems currently available. The architecture is open and well understood. It is important that the same guidelines used in the development of the Intel 8086(TM) microprocessor based personal computers be carried forward to the car information and communication system computer so that the hardware and software are updatable in the future. It is also important to take full advantage of the vast knowledge of hardware and software designers so that the two are closely aligned in their internal architecture. The way in which the computer communicates to the display and the way software is written must be the same. A preemptive multi-tasking operating system which runs on top of DOS will allow all of the features of the car information and communication system to run without collisions.

HARDWARE and SOFTWARE CONSIDERATIONS

As mentioned previously, the main CPU would be the Intel80386(TM)SL microprocessor because the computing horsepower is necessary to support the operating system, the applied geography and the other applications that run on the system. Just as in the personal computer, a BIOS (Basic Input/Output System) exists which includes all hardware dependent input and output drivers, including navigation sensor input, mass storage access, and display output. A BIOS that is stored on FLASH memory devices internal to the computer allows for updating of new display technology, a new type of CD player, or new sensor technology. The BIOS plus other critical navigation software for such algorithms as dead reckoning, map matching, GPS, and map display should remain embedded and not accessible by the driver. Similarly bootcode, navigation calibration constants, and any car-specific configuration information and/or critical system operating software must likewise be embedded so that it is accessible only by individuals licensed by the car companies.

The operating software and DOS could also be stored in other accessible FLASH memory for easy updating by the driver. Dual FLASH memory card ports are recommended in order to run an application program while data is being captured on another card (Figure 6).

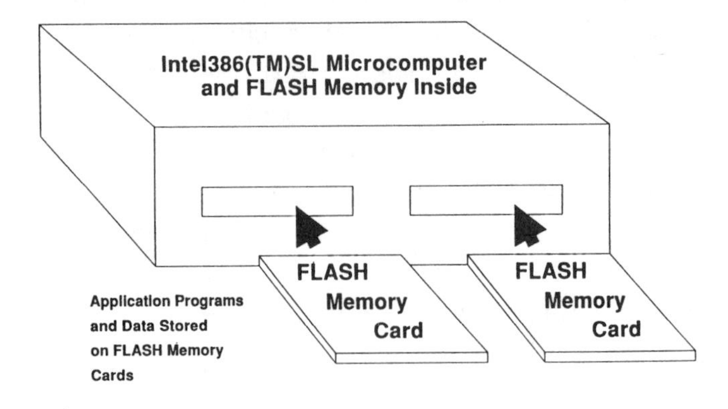

FIGURE 6: CAR INFORMATION and COMMUNICATION SYSTEM COMPUTER

CONCLUSION

Autonomous car navigation systems have been on the market since the mid 1980s. Lack of standards, map geographic coverage and local traffic information integration have prevented widespread acceptance of these systems. Even as these limitations are overcome, the user friendliness and utility of these new interactive systems called car information and communication systems, will be improved through standardization of a open computer architecture which exists in the personal computer. The car information and communication system will control the operation of the cellular telephone, entertainment system, climate control and reconfigurable display. It will be linked to other electronic systems in the car like powertrain, chassis and multiplexed control nodes. The computer architecture recommended is the Intel386(TM)SL microprocessor designed as a highly integrated personal computer chip set with reduced power and space requirements, offering the same performance of a desktop personal computer. Electrically reprogrammable FLASH memory is useful both embedded inside the computer for key algorithms and programs and externally in the form of a FLASH memory card which could contain application programs and data. By maintaining close hardware and software compatibility with the personal computer, the car information and communication system will serve as a useful link between the home and the office, as well as greatly increasing the use by the driver over the current autonomous navigation systems.

ACKNOWLEDGEMENTS

The author would like to thank James L. Buxton and Michael G. Sheldrick from ETAK Corp. for their guidance and contribution of ideas for this paper.

REFERENCES

1. Automobile Navigation and Intelligent Vehicle Highway Systems, R. L. French, SAE #89001, Feb., 1991.

2. An Advanced Driver Information System for North America, James H. Rillings and Robert J. Betsold, SAE # 901127, Oct., 1990.

3. Required Elements of Integrated Vehicle Control Systems, Edward H. Schmidt, C. David Wright, and John S. Zwerner, SAE # 901170, Oct., 1990.

4. Evolution of the Office on Wheels Towards the Intelligent Vehicle/Highway System, Daniel L. Frank and Oliver T. McCarter, SAE # 901162, Oct., 1990.

5. An In-Vehicle Navigation and Information System Utilizing Defined Software Services, J. B. Alegiani, James L. Buxton, and Stanley K. Honey, IEEE CH2789, June 1989.

950010

Development of a Configurable Electronic Engine Control Computer

Dave Steinmeyer
BKM, Inc.

ABSTRACT

To meet the need for a flexible electronic controller that is easily configured to meet different operating requirements, a general purpose Electronic Control Unit (ECU) has been developed. The ECU reads engine operating parameters from the sensors and generates output signals to the fuel injectors in response to those parameters as determined by the control software. Multiple signal path options allow a generic printed circuit board to be configured for different applications by populating alternate components as needed for the specific input and output requirements. Unused component mounting pads are left blank. This ECU has been applied to a variety of advanced electro-hydraulic fuel injection systems for gasoline, diesel, compressed natural gas, propane, and dual-fuel applications.

The configuration flexibility and signal processing power of the ECU make it an appropriate choice for both small volume development projects and OEM or after-market use. Various engine control software strategies including speed and load governing, overspeed protection, and skip-fire have been demonstrated. Generic control software is customized to satisfy the specific requirements of each engine application.

Engine calibration parameters are displayed and can be modified, via a personal computer, during engine tests. These include fuel delivery, timing, and sensor calibration maps. Once the parameters are proven to be satisfactory, a permanent Erasable Programmable Read Only Memory (EPROM) is programmed using the PC. This re-calibration can be performed in the field by an application engineer or test technician.

The ECU was designed for the heavy duty engine environment and to permit two different configurations - standard air cooling or optional fuel cooling. Environmental tests have been performed to verify reliability - shock, vibration, high and low temperatures, electromagnetic compatibility, liquid splash, salt spray, and power supply transient voltages.

BACKGROUND AND DESIGN CRITERIA

The subject Electronic Control Unit (ECU) was designed in response to the need for a microprocessor-based engine controller for the Servojet Fuel Injection (FI) system. This electro-hydraulic system was originally developed for diesel engine applications. Later versions have also been applied to gasoline, compressed natural gas (CNG), and dual fuel (diesel/CNG) engines. (ref. 2, 3, 4, 5, 6, 7)

Early Servojet FI systems used specially modified Ford EEC-IV engine controller hardware and custom software (ref. 7). However, as the demand for new applications grew, it became impractical to use the EEC-IV without control over the hardware configuration. The use of hybrid signal conditioning modules in the EEC-IV also made signal interfacing more difficult. In addition, the EEC-IV contained only a limited number of high-current injector solenoid driver output channels. The Servojet systems generally require one solenoid driver per cylinder - which dictated the addition of external drivers. This was acceptable for one-of-a-kind laboratory test systems but not for vehicle installations intended for higher volume customer use.

Therefore, the decision was made to design an ECU that would retain a high degree of compatibility with the existing software and incorporate the necessary hardware features. The EEC-IV contains an Intel 8061 microprocessor that was developed for and proprietary to Ford. The Intel 8097 microprocessor, which has a software command set that is very similar to the 8061, was chosen for the new ECU . This would allow relatively easy adaptation of software modules written for the 8061. All ECU control software is written in assembly language.

Some important features of the 8097 microprocessor, that make it especially well suited to engine control, are:

- High speed Output (HSO) unit: This triggers output events at specific times with minimal Central Processing Unit (CPU) overhead. Up to eight events can be pending at one time and interrupts are generated whenever any of these events are triggered. The Content Addressable Memory (CAM) file (ref. 7,8) can be loaded with the desired output action (I/O line number and output state definition - hi or low) plus a time tag expressed in clock counts. Nothing happens on that line until the time tag matches the contents of a separate timer that is continuously incremented. At that instant, the designated action occurs without further CPU intervention. In practice, this technique allows extremely precise timing of output pulses (2.0 microseconds at a 12 MHz crystal frequency).

- High Speed Input (HSI) unit: This is used to record the time at which an input event occurs with respect to a timer. This is primarily used to time tag the crankshaft position signal from a Hall effect sensor.

- Built-in Analog to Digital (A/D) converter: Eight analog input channels with 10 bit resolution are incorporated into the 8097.

The key ECU design requirements included:

- analog input channels for engine sensors
- injector solenoid driver output channels
- solenoid driver output channel for Pulse Width Modulated (PWM) control of rail pressure or electro-mechanical actuators
- high speed digital input channels for crankshaft position and digital pressure sensors
- high speed logic outputs for auxiliary injectors or ignition signals
- spare low speed logic outputs for relays, lamps, etc.

The current generation ECU meets these requirements and allows rapid implementation of alternative component configurations using a common Printed Circuit Board (PCB).

ECU SPECIFICATIONS

Microprocessors:
8097 or 80C196 main processor (CPU)
80C51 or 87C51 output multiplexer (MUX)

Clock Frequency:
12.00 MHz CPU (optional 16.00 MHz)
16.00 MHz MUX

Memory:
32 kByte strategy EPROM or ROM
32 kByte Static RAM
8 kByte MUX ROM or EPROM
128 Byte serial EEPROM (2 kByte optional)

Analog Inputs:
8 Channels, 0-5 volt, 10 bit resolution
A/D conversion time, 22 μSec

Digital Inputs:
3 high-speed inputs, TTL voltage levels

Outputs:
8 Solenoid drivers (4 A peak, 1 A holding current)
1 Pulse Width Modulated driver (4A/1A or FET)
8 high speed logic (400 mA current sink, 32V max)
5 low speed logic (400 mA current sink, 32V max)
1 regulated sensor excitation (+5 Volt DC, 500mA)
1 bipolar stepper motor output (500 mA per coil)

Supply Voltage:
8.0 to 16.0 Volts DC continuous
Load dump and transient protection
Supply Current: 250 mA standby plus up to 1.0 amp average current draw per solenoid driver.

Environmental:

Storage Temperature:	-40 to +105 C
Operating (air-cooled):	-20 to +60 C
Operating (fuel-cooled):	-20 to +105 C
Altitude (operating):	SL to 5500 meters
Mechanical Shock:	50 g peak, 10 mSec
Mechanical Vibration:	4 g at 5 to 200 Hz
Humidity:	98%
Material Compatibility:	SAE J1211
EMI:	SAE J1113

Weight: 1.93 kg

ENGINE CONTROL SYSTEM OVERVIEW

A typical Engine Control System (ECS) consists of the following hardware items:

Electronic Control Unit
Crankshaft Position Input Pulse Sensor (PIP)
Manifold Air Pressure Sensor (MAP)
Air Charge Temperature Sensor)
Fuel Pressure Transducer
Engine Coolant Temperature Sensor
Other engine sensors and actuators, as required
Wiring harness assembly

The ECU reads engine operating parameters from the sensors and generates output signals to the fuel injectors in response to those parameters as determined by the control software. Depending on the engine type and the specific fuel used, other outputs - such as ignition, idle speed control, fuel pressure control, etc. - may also be generated by the ECU.

The ECU is capable of controlling a wide variety of engine types from 1 to 8 cylinders. Twelve-cylinder engines have been controlled with dual-ECU systems. The maximum speed capability depends on number of cylinders and the complexity of the control software. It has been used to control an 800 RPM 12-cylinder locomotive engine, an 8-cylinder, 4500 RPM automotive engine, an 8000 RPM aircraft rotary (Wankel) and a 12,000 RPM single-cylinder two-stroke. Applications have ranged from heavy duty truck engines to high-altitude (25,000 meters) unmanned research aircraft.

Various engine control software strategies have been demonstrated. The most common strategies may be broadly classified as follows:

- Injection & Ignition Timing
- Starting
- Idle Control
- Torque Shaping
- Fuel / Air Ratio Control
- Speed Governing
- Fuel Pressure Regulation
- Altitude Compensation

Generic control software is customized for the specific requirements of each engine application.

MULTIPLE SIGNAL PATH OPTIONS

Analog signal conditioning features: The analog signal conditioning section is configured for maximum flexibility. A typical analog input circuit is shown in Figure 1. All channels contain filter capacitors, in-line current limiting resistors, and optional pull up or pull down resistor networks. While most pull up resistors are connected to 5 volts, one channel is pulled up to 12 volts allowing the system battery voltage (after scaling) to be sensed directly. This feature is used when the energize time of the injector solenoid needs to be adjusted to compensate for fluctuations in the battery voltage. During cranking the supply voltage may drop as low as 6.0 volts.

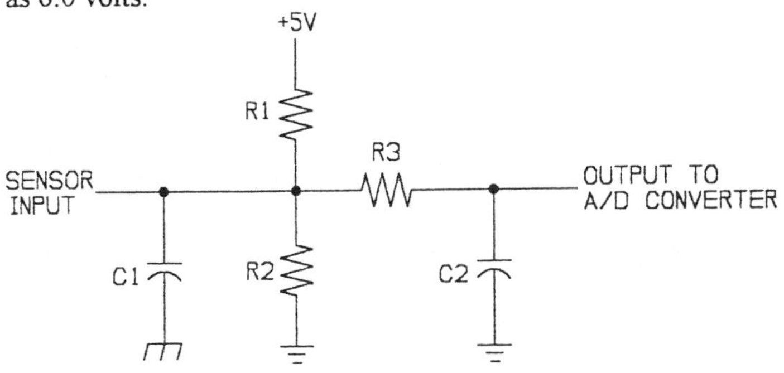

Figure 1 - Analog Signal Conditioning

Various channels may require a pull-up (R1) or pull-down (R2) resistor on the input to insure that a fail-safe condition exists in the event of a sensor or wiring harness failure. For example, if a high input voltage causes the software to respond by reducing engine power, then R1 is installed and R2 is deleted.

HSI/HSO: One of the Input/Output (I/O) lines from the 8097 may be used as either a High Speed Input (HSI) or a High Speed Output (HSO). The signal conditioning on this line therefore can be configured as either a filtered input or an open collector output (see Figure 2). As an output, this channel may be pulled up to 5 volts, 12 volts, or left as an open collector.

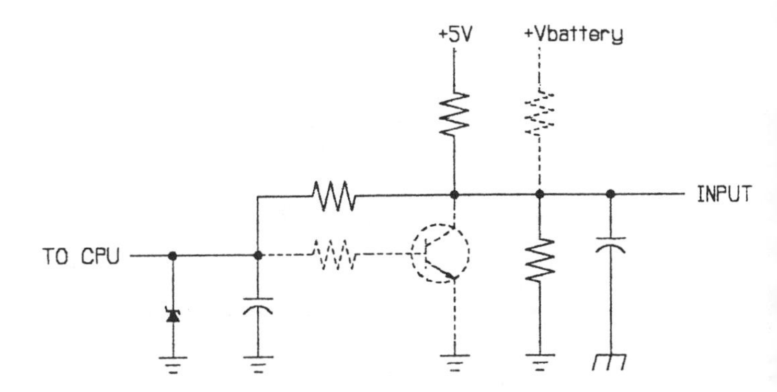

Figure 2a - High Speed Input/Output configured as an input

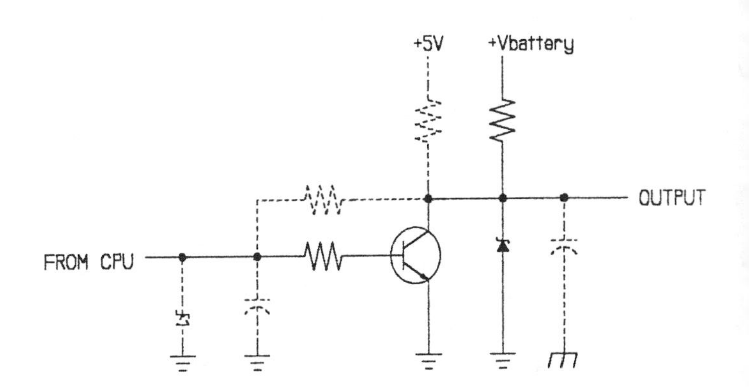

Figure 2b - High Speed Input/Output configured as an output

Solenoid Driver or MOSFET option: The Servojet fuel injector solenoids have low resistance (1.6 ohms) for quick response and because they remain energized for a relatively long time, require a current limiting driver. A Motorola MC3484S4-2 driver switches full supply voltage to the solenoid until a 4.0 amp current level is reached. The driver then "folds back" the current to a 1.0 amp holding level. The PWM driver, on the other hand, generally operates at a higher frequency (typically 200 to 400 Hz). At this frequency, a current limiting driver is not necessary since there is not enough time for the current to rise to the foldback level (4.0 A) during each cycle. The PWM output is configured so that either a high current Metal Oxide Semiconductor Field Effect Transistor (MOSFET) or a current limiting solenoid driver can be installed (Figure 3).

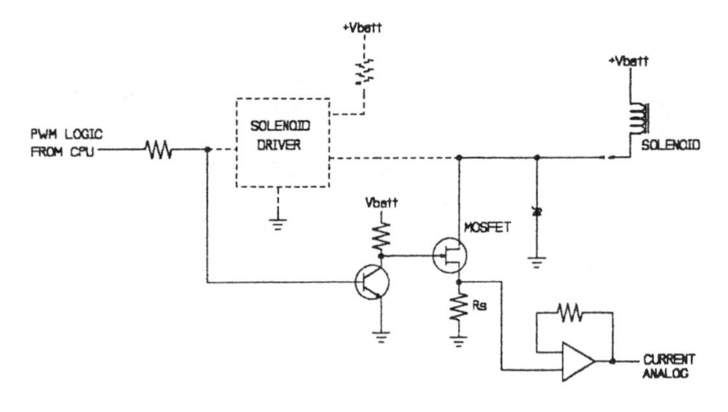

Figure 3 - PWM output circuit

PWM channel current sense feedback amplifier: The MOSFET driver channel (Figure 3) contains an in-line current sense resistor (R_S). The current sense voltage can be connected to an operational amplifier (OP AMP) that amplifies that millivolt signal to the 0 to 5 volt level needed for the analog input channels. This amplified current signal is routed back into one of the A/D input channels. Current feedback permits the software to maintain a constant average output current by adjusting the duty cycle to compensate for changes in sensed current as the output resistance varies (usually due to heating of the solenoid coil).

Auxiliary outputs to open collector NPN or solenoid driver: Often, only 4 or 6 solenoid driver output channels are used for fuel injectors but additional high current outputs are needed to drive relays, solenoid valves, etc. In those cases, auxiliary low speed logic output channels may be connected by jumper resistor to the unused solenoid driver inputs (See Figure 4).

Stepper Motor Driver Option: A number of throttled engine applications use a small stepper motor to control an idle air bypass valve. This valve bleeds air past the throttle, at idle, to compensate for transient speed changes due to engagement of the air conditioning compressor or other accessory loads. An integrated stepper motor driver has been designed into the ECU for this purpose. Three digital signal lines control the stepper driver analog outputs. They are CLOCK (1 pulse per step), DIRECTION (in/out), and INHIBIT (de-energizes all stepper driver outputs). The stepper outputs are jumpered to four ECU signal lines that can alternately be connected to open collector transistor outputs when the idle speed control is not used.

Multiplexer: The 8097 microprocessor has only 6 high speed CAM outputs. In many engine applications, when more outputs are required, an optional 8051 microprocessor is installed to multiplex precisely timed high speed signals from the 8097 into 16 outputs. This configuration permits total flexibility in output timing and sequencing. For example, the six 8097 CAM lines generate 8 solenoid outputs, plus 8 logic signals and 1 PWM output. In addition, the output sequence may be varied at will to achieve various "skip-fire" control strategies. The 8051 serial data lines may be accessed through unused auxiliary I/O lines.

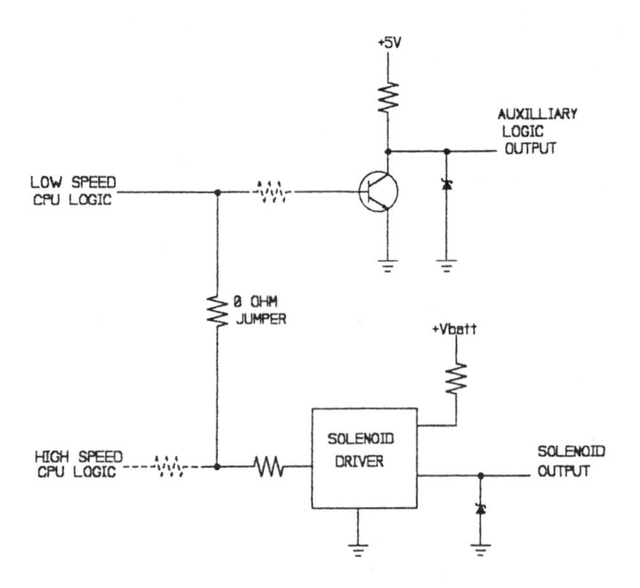

Figure 4 - Auxiliary output channel

CALIBRATION SOFTWARE

A separate piece of support equipment, the Calibration Interface, connects to the ECU and permits control calibration parameters to be displayed or modified during engine operation. This unit is used primarily by application engineers and consists of an MS-DOS personal computer, a serial interface adapter, and PC software. Once the calibration data is correct, a new EPROM may be programmed and installed into the ECU.

The PC calibration software is called "BKMPanel". Several typical display screens are shown in Figures 5a - 5c. The program is menu driven with the major menu selections listed across the top of the screen (FILE handling, COMMunications, PROM programming routines, data FORMs, and LOOK_AT screens). Under each menu, a list of options or available screens appears. Selecting a screen, will display a set of parameters as shown in Figure 5.

For example, the Governor screen, shown in Figure 5a, presents speed or load governing parameters. QCOM is the commanded fuel quantity per injection in cubic millimeters. QLIM is the maximum fuel quantity limit. RPMCMD is the commanded engine speed calculated from a throttle pot analog input. RPMERR is the speed error in RPM. PGAIN, IGAIN, and DGAIN are the proportional, integral, and derivative (PID) control gains respectively.

The INPUT screen of Figure 5b presents the raw analog input values (IACT, IECT,etc.) and the actual values in engineering units corresponding to the raw inputs (ACT, ECT, etc.). In this case, ACT and ECT are the air charge and engine coolant temperatures in degrees C.

FILE COMM PROM FORMS FORMS LOOK_AT

GOVERNOR

QCOM	17	QMIN	4	RPMERR	-4
QCOMM	18	QPUMP	4.5	RPMMIN	90
QINT	2.8	QPILOT	4.4	DGAIN	0.980
QPROP	1.0	GASDE	76	IGAIN	0.500
QLIM	68.0	SKIP_LIM	700	PGAIN	0.250
QLOLIM	23	RPMAVE	900		
QHILIM	50	RPMCMD	904		

Figure 5a - Governor Calibration Screen

FILE COMM PROM FORMS FORMS LOOK_AT

INPUTS

ACT	49	ECT	88
IACT	678	IECT	412
CNGT	28	VOLTAGE	13.75
ICNGT	801	IVOLTS	989
CNGP	280	RPM	1370
ICNGP	1002	DT	27
MAP	80	IPEDAL	221
IMAP	799		

Figure 5b - Analog Input Calibration Screen

FILE COMM PROM FORMS FORMS LOOK_AT

SUMMARY

RPM	890
GAS_DE	54
ACT	34
QCOM	87
ECT	78
QPUMP	23
CNGP	240
T_DWELL	15
MAP	101
AIR_MASS	907
A_F_RATIO	15.6
VOLTAGE	13.76

Figure 5c - Summary Calibration Screen

ECU ENCLOSURE DESIGN

In keeping with the overall design theme of maximum flexibility, the ECU enclosure is designed to permit two different configurations - standard air cooling or optional fuel cooling.

Since the solenoid drivers generate excess heat while they are limiting current, the enclosure must function as a heat sink. In most applications, convection air cooling is adequate. However, under severe conditions - especially at sustained high ambient temperature and engine RPM - liquid cooling may be necessary to maintain the drivers and other ECU components at acceptable temperatures. The diecast aluminum enclosure contains a cast-in boss that may be drilled with cooling passages (Figure 6). Liquid fuels or a water/glycol mix may be used as a coolant.

The enclosure is environmentally sealed using o-rings in the top and bottom covers and a rubber gasket between the main connector and the enclosure body.

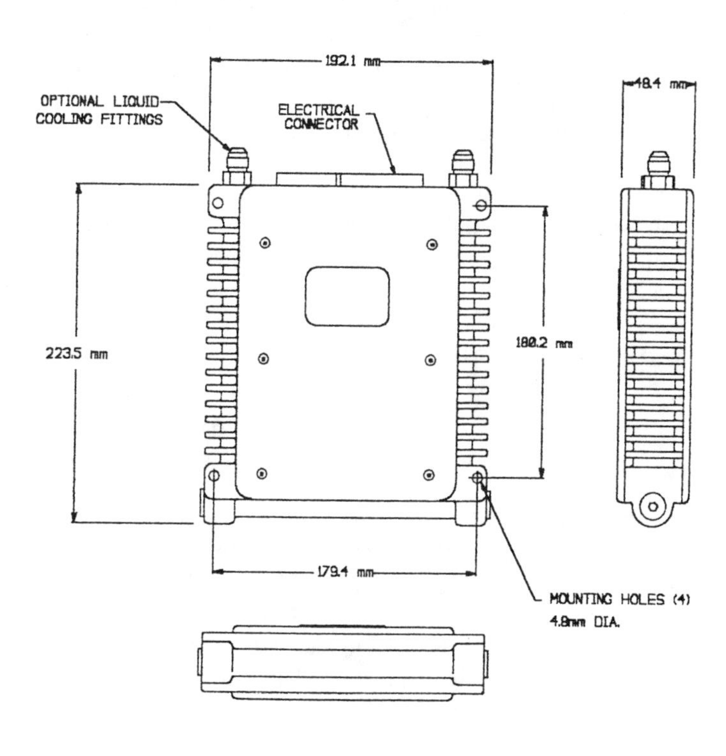

Figure 6 - ECU Enclosure

ECU RELIABILITY PROGRAM

1. Environmental Validation Tests: The ECU has been subjected to a series of environmental tests including shock, vibration, high and low temperatures, Electromagnetic Compatibility (EMC), liquid splash, corrosion (salt spray), and power supply transients.

Lessons learned during these tests suggested several design improvements that have been incorporated into the latest revision. For example, filtering of the battery voltage input was improved after it was discovered that inductive switching transients could result in a CPU reset. The improved filtering has eliminated the resets.

2. Failure Mode and Effects Analysis (FMEA): A FMEA was done for all ECU components. An evaluation was made of each postulated part failure including associated secondary failures (parts which fail as the result of primary part failure) to determine the effects on the operation of the engine.

Software effects were integral to this analysis and found to have a major impact on the fail-safe performance requirements that were established early in the program. In several cases, undesirable failure modes were eliminated using software error checking techniques.

CONCLUSIONS

The described engine control computer has been shown to be a flexible tool for development of advanced fuel injection systems. It is quickly adaptable to a wide variety of system input and output requirements and has demonstrated excellent performance and reliability.

System development and engine calibration are simplified using available personal computer hardware and proprietary calibration software to optimize operating parameters.

REFERENCES

1. Ward, David D., "Integrity of Automotive Electronic Systems. A View From Europe", SAE 940371

2. N.J. Beck, et.al. "High Pressure Fuel Injection - A Rational Approach to Diesel Engine Efficiency, Emissions, and Economics", SAE 830863

3. R.L. Barkhimer, N.J. Beck, and W.E. Weseloh, "Development of a Durable, Reliable and Fast Responding Solenoid Valve", SAE 831326

4. N. John Beck, et.al. "Direct Digital Control of Electronic Unit Injectors", SAE 840273

5. N.J. Beck, et. al. "Electronic Fuel Injection for Two-Stroke Cycle Gasoline Engines", SAE 861242

6. N.J. Beck, et. al. "Optimized E.F.I. for Natural Gas Fueled Engines", SAE 911650

7. N.J. Beck, et.al. "Electronic Fuel Injection for Dual Fuel Diesel Methane", SAE 891652

8. William Weseloh, "EEC IV Full Authority Diesel Fuel Injection Control" SAE 861098

9. A. Stephen Mihalik, Jr. and Ira Horden, "A New Approach to Engine Controls", SAE 831021

10. 16-Bit Embedded Controllers", Intel publication 270646-002

950430

Power Integrated Circuits for Electronic Engine Control

Keith Wellnitz and Jeff Kanner
Motorola Automotive Operations

ABSTRACT

This paper describes a series of power ICs designed specifically for powertrain control applications. The series includes low-side drivers and high-side drivers. The drivers are capable of fault detection and reporting to the Micro-Control Unit (MCU). Reported faults include short circuit detection, thermal limit, over-voltage protection, "on" open load detection, and "off" open load detection.

INTRODUCTION

Powertrain Control Modules (PCMs) use microcontrollers to control the system. The PCM environment requires Integrated Circuits (ICs) withstand extreme voltage transients. Without protection, these voltages are destructive to the MCU. MCUs need power ICs as protection from these voltages.

In addition, PCM modules must have extensive fault diagnostic capability to ensure reliability and meet emission requirements. These faults are detected and then reported to the motorist to indicate a potential need for servicing. Reliable designs off-load diagnostics from the MCU to the power IC to eliminate single point failures. Therefore, the power IC plays a critical role in the PCM system.

The ICs are fabricated using a mixed-mode process containing bipolar, MOS, and power FET technology. The power FET drives the load and is clamped to handle inductive flyback voltages. Bipolar and MOS devices allow both analog and digital functions. The choice of device technology depends on the type of control elements that must be integrated. Operational amplifiers, comparators and regulators work best with bipolar devices. CMOS devices handle logic, active filters, and current mirrors. This series of PCM ICs fully utilizes the diversity of available circuit technology.

QUAD INTEGRATED DRIVER

The Quad Integrated Driver (QID) is a four output low side switch with protection and diagnostics. The QID is designed specifically for PCM applications and utilizes the *SMARTMOS™* process. The QID is capable of controlling fuel injectors, transmission solenoids, canister purge solenoids, fuel pump relays, and incandescent bulbs. Each of the four power outputs on the QID is comprised of a DMOS transistor. The control inputs and open drain fault reporting are CMOS compatible. Fault diagnostics include current limit, thermal limit, "on" and "off" open load detection, and over-voltage protection. The Rdson of each output is 200 mΩ The IC is packaged in a high power 15 pin SIP.

A block diagram of the QID is shown in Figure 1. The key circuit blocks are Input, VREG, Output, and Fault Diagnostics. A description of each block follows.

INPUT - The input circuitry controls the gates of the power DMOS devices. It is CMOS compatible to receive direct MCU communication. There are 250 mV of hysteresis on each pin to provide double pulse suppression. The patented hysteresis cell takes advantage of unique characteristics of MOS devices and is insensitive to temperature variation. All inputs have active 10 μA pull-down current sources. This ensures that the power transistors do not engage the loads during an intermittent or lost connection. The input also contains a dual-select pin. This pin allows a fail-safe MCU to take control of outputs 1+2 and/or 3+4 in the event that the primary MCU is not functioning properly.

VREG - Battery is the only source of power available in automobiles. Most ICs are unable to operate from direct battery and require pre-regulation. The QID needs no pre-regulation because of the internal VREG circuit. VREG converts the battery (8-28 volts) to usable voltages and monitors the battery for an over-voltage condition. VREG uses high voltage analog transistors to withstand over-voltage conditions. Total quiescent current is less than 5 mA.

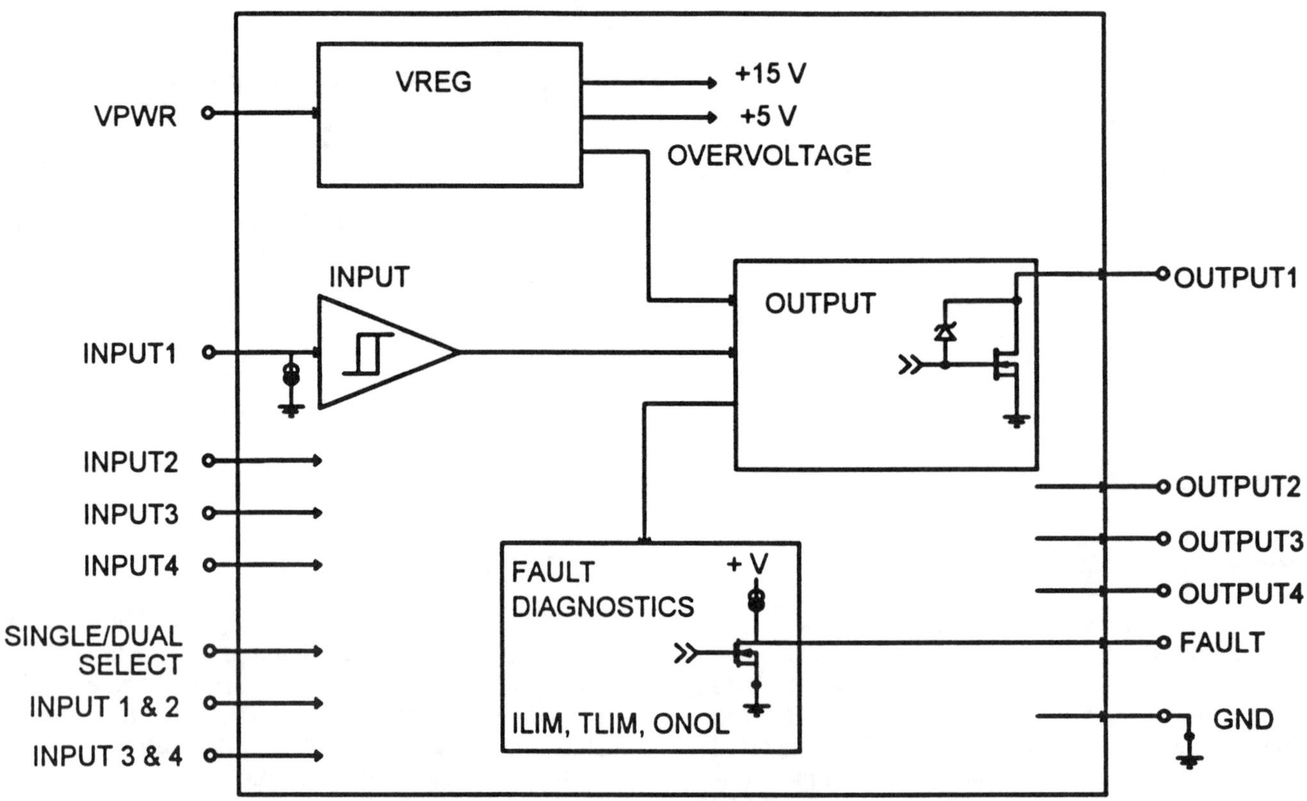

Figure 1: Block diagram of the Quad Integrated Driver

VREG contains a 5 volt regulator that powers the internal CMOS logic and bias circuitry. Its precision eliminates potential communication errors with the MCU due to supply mismatch. VREG also has a high voltage buffered shunt regulator. It produces a voltage one Vce_{sat} below the battery up to a 15 volt clamp. The 15 volt regulator drives the gate of the power FET. The ability to drive the gate to battery voltage levels lowers the RDS_{on} improving the power performance of the IC.

Over-voltage shutdown protects the QID during load dump. Since load dump overdrives the battery line, the power FET's drain experiences 80 volts. If a power device is on, excessive power dissipation causes a thermal overload. This extreme temperature condition puts unnecessary stress on the power device. Instead, VREG's over-voltage comparator trips and generates a fault flag. The flag signals the gate drive circuitry to turn off all power devices.

OUTPUT - Federal regulators place strict limitations on the amount of noise generated by automotive electronics. Fuel injection systems and full lock-up torque converters are prone to generate noise because the injectors and solenoids are pulse width modulated. Rapid switching causes high current slew rates at the output. The QID output circuitry controls gate charging current and limits the output current slew rate (dI/dt). Figure 2 shows the MCU's input voltage and the QID's output voltage and current. Note the drain current switches at less than 1 A/µs. External noise is mitigated.

Load dump, ESD, and the unclamped injector flyback voltage exceed the 80 volt breakdown of the DMOS device. To clamp the output to 60 volts, zener diodes are connected from drain to gate. When a disabled output is pulled to 60 volts, the zener clamp sources current into the gate enabling the output. This clamp enhances the robustness of the IC by turning on the DMOS and preventing avalanche breakdown. Avalanche breakdown is less robust because it increases the power density at the edges of the device. The clamp forces current to flow uniformly through the silicon and lowers the power density. At 25 °C, the energy capability of each output of the QID exceeds 100 Joules.

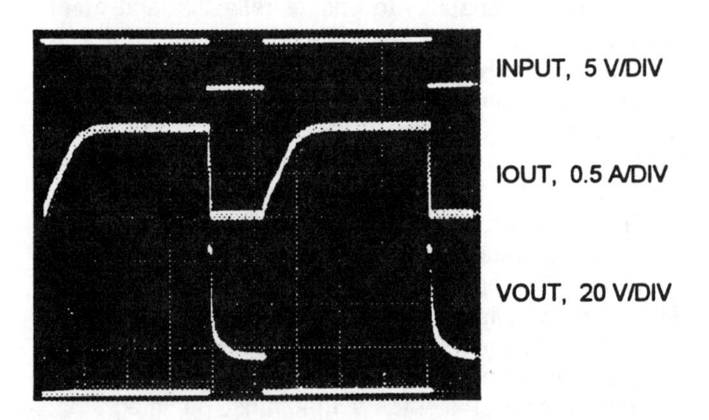

Figure 2: Oscilloscope photograph of the MCU input voltage and the QID's output current and voltage

The QID's DMOS device has a nominal RDS_{on} of 200 mΩ. The IC is packaged in a 15 pin SIP. Junction to ambient thermal resistance (Rθja) is 15 °C/W when mounted in still air with no heat sinking [1]. Proper heat sinking improves Rθja to 5 °C/W.

FAULT DIAGNOSTICS - Real time current limit (ILIM) protects the QID's outputs and external board traces during a faulted condition. Analog ILIM actively regulates the output DMOS to limit the current as shown in Figure 3. ILIM is implemented using SENSEFET™ technology to sample load current.

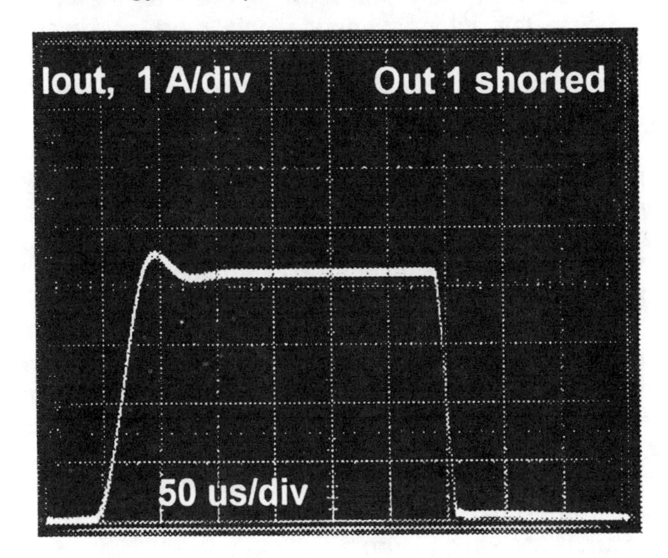

Figure 3: Analog current limit actively regulates the output current

The IC must protect itself from both hard and soft shorts. Hard shorts are direct shorts to battery. The ILIM circuitry detects hard shorts and turns off the outputs. Soft shorts are below current limit but exceed the power dissipating capability of the device. Soft shorts are difficult to detect. Soft shorts require over temperature sensing to protect the IC.

The QID includes over temperature detection and protection (TLIM). Faults are detected for each output separately and then the faulted devices are turned off. In a multiple output power IC it is highly desirable to turn off only the output device experiencing the faulted condition. Multiple output devices require independent local temperature sensing in place of a global temperature sense.

As shown in Figure 4, the four outputs of the QID turn off when the thermal limit of 170 °C is exceeded. All of the outputs were shorted to a 14 volt supply at 25 °C. A total current of 16 amperes initially flowed through the device. Note that each output turns off independently. Variations in turn off time of each output results from differences in current limit and thermal efficiencies. The shut-off temperature of each output is within 2 °C.

A PCM contains both Pulse Width Modulated (PWM) and DC switched loads. The injectors and torque converters are PWM loads. The fuel pump relay, canister purge solenoid, and incandescent bulbs are DC switched. For PWM applications, "off" open load detection is sufficient to verify the presence of the load. "Off" open load circuitry detects an open circuit between load and output when the input is off. "On" open load circuitry detects the same fault when the input is on. For DC switched applications, the activated load must be monitored to detect faults that occur during start-up.

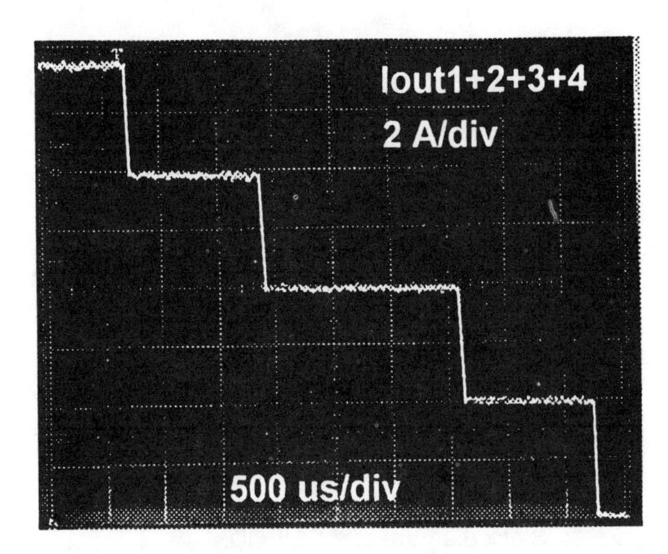

Figure 4: Independent thermal shutdown protects each output of the QID from a simultaneous short of all outputs to battery

For relay, solenoids, and lamps, the QID has "on" open load detection as a diagnostic feature. "On" open load circuitry must detect millivolt changes in the drain voltage. This requires precision analog circuitry. The QID uses a comparator network capable of detecting 10 mV changes on each output.

Figure 5 shows a die photograph of the QID. Each output has independent "off" open load, "on" shorted load, "on" open load, and over temperature detection and protection.

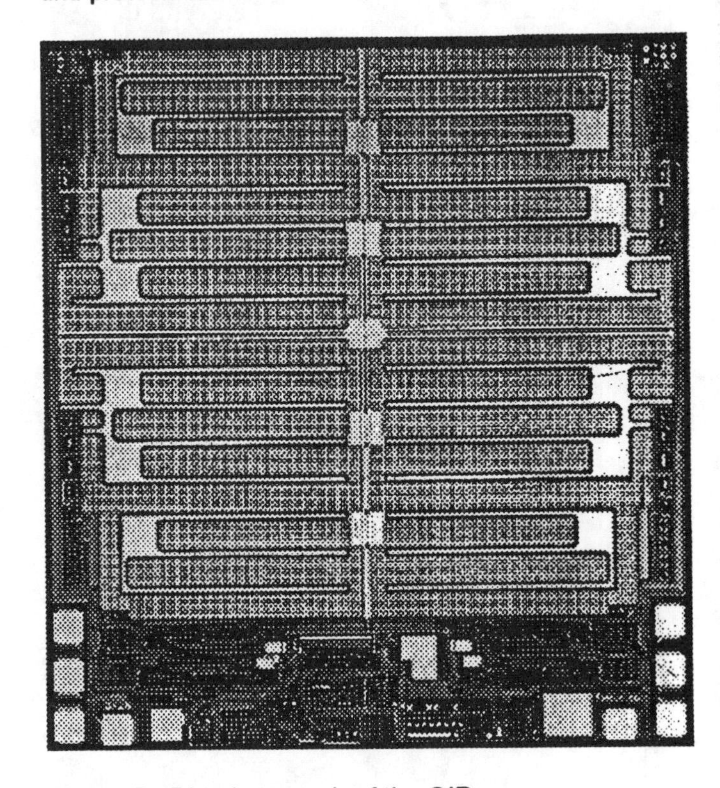

Figure 5: Die photograph of the QID

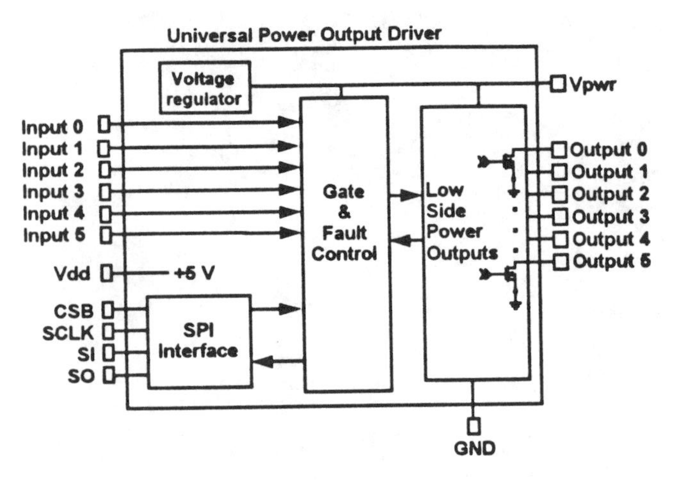

Figure 6: Block diagram of the UPOD

UNIVERSAL POWER OUTPUT DRIVER

The Universal Power Output Driver (UPOD) is a six output low side switch with fault protection, diagnostics, and a serial peripheral interface (SPI). Like the QID, the UPOD is designed for a variety of PCM applications from injectors to bulbs. Fault diagnostics include current limit, thermal detection, "on" and "off" open load detection, and over-voltage protection. The Rdson of each output is 200 mΩ. The IC is packaged in a high power 23 pin SIP.

High speed communications and increased flexibility are the trend in PCM ICs. The increase of ICs and diagnostic data in electronic modules increases the information available to the MCU. In order for the MCU to process this information, the data must be readily available in a high speed format. Parallel outputs are slow and require a separate MCU input pin for each output.

The UPOD utilizes an SPI to increase the communication with the MCU. The SPI is a two way communication port allowing the IC to receive instructions and transmit diagnostic feedback. With the SPI, the UPOD sends diagnostic information to the MCU in 8 bit words at frequencies up to 1.8 MHz.

Parallel control of the outputs is preferred for PWM applications. For DC switched applications, controlling the output through the SPI is sufficient. The outputs of the UPOD can be controlled by either the parallel inputs or an input command received on the SPI. Controlling non-PWM output via the SPI reduces the number of parallel ports required on the MCU. The UPOD block diagram is shown in Figure 6.

A die photograph of the UPOD is shown in Figure 7. The digital, analog, and power portions of this IC are labeled. The power DMOS outputs occupy sixty percent of the total die area. Even with a serial interface, the digital circuits require the least area because of the small geometry of CMOS.

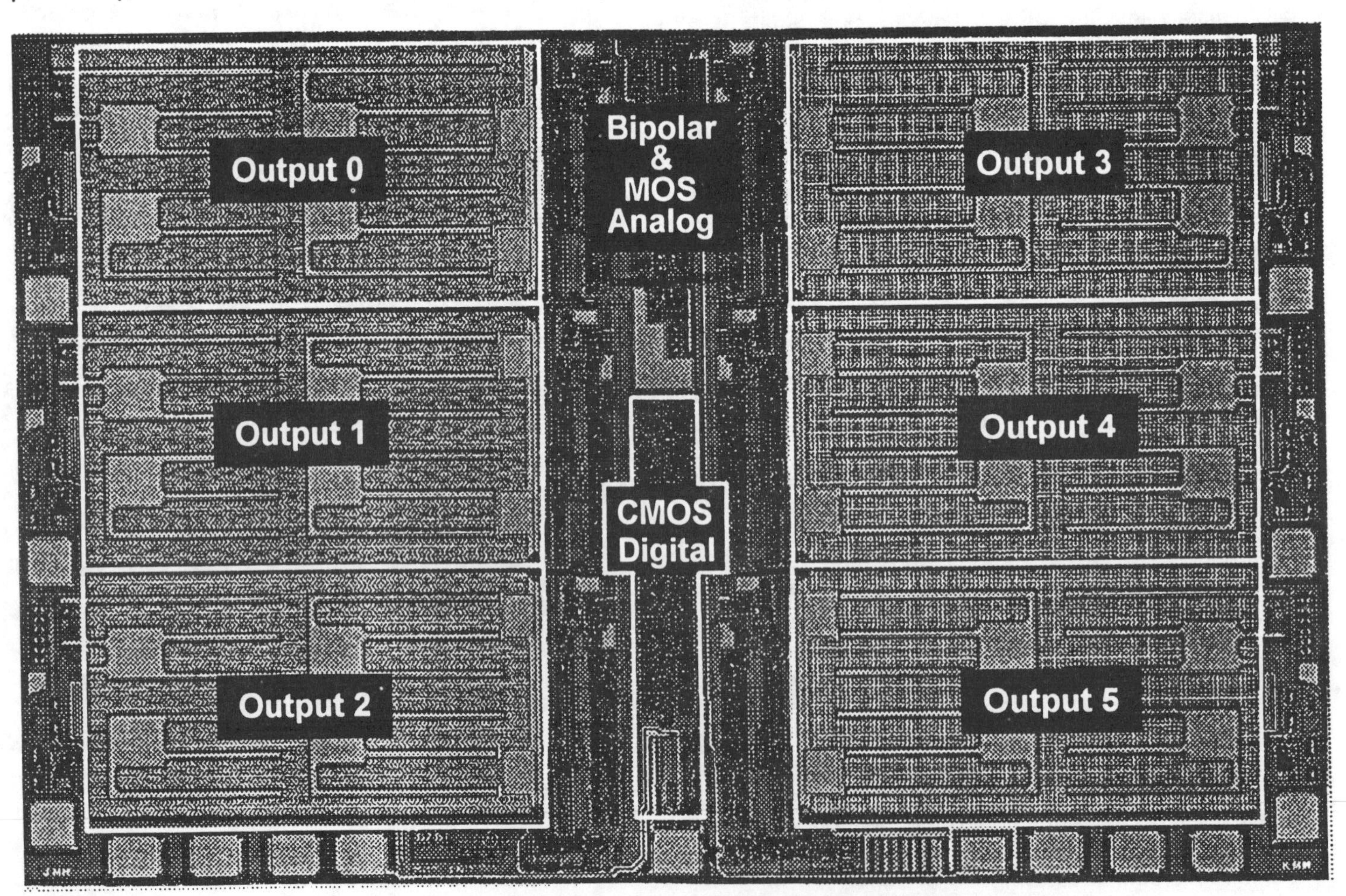

Figure 7: Die photograph of the UPOD

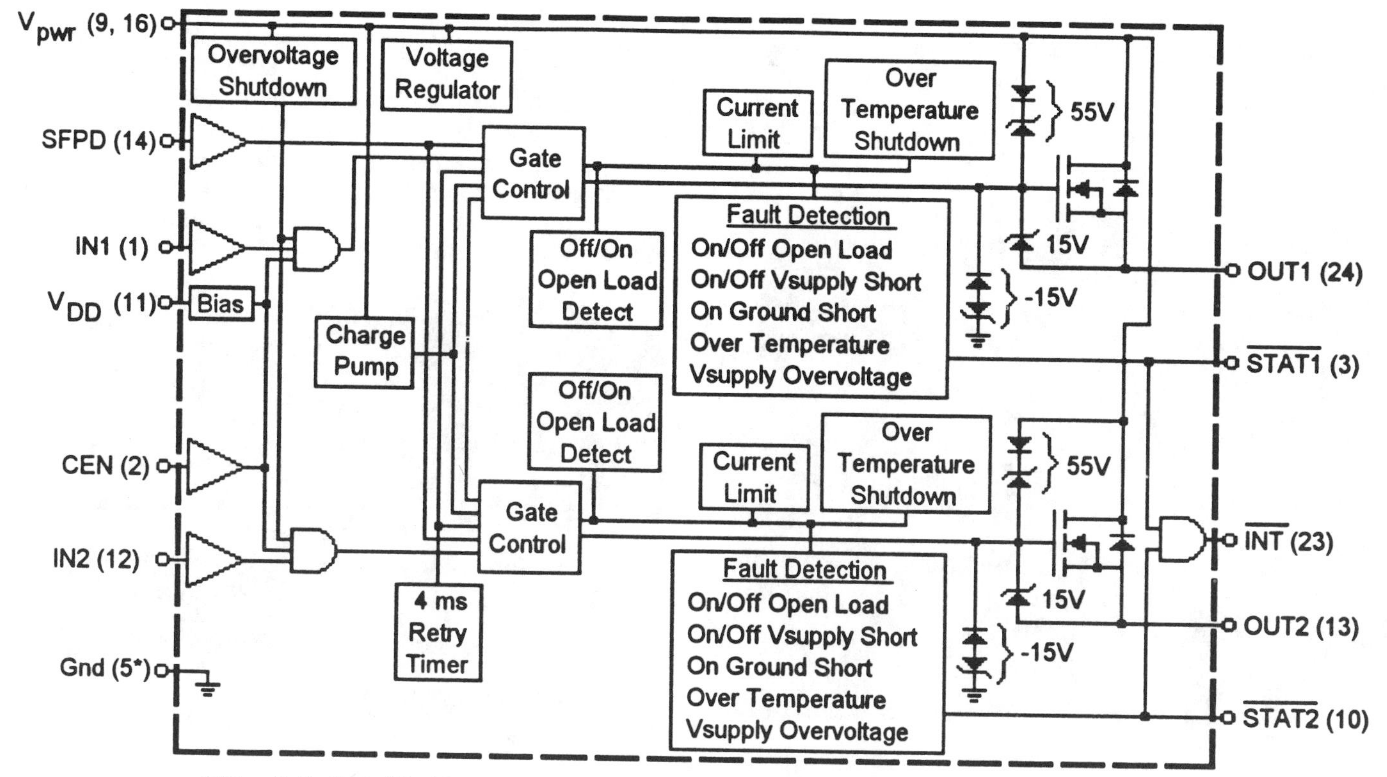

* Pins 5, 6, 7, 8, 17, 18, 19, & 20 should be grounded to provide thermal heatsinking.

Figure 8: Block diagram of the DHSSA

DUAL HIGH SIDE DRIVERS

The DHSSA and DHSSB are Dual High Side Switches (DHSSs) with protection and diagnostics. The DHSSA is designed to control transmission solenoids, relays, and automotive lamps. The DHSSB is capable of driving small relays and solenoids. Faults protection and detection applies to "on" and "off" open loads, "on" shorted loads, current limit, over-voltage shutdown, and output over-temperature. The outputs are clamped 15 volts below ground for inductive flyback clamping.

The block diagram for the DHSSA high side switch is shown in Figure 8. A description of the key features of both DHSS follows.

ULTRA-LOW NOISE CHARGE PUMP - Either DHSS's charge pump radiates less noise than any other integrated High Side Switch (HSS). The charge pump provides the gate voltage necessary for the output DMOS switches. The charge pump is a voltage tripler with VPWR (battery) as the input. The tripler allows operation with a battery potential from 5.5 to 28 volts.

The three sources of noise in an HSS are the storage capacitor, VPWR and OUTPUT pins. Noise generated in a charge pump is typically routed beyond the module via the VPWR and OUTPUT wires. This noise will manifest itself in the form of conducted emissions. The DHSSs' charge pump and output control circuits are current source driven. This limits the current transients and the resulting conducted emissions.

The major source of noise in any HSS is the VPWR input because the charge pump runs directly off VPWR. The noise spectrum from the DHSS at the VPWR circuit board input is shown in Figure 9. Above 10 MHz, the noise is less than -60 dB. This is a 20 dB improvement over the industry standard HSS.

The internal charge pump is self oscillating, modulating the frequency based upon the current loading. This circuit further eliminates current spikes on VPWR by insuring that one phase of the charge pump is always charging. The dynamic response of this circuit eliminates the requirement for an external storage capacitor. This significantly reduces radiated emissions within the PCM.

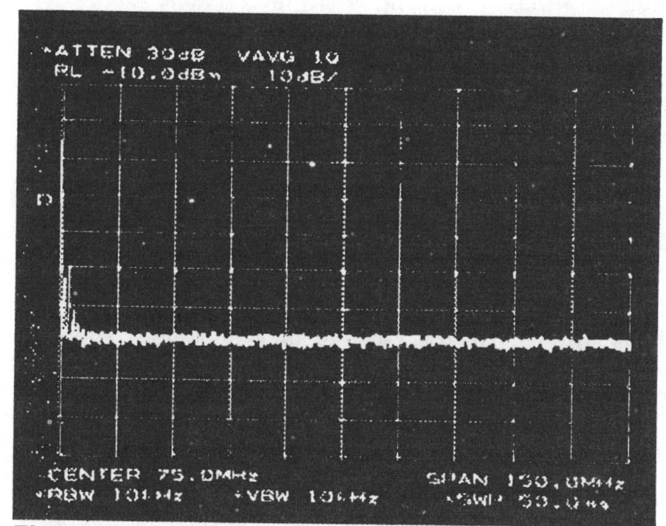

Figure 9: DHSS noise at the VPWR connector of the circuit board

The DHSS OUTPUT is coupled to the charge pump via internal current sources and the gate capacitance of the output DMOS. Figure 10 charts the OUTPUT noise spectrum for the ultra-low noise DHSS. In the AM, IF, and audio ranges (<15 MHz), the DHSS OUTPUT is 50 dB quieter than the industry standard and 20 dB quieter than a "Very Low Noise" HSS [2].

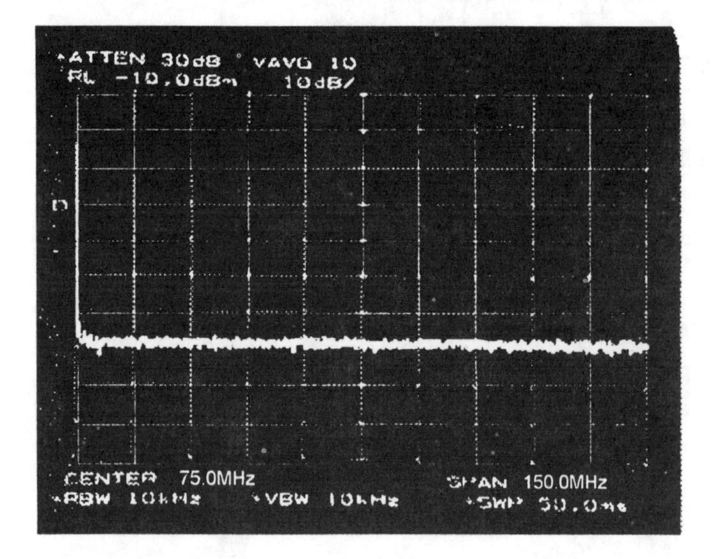

Figure 10: DHSS noise at the OUTPUT connector of the circuit board

OVER CURRENT PROTECTION - The DHSS offer three layers of protection from over current conditions. The three layers are analog current limit, short fault protection, and thermal limit.

Analog current limit actively regulates the output current. Limiting the maximum current protects the printed circuit board and the IC. During current limit, the output control circuitry senses the presence of excessive current and reduces the gate voltage on the output DMOS transistor. Reducing the gate voltage forces the DMOS transistor to operate in the MOS saturation region where the output current is a function of gate voltage. During current limit, the voltage drop from VPWR to OUTPUT is limited only by the external power source and load impedance. The resulting voltage across the output DMOS is larger than nominal and produces an increase in power dissipation. The increase in power produces a rise in the junction temperature of the IC. Left unchecked, the temperature rise is destructive.

The second layer of protection, Short Fault Protection (SFP), reduces the IC's temperature during analog current limit. SFP limits the duration of current limit to 50 μs to validate the fault condition. Then the output turns off for 4 ms to allow the IC to cool. The output is then reactivated for another 50 μs to monitor for fault condition removal. This cycle repeats itself until normal operation is achieved or until the output is commanded off via the Input. Figure 11 is a photograph of VPWR current for a DHSS with a short on Output 2. Figure 11 demonstrates current limit and SFP on Output 2 while Output 1 continues to operate normally.

The third and most significant layer of protection is Thermal Limit (TLIM). TLIM actively limits the temperature of the IC. TLIM deactivates the output when the junction temperature of the IC exceeds the over-temperature threshold, 180 °C. The over-temperature threshold is above the maximum operating temperature to allow the full range of operation. Seven degrees of hysteresis is provided to limit thermal oscillations. The output reactivates when the junction temperature drops seven degrees below the over-temperature threshold.

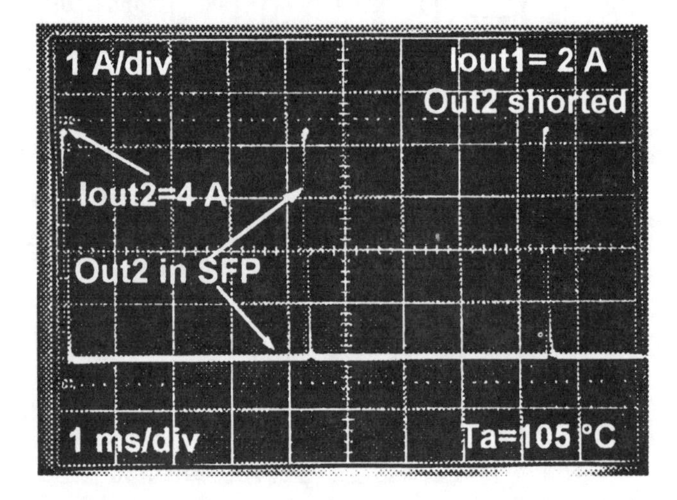

Figure 11: VPWR current with a short on Output 2 and normal operation of Output 1

OUTPUT SPECIFIC TLIM - Output specific TLIM is essential for multiple output high side switches to insure that only the faulted output is disabled. Each DHSS output contains a dedicated TLIM circuit. Mismatch between over-temperature thresholds on the same device is less than 2 °C [3]. This matching ensures that the device deactivates the faulted output before causing an adjacent output to fail. Figure 12 is a photograph of output specific TLIM in the 140 mΩ DHSS. At 105 °C, Output 1 controls 2 amperes while Output 2 controls a 3 ampere load. Every 46 ms, Output 2 exceeds its over-temperature threshold and TLIM turns off the output. After cooling off, Output 2 turns on. Output 1 is never deactivated by the fault on the adjacent output.

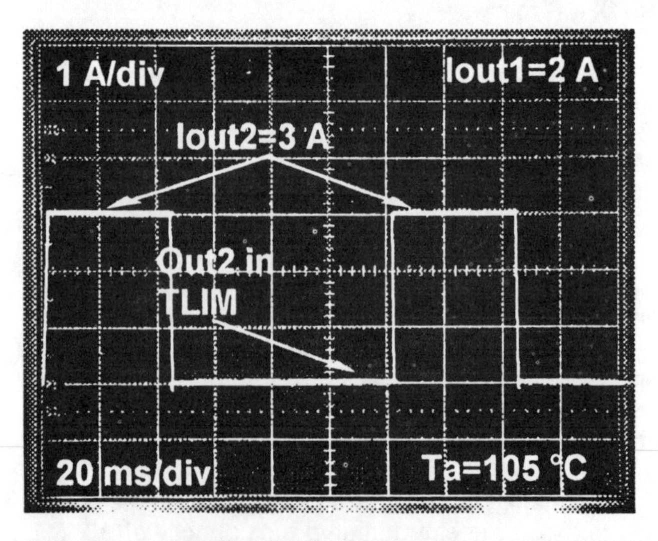

Figure 12: VPWR current with excessive power in Output 2 and normal operation of Output 1

The slow heating demonstrated in Figure 12 is a worst case test scenario. This slow heating of Output 2 allows the heat to propagate across the IC raising the temperature of Output 1. It would be easier to disable only the faulted output if a hard short were used for the fault condition. A hard short would have higher instantaneous power dissipation forcing a TLIM on Output 2 before the heat can propagate across the IC. As a result, the thermal gradient would be higher and easier to manage.

SHORT FAULT PROTECTION DISABLE - Short Fault Protection Disable (SFPD) is a selectable feature that extends the duration of current limit operation. SFPD is a CMOS input that disables the SFP circuitry. With SFPD active, a faulted output remains in current limit until attaining a normal current level or until TLIM deactivates the output.

Figure 13 is a photograph of the in-rush current into a brake lamp when driven directly from a supply. The current exceeds the DHSS current limit threshold, 4 amperes, for 30 ms. For the DHSS driving the same lamp with SFPD low, SFP activates the output 50 μs out of every 4 ms. The lamp would not incandesce with such a small duty cycle and on time. The MCU could incandesce the lamp by toggling the input every 50 μs. However, switching losses would increase the power dissipation and the MCU overhead requirements would degrade the system chronometrics.

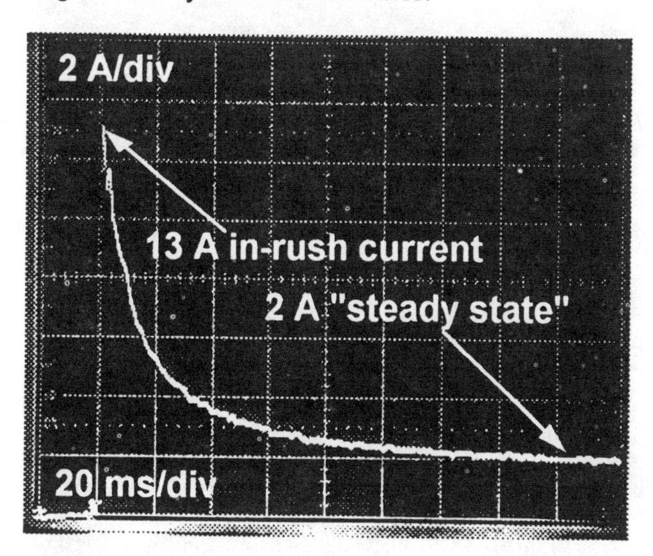

Figure 13: Brake lamp shorted to a voltage source

The SFPD feature permits incandescent lamp applications with in-rush currents in excess of the current limit threshold. To illuminate the lamp, the transient thermal response must keep the junction temperature below the over-temperature threshold. Figure 14 is a photograph of the in-rush current into a brake lamp driven by the DHSS with SFPD active. Having sufficient thermal capacity, the DHSS limits the current to 4 amperes until the bulb incandesces.

LOW OVERHEAD FAULT REPORTING - Fault reporting on the DHSS is software efficient. Monitoring the output status requires no background loops or word compares which are typical of SPI operated devices [3].

A fault on either output will induce a high to low transition on the INTB output. INTB is an open drain output with a current source pull-up of 40 μA. INTB is designed to be wired-or with other ICs to drive the MCU interrupt request input. INTB eliminates the need for the MCU to poll the DHSS for fault status. Because normal operation is the dominant mode of operation, a substantial saving of MCU "horsepower" is realized. When a fault occurs, individual status outputs, STATB1 and STATB2, indicate which output is faulted.

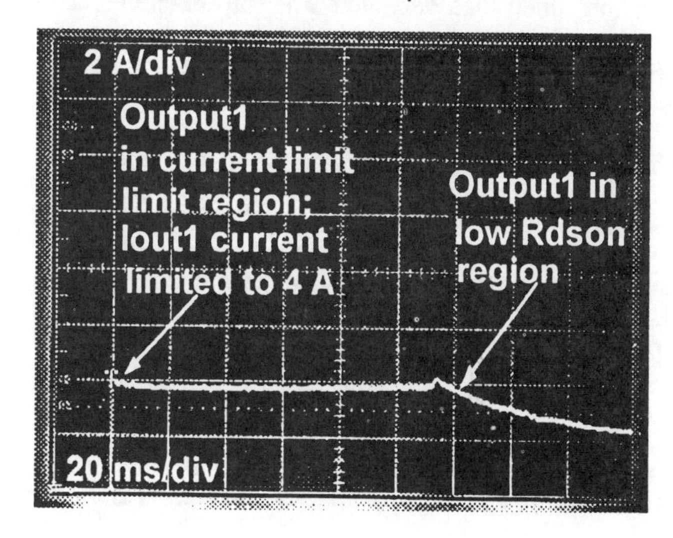

Figure 14: Brake lamp controlled by DHSS with SFPD active

OPEN LOAD DETECTION IMPEDANCE - The load impedance that is reported as an "on" open is a function of the "on" open detection current (I_{oon}), VPWR, and R_{DSon}. This relationship is shown in Equation (1). Figure 15 plots the load resistance that is reported as an open by the DHSSA.

$$R_{on\text{-}open} \geq (VPWR/I_{oon}) - R_{DSon} \quad (1)$$
$$R_{on\text{-}open} \approx (VPWR/I_{oon})$$

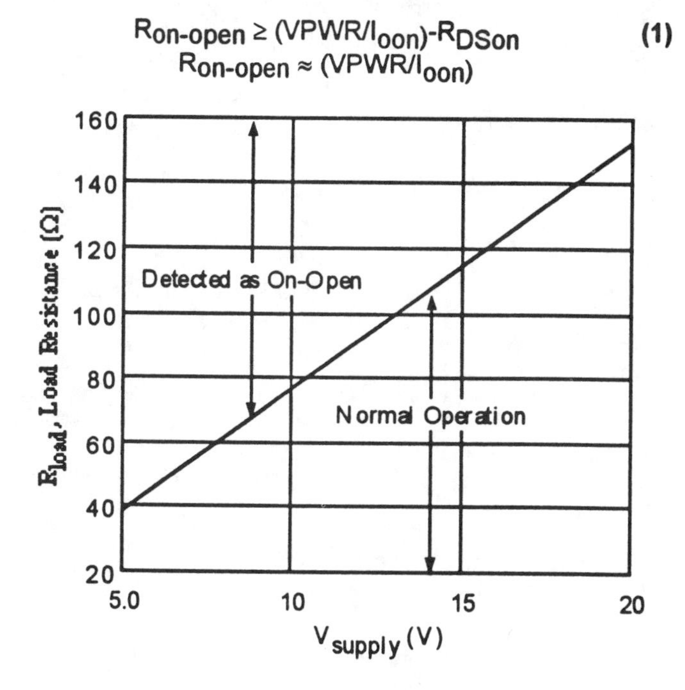

Figure 15: Load resistance reported as an "on" open for the DHSSA

The load impedance that is reported as an "off" open is a function of the output pull-up resistor (R_{opu}), VPWR, and the output fault threshold (V_{tho}). This dependence is shown in Equation (2). Figure 16 plots the load resistance that is reported as an open by the DHSSA with R_{opu} of 20 kΩ.

$$R_{off\text{-}open} = (VPWR - V_{tho})/(V_{tho}/R_{opu}) \quad \textbf{(2)}$$

"On" open load detection provides better resolution than "off" open load detection. The "on" open detection circuitry will detect loads in the 80 to 120 Ω range for VPWR in the range of 10 to 15 V. This impedance region is close to but above the normal load range of 12 to 40 Ω. Loads from 80 to 130 kΩ are detected as "off" opens for the same VPWR range. For "off" opens, the load impedance must change by two orders of magnitude before a fault is reported. "SLEEP STATE" - The DHSSs are designed to virtually eliminate VPWR current during periods of non-use. This circuit is called "Sleep State" (SS). SS is used in battery powered applications where the DHSS may not be active for extended periods of time and prolonged battery life is important.

The DHSSB controls 0.6 A per output in a 105 °C ambient (Tj ≤ 150 °C). Each output has 380 mΩ of resistance at 25 °C. The DHSSA delivers 1.6 A per output in a 105 °C ambient (Tj ≤ 150 °C). Each output has 140 mΩ of resistance at 25 °C. The die photograph of the DHSSA is shown in Figure 17.

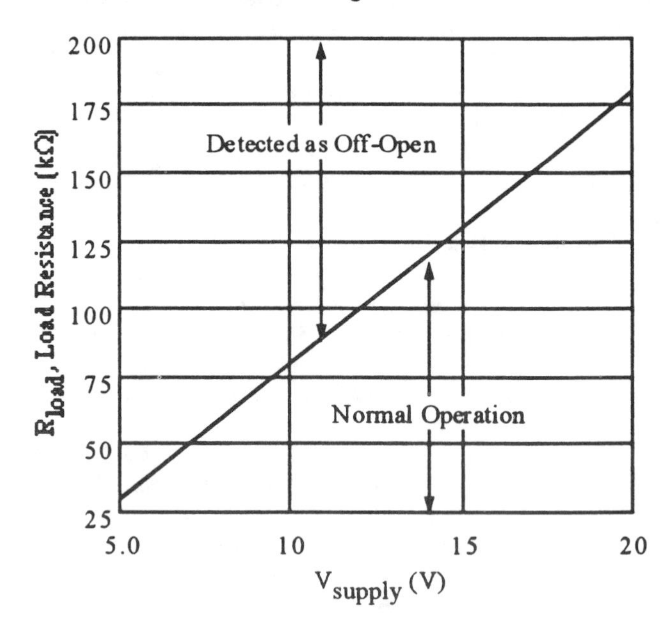

Figure 16: Load resistance reported as an "off" open for the DHSSA

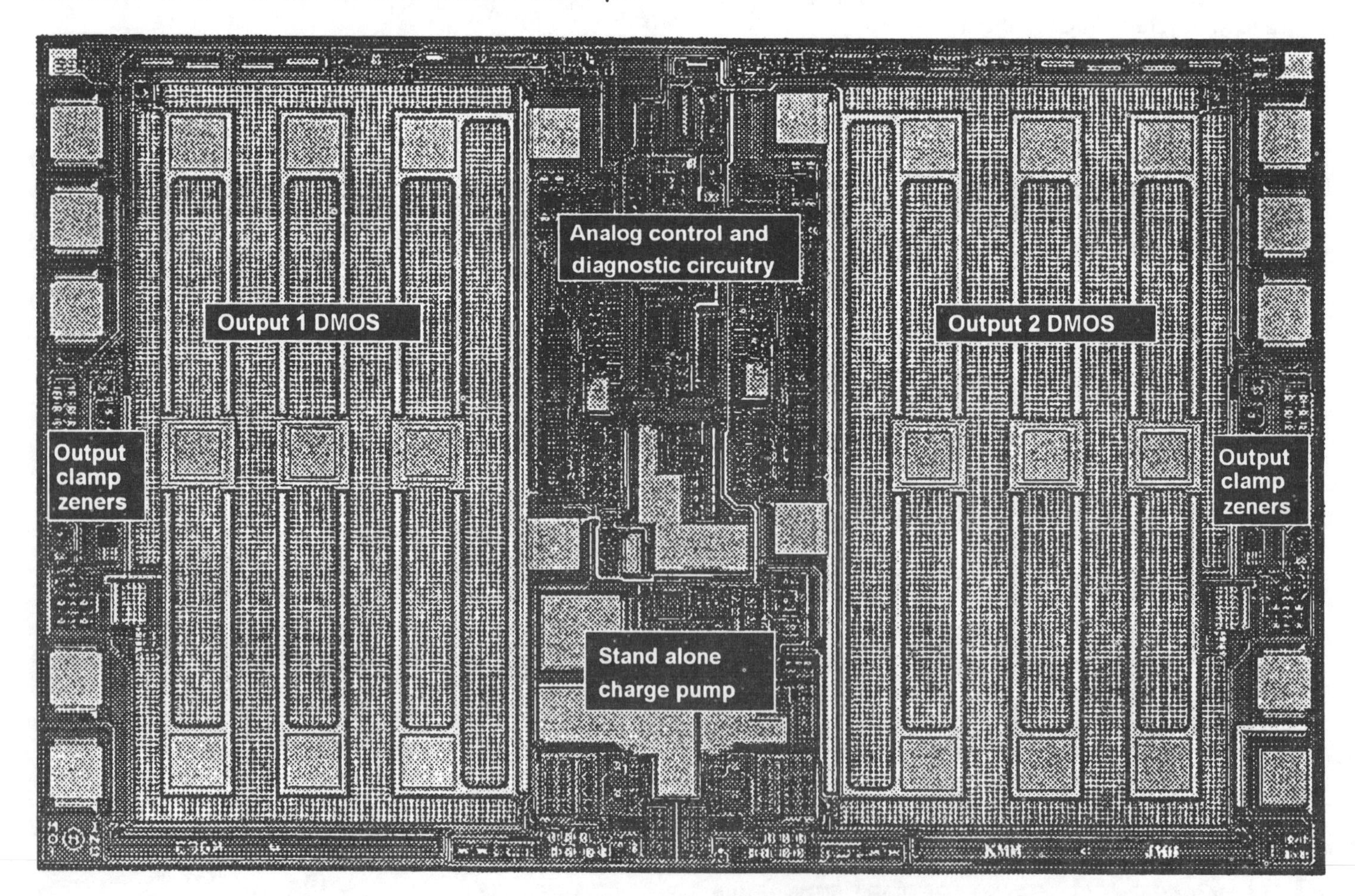

Figure 17: DHSSA die photograph

SUMMARY

The Powertrain Control Module environment requires power ICs withstand extreme voltage transients and have diagnostic capability. The QID, UPOD, DHSSA, and DHSSB utilize a mixed-mode process that enables them to meet rigorous PCM demands. Outstanding features of the ICs are reviewed in Table 1. All four of the ICs currently operate in PCM modules.

Integrated Circuit	PCM Application	Number of Outputs	RDSon per Output at Tj=25°C	Features
QID	fuel injectors transmission solenoids canister purge solenoids fuel pump relay incandescent bulbs	4	200 mΩ	current limit independent thermal limit on and off open load detection over-voltage protection high energy output clamp
UPOD	fuel injectors transmission solenoids canister purge solenoids fuel pump relay incandescent bulbs	6	200 mΩ	current limit independent thermal limit on and off open load detection over-voltage protection high energy output clamp two way SPI communication
DHSSA	fuel injectors transmission solenoids canister purge solenoids fuel pump relay incandescent bulbs	2	140 mΩ	current limit independent thermal limit on and off open load detection over-voltage protection high energy output clamp low noise charge pump
DHSSB	engine relays transmission relays	2	380 mΩ	current limit independent thermal limit on and off open load detection over-voltage protection high energy output clamp low noise charge pump

Table 1: Outstanding characteristics of the QID, UPOD, DHSAA, and DHSSA Powertrain Control Module ICs

REFERENCES

[1] Laurie Carney, Randy Frank et. al., "Recent Developments in Surface Mount Power IC Packages", Power Conversion and Intelligent Motion Proceedings, October, 1993, p. 300.

[2] Mario Paparo, William Javurek, et. al., "A Very Low Noise, μP Interfaced, Multiple High Side Driver with Self Diagnostics for Car Radio Application" presented at the SAE International Congress & Exposition, Detroit Michigan, February 28-March 3, 1994

[3] Jeff Kanner and Keith Wellnitz, "Power IC for Controlling Electronic Fuel Injectors" presented at the PCIM/POWER QUALITY '94 with Mass Transit System Compatibility '94 CONFERENCE & EXHIBIT, Dallas/Ft. Worth, Texas, September 17-22, 1994

[4] Ben Davis, Keith Wellnitz, and Randall Wollschlager, "A New Automotive SMARTMOS™ Octal Serial Switch", presented at IEEE Workshop on Power Electronics in Transportation, Dearborn, Michigan, October 22-23, 1992

960047

Programming of the Engine Control Unit by the C Language

Minoru Yoshida, Osamu Tada, and Joe Hashime
Daihatsu Motor Co., Ltd.

ABSTRACT

An engine control system turns complicatedly year by year to satisfy the requirements of the low pollution, low fuel consumption and high performance.

Though assembly language has been used for programming of the Engine Control Unit(ECU) so far ,we adopted the C language to improve the productivity of the software. Therefore it becomes possible to develop the high reliability program in a short period.

This paper introduces about the overview of ECU which employs high speed 16- bit microcomputer programmed by the C language.

SYSTEM OUTLINE

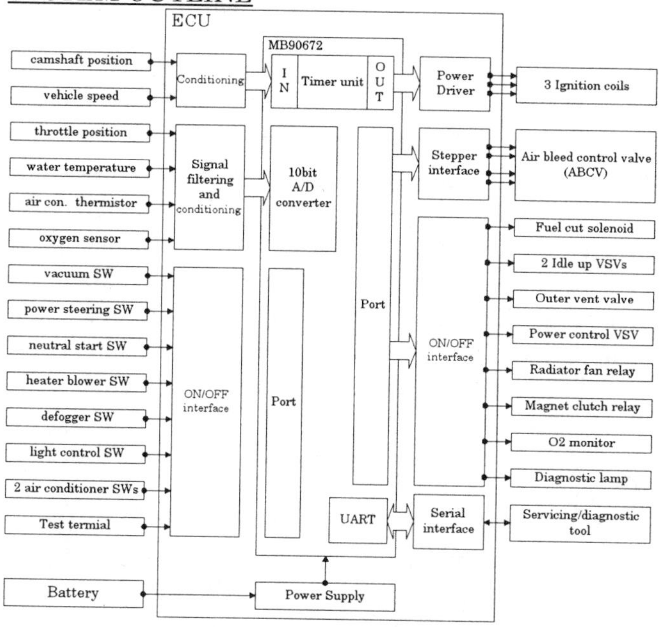

Figure1 ECU block diagram

Fig 1 shows the block diagram of the Engine Control Unit(ECU) which we have developed for our new mini-vehicle.

Mainly the ECU controls air-fuel ratio and 3 cylinders ignition system, further more carries out fuel cut control, idle up control, evaporating fuel control, air conditioner control, diagnosis, and so on.

Table 1 Specification of microcomputer

Internal bus frequency	16MHz
ROM	32Kbyte
RAM	1.64Kbyte
UART	2ch
A/D converter(10/8-bit)	8ch
24-bit free run timer	1ch
Input capture function	4ch
Output compare function	8ch
8-bit PPG timer	2ch
16-bit re-loadable timer	2ch

We have used many 8 bit microcomputers for our ECUs, and have used assembly language for programming. But at high engine revolution, the performance of 8-bit microcomputer is not enough because of its low bus frequency. To compensate this weak point, programmers have to write a very complex program.

In this development, we have used the following software techniques.

1) Adoption of the C language.
2) Adoption of the architecture which has layered structure.
3) Module programming.

The purposes which those techniques were used for are

1) Reduction of the development period.
2) Portability of the software.
3) Flexibility of the software.

However, in proportion to use these techniques, the quantity of the required memory increases, then the processing time becomes slow down. To solve these problems, we have employed the new 16-bit microcomputer.

Table 1 shows the specification of microcomputer. The system's core is the FUJITSU MB90672 integrated microcomputer that features a 16bit/16MHz CPU combined with powerful on board peripheral subsystems.

SOFTWARE ARCHITECTURE

Application layer

Logical value Throttle position, temperature,etc.
Switches ON/OFF status
Output devices ON/OFF command
Engine revolution
Vehicle speed
Spark advance
Diagnostic information

Input/Output layer

Interrupt layer

Analog sensors voltage registance
Switches port high/low status
Output devices port high/low

Hardware definition layer

Pulse sensor
Ignition pulse
Communication

ECU Hardware

VEHICLE
Sensor,Actuator,Communication

Figure 2 Software architecture

Fig 2 shows an example of software architecture, and it is similar to ISO/OSI model. The software architecture consists of four layers.

1) Hardware definition layer.
2) Input/Output layer.
3) Interrupt layer.
4) Application layer.

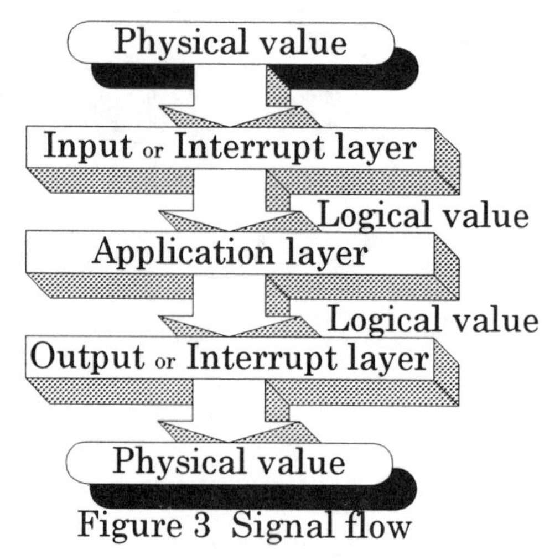

Figure 3 Signal flow

The signal flow of this software is shown in Fig 3. This diagram shows the fundamental flow to describe the function of engine control. The physical values (-i.e. A/D conversion value) are transferred in the input or interrupt layer, and changed into the logical values(-i.e. water temperature). In application layer, the application module calculates by using the logical values, and then transfers to output or interrupt layer, then the signal is generated from the port or timer unit of the microcomputer. By separating application layer from the process of input and output, it becames possible to write the application layer doesn't depend on hardware.

HARDWARE DEFINITION LAYER

This layer is not a program, but a definition body, and consists of two components.

1)IODEF.C,IODEF.H
(microcomputer internal I/O register is defined.)

2)PORTDEF.H
(the connection information of the port is defined.)

As it was divided into two components, the connection information of the port is often and sometimes changed in the middle of the development of ECU, but the definition of internal I/O register definition doesn't need to be modified so far as the same microcomputer is used.

As shown in the Fig 4,5, the registers and flags of the internal I/O are defined by using the unions and the structures(bit-fields) in IODEF.C and IODEF.H. A characteristic name is given to each flag by the macro substitution sentence (#define ...). It is made possible to access both by bit unit and by register unit.

As shown in the Fig 6, in the same way, the connection information of the port is defined by using the unions and the structures(bit-fields) in PORTDEF.H. Moreover, input or output, a logic level, initial value (only at output), and so on of each port are described.

As shown in the Fig 7, the unions and the structures are defined in another file STRUCT.H.

```
/*-------------------------------------------------------------------*/
/*                        Parallel Port                              */
/*-------------------------------------------------------------------*/
__io union flag8 PDR0;
__io union flag8 PDR1;
        -------
        -------
__io union flag8 ADER;
```

Figure 4 IODEF.C (A part is shown.)

```
extern __io union flag8 PDR0;
extern __io union flag8 PDR1;
        -------
        -------
extern __io union flag8 ADER;
#define  ADER_ADE0    ADER.bit.bit0
#define  ADER_ADE1    ADER.bit.bit1
        -------
        -------
```

Figure 5 IODEF.H (A part is shown.)

```
/*-------------------------------------------------------------------*/
/*         PORT 1 DEFINITION                                         */
/*-------------------------------------------------------------------*/
#define  P_IG        PDR1.bit.bit0
#define  P_ACSW      PDR1.bit.bit1
#define  P_ACEN      PDR1.bit.bit2
#define  P_DSW12     PDR1.bit.bit3
#define  P_DSW3      PDR1.bit.bit4
#define  P_PN        PDR1.bit.bit5
#define  P_VSW       PDR1.bit.bit6
#define  P_PST       PDR1.bit.bit7

#define  I_PORT1     0xFF
#define  O_PORT1     0
#define  QI_PORT1    BIT0+BIT1+BIT2+BIT3
#define  QO_PORT1    0
#define  DIR_PORT1   0
#define  INIT_PORT1  BIT4+BIT5+BIT6+BIT7
```

Figure 6 PORTDEF.H (A part is shown.)

```
struct flag8_s {                        /* 8 bits flag structures */
   unsigned char   bit0 : 1;            /* bit0 */
   unsigned char   bit1 : 1;            /* bit1 */
   unsigned char   bit2 : 1;            /* bit2 */
   unsigned char   bit3 : 1;            /* bit3 */
   unsigned char   bit4 : 1;            /* bit4 */
   unsigned char   bit5 : 1;            /* bit5 */
   unsigned char   bit6 : 1;            /* bit6 */
   unsigned char   bit7 : 1;            /* bit7 */
};
union flag8 {                           /* 8 bits flag union*/
   nsigned char    byte;                /* for byte access*/
   struct flag8_s  bit;                 /* for bit access*/
};
```

Figure 7 STRUCT.H (A part is shown.)

INPUT/OUTPUT LAYER

Many modules are contained in this layer to cope with the various kinds of input and output. This layer consists of four modules as follows.

- Analog input module.
- Pulse input module.
- Switch input module.
- Output module.

ANALOG INPUT MODULE

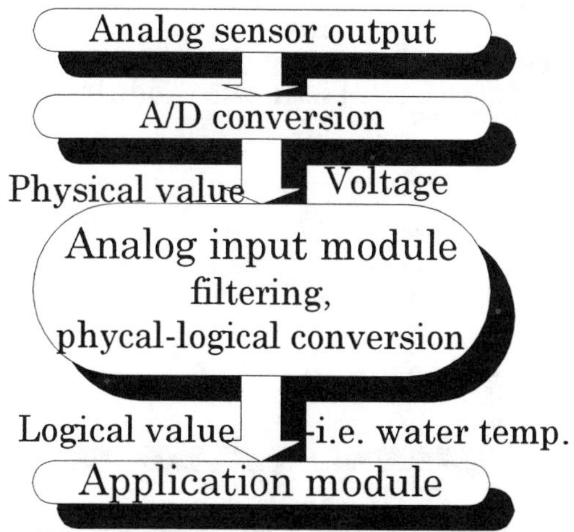

Figure 8 Analog input module

The concept of the analog input module is shown in Fig 8. A signal from the analog sensor is converted into the physical value by the A/D converter. In the input module, the physical value is averaged (if needed) and converted into the logical values. The signal from non-liner type sensor are converted in the method of the interpolation by searching the conversion table. An example of the conversion table for a non-liner type sensor(-i.e. water temperature sensor) is shown in Fig 9.

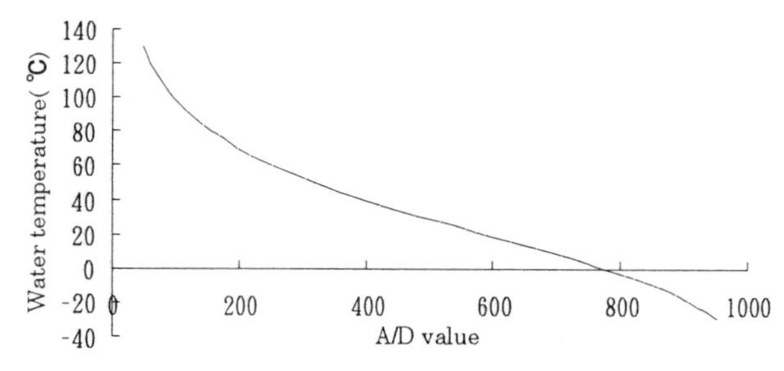

Figure 9 Non-liner sensor conversion table

It isn't decided by the characteristic of the sensor, and the application module can be computed by using the logical information. When the characteristic of the sensor is changed, only this table has to be modified.

PULSE INPUT MODULE

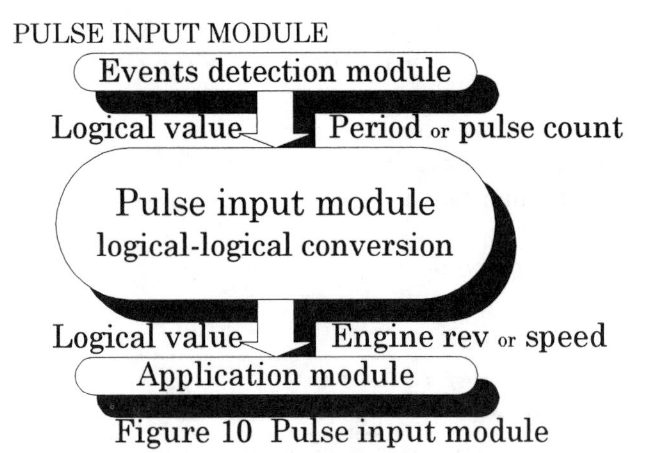

Figure 10 Pulse input module

The concept of the pulse input module is shown in Fig 10. The periodic or pulse count information from the events detection module in the interrupt layer is changed into the information about engine revolution or vehicle speed by the pulse input module.

SWITCH INPUT MODULE

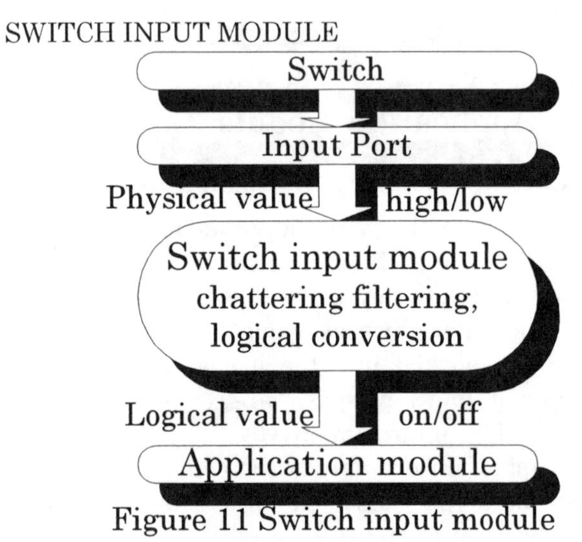

Figure 11 Switch input module

The concept of the switch input module is shown in Fig 11. A signal from the vehicle switch is converted into the physical value by the input port. In the input module, the signal passes chattering filter, and if the signal logic is negative, logic is reversed. Application module can get not the HIGH/LOW status of port but the ON/OFF information of the switch. This module absorbs those changes even if the position or the logic of the port is changed, by the change of the circuit of the ECU.

OUTPUT MODULE

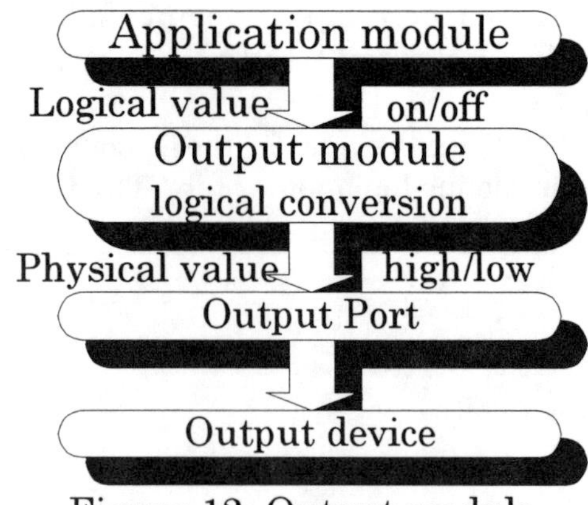

Figure 12 Output module

The concept of the output module is shown in Fig 12. Application module issues an ON/OFF command to output devices. In the output module, the command is converted into the port HIGH/LOW signal. If the port logic is negative, logic is reversed. The changes in the output circuits are absorbed by this module.

INTERRUPT LAYER

The interrupt layer consists of two modules.

·Events detection module.

(-i.e. camshaft position, vehicle speed etc.)

·Ignition management module.

These modules must be processed in real time, therefore the interrupt layer is in charge of the input and output function.

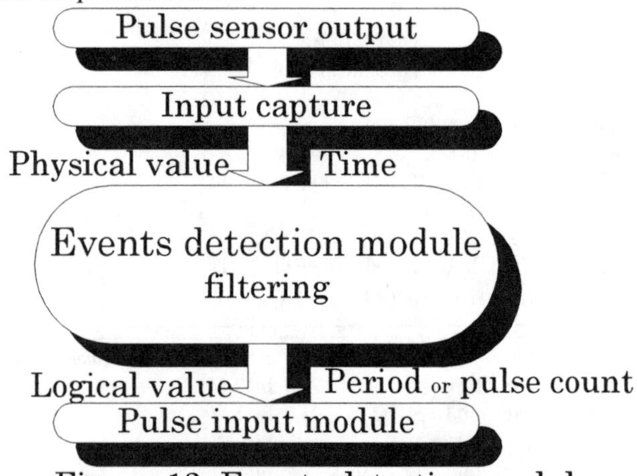

Figure 13 Events detection module

The concept of the events detection module is shown in Fig 13. The input capture of timer unit receives the pulse from sensor such as camshaft position sensor, and then in the events detection module, the pulse is changed into the information of the period or pulse count.

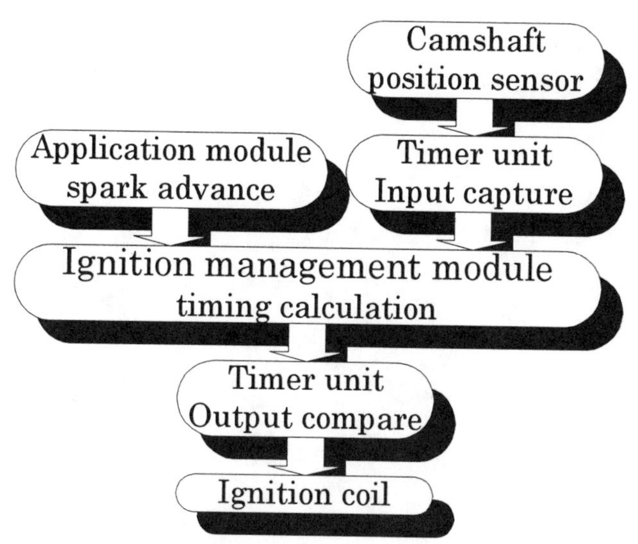

Figure 14 Ignition management module

The concept of the ignition management module is shown in Fig 14. So, the camshaft position (input), and the ignition timing (output) are complicatedly time-related, the structure of this module is very complex. The pulse from the camshaft position sensor is inputted into the timer unit (input capture). The time when the pulse was came in is known. The engine revolution position is recognized as a time information. From application module, the spark advance required by the engine is known as the angle information. From the information of the spark advance(angle) and the engine revolution period(time), the interrupt module converts the angle information into the time information. The calculation must be carried out with the dimension of the time for the spark advance management by the timer unit .From the two time information, the revolution position and the spark advance, the interrupt module calculates an appropriate time information for the ignition control and sends the ignition timing to the timer unit(output compare).

APPLICATION LAYER

In this layer, the module to realize a requirement function is described. This layer formed the functional center of ECU. The capacity of the object-code occupies more than whole 50%, too. One module copes with it toward one function, and each module is independent, and each module doesn't directly access the hardware.

REDUCTION OF THE DEVELOPMENT PERIOD

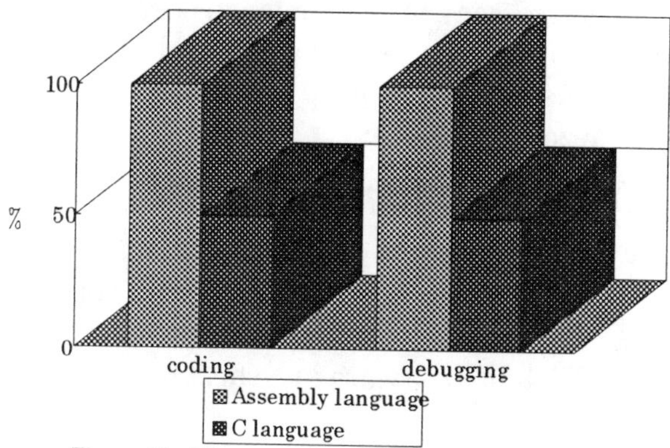

Figure 15 The comparison of the development period.

The main purpose to program by the C language is reduction of the development period. As a result of using a C language for programming, we managed to write the program easy to be debugged, easy to be modified and easy to be understood. The period to learn the assembly language which was characteristic of the microcomputer became unnecessary, and the period which it took to describe a program decreased about 50% of before, and as it was the same, the occurrence rate of the bug reduced, too. In Fig 15, the comparison of the development period is shown.(When assembly language is used, the period is 100%)

PORTABILITY and FLEXIBILITY

We have created the unique software structure, and the layer doesn't depend on hardware. This layer (named application layer) has only to be modified when a required function is changed, but there is no necessity for another changes.

Even if the change of the little scale occurs in ECU, it is possible to cope with the change only to modify the hardware definition module. The input/output module has only to be modified for the change in the comparatively large scale such as the addition of the sensor. It doesn't need to modify the application module that isn't influenced by the change.

If the necessity to use a different microcomputer with the same control contents arises, the hardware definition module, a part of the input/output module and a part of the interrupt module has to be modified. But, so far as the C language is used, it is no necessary to modify the application module.

If the necessity to use the same microcomputer with another control contents arises, though it is necessary to modify, it is possible to use the hardware definition module and the input/output module.

If the necessity to use a different microcomputer with another control contents arises, the some modules which have already been developed are possible to use. It doesn't need to develop the software at all from zero.

CONCLUSION

We have already developed two kinds of ECU in such a method. Compared with the software by using the assembly language, we have managed to develop the high reliability software in a short period.

REFERENCES

(1)G.Mercalli, P.Mortra, “A New Automotive Real Time Engine Control System with 32 Bit Micro-controller”. XXV FISITA CONGRESS, Vehicle systems and components, 17-21 October 1994 Beijing.

(2)K.Sato, O.Shiroshita, K.Asami, J.Nakano, “A Study on a Method for Developing Control Programs by using C-Language”. Journal of the Society of Automotive Engineers of Japan, Vol.48 August 1994.

960332

Single Chip Power System for Engine Controllers

William E. Edwards, Eric Danstrom, and Mitchell Belser
SGS-Thomson Microelectronics Inc.

Abstract

The U552 is a single chip, switched mode power supply (SMPS) for automotive engine controllers. It is a complete power supply for microprocessor based systems and includes several support functions to provide improved efficiency, higher reliability, size reduction, decreased parts count and significant cost savings over other implementations. The current mode SMPS runs at 100KHz via an on chip oscillator with an accuracy of +/-10%. Noise suppression and supply rejection are achieved by powering all linear and digital blocks from an internally pre-regulated 5V supply. A precision reference provides 2% output voltage accuracy from -40 to +150 degrees C. Connected directly to the battery, a low quiescent current, stand-by regulator powers SRAM containing on-board diagnostics (OBD). Two 100mA, linear tracking regulators are on chip to drive sensors such as manifold absolute pressure (MAP) and throttle position. These regulators will track a reference voltage such as the A/D's power supply within +/-10mV over all temperature and load conditions eliminating system gain errors and provides fault isolation between sensors. The combination of enable/disable hysteresis and power-on reset with a 20mS delay ensures well orchestrated, glitch-free power sequencing of the microprocessor. To meet all automotive requirements (temperature, voltage, etc.), this system on a chip maintains full performance during 50V load dump, provides 4KV ESD protection to all pins and has short circuit protection on all outputs.

Introduction

The electronic content of automobiles is continually increasing. This increase is driven by four primary factors: more stringent emission controls, engine performance, safety and increased consumer expectations of reliability and functionality. Subsequently, more powerful processors are required to provide the real-time computational capacity necessary to monitor and control the ignition system, emissions sensors, external actuators and diagnostics. This increase in microprocessor capability requires similar performance improvements in the controller's power supply. These include higher load currents and increased electrical isolation from the noisy and harsh environment of the battery and alternator. Engine controller power management systems must also satisfy requirements from sales, system integration & assembly and reliability. This translates into a power management system with a very high level of integration to achieve the goals of low cost, minimal parts count and size while delivering improved electrical performance and reliability.

To meet these new requirements, a new combinational SMPS has been developed. A combinational SMPS is an evolutionary development where the SMPS controller, power switch and other power or support functions are realized on a single device. To integrate the previously incompatible circuit functions of power

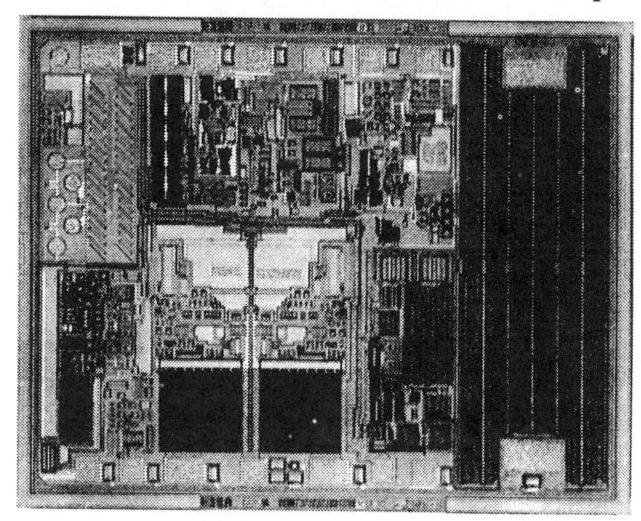

figure 1

switching, precise linear blocks and combinational logic, the U552 (figure 1) was developed in SGS-Thomson's BCD process (Bipolar, CMOS and DMOS) .

Process highlights include 60V vertical DMOS @ .5 ohms - sq. mm, 70V PMOS FETs, vertical NPN and lateral PNP bipolar transistors, 2.5um CMOS and a single or double metal interconnect system.

The U552 (figure 2) has eight (8) functional blocks:

1. SMPS
2. Stand-by regulator
3. Tracking regulator #1
4. Tracking regulator #2
5. Enable/Disable block
6. Power on reset (POR)
7. 2.5V precision reference
8. 5V pre-regulator.

Circuit Description

SMPS

The current-mode SMPS can be used in either the buck-boost or flyback topologies and is similar in function to the industry standard UC3842.

The Set/Reset circuit (SR) performs four (4) functions:

1. SR flip-flop
2. 50KHz clock signal to a 100KHz pulse train conversion
3. Pulse blanking during transformer flyback
4. Under-voltage lockout.

Setting and resetting the SR flip-flop turns the switch on and off thereby supplying the magnetics with current to charge the output capacitor. Pulse width modulation (PWM) is generated by the feedback loop of the comparator and the error amplifier sensing the DMOS switch drain voltage. During switch turn-on, information from the feedback loop is invalid and must be blanked to obtain proper operation of the SMPS. The under-voltage lockout circuit uses a "broken bandgap" architecture to monitor the battery. The SMPS is disabled whenever the battery voltage falls below 5V. The under-voltage lockout comparator employs hysteresis to avoid oscillations during periods of marginal battery voltage such as cold weather cranking or battery failure. These features ensure the SMPS has adequate voltage to operate thereby eliminating indeterminate output voltage states and possible microprocessor errors caused by low battery.

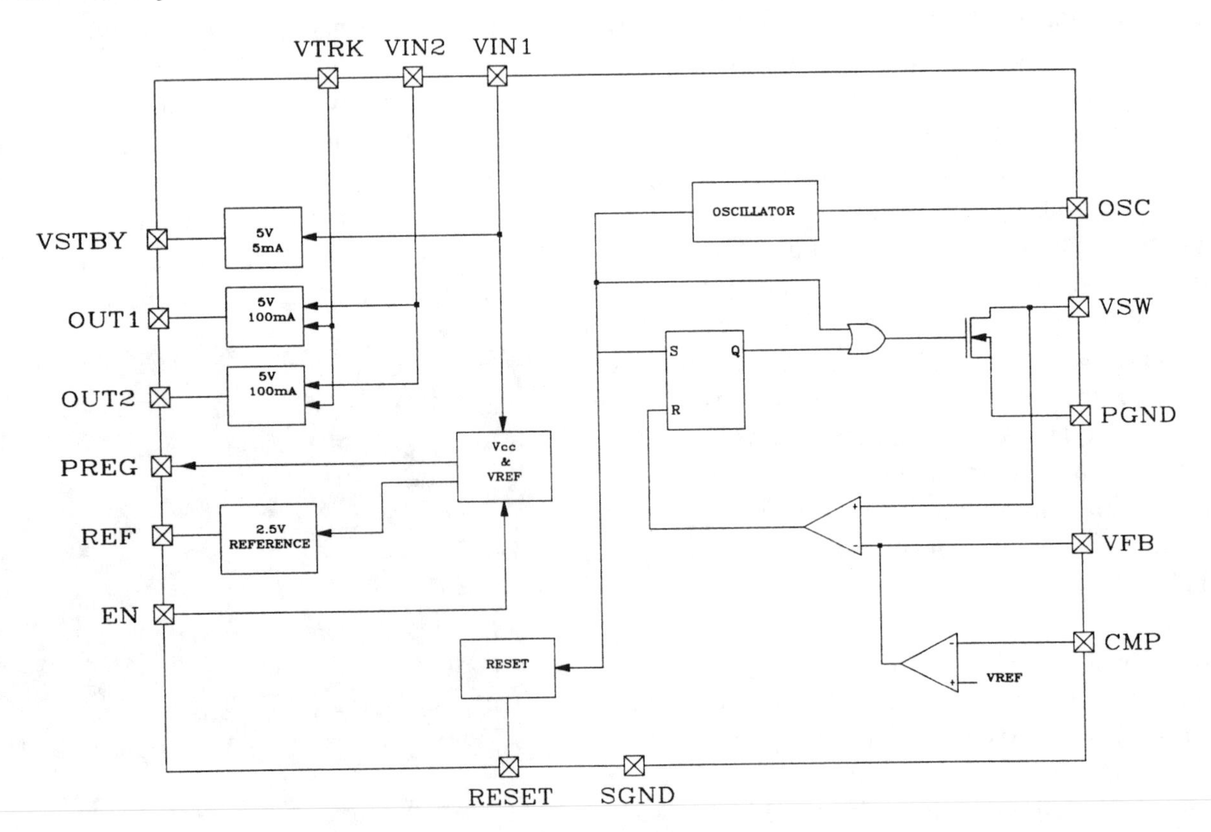

figure 2

The oscillator uses an LM555 topology (figure 3). An off-chip resistor and capacitor set the frequency of oscillation with an accuracy of +/-10% from -40 to +150 degrees C. The 100KHz sawtooth signal generated by the charging and discharging of the capacitor drives a D flip-flop to produce a 50KHz square wave with an associated period of 20uS. The 0 to 5V square wave forms the on-chip clock signal used to drive the SMPS and the power on reset delay circuit.

The Gate Drive circuit buffers the output of the Set/Reset circuit and provides the necessary drive to charge and discharge the gate of the DMOS switch. It uses active clamping from gate to source on power MOS circuit elements to protect the gate oxide and allow continuous operation during load dump of up to 50V.

The 6A DMOS switch is connected in the common source configuration and has a maximum on resistance of 0.2 ohms. The standard current sense resistor is eliminated; output current is directly sensed at the drain of the DMOS transistor. Resistorless load current sensing combined with no DC input current (only AC current to charge and discharge the gate of the DMOS switch) provide a significant increase in conversion efficiency over bipolar topologies.

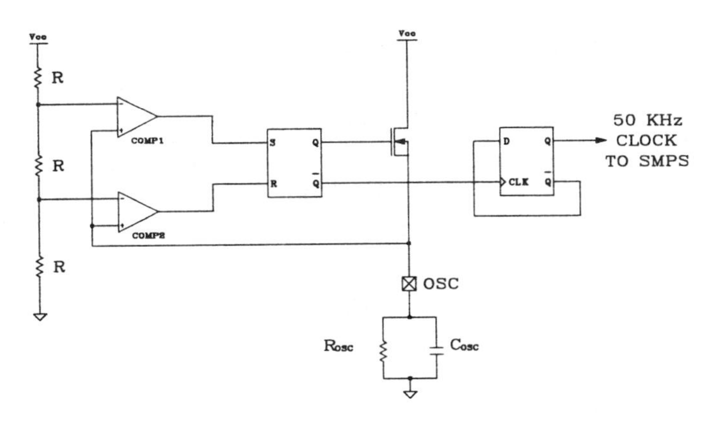

Figure 3

Common Circuit Blocks

The pre-regulator (Preg) runs off the battery and provides a regulated 5V supply for all the chip except gate drive, DMOS switch, stand-by regulator, and the tracking regulators. Requiring only 0.1uF of load capacitance, the pre-regulator provides supply rejection in excess of 65dB and protects the circuits from load dump. This results in reduced susceptibility to transients produced by the charging system and simplifies circuit design. To further increase system efficiency, a novel dual output stage topology is used (figure 4). The output stages are powered by the battery and the second winding of the transformer's secondary (Vin2) respectively; only one output is active at any given time. Under normal operation, the output stage supplied by the secondary winding is used. The battery supplied output stage is used whenever the secondary voltage is less than 8V.

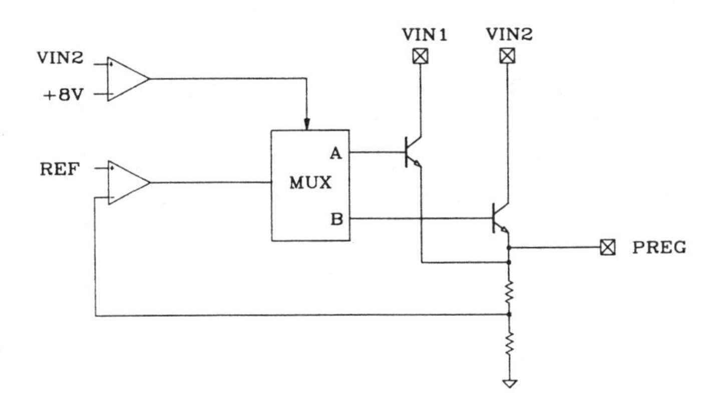

Figure 4

The power on reset circuit (figure 5) uses an n channel FET to hold the microprocessor in reset. A maximum of 0.4V is maintained at the reset pin while sinking 1mA. The reset comparator (rcomp) incorporates both offset and hysteresis to set the reset trip point at SMPS = 4.5V and eliminate oscillations during ramp up/down of the SMPS. The microprocessor is enabled 20mS after the output of the SMPS is valid (>4.75V). The reset delay is generated by a 10 bit counter and combinational logic. When the SMPS exceeds 4.75V the counter is enabled and clocked at 50KHz. After counting to 1000, the output n channel FET is turned off releasing the microprocessor's reset pin.

The precision reference is a band gap circuit. Based on the Brokaw cell it has +/-2% accuracy and provides a temperature invariant reference for the SMPS, reset threshold, the 2.5V external reference and the pre-regulator's output stage switching network.

Ancillary Circuits

The stand-by regulator (VSTBY) is a low quiescent current regulator. Running off the battery, VSTBY powers the SRAM that stores OBD data for such events as engine misfires, anti-lock brake (ABS) faults and California Air Resources Board (CARB). This regulator has a 1V drop out at 5mA load current and only requires 0.1uF for stability. When the SMPS is disabled, total current drain from the battery is typically 400uA.

There are two (2) identical tracking regulators included on chip for driving external sensors such as manifold absolute pressure (MAP), Barometric Absolute Pressure (BAP), lambda (O2) sensor, miscellaneous temperature sensors and throttle position. Output voltage accuracy is Vtrack +/-10mV. With only 0.1uF of load capacitance, each regulator can supply 0 to 100mA. In the system, the tracking regulator is referenced to the A/D's supply voltage. This provides sensor isolation and eliminates signal point failures. Because of the regulator's accuracy, system calibration to remove gain errors is not required .

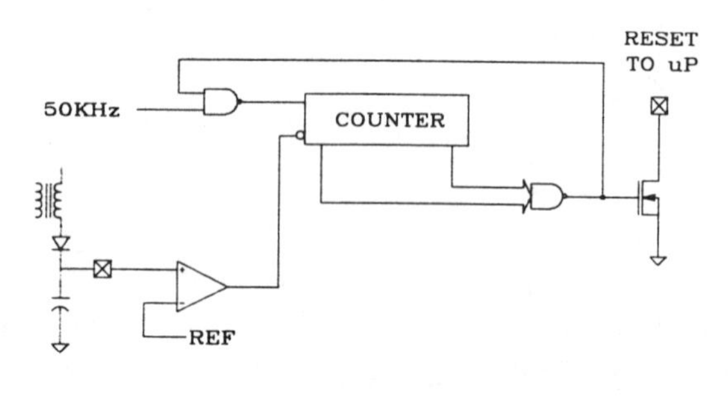

figure 5

A 2.5V precision reference is provided for the analog to digital converter (A/D) for the absolute measurement of voltage (battery, fuel sensor etc.) and current (ignition coil, injectors etc.). The reference is derived from the bandgap by a high gain, low offset operational amplifier configured with a non-inverting gain of 2.

Technology Comparison

While any individual parameter of the U552's performance can and has been achieved elsewhere either discretely or at the board level, the full benefit of this device can only be understood when compared to other approaches in the context of the application . The criteria by which the different approaches are judged should include:

- Electrical performance
- Thermal performance
- Total cost of the function/ease of integration
  - assembly, heat sink, test, parts count
- Reliability
  - Number of interconnects and parts
- Overall size.

Below is a comparison of three power supplies; a discrete SMPS, a linear regulator and a monolithic SMPS (U552).

Discreet Implementations

Function	Component	Pins
SMPS	controller	14
	power switch	3
	sense resistor	2
Stand-by regulator	3 terminal regulator	3
Tracking reg. (2)	dual power op-amp	8
	(2) capacitors	4
	(6) resistors	12
Power-on reset	comparator	6
	resistor	2
	capacitor	2
Total	*16 components*	*56 pins*

note: magnetics and support components same as U552

Function	Component	Pins
Linear regulator	3 terminal regulator	3
Stand-by regulator	3 terminal regulator	3
Tracking reg. (2)	dual power op-amp	8
	(2) capacitors	4
	(6) resistors	12
Power-on rese	comparator	6
	resistor	2
	capacitor	2
Total	*14 components*	*40 pins*

note: no magnetics
increased heat sink for linear regulator

Function	Component	Pins
SMPS	U552	20
Stand-by regulator		
Tracking reg. (2)		
Power-on reset		
Total	*1 component*	*20 pins*

The above comparison illustrates the advantages of a combinational SMPS over discreet implementations, namely reduced parts count, fewer interconnections and the subsequent improvements in reliability and assembly costs. The benefits this approach provides to the vehicle management system (sensors, power distribution, signal measurement and software both operational algorithms and calibration routines) are due to the circuit architecture and the high level of device matching and tracking inherent in integrated circuits. Key to the architecture is the use of a single, precision reference and the highly accurate tracking regulators/sensor drivers (figure 6). The band gap sets the accuracy and temperature stability for the SMPS, 2.5V reference and over/under voltage thresholds. Sensors for temperature, MAP, BAP and throttle position are powered by the tracking regulators which use the same reference as the A/D converter. This configuration ensures accuracy, temperature stability and temperature tracking of the entire vehicle management system. In contrast a discrete implementation (figure 7), requires either highly accurate and expensive components, trimming or software to compensate for gain and temperature variations between multiple components.

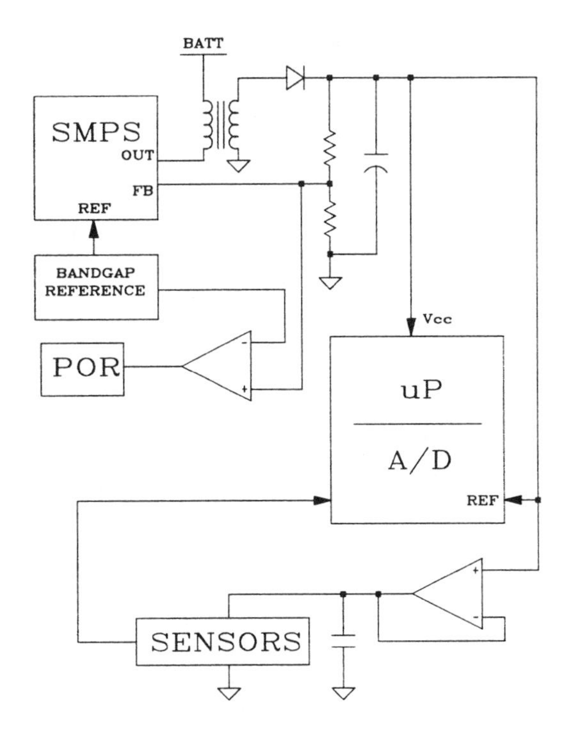

figure 6

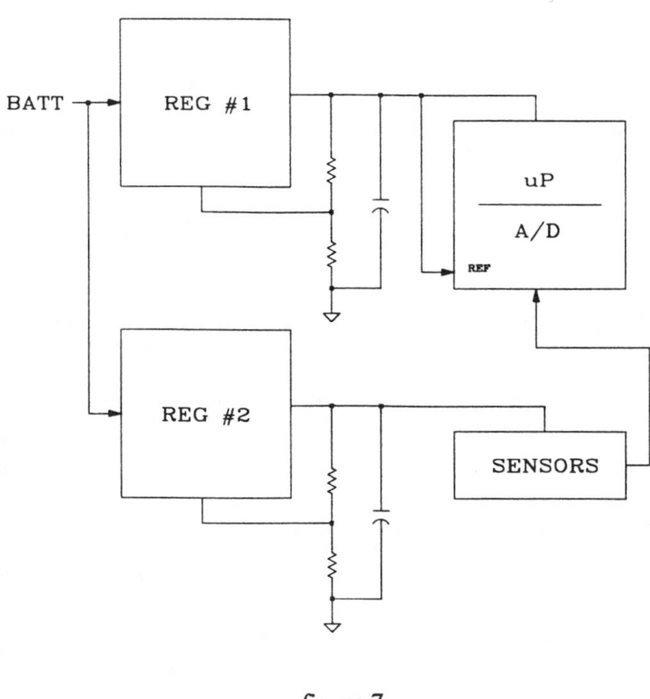

figure 7

Summary

In keeping with the trend of increased integration and electrical performance, the single chip power system presented is natural step in power management system development. Balancing cost, system partitioning, electrical performance and reliability, the U552 is an attempt to meet the many and sometimes conflicting requirements of the automotive market. Made possible by the advancements in mixed-power IC technologies and customer/vendor cooperation, the device is the next step, but not the last, in integrated power management systems.

See next page for test waveforms.

The author wishes to thank A. Hennigan and D. Swanson.

Test waveforms

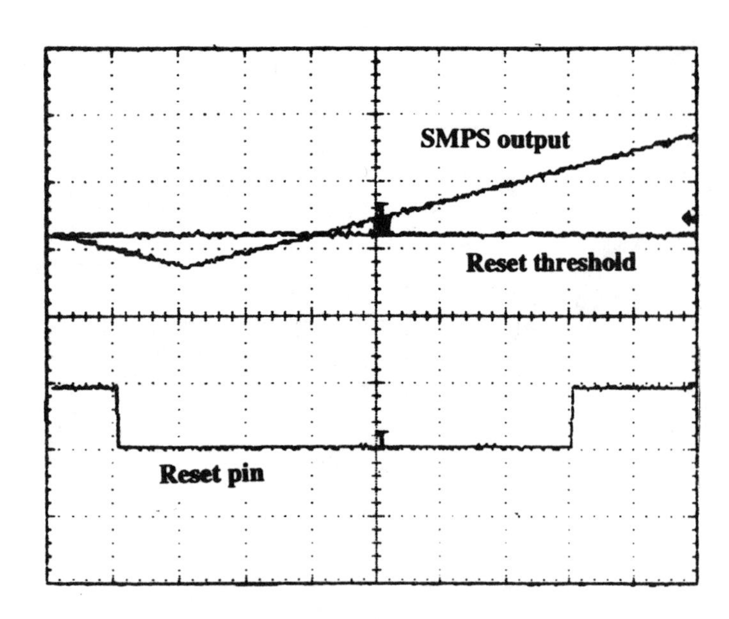

figure 8
Power on reset with 20 mS delay and hysteresis

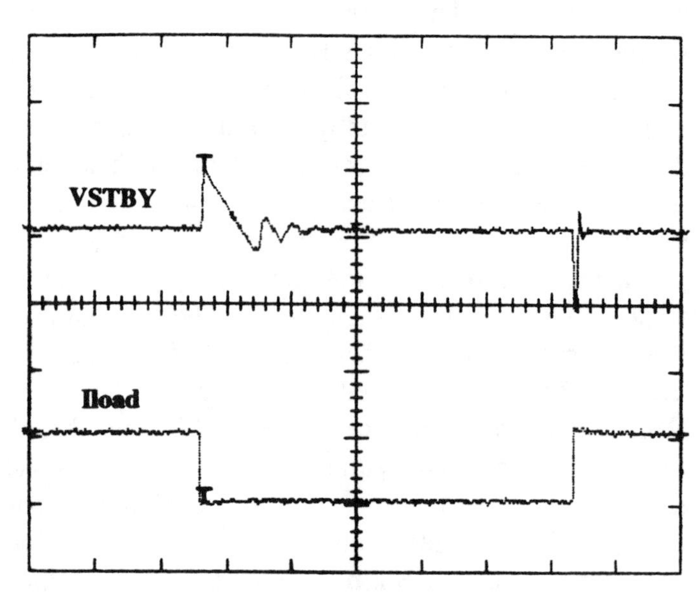

figure 10
Standby regulator
Transient response Iload 0-->5mA-->0
Cload =0.1uF

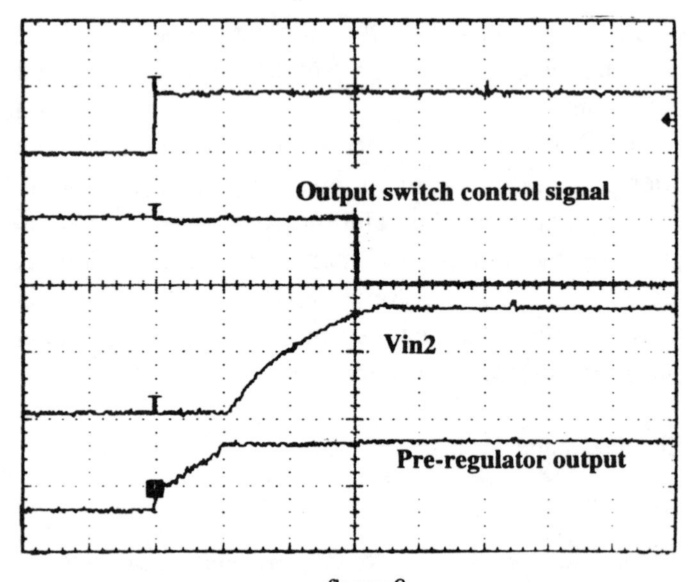

figure 9
Pre-regulator output stage switching

980831

Development of a Shift By Wire Synchronized 5-Speed Manual Transmission

Eric A. Bansbach
New Venture Gear, Inc.

ABSTRACT

This paper describes the modification of a rear wheel drive manual transmission for the purpose of shifting using electromechanical actuators under microprocessor control. This Shift By Wire (SBW) technology has applications in automated manual transmission systems and shift assist systems. The system is optimized for shift comfort and shift speed. A test vehicle was outfitted with the system and evaluated. The development concerns for this type of program are discussed and prototype results of the undertaken program are presented.

BACKGROUND

Electronics have been integrated into many automotive systems ever since the benefits of these systems were realized. In the drivetrain area, electronics have greatly enhanced engine control systems, automatic transmission systems (hydraulic transmissions) and four wheel drive systems.

Past history has also shown the application of electronics into clutch systems. Many of these systems initially were unsuccessful due to the lack of power in the controls area. Since the advent of the microprocessor and other advanced electronic technologies, many systems that were formerly not technologically feasible, are now realizable. Automatic clutch systems are now production items on several European small car platforms.

The manual transmission has also been the target of the electronics revolution. While purely mechanical manual transmissions are still dominant, some automated manual transmissions are making appearances into the marketplace. Most of this development has centered in Europe where the driving factors are most prevalent. These factors include fuel economy, emissions and transmission cost.

SBW SYSTEM OBJECTIVES

The goal of this transmission development program was to integrate into a vehicle, a transmission whose shift actuation was entirely electronically controlled. There was to be no mechanical link between the driver and the gearbox (clutch system not included). Shifting of the gearbox was to be commanded by an electronic shift lever accessible to the driver, or alternately through pushbuttons on the steering wheel which commanded an up or down shift. The advantages of such a system are reduced noise and vibration in the passenger compartment, greater control of gear engagement, increased protection from driver error, easier tailoring of shift lever feel and effort, easier location of the shift lever, and elimination of the shift tower components. This technology is also the basis for developing an automated manual transmission (SBW with clutch control).

The transmission used was a New Venture Gear model NV3500 manual transmission. This is a single rail gearbox with all forward speeds synchronized. The target production vehicles for this gearbox include rear wheel drive cars, light trucks and utility vehicles. The test vehicle used for this program was a Chevrolet 6 cylinder, 2WD pickup truck. The single rail design of the 3500 allowed for an easier method for actuator attachment. A dual motion (linear/rotary) actuator is connected directly to the shift rail.

SYSTEM OVERVIEW

The complete system requires an electronics control module, a dual-motion motor actuator, a variety of sensors, electronic shift lever (or pushbuttons on the steering wheel), vehicle displays, RS-232 serial interface (to remote computer) and numerous vehicle inputs.

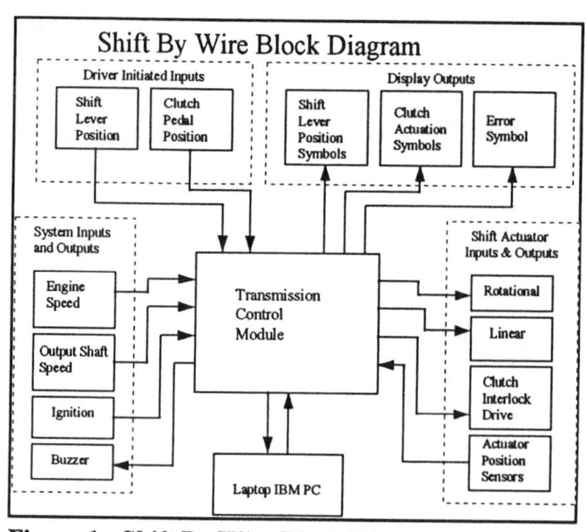

Figure 1 - Shift By Wire Block Diagram

CONTROL SYSTEM

The controls for the system addresses actuator control, shift control, shift force calculations and protection routines.

ACTUATOR CONTROL

The actuator used in this program is a dual motion linear/rotary motor actuator. The rotary force output is rated for a relatively low level since the gearbox rail crossover motion (motion between the shift planes) is of a low force level. The linear component of the actuator has a higher force potential to accommodate the higher forces required during gear synchronization and engagement.

SHIFT CONTROL

An electronically outfitted joystick shift lever is the primary shift command mechanism, and is used by the driver to select the desire gear. The shift lever is mechanically isolated from the transmission, allowing the lever to be placed anywhere within the vehicle compartment. By placing position sensors within the lever assembly, full electronic tracking of the lever position is monitored by the ECM. Mechanical gating on the shift lever limits the movement to that of the double "H" pattern typical on today's manual transmissions. The shift lever will be free of the sometimes high shift forces and also free from miscellaneous vibrations that permeate through the shift tower into the shift lever such as nibble, double bump, clash and other general powertrain vibrations. By imitation the same type shift method as a mechanical manual transmission (double H pattern shift lever) allows the driver to maintain the same apparent control as the standard manual transmission with no driving adjustments required.

A secondary gear selection option is through pushbuttons on the steering wheel. One set of buttons initiates a shift up to the next higher gear, and a second set initiates a shift to the next lower gear. A separate Neutral/Reverse switch assembly contains two switches. Pressing the neutral switch will engage neutral when depressed once, and pressing the reverse switch twice will engage reverse.

The clutch pedal operates in the same manner as with a mechanical manual transmission. For the driver to shift to the next desired gear, the clutch pedal must be fully depressed. If the shift lever is moved into a new gear with the clutch not disengaged, the gearbox will remain in the original gear with the dashboard display indicating this. This is called preselecting the next gear shift (see Shifting Preselect below). The system then waits for the clutch pedal to be depressed. Disengaging the clutch then initiates the gear shift. Once the clutch is disengaged, the shift will continue per the command from the shift lever. The clutch must remain disengaged until the shift is complete. Failure to do so will cause the clutch interlock to intervene (see below) for protection of the clutch.

From the driver's perspective, important aspects of performing a complete shift (from one gear through neutral into the next selected gear) include shift time, shift perception and NVH. For this purely SBW system to be acceptable to the driver, the shift times must be equal to or faster than a normal driver operated mechanical shift. That is, when the driver moves the shift lever, the gearbox must follow, without delay, according to the lever's motion.

A target of 400 milliseconds maximum (worst case) was initially set to accomplish this feat. This total time was broken down into the main subcomponents:

Maximum Shift Breakdown Times	
Shift Out of Gear to Neutral:	70 msec
Crossover (If Needed):	50 msec
Neutral to Synchronization Start:	20 msec
Synchronization of Selected Gear:	200msec
Engagement of Selected Gear:	60 msec

Figure 2 - Table Showing Breakdown of Worst Case, Target Shift Sequence Times.

The synchronization time of 200ms is a worst case figure, and is seen during a 1st to 2nd gear shift during hard acceleration. Most synchronization times were faster than 200 milliseconds.

Shift perception includes not only the perceived time for a shift, but also the pleasibility of the shift. Synchronization times much faster can be achieved with full applied synchronization forces, but consequences may include various harsh engagement sounds and synchronizer abuse. The vehicle conditions present during the shift determine the shift force applied.

PROTECTION ROUTINES

With such an automated system, protection measures are easily implemented into the control software. The primary concern for a manually initiated shift system is shifting into gears when conditions are not safe, either for the driver or the mechanics of the gearbox. With powerful actuators, the driver has little recourse once a shift has been commanded.

Down shifts are a major cause of gearbox component (including clutch) abuse and excessive wear. For an example, a downshift from 5th gear to 1st at 55mph would cause clutch failure, and possibly overheating of the synchronizers. The SBW ECM monitors vehicle speeds and selected gears to ensure that unsafe shifts are never initiated, thereby overriding shift lever commands.

For this system, the ECM will only allow engagement of the following gears when the transmission output shaft is within the following ranges.

Safe Gear Engagement Range (Output Shaft RPMs)

1st Gear:	0 to 1250 rpms
2nd Gear:	0 to 2150 rpms
3rd Gear:	575 to 3575 rpms
4th Gear:	800 to 5000 rpms
5th Gear:	1100 to 6850 rpms
Reverse Gear:	0 to 300 rpms

Figure 3 - Chart of Output Shaft RPMs For Safe Gear Engagement.

The electronic controls also allow for a reverse gear blockout option without the need for additional mechanical components. The traditional spring and cam components are no longer required. This function is accomplished by the ECM monitoring the vehicle speed and shaft speeds of the transmission. If the vehicle is moving forward with a rear output shaft of greater than 300 rpm, then the control logic does not allow the actuator to engage the reverse gear.

Some transaxles and transmissions do not have synchronizers on the reverse gear, mainly as a cost saving measure. Consequently, when shifting quickly from a rolling stop into reverse, clash will occur due to the output shaft still rotating slightly or due to the input shaft still rotating slightly. The SBW controls track both shaft speeds (transmission input and output) to ensure the shift into reverse is not made during these clash conditions.

ADDITIONAL DEVELOPMENT CONSIDERATIONS

SHIFT FORCE

The ECM (Electronic Control Module) calculates the optimum shift force required for the actuators to synchronize the gear within specified times. The standard formula used in the calculation is:

$$F_s = S1 - S2 \frac{I_r \times \sin\phi \times (0.1047 \, rad/sec\text{-}rpm)}{t_s \times \mu \times R_c} K_t \qquad (1)$$

where;
F_s = Synchronization Force
I_r = Reflected Inertia of Selected Gear
μ = Dynamic Coefficient of Friction
Φ = Synchronizer Cone Angle
R_c = Synchronizer Mean Cone Radius
S1 = Selected Gear Speed Before Synchronization (Input Shaft Speed Divided By Gear Ratio)
S2 = Selected Gear Speed After Synchronization (Output Shaft Speed)
K_t = Temperature Compensation Term
t_s = Desired Synchronization Time

For the NV3500 manual transmission used in this program, the synchronizer cone angle Φ, is 6.5 degrees. R_c for the five synchronized gears is 41.61mm. Depending upon the vehicle condition and the desired operating mode, the value of t_s can vary from the maximum 200 milliseconds down to the minimum which is set by a combination of the synchronizer material properties and shift engagement NVH.

K_t is a linear variable that is inversely proportional to the oil temperature in the gearbox. The viscosity of the oil increases with decreasing temperature, thereby slowing the synchronization times.

To obtain a formula that results in the total shift time given the programmed shift force, equation (1) can be solved for t_s as follows:

$$t_s = \frac{I_r \times \sin\Phi \times (S2 - S1) \times K_t}{\mu \times R_c \times F_s} \qquad (2)$$

When optimizing shift times for speed, this equation can be used with the programmed shift force F_s to calculate the total synchronization times.

Figure 4 shows a typical shift force used to shift the gearbox into neutral from an in-gear position. Note that this shift force is a constant figure over the entire distance from the in-gear position to neutral. This is due to the light load seen during this type of shift.

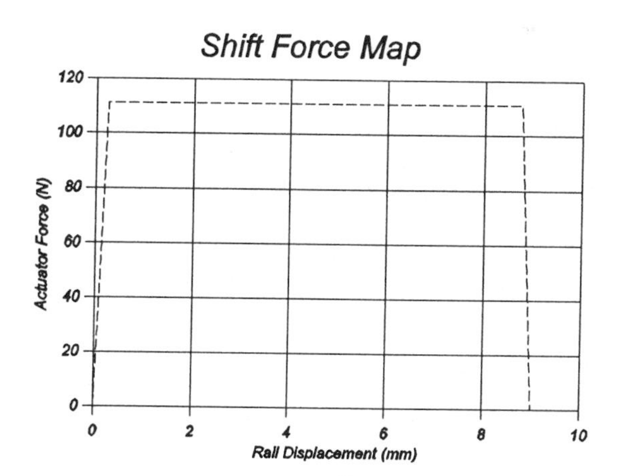

Figure 4 - Actuator Shift Force Map Detailing Applied Forces During a In-Gear to Neutral Shift.

Figure 5 shows the shift force map used when shifting from neutral to an in-gear position.

While the force level changes according to the vehicle and gearbox conditions, the shape of the curve is consistent for all synchronized shifts. During the first

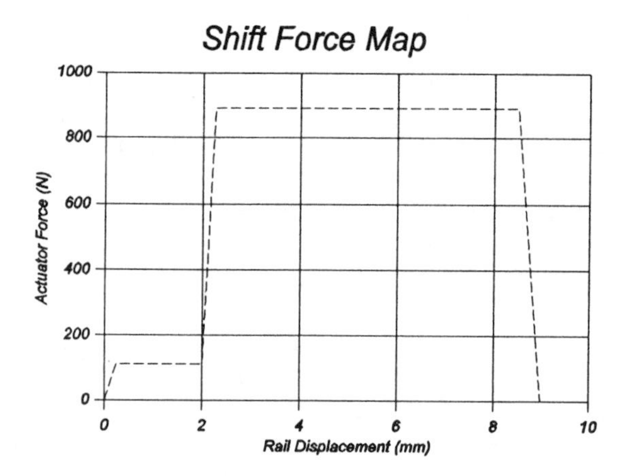

Figure 5 - Actuator Shift Force Map Detailing Applied Forces During a Neutral to In-Gear Shift.

2 mm of rail travel, the actuator moves the shift sleeve and synchronizers to take up synchronizer clearance. At this time, the calculated synchronizer force is used throughout the completion of the shift. The force is scaled back near the end of travel to avoid running into the end of travel on the shift rail. The synchronization force seen in Fig. 4 (between 2mm and 8.25mm), is calculated by the microcontroller before each shift. The proportional amount of full power available from the actuator is then applied to the actuator using simple PWM (pulse width modulation) of the actuator drive. Full actuator power (100% duty cycle) is never used on this system.

SYNCHRONIZERS

The synchronizer design must be carefully reviewed in any automated shifting scheme. This is especially important for add-on shift by wire or automated manual applications. While a gearbox for normal mechanical shifting is designed for specified typical and worst case shift forces created by the driver and shift lever, the addition of hydraulic or electromechanical actuators may introduce shift forces above and beyond those seen by the original design. Failure to properly analyze all potential shift forces can lead to overheating of the friction material.

The ideal shift times for a shift by wire, or automated manual transmission would be as small as possible, <50ms for example. This would result in a system where there would be no perceived delay in how the transmission responds to driver commands given through the shift lever (or through pushbuttons on the steering wheel). In fully automated manual transmission systems, torque interruptions are very noticeable, especially compared with a conventional automatic transmission. To approach these somewhat ideal shift times, the synchronization times (only a part of the total shift time) would need to be very small. Using a powerful actuator (hydraulic) might accomplish this goal, but with higher shift loads.

ACTUATORS: ELECTROMECHANICAL VS. HYDRAULIC

The questions of using motor actuators or hydraulically powered actuators may forever remain a subject of debate. There are still two schools of thought and each side's arguments often seem to conflict with each other. The issues of contention are cost, size, weight, average power consumption and system preference.

To properly compare hydraulic systems with electromechanical systems, one must review all components within the system, and system effects. Hydraulic systems require not only the actuators, but the pump system and reservoir (accumulator) hardware. Traditionally, the hydraulic systems have taken up more space within the engine compartment than a electromechanical system. The actuators themselves however are smaller. If space around the gearbox is tight, then the reservoir and pump can be located remotely thus allowing the larger system to fit in a more confined area.

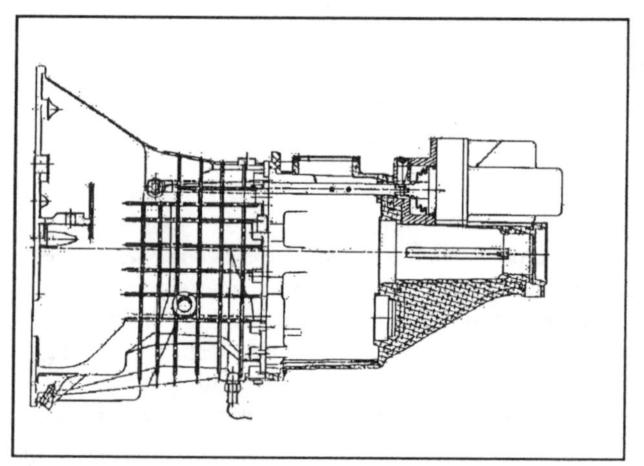

Figure 6 - Shift By Wire Transmission Side View. Actuator Mounted At The Rear Above The Output Shaft.

Hydraulic actuators may also, for a production system, utilize the hydraulic circuit already existing in a vehicle, such as the clutch or brake fluid circuit. While the vehicle's pump system may increase in size, the added cost is less expensive than adding an additional pump system to the gearbox. While this is an alternative, this decision must be made by the car manufacturer.

The weight of a hydraulic system may be more than the addition of one or two actuators. This is again, dependent upon the availability of an existing hydraulic circuit, and the willingness of the OEM to let the transmission subsystem utilize it.

Hydraulic actuators have traditionally been less expensive than motor actuators, however this is one area where new technology developments are making these two costs comparable. Finally, there has been some movement within the automotive industry to tend away from using hydraulic circuits.

After careful consideration, this SBW development program used electromechanical actuators due to the lack of a sufficient hydraulic source already in the vehicle (clutch supply not sufficient for both systems), pump noise (although only a demo vehicle, this noise may still be objectionable to a customer during a test ride) and finally, to maintain a fluidless add-on system.

ELECTRONIC CONTROL MODULE (ECM)

The design of the control module is straightforward, and therefore its functionality will only be discussed briefly. The ECM is a microcontroller based module that contains the appropriate A/D channels, timers counters, digital inputs and PWMs (Pulse Width Modulators). On board MOSFET H bridge drivers provide bi-directional control of the linear and rotary portions of the actuator. An Serial Communication Interface (SCI) is provided for communication to a laptop computer which allows for easy in-vehicle testing, calibration and parameter modification.

CLUTCH INTERLOCK

A final safety device has been included into the system to prevent inadvertent and concurrent clutch engagement and shift engagement. This device is a clutch solenoid, that is inserted between the clutch pedal and the clutch. Since the driver maintains control of the clutch, the module must ensure that driver errors do not compromise system safety. Two clutch switches are used to determine when the clutch pedal is approximately half depressed and also when the clutch is nearly fully depressed.

The ECM will only allow the gear shift to initiate when the clutch pedal is fully depressed. If gear engagement has been initiated and the clutch pedal is then released, the solenoid will either prevent the clutch from being released, or will slow the release of the clutch to allow the shift to complete. This valve contains a check valve to allow the clutch to be depressed even when the solenoid is activated. It will not interfere with the driver's operation of the clutch.

SHIFTING PRESELECT

An added feature of this system is called "preselecting" which allows the driver to choose the next gear to be engaged before the shift is actually made. The option was added to aid in implementing an automated manual transmission. The preselected gear is chosen by shifting the gear shift lever to the desired position without the clutch pedal being depressed. This preselection can be performed at any time, while in any gear, and with the vehicle going at any speed.

Once the new gear is selected, the dashboard display will continue to display the currently engaged gear, while indicating the new selected gear using a flashing indicator. The actual shift is then initiated by the disengaging of the clutch. When the clutch pedal is depressed half way, the ECM instructs the actuators to shift out of gear to neutral. When the clutch pedal reaches the fully depressed state, the completion of the shift is made by engaging the preselected gear. Throughout this sequence, the ECM is constantly monitoring the vehicle and gearbox for safe operating conditions (vehicle speed, shaft speed). An audible beep indicates to the driver the completion of the shift, which tells the driver that the clutch can be engaged.

The purpose of starting the shift sequence when the clutch pedal is only half depressed is to give the appearance of a faster perceived shift. Shifting out of gear as the clutch is being released does not present a danger to either the driver or gearbox.

THROTTLE / ENGINE CONTROL

It should be remembered that this system is a shift by wire system, and therefore operates functionally as a mechanical manual transmission. There is no electronic engine or electronic clutch control associated with this system. Therefore, when shifts are made, not only does the driver control the clutch, but the driver must also operate the accelerator pedal during the shifts. Upon release of the clutch, the driver must still lift up on the throttle to prevent engine flare.

The next step for this type of program is to integrate an automatic clutch system in with this SBW system. This next phase would indeed include engine control

RESULTS

The system was initially installed onto a spin stand and evaluated both statically and then dynamically. In-vehicle testing followed. The results are presented in the following three sections.

BENCHTOP TESTING

The purpose of the benchtop static testing was to ensure proper positional accuracy of the actuators within the gearbox. This proved the capability of the microcontroller to move the actuator within the designated shift planes and crossover paths between the planes. For this test, the input and output shafts were not moving and therefore any recorded shift times would be irrelevant since no time would be needed for synchronization.

For dynamic testing on the spin stand, the output shaft was driven by a motor to simulate the speeds as seen by a vehicle on the road. A clutch plate was mounted on the input shaft of the transmission to provide the proper inertia to be overcome during the synchronization process. The expected shift times and shift curves matched those measured in the vehicle during road testing.

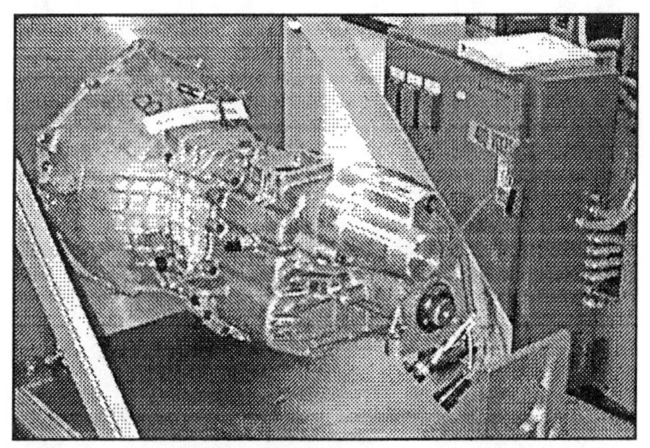

Figure 7 - NVG 3500 RWD Transmission On Spin Stand With Actuator Mounted Above Output Shaft

The final purpose for performing thorough dynamic spin stand testing is to ensure the gearbox has functional and safe operation before it is operated in a moving vehicle.

IN VEHICLE TESTING

With the SBW gearbox installed into the test vehicle, instrumentation was added to properly measure the vehicle shift parameters. Both data acquisition equipment and internal ECM diagnostics were used for measurements. ECM data was sent serially to a laptop computer for immediate analysis and storage. Specific measurements included total shift time, total synchronization time and analog plots of shift rail position.

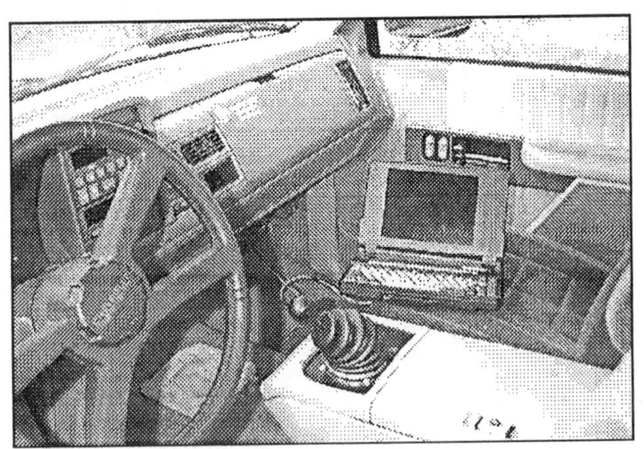

Figure 8 - Interior of Test Vehicle Showing Electronic Shift Lever and Laptop Interface.

Re-calibration of the shift points (location map of the shift plane and in-gear position of the shift rail) was used to confirm the calibration performed on the test stand.

SHIFT PLOTS

Measurements were made of the shift rail position during various stages of testing while in the test vehicle. Rail position provides clear information regarding total shift time, synchronization time and engagement times.

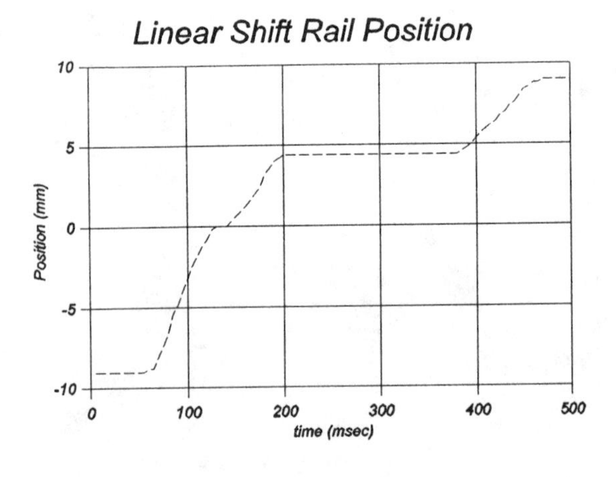

Figure 9 - Worst Case Shift Time 2nd to 3rd Gear With Programmed 200ms Synchronization Time

Figure 9 provides a plot of a shift made while testing, and whose synchronization time was predetermined to be 200 milliseconds. The long synchronization time was used initially to prove the mathematics for determining synchronization times, and to verify that all combinations of shifts could be made within the 200ms maximum target time.

For the plots of Linear Shift Rail Position in figures 9, 10 and 11, the total displacement for the shift rail was 18 mm, or +/-9mm from neutral. The position of the rail while in 1st gear (and 3rd and 5th) is set to be 9mm, while the gate position for 2nd gear (and 3rd and R) is set to be -9mm, all measured from neutral, 0mm.

Figure 9 starts with the gearbox in second gear (-9mm) and travels to 0mm (neutral). At this point the linear motion temporarily slows while the crossover from the 1-2 shift plane to the 3-4 shift plane is made. Once in the 3-4 plane, the linear motion continues with the synchronizer clearances being taken up. At approximately the 200 millisecond grid line, the linear motion once again slows as synchronization takes place. The blocking ring prevents the sleeve from continuing through to the gear until the two shaft speeds (input and output shafts) are equal (just before the 400ms grid line). Once this occurs, the linear motion continues until the gear is fully engaged.

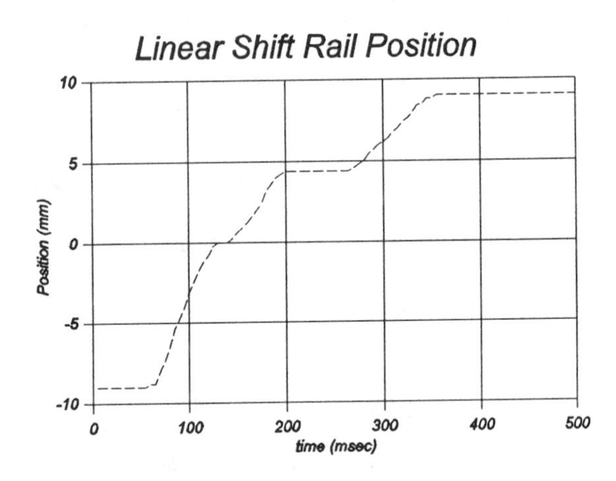

Figure 10 - Shift Time 2nd to 3rd Gear with Optimized Synchronization Time.

In figure 10 above, the synchronization times have been reduced by increasing the synchronizing force. These forces are still below the maximum allowable level as determined by the synchronizer friction material, thus allowing for further improvements. The total shift times for a 1-2 and 2-3 shift are usually the longest during hard accelerations.

Performing 3-4 and 4-5 shifts were typically faster since the difference in shaft speeds was less than the 1-2 and 2-3 shifts. Less speed difference translates, per (1) and (2), into faster synchronization times.

Finally, the shift performed in figure 11 is a 1-2 shift optimized for both shift speed and comfort (no harsh engagement sounds during the shift). The total shift for this condition was slightly less than 200 milliseconds. Notice that there is no slowing of the linear motion as the rail position passes through neutral (0mm). This is a result of a 1-2 gear shift remaining in the same shift plane, that is, there is no crossover motion required of the actuator.

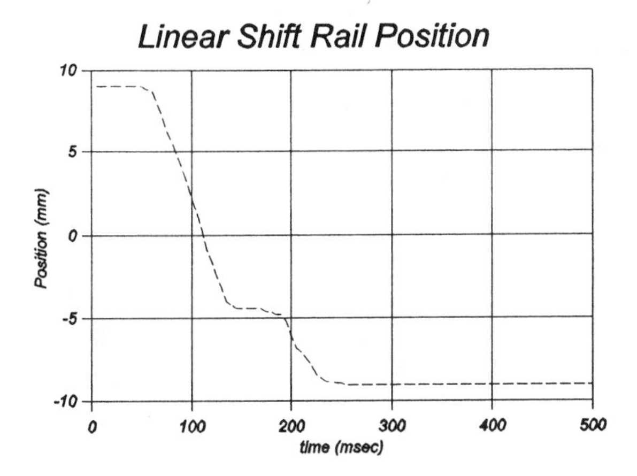

Figure 11 - 1st to 2nd Shift Optimized For Speed and Comfort.

One factor not presented graphically here, was the affect of temperature. All of the above measurements were made in warmer ambient temperatures and after the transmission had sufficient time to warm up. A temperature compensation scheme is built into the ECM which adjusts actuator forces to provide consistent shift times.

FUTURE DEVELOPMENTS

Given the results presented here, there are many areas in which further development can improve this type of system, primarily with respect to shift time.

For most automated manual applications, smaller gearboxes will be used than the one used for this SBW program. While this gearbox was targeted for a light truck, utility truck application, most automated manual applications are centered around the small and medium sized European car platforms. These vehicles are smaller than the US small and medium sized cars. This reduction in gearbox size means lower gear, shaft and clutch inertias, which translate into lighter synchronization loads. Additionally, actuator sizes may be reduced.

Mechanical modifications can be made to the gearbox to improve on shift times. These include using dual cone synchronizers and more efficient shift systems. Additionally, improved actuator design using the experience received in this program will help balance actuator size, speed and weight for optimum performance.

CONCLUSION

The development of a shift by wire transmission has been presented. Of primary concern to the end user of such systems in the total shift time and shift quality. The targets identified in this paper were met, although more stringent timing is required for automated manual transmission systems. Total shift times of 200ms were achieved.

Implementation of this system into a test vehicle was successful and proved both control schemes and shifting algorithms. Finally, the actuator design and specification for an automated shift system was completed.

ACKNOWLEDGMENTS

The author thanks Sanjeev Varma, Sankar Mohan, Chris Phelan, Randy Adler and Dan Miller for their technical assistance and advice.

REFERENCES

1] Socin, R.J. and Walters, L.K.,"Manual Transmission Synchronizers," SAE Paper No. 680008.

981341

Diagnostics in the Automotive Industry: Real-Time Information for Bottom Line Results

Gene Stovall
Cutler-Hammer

ABSTRACT

In the past, attempts by the automotive industry to implement various diagnostic systems into its process proved ineffective, costly to maintain and limited in their ability to provide specific information. Today, the industry is combining PC-based control, the DeviceNet network and intelligent field devices, to access diagnostic information that is enabling it to achieve new levels of productivity and cost-savings.

The integration of these three state-of-the-art technologies is delivering improved machine uptime; improved product quality; and reduced machine commissioning time, which translates to getting a product to market faster.

With the implementation of best-in-class intelligent components to provide device and process diagnostics in a PC-based control, open communication network environment, automotive manufacturers have the control solution to monitor status and performance on a real-time basis. This capacity is proving itself as a strategic enabler to proactively improve productivity.

The measurable results, including low-cost production and greater control, that diagnostics provide are illustrated by their application at Ford Cleveland and Chrysler Kokomo. The quantifiable efficiencies and economies give the industry a genuine competitive advantage.

INTRODUCTION

The technological synergies achieved by combining an open standard device-level network, best-in-class intelligent devices and PC-based control are the key factors that will enable users to achieve uptime and productivity gains.

This paper will examine:

- the attributes of an open automation system, device-level networks including DeviceNet, DeviceNet compatible intelligent devices and PC-based control;
- how the performance capabilities and functionality of best-in-class intelligent components play a critical role in providing superior diagnostics; and
- how this integrated open control solution provides diagnostics and information in addition to control.

The benefits of device and process diagnostics in an open automation control environment will be illustrated by their performance in several automotive production facilities.

OPEN NETWORK STANDARDS

Open standard systems can directly impact performance, price and productivity issues critical to achieving competitive advantage. The system is characterized by its open, modular, scalable, economical and maintainable aspects.

An open system allows the integration of off-the-shelf hardware and software components into a standard environment. Since there are times when a module of a control system needs to be replaced to provide additional benefits, such as additional capability or lower cost, the users want the ability to make that replacement easily.

Since creating an agile control system is a key goal, the ability to increase or decrease the functionality of a system as changes in the process are made is a paramount concern. A control system that is scalable or allows control modules to be added or removed and provides appropriate control capability to match application needs enables efficient reconfiguration. This typically involves adjusting system parameters and adding or removing devices. Open systems allow for control system changes to be made cost effectively since they allow incremental upgrades and easy integration of components.

DEVICE-LEVEL NETWORKS

Device-level networks enable users to benefit from the additional functionality and diagnostics that intelligent devices offer. This affords users cost and performance advantages unattainable with hardwired I/O. Instead of hardwiring each device, intelligent devices can be plugged into a single cable.

Open, device-level networks create opportunities for users to select devices to upgrade their systems using best-in-class replacement products. Device-level networks provide added functionality and diagnostics because they offer a low-cost communications link for industrial devices (i.e. limit switches, photoelectric sensors, motor starters, process sensors, bar code readers, variable frequency drives, panel displays and operator interfaces) to a network without expensive hardwiring. The direct connectivity has the potential to provide improved communication between devices for truly distributed control, as well as providing improved control diagnostics that are not easily accessible or available through hardwired I/O interfaces.

DEVICENET: THE OPEN NETWORK STANDARD OF CHOICE - Several control suppliers began an innovative vendor collaboration to develop the DeviceNet protocol as a robust, truly open network standard and established the Open DeviceNet Vendor Association (ODVA). The organization is dedicated to the development of the DeviceNet protocol, a truly open, real-time device-level network that enables devices from multiple vendors to plug into a single twisted pair cable and operate as a single system.

The DeviceNet communication link is based on a broadcast-oriented, communications protocol—the Controller Area Network (CAN). The CAN communications chip, which provides high-speed, real-time communication, was originally developed to replace expensive automobile wiring harnesses with low-cost network cable.

All DeviceNet products use the same CAN chips, which are available from nine vendors worldwide. The cost of the chips, which continues to be driven down by their volume application in the automotive market, is typically 5-10 times less than chips for other networks. This is critical because it enables best-in-class intelligent components to cost effectively add intelligence and add value, without significantly increasing the price of the product.

DeviceNet networks reduce the number of wires connecting all devices to one cable. Compatible devices simply plug into the cable rather than having to be hardwired to the controller. This can cut installation, commissioning and debugging costs in half. The networks support multiple controllers, distribute intelligence and control, increase system control through their ability to share data over one cable, improve maintenance and fault detection, and increase overall reliability and uptime of the entire connected system.

Unlike other "bit level" networks, which simply transfer the status of I/O using a source/destination message system, DeviceNet has the ability to communicate explicit diagnostic information about connected field devices. The network has the potential for a producer/consumer message delivery system that would enable devices to communicate directly with each other rather than relying on a high-level controller to orchestrate communication necessary for device interaction. For example, a switch could be programmed to respond to a sensor message that an event has occurred and open or close without controller direction.

The producer/consumer system would enable multiple nodes to accept the same message/data at the same time from one producer, facilitate node synchronization, provide for an efficient use of bandwidth, offer all the functions of source/destination messaging, support multicast, change of state and cyclic data production, and accommodate explicit and I/O messages on the same wire.

Users gain additional performance advantages with the combining of DeviceNet with Ethernet in a two-layer architecture approach, which creates three levels for information handling within a facility with one Ethernet network for the plant-level data control and another one for the rest of the plant's information. While Ethernet has been used extensively for networked computing in the business environment, in this configuration it can now be used for machine controller interlocking.

By combining DeviceNet and Ethernet, manufacturers now have the availability of fast control information for distributed control systems and the ability to coordinate plant-wide control. This provides the enterprise with easy data transfer as well as streamlined data access. In addition, it provides a low-cost, proven control system, which, with the recent advances in Ethernet, is very deterministic.

OPEN CONTROL

PC-based control simplifies a manufacturing application on a single processor platform. One PC can integrate four software packages—logic programming, network configuration, real-time control and human-machine interface (HMI).

With a PC, parallel and serial ports, networking facilities, input devices, and add-in hardware interfaces are all standardized. PCs can support more peripheral devices (CD-ROM drives, sound cards, voice recognition, optical scanning etc.). In addition, the Graphical User Interface, represented by Windows, is familiar, permitting the use of third-party tools and enabling users to choose from a multitude of off-the-shelf Windows applications to analyze or manipulate control system data while the control system is running.

Adoption of the PC platform reduces reliance on one supplier or group of suppliers, allowing users to pursue the most cost-effective solution. PCs also offer a litany of reliability features; are available worldwide on short notice from many vendors; preserve software investments; and offer compatibility with standard operating systems, networks, and user interfaces. Further, hardware can be upgraded as needed without an adverse impact on the software investment.

For users, open control means:

- lower total lifecycle cost;
- independence from proprietary control systems;
- easy to integrate control functionality;
- ease of integration of control, HMI, and programming;
- easier access to information by all control and business applications;
- software portability and preservation of investment; and
- systems positioned for the future.

These characteristics mean open control offers users a viable way to achieve cost reduction in the context of continuous improvement. Its implementation enhances profitability and competitive advantage because it directly impacts installation, maintenance, troubleshooting, and downtime costs. Further, it gives users the opportunity to make value-added control system improvements on an ongoing basis.

DIAGNOSTICS: THE CRITICAL KEY - In an effort to keep machinery running smoothly, the automotive industry has attempted to implement diagnostic systems into its processes. The diagnostic techniques range from writing ladder logic code in PLCs to external computer systems that monitor the machines. These methods are not effective and are costly to maintain. The software in these systems must be updated when changes are made on the machine; all too often, this is not done.

An open control device network solution, which is designed to manage device and diagnostic information better and faster, plays a key role in significantly improving both the quantity and quality of enterprise-wide information exchange. The network enables rapid, predictive fault detection and identification without adding significant cost to the system.

With increased access to more and more process information through diagnostics, it is important for users to understand what kind of diagnostic information can be captured so they can leverage it to improve processes and productivity.

Device diagnostics - An open device-level network enables intelligent devices to report faults to the controller for fast action. The location of the fault on the network is displayed on the operator interface for rapid troubleshooting. The open controller also provides the ideal platform to easily manipulate the data and display it to the operator. Limit switches, photoelectric sensors and proximity sensors all can send fault information to the human-machine operating device to instruct the operator and maintenance personnel about operational status.

Process diagnostics - With a open device level network, manufacturers have the ability to monitor the "health" of a process, which enables them to eliminate factors that can cause downtime. For example, intelligent motor starters can pass information like motor current and percent of thermal overload, which can be used to predict machine tool wear and problems with a motor running hot before it actually breaks down. Thus, there is no longer a need to create extensive diagnostic programs to improve machine process.

As open network communications are implemented in field devices, customers have the ability to choose those devices from specific manufacturers that present the optimum value for the application. This may be the device with the most diagnostics, most additional functionality (timers, counters, etc.), lowest price, highest performance, etc. Thus, a "best-in-class" buying decision can be made for almost every component of the system.

With the combination of an open network standard and control, users also have a choice of control software and a choice of intelligent field devices. A truly open device-level network's open architecture gives users the flexibility to make efficient, cost-effective improvements and take the incremental steps in a migration to truly distributed control. As improvements in control methods yield increasingly smaller returns, the combination of open networks, open control and best-in-class intelligent field devices provide manufacturers with the capacity to achieve increasing control system functionality with diagnostics.

SYNERGIES AT WORK

FORD CLEVELAND'S IMPLEMENTATION - The ability of DeviceNet and PC-based control to enable users to optimize diagnostics is demonstrated at the Ford Cleveland Engine 1 facility. Ford Cleveland Engine, seeking ways to increase the productivity of the plant's machinery by installing new technology that production personnel would find easy to use, installed a beta DeviceNet system.

The machine Ford selected performs an air leak test on the cylinder head of a 5.0 liter V-8 engine. Installed during the late 1960s, the equipment was controlled by relays. The goals established for the beta site were to standardize on an open platform, increase accessibility to the machine, improve uptime with diagnostics, reduce system commissioning time, and gain the buy in of production personnel for the new control system.

The system includes:

- 40 sensors interfaced to DeviceNet via a smart adapter module;
- 10 DeviceNet I/O blocks connected to solenoids and lights;
- Starters monitoring motor current and percent of thermal overload on the motors;
- Flat panel touch screen iPC at main control panel;
- Flat panel touch screen iPC at operator station; and
- soft logic, network management and operator interface software.

The software was installed on an iPC in the main control panel. This computer is used primarily for monitoring and making changes in the control program. A remote pedestal provides a flat-screen computer running HMI software, which mimics the operator console on the machine. This second iPC is used as the operator's window to the process, replacing an old light and button mimic display. The operator now uses the touch screen on the pedestal iPC to start and stop the process. The pedestal is also used to present the operator with the diagnostic screen that helps keep the process running. The two computers communicate via a communication link.

Diagnostics At Work - At the Ford installation, the production operator has used the system diagnostics to help electricians troubleshoot a problem with a limit switch. While modifications were being made to the new system, Ford activated the old control system. During that time, the production line experienced a problem.

He explains, "After the electricians were called and spent 20 minutes unsuccessfully trying to identify the problem, we switched the machine over to the new control system and the diagnostics immediately identified a problem with a limit switch on the machine.

In another instance, mechanical changes were being made on the machine. When they were completed, the machine did not run correctly. The production operator used the diagnostic screens to identify that a proximity sensor was out of range. The machine repairman, who had moved the sensor to reach into the machine, had not repositioned it correctly. He reports, "A problem that could have shut down the machine for about half an hour, was fixed in minutes because the system told us exactly what was wrong. It eliminated the possibility of a broken wire or failed electrical component, this in itself saved us money."

In both cases, the DeviceNet system pinpointed the problems quickly and accurately. Additionally, engineering, production and maintenance personnel responded very positively to the new system. The new control system has increased machine accessibility, improved uptime with diagnostics and provided usable troubleshooting information from intelligent devices.

CHRYSLER KOKOMO'S IMPLEMENTATION - At Chrysler's transmission plant in Kokomo, Indiana, a 68-station assembly carousel producing transmissions for RWD light trucks is one of the largest production lines ever to use purely PC-based control. It is also believed to incorporate more PCs than any production line anywhere. (And a year from now it will be twice as large after new PC controlled final test, button-up, and clutch line operations go into service.)

Networked at every manual and automatic station is an iPC that serves as both controller and HMI. The touchscreen iPCs also perform zone control and host tasks. Each computer is fitted with a 166 MHz Pentium card, Ethernet card, 3.2 GB hard drive, and 128 MB of battery-backed RAM. All the iPCs are nodes on a Microsoft Windows NT TCP/IP network.

The station and zone controllers are loaded with Windows NT and iRMX-based real-time OS software. The zone controllers serve as concentrators for historic data storage and provide station program backup and some transport coordination for the synchronous line. The host is primarily concerned with data collection, production counts, SPC functions, error roofing supervision, and program backup.

Depending on the station, other software loaded includes Microsoft Excel for Tool & Operation sheets, Autodesk AutoCad Lite for prints, Datalogic barcoding, Phoenix Contact Interbus-S, Ann Arbor Automation NetConnect for host communications, Sizemetric press force gaging, and Cincinnati Air Test air pressure testing.

Networks Assists Diagnostics - The carousel marks the first large Chrysler implementation of networked I/O to field devices. Each station PC has an Interbus-S network, as does each of the carousel's six zone PCs. Some station devices are on the station network, others on the zone network. The I/O does not communicate with any smart devices at the bit level at present.

Networked I/O has meant that the thousands of red wires of PLC systems do not exist, just a single cable in and out of I/O blocks mounted close to the equipment controlled.

Flow Chart Programming Speeds Troubleshooting - The entire carousel, excepting several nut runners, is programmed in Flow Chart. This language speeds troubleshooting because—unlike ladder logic— the sequence of operations is inherently provided when viewing the control program. Further, the flow language used is limited in how much logic can be squeezed into a block.

Swap Out Rather than Repair - Chrysler's philosophy is not to diagnose a malfunctioning PC online, but rather to swap out the unit and use an automated process to download programs from the network into the replacement. The goal is a seven-minute PC swap out and reload.

To provide the highest reliability for the carousel's 11 automatic station PCs and zone controllers, Chrysler will shortly add a second PC, RAID controller (redundant array of independent disks), and switch at each location. By providing PC redundancy, the zone and automatic station PCs should never fail. RAID will also permit hot swap out of malfunctioning hard drives.

The automated process to download programs over Ethernet relies on other carousel PCs as sources. All station PCs within a zone back up their information to the zone controller. The six zone controllers in turn back up each other's zone and station programs. All zone and station programs are also backed up to the host. Finally, all PC programs are written to CDs and floppy disks for alternative hand loading at the PCs. Chrysler will shortly add a read/write CD drive to the host to provide offline hard media.

Information: Accessible and User-Friendly - The fact that the carousel is under PC rather than PLC control is transparent to operators. Alarms and diagnostic messages are very similar to those commonly provided with modern PLC control systems. But the various troubleshooting helps that the new open control system gives operators and maintenance personnel is definitely noticed and liked.

Perhaps most often called up are the Excel-based tool-and-operation sheets, which are available as menu picks on the HMI. New operators especially like having these references readily available. T&O sheets, which list the operations for each station and the perishable tooling used, were always a problem in paper form to keep up-to-date and store. Since T&O sheets are prepared in Engineering in Excel, revisions can be simply put on a floppy and carried to the PC for loading. Any change made on one shift immediately carries over to all shifts.

In the same manner, the AutoCad Lite prints for each station are available on the station PC. If trouble occurs with an electrical or electromechanical component, such as a proximity switch, Chrysler's internal reorder number and the manufacturer's part number are provided by simply double-clicking the failed part. Perhaps most unusual, photographs of the station can be called up, and the photos will highlight any component in failure.

SUMMARY

As the automotive industry and other manufacturers are challenged to find ways to increase productivity and economic efficiencies, they will look to implement new technology that not only produces bottom line results, but will also be accepted by production personnel. The device and process diagnostics that are available with open automation systems, technological advances in controls, and intelligent plant floor devices provide the solution to meeting these needs today.

In the future, with advances in the implementation enhanced device and process diagnostics, plus the addition of discrete event diagnostics, distributed configurations of intelligent devices on an open network, the next generation of control will be able to communicate with each other on a peer-to-peer level, bypassing host devices. Intelligent devices and device networks can deliver a wealth of data. Diagnostics and troubleshooting functions will contribute to ongoing system performance improvements and reduced downtime. In addition, embedded value-added functions will facilitate users' ability to improve processes by providing documentation in critical areas like product quality, yield, profitability and environmental performance. Moreover, the technology's inherent flexibility and ever-increasing agility offer users like the automotive industry, the assurance that their investment will enable them to achieve ongoing improvements as well as position them for adoption of next generation control.

981459

Fitting Automotive Microprocessor Control Look-Up tables to a Response Surface Model using Optimisation Methods

R. J. Lygoe
Ford Motor Co. Ltd.

ABSTRACT

Modern automotive microprocessor control systems provide a characterisation of the spark timing to control knock. This representation is normally provided in the form of look-up tables and in this study is referred to as Borderline Spark. Five and six factor augmented Box-Behnken response surface experiments have been designed to characterise Borderline spark as a function of the relevant control system input variables. Least squares regression techniques were used to generate a Borderline Spark response surface model. Confirmation experiments on two different engines showed that the response surface model provided an accurate prediction over most of the operating range. Symbolic integrals together with Sequential Quadratic Programming have been used to achieve the optimal fit of the model to the look-up tables using MATLAB. Furthermore this software has been used to define look-up table sizes based on a user-specified interpolation error between the look-up table and response model surfaces.

INTRODUCTION

The purpose of this paper is to describe:

a) the experimental process used to map the Borderline spark (BDL) response surface

b) the development of an empirical model of the BDL response surface which is suitable to define the contents of the powertrain control software[1] for BDL

c) an optimisation tool which fits the empirical model to the powertrain control strategy for BDL in some optimal manner

It is beneficial to use a designed experiment to characterise the BDL response surface because:

i) the smallest amount of data is required to achieve a good representation of the response surface, thus calibration process efficiencies can be realised. Furthermore consideration in the experiment can be given to control factors which are time-consuming to set so as to minimise the duration of data collection. A designed experiment allows characterisation of non-linear behaviour and interactions, which would be comparatively inefficient to achieve using one-at-a-time experimentation.

ii) the nature of a designed experiment lends itself to model identification (including any significant interactions) using regression analysis. The model so identified can then be compared to the existing control strategy model to understand any inadequacies to seed strategy model development.

iii) an experiment can be designed to allow sequential assembly. This is important because should the model not validate acceptably, it is possible to augment the original designed experiment to allow higher order model terms to be identified. This is consistent with the Deming Plan-Do-Study-Act [1] cycle.

DEVELOPMENT OF A RESPONSE SURFACE MODEL

AN EXPERIMENTAL PROCESS TO GENERATE A BDL RESPONSE SURFACE MODEL

The link between graduating functions and the Taylor series expansion - In general a polynomial in the coded* regressors $x_1, x_2, \dots, x_n$ is a function which is the linear combination of powers and products of the x's. A term is said to be of order j if it contains the product of j of the x's (some of which may be repeated). Thus terms such as x_1^3, $x_1x_2x_3$ or $x_1^2x_2$ are all said to be of order 3. A polynomial is said to be of order d if the terms of highest order in it are of degree d also.

[1] referred to from herein as strategy

For example, with two regressors, the full second order polynomial can be written:

$$g(x,\beta) = \beta_0 + \beta_1 x_1 + \beta_2 x_2 + \beta_{12} x_1 x_2 + \beta_{11} x_1^2 + \beta_{22} x_2^2 \quad \text{EQ (1)}$$

* there are three principle reasons for using coded variables:

- coding improves the conditioning of the regression matrix
- coding allows the magnitude of the regression coefficients to be compared
- coding allows standard layouts for experimental design and facilitates easy calculation of the contrasts

A polynomial expression of degree d can be thought of as a Taylor series expansion of the true underlying theoretical function $f(x)$ truncated after terms of dth order. Consequently, the following usually applies:

1. The higher the degree of the approximating polynomial, the more closely the Taylor series can approximate the true function.
2. The smaller the region over which the approximation is applied, the better the quality of approximation with a polynomial of degree d.

In practice a polynomial of at most second order, over a limited region of x space, can be made to adequately represent the true function, especially if the right kind of transformation is made to either the regressor or response variables. There is evidence within Ford [2] that the BDL response surface can be adequately represented by a polynomial of at least second order throughout.

Least squares estimation of the model terms - Estimates of the model terms (β_i) can be readily derived using the principle of least squares. There are many sources of literature for least squares regression analysis, for example [3], the details of which will not be given here. Stepwise least squares regression techniques were used to estimate the model coefficients. The heuristic stepwise search attempted to maximise the PRESS r^2 statistic [4].

The motivation behind using the simplest model - Given that the least square estimates of the model coefficients have been obtained and that the available model diagnostics [3] indicate that this model is valid, a prediction for the response at any level of the regressors within the design space can be calculated using the following matrix equation:

$$\hat{y}_i = \mathbf{x}'\hat{\mathbf{b}} \quad \text{EQ (2)}$$

where $\hat{y}_i$ is the predicted response variable (scalar), $\mathbf{x}'$ is a (1 x p) vector of regressor variables and $\hat{\mathbf{b}}$ is a (p x 1) vector of regression coefficients

It can also be shown [3] that the average error variance ($\overline{Var(\hat{y}_i)}$) associated with the $\hat{y}_i$ is given by:

$$\overline{Var(\hat{y}_i)} = \frac{p\sigma^2}{n} \quad \text{EQ (3)}$$

Where p is the number of regression coefficients, n the number of observations and σ^2 the error variance. EQ (3) implies that there is a penalty associated with adding additional terms to the model in the sense that as p increases then $\overline{Var(\hat{y}_i)}$ increases. This provides strong motivation to use the simplest model to describe the variation in y; i.e. the model with the smallest number of parameters which adequately describes the behaviour of the response variable over the design space. For this reason the second order model was chosen for BDL, with the caveat that the order of the model could always be increased if justified, by a process of experimental design augmentation.

Selection of an experimental design - From [5], [6], [7], the analysis of response surfaces is greatly facilitated by the use of an appropriate experimental design. The literature lists desirable properties of response surface designs which have been considered in the context of this study. Given these requirements a suitable class of designs is due to Box and Behnken [8]. The Box-Behnken design does not require that data be gathered at the extreme setting of the factors (i.e. x = {1, 1, 1, ..., 1) or x = {-1, -1, ..., -1}). In the context of the BDL response surface investigation, this is a major practical advantage[2] offered by this class of design, while still providing a good distribution of points throughout the region of interest. Box-Behnken designs are also relatively compact in terms of the number of runs. For example a 6 factor Box-Behnken design with 6 replicates of the centre point contains 54 runs. In circumstances where it is difficult to change or set levels of some or all of the factors (for example air charge temperature (ACT) and engine coolant temperature (ECT)), it makes sense to utilise designs which require only three levels of each variable to generate a full second order response model. Further discussion of the properties is given in [9], but overall, the Box-Behnken designs are well suited to the BDL investigation.

---

[2] At the extreme factor settings (high engine speed/load/ECT/ACT there is an increased tendency to knock, which is more difficult to detect accurately (due to greater background noise) using subjective audible measurement. Therefore at these engine operating conditions there is an increased risk of engine damage from heavy knock or preignition through inadvertently over-advancing the spark timing when trying to identify knock.

Assignment of model terms - The second order BDL response surface model is specified in EQ (4) below:

BDL (deg. BTDC) =

$$\beta_0 + \beta_N N_c + \beta_L L_c + \beta_{LAM} LAM_c + \beta_{EGR} EGR_c + \beta_{ECT} ECT_c + \beta_{ACT} ACT_c + \beta_{NN} N_c^2 + \beta_{LL} L_c^2 + \beta_{LAMLAM} LAM_c^2 + \beta_{EGREGR} EGR_c^2 + \beta_{ECTECT} ECT_c^2 + \beta_{ACTACT} ACT_c^2 + \beta_{NL} N_c L_c + \beta_{NLAM} N_c LAM_c + \beta_{NEGR} N_c EGR_c + \beta_{NECT} N_c ECT_c + \beta_{NACT} N_c ACT_c + \beta_{LLAM} L_c LAM_c + \beta_{LEGR} L_c EGR_c + \beta_{LECT} L_c ECT_c + \beta_{LACT} L_c ACT_c + \beta_{LAMEGR} LAM_c EGR_c + \beta_{LAMECT} LAM_c ECT_c + \beta_{LAMACT} LAM_c ACT_c + \beta_{EGRECT} EGR_c ECT_c + \beta_{EGRACT} EGR_c ACT_c + \beta_{ECTACT} ECT_c ACT_c$$

EQ (4)

where: the βs are the regression coefficients and the coded regressor variables, e.g. N_c, L_c, LAM_c, EGR_c, ECT_c, ACT_c, are the variables provided by the powertrain control strategy.

In order to fit the response surface model to the look-up tables, the appropriate model terms must be assigned to the corresponding look-up tables, as shown below in Table 1:

BDL look-up table name(s)	2nd order BDL response surface model term allocation
Base table	$\beta_0 + \beta_{NN} N_c^2 + \beta_{LL} L_c^2 + \beta_N N_c + \beta_L L_c + \beta_{NL} N_c L_c$
A/F offset table	$\beta_{LAM} LAM_c + \beta_{LAMLAM} LAM_c^2 + \beta_{NLAM} N_c LAM_c$
EGR offset table	$EGR \times (\beta_{EGR} + \beta_{NEGR} N_c + \beta_{LEGR} L_c)$
ECT offset table	$ECT_c \times (\beta_{ECT} + \beta_{NECT} N_c + \beta_{LECT} L_c)$
ACT offset table	$ACT_c \times (\beta_{ACT} + \beta_{NACT} N_c + \beta_{LACT} L_c)$

Table 1: Assignment of the BDL 2nd order response surface model terms to the BDL look-up tables

Of the 28 response surface model terms only 18 can be represented by the existing look-up tables available. If any of the remaining 10 terms prove to be statistically significant:

1. if the magnitude of the term (as indicated by the size of the regression coefficient) is not of engineering significance, then it can be neglected. For example one may decide that if one of these 10 terms were statistically significant but had a regression coefficient of 0.5, then it can be neglected.

2. if the magnitude of the term is of engineering significance, then augmenting the existing look-up tables to take account of such term(s) may be justified.

DEVELOPMENT OF AN EMPIRICAL MODEL

Data collection - Data was collected on an engine installed in an engine dynamometer test cell with full vehicle induction and exhaust systems fitted. The test engine had previously been broken-in and power-checked to ensure it was functionally representative. The data was recorded from the powertrain control system using a computer-based data acquisition system. BDL is measured using an engine listening system. This comprised two surface mounted broad band accelerometers (one on the engine cylinder head and one on the block) to detect sudden changes in cylinder head and block vibrations caused by knock. A filtering unit was used to attenuate as much background noise as possible, before transmitting the resulting knocking sound output through a loudspeaker located in the engine dynamometer test cell control room. It was necessary to establish that the engine dynamometer test cell operators on both day and late shifts could repeatably rate audible knock using the engine listening device and both produce similar results at the same engine operating conditions. This is important because the detection of audible knock is a subjective measurement which can vary widely between different people. This process was supervised by an experienced engineer at a variety of engine operating conditions so that the definition of BDL was clearly understood and agreed with the test operators. Even so, BDL measurement repeatability is approximately ± 2 deg. BTDC.

Confirmation of BDL response surface models from initial experiments - Initial Box-Behnken experiments were carried out on 1.8, 1.4, 1.25 and 2.3 litre, 4 cylinder engine designs to verify that the proposed second order response surface model (EQ (4)) provided a satisfactory fit to the data. Valid BDL response surface models were developed [9] and it was then necessary to confirm that they were able to predict independent data satisfactorily. This independent or confirmation data consisted of a mixture of repeated and interpolated points compared to the original experiment. If the model is correct and the prediction is for an individual in the same population from which the data were obtained, then the data will appear in the Prediction interval. Thus Prediction intervals provide a useful measure that the model confirms with additional data. With all four engines a significant quantity of the confirmation data lay outside the prediction interval. It was therefore possible to conclude that the model was not adequate at these confirmation data points. Model inadequacies were further evident when the BDL response surface model prediction was plotted against the confirmation data at wide-open-throttle (WOT), an example of which is shown below in figure 1. The most crucial operating region as far as BDL is concerned is at or near wide-open throttle (WOT) on modern four-valve engines. This is because at such engine operating conditions, BDL often defines the maximum spark that can be tolerated and therefore the maximum torque output; also because it is at or near WOT that the engine is most likely to experience knock damage (due to higher cylinder temperatures and pressures) from BDL timing that is too advanced.

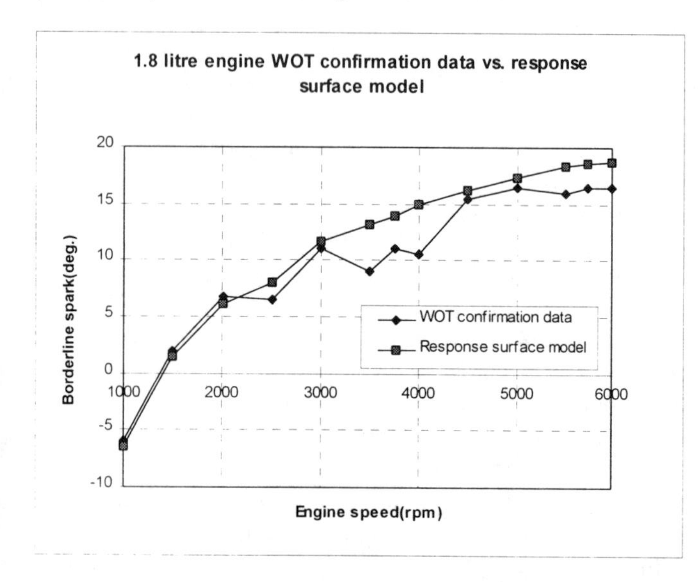

Figure 1 1.8 litre engine WOT BDL confirmation data vs. BDL response surface model prediction

There was evidence that the WOT BDL data exhibited third order behaviour with engine speed, as shown in figure 1.

It is possible to establish if third order response surface model terms are statistically significant [5], from analysis of the so-called 'alias' or 'bias' matrix. This matrix indicates how the coefficients of the second order model terms are biased by the coefficients of third order model terms. Having constructed the bias matrix, it was observed that the biased second order terms were confounded with linearly dependent groups of third order terms. Providing only one of the third order terms from each of these 'confounded groups' was selected, it was then possible to add these third order terms to the second order model. Least squares regression was then used on such an augmented model to determine if any of the third order model terms were statistically significant. From the results shown in Table 2 below, it can be seen that the $N_c^2 L_c$ was significant, based on rejecting terms with a probability of significance > 0.05. Thus an experimental design more elaborate than the Box-Behnken response surface design, was justified to all allow such third order terms to be estimated.

Extra Sum of Squares Method for Individual Coefficient Analysis				
Missing Coefficient	R Squared	Adjusted R Squared	F_o	Probability
N	0.870	0.783	112.2	8.34E-14
L	0.911	0.851	62.8	4.40E-10
LAM	0.952	0.919	13.6	6.07E-4
ECT	0.952	0.919	13.3	6.94E-4
N2	0.910	0.849	64.3	3.23E-10
L2	0.957	0.927	7.4	9.10E-3
LAM2	0.949	0.914	17.3	1.43E-4
NL	0.951	0.919	13.7	5.87E-4
LECT	0.955	0.925	9.5	3.48E-3
LAMACT	0.959	0.931	4.9	0.03
N2L	**0.957**	**0.928**	**7.0**	**0.01**

Table 2: Results from testing the addition of selected third order terms to the original second order model for significance using the extra-sum-of-squares method

Development of an augmented designed experiment - The aliasing that occurs with including third order model terms in the Box-Behnken response surface design can only be elucidated if there are further experimental runs. In otherwords the Box-Behnken response surface design had to be extended. One approach is to augment the Box-Behnken response surface design with a ¼ fraction, two level experiment, which for a five (six) factor experiment is an additional 8 (16) runs. The resulting augmented experiment for five factors consists of 54 data points (6 of which are centre points) and for six factors it consists of 72 runs (8 centre points).

The augmented second order BDL response surface model is specified in EQ (5) below.

BDL (deg. BTDC) =

$$\beta_0 + \beta_N N_c + \beta_L L_c + \beta_{LAM} LAM_c + \beta_{EGR} EGR_c + \beta_{ECT} ECT_c + \beta_{ACT} ACT_c + \beta_{NN} N_c^2 + \beta_{LL} L_c^2 + \beta_{LAMLAM} LAM_c^2 + \beta_{EGREGR} EGR_c^2 + \beta_{ECTECT} ECT_c^2 + \beta_{ACTACT} ACT_c^2 + \beta_{NL} N_c L_c + \beta_{NLAM} N_c LAM_c + \beta_{NEGR} N_c EGR_c + \beta_{NECT} N_c ECT_c + \beta_{NACT} N_c ACT_c + \beta_{LLAM} L_c LAM_c + \beta_{LEGR} L_c EGR_c + \beta_{LECT} L_c ECT_c + \beta_{LACT} L_c ACT_c + \beta_{LAMEGR} LAM_c EGR_c + \beta_{LAMECT} LAM_c ECT_c + \beta_{LAMACT} LAM_c ACT_c + \beta_{EGRECT} EGR_c ECT_c + \beta_{EGRACT} EGR_c ACT_c + \beta_{ECTACT} ECT_c ACT_c + \beta_{NNN} N_c^3 + \beta_{NNL} N_c^2 L_c + \beta_{NNLAM} N_c^2 LAM_c + \beta_{LLL} L_c^3 + \beta_{LLN} L_c^2 N_c + \beta_{LLLAM} L_c^2 LAM_c + \beta_{LAMLAMLAM} LAM_c^3 + \beta_{LAMLAMN} LAM_c^2 N_c + \beta_{LAMLL} LAM_c^2 L_c + \beta_{NLLAM} N_c L_c LAM_c \quad \text{EQ (5)}$$

The corresponding model term assignment is shown below in Table 3.

BDL look-up table name(s)	Augmented response surface model term allocation
Base table	$\beta_0 + \beta_{NNN} N_c^3 + \beta_{NN} N_c^2 + \beta_{LLL} L_c^3 + \beta_{LL} L_c^2 + \beta_N N_c + \beta_L L_c + \beta_{NL} N_c L_c + \beta_{NNL} N_c^2 L_c + \beta_{NLL} N_c L_c^2$
A/F offset table	$\beta_{LAM} LAM_c + \beta_{LAMLAM} LAM_c^2 + \beta_{LAMLAMLAM} LAM_c^3 + \beta_{NLAM} N_c LAM_c + \beta_{NNLAM} N_c^2 LAM_c + \beta_{NLAMLAM} N_c LAM_c^2$
EGR offset table	$EGR \times (\beta_{EGR} + \beta_{NEGR} N_c + \beta_{LEGR} L_c)$
ECT offset table	$ECT_c \times (\beta_{ECT} + \beta_{NECT} N_c + \beta_{LECT} L_c)$
ACT offset table	$ACT_c \times (\beta_{ACT} + \beta_{NACT} N_c + \beta_{LACT} L_c)$

Table 3: Assignment of the BDL augmented 2nd order response surface model terms to the BDL look-up tables

For the augmented 2nd order response surface model terms which cannot be represented by the existing look-up tables, the same two scenarios apply as with the original 2nd order response surface model. In this case only 25 of the 38 augmented response surface model terms can be represented by the existing look-up tables available.

Augmented Box-Behnken experiments were carried out on 1.0 and 1.4 litre, 4 cylinder engine designs. Least squares regression was used as before to develop augmented 2nd order response surface models. Confirmation runs were then carried out on both engines to confirm the quality of the model predictions. The Prediction interval plots shown below in figures 2a) and 2b), indicate that the majority of confirmation data points fall within the Prediction interval and thus the models were adequate.

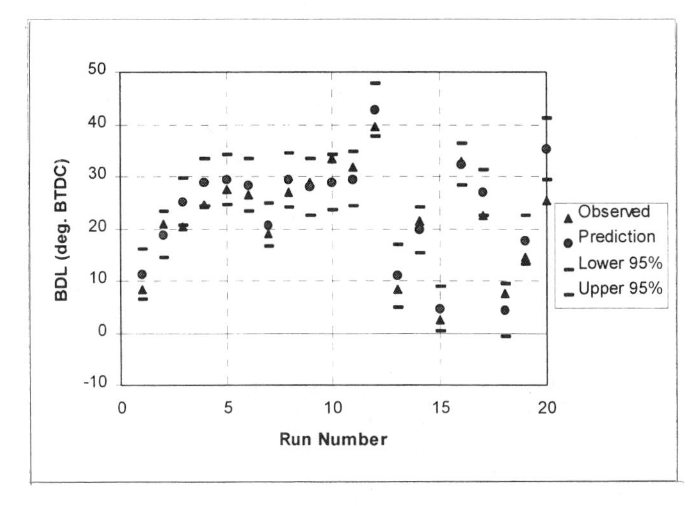

Figure 2a) 1.0 litre engine confirmation data ('Observed') vs. augmented 2nd order response surface model prediction

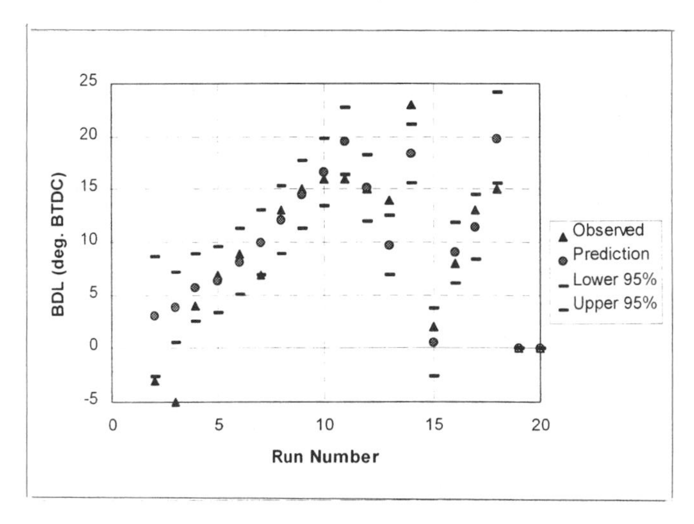

Figure 2b) 1.4 litre engine confirmation data ('Observed') vs. augmented 2nd order response surface model prediction

Furthermore the augmented 2nd order response surface model characteristic at WOT, as shown below in figure 3, was a significant improvement over the 2nd order response surface model.

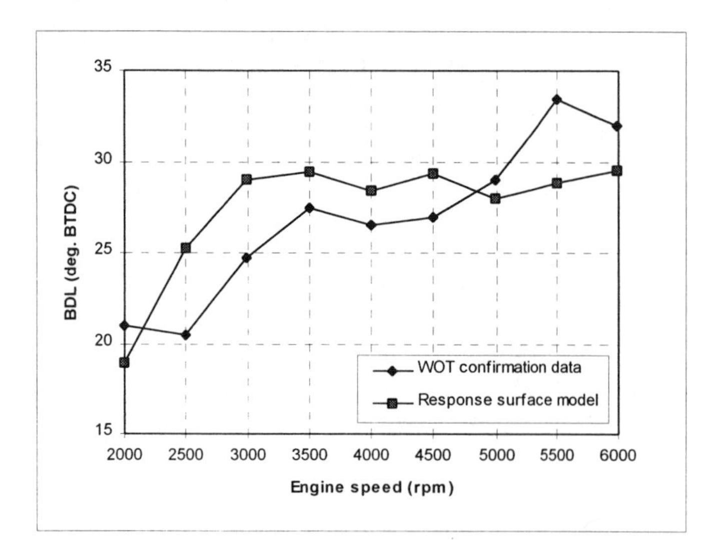

Figure 3 1.0 litre engine WOT BDL confirmation data vs. augmented 2nd order response surface model prediction

FITTING THE LOOK-UP TABLES TO THE EMPIRICAL MODEL

BACKGROUND - It has previously been demonstrated with a one-dimensional look-up table that it is possible to achieve the best fit of a look-up table to a model. This was achieved by minimising the integral squared difference between the look-up table and model by adjusting the position of the look-up table breakpoints. For the purposes of this study it was assumed that the number of look-up table rows (columns) equals the number of row (column) breakpoints.

FORMULATION OF THE OPTIMISATION PROBLEM FOR A TWO-DIMENSIONAL LOOK-UP TABLE - From the bilinear interpolation formula for a two-dimensional look-up table [10], any point, z, can be represented as:

$$z(x,y) = c + a_1 x + a_2 y + a_3 xy \qquad \text{EQ (6)}$$

which, for the Base engine speed (N) versus engine load (L) look-up table

$$z(N_{c_i}, L_{c_j}) = c + a_1 N_{c_i} + a_2 L_{c_j} + a_3 N_{c_i} L_{c_j}$$

EQ (7)

where a_1, a_2, a_3 are the coefficients for each look-up table cell, and the suffices 'i' and 'j' are the look-up table coded breakpoints.

The integral squared difference between that part of the response surface model assigned to the Base table (from hereon referred to as the Base table response surface model) and the Base table surface is a double integral. This was achieved using the MATLAB Symbolic Math Toolbox [11]. The double integral is shown below, using the 1.0 litre engine response surface model from EQ (8),

1.0 litre engine BDL response surface model (deg. BTDC) =

$$\begin{aligned}&31.75 + 22.26N_c - 10.73L_c - 18.93LAM_c - 5.13ECT_c - \\ &3.84ACT_c - 8.13N_c^2 - 1.3L_c^2 + 2.3LAM_c^2 - 1.03ACT_c^2 - \\ &1.92N_cLAM_c + 1.45N_cECT_c + 3.83L_cLAM_c + \\ &1.7L_cECT_c + 1L_cACT_c + 2.55LAM_cECT_c + \\ &0.61ECT_cACT_c - 7.34N_c^3 + 9.84LAM_c^3 + 3.16N_c^2L_c + \\ &6.87N_c^2LAM_c + 3.89L_c^2LAM_c - 2.73LAM_c^2L_c\end{aligned}$$

EQ (8)

Integral squared difference =

$$\int_{j=1}^{j=r}\int_{i=1}^{i=p}\left(\begin{array}{l}(31.75 + 22.26N_{c_i} - 10.73L_{c_j} \\ -8.13N_{c_i}^2 - 1.3L_{c_j}^2 - 7.34N_{c_i}^3 \\ +3.16N_{c_i}^2L_{c_j}) \\ -(c + a_1N_{c_i} + a_2L_{c_j} + a_3N_{c_i}L_{c_j})\end{array}\right)^2 dN_{c_i}dL_{c_j}$$

EQ (9)

where the Base table response surface model, $z\left(N_{c_i}, L_{c_j}\right)$,

$$z\left(N_{c_i}, L_{c_j}\right) = 31.75 + 22.26N_{c_i} - 10.73L_{c_j} - 8.13N_{c_i}^2 - 1.3L_{c_j}^2 - 7.34N_{c_i}^3 + 3.16N_{c_i}^2L_{c_j}$$

EQ (10)

and p and r are the default number of engine speed breakpoints minus 2 and the default number of engine load breakpoints minus 2, respectively. The 'minus 2' is because the end breakpoints in both dimensions of the look-up table are fixed.

The optimisation problem is defined as:

minimise:

$$\int_{j=1}^{j=r}\int_{i=1}^{i=p}\begin{pmatrix}(31.75+22.26N_{c_i}-10.73L_{c_j} \\ -8.13N_{c_i}^{\ 2}-1.3L_{c_j}^{\ 2}-7.34N_{c_i}^{\ 3} \\ +3.16N_{c_i}^{\ 2}L_{c_j}) \\ -(c+a_1N_{c_i}+a_2L_{c_j}+a_3N_{c_i}L_{c_j})\end{pmatrix}^2 dN_{c_i}dL_{c_j}$$

by varying $i_1, i_2, ... i_p$ and $j_1, j_2, ... j_r$

subject to: $i < i+1,\ j < j+1$

EQ (11)

IMPLEMENTATION IN MATLAB - The only constrained optimisation routine provided by the MATLAB Optimisation Toolbox is the '**constr**' routine which uses a Sequential Quadratic Programming (SQP) method. SQP methods represent the latest in non-linear programming methods [12]. Schittkowski [13] has implemented a version which out performs every other tested method over a large number of test problems.

Initial look-up table breakpoint values were determined visually by plotting the Base table response surface model in both engine speed and engine load dimensions and then placing more (less) breakpoints where the curvature (or rate of change of gradient) is greater (less). The plots and the initial breakpoint position selections are shown in figures 4 and 5.

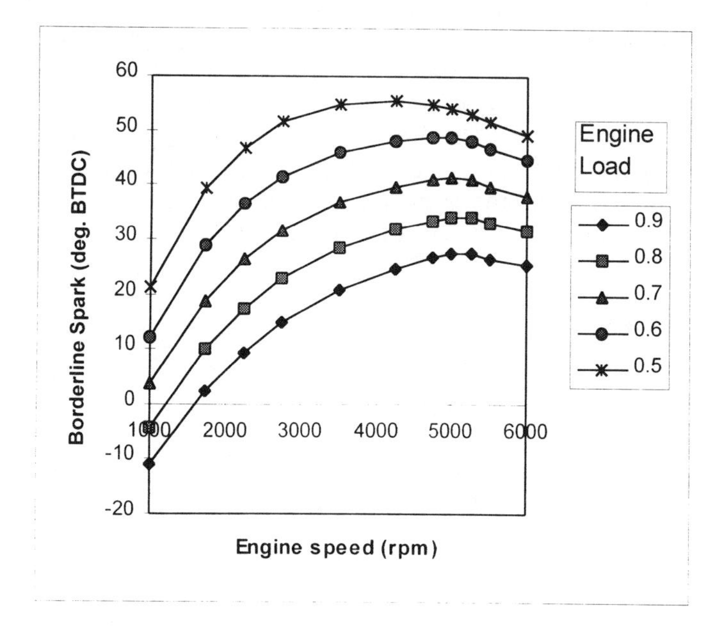

Figure 4 Initial breakpoint position selections for engine speed

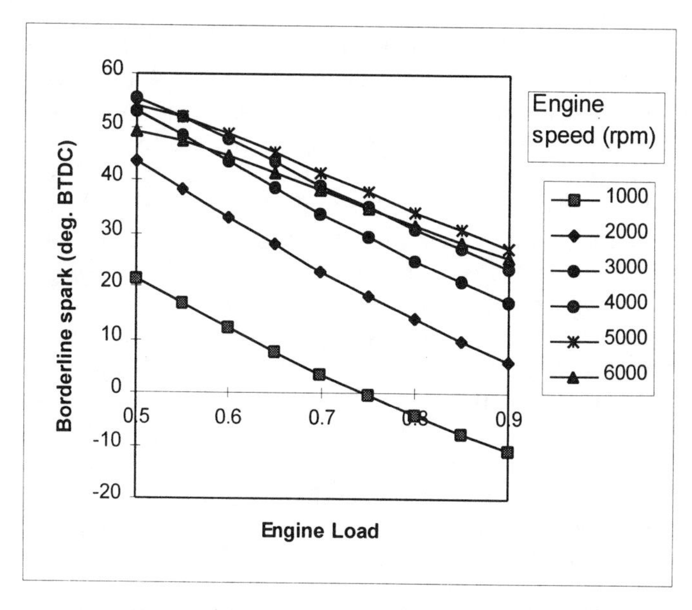

Figure 5 Initial breakpoint position selections for engine load

Gaussian elimination was used to solve for the matrix of coefficients, $\boldsymbol{a}$, for each look-up table cell, from EQ (7):

$$\mathbf{a} = (\mathbf{1}+\mathbf{N}+\mathbf{L}+\mathbf{NL})^{-1}\mathbf{z} \qquad \text{EQ (12)}$$

where $\mathbf{z}$ represents the Base table response surface model matrix

From EQ (11), constraints were necessary to prevent the coefficient equation, EQ (12), from becoming singular and therefore impossible to solve. The constraints were specified in the form of a small numerical difference between adjacent coded breakpoints using the '**diff**' MATLAB function. In MATLAB the constraint matrix, $\mathbf{g}$ is specified such that:

$$\mathbf{g} < 0 \qquad \text{EQ (13)}$$

DISCUSSION OF RESULTS - Firstly the symbolic equation of the integral squared difference EQ (11) was formed in MATLAB, based on the default Base table size (9 engine speed columns, 7 engine load rows). The constrained optimisation routine, '**constr**', was then called with the previously defined initial look-up table breakpoint values. Convergence to the following solution was achieved.

opt_speed_brkpts (coded) =

[-0.81, -0.58, -0.15, 0.07, 0.26, 0.43, 0.58, 0.73, 0.87]

opt_load_brkpts (coded) =

[-0.79, -0.57, -0.34. -0.10, 0.15, 0.41, 0.69]

This was proved to be a global solution by re-running the optimisation process from different initial look-up table breakpoint values to achieve the same result. A three-dimensional plot of the look-up table based on the optimal breakpoint positions, displaying each of the look-up table cells is shown in figure 6. In the engine speed dimension it can be seen that the breakpoints have been placed where the rate of change of gradient is changing. In the engine load dimension the rate of change of gradient is constant and so the breakpoints are equi-spaced. These results can be predicted from analysis of the response surface model equation, EQ (10). The highest order engine load term is quadratic, whose second derivative (or rate of change of gradient) is therefore constant. However there is a cubic engine speed term whose second derivative is dependent on engine speed.

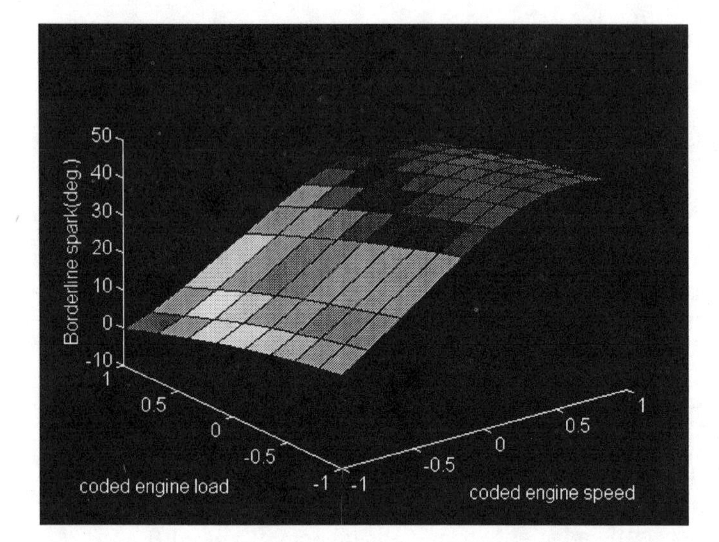

Figure 6 Base table surface based on optimal breakpoint positions for the default look-up table size.

The maximum interpolation error (MIE) between the response surface model and the look-up table can also be defined. This maximum interpolation error is of more relevance to engineers calibrating BDL, than the integral squared difference between the response surface model and the look-up table. To determine the maximum interpolation error, each Base table cell was divided up into a 10 by 10 sub-grid. For each one of the sub-grid cells the bilinear interpolation equation, EQ (7), and the Base table response surface model equation, EQ (11) were re-evaluated. The maximum interpolation error across all sub-grid cells for all look-up table cells was then determined. For the default table size the maximum interpolation error was 0.16 degrees BTDC BDL. A three-dimensional plot of the interpolation error between the Base table and response surface model is shown in figure 7. It indicates how well the Base table fits the response surface model across the range of coded engine speed and coded engine load.

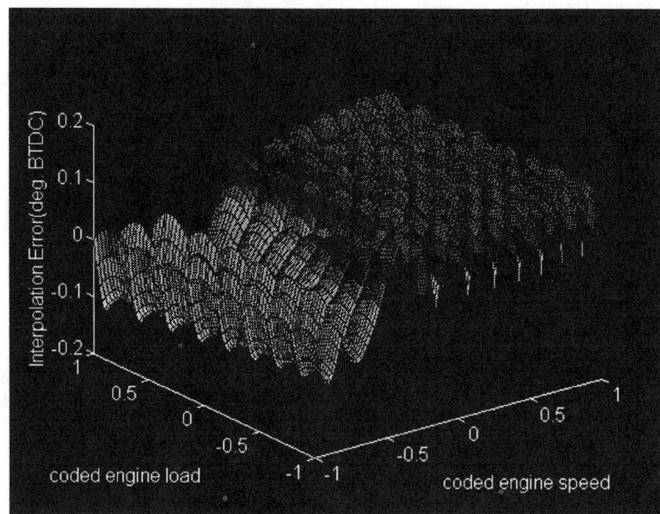

Figure 7 Three dimensional plot of interpolation error

A similar process of determining the optimal breakpoint positions and the consequent maximum interpolation error was carried out for each of the other look-up tables in the strategy (as described Table 1). This resulted in a total maximum interpolation error between the full BDL response surface model and all the look-up tables specified in Table 1 of 0.54 degrees BTDC BDL. If the total maximum interpolation error that is acceptable is specified in advance then the look-up table sizes can be defined. It is interesting to compare the default look-up tables provided by the Ford powertrain control strategy and those required from a process of fitting the response surface model assuming a user-specified total (across all the look-up tables) maximum interpolation error of approximately 1 degree BTDC BDL. (Also assumed is that the optimal look-up table breakpoint positions have already been determined). It can be seen that the total number of look-up table cells required to satisfy a 1.09 degree BTDC BDL maximum interpolation error is 47 compared to the 239 look-up table cells required to satisfy a 0.54 degree BTDC BDL maximum interpolation error. In otherwords a saving in look-up table cells of 80% results if the maximum interpolation error is doubled.

Look-up table name	Default look-up table size (rows by cols.)	MIE (deg. BTDC BDL spark)	Revised look-up table size (rows by cols.)	MIE (deg. BTDC BDL spark)
Base table	9 by 11	0.16	3 by 5	0.42
A/F offset table	8 by 10	0.38	4 by 6	0.67
ECT offset table	5 by 6	0	2 by 2	
ACT offset table	5 by 6	0	2 by 2	
	Total cells = 239	Total MIE = 0.54	Total cells = 47	Total MIE = 1.09

Table 2: Look-up table size comparison between the existing Ford powertrain control system and those required to satisfy a maximum interpolation error of approximately 1 degree BTDC BDL

It should be noted that the following terms in the full BDL response surface model, shown below in Table 3, cannot be represented in the look-up tables that currently exist.

Augmented 2nd order BDL response surface model term	Regression Coefficient
$L_C LAM_C$	3.83
$L_C^2 LAM_C$	3.89
$LAM_C^2 L_C$	2.73
$LAM_C ECT_C$	2.55
$ECT_C ACT_C$	0.61

Table 3: Full BDL response surface model terms not represented by the existing look-up tables

These terms would require additional look-up tables to characterise them. Alternatively the terms could be neglected resulting in an error in the BDL calibration represented by the look-up tables. The magnitude of this error is ± the regression coefficient.

CONCLUSION

Based on this study a new Borderline spark calibration process has been defined, which can be applied to many other powertrain calibration processes. A generic form is shown in Attachment 1: figure 8 below. In broad terms, a methodology has been proposed to define the powertrain control strategy to best represent engine operating behaviour.

For Borderline spark the use of a designed experiment has reduced the duration of the data collection required by 75% - this represents a reduction in time from six weeks to one and a half weeks.

Based on historical data, a second order model was proposed for Borderline spark and five and six factor Box-Behnken response surface experiments were designed to allow such a model to be generated. Stepwise least squares techniques were used to determine the model coefficients. Statistical methods were employed to ensure that a minimal set of significant effects were employed for the purposes of prediction. Initially four engines were tested and response surface models developed using this approach. With all four engines the models developed did not confirm adequately. Furthermore there was some evidence of third order Borderline spark behaviour with engine speed.

Analysis of the alias matrix formed by adding third order model terms to the Box-Behnken response surface experiment, and least squares regression analysis, allowed the presence of significant third order model terms to be detected. Consequently the Box-Behnken response surface experiment was augmented with a ¼ fraction two level experiment.

The powertrain control strategy contains look-up tables were then fitted to the response surface model. Using MATLAB, symbolic integrals together with Sequential Quadratic Programming were successfully applied to the 1.0 litre engine data to minimise the integral squared error between the look-up table and response model surfaces. One of the MATLAB routines developed was used to demonstrate that the look up table size can be defined to satisfy a user-specified maximum interpolation error between the look-up table and response model surfaces. Using the smallest acceptable size for each of the look-up tables is the most operationally efficient for the powertrain controller.

ACKNOWLEDGMENTS

My thanks to development engineers: Martin Moore, Peter Wintle, Karl Moore, Mel Coldwell and Steve

Valentine of the Engine Performance and Economy department, Ford R&E Centre, Dunton, UK for assistance with the Borderline spark experimental data collection. I am very grateful for the statistical support in the form of experimental design and least squares regression analysis provided by Mark Cary and David Hilton. I also thank Ian Noell of Cambridge Controls Ltd. for his patience in providing me with MATLAB training.

REFERENCES

1. Deming, W. E. T, *Out of the Crisis*, Cambridge University Press, Cambridge, 1986.
2. Ulrey, J., Ford Mapping database (approx. 1000 maps of different engines)
3. Montgomery, D., Peck, E. A., *Introduction to Linear Regression Analysis*, Wiley & Sons, 1992.
4. Hilton, D. H., CMT Regression Add-In for Excel, Ford Mo. Co. Ltd., 1996
5. Box, G. E. P, Draper, N. R., *Empirical Model-Building and Response Surfaces*, John Wiley & Sons, 1992
6. Khuri, A. I., Cornell, J. A., *Response Surfaces: Design and Analyses*, Marcel Dekker, 1987
7. Montgomery, D. C., *Design and Analysis of Experiments*, John Wiley and Sons, 1991
8. Box, G. E. P., Behnken, D. W.: Some new three level designs for the study of quantitative variables. *Technometrics*, 2, 455-475. Corrections, 3, 1961, p.576.
9. Lygoe, R. J., The Application of Response Surface and Optimisation Methods to Borderline Spark Calibration, MSc Advanced Automotive Engineering Project, Loughborough, 1997
10. Press, W. H., Teukolsky, S. A., Vetterling, W. T., Flannery, B. P.: Numerical Recipes in C, Cambridge University Press, 1992
11. Moler, C., Costa, P. J., MATLAB Symbolic Math Toolbox v1.0a, The MathWorks Inc., 1994
12. Grace A. W.: The MATLAB Optimization Toolbox User's Guide, The Mathworks Inc., 1994.
13. Schittkowski, K. W.: NLQPL: A FORTRAN-subroutine Solving Constrained Nonlinear Programming Problems, Operations Research, Vol. 5, 485-500, 1985

DEFINITIONS, ACRONYMS, ABBREVIATIONS

ACT	Air Charge Temperature
A/F	Air Fuel ratio
BDL	Borderline spark
BTDC	Before Top Dead Centre
ECT	Engine Coolant Temperature
EGR	Exhaust Gas Recirculation
MIE	Maximum Interpolation Error
PRESS	Predicted Error Sum of Squares
SQP	Sequential Quadratic Programming
WOT	Wide Open Throttle

ATTACHMENT 1

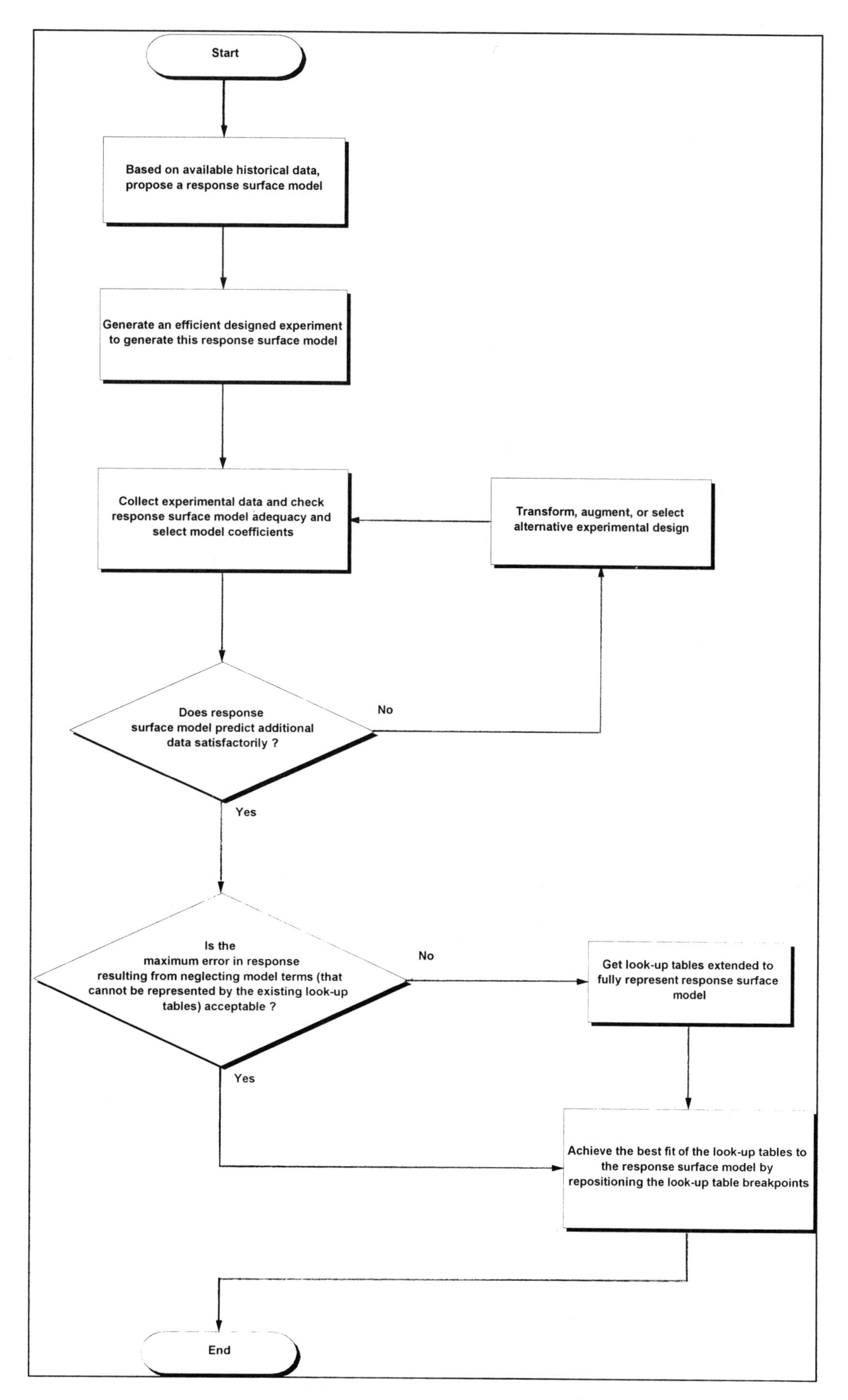

Figure 8 A generic form of the Borderline spark calibration process

WHAT THE FUTURE HOLDS

Future Developments in Automotive Microcontrollers

Ross Bannatyne
Transportation Systems Group
Motorola Inc.

The object of this paper is to discuss the changes that will occur in the use and implementation of automotive microcontrollers in the next generations of systems. The functions that will be required to be maintained by the microcontroller and the types of technology that will enable these functions are addressed.

MICROCONTROLLER PENETRATION

The next ten years will see as steady a growth in automotive electronic systems as the last ten. Driving this growth will be consumer demand for enhanced safety features, entertainment systems, and added convenience functions, as well as government edicts on emissions controls.

Figure 1 illustrates the implementation growth of microcontrollers in automobiles. Three categories are shown: low-end, mid-range, and luxury vehicles. Typically, systems that are first introduced on luxury vehicles migrate eventually down into lower-priced vehicles. The length of time for this migration depends on how quickly the costs of the systems can be reduced and how valuable the systems are perceived to be by consumers.

The electronic control systems currently used in vehicles are each being enhanced and expanded. For instance, it is expected that the two frontal airbags that are common today will be joined by further frontal airbags (feet, knees), several sidebags, rear passenger airbags, and advanced sensing systems for detecting occupant position. Similarly, braking, steering, suspension, powertrain, and body-control functions will be developed further by implementing more advanced electronic systems.

In addition to enhancing and expanding current systems, many altogether new systems will be added. In the next few years, systems such as GPS (global positioning system)-based navigation systems, stability management systems, "by-wire" braking and steering systems, collision warning and avoidance systems, voice recognition, Internet access, and night-vision enhancement will be introduced—and these examples are just the beginning!

The networking of new and existing systems is another factor supporting increased use of electronic control systems in vehicles. There are many benefits of networks in the automobile. One benefit it that data can be shared in real time. From a control systems standpoint, this means that more intelligent systems are made possible. For example, an integrated chassis-control system layer may be implemented by coordinating the data generated by the braking, steering, and suspension systems. Another benefit of networks is that "second guessing"—the practice of using data from one system to check the plausibility of the results of another independent system—becomes easier. Under certain conditions, data from one system could be used as a back-up for another system. For example, the wheel speed and vehicle directional information used in a stability-management system could be used to supplement the navigation system, especially in the event that GPS is lost.

MICROCONTROLLER FUNCTIONS

To deal with the new range of system requirements, the automotive microcontroller will be developed further to facilitate new functions.

Communications. The implementation of communications between electronic control units (ECUs) is a growing trend. In vehicles, multiplexed communications systems were developed originally to reduce weight, interconnections, cost, and complexity. It soon became apparent, however, that by providing the opportunity to share data in real time from different ECUs, vehicular systems could be enhanced greatly.

Unfortunately, a single communications protocol that cost-effectively addresses every automotive application does not exist. Instead a number of different communications subsystems are integrated together in a modern vehicle (Figure 2).

Figure 2 shows five distinct communications systems that are implemented on a vehicle. It is likely that there will be a number of additional "sub-bus" networks for sensors and actuators. Gateways exist between these networks to share information across "boundaries." The chassis control functions are grouped on a redundant network which meets fault-tolerant criteria: only safety critical information is allowed on these buses. Likewise, a highly robust independent network is provided for the airbag system.

Many new automotive microcontrollers will have more silicon devoted to communications capabilities than CPUs. Already, microcontrollers such as the M68HC912DG128 are being offered with two independent CAN (Controller Area Network) modules, along with several more synchronous and asynchronous communications systems. These communication interfaces are as autonomous as possible, so that the CPU need not devote a great deal of overhead to managing communications.

THE NETWORKED VEHICLE

Algorithm Complexity. The increasing complexity of automotive electronic systems has had a dramatic effect on the throughput requirements and peripheral integration of automotive microcontrollers. Algorithms are now required to handle the inputs from many sensors and communications systems, to execute real-time control cycles, and to control the outputs of many actuators.

Figure 3 illustrates the effect that this growing complexity has had on the physical characteristics of microcontrollers. Three generations of powertrain microcontrollers are shown. Over these three generations, the microcontroller has become about 100 times more powerful in terms of CPU throughput; the program memory (holding the algorithm) has grown 40 times larger; and the number of transistors on the chip has increased by a factor of 300. The powertrain application is by no means unique. Many controller systems in the vehicle have kept pace with these developments. For example, 32-bit RISC processors are being used for new generations of airbag and antilock braking systems (ABS).

Algorithm complexity is also leading to the widespread implementation of operating systems. Although most operating systems are still developed in-house by the application specialists, the industry will migrate very quickly toward standardization of the operating system and network management. The OSEK/VDX operating system has been adopted by many as the open standard. (OSEK is a German acronym for Offene System und deren Schnittstellen fur die Elektronik im Kraftfahzeug.) This standard was developed specifically to decouple the application code (algorithm) from the network management tasks and avoid incompatibility problems between the application code and the hardware. It includes a standardized application programming interface, behavior, and protocol. Implementation of OSEK/VDX should facilitate reusability and portability of software and also predictable system behavior. It is expected that many automotive microcontrollers in the near future will be implemented with an OSEK operating system.

Smart Nodes. As electronic control systems and communications networks grow, leading to an increased number of microcontrollers in the vehicle, there will be a growth in the number of sensors and actuators required. In conventional systems, sensors and actuators are connected directly to the appropriate electronic control unit, usually by a twisted pair. As the number of sensors and actuators increases, there are several problems that must be faced. For example, handling their interrupts or sampling their outputs and providing different interfaces for each sensor/actuator becomes problematic, as does processing and sharing the (usually analog) information that is exchanged between the control unit and the sensor/actuator.

As the cost of electronics comes down and the need for simplification of such systems increases, it will become more common to implement smart sensors and actuators. A smart sensor/actuator need not include a microcontroller (but most will have a small CPU), and will be connected to the ECU on a bus system. Much of the growth in automotive microcontrollers, shown in Figure 1, will be a result of smart nodes.

Today's smart nodes are composed usually of a package containing a microcontroller die, sensor die, and sometimes an analog interfacing die (depending on how much functionality is included on the sensor die). It is anticipated that it will be cost-effective in the not-too-distant future to integrate MicroElectroMechanical sensors (MEMs) and microcontrollers on a single monolithic silicon chip. Such a concept for a pressure sensor is shown in Figure 4. The sensing element, analog interface, and microcontroller functionality are all contained on a single silicon die.

Safety Critical Operation. Microcontrollers have been at the heart of safety critical systems for many years. Almost all of the safety critical automotive systems in which they have been used have provided a fail-safe function. In the near future, there will be an added requirement for fault-tolerant microcontroller-based systems.

There is an important difference between fail-safe systems and fault-tolerant systems. Today's ABS systems are fail-safe; that is, if an electrical system error is detected, the ECU switches to a safe "off" mode, allowing the foundation hydraulic brakes to operate without the faulty ABS system interfering. A fault-tolerant system, on the other hand, would not only have to recognize that an electrical fault had occurred, but would have to continue to operate safely with the existing known fault.

ABS systems use redundancy to facilitate a fail-safe system. Typically, the CPU at the heart of the system supervises the continual testing of all the major system components. The CPU can validate these components only if the CPU itself is known to be "sane". Hence a second, redundant CPU is used the validate the sanity of the first CPU.

A redundant CPU can be implemented either as a second stand-alone microcontroller or as an error-detection CPU with comparison logic on the same microcontroller. This type of dual configuration is shown in Figure 5. If the two CPUs disagree over the result of an instruction execution, the "fault" signal will be enabled and an interrupt will initiate a sequence of events to switch the control unit into the safe "off" mode. Dual CPU microcontrollers will increase in popularity in automotive safety critical applications such as steering and airbags, as well as ABS.

A fail-safe system will not suffice for emerging automotive applications such as brake-by-wire and steer-by-wire. These will require a fault-tolerant solution, as a hydraulic back-up mode will not exist. If a fault is detected in such a system, it must continue to operate safely in a "limp home" mode. Simple redundancy of components and a voting algorithm can be used to satisfy this requirement, however, this is normally a very expensive solution (as voting requires at least three CPUs). A further solution is to use time-triggered communications protocols (such as TTP/C), an area in which an increasing amount of work has been done. A cost-effective microcontroller-based control system that provides fault tolerance will require control modules (like a TTP/C controller) integrated with microcontrollers to support the new fault-tolerant communications systems.

MICROCONTROLLER TECHNOLOGY

Because of advanced system requirements, such as increased communications, algorithm complexity, and safety critical requirements, microcontroller technology requirements will be impacted.

CPU Trends. As algorithm complexity increases, increasing throughput requirement, very high-performance, low-cost CPUs will be required. In addition to providing a very high throughput and fast interrupt handling capabilities, these CPUs must also be conducive to generating efficient dense code when a C compiler is used. Currently, the CPU is usually a relatively small area of the die with respect to memory arrays. If a CPU is designed in a high-level language friendly way, a significant amount of memory (and hence cost) can be saved, as more efficient code is generated by a compiler.

In automotive microcontrollers, there is also a trend to use high-performance RISC CPUs in preference to the traditional CISC units. This trend will continue. Traditionally, the factors that have set RISC apart from CISC are (1) RISC CPUs are able to execute an instruction in a single clock cycle, and (2) RISC machines do not use microcode to decode instructions, but are *hardwired.*

It has been argued that the vast majority of most software consists of very simple instructions. The philosophy of RISC is to produce processors that can execute these simple instructions (such as ADD, SUB, SHIFT, etc.) in one clock cycle. More complex instructions, such as MUL and DIV, were not available on early RISC processors.

When an opcode is generated by a line of software in a CISC machine, this opcode is basically an address for a *microcode memory* (also sometimes called a control store), pointing to a certain string of control bits that are applied to the execution unit (via a small amount of combinational control logic). These bits include many control signals to ensure that the execution unit performs the desired function. This microcode memory is a regularly structured array that is straightforward to design, and offers flexibility to the designer, allowing him or her to design the optimum microcode.

A RISC processor does not have a microcode memory. Therefore, the opcode that is generated in software is applied directly to a larger array of combinational control logic in order to generate all the appropriate signals to operate the execution unit. This is a more complex design, which results in a smaller silicon size (as there is no microcode memory) and usually faster operating speed.

The distinction of RISC as being capable of single clock instructions is becoming more vague, as many RISC CPUs now include complex instructions that can take several cycles to execute. RISC processors allow operations to be performed only directly between registers in the CPU. This means that if there are two bytes in memory that are to be ANDed together, they must first be loaded into CPU registers. CISC machines would normally allow the ANDing of a user register with a location outside the programmer's model.

Memory Trends. The standard microcontroller memory types that have become established in automotive systems are ROM, EPROM, and Flash EEPROM for program store; RAM for stack and scratch-pad memory; and byte erasable EEPROM for storage of calibration and security data. As Flash EEPROM is becoming more cost-effective, it will replace ROM ultimately as the favored memory solution.

Automobile manufacturers often wish to revise software in the field. Unless there is a method of reprogramming the memory array remotely, this typically requires that the sealed ECU be removed and replaced—a time-consuming and expensive process for the automobile manufacturer, not to mention a greater inconvenience to the owner. Flash EEPROM provides the technology that allows such field revisions.

Flash EEPROM can be used for software upgrades as well as software revisions. It is now becoming more common for ECU suppliers to build generic platforms which may have slightly different features implemented in software. For example, some automobiles adjust automatically the side-view mirrors (which turn to point to the rear wheels) when the vehicle is placed in reverse gear. This feature is implemented in software and feasibly could be offered as a software "upgrade" in the after-sales market.

Memory sizes for most automotive applications will continue to grow, leaving the CPU on the microcontroller as a relatively small portion of the chip. Figure 6 illustrates a die photograph of the M68HC912DG128, set to be used in several automotive applications. The relative sizes of the memory modules and other microcontroller functions can be seen clearly in the photograph. The microcontroller shown has 128K of Flash EEPROM memory in four 32K arrays. Each array can be bulk erased separately. Having only two arrays of 64K would reduce the die size, but erasing the memory would not be as flexible. In addition, a 4K RAM is included and 2K of byte erasable EEPROM. Typically, a ratio of 32:1 is considered appropriate between program memory and RAM on modern microcontrollers. Certain applications that are RAM intensive may require more, although modern design techniques are usually helpful in estimating exact code-size requirements early in the design cycle.

The memory system in future automotive microcontrollers may be enhanced in certain cases to allow simplified memory verification. In a safety critical system, it may be desirable to check that memory contents are stable and have not been corrupted during the course of operation. There are several techniques that can be implemented to validate memory contents, each having its own merits and problems. The most straightforward technique is to implement a parity bit on the entire memory array. Whenever a byte is written to memory, a parity generator adds an extra bit, and whenever a byte is read from memory, a parity checker will ensure that a parity error has not occurred. Another scheme that has been widely used is to perform signature analysis on blocks of memory using a Linear Feedback Shift Register.

Packaging Trends. It can no longer be assumed that automotive microcontroller packaging requirements are conventional PCB-mounted plastic/ceramic units. Processing is becoming distributed around the vehicle as dictated by the locations of sensors and actuators, and packaging requirements are changing accordingly. A mechatronic approach is now being taken for microcontroller packaging.

Mechatronics is a discipline that has arisen from a need to look at systems as a whole, rather than component parts such as electrical/electronic engineering and mechanical engineering. Mechatronics aims to bring about completely optimized systems by integrating the individual components of design into a process. (In contrast, the traditional approach often yields a collection of electrical, mechanical, and hydraulic subsystems interfaced together with a control unit.) There is much opportunity for mechatronic technology to improve today's automobile systems.

Figure 7 shows two approaches to an electronic motor controller. On the left, is an electronic control circuit that uses a number of different interconnect technologies including rivets, solder, surface mount devices, and wire bonding to a silicon die. On the right is an example of a mechatronic solution where an electronic microsystem has been constructed using a multichip module and "connectorized" for robust and efficient interfacing with a motor housing. This type of mechatronic system would usually have many fewer interconnections than the traditional solution.

Both motor control circuits are connected to a bus via a plug. On a modern automobile, this interface is likely to be a communications bus, linking several motor control circuits (such as window lifters, seat positioners, and mirrors).

Noise Reduction. ECUs and all electronic components contained within are tested thoroughly to measure electromagnetic compatibility (EMC) performance. EMC is becoming a bigger issue for automobile manufacturers as the operating speed of electronic components is becoming faster (higher frequencies lead to increased electromagnetic emissions), and as the number of ECUs that would potentially affect one another's operation is increasing. The worst case would be the result of the operation of an ECU being corrupted by radio-frequency emissions (from another ECU or external source from the vehicle), which could be a potential compromise on safety.

EMC can be optimized by careful design of the integrated circuit and the printed circuit board. A system is considered electromagnetically compatible if it satisfies three criteria:

1. It does not cause interference with other systems.
2. It is not susceptible to emissions from other systems.
3. It does not generate interference with itself.

In recent times, the automobile manufacturers have been putting more pressure on the ECU suppliers to produce units with better EMC performance. In turn, this has put pressure on the semiconductor suppliers to produce more robust microcontrollers.

At the integrated circuit design level, there are many considerations that can enhance the EMC performance of the design:

- Using fewer clocks and turning off clocks when not in use
- Reducing output power buffer drive
- Using multiple power and ground pins and reducing internal trace impedance on these pins
- Eliminating integrated charge pump circuitry
- Positioning high frequency signals next to a ground bus

All of these steps are now taken to improve EMC.

Power Consumption. Until just recently, power consumption was never considered to be a priority requirement for automotive ECUs. This situation has changed because of the number of systems that are required to operate when the ignition is turned off. These systems would drain the battery quickly were special attention not given to their power consumption. The door modules are a good example. These ECUs must exist permanently in a "ready" mode in order to recognize a signal from a remote keyless entry (RKE) device.

All automotive MCUs are now optimized for power consumption. This is done mainly by switching off clock sources inside the chip when they are unused. This also reduces emitted noise.

Power consumption is also a consideration for airbag ECUs. They must function in the event that a crash situation disconnects the electrical power supply to the ECU. A large capacitor is used normally to ensure that under these circumstances there is enough energy available to fire the bag(s). By careful design attention to power-consumption requirements, the size of this capacitor may be reduced, thus reducing ECU cost. Airbag microcontrollers are often selected primarily by measuring maximum MCU performance while operating at a speed defined by a given power-consumption limit.

Integration. All microcontrollers have followed an integration roadmap whereby feature sizes have dropped from several microns to around 0.5 μm effective gate width in today's state-of-the-art systems. Feature size shrinkage is set to continue and will result in lower costs, higher operating speeds, and an increased functionality of the chip. One of the challenges that offsets these benefits is that of reduced operating voltage requirements (the smaller gate oxide cannot withstand the higher voltage). In order to function at 0.5 μm, a 3.3 V power supply is required for the CPU. To function below 0.5 μm, a further requirement for a 1.8 V power supply will be imposed. This means that in order to take advantage of advances in the integration technology of the microcontroller, the system will require additional power supply capabilities.

Previously, the development of monolithic smart sensor devices that integrate sensing elements together with microcontrollers were discussed. For a number of years, mixed-signal devices that include high-voltage/high-current capabilities have been produced for automotive applications. Mixed-signal-capable microcontrollers will continue to be developed where it is cost effective to do so, and where it makes sense from a system-partitioning perspective.

Partitioning functions in the ECU and satellite modules (sensors, actuators, other ECUs) is not always a straightforward task. The problem is multidimensional because functions may be implemented in different semiconductor technologies, as well as in hardware or in software. There also may be other considerations, such as the physical location of the function. The first step is usually to isolate each function to be performed (like sampling a sensor output or driving a motor), then examine each possible implementation. When all functions are identified and their interrelationships understood, the lowest-cost/most efficient solution may be determined.

While repartitioning an existing system, it is important to always look at basic functions rather than existing solutions. New technologies may be available that will allow a better implementation of the function. By using a new type of sensor that does not require costly interfacing, a cheaper solution may result. For example, it may be possible for the sensor to have a digital interface integrated which will allow the information to be shared on a common bus with other systems. In all cases, the best solutions will result when there is close cooperation between semiconductor integration experts and systems experts.

The semiconductor vendor is positioned uniquely to assist in partitioning the system. It is possible to integrate power functions onto microcontrollers, digital functions onto analog chips, and signal conditioning and digital interfaces with MEMs sensors (e.g., accelerometers, pressure, etc.). In addition, many other combinations of solution are possible. Software is an important element that must also be considered. As microcontroller bus speeds have increased enormously, it may no longer be necessary to use complex hardware to off-load interrupts; the processor may be more than capable of handling these due to its vastly increased bandwidth. Conversely, a great deal of bandwidth may be consumed by software which could be replaced by minimal hardware. Figure 8 gives an example of a "system chip" with various types of technology implemented on the same silicon.

The most highly integrated product is not necessarily the best product. When reduced cost is the sole motivation for integration, there will be a point on the cost vs. integration curve that determines the optimum level of integration. However, other motivations for increased integration may exist, such as reduced chip-count/size, reduced power consumption, general simplification, and increased reliability.

More highly integrated products also tend to have less generic appeal. This translates to reduced economies of scale, and usually slightly higher costs. Often, however, the additional cost is offset by a more competitive product in the market.

ELECTRONIC DESIGN AUTOMATION

Future automotive microcontrollers will be impacted by changes in the design methodology that is being adopted in the automotive electronics industry. There is a trend to take a systems-engineering approach, an integrated design methodology incorporating electronic, mechanical, hydraulic, and design for manufacturing considerations. Simulation, code generation, optimization, and fast prototyping tools are now used more commonly in order to reduce time to market, to generate more highly integrated designs, and to reduce costs and increase reliability.

The modern automotive microcontroller must now be supported early in the cycle with software models that can be integrated into simulation environments. A conventional system design is illustrated in Figure 9.

The conventional system design process starts with abstracting a software and hardware specification from the systems specification. Both hardware and software are developed independently and are then integrated and tested/debugged. If a functional problem exists at this stage (as it often does), it is usually very difficult to redesign hardware, so it is most common to develop software fixes insofar as possible. The main problem with this methodology is that both the hardware and the software are being debugged at the same time, thus it is more difficult to determine the source of faults. It is estimated that over 50 percent of the entire development cycle is spent at this stage.

Often, a new microcontroller design is undertaken as part of the "develop hardware" effort. This design will usually take an existing CPU and existing peripheral functions (i.e., timers, serial communications, etc.) and integrate these together with the required memory arrays for the particular system. These memory arrays are based on the estimated software size, which is being developed independently, and so they are rarely efficient. Quite often, a new "custom" peripheral is also developed to be integrated with the microcontroller. This could be a "knock" detect module for a powertrain system or a wheel speed interface for an ABS system.

To resolve the problems of optimizing a systems design, a new approach has been developed in which software-based toolsets allow alternative implementations to be evaluated rapidly. The new approach allows trade-offs to be made quickly and an efficient design to be generated with confidence that few or no problems will surface later. The new systems design process that is being adopted is illustrated in Figure 10.

The control algorithm development is now usually a model-based operation. The algorithm is simulated on a workstation in an environment that simulates the behavior of the physical system. The algorithm can be tweaked and simulated in the model on an iterative basis until the algorithm development has been completed.

The control algorithm model, usually mathematically based, is then processed using an Automatic Code Generation (ACG) tool which will provide a high-level language file representative of the control strategy. This high-level language can be compiled to run on any execution-unit-based system (such as a microcontroller). In order to verify the code, it is tested in a hardware environment of inputs (sensors, serial communications links) and outputs (actuators) in order to verify the operation of the control algorithm. After the model-based algorithm development is completed, it is time to partition the hardware and software.

Software tools are used again at the partitioning stage to simulate the system operating with different configurations of hardware and software. The goal is to understand the trade-offs involved in implementing different functions as either software routines or as a dedicated hardware block. For example, a routine that constantly services certain low-level interrupts that occur very frequently might best be implemented as an autonomous (or "smart") peripheral on a microcontroller. The algorithm is run on a model of a CPU and peripheral modules can also be implemented as behavioral models.

The outcome of the hardware/software codesign stage is to determine the final specification for the hardware. For the most part, the software will already exist and have been verified at this stage. The hardware requirements that are identified at this stage will be optimized to meet the system requirements exactly, and a new microcontroller can be developed which suits the specifications exactly. It is estimated that the overall development time required using the model-based development methodology can be about half the time required for the conventional systems design methodology.

The impact on the microcontroller is that quality software models will be required not only of the CPU but of the peripheral modules which could be used in the system. The models will be required very early in the design cycle and it is likely that there will be many different types of models required for use with different toolsets (e.g., Verilog, VHDL, C++). These models will someday be regarded as essential to the design process as a databook.

ADVANCED AUTOMOTIVE MICROCONTROLLERS

Probably the best example of a "next generation" modern automotive microcontroller is the MPC555, which was designed for powertrain control and Intelligent Transportation Systems (ITS) applications. A block diagram of the MPC555 is shown in Fig. 11.

The MPC555 is based on the PowerPC architecture and is composed of over 6.7 million transistors, having over 300 times the complexity of a microcontroller used in a comparable application a decade ago. The 32-bit CPU includes multiple execution units and a floating point. It supports a Harvard architecture with separate load/store and instruction buses for simultaneous instruction fetching and data handling.

The chip is well-equipped with peripherals to interface with the rest of the system. There are 32 analog inputs as well as 48 Timer Processor Unit (TPU) timer-controlled I/O channels. Two CAN (Controller Area Network) serial communications interfaces are also included to provide multiplexed communications with other vehicular systems. The program memory is 448 Kbytes of Flash EEPROM with 26 Kbytes of RAM.

Certain I/O structures have been added to the chip to accommodate 5 V signals around the chip. Although the MPC555 has been developed for a 0.35 μm manufacturing process, it is expected that the technology of other system components will develop more slowly and will still operate with a 5 V power supply and signaling level.

CONCLUSION

This discussion has centered on the functions that next-generation automotive microcontrollers will be required to handle, as well as the technology developments that are being established to satisfy the processing and ancillary performance needs that will arise.

The growth in complexity of microcontroller-based automotive systems has been enormous and is expected to continue. The next major challenges in the development of these systems will be to optimize the efficiency of the controller and associated software by model-based development techniques, provide open architectures, and ensure reusability of hardware/software. As these challenges are met, it will ensure that the perennial requirements of the industry are met: to reduce cost, increase performance, and reduce time to market.

BIBLIOGRAPHY

R. Frank, *Understanding Smart Sensors*, Artech House, 1996.

B. Hedenetz and Belschner, R., Daimler-Benz AG, "Brake-by-wire without Mechanical Backup by Using a TTP-Communication Network," SAE Paper No. 981109, Society of Automotive Engineers, Warrendale, Pa., 1998.

R. K. Jurgen, Ed., *Automotive Electronics Handbook*, McGraw-Hill, 1995.

K. Klein, "Measurement and Modeling of Boron Diffusion in Si and Strained Si (1-x), Ge (x) Epitaxial Layers During Thermal Annealing," *Journal of Applied Physics*, November 1993.

I. Kobeissi, "Noise Reduction Techniques in MCU-Based Systems and PCBs," Motorola Application Note, Austin, Texas, 1997.

C. Kuttner, "Hardware-Software Codesign Using Processor Synthesis," IEEE Design and Test of Computers, Fall 1996.

R. Q. Riley, "Alternative Cars in the 21st Century—A New Personal Transportation Paradigm," R-139, Society of Automotive Engineers, Warrendale, Pa., 1994.

W. Wray, Greenfield, J., and Bannatyne, R., *Using Microprocessors and Microcomputers*, Prentice Hall, 1998 (ISBN 0-13-840406-2).

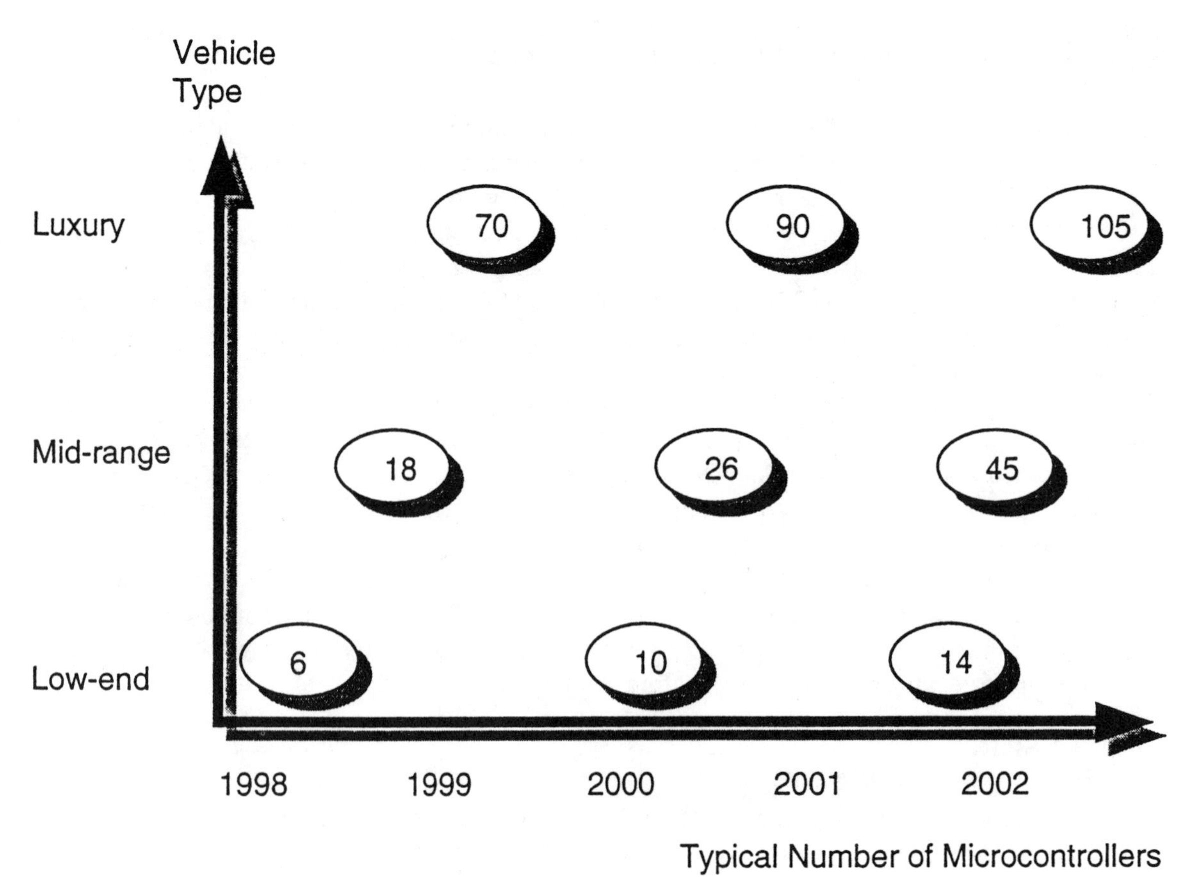

Figure 1. Microcontroller implementation growth in automobiles.

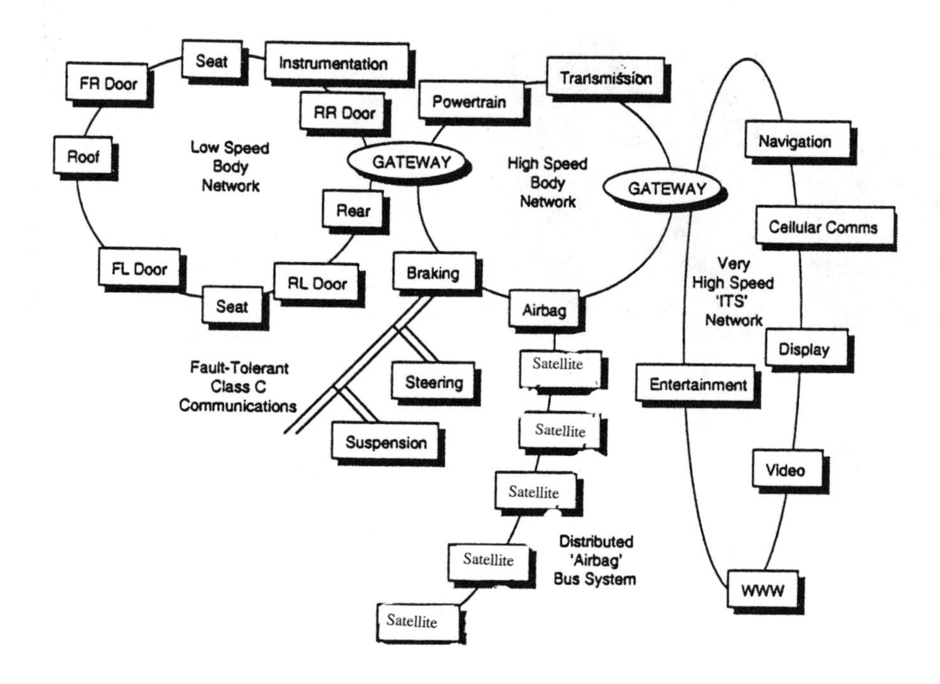

Figure 2. The networked vehicle.

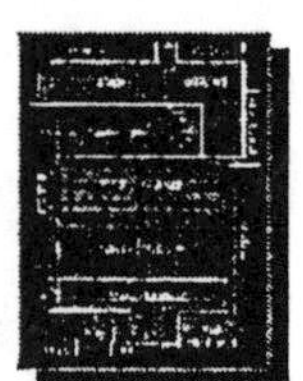

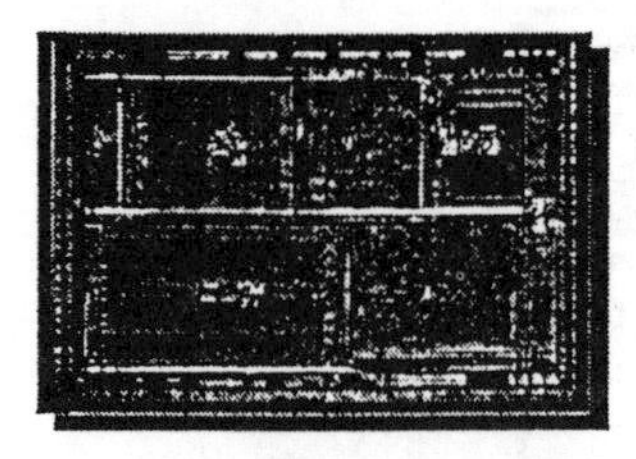

1982
- M68HC11
- 8-Bit CPU
- 4MHz operation
- 12K program store
- 20,000 transistors

1990
- M68300 based
- 32-Bit CPU
- 20MHz operation
- 64K program store
- 200,000 transistors

2000
- MPC500 based
- 32-Bit RISC CPU (multiple execution units)
- 50MHz operation
- 512K program store
- >6M transistors

Figure 3. Automotive microcontroller complexity.

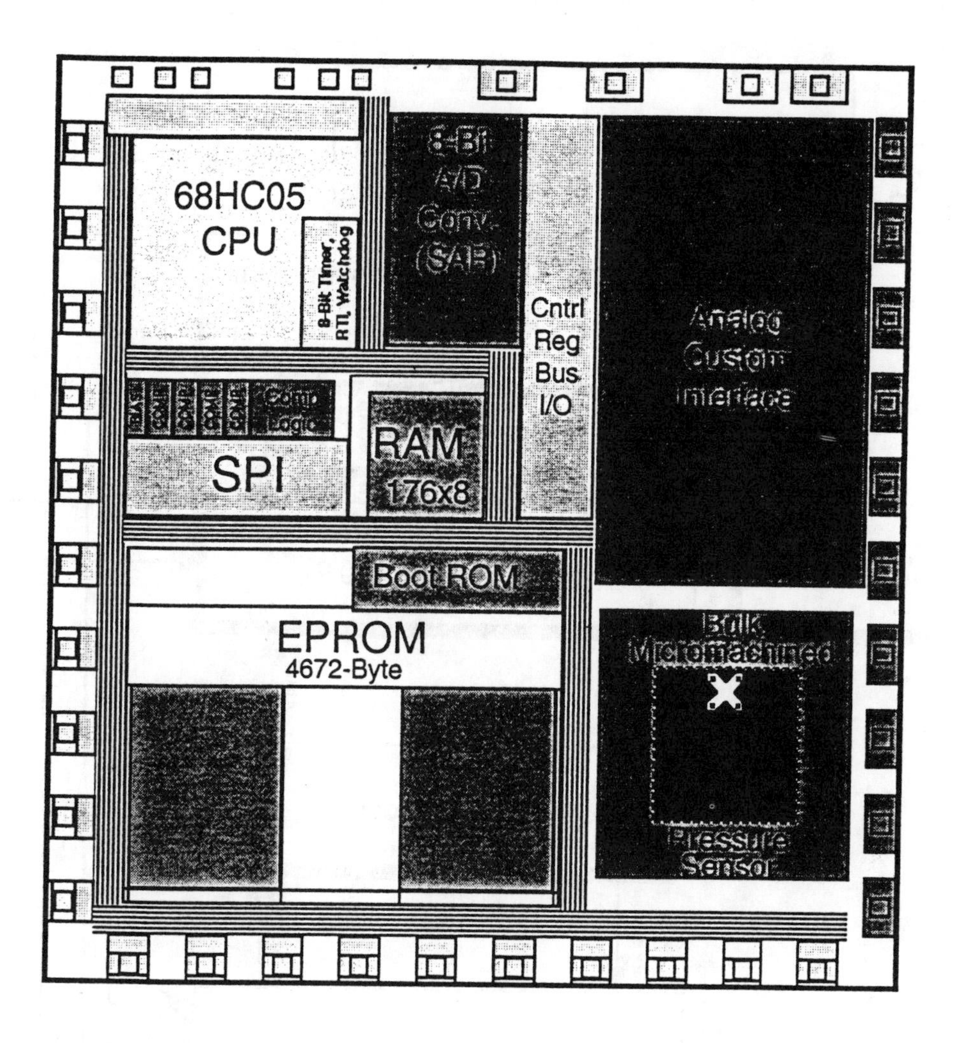

Figure 4. Integrated single chip smart pressure sensor.

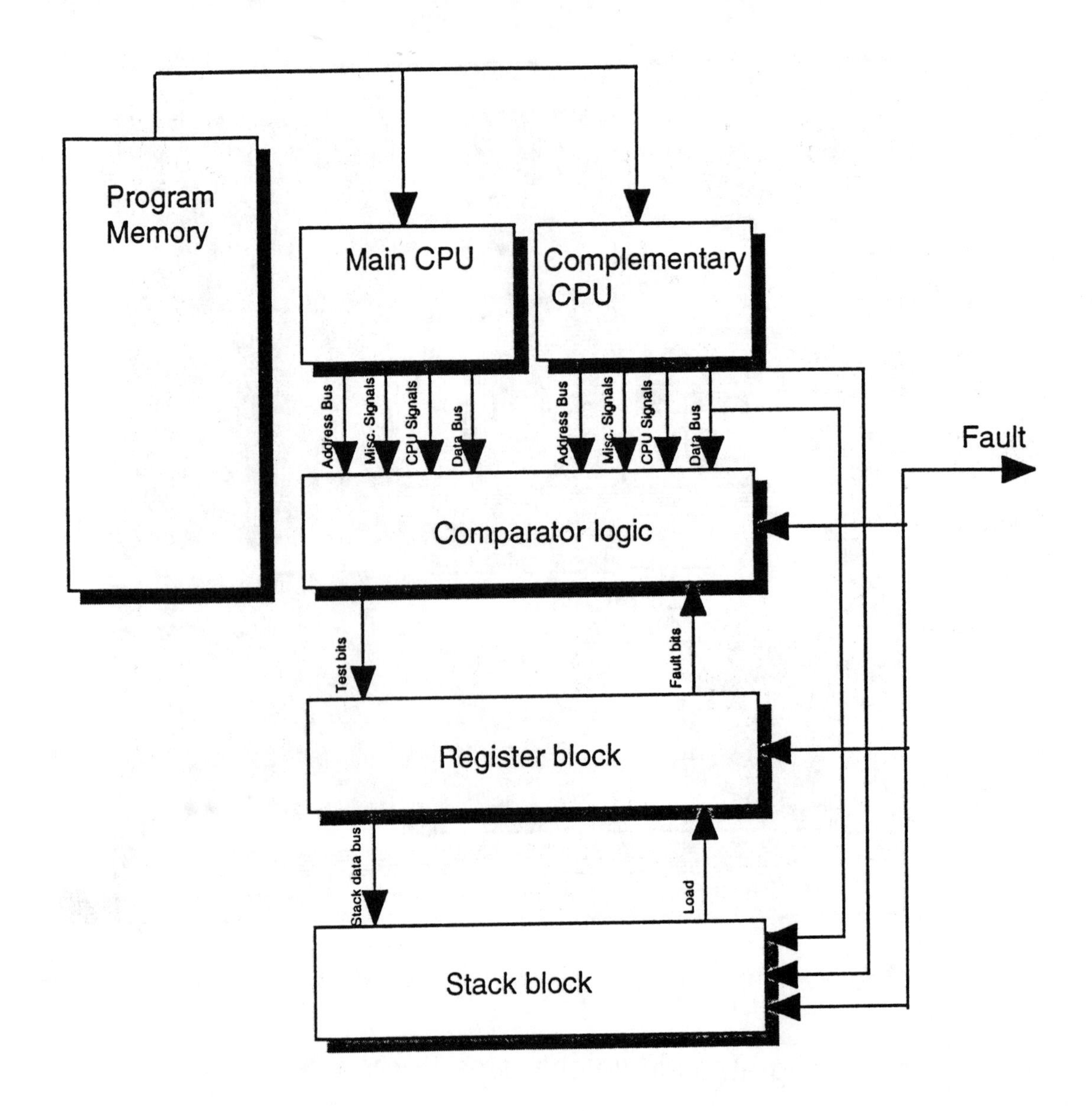

Figure 5. Redundant CPU configuration.

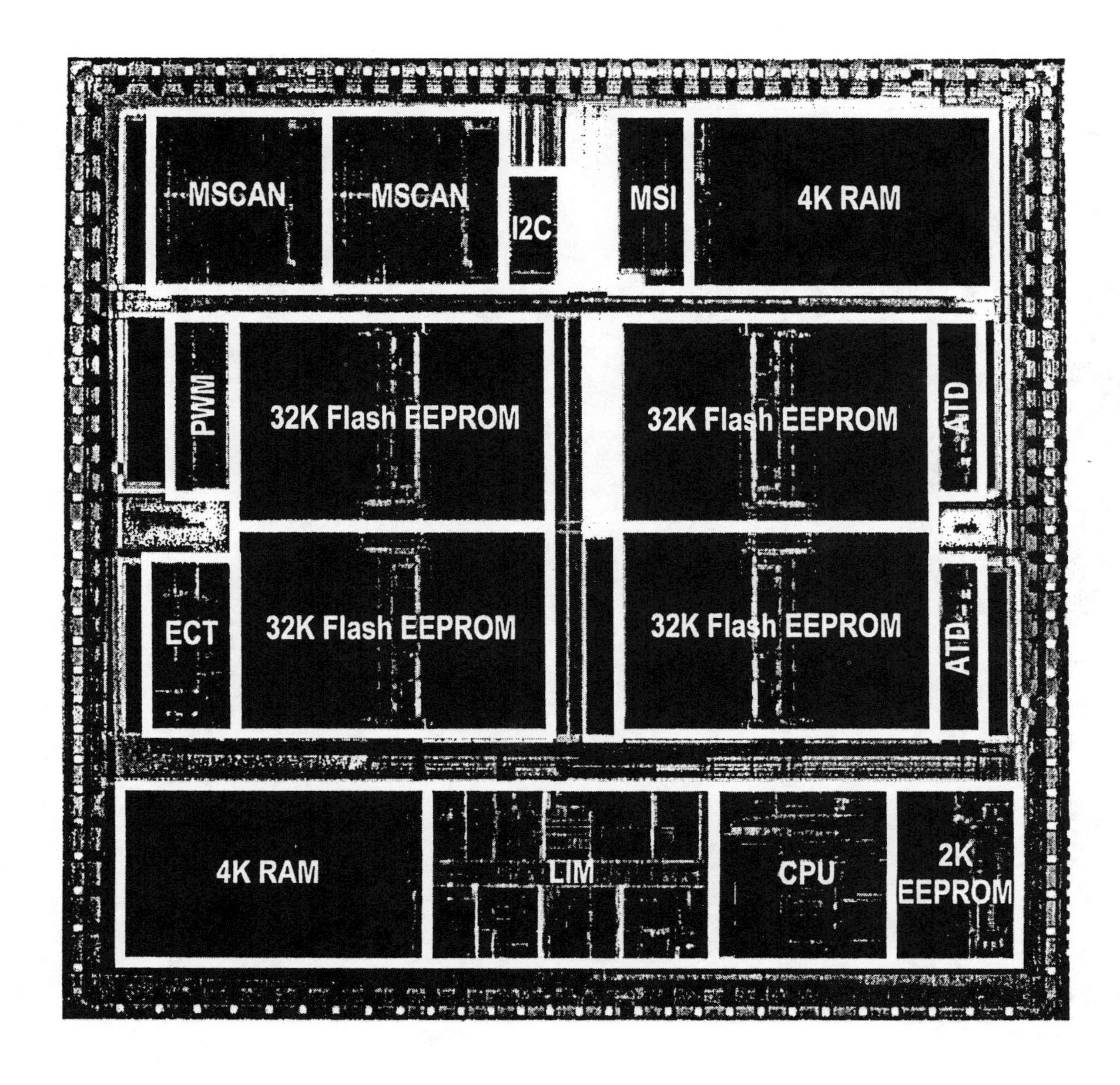

Figure 6. M68HC912DG128 microcontroller.

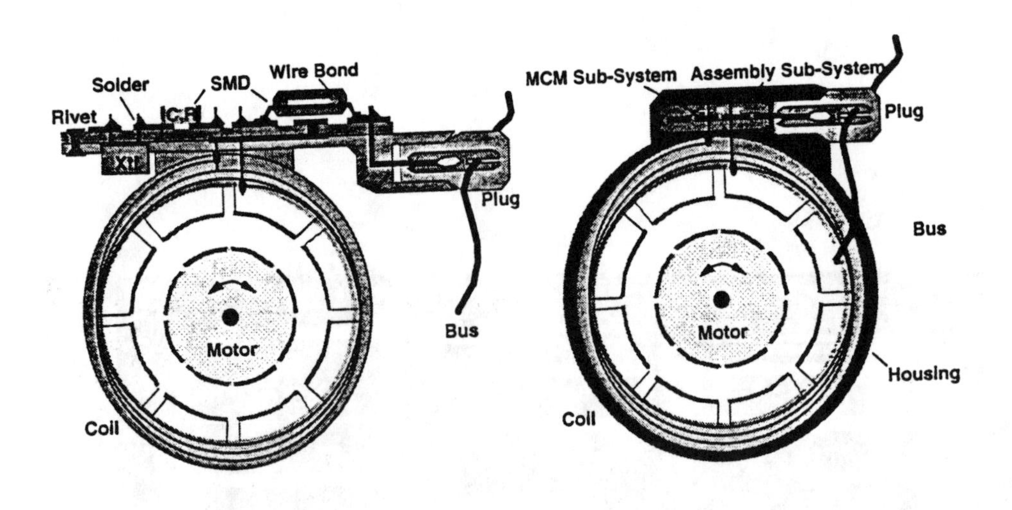

Figure 7. Mechatronic motor solution.

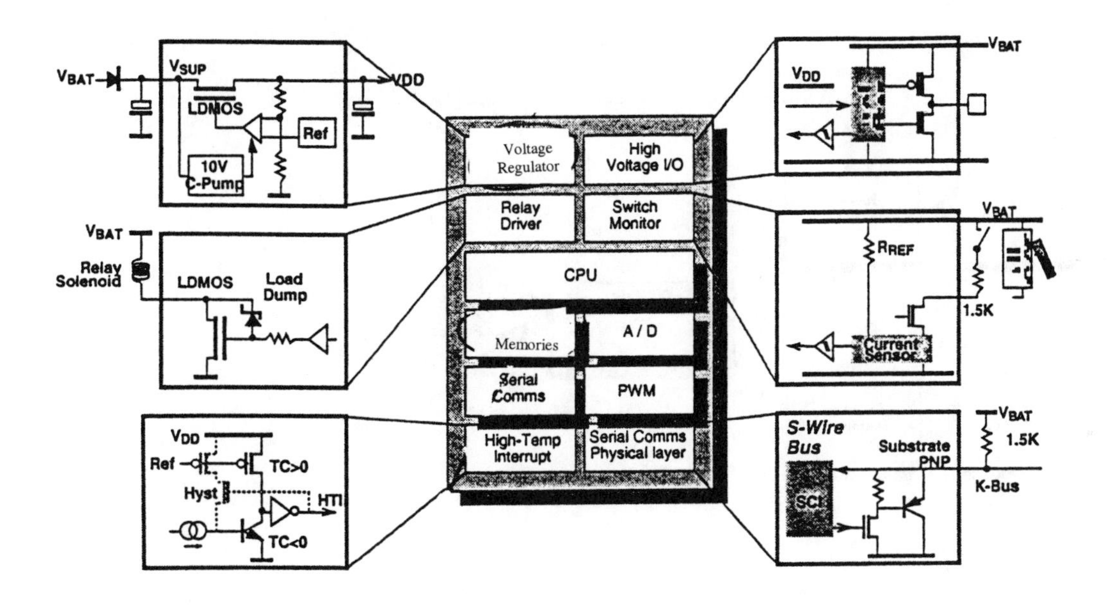

Figure 8. System chip.

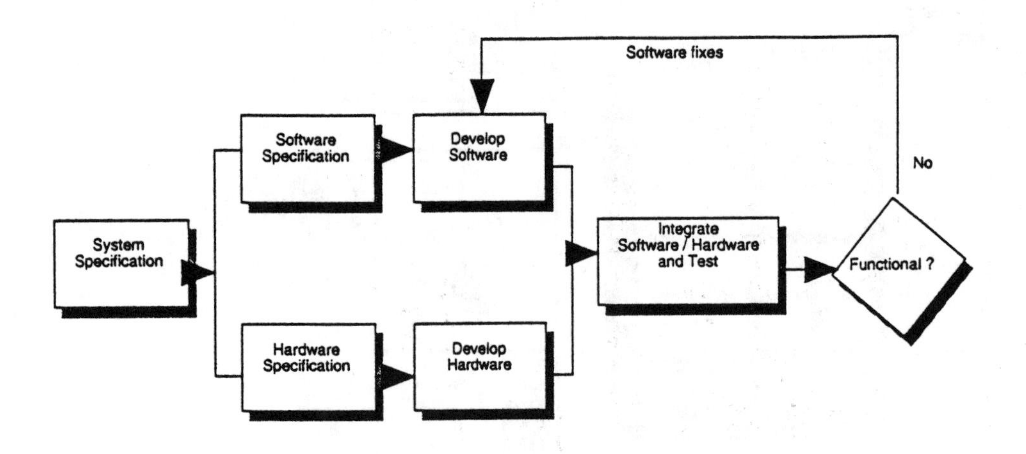

Figure 9. Conventional system design.

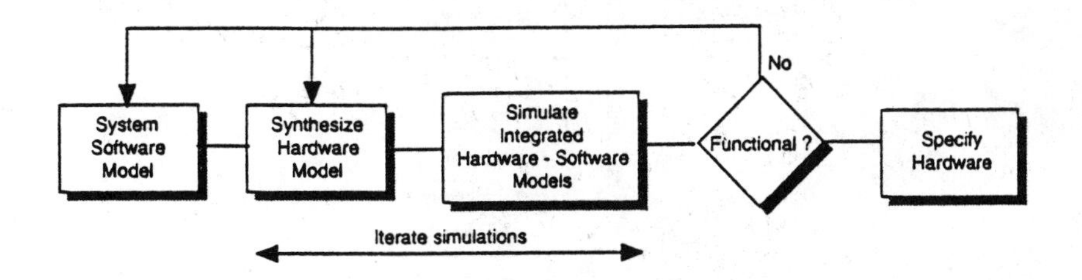

Figure 10. Modern systems design methodology.

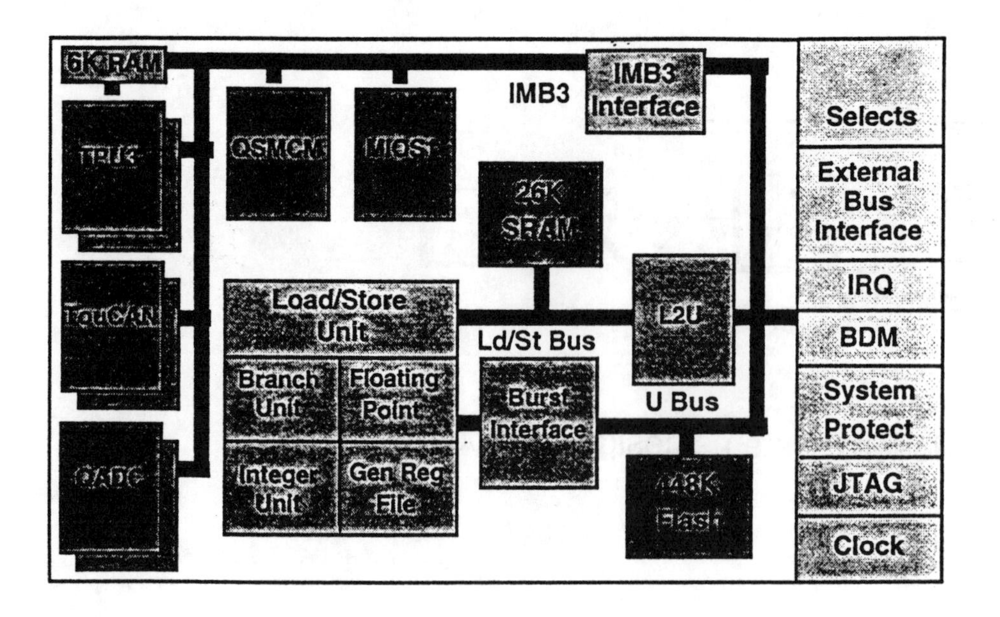

Figure 11. MPC555.